CAMBRIDGE LIBRARY COLLECTION

Books of enduring scholarly value

Life Sciences

Until the nineteenth century, the various subjects now known as the life sciences were regarded either as arcane studies which had little impact on ordinary daily life, or as a genteel hobby for the leisured classes. The increasing academic rigour and systematisation brought to the study of botany, zoology and other disciplines, and their adoption in university curricula, are reflected in the books reissued in this series.

Memoirs of Hugh Edwin Strickland

Published in 1858, this memoir recounts the life and work of the natural historian and geologist Hugh Edwin Strickland (1811–53). Written by his father-in-law, the Scottish naturalist Sir William Jardine (1800–74), the book covers Strickland's early childhood, his education at Oxford, his involvement in and influence upon the establishment of the Ray Society and his notable academic pursuits in natural history before his life was tragically cut short by a freak railway accident in 1853, when he was just forty-two. The reader will gain an insight into Strickland's character, his scientific acquaintances, including Henslow and Darwin, and his wide-ranging interests in the area of natural history, including geology, zoology, palaeontology and especially ornithology, demonstrated by his study *The Dodo and its Kindred* (1848). Drawing upon revealing and informative extracts from Strickland's journals throughout, the book also contains a wide selection of Strickland's shorter scientific writings.

Cambridge University Press has long been a pioneer in the reissuing of out-of-print titles from its own backlist, producing digital reprints of books that are still sought after by scholars and students but could not be reprinted economically using traditional technology. The Cambridge Library Collection extends this activity to a wider range of books which are still of importance to researchers and professionals, either for the source material they contain, or as landmarks in the history of their academic discipline.

Drawing from the world-renowned collections in the Cambridge University Library, and guided by the advice of experts in each subject area, Cambridge University Press is using state-of-the-art scanning machines in its own Printing House to capture the content of each book selected for inclusion. The files are processed to give a consistently clear, crisp image, and the books finished to the high quality standard for which the Press is recognised around the world. The latest print-on-demand technology ensures that the books will remain available indefinitely, and that orders for single or multiple copies can quickly be supplied.

The Cambridge Library Collection will bring back to life books of enduring scholarly value (including out-of-copyright works originally issued by other publishers) across a wide range of disciplines in the humanities and social sciences and in science and technology.

Memoirs of Hugh Edwin Strickland

William Jardine

CAMBRIDGE UNIVERSITY PRESS

Cambridge, New York, Melbourne, Madrid, Cape Town,
Singapore, São Paolo, Delhi, Tokyo, Mexico City

Published in the United States of America by Cambridge University Press, New York

www.cambridge.org
Information on this title: www.cambridge.org/9781108037693

This edition first published 1858
This digitally printed version 2011

ISBN 978-1-108-03769-3 Paperback

The original edition of this book contains a number of colour plates, which have been reproduced in black and white. Colour versions of these images can be found online at www.cambridge.org/9781108037693

MEMOIRS

OF

HUGH EDWIN STRICKLAND, M.A.

HUGH EDWIN STRICKLAND.

1837.

M & N. HANHART, IMPᵗ

MEMOIRS

OF

HUGH EDWIN STRICKLAND, M.A.,

FELLOW OF THE ROYAL, LINNEAN, GEOLOGICAL AND ROYAL GEOGRAPHICAL SOCIETIES, ETC. ETC.

DEPUTY READER OF GEOLOGY IN OXFORD.

BY

SIR WILLIAM JARDINE, BART.,

F.R.S.E., &c.

STRICKLANDIA ACUMINATA, BUCKMAN.

LONDON:

JOHN VAN VOORST, PATERNOSTER ROW.

MDCCCLVIII.

PRINTED BY RICHARD TAYLOR AND WILLIAM FRANCIS,
RED LION COURT, FLEET STREET.

CONTENTS.

PART I.

CHAPTER V.—1836.

CHAPTER VI.—1836 TO 1842.

CHAPTER VII.—1842 TO 1846.

CHAPTER VIII.—1846 TO 1853.

CHAPTER IX.—1853.

LIST OF PLATES.—PART I.

WOODCUTS.

MEMOIR

OF THE LATE

HUGH EDWIN STRICKLAND.

RIGHTON IN 1857.

CHAPTER I.—1811 TO 1828.

INTRODUCTION.—BIRTH AND CHILDHOOD.—AUTOBIOGRAPHY.—FIRST LETTERS.—EARLY JOURNAL.—DR. ARNOLD'S OPINION.—ENTERED AT OXFORD.

THERE is perhaps no subject more difficult to write well than that of Biography, especially if the sketch or memoir be that of a relative or dear friend or companion, where recollection of times pleasantly spent and subjects kindly discussed gives rise to partiality and to ideas of ability, and fixes an importance to events which the unconcerned cannot view with similar feelings, and often leads the author to draw an exaggerated picture, or enter into long and tiresome details; thus it is that we have a fear of either introducing incidents too trivial for the general reader, or of appearing careless and cold-hearted to those most immediately interested. Hugh Strickland's life was one of peace; the events beyond the family or domestic circle were those that occurred in the paths of science, or in the pursuit and study of the works of God; there could be little therefore of excitement or of wild adventure to hurry on the reader; at the same time, nearly the whole range of the progress of Natural Science was searched into, and there were few scientific subjects started during the last fifteen years of his life in which he did not more or less share the interest, or take some public part. He had just reached that period of his career when his ability was publicly recognized, and opportunities had been given to him which would have confirmed the general choice, and enabled him in all probability to have taken a foremost place, and to have filled some blank which increasing years must soon make among our able scientific leaders.

To have gained this eminent position, many memoirs had been published, evidencing great care and research. There had been much and varied correspondence both at home and abroad, and it is upon these we would chiefly draw as our sources of information. The fondness of a mother has supplied the facts and incidents of his early youth; but even when only fifteen years of age he had begun to tell his own story, and short extracts from a little journal kept then, give us far more insight than what hundreds of nurses or tutors could remember or relate. When come to man's estate and visiting other countries, a minute journal is again kept, and though much has been done in Asia Minor through which he travelled, and more might now be told, yet this record is interesting and valuable from its proved accuracy. This has been given nearly as it was written. Soon after his return from the East, my own correspondence and acquaintance with him commenced, and from that time until the end it was maintained, touching upon or discussing many of the scientific projects of the time; and therefore it is that we are able to tell the story of some of them, adding his views from letters to ourselves and others. But here we have often occasion to complain: all Strickland's interesting correspondence from others has been preserved, but it has been difficult often to obtain his own;—not that any of it has been wished to be withheld, but perhaps some did not look upon it of the same value at the time as we do now, while the acknowledged carelessness of others who follow after scientific pursuits is remarkable. Some take no care of their correspondence at all, and permit letters to tumble about and be lost, which probably required much research and time to write; while two scientific gentlemen replied to a request for permission to see letters to them, the one "that he never preserved letters," the other, "that he burnt them all annually."

Next to a man's correspondence, his published writings are the best sources for insight into his scientific character; but rather than give extracts only from the many printed papers and memoirs, as they occurred in time, and as was originally intended, we have preferred collecting and printing them entire, as a separate part of this volume; in this Memoir, therefore, they are merely referred to. Those published conjointly with Sir Roderick Murchison and Mr. Hamilton we have received the sanction of the co-authors to reprint, and they have kindly, at the same time, added notes of the farther progress or information obtained.

Throughout the volume we trust that we have printed nothing that can give offence, or betray the sacredness of private correspondence. If we have printed too much, or recorded some events of little public importance, it must be borne in mind that while the volume is intended as a record of his public and scientific life, it is also given to his family and friends, as a memorial of the private character and affections of one whom they admired and loved.

Hugh Edwin Strickland was the second son of Henry Eustasius Strickland, a younger son of the late Sir George Strickland, Baronet, of Boynton, and Mary the eldest daughter of Edmund Cartwright, D.D., F.R.S.

He was born on the 2nd of March, 1811, at Righton, in the East Riding of Yorkshire, a small village situate between Bridlington and Scarborough, looking down upon the beautiful Bay of Filey. His father occupied a farm of considerable extent in the vicinity. The locality was retired, and the bold features of the neighbouring scenery could not fail to make an impression on the mind even of a child, and there can be little doubt that the constant sight of the Cliffs of Specton and Filey, with very frequent rambles on that interesting coast, formed the basis of his subsequent application to the study of geology. In the spring of 1820, being now nine years old, a private tutor was engaged to instruct him at home, and from that period the beginning of his classical education may be dated. Although the regularity of the proposed plan of education was interrupted by accidental circumstances, it was still continued under his father's roof until the year 1827, when he became a pupil of Dr. Arnold, previous to his being appointed head-master of the Rugby School.

As a child he is recollected cheerful and happy, but had a thoughtful expression of countenance, which frequently attracted attention. He had at this early age an instinctive distaste for all exaggeration, and especially in relating any event or circumstance he made a point of being strictly accurate. This fact, and minute attention to details, may be traced through, and give a value to all his writings, and descriptions; and he did not trust to memory alone: he was in the habit of keeping memoranda of what he saw or read, and began to act on this practice, which was afterwards constantly maintained, from the time that he entered upon his public studies with Dr. Arnold. The little journal which commences this early record of his daily proceedings has been fortunately preserved, and innocently commences with the following sentence:—"Before I begin my regular journal, it may not be amiss to give a sketch of my life up to the present time." This "sketch" gives an insight into the young character and feelings, more than could be otherwise collected or remembered; it is the simple story of a boy ripe in observation beyond his years, and yet untainted by communication with the world, or by the example of thoughtless or vicious companions, and though partly repeating what we have already told, we give it as the best and most creditable testimony of the events of his boyhood, and from the same motives will continue afterwards extracts from the journal which follows it.

"I was born on the 2nd of March, 1811, at Righton, in the East Riding of Yorkshire. My earliest recollections commence when I was staying at Hildenley, near Malton, with two of my father's sisters, while my father, mother, brother and sister were on a visit to the south; this was in the end of 1814. I remember the party returning, and taking me home to Righton. I also remember my fourth birthday, which came soon after. About this time my above-mentioned aunts, whose names were Julia and Charlotte, left Hildenley, and bought a house in Gloucestershire, near Tewkesbury, having about 300 acres of good land belonging to it, whither they went, and have lived there ever since.

"Towards the end of the next year, when I was in my sixth year, I accompanied my mother on a visit to three of her aunts, who lived at Markham, in Nottinghamshire. We also went to Bracebridge, near Lincoln, to visit my uncle and aunt, Mr. and Mrs. Penrose, and then returned home. Nothing further of importance happened till the 12th of February, 1818, when my beloved brother Henry died, at a school at Ripon, where he had gone about a week before. His death was a severe shock to the whole family.

"In the beginning of 1820 a tutor was procured for me, as I began to get too old for my mother to teach. His name was John Monkhouse; he came out of Westmoreland, and stayed with us a year. Soon after this, my father was invited by his brother-in-law, Mr. Freeman, to change his residence to Oxfordshire, in order to be near his sister, Mrs. Freeman, who was in a declining state of health. We accordingly set about packing-up and preparing for our departure, which, after several journeys taken by my father, finally took place on the 23rd of June, to the great grief of myself and some other members of the family, to whom the place had become endeared by long acquaintance.

"Soon after we came to Henley Park, my cousin George Cartwright came to participate with me in our studies. We then passed half a year very happily, when Mr. Monkhouse leaving us, George also went home, and I was sent every day to Mr. Fanshawe's, who took-in a few pupils. This served to give me some insight into the ways of the world, of which I had before very little idea. In February 1822, it being supposed that my spine was out of order, I went with my mother to London to consult Dr. Harrison on the subject, who recommended that I should lie down, according to his plan, which I accordingly did, and remained five months upon my back, after which I began to get up gradually, beginning with five minutes a day, and gradually increasing the time. In August 1822, our family went to Southend, in Essex, taking me with them, where we stayed six weeks, and then returned to Henley Park. Shortly after this, Henry Monkhouse, a brother of my former tutor, came to act in that capacity to me. He had gone to Cambridge not long before, and it was agreed that he should spend his vacations with us, and that while he was at Cambridge, I should write letters to him, containing exercises, &c. In this way passed 1823, in which year I spent every night under the same roof, and which I am not aware I ever did before, and most probably never shall do again. In January 1824, Mr. Monkhouse left us as usual, and being desirous of spending his next vacation at home, it was desirable that some other mode of instruction should be found for me. Upon this I commenced going three times a week to Mr. Parr, the clergyman of Henley, accompanied by my brother Algernon, who was now eight years old. This continued, with the exception of a fortnight in March, when I had the measles, till the 8th of April, when we discontinued our attendance. On the 23rd of the same month, I set off with my father to go into Yorkshire, where I had not been since 1820. We stayed at Boynton, the seat of my uncle, Sir W. Strickland, about a month, in which time I twice had the pleasure of visiting my former home at Righton. On the 14th of June a younger brother of Mr. Henry Monkhouse came, and remained with us till December, when he went to Oxford.—June 17th. Went with my father, mother, and two sisters to London, where we spent three weeks at the town house of my two aunts Strickland, who had come from Gloucestershire to spend a little time in town. Upon our return I commenced lessons with Mr. William Monkhouse, and continued in this manner till he went. I still continued to lie down on a sofa about an hour every day, as I had done ever since 1822, but soon after left it off altogether. In January 1825, we all had the scarlet fever, upon recovering from which, Mr. Henry Monkhouse, who had now left Cambridge altogether, came again to us, and remained with us till September 1826.

"A taste for natural history which I had acquired, and which still continued unabated, was increased by a trip which our party, together with Lady Strickland, her son Arthur, and her daughter Emma, took to Little-Hampton, in Sussex, where we spent seven weeks very agreeably, in collecting shells, &c. In July I went with my father to Apperley, in Gloucestershire, where we stayed a fortnight with my aunts, and then returned home, when I resumed my studies, along with my two brothers. At the end of this year, I was informed that it was settled that I should go to Oxford in three or four years' time. This was a deathblow to all my hopes, as I had always entertained a dread of going to college, as being an almost inevitable cause of my taking root in England, whereas I had always cherished a hope of being a sailor, or at least of engaging in some *locomotive* profession, which should oblige me to roam about and see the world; and this hope, though repeatedly smothered, had never, till this final decision of my fate, been totally extinguished. And much did I grieve to see my cousin and former playfellow, George Cartwright, enter the navy; nay, to see my own brother, and of late my almost only companion, Algernon, destined for the same profession, while I am doomed to spend the rest of my days in the capacity, or rather incapacity, of an English country gentleman.

"But to continue my narrative: in March 1826, I commenced attending a series of lectures on chemistry, which were delivered at Henley by Dr. Venables, and continued till the end of April. In the next summer Mr. Monkhouse went home to see his friends in Cumberland, and I was to have accompanied him, but was disappointed, as well on account of a severe illness which my mother had, as that I was ill with an inflammation of the lungs at the same time, which I had caught with listening to the nightingales (which were unusually plentiful that year) on the cold nights in April. At length, after a long and dangerous illness, my mother was recommended to take change of air (*i. e.* of place). Upon this, she and I went to Apperley, where we stayed a fortnight, when my mother went to Cheltenham, and I returned home with my father, who had come to us. I find our future plans as yet quite undetermined, as my father wishes to settle in Gloucestershire, but has not yet found any house likely to suit him; so, where we shall spend our next new-year's-day, God only knows. My father has informed me that he thinks of sending me next Easter to Mr. Arnold, a brother-in-law of my uncle Mr. Penrose's, who takes a few pupils, for a year or so; and after that, my father and mother think of going abroad for a time, and of taking me with them; but all this must be regulated by circumstances. In this state of uncertainty I commence my *Journal*, in my own bed-room at Henley Park; but where, or when, or whether I ever shall finish it, who can tell?"

The restraint incident to a reclining position, mentioned in the preceding sketch, was rendered necessary by an accident arising from a boyish exertion of strength, which sprained his back, and his health being evidently affected in consequence, he was taken to London, and immediately placed under the care of the late Dr. Harrison, an old friend of the family, whose treatment proved ultimately successful. No bad effects from this accident seem to have continued in after-life; he was an excellent walker, and could stand fatigue well. This long confinement, of nearly twelve months before his health was restored, must have been very irksome to a comparative child, but it was made as agreeable as possible; a taste for reading was formed and encouraged, and a bias was thus given which told on the future bearings of his mind. "It will be easy thus to perceive," his mother writes, "the causes that seemed to encourage so marked a bias as he had already shown in favour of the physical sciences. The country also around our new residence at Henley, of a very different aspect in its beautiful woods and meadows from that of Flamborough and Filey, abounded in subjects of interest sufficient to gratify the most determined naturalist, and it was no wonder that, feeling himself restored to health and strength, he should eagerly return to those pursuits with delight, increased from the variety and novelty of a fresh field of observation. An additional encouragement had also arisen in the assistance of his brother Algernon, who had now arrived at an age to bring to memory the happy companionship of his much-loved brother Henry, and collections were formed and plans of classification devised by those yet juvenile and simple naturalists."

Very seldom from home, there was yet not much occasion for correspondence; the following is the earliest letter of his writing extant: with others, it is carefully preserved by his mother. Written when little beyond twelve years of age, the last sentence is eminently characteristic of the importance of a boy, who thinks he is very busy, and is able to express his thoughts. As a whole, and with those which follow, it would put to shame some of the young gentlemen of more advanced age, who now aspire to official appointments and the service of their country.

"Henley Park, Nov. 9, 1823.

"DEAR MAMMA,—I received your letter this morning by the box, and am much obliged to you for it. I had also a letter from Mr. John Monkhouse this morning, and he says that in his part of the world all the corn is not cut, and that he saw hay in cock on the 1st of November. In consequence of the wet weather I could not have my bonfire on the 5th inst., but the next day being fine I made a nice one. Algy and I had colds in the beginning of the week, *but in consequence of being out so late on the night of the bonfire, we are now quite well.* There is such a riot among the children I can say no more, but that I am, dear mamma, your affect. son."

In the spring of 1824, whilst on a visit in Yorkshire, to his uncle Sir William Strickland, he took the opportunity of going over to Righton, where he seems to have experienced the not unfrequent disappointment of those who anticipate peculiar pleasure on revisiting the scenes of their early childhood. "Poor Righton is altered much, and chiefly for the worse; many things are fallen down, and but few are built up. Speeton Beacon, and the aged Patriarch of Thorn-tree flat, still exist, though probably in a mutilated state. Horses and cattle run at liberty about the plantation behind the stables and the yard, which is overgrown with weeds. The plantation at the back of the kitchen-garden has got up surprisingly; it was with difficulty I could force my way through the trees to the seat and table, both of which I found buried among the trees."

In the autumn of 1826 he spent some weeks with his uncle and aunt, Mr. and Mrs. Penrose, at Bracebridge near Lincoln*, and again writes to his mother:—

"I have now accumulated sufficient news to fill a letter. Mrs. Markham is deeply involved in the affairs of France, which are to be cleared up and publicly declared in December. A *rival* work is advertised under the title of 'History of Spain, on the plan of Mrs. Markham's History of England.' Several of the engravings for 'France' have arrived; they are on wood, and are well executed. You ask me, how I employ my time? I am usually employed in learning Greek, &c., which my uncle is so good as to hear me say after dinner; then in the middle of the day I either go to Lincoln with my uncle and aunt, both, or neither, or I take a walk somewhere to potter after fossils, &c. In the evening we generally read aloud to my great-aunt Frances some of the Waverley novels, or I read Parkinson's Organic Remains to myself. A gentleman and his wife in this neighbourhood live at an artificio-naturali-cultivato-rurali-romantic sort of a place, called the Jungle; the house is faced with irregular and deformed bricks, which have melted and run together in the fire; it is overgrown with ivy, and somewhat resembles our root-house; within it is fitted up like other houses—in short, it is a complete modern antique.

"The grounds are wild and jungle-like, being stocked with kangaroos, two kinds of deer, gold pheasants, &c. To-day I escorted my aunt to Lincoln; in the evening she will go to the ball, which with some difficulty I have escaped from. I went to the top of the central or highest tower of the Minster, where I imitated the example of hundreds, and cut my foot and name on the leads, and then descended to the ground (a distance of 283 feet perpendicular). I ascended another, and paid Great Tom a visit, where I *underwent* the usual ceremony of creeping under him, and standing upright under him. I thought I should never come to the end of these eternal corkscrew staircases; however, I am far less tired than I expected. By-the-by, my aunt has lived here twelve years and has never yet been up any of the towers, not even Great Tom's!! I tried to persuade her to come with me to-day, but she said, what was fair enough to be sure,

* Mrs. Penrose, the daughter of Edmund Cartwright, D.D., was the authoress of the works now so well known to the public as Mrs. Markham's Histories of England, France, &c.

that if I would not accompany her to the *Ball*, it was not likely she would come with me to the *Bell*. However, to me at least, the latter was best worth seeing of the two.—I have got more fossils already than I can take home when I come, so I suppose I must give my aunt the more bulky ones."

In a letter to his father, he says—

"Pray be so good as to inform me whether you have finished your shell-cabinet. I have found hardly any land shells here, there being no woods or mossy banks; of freshwater shells nothing particular, except the large Cardium which we found in the Thames."—"The greatest curiosity in Lincoln, *next to the Minster* (for she lives close to it), is a lady of the name of Dixon, who works most beautiful pictures with little bits of coloured cloth, sewn on to canvas; it is, in fact, a mosaic of cloth instead of stones, and the effect is so good, that at a short distance it is like life itself, as it has not the flat and shining appearance of oil painting."—"You cannot think what nice genteel-looking books Mrs. Markham's Histories of England and France are, now that the engravings are on wood."

On the 1st of January, 1827, the little Journal we have mentioned, prefaced by the "Sketch of my Life," was commenced. It embraces one entire year, and contains entries of every day's transactions. Its insertion here is essential to the delineation of Strickland's early disposition, and it besides carries on the family narrative of removal from Henley to Cracombe House, the routine of study under Dr. Arnold, and the holiday occupations.

"Monday, January 1st, 1827.—Dined with father and my sister Frances at Mr. Combe's; present, Mr. and Mrs., three Miss Bussels, &c.

"Tuesday, 2nd.—I finished making the trays for his new shell-cabinet, which is therefore completed and ready for packing.

"Weddingsday, 3rd.—Sarah Bell, who has lived with us in the capacity of cook nearly ten years, was this day married (after a long courtship) to William Sever, who has been one of my father's farm-servants about four years. They were married at Henley, and then set off to London, where they will take a cook's shop at No. 6, George Street, Oxford Street.

"Thursday, 4th.—Went to Henley with father, who spoke to Mr. Lambe about my taking some lessons of him in algebra and mathematics: agreed that I should go to him tomorrow.—Wrote to H. Monkhouse.

"Friday, 5th.—Went to Lambe, who tried me in the 1st book of Euclid, &c.—The ice bore this day for the first time this winter.—Duke of York died.

"Saturday, 6th.—Father and I began packing up the shells in the cabinet.

"Sunday, 7th.—Received a letter from Arthur Strickland, from Tenby.

"Monday, 8th.—I have lately been reading Ellis's Tour through Hawaii or Owhyhee; it is a very interesting book, and well written.

"Tuesday, 9th.—Went to Mr. Lambe's.

"Wednesday, 10th.—Father went to London.—Stayed in all day, copying Ellis's Map of Hawaii, with the Pantograph.

"Thursday, 25th.—Went to Mr. Lambe's; coming back, conversed with a gipsy, who gave me the following words of his language:—tree, *roke*; dog, *jùgnor*; water, *pàhni*; man, *gòrjo*; woman, *gorjah*.

"Friday, 26th.—Aunt Frances and Canon Riego went to London, taking with them Julia, who is going back to school next Monday.

"Saturday, 27th.—Finished writing the list of British shells with their habitats, which I flatter myself is very complete, as it contains many new species, with a vast number of habitats. This work has occupied me (with the exception of three months when I was in Lincolnshire) since last April, and has been almost entirely composed between the hours of six and nine in the morning.

"Thursday, Feb. 1st.—Despatched a letter to the 'Mechanics' Magazine,' signed 'Boreas,' and describing a plan I had invented for uniting a wind-gauge to a weather-cock, and exhibiting the results of both upon two dial-plates.

"Sunday, 4th.—Went to Henley Church; text, Col. iii. 17.

"Monday, 5th.—Wrote a letter to the 'Mechanics' Magazine,' signed 'Eboracus,' upon the particles of water and alcohol, and upon producing light by blowing inhaled vapour upon a candle. Cleaned the water-clock in mamma's room, and set it a-going.

"Tuesday, 6th.—Stayed in, copying the Hieroglyphical alphabet in the last No. of the 'Edinburgh Review,' which makes out that the Egyptian hieroglyphics are for the most part letters, and not symbols; that is, that the Egyptian name for the thing represented, begins with the letter which that thing stands for: thus, Strickland would be expressed by a spear, a tree, an ibex, a king, a lute, an altar, a nose, and a duck.

"Sunday, 25th.—Went to Fawley Church; text, Cor. xiii. 1.—I am rather surprised at neither of my communications being either inserted or noticed in the 'Mechanics' Magazine' of this morning, as they have been acknowledged as received some time; but I suppose it is owing to the press of matter, that the editor cannot find room yet.

"Friday, March 2nd.—This is my sixteenth birthday, but I have not time to be as sentimental as I could wish upon the subject, since it is past eleven, and I am very tired, having been busy the whole day packing, cording and nailing boxes of pictures, birds, bottles, &c. Five thousand eight hundred and forty-four days, or sixteen years, have passed over my head, three years of which are entirely lost in oblivion; as to the remainder, though I have frequently, upon any slight vexation, thought myself the most miserable of human beings, yet I have reason to believe that I have, upon the whole, been as happy as most persons of my age, and happier than many. Rather more than nine years of my life were spent in Yorkshire, and nearly seven more have glided away from us here; but where our family will take up its abode next, neither I, nor, I believe, any one else has the smallest idea, except that it will be equally, or more remote from my *first* home, viz. in Gloucestershire. And now I am going to conclude the last out of six birthdays which I have spent under this roof; but whether I ever shall, or if so, where live to see my next, thou, O Lord, only knowest! and under thy protection I resign myself to repose, under the hope of opening my eyes tomorrow to the beginning of my seventeenth year.

"Monday, 5th.—Stuffed Algernon's bullfinch, which died a day or two since, having been in his custody about a year and a half.

"Wednesday, 7th.—I have lately been reading Mollier's Travels in Columbia; it is a pretty good book, considering it was written by a Frenchman, and gives a good idea of the country. He went from Carthagena to Bogota and Popayan, and from thence by sea to Panama, where he crossed the Isthmus, and came home.

"Thursday, 8th.—Measured two spruce firs in the woods beyond Pining, one of which by actual admeasurement was 81 feet in height, but by taking its altitude with a quadrant, was 73 feet: the other I did not measure, but by the quadrant was 80 feet, and allowing for the same error would be about 88 feet high.

"Friday, 9th.—A young man named Marklew came from Henley, to make proposals to Ruth Mainprise, our nursery-maid; it seems he first saw her at Sarah Bell's wedding, on January the 3rd, where she was present; so if the maids get husbands so fast, I suspect we shall soon find ourselves without any.

"Saturday, 10th.—Two engineers came down from London to see and examine my father's steam-engine, and the machinery attached to it, to enable Mr. Freeman our landlord to decide about buying it of my father, as it is attached to the premises; and he thinking the expense too great, has refused to buy it, and accordingly my father will pull it down and sell it as well as he can.

"Sunday, 25th.—Went to Fawley Church. Mr. Fanshawe made me a present of an 8vo edition of Tacitus' works in four volumes. This day being Lady-day, my father nominally resigns his farm, as he only continues further upon sufferance.

"Wednesday, 28th.—I hear that Captain Parry has just sailed on his fourth voyage. In this voyage he goes past Spitzbergen, direct to the North Pole. Oh! what would I give to be along with him? but as that is not to be, all that I can do is to pray that he may succeed in his undertaking, and return safe.

"Monday, April 9th.—Two-thirds of our goods have this day been stowed on board the boat; but it is doubtful, though her burden is forty-four tons, whether she will carry the remainder.

"Tuesday, 10th.—The boat has taken the greatest part of our goods, and as she will hold no more will depart tomorrow morning. I have laid myself on my bed for the last time at Henley Park; this evening we removed to the Bell at Henley, where we shall remain until all affairs in this neighbourhood are settled, and then take our final leave of it.

"Tuesday, 17th.—I hear that Mr. Arnold has gone to Rome, and consequently my presence will not be required at Laleham (where he lives) till the 1st of June.

"Wednesday, 18th.—I this day saw the body of Margaret Myers, who with her father and mother followed our fortunes out of Yorkshire; last autumn she caught a cold which settled on her lungs, and since that time she has been getting weaker and weaker, till last Monday she breathed her last, aged thirteen. I saw her stretched out in the bed in which she died, with her coffin beside her; her cheek was pale and cold, and her eyes closed; I had never seen a dead body before; and it was the more affecting as I had known her well, as long as I have any distinct recollection of anything, for many a game of play have I had, and many a house have I built among the haystacks in my younger days with Margaret Myers; but these days are past; tomorrow will see her restored to her mother-earth, and all that I can say is, 'Alas, poor Margaret!' she was a pious, good girl, and I hope is now happy.

"Saturday, 28th.—I was much surprised at seeing 'Boreas's' communication in the 'Mechanics' Magazine' of this morning, as from the length of time since it was sent, I had given up all expectation of ever seeing it in print*.

"Sunday, 29th.—Went to Henley Church in the morning; Mr.

* Mechanics' Magazine, vol. vii. p. 264.

Richards preached from John x. 11: and in the evening to Fawley; Mr. Peachell from Romans v. 19.

"Wednesday, May 2nd.—A large concourse of persons assembled at the Park; and, as is usual at sales, many valuable things went for almost nothing, while others of less value fetched a greater price than they were worth: the sale continued till it was quite dark, when about forty lots remaining unsold, it was settled that they, together with the remainder of the farm, should be sold by auction on Wednesday week, the 16th instant.

"Thursday, 3rd.—Henley Park begins to look forlorn in the extreme: the floors dirty and strewed with orange-peel, the rooms half or quite empty, the furniture discommoded and piled one upon another; the house and garden full of men, and straw, and waggons, and carts, and barrows, all employed in the removal of the effects sold yesterday: such is the state of a house in which I have passed many happy days, perhaps more than I ever shall again enjoy.

"Friday, 11th.—Spent the day at the British Museum.

"Saturday, 12th.—Went to see the Chinese ladies exhibiting at 94 Pall Mall. Their countenances and dress exactly resemble the figures upon Chinese paper; their finger-nails were about two inches long, and they were accompanied by a man, who was also dressed in the costume of his country. From thence we took a boat and rowed down to Rotherhithe, to see the Tunnel constructing at that place for the conveyance of wheel carriages under the Thames. The first object which presents itself is a large shaft or well, walled round with brick, 75 feet deep, and perhaps 40 feet wide. The central part of this shaft is at present filled with staircases to descend by, and machinery, comprising a steam-engine, working two large pumps to clear it of water; it also draws the carts laden with the material to the bottom of the shaft, and then raises them to the top, where they are emptied and let down again. Descending to the bottom by a staircase, you see two long passages lighted by gas, and placed side by side; each of which is wide enough for one carriage, besides a footpath by the side. The length of the Tunnel at present is about 500 feet (not more than a quarter of the whole length), only 300 of which is open to the public, so that one cannot see the mode of working on account of the distance to the end of the Tunnel; however, we returned highly gratified with our day's work.

"Monday, 14th.—Went to Exeter Change, and afterwards returned home to Henley.

"Monday, 21st.—All accounts being now settled, and business concluded, our whole party took our final leave of the neighbourhood this day; we went as far as Witney, where we slept.

"Tuesday, 22nd.—At Apperley Court, near Tewkesbury. We arrived here this day, and shall stay here some time, and then take a lodging somewhere in the neighbourhood for the present.

"Saturday, 26th.—Went to a Roman pavement lately discovered at Coomb-hill; it consists of a long plain tessellated pavement, about 30 yards long and 2½ wide, with abundance of ruins, and traces of walls, &c., but no figured pavements have as yet been discovered.

"Sunday, June 2nd.—At the Rev. T. Arnold's, Laleham, near Staines. Came here this afternoon, and was introduced to Mr. Arnold, whom I had never seen before; he only returned yesterday, and none of his pupils besides myself have yet arrived.

"Monday, 18th.—All Mr. Arnold's pupils having arrived, and all things having fallen into regular routine, I will describe my companions, as well as the plan which we pursue. Of the seven young men or boys, to which number Mr. Arnold limits himself, four are new-comers at the present vacation, and three have been with him before. Of the latter class, the one that has been here the longest, and consequently has the precedence, is named Henry Hobhouse; next to him comes John Pendeill, then Arthur William Tooke, then *egomet ipse*, then Nash Vaughan Edwards, then Charles Lloyd, and lastly, Charles Robertson.

"Mr. Arnold sits in his own room, and each of us in turn go in to say our lessons to him twice a day, according to the above-mentioned precedence, so that when the last has been in the first time, it is time for the first to go in for the second time. The first lecture, as it is called, consists, on Monday, Tuesday, Wednesday and Thursday, of Herodotus, and the second of Livy; on Friday and Saturday, there is only one lecture; on the former of those days I do Sophocles, and on the latter am examined in Hallam's Middle Ages. We get up about 7 A.M., dine at 5½ P.M., tea 8 P.M., go to bed, or supposed to do so, 10 P.M. We have moreover three themes a week, which are shown up on Saturday; and twice a week go in with Greek Testament, between dinner and tea.

"Saturday, July 7th.—Received a letter from Fanny, with the important news that father has taken a house, named Cracombe, situate between Pershore and Evesham, in Worcestershire, and about four miles from either place. It is placed half-way down a hill, near the river Avon; it will be entered upon next Michaelmas, when all our goods, which are now lying at Apperley, will have to be removed to that place, a distance of about eighteen miles.

"Wednesday, 8th.—This day there was a general dispersion of our party, for the midsummer holidays: I myself came to Cheltenham, where I slept.

"Monday, 13th.—Went with father to see our future residence, Cracombe; the house is now in the occupation of the proprietor, a Mr. Perrott and his family, who will remove to the adjoining village of Fladbury, and let the house to us. The house is very large, though only two stories high, insomuch that there will be plenty of room for all our devices, and still several rooms will have to be shut up. My father has taken it with about 250 acres of land, for seven years. The situation is very beautiful. There are large shrubberies, and an excellent kitchen-garden, all which, together with the house, are in great disorder, and stand in need of a thorough repair.

"Wednesday, 15th.—Returned to Apperley Court much pleased with the appearance of our new abode.

"Saturday, 18th.—Took an hornet's nest.

"Wednesday, 22nd.—Set out with father on a short tour. Went as far as Ross; country beautiful. Soil red marl, which becomes more sandy and harder the further westward one goes, so that at Ross, the houses are almost entirely of red sandstone.

"Thursday, 23rd.—Went to see Goodrich Castle, a most beautiful ruin; from thence to Monmouth; from thence to see the Buckstone, a magnificent rocking-stone on the top of a hill, calculated to weigh 50 tons. From thence past numerous iron-works and collieries through the Forest of Dean to Purton Passage, where we slept. During this day I passed through some of the most beautiful country I ever saw.

"Friday, 24th.—Slept at Newnham; went to Westbury Cliff for fossils, did not find many.

"September 24th, Monday.—The party having arrived, affairs have relapsed into their old channel.—There is one new comer, who increases our party to eight: his name is James Lewis.

"October 1st.—Wrote to father.

"Tuesday, 9th.—Received a letter from Fanny: the party have at last removed from Apperley Court to Cracombe.

"November 20th, Tuesday.—I have at last given up bathing, after having continued it constantly three or four times a week through the summer. The time flies on steadily and swiftly, and I care not how swiftly, for I want sadly to go to our new home, and to give it a more careful examination than I could do in the short visit I gave it last summer.

"December 1st, Friday.—Our chief amusements consist in sailing, rowing, or punting, or in other recreations connected with the river. I see in the paper among the list of killed at the battle of Navarino, Charles, the son of my brother Algernon's schoolmaster, Mr. Bussel, who had gone to sea hardly a twelvemonth ago.

"Wednesday, 19th.—Set off at 6½ A.M. with my brother and sister, and passing through Woodstock, and Moreton-in-the-Marsh, arrived at Evesham at 5 P.M., and from thence came to Cracombe, where I found all the party well, though still many degrees removed from a comfortable settling, since they have been retarded from establishing themselves comfortably by the great want of repairs throughout the house.

"Friday, 21st.—Went out in my father's boat to explore the river Avon, which is very deep with a muddy bottom, and winds amazingly.

"Sunday, 23rd.—Went to Fladbury Church, which is about half a mile distant from Cracombe. It is a very large and handsome church for a village, and is in excellent repair, which is chiefly owing to the liberality of the present incumbent, Mr. Stafford Smith.

"Tuesday, 25th.—Christmas day; went to Fladbury Church.

"Thursday, 27th.—Commenced clearing the dead trees and branches out of the garden, which promises to be a long job, as the garden and shrubbery are very large, and in a state of extreme neglect.

"Here then I have arrived at the termination of another year; a year which has seen our family take its departure from one home, and settle itself in another which promises to be equally or more agreeable. And now, having a few pages to spare to the end of the volume, I do not see that I can better employ them than by a few general observations as well as remarks upon particular passages which occur upon a reviewal of the foregoing pages, which being written from time to time by snatches, contain many things which require explanations which a more attentive perusal suggests.

"In the first place, it will be seen that the handwriting is very various in different places; nor will this seem extraordinary when it is considered that this book was written frequently where good pens were not attainable, or often with nothing but my knee to write upon, or late at night, when overcome with sleep, &c.

"In the second place, it will be observed that I have been much less diffuse of late than at the beginning of the year, which is occasioned, first, by the monotonous life I lead at Laleham, which furnishes but few incidents worthy of being recorded; secondly, by my seldom having time to record those incidents, which caused them to be sometimes postponed till they were forgotten; and thirdly, by my taking less interest in it than at first, and consequently not taking so much trouble to find out or invent subjects to dilate upon."

Throughout the whole of this record of the doings of the boy, as well as in the simple letters quoted before, we can trace strongly the beginnings of many of those various subjects that afterwards so much engaged his attention. They exhibit a remarkable advance of mind, and at the age of sixteen he had ventured to put his ideas into writing for the press, and to send his first contribution to science to one of the monthly periodicals*.

But these early pursuits, for they seemed already inseparable from his thoughts, made him appear among youths of equal age as not so forward in the world. After half a year's trial of the school at Laleham, his father requested Dr. Arnold to give him his opinion of his son's capabilities. He returned the following friendly and kind reply:—

"Laleham, Dec. 14, 1827.

.... "With respect to your son's abilities and comparative proficiency, both are certainly much above par; but I do not think he is fond of history, or moral and grammatical studies, as opposed to natural history, to warrant the expectation of his arriving at any high distinction in them. There is no doubt at all of his passing his examination very respectably; but as things now are at Oxford, I apprehend there must be very high powers of mind, and a strong interest in the subjects studied, to give a young man a place in the highest class of *Literæ Humaniores*. At the same time, it is very likely that a further experience of him may agreeably disappoint me, and that as he advances more, which he has done, and is sure to do still further, he may acquire a more lively interest in his studies, such as will lead him to combine what he reads with me with his own miscellaneous reading, and to retain it more vividly in his memory by its being more naturally associated with his own mind..... His reserve is very great, and makes him appear to great disadvantage; but his conduct in all essential points is most excellent, and he has worked regularly and well."

Dr. Arnold's opinion is deserving of attention. Strickland's habits of mind and manners at this period were the natural effects of his bringing up, which his long confinement from the accident to his back, and intercourse then chiefly with grown-up people, materially confirmed. The reserve alluded to had been increased by the facility afforded of encouraging tastes intended to have been kept within the bounds of recreation and amusement. Before he was placed at Laleham, he had comparatively rarely associated with

* "Plan for uniting a Windgauge with a Weathercock," by Boreas.—Mechanics' Magazine, vii. p. 264.

young people of his own age. With older people he was much more at ease; and when he found himself within a circle of youths to whom he was perhaps more than equal in general knowledge and miscellaneous reading, yet not so in their walks of study or amusements, it is very probable that he felt his position, and appeared unsocial and reserved, when it was very far from his wish to be so. The short period, however, passed at Laleham eventually proved of inestimable value by the fresh views it had opened to him of a higher order of intellectual culture. He had learned the standard by which a young man's powers are to be tried at the Universities; but above all, by enlarging his sphere of study, he had been taught to guard against that narrowness of mind so frequently induced by too close a cultivation of any absorbing object.

In the following summer, Dr. Arnold being appointed to Rugby, and his private pupils leaving him, Strickland was matriculated at Oxford; but as he could not be entered at College before the ensuing spring, the few remaining months were passed with the Rev. Joseph Jowett at Silk Willoughby in Lincolnshire, and in February 1829 he became a student at Oriel College.

SPEETON CLIFFS.

CRACOMBE HOUSE.

CHAPTER II.—1828 TO 1835.

OXFORD.—COLLEGE LIFE.—BUCKLAND'S LECTURES.—VISITS PARIS.—READS WITH A TUTOR IN ISLE OF WIGHT.—DEGREE AT OXFORD.—RETURN AND RESIDENCE AT CRACOMBE.—MR. MURCHISON.—BOUNDARY OF THE LIAS.—MR. HAMILTON.—PROPOSAL TO VISIT ASIA MINOR.—DEPARTURE.

ON the 29th May, 1828, Strickland was matriculated to Oriel College, Oxford. The usual college amusements did not at first gain his attention, his scientific pursuits took the lead; and on reporting to his father soon after being placed, he writes that he had spent two or three days at the museum, "to which a subscription of two shillings gives admission for a year. It is already a considerable collection, and is daily increasing; it has been quite fresh arranged by Mr. Duncan, and little remains of the Ashmolean Museum."

Having entered upon the usual course of study preparatory to passing his first examination, he at the same time attended the geological lectures of Dr. Buckland. "They are so rational and interesting, and agree so well with all the opinions I had before formed on the subject, that I am quite delighted with them, while it is worth one's while if it were merely for the sake of examining his immense collection of organic remains*." A note-book was kept of these lectures, which is now curious to look over, as many of the Professor's early views and ideas are taken down. By the month of June in the same year, however, he seems also to have entered heartily into the spirit of the place, its games and boating. He exults in the success of the men of Oxford over those of Cambridge, and writes of a boat-race with a degree of enthusiasm unusual to him in matters of this kind:—

"The gaiety of the assembled multitude, the numerous associations which connected me with the place, and the splendid termination of the whole, wound me up to a state of rapture such as I never remember. Throughout the day, to the last moment, it was fully expected that Cambridge would beat, as they had a crew in better condition than the Oxford men, a quite new boat, and the first choice which side of the river to take, so that being left in the rear by fifty yards, proves plainly how superior we are to them in rowing. In cricket, too, there has been a match today, and this evening we beat by 116 runs."

In November of the same year Strickland passed his first examination, his "Little-go," and thus reports his opinion of the proceedings:—

"About half-past twelve yesterday I went into the schools. I was the first called up, and was examined about twenty minutes. I was not nearly so frightened as I expected to be, and was quite surprised at myself, as I expected I should not be able to answer a word, instead of which I found myself very much at my ease. I was then set to do various translations, &c., which occupied about two hours and a half, when the examiner said that all I had done was very satisfactory, and I was at liberty to go. After all, this same *Little-go* is a great farce; only, few like to consider it so till they have passed it†."

During the long vacation of this year he accompanied his father to Paris. His notes and journal of the tour, which

* Letter to Mrs. Strickland, May 9, 1829.

† *Ibid.* November 1829.

included about a month, indicate great activity on the part of the travellers, as well as his own industry and habit of observation; but what is most worthy of remark, is the vigilant attention with which he scanned, and seemed, as it were, to seize upon, the geological features of the country he passed over.

At the usual time he returned to Oxford and pursued his studies. Strickland was now reading for honours, and his parents were naturally anxious about his progress and the degree he would be able to take. In reply to their inquiries, he writes in February 1831,—

"You desired that I would let you know as soon as I could form an opinion, as to what I might be able to accomplish in the way of reading for a class. Considering how much is required for the several classes, and how much I consider myself capable of performing in the time that remains, I cannot help thinking that I shall be in danger of slipping between two stools, and failing in both. For all persons, who like me do not really possess a mathematical head, ought to devote at least three hours a day to mathematics, and this makes a most material deduction from the eight or nine, which is as much as most people can employ in reading. Not that by any means I despise mathematics, for it is a science which I should be very happy to pursue farther; but as far as regards taking it up in the schools, I think it would consume so much of my time as would be injurious in the other branches, and as a single class in classics is more desirable than in mathematics, I think it would be best for the present to attend to the former only. If I were to do this, I think I might possibly accomplish a second; but you must know that this is not so easy as it was, for they have now made four classes where there were only three, so that the second class is now above the middle, instead of being in it*."

During this term he attended Dr. Kidd's anatomical lectures, and kept his accustomed note-book of the course. In the autumn following, he retired to read with a tutor. The place selected for this purpose was the Isle of Wight, unexceptionable for beauty of scenery and retirement, and then much less frequented and less public than now; but for a person of Strickland's pursuits it was placing a strong temptation within reach, for there is no finer geological field in

* Letter to Mr. Strickland, February 1831.

the world. Not that we think secluded study the best, for we maintain that recreation and variety intermingled, are essential to keep the mind in the full vigour of its powers; but the whole of the island is so seductive, that we cannot wonder at every spare moment being given up to it; and such was really the case, for in a little note-book* we find sketches and sections, and notes, particularly of the St. Helen's, Whitecliff Bay and Alum Bay beds, which show that the whole had been gone over and carefully examined. They were, in fact, carefully compared with Professor Webster's letters printed in Sir Henry Englefield's work; and had even at this early time of his geological study given him much thought, for he observes that he could not reconcile the beds at the one end of the island as counterparts of those at the other. But he worked at his books also, for he writes:—

"I cannot wonder at the anxiety you express in your letters, and I can assure you I am not without anxiety on the same subject, being conscious that my powers are but moderate, and therefore require a corresponding degree of exertion; but as trying for honours is always to a certain degree a lottery, it is at least worth the attempt; and certainly the spell I have had with a private tutor has been a great advantage to me, as I have now read through all my books for the first time."

On the 6th of May, 1831, Strickland passed his public examination in "Literis Humanioribus," and in the following spring took his degree for M.A. Upon thus completing his studies at the University he returned to his home, and entered keenly into his favourite pursuits. His father still resided at Cracombe House, in the Vale of Evesham, where the interesting character of the vicinity and surrounding country afforded almost inexhaustible materials to the geologist. The riches of the Great Severn Valley also began to be opened up by

* These note-books were continuously kept by him, and are now of the greatest use in working out this memoir. They sometimes contain very full and extended notes, the groundwork of his published papers; in others, memoranda to be afterwards referred to, of things seen and to be seen, of books read and to be read, subjects of all kinds, geology and zoology, statistics and antiquities, mechanics and literature, &c.

railway cuttings, and attention to be more widely drawn to the unworked-out strata of the Keuper and New Red Sandstone beds. His geological collection may perhaps date its real and systematic commencement about this time; it now rapidly increased, and its arrangement naturally brought him into correspondence and acquaintance with men of kindred tastes and acknowledged reputations both at home and abroad. At the same time other branches of natural history occupied a great share of his attention, and the foundation of his large ornithological collection was now also laid.

Among the men of science with whom Strickland at this time became acquainted, through the particular direction of his studies, was Mr. Murchison (now Sir Roderick), whose uniform kindness and advice were continued through a long and intimate acquaintance, first as to a young man and pupil entering upon active life, and afterwards as a fellow-labourer in science, who became alike the counselled and the counsellor. This intercourse and friendship was incident to Strickland's knowledge of the geology of Worcestershire and the adjoining districts, and to the information he had furnished to Mr. Greenough, when constructing the second edition of his geological map; and in 1834 a request was made to him by Mr. Murchison, through the introduction of Sir Charles Hastings, that he would lay down the boundary between the Lias and the New Red Sandstone for the Ordnance Map then in progress*.

It was to the same circumstances also that he owed his introduction and acquaintance with Mr. Hamilton, who was to become his friend and travelling companion, and through whose influence he undertook his Eastern travel, and obtained so much important information. In April 1835 Mr. Murchison offered a short visit to Cracombe. "I decidedly wish to see your gravel and alluvial case, either outward or home-

* The first geological map of Worcestershire was constructed by him and Mr. Edwin Lees for Sir Charles Hastings' 'Illustrations of the Natural History of Worcestershire;' and a most remarkable production it is, considering that the 'Silurian System' had not appeared, and that most geologists were "all abroad" on the geology of the Malvern and Abberley range.—Trans. Malvern Field Club, 1855, p. 12.

ward bound, and by reference to your letter I suppose the *first* will suit you best*." "I must mention that I am accompanied by my young friend, our Geological Secretary, W. Hamilton, who contemplates an expedition to Asia Minor, and is going to take a lesson from me in this tour."

This visit was paid, and the parties remained a night at Cracombe. The expedition to Asia Minor was, no doubt, then discussed, for in the course of a month afterwards matters were arranged; and in writing to his brother Algernon, then at sea, before setting out, he thus explains his motives and expresses his feelings when undertaking so long a journey:—

"18th May, 1835.

"I have lately become acquainted with a gentleman of the name of Hamilton, who intends going to Asia Minor, to explore the interior of that country, and he has asked me to accompany him. I told him I could not decide till I had consulted my father and mother, and they were at first unwilling to part with me, but having given the question a fair consideration, they saw so many advantages in my going as to overcome their reluctance. I am at present, therefore, very busy making preparations and drawing my plans to a focus. We propose to remain at Smyrna some months to learn the Turkish language, and become acquainted with the customs of the country, and then we shall decide what route to take. We wish principally to explore the eastern and central parts of Asia Minor, much of which has not been visited by Europeans in modern times. We expect to be out the greater part of two years, that is, if we find enough to occupy us, and if we continue to enjoy good health. If we are tolerably fortunate it can hardly fail to be a very interesting expedition, and I look forward to it with great pleasure, though mixed with a good deal of regret at leaving my dear friends and dear home for so long a time; I shall also feel great reluctance at increasing by my absence the anxiety which my father and mother already feel on various accounts; for though their wish for my advantage and improvement has induced them to consent to my going abroad, I am afraid they will much regret my departure, as I have so long been a member of the home circle. This regret I must endeavour to lighten by writing home as often as possible. I am sorry also to be still further from you than we are already, for it can only be at very distant intervals that we can hope to hear of each other. However, I hope to return about the same time as

* Letter, London, April 7, 1835. The "gravel and alluvial case" was the discovery of marine shells by Mr. Jabez Allies, in the gravel of Kempsey, Painswick Hill, &c.; these will be noticed more particularly afterwards.

you will, so we must look forward with confidence, and pray for a happy meeting in two years at Cracombe."

In July 1835 the party left London and proceeded directly to Paris. From the time of starting, a journal of notes and all transactions was kept by both travellers. Those of Mr. Hamilton have already been published*, while from Strickland's notes, various papers were from time to time read before the Geological and other Societies, and several of them were printed in their Transactions; but the Journal of the entire route, with its illustrations, &c., remains in note-books as brought home. At the same time he wrote at length to his parents of all his proceedings. These letters are freely written, and are very interesting; they are the open letters of a son to his parents, often alluding to family and private concerns; and as printing them entire would at the same time only be to repeat facts afterwards detailed more minutely, we think it better to omit these altogether, and to print the journal as complete as possible.

* Researches in Asia Minor, Pontus and Armenia, by W. J. Hamilton, Sec. Geol. Soc. 1842.

BREDON HILL.

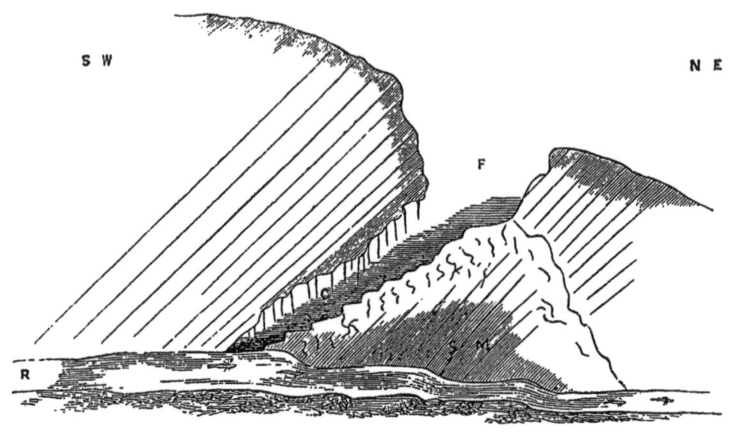

MADDALENA CAVERN *.

CHAPTER III.—1835.

JOURNAL.—LEAVE LONDON.—PARIS.—FRENCH SAVANS: ELIE DE BEAUMONT, BOUÉ, VON BUCH, &c.—VALLEY OF THE LOIRE.—LIMAGNE D'AUVERGNE.—PUY DE DOME.—MONT DOR.—PIC DE SANCY.—LE PUY.—M. BERTRAND.—VALLEY OF THE RHONE.—TURIN.—MANTUA.—PALAZZO GAZOLA.—VENICE.—TRIESTE.—CAVERN OF ADELSBERG.—CAVE OF MADDALENA.—PROTEUS ANGUINUS.—IDRIA.

"July 4th, 1835.—Left London at 5 A.M., in the Lord Melville steamer, in company with W. J. Hamilton, Esq. We arrived at Calais at 5 P.M., and at 6 departed for Paris, which we reached at 4 A.M., the 6th July.

"We afterwards sallied out in quest of some of the geologists and other savans, to whom we had letters. We proceeded first to the Museum of the Jardin des Plantes, and after a rapid survey of the collection of fossils, went to the Ecole des Mines to call on M. Elie de Beaumont, who is one of the professors at that establishment. Not finding him, we proceeded to the Serbonne, to call on M. Constant Prévost, but found he had gone into the country. His abode is *au quatrième*, at the very top of a staircase in the Serbonne, exactly resembling those of the Oxford colleges. Nothing is more remarkable

* *S. M.* Subsided Mass. *C.* Cavern. *F.* Funnel. *R.* River, with *Proteus anguinus.*

than the unpretending abodes in which French gentlemen, even of the highest distinction, are satisfied to reside. We next visited M. Boué, whom we found in an equally humble, and yet equally exalted residence, surrounded with books, specimens, and other objects of study. He received us most affably, and I was by no means sorry to find that he spoke English very fluently, having passed some time in examining the geology of England and Scotland. We then called on M. von Buch, and had not been there many minutes before M. Elie de Beaumont accidentally entered. The conversation of course turned upon geology, and both these eminent men expressed great interest in our intended expedition to Asia Minor, each of them recommending to our attention those departments of the subject in which they were most interested; M. Elie de Beaumont laying great stress on the direction and age of mountain chains, and M. von Buch on volcanic phenomena and craters of elevation.

"July 7th.—The forenoon was passed in seeking for various persons, books and things, without much success; and after an early dinner we visited the gypsum quarries at Montmartre, where we succeeded in finding the bed of small oysters which caps the freshwater marls, and which I had not noticed when I visited the same place five years before. In the evening we called on M. Virlet, who, with M. Boblaye, published the volume on the geology of the Morea, which forms a part of the scientific survey of Greece, now publishing by order of the French Government. M. Virlet is an intelligent man, who speaks French more distinctly and intelligibly than the generality of his countrymen. He gave us some useful information, both on geology and other subjects connected with Greece and Turkey.

"July 8th.—I paid a second visit to M. Boué, who among other things advised me to visit a section near Passy, where the calcaire grossier contains a subordinate bed of lignite, proving that the lower freshwater of Cuvier is in reality contemporary with the calcaire grossier, and not a distinct formation as Cuvier supposed. In the afternoon we went to Passy, but failed in finding the lignite.

"July 9th.—Went to the Museum at the Jardin des Plantes, and took a hasty glance at the bird-gallery above, where I had just time enough to be tantalized by the glorious scene, without being able to enjoy it. It by far exceeds all other ornithological collections which I ever saw, not only in extent, but in beauty of arrangement. Every specimen has a neat label attached to it, stating the name and synonyms, the habitat, sex, and donor. Is there any room to hope that the managers of English public museums will ever see the necessity of imitating this laudable practice? There is, however, one defect in the labelling, which indeed belongs to all French, and I am sorry to say, some English works on natural history: the absurd, unscientific, and unnecessary vernacular names are placed foremost on the labels, and the Latin or systematic ones are only added as accessories, instead of being used exclusively, as they ought to be.

"In the afternoon I called on M. Deshayes, and gave a glance at a few of the drawers which contain his magnificent collection of shells. It is arranged zoologically, but the species of each genus are placed in geological order, the recent ones first, and then the fossil in the order of the formations to which they belong. M. Deshayes showed me some of the secondary fossils figured in Boblaye's work on Greece, which are of very rare occurrence in that country.

"July 10th.—Went to the Jardin des Plantes, and called on M. Cordier, to whom M. Virlet had given us a letter; M. Cordier directed M. D'Orbigny to show me the collection of rocks which M. Virlet had collected in the Morea. They were placed for the present in one of the vast warehouses attached to the establishment, in which are very extensive collections made by different persons, and placed here till room can be found for them in the general arrangement. They are erecting a very splendid building on the east side of the Jardin to be exclusively devoted to geology, and when finished, the geological collections now in the Museum will be moved thither, and the latter building confined to zoology. M. Virlet has evidently bestowed much labour in forming his collection of specimens in the Morea, which are all labelled and catalogued in a superior style. I had only time to look at the secondary rocks, the variety of which was very great, and the most part of them quite unlike the secondary rocks of Northern Europe, but rather resembling the primary and transition rocks. This is probably owing to their having undergone great alterations by heat, and to the same cause may be attributed the very great scarcity of organic remains in these deposits.

"All our arrangements being completed, we at last took our departure from Paris at 1 P.M. The *road* offered nothing worthy of notice till we reached Fromanteau*, beyond which it descends into a wide valley by an inclined plane on the top of an artificial bank; the river Seine flows on the east side of this valley. Beyond is a bare and open country, with occasionally rows of lopped elm-trees, like those in

* In the valley south of Fromanteau examined a gravel-pit by the roadside, in the flat, on the west of the Seine. The gravel is mostly chalk-flints and rolled fragments of tertiary limestone, &c. It is minutely stratified in layers of clean washed sand and gravel of various degrees of fineness, and much resembled the gravel of the Avon. It contained some large erratic blocks of sandstone, apparently the same as at Fontainebleau, some of them nearly a ton weight. No shells or bones were noticed in it, but it is a very likely place for either. This diluvium is quite different from that which occurs on the *hills* in the neighbourhood.

Worcestershire. A few miles miles beyond Essone the road suddenly opens on a view of the Seine, which makes a fine sweep, and diversifies the monotonous prospect; this soon disappears, and the corn-fields and elms return in all their perfection of dulness. We met several carts loaded with wool; the dust and monotony was most agreeably and suddenly relieved by our entering the forest of Fontainebleau, where, instead of endless and open corn-fields, we found ourselves surrounded by magnificent timber trees, chiefly beech and oak. Our postillion seemed to possess more love of natural beauties than is usual with persons of his profession, and his eyes sparkled with pleasure as he called our attention to the scenery of the forest: a herd of Alderney cows which enjoy the right of pasturage were returning home for the evening, and he took great pride in showing us one which belonged to him, and which he seemed to think the flower of the flock; at his recommendation we turned out of the main road to a place called Belle Croix in the heart of the forest. Near here the romantic sandstone rock is quarried for paving-stones to supply Paris, and in one of these quarries occur the remarkable siliceous sand-crystals, which are often seen in collections. After visiting this place we returned to the main road, which was still surrounded by forest, and reached Fontainebleau at 9; we resolved to stay and have some refreshment. I found that I was fast leaving the realms of comfort and cleanliness: having called for some water to wash my hands, I was shown into the kitchen, and a copper saucepan of cold water was placed for me on the table, close to where a man-cook was employed in trussing a fowl. Having accomplished my ablutions, we supped and proceeded. We reached Nemours in the dark, where there is a curious old castle, the appearance of which when seen by moonlight was solemn and majestic: I think it was near Neuvy-sur-Loire that I first saw a woman spinning with a distaff, a practice which is universal further south, where a woman is rarely to be seen tending her sheep or pigs, without at the same time employing her fingers in spinning or knitting. We had now entered the valley of the Loire, which is here in many places a quarter of a mile wide, but very shallow, with extensive beds of clean washed sand and gravel. From Fontainebleau the tertiary strata continue to Neuvy-sur-Loire, where the chalk appears. Beyond Neuvy the road descends towards the level of the Loire, and a bed of blue clay appears in a pit close to the road on the left. This clay contains some nodular masses of iron claystone in which a few fossils occur, one of which resembles *Rostellaria parkinsoni.* The clay itself appears to contain no fossils, but from its situation with respect to the chalk, and from the Rostellaria which occurs in it, I conclude it to be gault. The chalk appears to follow the top of the hill on the east parallel to the Loire, and after passing Cosne, the road reascends the chalk, and follows it to within about two miles of Pouilly, where on both sides of a valley near a village called Liberquey (or something like it), the Kimmeridge clay appears with vast numbers of *Gryphæa virgula*. On the south side of the valley towards the top are some beds of whitish stone, in which we found a Trigonia and *G. virgula*. At Cosne there is a fine view of some hills on the opposite bank of the Loire, including the château of Sancerre, on the summit of one of them. On descending to Pouilly there is a very extended prospect, and the country is much superior to the northern parts of France. Vineyards are now become almost universal on the sides of the hills. Beyond Pouilly the ground is flat, and no rock appears till about five miles, when a quarry occurs on the left of the road, of a very coarse soft pisolite of a pure white, containing hardly any fossils. Oolitic beds extend from hence to La Charité, beyond which place a whitish limestone forms the hills for some miles. This limestone has few or no fossils, and no trace of oolitic structure.

"On descending to Pougues, a town in the bottom of a beautiful and extensive valley surrounded with vineyards, and having also some mineral waters, a blue marly bed appears, in which I found a striated Modiola. On reascending the hill of Pougues, which commands a very rich and beautiful prospect, the same whitish limestone forms the hills for some miles. It is purely sedimentary, and has no traces of oolitic structure. It is not abundant in fossils, but contains a discoid Ammonite, an Avicula, a peculiar Echinus, and two or three other fossils. The road is mended with weathered pebbles of this stone picked off the land, which are more abundant in fossils than the stone which occurs in the quarries. The same limestone appears to form the hill in the neighbourhood of Nevers.

"We stopped for the night at Nevers, a town on the Loire of considerable trade; there are some iron mines in the neighbourhood: it is singular that at Nevers, which is in the very centre of France, vast numbers of anchors are made, which are transported to various parts of the coast by land or water-carriage. The hill south of Nevers is composed of a coarse oolitic brown limestone resembling the cornbrash. It contains *Pecten vimineus* and another large species, besides several other fossils. This formation apparently continues to about a mile north of Magny, where a good section is exposed of a stratum of blue clay on each side of the road. This clay is apparently lias (in which case the oolite last mentioned would be lower oolite), but it may possibly be Oxford clay, and the incumbent oolite south of Nevers would then (as *Pecten vimineus* seems to show) be coral rag. The blue clay aforesaid contains many fossils, including a slender Belemnite,

Crassina lurida, an Arca, two Nuculæ, Avicula, and a muricated Turbo with a very tertiary aspect. From Nevers to St. Pierre le Montier the country was more like England than any part of France I have yet seen: the enclosures are small, with a fair proportion of pasturage, and the white-washed cottages and thatched roofs almost made me believe myself in England again. As we approached Moulins, we were struck by the singular bonnets worn by the women: they are made of straw, curved upwards behind, exposing the back of the head. At Moulins we crossed the Allier, a river running into the Loire, and continued to travel parallel to its general course for some miles. At first marly freshwater deposits appear, which are succeeded as you ascend the hills by vast deposits of tertiary sand and gravel, composed almost wholly of white quartz pebbles. Chatel le Neuve is on a high hill, composed of this gravel, which contains so little argillaceous matter as to be in some places almost destitute of vegetation. About four miles beyond Chatel le Neuve, the road descends into a deep valley, and near a house on the left was a pit in the side of the hill, where about twenty feet of greenish tertiary clay and marl appeared to be covered by the gravel, of which the upper part of the hill consists. From this to St. Pourçain are marly beds, and at the latter place we first found the curious *indusial* limestone. It occurs in a bank by the side of the brook, composed chiefly of blue marls without fossils, but containing five or six strata of limestone. The latter occurs in beds more or less regular, from 4 to 12 or 16 inches thick; but it often assumes the form of rounded nodules, coated with concentric layers of travertino. The limestone is entirely composed of indusiæ of Phryganeæ, coated with shells of a species of Paludina. No other shell was noticed except a Helix. The beds of this limestone were from 3 to 6 feet apart, and interstratified with the marl. A stratum of brown oolite also occurs 4 inches thick, the ova very small and beautifully round and regular.

"We were at last rewarded with a sight of the majestic Puy de Dôme in the distance: we were now in the Limagne d'Auvergne, a comparatively low district, which is considered one of the most fertile parts of France. The crops of wheat were exceedingly heavy, though much less forward; very little of the grain was ready to cut, and the greater part quite green. For many miles the road is lined with rows of large walnut trees, grown for timber and for the sake of the oil which is extracted from the walnuts; the oil is used instead of olive oil, but much inferior. Darkness overtook us at Riom, and we stayed there for the night.

"We left Riom and reached Clermont at half-past one, and took up our abode at the Hotel de la Poste, which, for fleas, dirt, and other discomforts, beats any I ever entered; we comforted ourselves that it was a very good and salutary preparation for our Asiatic expedition. The morning had been intensely sultry, and we had not arrived many minutes before symptoms appeared of a thunder-storm; it was preceded by a violent gust of wind, raising such clouds of dust as completely to hide from our view objects thirty yards from us; the rain soon fell in torrents, but it was many minutes before the falling drops had cleared the air of dust; the lightning increased every moment; it lasted about half an hour; we found afterwards that we had arrived at Clermont just in time to escape a storm such as had not been known for many years. As the town stands on a hill, the water soon ran off, and left the streets in an unusual state of cleanliness. We issued forth to call on M. Lecoq, a geological *savant* to whom we had letters of introduction. M. Lecoq is a druggist, and a most sensible, obliging and well-informed man. He is also Professor of Botany in Clermont, and it so happened that he had arranged to take a botanical walk with his pupils that afternoon. We were very happy to join his party, and he took us to two small hills in the neighbourhood called the *Puy de la Crouelle* and the *Puy de la Poix*.

"The Puy de la Crouelle near Clermont consists of peperino, *i.e.* volcanic ashes apparently deposited by water, being obscurely stratified, and much disturbed. It contains fragments of scoria and of tertiary limestone, of which the base of the hill is composed; but whether the peperino has been forced up through the limestone, or rests on it like an outlier, is not easy to decide, but I am inclined to the latter opinion.

"The Puy de la Poix is a very small hillock in the midst of the lacustrine beds. It consists of a peperino, but darker-coloured and more compact than the last-mentioned. A deep trench has been cut in the north side of the hill, and from a small cavity at the bottom bitumen slowly oozes, giving out bubbles of sulphuretted hydrogen. Near this part of the hillock is a very large mass of lacustrine limestone, much shattered, which has been caught up in the peperino, but does not seem to have been altered by the action of heat.

"17th.—We have heard this morning dreadful accounts of the effects of the storm; many mills have been washed away, and several people drowned. We went to the petrifying spring close to the town of Clermont. The water is so strongly charged with carbonate of lime as to have formed a large natural arch over a stream, and is now forming another. Great numbers of medallions and other incrustations are formed here. The objects to be incrusted are enclosed in a wooden chamber, and the water is made to fall on them from the top. Great damage had been done yesterday by the floods, as the place is close by the brook. The chambers had become filled with mud, and men were employed digging out half-petrified branches of flowers, stuffed birds, cabbages, and other rubbish. Among other curiosities we were shown a stuffed cow, with a man by its side, which had undergone the petrifying operation. In the evening we walked to the hill of Gergovia, on which stood the ancient camp of the Gauls, which was attacked by Julius Cæsar. No traces exist of the ancient camp, except an embankment which runs round the brow of the hill. An estate on the east side of the hill still bears the name of Gergovia.

"In going from Clermont to Gergovia, you first ascend a *coulée* of lava on which stands the village of Beaumont, and having passed that, the freshwater marls recur, and continue to about half-way up the hill of Gergovia, where they are traversed by a dike of ancient basalt. Above this the marls contain vast quantities of volcanic ashes, so as sometimes to become regular peperino like that of the Puy de la Crouelle. The stratification of this peperino is quite distinct, and it is evidently of an aqueous deposition, and forms part of the lacustrine deposit.

"In ascending the hill west of Clermont by the road to Limoges, you first come to granite in a state of decomposition, which has been overlaid by a current of ancient basalt. The following is a section taken at the point of contact.

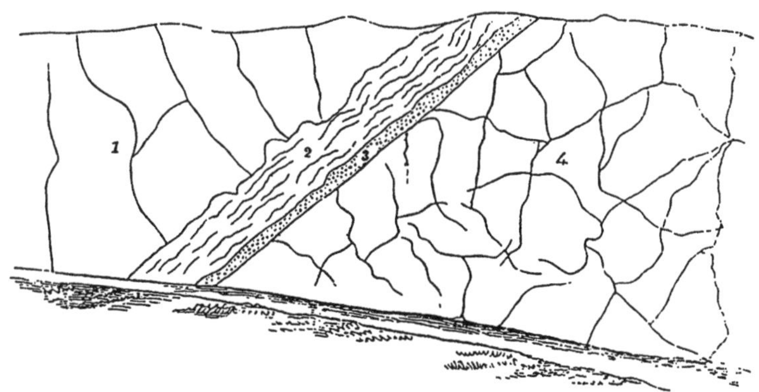

CLERMONT SECTION.

1. Hard but cellular trap.
2. Fragments of scoriaceous lava and clay.
3. Sandy clay.
4. Decomposed granite.

"Above this is a quarry, where the basalt is columnar, the columns perpendicular. The road then enters on a *coulée* of lava which has flowed from the Puy de Pariou. In one place, where the road crosses a bridge, there is a vast quantity of small black cinders in the ravine which separates the lava *coulée* from the granite.

"The Puy de Pariou, like most of the other Puys, consists of sand and scoria of a dull reddish colour. The crater is very perfect, and part of an older crater encircles it on the north. All the scoriaceous Puys are covered with vegetation, but the domitic Puys, from the greater hardness of their materials, exhibit more abrupt descent and projecting masses of naked rock.

"18th.—This morning we set off to visit some volcanic mountains in the neighbourhood. We proceeded by the high road to Limoges, which for at least three miles from Clermont winds up a steep ascent till it reaches the elevation of 3300 feet above the sea, and has then gained the summit of a platform or table-land running north and south, and forming the western boundary of the Limagne d'Auvergne. On this platform are situate the volcanic mountains which are so celebrated. They are scattered in an irregular line from north to south, and are thickly covered with herbage, and in some parts with low shrubs, supplying food and shelter to large herds of cattle. Notwithstanding their present peaceful aspect, the most cursory observation will show that these mountains have been the scene of the most destructive catastrophes. Most of the mountains rise suddenly from the plain in the form of a cone, and many of them have deep circular craters on the summit. With four or five exceptions these hills consist entirely of ashes and cinders. Vast streams of lava have burst forth in many places, and may be traced with the greatest ease winding their way into the valleys below, their rugged and cinder-covered surface presenting a marked contrast to the smooth and fertile soil over which they have flowed. We first ascended the Puy de Pariou, remarkable for the size and regularity of its crater. From the Puy de Pariou we crossed an open heathy country encircled by numerous volcanic Puys, to the Puy Chopine, which, being composed of rock instead of loose ashes, is much steeper and more abrupt than the other cone, and, from the rapid decomposition of some of its rocks, it presents some extensive scars on the east and south. We, however, accomplished the ascent, and remained at the top a short time, enjoying the view and geologizing. I could not find the granitic platform assuming that great regularity which Scrope describes. He says that the line of contact of the granite and the subjacent basalt is straight, but it appeared to us that large portions of granite existed below the general line of the tabular mass. We also noticed some domite on the south-east face, with a dike of trap within it; but Scrope asserts that the domite only occurs on the west half of the mountain.

"We proceeded to the Puy de Nugère, to observe the great stream of lava which has burst from its crater, and flowed in a sort of cascade down its side. In its course it has met with a projecting hillock of granite, which it has completely encircled, and which presents the phænomenon of an island of fertile corn-fields surrounded by the black and rugged lava which must ever remain untouched by the agriculturist.

"The lava is here quarried to a great extent, as most of the towns in the vicinity are built of it, and it is even sent to Paris to make '*Trottoirs.*' Here, as in most lava-currents, the edge of the lava is rounded and steep, leaving a small ravine between it and the granite. The high road to Volvic follows the bottom of this ravine for some distance, and presents a wall of granite on one side and lava on the other. The lava, as exhibited in numerous quarries, is very rugged and scoriaceous for eight or ten feet, which is cleared away to reach the more solid stone below. The latter still contains numerous fine cells, which render it more easy to work than solid lava. I did not see the bottom of the lava in any of the quarries, but in the lava of the Puy de la Bannière, the lower part is highly scoriaceous for some feet where it rests on the granite, and the lowest part of all consists of loose cinders with imbedded fragments of granite. The upper part of this lava is solid. This is exhibited about 200 yards south of Chateau de Tournoël, by the side of the level path leading to it through the woods.

"The ancient baronial castle of Tournoël stands in a commanding situation on a steep hill overlooking the low grounds of Limagne. This castle is in very good preservation, and appears to have been recently inhabited, as some of the painting on the walls is very fresh, and the shrine and table still remain in the chapel. The low circular tower at the entrance is apparently very ancient; its wall is covered alternately with rows of projecting hemispheres carved on the stone.

"On descending from this castle, the granite, which consists in great measure of felspar, is greatly decomposed, and a conglomerate of the lacustrine age appears below, in which the decomposed fragments have been reintegrated into a soft stone, hardly distinguishable from the original granite, except that it occurs in strata with interposed beds of rolled pebbles.

"19th.—Paid another visit to the Clermont Museum, where I stayed some time trying to understand and remember the various puzzling volcanic rocks which occur in this neighbourhood. In the evening we walked to Royat, a small village about two miles from Clermont, in a most romantic valley, or rather ravine, the rugged sides of which are thickly covered with chestnut trees. There is a very curious old church which looks more like an old Norman castle, consisting chiefly of two low square towers with circular arches. In the bottom of the valley runs a stream which turns a great number of mills. It was here that the storm of Thursday last was most severely felt. Some of the mills were entirely swept away by the torrent, and all of them sustained more or less injury. The sides of the stream presented a grievous scene of destruction. The gardens within reach had been in great part destroyed, and were scattered with trees, fragments of mill-work, or large blocks of stone. Vast crowds of spectators were attracted to the spot. Clermont is a considerable town, built on a low hill, and surrounded on three sides by mountainous ranges, above which the magnificent Puy de Dôme is at all times conspicuous.

"20th.—We set off at 5 A.M. to visit Puy de Dôme, the highest of the volcanic mountains in this neighbourhood; we left Clermont by the same tedious ascent which we passed two days ago. We first ascended the Petit Puy de Dôme, an adjunct of its colossal namesake, very irregular in form, but containing a very perfect crater. We then commenced the ascent of the larger mountain, one of the steepest of the Puys, owing to the hardness of the rock which composes it, termed *domite*, whereas most of the other Puys, being composed of loose ashes, assume a more gentle inclination. We at last reached the top, where we stayed some time enjoying the magnificence of the scene, and taking a lesson in geography by the help of our map. Having descended, we proceeded to Pont Gibaud, along the south side of the Cheire of Côme. This is the most remarkable *coulée* or stream of lava in the district. None of the others have preserved their freshness so well as this, and none have displayed the characters of these destructive deluges on so grand a scale. The Cheire of Côme takes its rise from the Puy de Côme, a beautifully regular cone, with two craters on its summit. The lava-torrent has issued from its base, and soon after divided into two streams, leaving untouched between them a large tract of fertile country. We followed the southern margin of the most northerly current. To have shortened our walk by crossing the lava would have been impossible. It is one entire mass of naked rocks, bristling up with every variety of rugged points, and mostly destitute of any vegetable soil. Small trees and shrubs can alone flourish here, by insinuating their roots between the crevices of the lava. The lava has been so imperfectly fluid that its sides are 30 or 40 feet high, and descend with an inclination of about 45° to the plain below. It has flowed exactly in the manner of a thick viscid liquid, entering any depression that existed for a short distance, or turning aside whenever any rising ground offered an obstacle. From the top of the Puy de Dôme the two branches of this lava-current are very conspicuous, presenting two vast undulating districts of a dull grey colour, devoid of all cultivation, and yet surrounded on all sides by corn-fields and meadows.

"The Cheire de Côme (Cheire in the patois of Auvergne means a lava-current), after flowing for some miles over a comparatively level country, poured headlong into the valley of the river Sioule, at Pont Gibaud, and flowed for nearly a mile down the river, which has since made itself a channel between the lava on one side, and the granite of which the country consists on the other. After resting awhile at Pont Gibaud, we set out along the bank of the Sioule, to visit some lead mines at Pranal. The Sioule runs through a deep romantic valley, covered with wood, and much resembles the valley of Waters-meet, near Linton, Devon. At Pranal is a cliff of basalt, in which some singular arch-like cavities have been formed. Below the basalt is gneiss, in which are some veins of lead which supply the mines. The latter consist of only some horizontal tunnels, but we could not get leave to enter them. We returned by a shorter but more fatiguing way over the hill, which rewarded us with a splendid view of the valley of the Sioule. We reached Clermont at 11 P.M.

"21st.—This morning I went to visit M. Bouillet, who has a most beautiful collection of the tertiary fossils of the neighbourhood, as well as of recent shells. The fossils all belong to land and freshwater species, and in his collection of recent shells he has entirely confined himself to those which inhabit land and fresh water. This part of his collection is in three divisions. 1st, the shells of Auvergne; 2nd, those of France in general; 3rd, those of other countries. The land and freshwater shell department is one of the most complete in France. M. Bouillet has lately discovered at the foot of Gergovia, a lacustrine deposit, probably of the pliocene era. Great numbers of Unio, of an extinct species, occur, and from his description it seems to be a case very similar to the fluviatile deposit of the Vale of Avon. He has not yet published an account of this discovery. I regret not having time to visit the place.

"22nd.—Left Clermont at 4 A.M., for Mont-Dore: M. Lecoq had recommended us to take the 'Petite Route' to Mont-Dore, but it had become impassable by the late storm; we were therefore compelled for the third time to ascend the hill west of Clermont, by the route to Limoges. When we arrived near the top, we turned to the left and followed the road towards Rochefort. As we wished, however, to visit some objects on the other road, we turned again to the left, by the village of Laschamps, and followed a bad road which crosses hillocks of loose black scoria and masses of lava. With little variety we reached the Puy de la Vache and the Puy de Lassolas, which are very curious. These two volcanoes stand close together, and are exactly similar in form and appearance. Each of them presents a vast crater, whose depth is equal to the height of the cone. These craters have been at one time nearly filled with lava, and the height it has risen in one of the craters is marked by a horizontal band of scoriaceous lava; when the melted lava had reached this height, the loose cindery sides of the crater were no longer able to sustain the weight, and one side was torn away, the lava rushing like a deluge over the plain below. The remains of the crater now resemble a semicircular theatre, from the centre of which, as from the pit, the lava still appears as if it were flowing. The sides of the volcano are composed of loose red scoria, mostly destitute of vegetation. This description is taken from the Puy de la Vache, and the Puy de Lassolas is its exact counterpart. The lava of the two volcanoes unites into one torrent, and forms an extensive '*Cheire*,' in the crags and crevices of which multitudes of curious plants and insects luxuriate. We now proceeded to Randanne, the country chateau of M. Montlosier, to whom we had a letter, but finding he had gone to Paris, we proceeded on our journey to Mont-Dore. The valley in which the village of Les Bains or Mont-Dore is situated, is very beautiful; at its upper end is the Pic de Sancy, the highest point of Mont-Dore, and at this part of the valley are some rugged and precipitous crags. The valley may be about two miles long, and one quarter wide at the bottom. Its sides are covered with patches of silver fir. Les Bains is a small village, lately become very fashionable in consequence of the erection of a large 'Etablissement Thermal,' on the site of the hot springs. In digging the foundation the remains of some Roman baths were discovered, consisting of some very massive columns and other stones. They appear to belong to the latest period of Roman art, and have more resemblance to Egyptian workmanship. The columns are sculptured in bas-relief, and some of them with figures of combatants, and others with pointed leaves disposed in a scale-like pattern. These very curious remains are set up in the middle of the street, and they can hardly fail of receiving injury*.

"23rd.—In compliance with the attraction which exalted situations constantly exert upon mankind, we set out to ascend the Pic de Sancy, which is 6217 feet above the sea. We first ascended to the head of the valley, which presents much beautiful scenery, and then commenced the ascent of the Pic by a winding path; the ascent is very easy, as these mountains are for the most part of rounded forms and covered

* The white-looking hill at the upper end of the valley of Dore, to the right of the valley D'Enfer, consists of beds of white pumiceous sandy tufa, more or less compact, remarkable for its regular horizontal stratification. This tufa alternates with, and is capped by, thick beds of trachyte. On ascending the hill west of Les Bains de Mont-Dore, a section by the side of the road, near the bottom of the hill, exposes series of white laminated marls, dipping to the west, much like the lacustrine marl of Clermont, but probably subordinate to the volcanic tufa which abounds in other parts of the valley.

with herbage. On the top of the Pic de Sancy is a stone on which are inscribed the height above the sea, and the longitude and latitude. It was a bright sunny day, though a dim haze in the distance prevented our seeing further than about thirty miles. It was therefore needless to seek for the Alps or Pyrenees, both of which in clear weather may be seen. The only striking objects in the distance were the Puy de Dôme, and its accompanying volcanoes on the one side, and the mountains of Cantal on the other. The latter are nearly equal in height to Mont-Dore, and are similar in geological composition, both being volcanic, but of much older date than the hills of the Puy de Dôme. We now descended on the south side, to see the lake Pavin, which lies at some distance. In our way we passed a large mass of snow lying in a ravine directly facing the sun: the nights are so cold in these elevated spots that the snow often remains in ravines the greater part of the summer. Lake Pavin is apparently an ancient crater, being quite circular and very deep; the water is clear as crystal; the sides of the crater are covered with wood, which gives this lake a highly beautiful appearance. We were obliged to return to Mont-Dore by the same route.

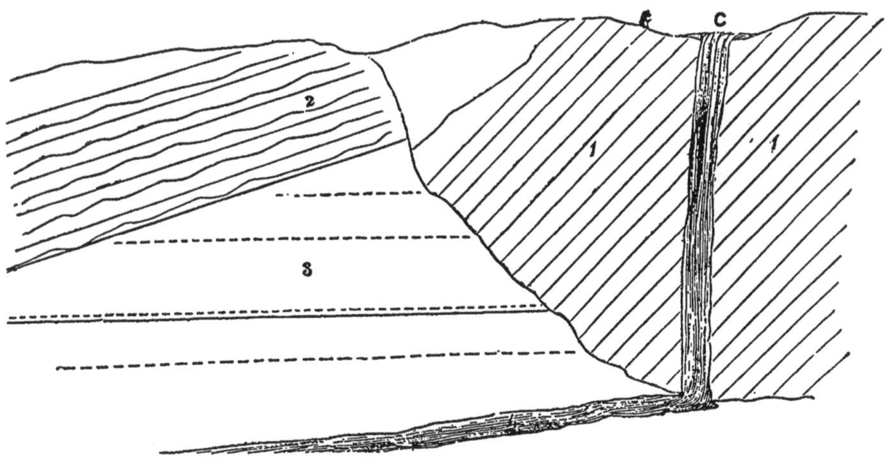

CASCADE OF QUEREILH.

1. Columnar Basalt. 2. Stratified Tufa.
3. Indistinctly stratified Tufa. 4. Rivulet. *C.* Cascade.

"24th.—Visited the cascade of Quereilh, which falls over some rocks of columnar basalt, in a sequestered nook clothed with fir-trees. We then crossed the river Dogne, and ascended the opposite side to lake Guéry, which presents two small cascades at the upper end, falling over basaltic columns. We ascended the hills on the north side of lake Guéry, and found ourselves opposite the two basaltic rocks called Sanadoires and Tuillières: we made our way to the former rock, which rises like a vast cone with a perpendicular cliff, perhaps 400 feet high on the south side, but is accessible on the north. On the summit of this rock the English had a fortress in the fourteenth century. The greater part of this rock is columnar, and consists of a kind of basalt, called *clinkstone,* from the sharp glassy sound it emits when struck. The columns which form the western side are very curious, from their great length, and the singular manner in which they radiate from the apex of the rock, which resembles in form a mighty bell; we descended this rock, and ascended its opposite neighbour, which equals or exceeds it in height. The columns of this rock are for the most part parallel and perpendicular, and the side which faces the Roche Sanadoire affords an admirable specimen of basaltic columns. The columns split across into thin flakes, which are quarried and used as slates for roofing, whence its name of La Tuillière. Having satisfied our curiosity, we returned to the hill west of lake Guéry, and then proceeded over a bare open country, grazed by vast herds of cattle, at the back of Puy Gros; whence we descended to the valley of Bourboule. We here met our landlord, old Cohadou Bertrand, at whose hotel at Mont-Dore we were staying. Though a man of some property, he still retains the dress and manners of an Auvergnat peasant. He was busy with his hay, and crossed the hay-field to show us the way to the cascade of Vernière, a small but picturesque waterfall at the bottom of a deep glen, densely clothed with fir-trees. We now ascended this valley, called La Scie, which contains some magnificent silver-fir-trees. Crossing the hill which divides this valley from that of Les Bains, we returned home at 7 P.M.

"25th.—Having prepared for our departure from Mont-Dore, we ascended to the cascade above the village, which falls in a clear sheet over a hanging cliff about 30 feet high. Many saxifrages and other curious plants grow in the damp rock behind the fall. Having gained the summit above the fall, we passed on till we reached another beautiful valley called Chaudefour. The rocks at the upper end assume a pointed form, and are called Les Aiguilles. Before we reached Chambon, we sat down on a projecting knoll to enjoy the splendid view presented by the upper end of the valley. The steep sides were covered with trees, and rugged or pointed crags projected through their foliage. Several ravines furrowed the mountains, in each of which a rivulet formed a succession of glittering cascades. A ravine on the opposite side of the valley added to the beauty of the scene, by the lovely red and yellow tints of the granite which composes its sides, forming an admirable contrast to the grey basaltic rocks that frown above it. We stopped at the village of Chambon, and then followed the lake of that name to the Puy Tartaret, a volcano apparently of very recent origin. It is composed of loose red cinders, and is almost bare of vegetation; it has been thrown up directly in the middle of the valley, and by damming up the river Couze, has formed the lake of Chambon. It has given out a stream of lava which has flowed down the narrow valley of the Couze for thirteen miles as far as Neschers, where there is a hot spring rising out of granite, but yet deposits abundance of carbonate of lime. The Puy Tartaret has probably burst out exactly in the middle of the valley of Chambon, in consequence of that being the weakest point of resistance, the granite hills on each side furnishing a pressure which was wanting on that point. Puy Tartaret is commonly described as having three craters, but to us it rather appeared that the highest point, which is nearly central, has been a more recent cone, thrown up in the middle of an older crater, portions of which still encircle it, and now form three cavities, which are not circular, but oblong or curved.

"We passed on the south side of lake Chambon, and fancied we could see on the opposite side, a lava *coulée* which had burst from the side of a granite hill facing Puy Tartaret. M. Croizet, however, says that this is only a landslip. We followed the road to St. Nectaire, passing over the lava-current, which is covered with hillocks of scoria and lava in fantastic forms. There are some hot-baths at St. Nectaire, the water of which contains a good deal of salt, and hence the *Plantago maritima* and other sea-plants grow on the rocks near the springs. In the evening we ascended a rock at the back of our hotel, to see a Druidical cromlech; it consists of a block of stone about 10 feet long, 6 wide, and 18 inches thick, supported on four stones; there has been a fifth, which has fallen.

"26th.—Set off in a wine-cart of the rudest construction, which carried all our goods. Followed the course of the Couze, which flows at the bottom of a narrow granite gorge several miles long. The stream of lava from Puy Tartaret has flowed down this ravine, and risen to the height of 30 or 40 feet above the present rivulet, as is shown by fragments of lava still clinging as it were to the sides of the granite. The valley of the Couze is remarkable for the great denudation of the lava *coulée* of Tartaret, which terminates just below the town of Neschers, thirteen miles from its source. The main body of the lava has been removed by the denuding action of the rivulet, which is often swelled into an impetuous torrent. The extent of this denudation is well displayed near Verrières, a small village, which in winter never sees the sun. We reached Neschers at 10, and called on the Abbé Croizet, the *curé* of the place, a great geologist, who has an admirable collection of fossil bones from the neighbourhood, both

Eocene from the lacustrine, and Miocene from the alluvial beds. Also some very perfect specimens of fish from the tripoli basin of Menat. He has found bones of Felis, and other genera supposed peculiar to the more recent tertiary, in the Eocene beds of Auvergne, especially near Gannat. He says that molars of Cervus differ from those of Taurus in this: the latter has the tube in the medial groove long enough to reach the surface or crown, but in Cervus these tubes are only incipient. He possesses 150 species of fossil mammalia.

"The alluvium on the top of the hill, north of Neschers, consists of rounded boulders of granite and various volcanic rocks, interstratified with volcanic or pumiceous sand. M. Croizet has found various bones in it; I could find no shells in it, and recommended him to search for them. After showing us his beautiful collection, the Abbé supplied us with hammers and chisels, and accompanied us to some quarries in the neighbourhood. In addition to his scientific pursuits, he spoke much of politics. After becoming deeply imbued with geology and good cheer we remounted our elegant vehicle, which conducted us to Issoire, where we stayed the night.

"27th.—Set off to visit Mont Perrier, a hill famous for the beds of volcanic detritus, which alternate at different heights with alluvial beds containing bones, as well as for other curious geological phænomena. Some of the beds of volcanic breccia, which form a steep craggy cliff above the village of Perrier, have been hollowed out into caves, some of which are still inhabited, and are entered by means of ladders. On my return we departed from Issoire, and after passing a small coal-field, of which we saw the steam-engines at a distance near St. Germain Lambron, we arrived at Brioude, where we passed the night.

"28th.—We started at 4 A.M. and passed through a very hilly country which furnished many extensive views; the road wound for many miles among volcanic hills; but much less distinctly defined than the beautiful circular cones of the Puy de Dôme. Basaltic columns are frequent in the neighbourhood, and in many places are set up for posts to prevent the carriages going over the bank. After we caught sight of the Chateau de Polignac, the road turns through the hills to the right, and is carried along the side of a steep and high hill till it descends to the town of Le Puy. On the right is a quarry, where is a vast mass of beautiful basaltic columns, straight, parallel and upright. These columns are broken into regular lengths, and used for building walls, in which the hexagonal ends of the columns have a singular appearance in the side of the wall. They are also used, when longer, for bridges over ditches, &c. The view of Le Puy from this point is very singular. The town is built on the southern base of an isolated hill, crowned with a precipitous rock, at the foot of which stands the cathedral, overlooking the town. On the left of this hill, in the bottom of the valley, rises a most singular rock, exactly the form of a sugar-loaf, its height being much greater than its width. The top is just large enough to support a church, dedicated to St. Michael, who seems to be the patron of all precipitous rocks. The church is very old, and has a high tower, which appears in momentary danger of falling. The women in this part of France wear small round black hats, exactly the size and shape of a soup-plate.

"We arrived at Le Puy, and put up at the hotel kept by Berjat, which is considered one of the best, but, like all the others in this part of France, is dirty in the extreme. Many Roman relics have been found in the precincts of the cathedral.

"We returned to Le Puy and called on M. Bertrand de Doué, a geological friend of Mr. Murchison's. He lives in a chateau, formerly an old abbey, in a beautiful situation on Mont Doué, overlooking the Loire, about three miles from Le Puy. M. Bertrand has a good collection of bones of Anthracotherium, Crocodile and Tortoise, from the freshwater limestone near Le Puy. A skull of a Hyæna in scoria from St. Privat d'Allier, vegetable remains from "arkose" at Chartreuse, near Brives, including a dicotyledonous leaf; also lignite and leaves from the 'Terrain de transport.'

"The following is the succession of beds in the neighbourhood of Puy, according to M. Bertrand:—

1. Granite.
2. Psaumite, or arkose, with vegetable remains, chiefly, but not solely, monocotyledonous, and referred to the secondary period. Brives.
3. Red clay and marl, unconformable to the last, with fossils. Lacustrine tertiary. Mont Doué.
4. White marl without fossils. Mont Doué.
5. Gypseous marl with occasional bones.
6. Limestone with bones and freshwater shells. Le Puy.
7. 'Terrain de transport,' unconformable to the last, with lignite, bones of quadrupeds, leaves, and at Cussac freshwater shells of living species. N.B. This is properly a pliocene lacustrine deposit.
8. Volcanic deposits.

"M. Bertrand accompanied us to the top of the hill above his house, which is steep and rocky, and planted with Scotch firs, which are cut for fuel at certain times, and are the only produce which can be raised on these barren hillsides. After taking an early dinner, we visited the Rocher range, a vast rock of lava rising like a column out of the side of a hill composed of granite. It is in fact the *cast* of a volcanic chimney or orifice in the granite which has been filled with lava, and the granite has decayed and been removed, leaving the harder lava projecting.

"The Museum at Le Puy contains a collection of volcanic and other rocks of the neighbourhood. Some good specimens of bones from the lacustrine beds at Puy. Some singular insects resembling Oniscus, found by M. Bertrand in the same. Bones of Hyæna and Deer, from Privat d'Allier, found below a mass of scoria, resting on an old bed of basalt. Some of the joints and vertebræ are still united, which remarkably distinguishes them from the bones commonly found in alluvium, and proves them to have been suddenly killed by the volcanic eruption which has covered them,—not by Hyænas, as M. Bertrand thinks. Lignite and leaves, and a fossil chrysalis from the 'Terrain de transport;' also bones of Horse at Taulhac, of Rhinoceros and Cervus at Cussac, with five species of shells, viz. *Cyclas fontinalis, Cypris faba, Cyclostoma maculatum, Lymneus ovatus,* and *Planorbis hispidus.*

"30th.—Called on M. Aymard, who has a good collection of land and freshwater shells, of which he gave me many new species. He has also a considerable collection of fossils and antiquities, especially of Gallic stone axes or '*Celts*,' of which great numbers are found in this part of France. We remained some hours looking at M. Aymard's curiosities, and then went to call on M. Robert to see his collection, but he was out. M. Bertrand called on us this morning to show us the Museum; it is supported by Government, and contains a multifarious collection of objects, among which are several local curiosities of considerable interest. There are several Roman inscriptions which have been found in the neighbourhood, and some curious old bas-reliefs found in the precincts of the cathedral, representing hunting pieces, supposed to be Roman, but which, from the cross-bow being represented in them, I should suspect belong to a later age. There is also a good collection of medals, chiefly Roman, found in the neighbourhood, and a good series of the minerals and fossils of the country, In the evening we walked to Espilly, to visit some curious basaltic columns called the *Orgues d'Espilly,* perpendicular, of great height, and overhanging the river.

"31st.—I went before breakfast to Cussac, a village about five miles from Le Puy, to visit a curious alluvial deposit, on the property of M. Robert, in which numerous fossil bones have been found. It consists of a succession of beds of volcanic sand, boulders more or less rolled, of granite and lava, and some beds of marly clay. It is chiefly in the clay that the bones occur. It is more decidedly an aqueous alluvium than the generality of volcanic tufa in the neighbourhood, being more regularly and minutely stratified, and the enclosed fragments more rolled. These beds lie against the side of a hill, conformable to its general slope, but unconformable to the strata of lacustrine marl on which they rest. The hill is capped with lava, which probably overlies the alluvium. I found in the marly beds of the alluvium a Cypris, and *Cyclas fontinalis*. Vast quantities of bones have been found by M. Robert, and are in his collection at Puy.

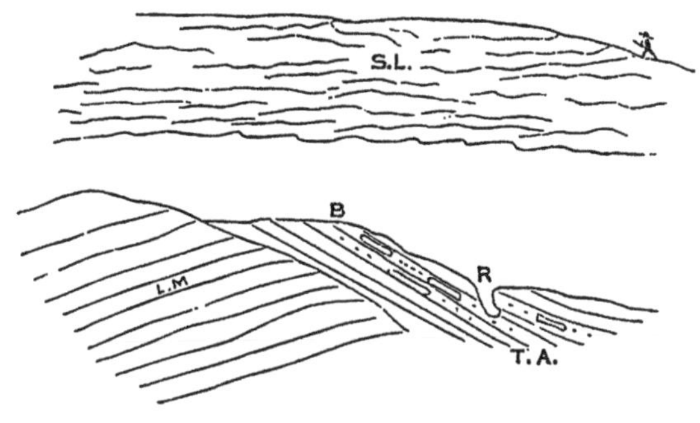

HILL NEAR CUSSAC.

S. L. Scoriaceous Lava.
L. M. Lacustrine Marl.
T. A. Tufaceous Alluvium.
B. Bone-bed.
R. Small ravine.

"In the afternoon we accompanied M. Robert and M. Aymard to a place called Paradis, near Espilly, where some fossil shells and insects occur in the lacustrine marl. When we got to Espilly a thunder-storm came on, and we took refuge under the arches of the bridge, where we found ample occupation in breaking the stones which had been washed down from the cliff at Paradis, insomuch that we remained there the whole time, without going further, and after getting a considerable number of fossils, returned to Le Puy.

"August 1st.—We set off at 5, to visit the valley of the river Sumene, near Mont Doué; we met by appointment our friend M. Bertrand, whose kindness led him to meet us at this early hour to show us some curious basaltic dikes*. Proceeded to the village of Blanozy, to examine some quarries of coarse sandstone, called 'arkose,' used for millstones. We then crossed the river Sumene, and made our way

* About a quarter of a mile south of the road from Brives to Blanozy, and east of Mont Doué, a remarkable dike of basalt rises through the lacustrine strata. It consists of two subcircular masses, connected by an ordinary dike, and rises a few feet through the tillage fields. The most northern mass is nearly circular, and consists of basaltic columns regularly radiating from its centre. The columnar structure of the southern mass is more irregular. The surface of the valley on this side of Mont-Dore is scattered over with large loose boulders of granite and clink-stone.

through some woods of Scotch fir. We redescended to the valley of the Sumene, and followed it to Ayzac, a wretched little hamlet, where the river passes through a black basaltic gorge, the rock of which is columnar, and contains crystals of olivine; we then went to the village of St. Pierre Ayzac, to examine the geology of the neighbourhood. On our return a storm came on, and we witnessed a custom very common on these occasions in this part of France, that of ringing the church bells. This custom has originated in some superstitious notion of the saints interfering to avert the storm, but which is defended on no less absurd grounds, viz. that the vibration of the air caused by the bells, prevents the storm from gathering. On this occasion the bells were rung by two boys, standing on the church roof close to the bells, which were suspended under an arch on the roof, as they sometimes are in England; this practice is chiefly kept up in harvest time, and the bellringer, who is well paid, is obliged to rise at all times of the night, when an occasion for it arises. When the storm was over we set out on our return, but were caught in another at St. Germain; we at last reached the mansion of M. Bertrand; his abode is intermediate between a farm-house and a country gentleman's mansion; some of the walls of the rooms are handsomely wainscoted. After passing a pleasant afternoon with M. Bertrand and his lady, both of whom have been in England, we took our leave with many good wishes on either side; but here a difficulty arose: M. Bertrand's chateau is at a great height, and the road from it very steep and rough, and the night was very dark; so he called his man Pierre, who got some long straw and tied it into several bundles, one of which he lighted, and led the way down the steep zigzag path; we followed as well as we could, and away went Pierre, waving his straw about to keep it burning, and suggesting the idea of a will-o'-the-wisp, whom we were following; however, the descent was at last achieved, and we gained the high road, which brought us home to Le Puy.

"2nd.—Left Le Puy; ascended the hill south of the town, and on looking back had a fine view of Le Puy, rising like a pyramid from the bottom of the valley, and surrounded by the grey and precipitous Roche Corneille; the town looked bright and gay in the sunshine, and contrasted well with the green vineyards beyond, which were studded with small square summer-houses. A little further we left the road for a few hundred yards on the left, to look at the deep valley through which the Loire winds before it enters the basin of Le Puy. A dull country succeeds, covered with small volcanic cones, and the scoria which they have emitted. We stopped for an hour at the small village of Costano, where seventy or eighty peasants were stopping at the auberge; they were migrating from the mountains to work at the harvest in the lower country. Their dress presented every variety of form, colour and material; they were of small stature, and much inferior in appearance to the peasants of Auvergne, who appear to be a very peculiar race, tall with very dark complexions, and features much like those of gipsies. The dress of the latter consists commonly of a coarse blue jacket and trowsers, and a black hat with a round-topped crown, and an immense brim, so large that it often requires a riband attached to the crown on each side to support it. The ploughs used in this part of France are of the most primitive description; they consist of a pointed block of wood with a piece of timber attached to it for a handle, and another in front which is fixed to the horns of two oxen. The man holds the plough with one hand and drives the oxen with the other. After leaving Costano, we traversed a bleak and dreary country to Pradelles, a village where we were compelled to pass the night at a wretched inn.

"3rd.—From Pradelles we ascended a steep and bad road till we had reached the height of about 3700 feet above the sea; to our right was the forest of Banzon. We were shown a solitary little auberge by the road-side, the keeper of which, with his wife, were executed a few years ago for having murdered (as is said) upwards of two hundred of his guests at different times before he was found out. We now followed a road by the side of the Allier, which is here but a small stream. At last we reached the dividing ridge which separates the waters flowing into the Bay of Biscay from those which fall into the Mediterranean, and descended the valley of the Ardêche, the view of which from the top of the valley is beautiful in the extreme. A succession of mountains with jagged outlines form each side of the valley, and a delicate blue haze, which pervaded the air and seemed to give it additional transparency, supplied to each *distance* in the picture a peculiar tint which distinguished it from the rest. The summits of the mountains on each side are occupied with forests of fir; parallel to which is a zone destitute of trees, but which the industry of the natives has rendered profitable by building walls along the steep mountain side, thus forming narrow terraces on which they grow corn or grass, which they irrigate by the numerous rivulets which spring from the rocks above. Lower down are thick groves of chestnut-trees, which always keep at a regular distance from the fir forest overhead. Lastly, at the bottom of the valley, we reached a climate adapted for the growth of mulberry-trees and silkworms; the last were just in the state for winding the silk from the cocoons. The road descends along the side of the valley to the village of Maire, a distance of about three miles. We thence proceeded to Thueyts, and ascended the volcano of Montpezat, which is of considerable height, and composed of loose red cinders and ashes, nearly destitute of vegetation. On the summit is a crater of very regular form. The village of Thueyts stands on a mass of lava which has flowed for some distance in the bottom of the valley. The river Ardêche has since eaten out a deep gorge about 150 feet deep, with a wall of granite on the one side, and of lava on the other. The lava is here in the form of hard black basalt, the lower part of which forms beautiful regular columns. At one place is a deep crevice in the basalt about 3 feet wide, in which a steep staircase has been made, called *L'échelle du Roi.* We descended by it to examine the columns at the bottom, where they rest on the granite. The same ravine lower down is called the *Gueule d'Enfer,* and affords a striking assemblage of precipitous and gloomy crags.

"A short distance before reaching Aubenas, the jurassic limestone commences, resting at a considerable angle against the gneiss. It is best displayed to the right of the road to Privas, on the side of the Coiron, where there is a section of 500 or 600 feet of this limestone. It much resembles the English lias, consisting of beds of blue limestone 6 or 8 inches thick, alternating with shale throughout the whole series with great regularity. It differs from the English lias in the greater compactness of the stone, and in the much greater proportion which the limestone bears to the shale. It contains Ammonites, Belemnites, &c., and the *Aptychus latus,* which I am not aware has before been found in so old a formation (supposing this to be the true lias).

"We now left Thueyts, and followed the bank of the Ardêche to the village of La Baume, where we passed the night. Here is a beautiful castle on a rock which appears to be very ancient, being of a simple style of architecture, and containing numerous arched dungeons and vaults. Among the ruins grow bushes of box, ilex, and many plants which denote a southern clime.

"4th.—Started to the village of Jaujac, three miles from La Baume, to see another lava-stream which has flowed from a volcano called the *Coupe de Jaujac.* The village is built on the top of the lava, the decomposition of which supplies a fine soil which grows magnificent chestnut-trees. Immediately below the village the river has cut a very deep gorge, and exposed beautiful cliffs of columnar basalt. We followed the Ardêche for a few miles further, and emerged from this valley near Aubenas, a handsome-looking town, which we passed on the right, standing on a hill. We now turned north-east, and passed on the left of the hill called the Coiron, ascending a very long and steep road; when we arrived on the summit I enjoyed a distant view of the Alps, which I now beheld for the first time. The view was very striking; the afternoon sun threw a strong light on their western sides, and made them appear much nearer than they really were, while their vast height and the peculiar blue haze which enveloped them, gave them the appearance of an enchanted land, rather than of a region belonging to this world. We descended to Privas, where we slept. The neighbourhood of this town, though warm and sheltered, is very barren, owing to the dry slaty rock which composes the country, and gives to the surrounding hills a grey and miserable aspect.

"5th.—Left Privas, and descended to the banks of the Rhone. This fine river is here so divided by islands, that the whole width of the stream can seldom be seen at once; at present it is very rapid and turbid, in consequence of the melting of the snow in the alpine heights where it takes its rise. The plain east of the Rhone is composed of a vast mass of alpine diluvium, *i. e.* gravel and coarse boulders, in which M. de Beaumont has discovered lignite and shells. A singular practice prevails in the valley of the Rhone. The farmers do not use barns, but make their corn into small clusters of ricks in the fields, and prepare a threshing-floor on the open ground near them. As soon as the corn is got in, the operation of threshing commences, apparently with the intention of threshing up their whole store at once, and keeping the corn in granaries. As the harvest was now over, we saw many of these parties of threshers. From three to five men stand in a circle, and strike exactly on the same point; they strike very fast, and as they were exposed to an intense sun, it must be very hard work. They keep gradually moving round the threshing-floor, which is of considerable size. We followed the bank of the Rhone, and crossed it at Valence, which is on the high road from Paris to Marseilles. Crossed the Isère at Romans; it is a considerable river, and falls into the Rhone at a little above Valence. We travelled all that night, following the valley of the Isère, and enjoying some beautiful scenery by moonlight.

"6th.—At 6 A.M. we reached Grenoble: we still followed the Isère. On the left is a range of steep cliffs, composed of limestone, stratified more or less horizontally. On the right of the valley is a range of mountains of greater height and more rugged aspect, being, in fact, a portion of the Alps themselves. Irregular patches of snow appear in many of the ravines near their summits. The valley abounds in mulberry-trees and vineyards. The vines are here trained in the manner Virgil describes, viz. attached to growing trees planted at regular distances. Virgil describes elms as being employed for this purpose, but those here made use of were maple. Luxuriant fields of maize contribute to increase the verdure of this beautiful valley. We reached the town of Chapareillan near the Sardinian frontier. Here we must have performed three days' quarantine, in consequence of the cholera existing in the South of France, had not the good mayor dated his endorsement of our passports three days back, so as to make it appear that we had remained that time at Chapareillan. We then crossed the frontier, where the officers of the Sardinian custom-house detected the trick, and called the mayor an impostor; yet, as the endorsements of our passports were all right to appearance, they very obligingly allowed us to pass. We had not time to visit Chambery, but turned eastward to Montmeillan, and then southward, and began to enter an alpine valley traversed by the Arc, a stream which falls into the Isère. We followed it as far as Aiguebelle, a neat little town, where we slept: the situation is very beautiful, the valley full of green fields of maize and hemp, and the steep sides covered with woods of beech and fir, above which some snowy and cloud-capped peaks appear. The houses here partly resemble those in Switzerland. The staircases are external, and the roofs project far beyond the walls. The latter, however, are of stone instead of wood, as in Switzerland.

"7th.—Started at 5 A.M., and travelled for upwards of sixty miles in the valley of the Arc, which from the melting of the snow is a turbid and impetuous torrent. Towards evening we reached the village of Lauslebourg, and now commenced the actual ascent of Mont Cenis. The glaciers at a distance appear to differ only in size from the numerous other patches of snow which surround them; one of them above Lauslebourg is nearly four miles in length. When we reached the summit, we saw beneath us a beautiful lake, surrounded with snowy crags, reflecting the light of the full-moon, which had now taken the place of daylight. We descended to the hotel on its banks, where we enjoyed ourselves over a good fire, though in the morning of the same day we had experienced a natural heat far greater than was agreeable.

"8th.—Left Mont Cenis, and commenced the descent to the valley of Suza; on this side the whole descent is accomplished at once, which makes it both longer and steeper than on the other side of Mont Cenis, where Lauslebourg, the village at the foot of the mountain, is already at a very considerable elevation. The road on the Italian side, steep as it is, is laid out with very great skill, and is carried in a sort of curved zigzag down the sides of a small valley or ravine, crossing repeatedly a rivulet which flows down it. The sudden turns are somewhat dangerous, but the whole distance is protected by substantial bulwarks. At one part of the road, where avalanches are frequent, a tunnel is now being constructed, which enters the hill and rejoins the road beyond the dangerous part. About fifty men are constantly employed in keeping the pass of Mont Cenis in repair, and the expense is paid by a toll of five francs per horse over the summit.

As the descent was rapid, the change of climate was equally so; and by the time we reached Suza, we experienced the full heat of an Italian summer. We here examined a triumphal arch of Roman construction, and in very good preservation; the workmanship is however inferior, the bas-reliefs over the arch, representing a sacrifice, being very awkwardly designed. We followed the bottom of the valley of Suza for some miles, and then entered on the level plains of Piedmont, and reached Turin in the afternoon.

"9th.—In the evening Hamilton dined with his friend Forster, the English envoy, and I walked up the hill to the south of Turin, where I had a fine view of the rich vale of Piedmont and the Alps which encircle it on the north. To the left rose the cone of Monte Viso, in front was Mont Cenis and numerous other peaks, and further to the right might be seen Monte Rosa, capped with perpetual snow.

"10th.—Visited the Museum, which contains some Roman antiquities, including some handsome mosaic pavements, and a good collection of Egyptian antiquities, some of the colossal statues being beautifully perfect. The collection of Natural History is very extensive, and is fully equal to that of Paris in the regularity in which it is arranged and labelled. The geological collection (with the exception of the tertiary shells, which are incorporated with the conchological collection) possesses less merit, the arrangement being incomplete and antiquated. The streets of Turin are very clean, regularly built and handsome. Most of them are furnished with piazzas on each side, a very useful protection from the heat of the sun; and almost all are supplied with a clear stream of water, always flowing down the middle, which keeps them clean and increases their coolness.

"11th.—At 11 P.M. we left Turin, preferring to travel in the cool of the night. At Poyrin, the second stage from Turin, we found, to our great dismay, that a portmanteau belonging to Hamilton was missing; the postillion and Hamilton immediately went back, and I remained to look after the carriage: Hamilton went all the way to Turin without success, and as it was now pretty clear the portmanteau had been stolen, it appeared useless to do more, and we were compelled to proceed, very disconsolate at the loss of a number of books and other articles of importance, as well as twelve hours' valuable time. We followed a straight, monotonous road, in a scorching hot sun, and the dust about 4 inches deep. This description will apply universally to summer travelling in Lombardy: we caught occasionally a glimpse of the Alps. The country is exceedingly fertile, producing luxuriant crops of maize, hemp and corn; the vineyards, when enclosed at all, are with hedges of acacia, the various tints of which, mingled with straggling vines, are very beautiful.

We travelled all night, passing through Asti and Alexandria, and at 8 A.M. on the 13th reached the frontier town of Broni. This was a day of trouble and vexation, owing to the regulations of the Italian States respecting the cholera at Genoa. We were now on the direct road through Parma to Ancona. Hearing of the *cordon sanitaire* on the frontier of the Duchy of Parma, we proceeded to the Austrian frontier at Sostegno, where no difficulty was made on the score of cholera, but the officer refused to let us pass, because our passports were not signed by the Austrian ambassador at Turin: we had been told at Turin this was not necessary, having been signed by the Austrian ambassador at Paris; however, it was useless to remonstrate, so we turned about and recrossed the ferry over the Po, which we had passed in the morning. As we were only a few miles from the Parmese frontier, we thought it best to try it, especially as we had procured a certificate from Turin that there was nothing of the kind in that town. With these credentials we were allowed to pass the frontier, but were again stopped at Castel San Giovanni, and compelled to return to Broni. We gave up the idea of going by Parma, and crossed the Po by a bridge of boats, and on reaching the frontier were permitted to pass: we now hoped our troubles were over, and pushed on as fast as we could. We passed Pavia at 9 P.M.; we again travelled all night, and reached Cremona on the morning of the 14th. We were now on the direct road to Bologna and Ancona, which we had left two days before, to avoid the Duchy of Parma; we were therefore rather annoyed at being told at the gate of Modena, that it was 'impossible' for us to enter, having come from Turin, a proscribed town; however, after some parley, the police-officer opened our passports, and they were duly signed, so we hastened along to Bologna, as far as the Papal frontier. But here we were again stopped, which upset all our plans; the Pope's officers were inexorable, and differed from most of those with whom we had hitherto been concerned, in being inaccessible either to bribery or persuasion; we were compelled to return, and a soldier was mounted behind our carriage to see us safe out of the territory.

"15th.—We had been arranging all our plans so as to be at Ancona about the 18th, and sail for Corfu in the monthly steamer. To do this was now impossible, and as it was reported the Austrians intended establishing a quarantine in a few days, there remained only one way, which was by Venice, and this we resolved to adopt. We returned to Modena, and in the evening we took a stroll on the ramparts, which were very gay in consequence of this being the feast of the Annunciation. About 10 P.M. we set out for Mantua, and again crossed the Po for the sixth time; we reached Mantua at 6 A.M., and as we were now less pressed for time, we spent a few hours in lionizing. The only object of interest we passed was Castel Franca, an old fortress, built of brick, with battlements of a singular form. The fortifications of Verona are strong, but not equal to those of Mantua. The greatest curiosity in the town is the amphitheatre, one of the most perfect in existence; it is surrounded by houses, but not in contact with it, and on one side is an open space, which affords a view of its colossal proportions. It has been surrounded with a facing of three tiers of arches, of which but a small portion, the width of four arches, now remains; between this external front and the part of the amphitheatre which now remains, there are three lobbies at different heights, from which were numerous passages, through side arches, into the interior of the amphitheatre. Of these side arches there are two tiers, so that the outer facing was higher by one row of arches than the main body of the building which now exists. The lower tier of arches are now used as shops and dwellings, which detract not a little from the effect. The whole of the interior of the amphitheatre has been restored in modern times, so that the seats are all perfect and nearly new, but the external arches, and the gloomy vaults and passages leading to the interior, are still in their original state.

"We next performed a pilgrimage to the tomb of Pepin, a damp vault which contains a stone trough full of water, and without an inscription. This trough once held a skeleton, now removed to St. Denis, and honoured as that of Pepin. From here we went to a private house called Palazzo Gazola, to see a very magnificent collection of the fossil fish from Monte Bolca in this neighbourhood. There are a vast number of species, all of them beautifully perfect, some of them nearly 3 feet long. Signor Gazola has a magnificent collection of fossil fish from Monte Bolca, including a vast variety of species and genera; also leaves of plants (chiefly marine) from the same place. Crustacea; bones of elephants from Romagnano; several other fossils, including a flat circular species of Patella? in reddish marble, and a collection of recent shells. Verona is, in fact, situated at the foot of the Alps, for immediately on the north of the town the hills begin to rise from the level plains of Lombardy: in the evening we walked up one of these hills, called St. Maltia from the church on the top; we collected several fossils and a young scorpion.

"The marble used at Verona for the amphitheatre and other buildings comes from St. Ambrosio, a few miles north-west of Verona. It contains Ammonites, but not many other fossils, and probably belongs to the oolitic period.

"Immediately north of Verona commences the nummulitic limestone, yellow, soft, of a tertiary aspect, but probably of the cretaceous age. It contains Nummulites, Cerithia and many other fossils. I regret not having time to reach the quarries of Santa Maltia, which are of this formation, about two miles north of Verona. The upper nummulitic beds are well shown in a deep road ascending from the Porta Tirolo to the church of Santa Maltia. The beds have a slight dip to the south.

"18th.—Left Verona, and stopped a few hours at Vicenza. The Olympic Theatre, built by Palladio, is very curious; it is built on the model of the Greek theatres, the seats being arranged in a semicircle; the scenery at the back of the stage represents three streets diverging from the stage, so that the audience can look straight up. The houses on each side of the streets are of carved wood. We ascended the bell-tower by a crazy wooden staircase of 340 steps, and enjoyed an extensive view of the plains of Lombardy, and the Alps rising suddenly out of them; to the south-east was Padua, and on the south the volcanic cones of the Euganean hills. We now resumed our journey, and came to Padua: we stopped at the Aigle d'Or, near the church of St. Antonio.

"19th.—Visited the University, said to contain 1800 students. It is merely a building for delivering lectures, the students lodging in the town. It contains a tolerable museum of natural history and fossils, chiefly from the neighbourhood, and a good collection of fish and shells from the Adriatic. In the afternoon we started for Venice; we passed over a country perfectly flat, and full of vineyards; a few sea-gulls at last announced our approach to the Adriatic. We reached Fusina, the place of embarcation, and were at last gratified in beholding Venice, crowded with domes and church towers, and looking like a city which had just been flooded, standing completely in the sea: several gondolas were waiting for passengers, and we soon deposited ourselves in one; the passage is about four miles, the water clear, and from 4 to 10 feet deep. By the time we reached Venice it was getting dusk, but sufficient light remained to see the ducal palace as we rowed past, one of the most striking buildings in Venice. We were landed at the door of the Hotel d'Europe.

"20th.—Commenced sight-seeing, in Venice no small undertaking.

"21st.—Walked about the streets in various directions, looking at the people, shops and buildings. The canals of Venice, with the exception of the Grand Canal, are not so important a feature as I expected to find them, though very numerous and intersecting the town in all directions; there is at least an equal number of streets, and there is no part of Venice to which one cannot walk on foot. The bridges over the canals are very numerous, but all of a single arch, and destitute of architectural merit. The Rialto is the only bridge over the Grand Canal; it is an arch of considerable span, with two rows of shops, and three footways over it. Every one who visits Venice should ascend the Campanile, or bell-tower of St. Mark. It is the highest building in Venice, and the view from it exceedingly interesting; giving a better idea of the geography of Venice and the surrounding islands than can be otherwise obtained. Two men constantly keep watch on the top of it, to give alarm of fire, and to point out the different objects to visitors. On the south-east is the straight horizon of the Adriatic; and on the east the distant mountains of Istria may be seen, uniting on the north to the Tyrolese Alps, which rise like a lofty barrier, and gradually lose themselves in the north-west. On the west are the flat plains of Lombardy, interrupted by the Euganean hills, a cluster of conical eminences south-west of Padua, which may be distinguished in front of them. In the foreground is the Lagune of Venice, studded with little clusters of buildings erected, some on small islands, and others on piles driven into the shoals.

"In the afternoon we proceeded to the Armenian monastery of St. Lazare, on a small island about two miles from Venice. Its inmates are all Armenians, who come from the East in their youth, and devote themselves to the study of languages. One of the brethren, who spoke very good English, received and conducted us over the establishment. The good monks are constantly occupied in translating the best European authors into the Armenian language; they have also edited a variety of grammars and dictionaries of the Oriental languages. They have a very curious library of MSS., chiefly Armenian, and some of them very ancient.

"In our way back we met some fishermen hauling a net into a boat, and stopped to examine the nature of their game. It consisted chiefly of small cuttle-fish, eaten by the Italians; they also eat all the larger kinds of snails. But what most surprised me was the variety of fungi which are eaten: I have seen in the markets large quantities of Agarics, and Boleti of several kinds mixed together, such as in England would be considered poisonous.

"24th.—We were occupied in preparing for our departure, but found time to visit the church of St. Giovanni e Paolo: I greatly admired some carving in wood, in a chapel on the north side of the church. We had intended visiting several of the manufactures of Venice, but were obliged to content ourselves with one, that of gold chains, which are celebrated for the extreme delicacy of their workmanship. We afterwards visited the Palazzo Treves, which contains many beautiful paintings by living Italian artists; but the gem of the Palazzo consists in a pair of colossal statues by Canova, representing Hector and Ajax. Hector is in an attitude of calm dignity, and is in my opinion much more pleasing than that of Ajax, who is in the act of drawing his sword to rush on his foe. The vigorous action of the muscles is admirably sculptured, but his countenance is rather expressive of rage than of warlike bravery. In the evening we deposited our goods on board the Trieste steamer, and at 9 P.M. took leave of Venice and her hundred isles.

"25th.—Just as it was becoming light we approached Trieste, and as this is a free port, we landed without having our luggage examined: we put up at the Black Eagle, a common sign in the Lombardo-Venetian kingdom since its occupation by Austria. Trieste is at the foot of a high ridge of limestone hills, which run parallel to the sea, and to the north of the town rise precipitously out of it. We passed the Lazaretto, near which I saw a very fine piece of fir-timber; it was 34 yards long, and about 10 inches in diameter at the small end, so that its whole length must have been at least 120 or 130 feet. We pursued our walk along the shore and ascended to Contobello, a village on the hill above the sea, three miles north of Trieste. The limestone here forms a nearly perpendicular cliff, parallel to the sea, and about one-third of a mile distant from it. On the slope between it and the sea are vineyards and olive-groves, the first I have seen. The day became very wet, which interfered with our geological observations*. However, it set all the snails on the move, and I was enabled to reap

* In ascending the hill north of Trieste, on the road to Prosecco, is a quarry of beds of hard calcareous sandstone, of a grey colour alternating with shale, in which I could find no animal remains, except a small fragment of an Echinus. Indistinct fragments of vegetable remains, apparently marine, are frequent between some of the beds of shale. In the sea-cliff between Trieste and a small village in a valley, these beds of stone and shale are very much contorted. In ascending from this village to another further north, with a church on a hill called Contobello, the calcareous grey sandstone above mentioned underlies a greenish micaceous sandstone, softer and more thickly bedded than the last, also without fossils. It dips with a very high inclination to the north-east, and has been overturned. On approaching the village, the dip of the sandstone and olive-green marls that accompany it amounts to about 80°, and it is succeeded suddenly and conformably by the nummulitic limestone which forms the top of the hill. This is well exposed by the sections of a new road, which has been cut in the face of the hill between the village and Trieste. I have since found that the shale rests *upon*, and is younger than the limestone, though at the bridge between Dollina and Balanos, a series of shales, like those above mentioned, clearly underlie the limestone, and pass into it at an angle of about 40° East. The nummulitic limestone, in appearance and texture, is much like the light-grey Derbyshire marble. It contains innumerable Nummulites, of two species, one with many, the other with few spires. Other fossils are very rare; in one part near the top, a small species of Spatangus occurs. This nummulitic limestone runs from north-west to south-east behind the town of Trieste.

a rich conchological harvest, and found many species of land *testacea* that were new to me.

"26th.—Set out to visit the castle and cave of St. Servolo, six miles from Trieste; in our way we visited the cathedral of Trieste, a small old church on a hill overlooking the town. It is chiefly remarkable for having been built on the site of a Roman temple dedicated to Bacchus. The tower of the church has been built on the portico of the temple, of which five or six columns, and the architrave, are still remaining; the interstices of the columns having been walled up, and the upper part of the tower continued on the top of the architrave. Several Roman inscriptions are built into the wall of the church. In a small enclosure near the church is a monument to Winckelman, the antiquarian, who was buried at Trieste. It has been repaired and adorned by the late Emperor of Austria. In the same enclosure are several very perfect Roman inscriptions, found in the neighbourhood of Trieste. We now pursued our walk to St. Servolo; we followed the low ground south of Trieste, between the hills and the sea. All the mountains are of a hard greyish limestone, similar in appearance to the English mountain limestone, though of the same age as the chalk. It resists decomposition to such a degree that the surface of these mountains consists of little but bare rocks, except where a few patches of firs contrive to find a footing. This grey rock gives the country a most desolate appearance. The valleys produce vines, olives and maize. We followed the valley to the village of Dollina, and then ascended the mountain on which St. Servolo stands. It is a bleak and desolate spot, near the summit of the mountain, with a precipitous rock rising above it, crowned with the ruins of an old castle. The staircase which led to the castle is now gone, and it is no easy task to climb up the rock. It is of small extent and much ruined, containing nothing of interest beyond the singularity of its position and the extensive view from it. A few hundred yards from the castle is the cave of St. Servolo. In this damp and dark abode the good saint dwelt for a year and a half, to purify himself from all worldly contaminations. The cave is now a sort of chapel, a shrine having been erected to the saint; it is of no great extent, but contains some good specimens of stalactites.

"We returned by a shorter route to Trieste, which we reached at 8 P.M. Trieste is now a flourishing town; the townspeople speak Italian, but the language of the peasants is Corinthian, or as they call it, *Craintsch*. It is a language of Slavonic root, resembling the Polish; further inland, German is spoken instead of Italian.

"27th.—Set out on a three days' excursion into the interior of the country; followed the great road to Vienna, and ascended the hill east of Trieste; when it reaches the top of the limestone platform, it runs for some miles along its surface. The country generally is very barren, masses of bare rock covering it in all directions. Numerous funnel-shaped cavities exist on the surface, from a few yards to a quarter of a mile in diameter. These cavities are no doubt owing to the subsidence of the ground into caves and fissures, which are very numerous in this limestone. One of the largest of these is near the village of Storia; it contains an open fissure at present. We here found several rare shells and plants among the rocks, including the beautiful red Cyclamen, cultivated in English gardens. We stopped for an hour at Prewald, a town at the foot of Mount Nanos. This mountain rises suddenly from the high table-land to a vast elevation, and we were prevented seeing its summit by the clouds. It is said to be about 6000 feet above the sea. Beyond Prewald the country becomes more verdant and wooded. We reached Adelsberg in the evening, and as it was becoming dusk, we set out to see the cave or grotto, about half a mile from the town. Close to the mouth of the cavern is a most striking and singular sight. The river Pinka running at the foot of the hill, makes a sudden turn, and instantaneously disappears beneath an overhanging rock. Immediately above this is the mouth of the cave, closed by a gate. On entering, we followed a passage for a few yards, and found ourselves in a lofty cavern of great width, in the lower part of which the river Pinka reappears, and after roaring and dashing over the rocks, is a second time engulfed, and appears no more in the cavern of Adelsberg. Up to the year 1819, this vast vaulted chamber constituted all that was known of the cavern of Adelsberg. But at that time an enterprising gentleman, the Ritter von Löwengreiff, contrived to cross the torrent of the Pinka, and to climb up an almost perpendicular cliff, about 50 feet high, immediately above the spot where the river is for the second time swallowed up. His exertions were rewarded by finding the mouth of a cave about a mile and a half in length, running directly into the mountain, and beset with stalactites throughout its whole length. A bridge has been built over the river, and a staircase up the cliff, so that visitors can see it without danger. The cave is of very regular width and height compared to its length, and seems clearly to have been formed by the waters of the Pinka, which, after wearing and dissolving the limestone of this cave, have since found means to escape through some fissure at a lower level. Such was the conclusion I came to from the appearance of the original rock of the cave, where it is not covered with stalactites. It has exactly the appearance of water-worn and weather-beaten rocks, or of ice which has been partially melted, leaving a somewhat honeycombed surface. The breadth and height of the cave vary from 20 to 50 feet. The quantity of stalactites is immense. The length of time which some of them must have required for their formation is very great. In numerous cases the stalactites have united with the floor, forming massive columns 4 or 5 feet in diameter. These are sometimes clustered like the columns of a Gothic church, and in other cases single. Every stage in the formation of these columns may be traced. The incipient stalactite is accompanied by a slight protuberance in the floor beneath; as the process advances the protuberance rises, and the stalactite descends; the one becomes a pillar 6 or 8 inches in diameter, and perhaps 10 feet high; the other takes the form of an inverted cone suspended over it; at last they meet and form a solid column, supporting the roof. Such is the process when the water has a clear fall, but when it only trickles over the rock, its effect is to form a glazing with a rippled surface. In general the two processes are combined, and the result is to produce that extraordinary diversity of form which the stalactites assume. In some cases, where a narrow crack exists in the rock above, a thin sheet of stalactite follows the line of the crack, and hangs from it with irregular contortions, though with a uniform width. In one place a stalactite hangs against the slope of the rock, and has assumed exactly the form and folds of a piece of drapery, the water constantly trickling along its edge and adding to its size. The resemblance is increased by a double stripe of a darker colour running parallel to its lower edge, about 3 inches from it. This curious stalactite, which is called *Vorhang* or Curtain, is about half an inch in thickness, and very delicate and transparent; it is perhaps the most curious specimen in the cave. To describe all the various forms would be endless. The most remarkable are known by German names, given from some real or fancied resemblance; such as the Lion, Waterfall, Pulpit, Garden of Cypresses, Organ, and the Statue of the Virgin and St. Stephen. Some of the larger columnar stalagmites, when struck, emit a deep musical note like a large bell, and the smaller stalactites produce various sharper notes, on which it would even be possible to play a tune. When we had advanced about half the length of the cave to a large apartment called the Riding-school, from a single columnar stalagmite standing in the middle of a clear space, our guides tried to make us return. Being, however, resolved to proceed till we could get no further, we insisted on going on, and were abundantly repaid. Much of the finest part of the cavern is beyond this point, and the termination called Mount Calvary is the most astonishing place of all. The cave here expands into a large chamber, formed by the falling-in of the roof, in consequence of which the floor is piled up with fragments into a considerable mound. Since this event vast quantities of columnar stalagmites of all sizes and heights have been formed, producing the effect of a museum crowded with marble statues and busts standing on their pedestals. At this spot, a mile and a half from the mouth, was found the skull and bones of a beast, allied to the Bear, which is now in the Museum at Trieste. By breaking up the stalagmite with my hammer, I extracted several fragments, and one or two joints of the foot of this beast, which still remained imbedded in the stalagmite. The only other part of the cave where bones have been found is in a chamber called the Tanz-Saal, near the entrance, in the floor of which many animal remains have been dug up. I have, however, no doubt that if search were made, this would be found a very productive bone-cavern. After remaining three hours in the cave of Adelsberg, we emerged from it into the equal obscurity of the night, and returned to our inn.

"As we had a great deal to do on the morrow, we resolved on visiting the cave of Maddalena the same night, much to the astonishment of our landlady: after supper we found our guides, who with lamps conducted us to the cave, through rain, mud and darkness. The guides turned suddenly from the high road to Vienna into a path through some bushes into a crater-shaped cavity, like those we had seen in coming from Trieste. At the bottom of this cavity, the mouth of the cave yawned before us, and we again entered the bowels of the earth. This cave is wider and higher than that of Adelsberg, but does not enter so far into the hill. The stalactites are numerous and splendid, and quite equal to the others. This cavern is terminated by a stream of water, said to be the same as the Pinka, which is swallowed up in the other cave. It is in this stream that the singular reptile *Proteus anguinus* is found; when the water is clear they are not unfrequently seen, but the stream was so muddy that none were visible, and after groping about with my scoop-net for some time, I was obliged to give up the pursuit. On our return to Adelsberg I procured one from the guide, who had three or four alive. They may be kept for a year or two, and require no food, though they will occasionally eat a worm. The only precaution necessary is to change the water often, and keep them from the light, which always renders them uneasy. Had I been on my way home I would have tried to keep my specimen alive, but situate as I was, my only alternative was to put my Proteus in spirits. The Maddalena cave* runs in the line of dip of the strata, and is entered from the top by a funnel-shaped cavity. Both the cave and funnel seem to have been formed by the subsidence of the strata, from being undermined by the river Pinka, which still flows at the bottom.

* See woodcut, p. xxviii.

"28th.—Started at 5 A.M. for the quicksilver mines of Idria. We followed the Vienna road for some miles, and passed a small castle called Planina, near which the river Pinka makes its exit beneath a mountain, after running nearly nine miles under ground. Idria is a large and flourishing village, situated in the bottom of a deep valley, surrounded on all sides with mountains clothed with fir-forests. The scenery around is of quite an alpine character, and extremely romantic. A torrent of a glass-green colour, owing to the melting of the snow, runs through the village, and brings down the logs of fir-timber, used there, or transported to Trieste. Idria owes its prosperity to its quicksilver mines; we employed the daylight that remained in visiting the works above ground, and after dark descended the mine. Some of the quicksilver is procured in a native state in the mine, but the greater part exists in a black slaty ore. The best ore is of a reddish colour, and very heavy. As soon as the ore comes from the mine it is screened, and passed through a succession of gratings, each finer than the last; through these gratings a constant stream of water flows. The ore is thus separated into different degrees of fineness, till a residue remains resembling fine black mud; into this state all the coarser fragments are reduced by stamping. The black ore is stamped by wooden rammers, worked by a water-wheel; a considerable quantity of mercury runs from it in a pure state, the remainder is laid by to be smelted. The smelting-house is removed to a distance from the village, from which such noxious fumes are exhaled into the air as to kill the cattle in the neighbouring fields; the process of smelting is therefore only carried on in winter, when the cattle are not there. The smelting-house is a kind of large still. The powdered ore is placed in shallow earthen dishes, and these are placed in a furnace. The quicksilver by this means is evaporated, and the fumes pass in succession through a number of large chambers, on the walls of which the quicksilver condenses; the vapour passes through five or six of these chambers, and then into the open air. The quicksilver drains from these chambers into a common receptacle, and is then packed in iron bottles or sheepskins, and shipped to all parts of the world, especially South America. We afterwards visited the mine itself, which presents a succession of passages and staircases, descending to the depth of about 600 feet from the surface. The quicksilver is found in only one or two strata of the shale of which the ground consists, and is worked much in the same manner as a coal-pit.

"East of Planina the limestone has an easterly dip, but whether this marks an anticlinal axis, or is only a local disturbance, I cannot say. The limestone continues with little change from thence to near Idria, where, on the hill east of that place, a conglomerate of slightly rolled fragments of limestone and red claystone in a calcareous paste takes the place of the limestone, and apparently rests conformably on it, dipping north-west. Soon after you begin to descend to Idria, a greenish splintery shale rests on the conglomerate, and seems to have been altered by heat; it much resembles hornstone. Various shales occur lower down, and fragments of trachyte? in one place. The limestone reappears at intervals down to the village of Idria. The quicksilver mines of Idria appear to be in clay-slate of the transition period, but this is merely a conjecture. At all events, they are probably below the vast formation of grey limestone which forms the mountains around. Their greatest depth, as stated by the guide, is 136 fathoms. According to his account, the succession of strata here is,—1st, the white limestone at the top of the hill; 2ndly, greenish schist, about 120 fathoms thick, and containing a little quicksilver in its lower part; 3rdly, limestone a few feet thick, without quicksilver; 4thly, black schist, 1 to 7 feet thick, with rich ore of quicksilver in veins, but commonly following the separations of the strata; 5thly, 'Liegende,' a sort of hard, bony, greenish clay-slate, without quicksilver, and which continues without change for at least 100 fathoms. All these beds seem to be conformable, dipping partly north-east, partly south-west.

"29th.—Commenced our return to Trieste by ascending a zigzag road on the west of Idria, and were tantalized by the fogs, which concealed much of the distant mountain scenery. We descended to the town of Wippach: at this place a singular phænomenon is presented, by a river nearly as large as the Worcestershire Avon, which suddenly rises in full force at the foot of a limestone cliff. It does not proceed from a cavity, but gushes out between the crevices of a mass of loose fragments of stone. This may be considered as the fountain-head of the river, as it is not known to enter the hill at any previous point. In this limestone tract, where caverns are frequent, these sudden appearances and engulfments of rivers are very common; and one river of Illyria, the Timavus, was celebrated by the ancients for this circumstance. We left Wippach, and after passing some bare rocky hills, rejoined the road to Adelsberg at Storia, and reached Trieste at 9 P.M.

"30th.—We were all day arranging our specimens and preparing for our departure from Trieste.

"31st.—Occupied in packing and sending our luggage on board. In the evening we visited the Museum of the Accademia Reale: from the hasty glance I was able to give before getting dark, I could see it was, for Trieste, a very good collection. It contains several bones from the cave at Adelsberg.

BAY OF VATHY.

CHAPTER IV.—1835-1836.

LEAVE TRIESTE. — CORFU. — ARGOSTOLI. — CURRENTS OF SEA-WATER FLOWING INLAND.—CRANII.—CYCLOPEAN WALLS.—LIXURI.—ITHACA. —BAY OF VATHY.—CORINTH.—ATHENS.—ISLAND OF PSYTTALIA.— PIRÆUS.—LAND AT SMYRNA.

"At 10 p.m. we embarked on board the Austrian packet 'Vigilante' for Corfu. This is a 'goletta,' and sails every fortnight from Trieste for Corfu and Patras. We set sail at daylight on the 1st of September, and enjoyed a beautiful view of the Tyrolese Alps. We sailed gently along the coast of Istria. Low hills line the coast, and higher mountains occupy the interior. We were shown the situation of Pola at a distance; it contains some beautiful Roman antiquities, which we regretted not being able to visit.

"2nd.—We coasted within ten to fifteen miles of the Dalmatian shore. This country seems to consist of little else than rocky mountains of a most desolate and arid appearance. They descend precipitously to the sea, and present nothing but bare grey rock, apparently the same as the Trieste limestone.

"3rd.—We were now more distant from the shore, which still presented the same character. The coast of Apulia was in sight most of the day. The voyage from Trieste to Corfu generally occupies six to ten days, but having a most favourable breeze, the morning of the 4th brought us opposite the Albanian coast. This is even more rugged than the Dalmatian coast, and seems well to merit Byron's epithet of 'Rugged nurse of savage men.'

"Towards the afternoon Corfu came in sight, and before dark we were running along the north coast of the island, whose green and rich vegetation was a pleasing contrast to the barren crags of Mount Chimarra on the Albanian coast. The moon was nearly full, and exhibited the beautiful mountains round the bay with a calm transparency exceeding anything I had witnessed in a British atmosphere. We anchored in the bay of Corfu about 5 A.M. on the 5th. Corfu and the Ionian Islands are on the safe side of the quarantine system, and hence every vessel that enters the harbour is considered unclean. We waited till six, when a boat approached, and the ship's papers were taken by a pair of tongs and carefully examined at arm's length, till the authorities were satisfied we had not the plague on board. A signal was then made, numbers of shore-boats now approached, and we soon found ourselves in the streets of Corfu.

"We here perceived a change in the dress and appearance of the inhabitants; we had now Greeks for Italians, and the effect produced on landing was of the most striking kind. The Greek dress is much more gay than that of the Italians. The small cap of red cloth is universally worn by the men; they sometimes wrap a piece of cloth round it, and it then becomes a regular turban. From the proximity of Corfu to the opposite coast, numerous Albanians are to be found in the island. Their dress is white like a Highland kilt, often richly adorned. They carry in their belt a dirk and pair of pistols, which add to the wildness and ferocity of their appearance. All the Greeks, no matter how humble their grade, wear moustaches, but shave all other parts of their beard; their hair is drawn backwards and left to accumulate behind.

"Corfu is a strongly fortified town, built on a promontory and protected by several fortresses. One of these, the citadel, is a most picturesque object: a conical and almost isolated rock of great height, strongly fortified and deemed impregnable. The great ornament of the town is the esplanade, used for practising the troops. At the upper end is a magnificent structure, the Governor's house, built by Sir Thomas Maitland. Hamilton being acquainted with Sir Howard Douglas, the present Lord High Commissioner, he called on him, and we were both invited to dine there in the evening. An English dinner and English company was a pleasure we had not enjoyed for a long time, and the kindness of Lady and Sir Howard Douglas will cause us to look back with pleasure on our stay at Corfu.

"Sunday, 6th.—Corfu gave us many proofs that we were approaching southern climes. We saw magnificent Cacti and Aloes used as hedges, and for the first time saw the growth of pomegranates, citrons, cotton and tobacco. The state of the ground also showed the fiery ordeal which it had undergone during the past summer, every blade of herbage having long since disappeared. This would have given the island a desolate appearance, were it not for the olive-trees and vineyards which cover every part of it, and render the scenery most enchanting.

"7th.—Set out to Garuni, celebrated for its scenery. Every part of the island not cultivated is thickly covered with olive-trees, for the most part very old, many of them as large as the pear-trees in Worcestershire, which they much resemble in their mode of growth. Many of the stems are of large size, and very fantastic in their form. It is a peculiarity in the olive, when it grows old, that the stem becomes full of deep oval cavities, which very often extend quite through the tree, and contribute greatly towards its picturesque appearance. The vineyards are at present loaded with ripe grapes of the finest quality. In every vineyard was a peasant armed with a gun, keeping watch over his fruit, in case of its being stolen; they remain night and day in little huts they build of boughs. They will, however, allow a person to gather a few bunches on asking their leave, and for the price of a few grains (a small copper coin the tenth part of a penny) will give you three or four bunches, such as the table of an English nobleman would be proud of. The road to Garuni continues rising till it reaches the top of the hill, which has been cut through for a short distance. On passing this gap a most splendid prospect bursts on the view. For the first time you see the sea on the west side of the island, expanding far and wide, with a beautiful little bay immediately below. On the left is the lofty mountain of Sta Decca. We descended to the village of Garuni, and ascended the hill on the opposite side, whence by a long descent we reached the sea, and rejoined the pass before mentioned. In going from Corfu towards Garuni, the road exposes sections of a breccia-like deposit resting against the flanks of Mount Decca. It consists of breccia of scaglia, of white marls, and of various decomposing and rubbly calcareous beds. I should have referred this formation to modern alluvium had I not found marine shells, apparently tertiary, in one or two places. It extends to a considerable height above the sea.

"On the side of the road east of Garuni is a deposit of grey marls with fragments of shells dipping east, and either running under or abutting against Mount Decca. Below this, descending towards Garuni, are yellow sands, seemingly underlying the marl last mentioned. On the south-west of Mount Decca is a mass of conglomerate, upset, forming a considerable hill with a cliff facing the east.

"The hill west of Garuni consists of stratified conglomerate, dipping about 40° East, and very hard and rocky. It consists of rolled pebbles of scaglia and flint. It presents an escarpment down to the sea on the west. In descending this escarpment the beds about the middle are more calcareous, and the fragments more regular, often resembling scaglia; but at the bottom the rolled flinty pebbles reappear, and at this part the beds dip in various directions.

"8th.—Set out to visit Mount St. Salvador, which is the highest mountain in Corfu; it rises suddenly from the water on the north side of the bay, to the height, it is said, of about 3000 feet. It consists of alpine limestone, dipping (with a few local exceptions) about 40° East. The western side of the mountain is thick-bedded; but on the eastern side, towards the village of Signies, the limestone is in strata of 6 inches to 2 feet thick. Fossils are very rare, but I found joints of Crinoidea and fragments of Ammonites. North of the village of Signies it is overlaid conformably by a series of finely laminated blue shales, containing in some of the beds vast numbers of small bivalves (Venus? or Astarte?) concentrically striated. The only other fossils contained in this formation were Aptychus, and dark-coloured fragments of fish-bones (or insects?). This bed probably runs across the island from north to south. We met with it near a well in ascending from the sea towards Signies. This formation contains many strata of flint, alternating with the shales. On it rest other beds of white limestone and flint, both nodular and stratified, some of it black, and exactly like English chalk-flint. These beds (which in the bottom of the valley of Signies are covered by a vast mass of alluvium full of angular flints, &c.) appear to dip conformably under the mountain north-east of the village; they are composed of white stratified rock. I should have added to the above list of fossils, some conical, irregularly radiated bodies, somewhat like Balani (but the nature of which *I could not determine*). The flints of the series of beds have a great analogy to those of the chalk, but the fossils seem decidedly more analogous to those of the oolitic period.

"The base of Mount St. Salvador is clothed with shrubs and bushes, through which we ascended to Signies, a desolate-looking village, half in ruins. Immediately above rises St. Salvador, on the top of which stands a monastery with apartments for a large number of monks; but the excessive inconvenience of the situation has at last prevailed over their desire of being nearer to heaven than other people, and the few who now remain attached to the establishment reside in the village below. These monks heard of our approach and posted up the mountain before us, and when we reached the top we found the lamps lighted in the chapel, and the old monk waiting to receive us. He was a most hermit-like figure, dirty and ragged, and very old, with a long white beard. The monastery presents nothing remarkable, except the zeal which could build and maintain it in so singular a situation. On the west the low coast of Italy and Otranto may be dimly distinguished. In the north-west are Fano and one or two other rocky islands, and in the north-east are the magnificent mountains of Chimarra on the Albanian coast, the 'infames scopulos Acrocerauniæ' of Horace. On turning to the south-east, the eye still ranges over the endless mountains of Albania. In the south the distant island of Santa Maura appears, and in the foreground is Corfu with its beautiful mountains, valleys and bays. When we had sufficiently enjoyed the prospect, we descended to the sea, and returned to Corfu in our boat.

"9th.—Visited the island of Vido, which lies opposite the town of Corfu, at about a mile distant. This little island was formerly thickly clothed with olive-trees, but when the French had possession of Corfu they rooted up these emblems of peace, and covered great part of the island with fortifications. These the English pulled down, and at enormous expense built fortifications of their own. One of the three regiments stationed at Corfu is always quartered here and employed at the works, which have been in progress ten years, and are yet unfinished. When these are complete, the poor little island of Vido will more resemble the work of art than of nature.

"The island of Vido consists of white scaglia, similar to that on which Corfu is built. It contains smooth Terebratulæ, and traces of other fossils.

"26th.—Left Corfu at 7 A.M. by the Ionian steamer 'Eptanisos,' after a stay of three weeks, during a fortnight of which I had been unable to enjoy the beautiful scenery of the island, or to examine its curiosities, in consequence of a slight fever, which confined me to my room. In the middle of the isthmus stands the fort of Santa Maura, which at present contains an English garrison. The town is a mile further on at the extremity of the isthmus, and boats are towed to it along a narrow canal. The marshy channel between it and the mainland is seldom more than two or three feet deep, which the people navigate in small flat-bottomed canoes, made of a hollow tree, which they call *Monoxyla*.

"As the steam-vessel remained here for about four hours we landed, and proceeded to visit some ancient remains about two miles from the town. We were accompanied by Mr. and Mrs. Bracebridge, who had come with us from Corfu, on their way to Zante and Athens. They had both been great travellers, had visited Syria, and many of the Eastern countries. They had just returned from visiting Norway, when they again started for the East. We passed some fine groves of olive-trees, and went up a beautiful valley on the right, ascending its opposite side, on the summit of which we found the walls we were in quest of. They are of small extent, and of the rudest style of Cyclopean architecture: the stones are not remarkable for their size; their disposition is also remarkably irregular. At one part is the ruin of a small tower or fortress, apparently of later date than the walls below. It commanded a beautiful view of the straits which separate Santa Maura from the mainland of Epirus. At the town of Santa Maura a conglomerate occurs, composed entirely of small limestone and flint-pebbles the size of peas. In ascending from the town to the Cyclopean walls, various rocks of the scaglia formation are met with, including a whitish marl. About 8 P.M. we reached the steamer and resumed our journey. The celebrated promontory from which Sappho leaped into the sea was pointed out to us.

"27th.—At 5 A.M. returning light showed us the town and bay of Argostoli in Cephalonia, where we had just cast anchor. Argostoli is a long straggling town, built on the west side of the bay, which is itself a branch of a much larger inlet. It is a place of considerable trade, principally in wine and currants. On account of the frequency of earthquakes, the houses are seldom built more than two stories high. We called on Major Parsons, the Governor of the island, who had the hospitality to find us an apartment in his own house. We afterwards went to visit the grand curiosity of the place; this is, a stream of water which runs *out* of the sea, and is swallowed up in a cavity of the land. This extraordinary circumstance occurs about a mile north of Argostoli, at the very extremity of the promontory which separates that town from the larger bay. It had been known to the peasants for many years, but it is only two years since it excited the attention of the English. It was first noticed by Mr. Stevens, a merchant of Argostoli, who has erected a mill on the spot. The whole coast here consists of rugged and weather-beaten rocks, between the crevices of which the water was seen to rush in at four or five different points. Mr. Stevens selected the principal of these, and stopped up the remainder. He then excavated the rocks for some distance, so as to allow a much wider passage for the current. He now erected his mill, which has ever since been in constant use, and has well repaid him for his enterprise. The following are the principal facts connected with this extraordinary retrograde current, which appeared to me so interesting a phænomenon that I took some pains to collect them. The channel leading to the mill-wheel is about 3 feet wide, and in the mean state of the tide, the fall is about 3 feet. The usual rise of the tide here is 6 inches, but after southerly winds it rises considerably more. After passing the wheel, the current flows for a few yards in a channel Mr. Stevens has excavated among the rocks, and is partly absorbed by swallow-holes in the bottom, and partly disappears under the rocks at the extremity. The water at the bottom of this excavation rises and falls with the tide, or what is more strictly the case, the greater influx of water at high tides raises its level in a proportionate degree. A small spring of fresh water enters it on the land-side, and when the sea-water is effectually stopped, renders the water at the bottom of the excavation quite fresh in the course of a day. A remarkable fact is stated by Mr. Stevens, that when the sea-water is stopped out, the influx of fresh water raises the water at the bottom of the excavation several inches up to a certain spot, where it stops. This may perhaps be explained by supposing, that, in consequence of its less specific gravity, the fresh water must form a higher column in order to overcome the obstacles it meets with in its subterranean course. The stream at the moment of its disappearance, seems to run immediately under the sea on the opposite side of the promontory, which is only a few yards distant. With the view of ascertaining the direction of the stream, Colonel Brown, Commander of the regiment of Riflemen stationed here, has made an excavation near the spot, in which the salt water has risen to the level of that below the mill, and a slight current is here perceptible. In addition to the apertures already mentioned, there is another, 300 or 400 yards nearer to Argostoli, into which a considerable body of water rushes; it is proposed to erect a mill at this spot also.

"In endeavouring to explain this extraordinary phænomenon, it might perhaps be supposed that the water flowed under the land nearly at the same level to a great distance, till it was gradually absorbed by the incumbent soil, and evaporated at the surface. This hypothesis might perhaps be tenable if the case occurred on a coast of a nearly level continent, where the water might possibly flow for many miles at a short distance from the surface. But occurring as it does in an island, and at the extremity of a steep and rocky peninsula not a mile in width from sea to sea, such a theory can hardly be maintained. The only other supposition which appears to me at all probable, is that an earthquake (a frequent occurrence in this island) has at some period opened a communication between the margin of the sea and the regions of volcanic fire. The water being thus converted into steam, must be supposed to escape through some volcanic vent, of which we know of none nearer than Etna or Vesuvius*.

"About a mile north of the town of Argostoli, near the Lazaretto, a

* An account of this subterranean stream was read to the Geological Society from a letter written at Argostoli, an abstract of which was published in the Society's Proceedings. See List of Scientific Writings.

tertiary bed exists, composed of a yellow calcareous porous rock, consisting of minute particles of limestone cemented together, without fossils, horizontally stratified, resting against the inclined beds of the scaglia formation, and used for building rough walls at Argostoli. Further northwards the beds of secondary limestone, which dip about 30° or 40° to the east, contain an abundance of fossils, including many small spiral univalves, with a very tertiary aspect, though there can be no doubt that they belong to the secondary formation.

"On the evening of the 27th we dined with the mess, where the officers invited us every day during our residence at Argostoli.

"28th.—Visited the Cyclopean walls, said to be those of the ancient city of Cranii, mentioned by Thucydides as one of the four cities of Cephalonia. After following the road to Samos for two miles, we turned up a stony lane to the right, which led to a grove of very ancient olive-trees; as we were admiring their grotesque forms, we suddenly found ourselves close to the walls of which we were in quest. The blocks which composed them are of such enormous size, and their grey weather-beaten hue so nearly resembles the stony masses which cover the surrounding ground, that at the first glance these walls might be taken for a natural ledge of rock. The eye, however, soon detects the mathematical precision with which these shapeless polygons have been fitted together, and the mind feels a sensation of awe at beholding these mysterious relics of an unknown people, whose very name seems to have been lost before the commencement of history. These walls run in a straight line for more than a mile along the eastern side of a hill. Notwithstanding the great labour bestowed on these walls, they must have been very inefficient as a means of defence, as their height seldom exceeds 5 or 6 feet. The accuracy with which these enormous blocks are fitted together is wonderful, and betokens great mathematical skill in measuring their sides and angles. There are not the slightest traces of habitations in the area enclosed by this wall and the precipices which encircle the hill. It is therefore probably only a fortified enclosure, into which the inhabitants took refuge when menaced by their enemies. The ancient city of Cranii, if really in this neighbourhood, was probably on the plain west of this fortification, at the head of the bay of Argostoli, where are some traces of tombs, and till lately an ancient harbour, the stone walls of which were destroyed by the late resident or governor, Colonel Napier, to form an embankment to drain the upper part of the bay. To return to the Cyclopean walls: small square towers of the same height as the walls, project at regular distances of about eighty yards. The large blocks which compose the walls are cut to a tolerably even surface on the *outside*; their inner surface is left rough and uneven as they came from the quarry. A wall formed of smaller blocks, and of less height, runs parallel to the outer one on the inside, at the distance of about three yards, and the interval is filled up with small stones to form a sort of rampart for the men to stand when defending the walls. On approaching the southern extremity, we found a gateway at the bottom of a small valley. A few yards to the south of this gateway there is a remarkable change in the style of architecture of the walls. The stones, instead of being regular polygons, become right-angled, and are placed horizontally. The only irregularity in their structure consists in their height being uneven, and hence they are occasionally *let in* to the subjacent blocks. This circumstance seems to denote that these fortifications had been extended at some later period, when the newer school of Cyclopean architecture had been introduced. Some distance further south is another gateway, and a little further the wall terminates. The hilly parts of all the Ionian Islands which we visited have a great uniformity of character, being covered with projecting masses of naked rock, a very hard white or grey limestone.

"29th.—Crossed the bay to Lixuri, a small town on or near the site of the ancient city Palæa. We found here a very interesting and extensive tertiary formation, containing a great variety of fossils, and we resolved to devote another day to its examination. We crossed over at an early hour, and spent a long and very hot day in examining the several strata and collecting their fossils. The peasants were very liberal in inviting us to partake of their grapes as we passed the vineyards, and the heat of the day made their donations very refreshing. Amidst the vineyards were occasional patches of cotton, which was now ripe, and the pendent heads were bursting, and displaying their snow-white fleeces. I sat down in one place where the ground was literally strewed with tertiary shells, and I employed some of the cotton which was growing within an arm's length, to wrap up these delicate shells*.

"October 1st.—At 1 P.M. we left Argostoli for the opposite side of the island. We crossed the shallow part of the bay by a bridge or causeway (for it was between both), and ascended the naked and steep hills on the other side. After proceeding a few miles we descended into a valley, nearly in the centre of the island. It contains two or three villages, and the monastery of St. Hieronymo. Opposite is the Black Mountain, said to be 6000 feet high, and the highest point in the island. It is composed of white limestone, but receives its name from the forest of pines which clothe the upper part. The road passes to the north of this mountain, and hence descends all the way to Samos.

* See Collected Papers for sections and a description of this deposit, p. 70.

Towards the lower part of the descent we had a beautiful view of the valley and bay of Samos, with the classic isle of Ithaca, rising abruptly from a sea of the most brilliant azure, beyond which were the distant mountains of Santa Maura. We reached Samos as it was getting dark, having come fourteen miles from Argostoli. Samos was once a flourishing city; it now consists of a few wretched huts inhabited by fishermen, and is one of the most unhealthy places in the Ionian Isles.

"2nd.—On rising this morning I was saluted by a beautiful specimen of a paper-nautilus which a fisherman had just caught, and had the wit to bring it to me instead of throwing it into the water. It was occupied by a species of Sepia, which by many is supposed to be a parasite, and not the natural inhabitant of the shell. This I am not able to decide, but can say that the shell was perfectly clean and unbroken, and had no appearance of having been a dead shell picked up by a parasitic animal. The animal when dead fell out of the shell with its own weight, and had apparently not the slightest muscular connexion. When alive it was semi-transparent, and exhibited the most delicate prismatic hues. After I had examined this beautiful and rare production, we went to see the remains of the ancient town of Samos. It stood on the western slope of a hill, the summit of which is surrounded with a Cyclopean wall, forming an acropolis to the city below. This citadel, like that of many other Greek towns, has been connected with the sea by two walls, between which stood the town. The walls which fortify the citadel itself, appear to be more ancient than those which connect it with the sea, the former consisting of irregularly formed blocks, and the latter of squared stones. In both, however, the stones are of enormous size, and the height of the walls is much greater than those at Cranii. The wall on the south side is in some places 20 feet high, and consists of regular and horizontal layers of stones descending the steep side of the hill; it has been traced for some distance along the ravine at the bottom, and then ascends the opposite side to a monastery, parts of the wall of which are formed out of the Cyclopean fortification. The slope below the monastery is thickly covered with evergreens, among which the red berries of the arbutus were very conspicuous. On this side are numerous ancient tombs, many of which have been opened by Lord Nugent, Mr. Rankin and other antiquarians. They are merely graves surrounded with flat stones, and covered over with others.

"The citadel has been occupied in later times by the Romans, who have heightened the ancient walls. They also built towers at several points, which are easily distinguished from the more ancient buildings by the small stones, tiles and mortar of which they consist. Coins, both Greek and Roman, are occasionally found, and we bought a few of some children who brought them to us. On the east side of the citadel is a small gateway, which seems to have been the sole entrance from the country. On the side of the hill below the citadel are the remains of several oblong enclosures, surrounded on two or three sides by walls of irregular Cyclopean architecture; they have probably been dwelling-houses. Interspersed with these are the remains of Roman houses, especially along the sea-shore, where are also the foundations of an ancient pier, visible through the clear water, and enclosing a very small port.

"The road from Argostoli to Samos crosses the secondary limestone formation. In descending from the highest points of this road on the north of the Black Mountain, down to Samos on the east coast, the successive strata of this formation are admirably displayed. They dip regularly about 25° to the east all the way to Samos. The strata are well defined, and may be seen running along the sides of the hills for miles on each side of the road. The thickness of the series here displayed must be immense, as the road crosses the strata for at least six or eight miles. There is very little change throughout the whole series. All is white limestone of various degrees of hardness, from splintery siliceous limestone to the state of hard chalk. Fossils are rare, but near the lower part of the series I found a bed of oysters with small plicæ, about one foot thick, which reappeared once or twice in ascending the series. Also towards the middle I found a few Nerineæ? or Turritellæ?, and in one place Hippurites.

"3rd.—Started early and crossed in a small boat to Aito, the landing-place of Ithaca, where I enjoyed a delightful bathe in a sea whose crystal clearness seemed to transport one, not only into another element, but almost into another world. We then proceeded to the Castle of Ulysses, an ancient fortress so called, built upon the summit of a steep rugged mountain, which must be at least 1000 feet high, and standing on the narrow isthmus which connects the two halves of the island of Ithaca. Its summit contains water-tanks and remains of dwellings of Cyclopean architecture, and is surrounded by a Cyclopean wall of the rudest and most ancient style. The stones are polygonal and tolerably well fitted; but the *outside*, which in the others we have seen is nearly smooth and even, is here very rough. This fortress being very ancient and the only one on the island, may possibly have existed in the time of Ulysses and have been occupied by that hero; but it is most probable that the ancient capital of Homer's Ithaca was on the slope between the foot of the mountain and the Gulf of Molo, protected by the acropolis above and enclosed by the two Cyclopean walls, which may be traced running down the side of the mountain towards the harbour.

"The view from the summit is beautiful in the extreme. On one side is the channel of Cephalonia, and on the other the magnificent bay of Molo; in the distance was the coast of Acarnania. We descended to the bay of Vathy, which projects on the south side from the larger bay. We passed another of these little circular inlets called the bay of Dexia. Vathy is a beautiful bay with a very narrow entrance, and is so called from the depth of its water, which is very considerable. The town of Vathy contains 5000 or 6000 inhabitants, and is the chief place of the island. We were met by the Resident, Major Eden, to whom Mr. Parsons of Cephalonia had written to introduce us. Vathy has some little trade in oil and wine, but its staple commodity is currants, or as here called, *uva passa*. The cultivation is on the increase, and the produce this year is 500,000 lbs., at 42 dollars the 1000 lbs. The whole island of Ithaca is one vast rock of limestone, sloping on every side abruptly into the sea. All the central parts of the island seem to belong to the secondary limestone formation. On the east side of Sehanus Bay it occurs in strata about 6 inches thick, of a red colour, and contains Ammonites and *Aptychus imbricatus*?. It is used at Vathy for paving the streets. It apparently reappears in the opposite island of Jotako.

"Oct. 6th.—Embarked from Vathy in a small boat for Patras. The day was lovely and bright, and the sea as clear as crystal. We were followed for some time by a shoal of large tunnies, which displayed their painted backs and silver bellies in the sun, and could be seen plunging down to the depths of many fathoms. We enjoyed the beautiful outlines exhibited by the mountains of Cephalonia, Ithaca, and Acarnania. We passed between the islet Oxia and the promontory of Serophæ, once one of the islands of the Echinades, but now united to *terra firma* by the alluvium from the river Achelöus.

"We landed at Patras at seven, and felt removed one step further from the civilized world by the consideration that we were now subject to seven more days' quarantine before we could return to Italy or any of the *clean* parts of Europe. In the afternoon we visited the remains of a Roman aqueduct a mile from the town, but too much ruined to exhibit much architectural beauty*. Patras suffered severely in the late revolution; the town was entirely destroyed, and the extensive currant vineyards burned and uprooted by the Turks. It is now gradually recovering, but exhibits on all sides proofs of the havoc of war. On the west side of the town I noticed many remains of Roman dwelling-houses, the solidity of which had survived the ruin which has destroyed the modern town. I here noticed an operation

* At Patras, near the aqueduct, the hillocks of subapennine sand contain several species of freshwater shells mixed with the marine; a fact not noticed by Boblaye.

a proof that we were approaching Eastern climes,—some men making *bricks with straw* and drying them in the sun, which was all the baking they received. They are made in a large mould which makes six at once. We spent the evening with Mr. Crow the British consul.

"8th.—Set off on our journey to Corinth at 7 A.M. The country is densely clothed with beautiful shrubs,—myrtle, oleander, judas-tree and two kinds of arbutus. Planes and wild olives are interspersed, and a species of Pinus resembling in its leaves the Weymouth pine; the assemblage of these formed a beautiful shrubbery of many miles in extent. On the left the Gulf of Corinth, its opposite side lined with magnificent mountains, presented much beautiful scenery, and as we approached Vostitza the country again became cultivated with currants, maize and cotton. Vostitza was formerly a place of considerable trade, but like Patras reduced to ruin in the revolution.

"9th.—Votitza to Avgos. In the evening we reached Avgos, a hamlet of about half a dozen mud cottages: we were lodged in a loft over a stable; the floor was so crazy that we expected to be precipitated among the mules that were munching their hay below. The only furniture was a mat, on which we placed our cloaks and slept soundly.

"10th.—Avgos to Corinth. After some miles of hot and dusty travelling, we caught a sight of the Acrocorinthos, that frowning cliff which gave rise to Homer's phrase—*ὀφρυόεντα Κορίνθον*. We entered on the plain of Corinth, unenclosed and destitute of trees, and thickly covered with a very high and strong species of thistle. Towards evening I entered Corinth, full of excitement at visiting a place of such renown. But the scene which presented itself was most humiliating to human pride. This city, once so celebrated for her power and magnificence, has in later times become an assemblage of mud cottages. The fortress of Acrocorinthos was occupied by the Greeks, the town by the Turks, who, after destroying everything within reach, were blockaded and reduced by famine. We scrambled over the broken walls and heaps of rubbish which render the streets almost impassable, and made our way to the small portion of the town that has been rebuilt.

"11th.—We spent the day in visiting Corinth. The town stands on a platform elevated perhaps 50 feet above the plain, which extends to the sea and to the ancient port of Lechæum. On the south the ground rises rapidly, and terminates in the steep crag of Acrocorinthos, a detached hill of 800 or 900 feet, crowned with a strong fortress built by the Venetians*. The sole relic of ancient art is a Doric

* The Acrocorinthos is of hard scaglia limestone, resting conformably on the west side on beds of red jasper and shale, much altered, and passing lower down (apparently) into a very decomposed igneous rock, called serpentine by Boblaye. On

temple; seven columns are still standing, remarkable for their shortness compared with their diameter, from which cause they are supposed to be very ancient. They are composed of single stones, of a soft tertiary sandstone, and have evidently been quarried on the spot, for about 30 yards further west I noticed a projecting mass of stone exactly similar, the greater part of which has been quarried, and only this small outlier remains. We climbed over the ruins of the Pasha's palace and of a mosque; these contain many fragments of sculptured marble intermixed with plebeian Turkish bricks. The only biped which we met among these ruins was a small owl, which we pursued in vain with stones and missiles.

"We then visited the amphitheatre east of the town, which many travellers have overlooked. It is an oval excavation in the surface of a level platform, and not visible till close to it. It has probably been a small valley of denudation which has been made more regular by art. The strata of stone at the sides being quite horizontal, have furnished ready-made seats, which only required to be in some places smoothed, and where broken supplied with masonry, to form a convenient amphitheatre. Some caves have been scooped out at the upper end under the more solid strata, and are inhabited by peasants. In the evening we ascended the Acrocorinthos, which is garrisoned by thirty-six Bavarian soldiers. No antiquities remain, except a few fragments of columns, one or two of which are of verd antique. On the highest point is a small mosque, round which is a wall containing some very large squared stones, the relics, probably, of some ancient fortification. The view from hence is magnificent, teeming with historic associations. The celebrated isthmus lay almost beneath our feet; beyond was the Saronic Gulf, with Salamis, Mount Egina and many small islets. Towards the south-west is an endless succession of mountains encircling Arcadia, and extending thence to Achaia. On the west extends the Gulf of Corinth like a vast inland lake; and on the north the heights of Parnassus and Cithæron, with Mount Geranion in front, and the humble relics of Corinth lie beneath our feet.

"12th.—Left Corinth and crossed the Isthmus to Calamaki on the Saronic Gulf, the ancient port Salamis. The Isthmus is much less

the north side of the cliff of Acrocorinthos is a slope composed of rolled masses of scaglia, evidently an ancient tertiary shingle beach.

The south-eastern part *at least* of the island of Salamis is of the scaglia formation. Boblaye's map marks the isthmus in the middle of the island as tertiary, but I went as far as the village of Ambelaki, and there it is secondary. The Cynosure is a long narrow reef, formed by some strata of the scaglia which jut out of the water at an angle of about 30°, and dip south-east. They probably belong to the upper parts of the formation, and contain numerous shells, especially a small Mytilus.

level than I expected to find it, and the operation of dragging vessels across it, occasionally practised by the ancients, must have been a work of immense labour. Its surface is undulating, and covered with stunted shrubs. We passed the remains of an ancient wall built across it, and then descended into a ravine, which conducted us into a small plain bordering the Saronic Gulf, on which the Isthmian Games were formerly celebrated. We could see no traces of the temple of Neptune mentioned by Gell and other travellers. A little further is Calamaki, a small port, where we engaged a vessel to take us to Piræus. While waiting to embark, a string of camels loaded with sacks of corn came down to the water's edge, the first we had seen.

"We embarked in a small boat with a very large sail, and at daylight, on the 13th, we turned the point of Salamis and entered the Bay of Piræus, where we landed. We despatched our goods in a cart, the first wheeled vehicle we had seen in Greece, and set off on foot to Athens. We had a fatiguing march, owing to the dust and sirocco wind which then prevailed. This is certainly the worst side to approach Athens. The hills of the Museum and the Pnyx completely hide the town, except the Acropolis, of which we caught a glimpse at intervals: we ascended a small elevation, and suddenly found ourselves close to Athens. The first *coup-d'œil* is somewhat disappointing. Many of the antiquities are buried in the mass of modern buildings. The Acropolis, however, crowned with the majestic Parthenon, and the beautiful temple of Theseus, which stands on a rising ground apart from the town, are highly interesting and striking objects. The town presents a confused mixture of ruined mud cottages and neat whitewashed houses. We issued forth to visit the temple of Theseus, which is too well known to require description. In the late war it received one or two cannon-shots, and was considerably peppered by musketry; but at a little distance these blemishes are not visible, and it resembles more the production of yesterday than an edifice which has braved the battle and the breeze during 2300 years. The white marble of which it is built has contracted a yellowish hue. I noticed in several of the columns that the stones of which they are composed are displaced alternately to the right and left about half an inch, which I attribute to the vibratory motion of earthquakes. The interior of the temple has been used as a church by the Greeks, but is now applied as a museum for the antiquities which are found from time to time, and are immediately taken possession of by the Government.

"Nothing is more curious than the extraordinary preservation of some of the ancient buildings at Athens, compared with the total destruction of others. The most perfect relic of antiquity is unquestionably the temple of Theseus. The Tower of Andronicus, commonly called the Temple of the Winds, has also withstood the effects of time in a wonderful manner. The roof, consisting of wedge-shaped stripes of stone radiating from the centre, is quite perfect, as are the bas-reliefs on the outside, representing the eight Winds, described by Vitruvius: the lower part of this very singular building is under-ground, in consequence of the accumulation of soil on the outside. Close to this building the columns of a temple have been lately discovered, 20 feet long, and standing upright, yet entirely *covered by soil.* The Arch of Adrian is another remarkably perfect structure; it was the boundary between the old town of Athens and a new portion built and adorned by Adrian. It bears on each side an inscription, defining the two portions of the city which it separated. Close to this are the remains of the temple of Jupiter Olympius: sixteen columns are still erect, but of the rest of the temple not a vestige remains, except the level area, which serves to mark its extent. These magnificent columns are of white marble, and nearly 60 feet high; several of them still support their architraves. On the top of one of these architraves is an edifice two stories high, built of bricks and rubbish, once inhabited by a hermit, a second Simon Stylites. The two elegant little columns above the theatre of Bacchus have had a wonderful escape from the risks to which they have been exposed. They have once supported tripods, gained in prizes in the contests of the theatre; they stand close together on the steep slope of the rock below the wall of the Acropolis, and immediately above the central point of the theatre, of which they almost formed a part. The latter edifice has so entirely disappeared, that its very site can with difficulty be traced. The theatre of Bacchus, in which Sophocles and Æschylus of old received their well-earned plaudits, was on the slope of the Acropolis, at the south-east angle. Of the building no traces remain; but the site is marked by the native rock, which is cut perpendicularly 10 or 12 feet deep, in a curved form, constituting part of the outer circle of the theatre. In the centre of this curve is a cave, made into a church by the Greeks, but now broken open and desecrated. At some ancient period it has had a solid floor of stalagmite, which has been cut through in chiselling the inside into a regular form. The choragic monument of Thrasyllus, which once stood in front of this cave, was thrown down in the late war, and its fragments are yet scattered about. The theatre of Herodes is of much later date than that of Bacchus, and is of Roman construction. Its massive arches are a fine specimen of the ponderous and durable style of architecture of which that people were so profuse. The hill of Areopagus, once crowded with temples and justice-halls, is now nothing but bare rock, surrounded on three sides by rugged precipices.

"The Pnyx appeared to me one of the most striking and interesting relics of ancient Athens that now exist. This was the place of assembly of the Athenian *Demos*, that turbulent mob whose caprice is so strongly satirized by Aristophanes: it bears proofs of the most remote antiquity, of a period long anterior to the invention of theatres, which afterwards were often used for the public meetings of the populace in Greek cities. It is semicircular, and bears a distant resemblance to a theatre, the straight side being occupied by the orators and public officers, and the area of the semicircle by the spectators. Like the ancient theatres, it is placed on the slope of a hill; but in the Pnyx this was rather a defect, for the audience were placed on the lower side, and the speaker above; while in all theatres the reverse is the case. In order to remedy this defect, a curved wall has been constructed at the lower side, and the space above it filled with soil. The pressure is on the *concave* side; and as there are no buttresses on the other, it would long since have fallen, were it not for the enormous size of the blocks; *one* I measured was 12 feet 9 inches by 7 feet; another is 7½ feet by 8. Their thickness is from 3½ to 5 feet. They have parallel lines scored round the circumference, apparently as guides to the workmen in levelling the face of the stone. Others of the stones are cut down to an even surface. This wall is the only specimen of the kind now extant at Athens, and is probably the work of the Pelasgi, who are said to have fortified the Acropolis in ancient times, though no trace of their workmanship in that fortress now remains. The *straight* side of the Pnyx is not less curious than the circular. It is formed by the native rock, which has been hewn into a perpendicular wall about 10 feet high, and consists of two wings, which form a very obtuse angle at the centre. At this part is the *bema* or pulpit, a cubical projection of the solid rock, surrounded with steps of the same material. Their state of preservation is beautiful, the angles exhibiting the greatest sharpness and perfection. This mass of rock was doubtless the source from which flowed those brilliant streams of Athenian eloquence which have descended to our days.

"Another instance of the apathy of the Greek Government for the fine arts occurred at the Stoa of Adrian, a beautiful Corinthian edifice. Close to this they lately built a large barrack: in digging the foundations several columns and capitals were found which belonged to the Stoa of Adrian. Instead of rescuing them from their situation, they were buried in again, and the barrack built over these venerable remains. The number of Greek churches in Athens is one of the most singular features of the place; they abound in every part, and must have been ten times more numerous than necessary for the population. The greater part of them are now in ruins. Most of them are very small, and are decorated on the walls with tawdry paintings of saints. In many places around the modern town the soil is almost made up of fragments of ancient pottery similar to the Etruscan, and often containing ornamental patterns. I searched several times in these places for coins, but without success.

"I have as yet said nothing of the Acropolis: it is an oblong rock, surrounded on three sides by steep precipices, on the brink of which stands the wall, enclosing an area of four or five acres on the summit. This wall is chiefly composed of Venetian and Turkish patchwork*, amidst which the solid masonry of Ancient Greece may in several places be observed: on the north side a number of massive portions of marble columns are built into the wall, and these are doubtless what Thucydides alludes to when he mentions, as a proof of the haste with which Themistocles rebuilt the walls of the Acropolis after the Persian war, that many columns and portions of temples were in his time to be seen in the wall. These columns are of marble, and 6 feet in diameter, and supposed to belong to the Hecatompedon, or ancient Parthenon, which was burnt by Xerxes; they have been fluted on the upper half, and plain on the lower. In the wall near them are also some large Doric triglyphs, which probably belonged to the same edifice; they are of a coarse kind of stone which comes from Piræus. The metopes between these triglyphs are, however, of marble.

"In the cliff below this spot is a cavern, which was dedicated to Agraulus. Ancient authors allude to a communication between this cave and the Acropolis above. Recent excavations have brought this matter to light, by discovering a staircase leading from the inner part of the cave through the solid rock to the area of the Acropolis above. The staircase being filled with stones and rubbish, had, till lately, escaped observation. On ascending at the west end of the Acropolis, you come to the Propylæa, a marble building, which extends nearly across that end of the fortress. At the south-west angle there stood till lately a Turkish battery, which was supposed to occupy the site of a temple dedicated to the *Wingless Victory*. On pulling down this battery last year, the pavement of this temple was found entire, and so

* The Acropolis, Areopagus, and other hills of Athens are of a reddish or greyish compact limestone, sometimes brecciated, much contorted and altered. It rests occasionally on a reddish shale, which, from its greater softness, has caused the abrupt cliffs and caves of the Acropolis and other hills.

In the Acropolis of Athens are many fragments of tertiary rock, which has been used for the fortifications and other common purposes. It contains vast numbers of *Cardium edule*, and other shells, which have in general perished, leaving only casts. I could not find it *in situ*: but I suspect that it comes from the east side of the Phaleric Bay.

many of the columns have been discovered among the rubbish, that the Government are now occupied in restoring it. It is only about 12 feet in length, but from its conspicuous situation must have been a very beautiful object. It stands at the very corner of the Acropolis, on the top of a wall or cliff upwards of 30 feet high. Colonel Leake supposed, erroneously, that there was a cave dedicated to Ceres under this temple; but a doorway, which he thought was an entrance of this cave, is only a niche. Some metopes have been found belonging to the temple of Victory, containing very beautiful bas-reliefs, and others belonging to the same temple have been for some years in the British Museum.

"On passing the Propylæa, the magnificent west front of the Parthenon bursts upon the spectator's view. This is the most perfect portion of the edifice, and the effect produced by the first *coup d'œil* is probably as great as in the days of its prosperity. Two hundred years ago the Parthenon was as perfect as the temple of Theseus is now; but some evil genius suggested to the Turks that this temple would make an excellent powder-magazine, and in 1689 a shell from a Venetian mortar exploded this receptacle, and annihilated the central parts of the temple. A small Turkish mosque has been built on the space cleared by the explosion, and its clumsy architecture forms a strange contrast with the classic perfection around it. Such is the grandeur and simplicity of this temple, that at first it is difficult to form a just idea of its size. At the distance of a few yards the columns appeared about 3 feet in diameter, and I found it necessary to go close, and even to touch them, before I could be convinced that their real dimensions were 6 feet. In the middle ages this temple was used as a Christian church, and traces of painted saints may be seen on the inner surface of the marble walls. The ancients also endeavoured to improve the virgin purity of the marble by painting it with ornamental patterns, traces of which may still be detected in many places. The wall of the *cella* is composed of square blocks of marble, closely fitted, and each of them is cramped together with a piece of iron. The iron, swelling from the effects of oxidation, has in numerous cases split the stone and driven out triangular fragments from the front surface. Not a particle of mortar exists in the edifice, nor, I believe, in any Grecian building anterior to the Roman invasion. The columns of the Parthenon consist of (if I remember right) eleven pieces of stone, 6 feet in diameter and rather more than 3 feet in height. The surface where these massive cylinders are united has been cut to a perfect plane, and even polished, so that they have united with the greatest accuracy. The junction is sometimes so intimate that the eye can with difficulty detect it. A small square hole is cut in the centre of these cylinders, probably to receive an iron plug to prevent any lateral displacement. The stones which compose the prostrate columns still lie in their original order, and by well-directed endeavours many of them might be restored to their former position.

"During my stay at Athens I paid a visit to the island of Salamis. I took a boat at the Piræus, and passing the island of Psyttalia, which is uninhabited, I landed on the Cynosure, a long narrow rocky ledge, mentioned in the account given by Herodotus of the battle of Salamis. From hence I had a good view of the scene of the conflict in the straits between the island and main. It is a good anchorage, and two or three Austrian and French ships of war were floating peaceably over the wrecks of the Persian galleys. On the opposite shore was the hill on which Xerxes sat to behold the battle. The remains of the triumphal column erected by the Athenians may yet be seen on the point of the Cynosure. From hence I walked along the shore to Ambelaki, a village on the isthmus which unites the two halves of the island. Salamis is steep, rocky and barren, and only admits of cultivation in a few limited districts. The ancient town of Salamis stood on the side of the bay north of Ambelaki; its site is marked by numerous foundations of houses and traces of the town walls. From this point I returned to the mainland in a fishing-boat, which landed me in the bay about a mile west of the Piræus. I here noticed that the sand, which is calcareous, has a tendency to harden into solid rock, forming a kind of coarse oolite, similar to that of Ascension Island.

"The island of Psyttalia, or Lipsocatalia, is of the same formation as the Piræus. It is heaved up in a north-east and south-west direction, dipping to the south-east. It is difficult to say whether this and the

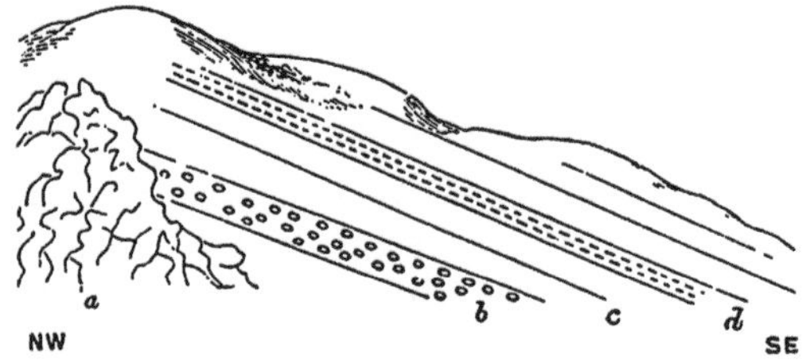

ISLAND OF PSYTTALIA.

a. Breccia of compact limestone.
b. Red sand and conglomerate of rolled pebbles of marble, schist, jasper, &c. In crevices of this red sand is salt, as much as 40 feet above the sea.
c. Whitish marl; passing upwards into compact scaglia limestone.
d. A stratum of limestone with many turreted shells, apparently Cerithia.

Piræus be a tertiary or secondary formation, but it is certainly anterior to the subapennine tertiary formation, which occurs at Patras, &c. &c. It is remarkable that Boblaye has not noticed it. The limestone of Psyttalia is compact and hard, identical in every respect with the scaglia of Greece and the Ionian Islands. Fossils are less abundant in Psyttalia than at the Piræus. Large bivalves occur in the massive limestone in the southern part of the island. At the northern part is the annexed section on the north side of the bay on the south-east side of the island.

"28th.—Took our leave of Athens and came down to the Piræus, where we spent the day with Captain Price, who accompanied us over the remains of the ancient fortifications and other public works. The harbour of Piræus was divided into three ports, one within the other, and each of them closed at the mouth by piers which advanced from either side and left a narrow passage in the middle. These piers, which now exist as reefs of rocks a few feet under water, render the entrance of the harbour somewhat dangerous. They are built in a considerable depth of water, and their construction must have been a work of great difficulty. The dockyard was on the west side of the harbour and was surrounded by a massive wall, many portions of which remain. At the north-west angle is a gateway between two round towers, the lower parts of which remain, and are of admirable workmanship. On the east side of the harbour the sea-wall commences which surrounds the whole peninsula of Munychia. It runs along the top of the low cliff, and may be traced nearly the whole way, for the rocks are here so durable that the sea has made very little progress since the time of Themistocles. We visited the so-called tomb of that general, which Colonel Leake shows to be wrongly appropriated. There are two or three sarcophagi cut in the solid rock, close to the water's edge, and when the south wind blows, the waves dash over and fill them. The massive joints of a column lie scattered around, honeycombed by exposure to the waves. We followed the sea-wall till it brought us to the circular harbour of Munychia. In some parts of the wall polygonal stones occur, but it is for the most part built of squared blocks of considerable size. The port of Munychia has also been closed by piers at its mouth. Foundations run sloping into the water in many parts of it, and seem intended as slips for launching vessels. From this port we crossed the isthmus and returned to Piræus. On the isthmus are the remains of a Doric temple, supposed to be that of Diana. The original temple has been reconstructed into a more modern edifice; the joints of the columns have been laid side by side underground by way of foundation to the walls. All this part of the peninsula has been densely covered with houses:

the ground-plan of houses, and even of streets, composed of large square blocks of stone, may be easily traced. The soil seems made up of fragments of tiles and pottery. I saw numerous circular cisterns sunk in the solid rock, and destined as receptacles for water, oil, or perhaps corn. These cisterns are also frequent near Athens. They may be compared to immense jars, about 20 feet deep and 10 feet diameter at the bottom, with an aperture of 2 or 3 feet in width at the top. Near to this are some chambers cut in the rock, of considerable size. They may either have been intended as tombs or as habitations, to which latter purpose they are still applied.

"29th.—We resumed our examination of the Piræus. We crossed the isthmus of Munychia and followed the sea-wall to the harbour of Phalerum, passing on our way an ancient fortress. The eastern extremity of the hill of Piræus, next the Phaleric Bay, consists of strata of sand and gravel, with Ostreæ, Pectens, species of Echini and Crustacea. This is surmounted by beds of white marl and compact white limestone, forming the hill of Piræus. On the east side of the Phaleric Bay are strata of rolled conglomerate, dipping towards the sea, but without organic remains as far as I could see.

"It appears that the tertiary stone with *Cardium edule,* &c. used in the Acropolis of Athens, must have been brought from a distance, as I could find no tertiary rock of so recent a character in the neighbourhood of Athens and the Piræus.

"In the afternoon we embarked in the Levant steamer, and at 7 A.M. of the 30th entered the harbour of Psyra. I was much struck with the appearance of the town, the houses being all white and flat-roofed. They surround the bay like the steps of a theatre, and in the centre rises a conical hill which is covered by the houses of the *old* town, the whole being crowned by a large cathedral. The modern town, which is now the most flourishing in the Greek empire, has been built within the last six years; before that, the inhabitants, from the fear of pirates, lived at the top of the hill, 500 or 600 feet above the sea, and throughout these islands the villages almost universally occupy similar situations: Psyra owes its rapid rise to the excellence of its harbour and central situation. We landed for a couple of hours, and walked up the hills, which are barren and rocky in the extreme. At noon we resumed our voyage, and found ourselves surrounded with the Cyclades. To the right of Psyra appeared the Isle of Ghioure, then the extremity of Negropont, Andros and Tenos. We passed between this island and Myconos, to the right of which might be seen Naxos, Delos, Antiparos and Paros. All these islands are similar in their general characters, rising abruptly from the sea to a great height, and being for the most part bare and rocky. We caught a glimpse of Samos before dark. In the morning we found we had passed the Straits of Scio, and were entering the Gulf of Smyrna. It rained all the morning, but I could not help remaining on deck, enjoying my first sight of the coast of Asia. My first impressions of the new continent were highly favourable. The mountains of Carabournou were thickly clothed with evergreens, and the plains at their base presented a bright green herbage, which was most refreshing to the eye, after the naked limestone hills and arid plains which formed the prevailing feature in Greece. We landed at Smyrna in a deluge of rain.

CYCLOPEAN WALLS.

TEMPLE OF AZANI.

CHAPTER V.—1836.

EXCURSION FROM SMYRNA WITH COLONEL MACKINTOSH.—EPHESUS.—MAGNESIA.—AIDIN.—VALE OF CAYSTER.—SMYRNA TO CONSTANTINOPLE.—SCUTARI.—KIRMASTEN.—RUINS OF POIMANEUS.—VALLEY OF THE RHYNDACUS.—ADRIANI.—RHYNDACUS.—MOIMOUL.—AZANI.—GHIEDIZ.—THE HERMUS.—AHAT-KIEUI.—ANCIENT BLAUNDUS.—KARADEWIT.—ADALA.—TOMB OF HALYATTES.—SARDIS.—RETURN TO SMYRNA.—HIERA.—TIRYUS.—TOMB OF AGAMEMNON.—MYCENE.—PATRAS.—ZANTE.—EMBARK FOR ENGLAND.

"30TH JANUARY, 1836.—After passing a long and somewhat tedious winter at Smyrna, during which the severity of the season prevented me from going more than a few miles from the town, I was at last greeted by a decided improvement in the weather. Hamilton being somewhat indisposed, I agreed to make a short excursion with Colonel Mackintosh. Our party consisted, besides ourselves, of Constantine a Greek servant, and an Armenian *suriji* or groom. We left Smyrna at 1, with the hope of reaching Trianda, a village on the road to Ephesus. After travelling two hours, the rain compelled us to leave the high road and make for the village of Sedikeni, where we passed the night at Constantine's house.

"31st.—Left Sedikeni at a quarter past 8, our prospects of good weather far from flattering. We passed over an undulating open country, with scattered shrubs and fir-trees. A mile beyond Trianda we entered on the plain of Bourghas-ovasi or Tourbali. This plain is thickly covered with the shrub called Agnus-castus (*Vitex agnus-castus*), the brown stems of which, being bare of leaves, gave it a black and gloomy aspect. The soil, however, seems excellent, and nothing is wanted but population to render it fertile; at present it only supports a few miserable villages. The chief of these is Tourbali, probably named from the ancient town of Metropolis, which is at no great distance. We passed the remains of an aqueduct leading to a hillock on our left. It is carried on arches for nearly a mile in length, and I could trace it along the side of the hill for some distance on our right. After crossing a stream, anciently called Phyrites, we followed the base of the hill on the right, till we reached the remains of Metropolis. The town stood on a hill, which forms part of the range of Mount Gallesus. On the east is the Bourghas-ovasi or Tourbali plain. Nothing remains of the town but the walls of the citadel, formed of massive blocks of marble, beautifully fitted and squared. Some long walls descending the hill are of the patchwork architecture of the middle ages. At the foot of the hill is a Roman building with arches, perhaps a bath. Numerous remains of columns and sculptures in the adjacent burial-ground prove that Metropolis, like all other Grecian cities, possessed her temples, porticoes and other ornamental edifices. It was admirably supplied with materials for architectural display, for the town stands on a rock of very pure white marble. After leaving Metropolis, we passed a small marshy lake covered with wild-fowl, anciently called the Pegasean Lake. We then turned into a valley on the right, accompanied by the river Phyrites, which had entered and reissued from this lake. After following it for four miles, the Cayster enters from the eastward, and receives the Phyrites. The classic Cayster is a stream from forty to fifty yards in width. The fortress of Ghatzi-cali, or Goat-castle, was on our right, seated on a lofty and precipitous peak, perhaps 1000 feet in height. It stands on a most commanding situation, and is visible from the plain of Ephesus and the whole upper valley of the Cayster. Beyond this we passed some lofty and romantic cliffs overhanging the road, the abode of numerous eagles, which were soaring over our heads. We crossed the Cayster by a bridge, perhaps of Roman construction, in the repairs of which many fragments from Ephesus have been employed. We now emerged from the valley into the plain of Ephesus, once an arm of the sea, with a small island at the upper end, which is now the hill of Aiasaluck. We at last reached Aiasaluck, a miserable village of about half a dozen huts, and were housed in a *cafinet*. A Turkish cafinet is a small hut containing a single room, with a flat roof, through which the rain was streaming on our arrival. A kind of boarded stage runs round the room, on which the Turks sit cross-legged during the day, smoking their pipes and warming themselves round a pan of charcoal. At night they roll themselves in their cloaks, and lie down to sleep on the same spot which they have occupied during the day. This appeared to be the sole employment of an old negro and two or three Turks during our stay at the cafinet of Aiasaluck. At the upper end of the room is a fireplace 3 feet from the ground, with a chimney above it. The appurtenances of the hearth consist of a lamp, and two or three small saucepans for making coffee. These cafinets recur every four or five miles on the high roads in this country; near them is commonly a well or fountain, and a large plane-tree, beneath which the caravans of men and camels repose during the intense heats of the summer.

"February 1st.—Before starting for Ephesus, which is about a mile distant, we visited the curiosities of Aiasaluck; it was the successor of Ephesus, and under the Byzantine empire rose to be a considerable town. It is now reduced to a wretched village, scarcely habitable in autumn, from the malaria produced by the marshes below. We first visited the Mosque of St. John, a large building, supposed to have been originally a Christian church. This is a point not easily decided. The building stands east and west, is of large size, and built of common stone, except the west front, which is of marble, and richly ornamented in the Arabesque style. Many Arabic inscriptions are carved round the door and window-cases. This marble facing, therefore, must be subsequent to the Turkish invasion; at the same time it is possible that the body of the mosque, which is of much plainer workmanship, may have been a Christian church. Within are some vast granite columns, supposed to have belonged to the temple of Diana. The capitals of these columns are in the Arabesque style, and hence M. Texier thinks that the whole edifice must be Mahometan; but it is possible that the Turks may have new-roofed the original church, and placed new capitals on the columns. The top of the hill has been surrounded with fortifications of the Greek empire. They are composed of blocks of marble taken from ancient Ephesus, intermixed with brickwork of great solidity. On the south side is an arched gateway, over which stood some years ago three handsome *bas-reliefs*, which are engraved in Tournefort's work. Two of these have been removed, and there now only remains the left-hand one, and the prostrate figure of Patroclus in the middle sculpture. I compared Tournefort's engraving, and noticed the following inaccuracies. He has omitted a sitting figure between the first and second figures from the left hand; also to give wings to the fifth figure in the left-hand sculpture; and to depict the chariot from which Patroclus has just fallen, the wheel of which is seen below his uplifted knee. Within the area of these old fortifications the Turks have in later times built a castle, on the motley walls of which are some inscriptions, two of which I copied: the characters of one of these are of a singular form, and belong to an age when many of the *small Roman* letters had been invented. The town of Aiasaluck has been supplied with water by a handsome aqueduct, leading across the plain from the hills on the east. This structure, built probably by some of the Greek emperors, is composed almost entirely of marble pillaged from the ruins of Ephesus. Numerous inscriptions are seen on the piers of this aqueduct, some of them at too great a height to be copied from the ground; and Dr. Chandler, who copied all within his reach, regrets that he could not procure a ladder. We conceived the idea of copying them by means of a small telescope, and found the plan succeed; but we visited it just before dark, and had only light enough to copy part of one.

"From Aiasaluck we proceeded to visit Ephesus. This renowned city was built upon and around a hill of grey marble named Prion, standing detached from the larger range of Corissus. The sea, which once reached the city-walls, has been removed to the distance of two or three miles by the alluvium brought down by the Cayster *.

"We were accompanied to Ephesus by a Greek, who acted as our cicerone. He lived at the village of Kirkadzai, or Kirkingecui, three or four miles over the hills east of Aiasaluck, which is said to contain 400 houses, all Greek. The few inhabitants of Aiasaluck are all Turks, except a Greek baker, whose son we found reading a Greek Testament given him by a missionary. We found the description of Ephesus given by Dr. Chandler very exact. The east end of Mount Prion is full of quarries and sepulchres cut out of the marble rock; one of these quarries has been converted into a Greek church, grown up with bushes, and approached by a narrow path. It is called the Church of the Seven Sleepers, on account of a legend of seven children who slept there for seven years, during which time they were miraculously fed by angels. We then followed the north base of Prion, past numerous tombs, vaults and foundations.

"We proceeded to the stadium, the walls of which are of that solid and well fitted masonry peculiar to the Greeks. The arch, which has formed the entrance, is interesting and well preserved, but has been deprived of a sculpture mentioned by Chandler over the inner front. A transverse street passed in front of the stadium, the pavement of which, formed of massive slabs of marble, still remains in many places. North of the stadium stood a large square building; the

* See Plate, commencement of chapter, p. lxxxviii.

vast massy piers which remain show that it has been an important edifice. On the west of the stadium is a square elevated platform, the site of some temple or other edifice. South of the stadium is the theatre celebrated as the scene of St. Paul's persecutions (Acts, xix.); the seats have been removed, but the two wings remain, though in a ruinous state. To the right of this are the ruins of a temple, the whole lying prostrate, and the beautiful ornaments of the frieze scattered and defaced. The columns are nearly 40 feet in length, and 4 feet 6 inches in diameter; they have each been a single stone, though now all broken. I noticed that some of the columns were unfinished, the flutings being roughly chiselled for a part of their length, while others were highly polished and finished. To the north-west of this temple are some massive and extensive ruins, close to a reedy marsh, once the port of Ephesus. These ruins were supposed by Tournefort to be the temple of Diana; they have none of the characters of a temple, and have more probably been some of the public offices of the city. They consist of walls, piers and arches, built of stone. Some of the upper arches are of brick, proving that these ruins are of the Roman age. These buildings stand upon extensive vaults, the entrance to which is by a long narrow descent, which I followed till I could see no longer, and regretted not having a lamp.

"We now turned up the valley between Mount Prion and Corissus. We first came to the *Odeum*, or Music Theatre, the circle of which remains, of large squared stones. In front of the *Odeum* is a large oblong, level space, which, from the many broken fluted columns scattered round its margin, I conjectured to have been a market-place or square, surrounded by an arcade. Beyond this is the Gymnasium, one of the best-preserved and most interesting ruins of Ephesus. The front consists of some large arches with wings at each end; behind this are walls enclosing a square space. We found the niche mentioned by Chandler, stuccoed inside, and painted with fish and waves. This building, which was a species of college, may be perhaps the 'School of Tyrannus,' in which St. Paul disputed daily. The ruins of Ephesus are rendered particularly impressive, from the death-like solitude which reigns in a spot once the gay and luxurious metropolis of Ionia; they are overgrown with bushes and shrubs, and I roused a fox from his lair amid the ruins of the Corinthian temple. The temple of Diana, one of the seven wonders of the world, is an object for which travellers have hitherto sought in vain. From a comparison of the notices given us by the ancients, it appears to have stood in the marsh to the north of Mount Prion, and its foundations are probably buried below the detritus brought down by the Cayster.

"I returned to Aiasaluck, deeply gratified with a place where the generality of visitors tell you there is nothing worth seeing.

"2nd.—Left Aiasaluck for Aidin. The rainy weather had now passed away, and during the remainder of our trip we enjoyed a beautiful clear sky. We followed a romantic ravine, thickly studded with shrubs: in about an hour we passed under a Roman aqueduct, or perhaps a bridge. Three arches remain below and six above, built of stone, and in remarkably good preservation. On the left of the arch is an imperfect inscription.

"Beyond this we continued to ascend the ravine by a paved road, which in many parts has been cut through the rock in the side of the ravine. From the summit of the pass I could see Mount Tartali, near Smyrna. We then descended into a picturesque country, with a mountain called Gumuch Dag a few miles distant on the south. We here entered the ravine of the river Lethæus; the sides of the valley were clothed with fir-trees and shrubs, and the ground was covered with a beautiful purple species of Anemone. We met several caravans of camels, for this pass is almost the only outlet from the valley of Mæander to the northwards. We visited the ruins of Magnesia, following the route mentioned by Mr. Arundel. In a short time we reached the commencement of a marshy plain, on the opposite side of which we saw the ruins. On the west side of this plain is a conical hill, projecting from the range of Gumuch Dag. In this hill I could see a large cavern, mentioned by Van Egmont, who passed near it in coming from Scala Nova. I regretted not being able to explore it, as it is doubtless the cave sacred to Apollo, mentioned by Pausanias. The town of Magnesia stood on the north slope of a hill overlooking the plain, and surrounded by picturesque hills and mountains.

"Crossing the Lethæus we came to a massive ruin, supposed to be a gymnasium. The eastern front has consisted of a series of arches, the piers of which remain. Parallel to this is another row of piers, which have also supported arches. The rest of the building consists of majestic masses of wall, composed of ponderous blocks of stone. A fig-tree had taken root beneath one of the blocks, and the force of vegetation had raised the ponderous mass at one end to the height of a foot or more. I then proceeded westwards to the walls of the city, which are very strong. There is a gateway with a small chamber on each side, perhaps part of a guard-house. I followed the walls to the summit of the hill, where there has been a square tower. On the south side is the plain of the Mæander, with that river winding through it with its proverbial curves, which are not quite so numerous, however, as represented on the maps. To the south-west is the Lake of Bafi, over the Gulf of Latmus, but cut off from the sea by the alluvium of the Mæander. From this hill the walls of Magnesia turn eastward, and follow the ridge, till they descend to the river Lethæus. I now descended to the stadium. The greater part of the marble seats are yet remaining, and, if cleared from the soil which has collected on them, would appear nearly in their original state. These seats are 2 feet 4 inches wide, and 11 inches high. The width of the stadium at the bottom I made 49 yards measured by pacing, and its length 218 yards. The theatre of Magnesia is very small. Little remains except the semicircular cavity in the side of the hill. The width of the bottom is about 39 yards. In the north-eastern part of the town is a large space enclosed by strong, well-built walls, nearly 20 feet high, the greater part of which are standing. This enclosure has probably been a citadel into which the inhabitants retreated at the approach of danger, though it is singular that it should have been placed in the plain instead of the hill above. Within this space are the ruins of the famous temple of Diana Leucophryne, and perhaps this circumstance induced the Magnesians to build their citadel in the plain, in order that they might rally round their goddess and defend her temple. This edifice, one of the most celebrated in Ionia, is now levelled with the ground, apparently from the effects of an earthquake. It presents a confused heap of ruined columns, friezes and capitals, the whole of the purest white marble. Nothing appears to have been removed, for Magnesia enjoys the advantage of not having been succeeded by any modern town, which might consume the materials of the old one. We saw some portions of the frieze, on which were beautiful bas-reliefs, though much defaced. On the east side of this enclosure is the entrance, with a row of arches on each side. We were compelled to hurry from Magnesia much sooner than I could have wished, but we were still five hours from Aidin. We followed the plain of the Mæander. On our left was the range of Messogis. After crossing several rivulets flowing from Messogis into the Mæander, we reached Aidin (called also Guzel-hissar) at half-past 8.

"3rd.—In the morning we called on one of the few Franks who reside in Aidin. This was a *Medico* or physician, a Greek from Athens, who had been settled in Aidin two or three years. He also officiated as Vice-Consul for France and Holland; and, as he had no other colleagues, might be considered as the representative of all the European powers. Signor Spiridaki was very hospitable, and insisted on our removing from the Khan, and passing the following night at his house. We now proceeded to explore the curiosities of Aidin. It is said to contain 300 Greek houses, and 7000 or 8000 Turkish ones. Like all Turkish towns, the streets are dirty, narrow, and ill-paved. In these towns the houses are commonly two storeys high, the upper one projecting 2 or 3 feet over the street. The lower windows commonly look into a small court, so that in the streets nothing is to be seen but dead walls. The shops being all concentrated together at one part denominated the Bazaar, the remainder of the town has a dull and lifeless appearance. Few women are to be seen, and those are enveloped from head to foot in a white sheet, with a black crape over their faces, and politely turn their backs on all male passengers.

"Aidin stands on the plain of the Mæander, at the foot of a steep cliff, above which is a flat table-land, on which stood the ancient town of Tralles: it has been a town of great extent; it is now a grove of olive-trees, amid which the Turks are constantly excavating for building-materials. The only ruin is an arcade of very gigantic proportions: three arches are standing; they have formed the western front of a large square building, probably a gymnasium, the rest of which is in complete ruin. On the inside of this arcade are the remains of a transverse brick arch, which has formed the roof of a corridor. The rest of the arcade is built of massive stone. The inside of these arches has been plastered and painted in fresco, with an ornament resembling a true-lover's knot. The eastern front of this building has been adorned with a colonnade, the columns flattened. Seven of these columns remain, but are broken off near the ground. The whole area of this building must be more than an acre: it is clearly of Roman construction. In repairing it at some later date, many sculptured marbles have been employed, and some inscriptions, which, being at a great height, I could decipher only by means of a telescope.

"The valley of the Mæander, so celebrated for its modern alluvial deposits, presents also some very remarkable ancient formations of the same kind. The lofty range of Messogis, which consists of micaceous schist, with associated marble and quartz, is flanked on the southern side by a lower range, composed of boulders and rolled fragments of those materials. These conglomerate beds are sometimes of a bright red colour, but more commonly grey or yellow. They are but slightly indurated, and, being deeply furrowed by ravines, form a landscape of the brightest picturesque beauty. These hills of conglomerate rise to the height of about 600 feet in the neighbourhood of the mountain-range, but spread out into flat plateaux towards the Mæander valley. The plateaux are elevated about 200 feet above the alluvial plain of the Mæander, on the sides of which they terminate by a precipitous descent. Historical as well as ocular evidence renders it probable that this vast alluvial plain has been occupied by the sea for a great distance up the country; it has therefore, in all probability, been that element which has encroached on the ancient deposits of alluvium, and caused the abrupt cliffs by which they are now terminated. The ancient town of Tralles occupied one of these plateaux of conglomerate, and was defended by a fort placed on a lofty peak of the same geological composition. The old buildings are of travertine with Limnææ. Whence comes it?

"On the site of Tralles is a small country-house belonging to Signor Paseali, now British Vice-Consul at Scala Nova, in the garden of which are some handsome Corinthian capitals and a sepulchral column, with the inscription given in Arundel's 'Seven Churches.' The walls of Tralles may be traced to a great distance. I followed them to the top of a pointed hill, on which has been a fort or small citadel. The view from this point is magnificent. To the south is the rich plain of the Mæander, bounded by a succession of mountains extending to the Taurus range. To the north the view is more confined, but highly picturesque, consisting of hills and ravines, broken into every variety of form, and clothed with olive- and pine-trees. In the background was the snow-covered range of Messogis. Immediately below, the little river Eudon murmured along, at the depth of 500 or 600 feet. In descending I came unexpectedly upon the remains of the theatre; it is of large dimensions, but the seats are all removed. The two wings are most massive pieces of masonry; they are built of rough stones, laid in mortar in the Roman style, and resemble at first sight a naturally stratified conglomerate. Their height must be about 60 feet, and their thickness 30 or 40 feet. They have been faced with large squared stones, a few of which remain, stuck in the face of these artificial cliffs. Immediately below the theatre is the stadium, of large size, surrounded by a grassy bank, the remnant of the seats and walls with which it was once adorned.

"I now returned to Aidin, and spent the evening at Signor Spiridaki's, who had asked some of his Greek friends to come and smoke their pipes with us. One of these was a native of Cæsarea, and, though he had lived fifteen years at Aidin, could only speak the Turkish language. The Greeks in the interior of this country are generally ignorant of their own language, and have copies of the Scriptures read in their churches in Turkish, but printed with the Greek characters. In the course of the evening two musicians entered, and played some Greek tunes; one of them performed on a fiddle, the other on a guitar, the strings of which are struck with a small piece of quill instead of the fingers.

"4th.—Took leave of our host, and took the road to Tireh, across Mount Messogis. It has been rarely passed by travellers, and has, I believe, been described only by Van Egmont and Paul Lucas. We soon began to ascend a high hill by a very steep path; we then descended, and followed for some distance a small river called Coom Chay. We passed two or three high ridges, whence we had extensive views of the plain of the Mæander, with the mountains of Latmus and Grius, the Lake of Bafi, and the small hillock which was in ancient times the island of Lade. Beyond this the sea was visible for a short distance, and the view in this direction was terminated by a distant island, probably Leros. To the east, at the distance of about five miles, was the high range of Mount Messogis, covered at this season with snow.

"In going from Aidin to Tireh the band of micaceous conglomerate is about six miles in width, forming some highly picturesque hills, some of which may be 800 feet high. The sand, gravel and conglomerate are commonly stratified, and for the most part approach to horizontality; they are sometimes hardened into puddingstone, but are generally of loose materials. Hence to Tireh is micaceous schist, with veins of quartz.

"We now arrived at the eastern side of another deep valley, with a stream running to the Mæander. In these solitary vales are a few small Turkish villages, whose inhabitants cultivate vineyards on the sides of the hills. We continued to ascend till we reached the region of snow, which was now rapidly dissolving, and at last arrived at the highest point of the pass, which I estimated at about 2000 feet above the sea. A new and splendid prospect now burst upon our view. The vale of the Cayster lay below us, a level plain like that of the Mæander, with mountains rising abruptly around it on all sides. Some small circular hillocks are detached over its surface, and on the south-west side a range of low hills runs parallel to the Messogis range, and forms with it a lateral vale, detached from that of the Cayster. Through this vale is the road to Ephesus. On the opposite side of the Caystrian vale we saw the range of Mount Tmolus, commencing with Mount Tartali near Smyrna, and stretching eastward till it united with Messogis at the head of the vale. Along the base of Tmolus are several villages and the town of Baindir. Immediately below us was the town of Tireh. We entered Tireh a little past 5, and called on the Aga. He is a son of Cara Osman Oglou of Magnesia, and nephew of the Pasha of Aidin, very young, and of gentlemanly and pleasing manners. He quartered us at the house of a Greek, where we lodged very comfortably.

"5th.—Tireh is a large, straggling, dirty Turkish town; it is supposed to occupy the site of the ancient Larissa. In front of some of the mosques I found some small granite columns, and a few capitals, but no remains of any consequence. We left Tireh at a quarter past 11 for Baindir, on the opposite side of the plain. The vicinity of these two towns and the fertility of the plain have induced a higher degree of cultivation than is commonly found in this country, and the vineyards and cornfields are well tilled, and enclosed. We crossed the Cayster by a bridge of six arches. At this place the river is fifty or sixty yards wide, a slow, muddy stream. We soon after passed a hillock, apparently of marble, with an ancient quarry on the top. There are many of these hillocks scattered over the plain, and on several we saw tumuli at the summit, or perhaps the remains of ancient fortresses. About 1 we passed a burial-ground, with a few blocks of ancient marble. We saw on a hill to the right what appeared to be some ancient ruins, and turned aside to visit them. They were the remains of a small Greek church, which, with several other walls and vaults, proved the existence of a village or small town during the middle ages. The church is built of coarse materials in a plain style. The piers have supported arches, one of which remains. This spot may perhaps be the site of Anagome, a place only known from the ancient map called the Pentinger Table. It answers more nearly to the position of Anagome than Baindir, which is too near Ephesus. Perhaps its name may have been *ἄνω κώμη* (*Upton*), or, the village on the hill. We now returned to the high road, and reached Baindir at 3. We passed a spot called *Old Baindir,* where I could see nothing but the foundations of a long wall, of no great antiquity. We passed some large herds of buffaloes, which were returning to Baindir from their pastures; they turned aside to their respective abodes by twos and threes at a time, like the pigs at Schlaugen Bad. Buffaloes are much used here for ploughing and for drawing *arabahs,* a kind of cart.

"Baindir stands on the south slope of the Tmolus range, and commands a fine view of the Caystrian plain. The hills above are clothed with olives to an unusual height, proving the fineness of the climate. The mountains of Tmolus immediately above Baindir abound with panthers, jackals and bears; hence the necessity of driving home the herds of cattle every night.

"6th.—Left Baindir for Smyrna at 7. I should have mentioned, I could find no vestiges of antiquity at Baindir, and I do not believe it to occupy an ancient site. After some miles, we left the plain of the Cayster, and entered a valley, which brought us out on the east side of the plain of Tourbali, which we had passed seven days before. At 11 reached a cafinet, near a spring. On the left of this is a square platform, the site of an old fortress, perhaps Roman; nothing but the foundations remain. On the right are two or three tumuli on a hill. We were told, that, at a place called Sorghoori, an hour distant in the mountains, are some ancient ruins. At 12, came to another

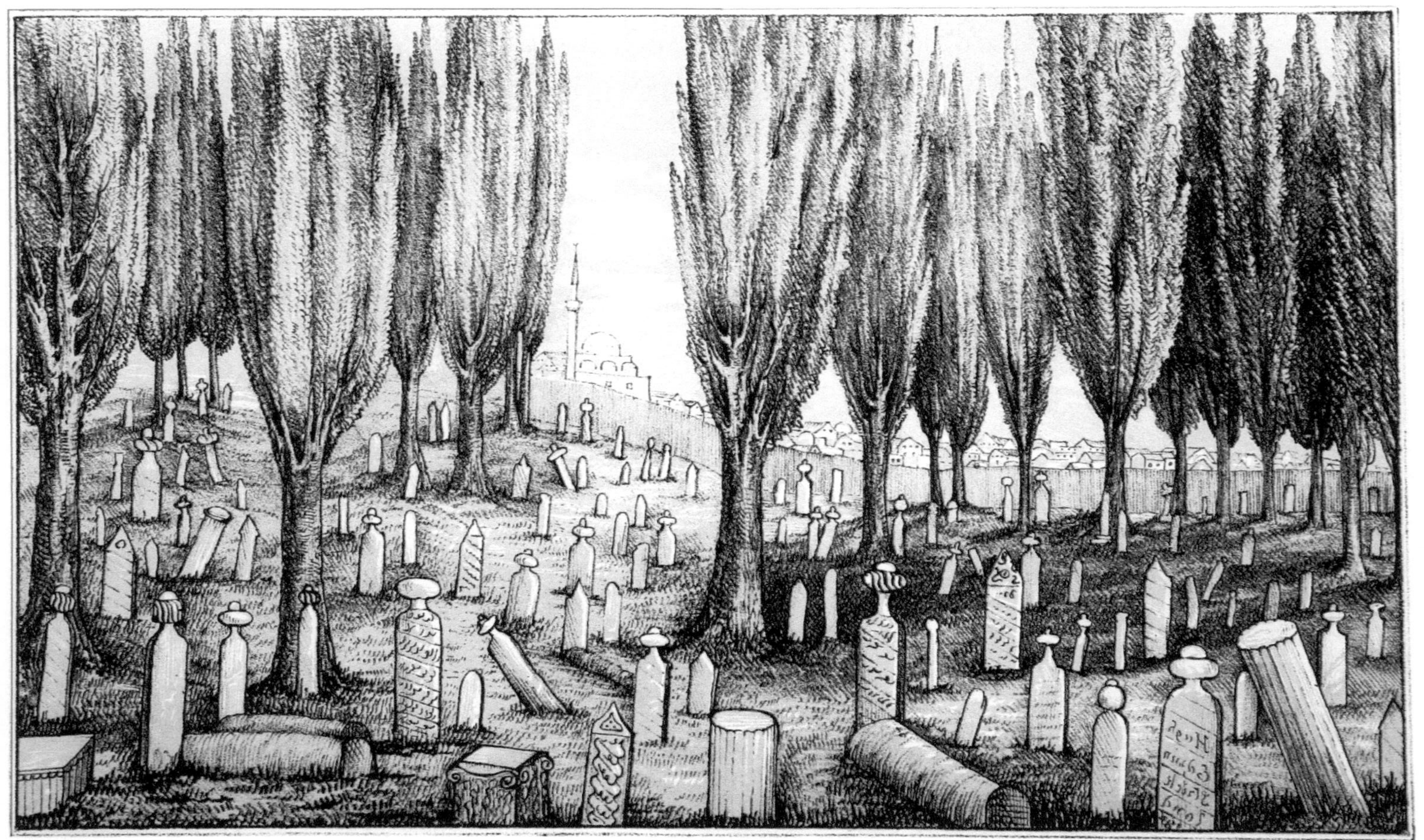

C.D.M.S. litho. W. West Imp.

Turkish Cemetry at SMYRNA. From the E.

cafinet, at the foot of a hill of white marble; close by is a burial-ground, with many ancient marbles, among which I found two inscriptions. It is, I think, hardly likely that these remains can have been brought from Metropolis, which must be five or six miles distant; and I am inclined to suspect the existence of some ancient site in this vicinity. The village of Fortena is near this spot on the right. We arrived at Smyrna about 7.

"15th.—Set out with Colonel Mackintosh, with the intention of visiting the village of Nymphi, on the road to Sardis. Colonel Mackintosh's object was to inspect the Pass of Cavaklederi in a military point of view, and I wished to see the geological and other features of the country. We followed the caravan road, which proceeds nearly east. Notwithstanding the wealth and commercial importance of Smyrna, there is nothing in the environs which, in England, would deserve the name of a road. For a short distance out of the town there is indeed a pavement, rough, slippery and full of holes, but this soon fails, and nothing remains but a mere track, worn in the native rock by the feet of the countless trains of camels, which for centuries past have daily traversed the same path. We proceeded along the plain which forms the termination of the bay of Smyrna. In this plain are the following villages, placed near the foot, or on the slope, of the surrounding mountains. On the south side, Cucklujah, Ishekli and Bounarbashi; on the north, Hadjilar, Narlecui and Bournabat. This plain is fertile and well cultivated, abounding with fig, olive and pomegranate trees. On the north is Mount Sipylus, and on the south Mount Tartali. At the head of the plain a transverse range connects these two ridges, and forms the true termination of the Gulf of Smyrna, though the sea is now about five miles distant. A remarkable conical hill stands in the centre of this range, and on the south side of it is the Pass of Cavaklederi. The ascent is steep and rugged. In the ravine on the right is the village of Cavakli. At the top of the pass the rock has been cut through to the depth of 20 or 30 feet, probably by the Romans. The road immediately begins to descend into the valley of Nymphi, similarly placed to that of Hadjilar, but opening to the east instead of to the west. It is drained by a small river, flowing eastward at first, then running round the termination of Mount Sipylus, and joining the Hermes. In many maps the drainage of this valley is made to flow north in two streams passing through defiles of Mount Sipylus which have no existence.

"On the north of this vale is the village of Nymphi, at the foot of some picturesque mountains, the northern base of the Tmolus range. In a recess in the side of the mountain some height above the village, we saw the ancient castle of Nymphæum, the rural retreat of the

Greek Emperors, but now in ruins. There are also near Nymphi some ancient mines of gold and silver, now, I believe, neglected. I much wished to visit them, but when we got within about two miles of the village we found our time was so short that we were compelled to return to Smyrna, having only been able to take a general view of the features of the country. A snow-storm also commenced, which rendered it cold and comfortless, and we were glad to reach Smyrna at half-past 6.

"20th.—We at last resolved on breaking up our long sojourn at Smyrna, and Hamilton arranged to start this evening for Constantinople, in the steamer. The morning being fine, I resolved to start for the great city by land, in company with Colonel Mackintosh, on Monday next. We had just arrived at this conclusion, when it began to rain in torrents. With this prospect before me, I altered my plan, and resolved on going by steam that evening, rather than run the risk of remaining at Smyrna another week. At half-past 7, Hamilton and I were on board the Crescent, a magnificent steamer; the other cabin passengers were two or three Turkish Effendis, and the Rev. F. Arundell, who was an agreeable acquisition to our party. The Turks were polite and good-humoured, but as we were mutually ignorant of each other's language, conversation was out of the question. The night being dark and rainy, we saw nothing till daylight, when we were off Cape Baba, the south-west extremity of the Troad. It is a rocky bluff, with a small town to the north of the point. Behind us was Mitylene, fast diminishing in the distance. The classic isle of Tenedos was in front,—a bleak-looking island, but celebrated for its wine. On the opposite side was the coast of the Troad, with low cliffs, and undulating hills behind, on which we could see the ruins of Alexandria Troas. Had the weather been fine, we should have seen Mount Ida, Lemnos, Samothrace, and even Mount Athos.

"22nd.—We ran along the north shore of the Sea of Marmora; the thickness of the weather only allowed us imperfect glimpses of the Island of Marmora, Calolimni, and Princes Island. At last we turned Point St. Stephano, and Constantinople with its countless minarets, came into view. We now entered the Golden Horn, an arm of the sea which divides Constantinople from the suburbs of Galata and Pera. We landed, and ascending the steep narrow streets of Pera, arrived at the Frank quarter, and took up our lodgings at the house of M. Giuseppino.

"23rd.—Mr. Cartwright, the Consul, sent his *cavasse*, or Turkish guard, to accompany us over the town. His name is Mustapha, a celebrated character. He is introduced in the novel of the 'Maid of Kars,' and his history is to be found in 'Keppel's Travels.' We first visited the College of the dancing dervishes; and as this was the day on which they perform their weekly dance, we had an opportunity of witnessing that strange exhibition. It takes place in a large octagonal room of light and elegant construction. In the centre is an area enclosed by a railing, round which the spectators seat themselves. We left our shoes at the door,—a ceremony always necessary on entering a Turkish mosque. Ten or twelve dervishes entered the middle area, and performed the chants, prayers and prostrations, which form the regular church services of the Mahometans. They then walked round the room, each bowing twice as he passed the carpet on which the head dervish takes his seat. They now resumed their places at equal distances round the room, with their arms folded on their breasts. Music now struck up in the gallery, the instruments being a drum and flute. The dervishes commenced a kind of waltz, whirling round with considerable rapidity with outstretched arms, and progressing slowly round the room. Their long petticoats expanded with the centrifugal force into an obtuse cone. The great skill in the performance consists in the perfect regularity with which they keep up the rotatory motion, which they continued for about half an hour. Some of them were very young, one apparently not more than twelve years old. Nothing can be more heathenish than this ceremony, which is more worthy of the South Sea Islanders than of the civilized Mahometans.

"We afterwards ascended the Tower of Galata, said to be of Genoese construction. It is circular, and of very great diameter, I suppose about 50 feet; it stands at one of the angles of the old walls of Galata, and is used as a look-out station, to give alarm in case of fire. It commands a magnificent prospect of Constantinople, on one side separated from Pera and Galata by the beautiful Golden Horn, and from Scutari by the equally beautiful Bosphorus, opening into the Sea of Marmora, beyond which are the mountains of Asia, with the snow-covered heights of Olympus in the extreme distance. The variegated masses of habitations are interspersed with mosques and minarets of dazzling whiteness, with gloomy cemeteries crowded with cypress-trees, with dome-covered bazaars, with palaces, barracks and other public buildings, and the deep blue sea which intersects this mass of human habitations is chequered with shipping, countless boats and flocks of wild fowl, which, being protected by the Turks, form as busy actors in the scene as the featherless bipeds which are crowded amongst them. The whole is a scene of the highest beauty and interest, and on which I could have gazed for hours.

"24th.—Took a walk for three or four miles along the shore of the

Bosphorus, which is lined nearly the whole way with villages and scattered houses. The Bosphorus averages perhaps a mile in width, and has the appearance of a large river; the land descends rapidly to the water's edge; and houses, gardens and groves cover the slope. In the evening we dined at Mr. Cartwright's, where we met Mr. Arundell, and Capt. Richards of the Volage, now at Therapia attending on the Embassy.

"25th.—Walked to the burial-ground north of Pera, which forms the chief promenade of the Franks. Busy writing letters.

"26th.—Being Friday we went to see the Sultan go to the Mosque. He is now residing at Dolma-Batchie, on the north side of Pera, and attends a small mosque on the hill called Yeni-Djamie. Eighty or a hundred persons were assembled to see the procession; the greater part were women. I was much struck with the freedom they enjoyed; they walk about alone or in parties in the town or environs, or enjoy a jolting over the rough roads in the springless, but gilded *arabahs*. At Smyrna the women appear in public much less than at Constantinople, and totally conceal their features with the *yashmak*, which is never seen in the latter place. There was nothing very striking in the procession, which consisted of a few guards, pages, nobles and several of the Sultan's horses, in the midst of which Mahmoud himself appeared.

"27th.—Crossed over the Golden Horn to Constantinople with Mustapha to see sights. We first visited the Sulimanieh mosque, the finest in Constantinople. The courtyard is surrounded with pillars of porphyry and verd antique, said to have come from Alexandria Troas. Some of the windows have painted glass, said to have come from Persia. The internal decorations are plain, and there is little of that profuse painting and gilding in which the Turks generally delight. The floor is covered with Turkey carpets. In this mosque are four large granite pillars similar to those in the mosque at Aiasaluck, and perhaps also brought from the temple of Diana at Ephesus. We now went to the Seraskier's Tower, a building of very singular architecture. In our way from hence we passed the Burnt Column erected by Constantine, long after the pure classic taste of Greece had ceased to exist. It is of porphyry or some reddish-brown kind of stone, in eight pieces, the joinings of which are concealed by projecting fillets of leaves, and it has been much cracked and injured by fires. We next visited two of the immense cisterns made under the Greek emperors to supply the town in case of siege, one of which is supported by thirty-two columns; these immense vaults are used for the purpose of spinning silk and twine. Hence we visited the Hippodrome or ancient race-course, a large oblong enclosure, containing the most interesting monuments of antiquity in all Constantinople. The first of these is an obelisk of Egyptian granite, the hieroglyphics on which are said to prove that it was erected by Shishak in honour of Sesostris. This beautiful relic, after standing perhaps 2000 years in Egypt, was brought to Rome, and then by Theodosius to Constantinople, where it still exhibits the angles as sharp, and the surface as polished, as at the day of its erection. It rests upon four blocks of solid bronze, which are supported by an enormous block of marble about 10 feet square and 6 high. On each side of this is a bas-relief representing the Emperor Theodosius and his queen with crowds of attendants, but in an inferior style of sculpture. This rests on a still larger block of marble, on two sides of which are inscriptions in Greek and Latin, in honour of Theodosius, and on the other two are bas-reliefs, one of which is a view of the hippodrome, with chariot-races, and the other represents a number of figures in the act of bringing the obelisk to the spot where it now stands. This relief is unfortunately much defaced, or one might perhaps detect the machinery by which this enormous mass of granite was placed on its pedestal: it stands much in need of being cleared from the soil, which has accumulated to the depth apparently of several feet. The same applies to another monument, possessing even greater interest,—the Brazen Pillar, composed of three serpents twisted together, which stands at the distance of a few yards from the obelisk; at present it is about 11 feet above the surface of the ground. It would be a most interesting object to clear it from the soil and to ascertain the form of its lower termination, which may be accompanied with some inscriptions that might throw light on its history. It is a beautiful piece of workmanship, hollow within, and about half an inch thick. Its largest diameter is at 3 feet from the ground; each serpent is here about 11 inches in thickness, diminishing in size downwards towards the tail. It may perhaps admit of doubt whether this column was captured from the Persians at the battle of Platæa; its chaste style and superior workmanship rather refer it to the classic days of Greece; it is, however, well authenticated that it existed for many centuries at Delphi, where it is said to have been used as the sacred tripod.

"At the west end of the Hippodrome is a lofty obelisk built of ordinary stone, and once faced with plates of bronze; it is much decayed, and threatens to fall; at the base is an inscription hardly legible. On the south side is the mosque of Sultan Achmet, a large structure with some columns of porphyry and verd antique in the court. Hence we proceeded past St. Sophia, contenting ourselves with a view of its exterior. We then descended to the Golden Horn.

"28th.—Crossed over to Scutari, where I had a new and enchanting view of Constantinople and Pera, heightened by the brilliancy of the sky and the water. I passed some magnificent barracks, built by Sultan Selim. Beyond this I followed the cliff to the village of Cadikeui, on the side of the ancient Chalcedon, of which nothing now remains. Beyond it I reached the Cape of Monde-Bornou, a favourite resort of the Franks in summer. To the south of it is a small bay, and opposite is another cape, which forms the entrance of the Gulf of Isnik. I now returned to Scutari, passing one of those enormous burial-grounds, a perfect forest of cypress-trees and tombstones, which form one of the most striking features of all Turkish towns, and of Constantinople in particular.

"29th.—Accompanied the Rev. F. Arundell to Constantinople, where we rambled through the bazaars, and revisited more at leisure some of the spots which I examined on Saturday. We then set out on a voyage of discovery into the northern regions of the town, and, being unencumbered by cicerones or cavasses, we groped through the streets at our leisure, stopping to examine everything of interest that came in our way. This part is less thickly inhabited, and much of it is in a state of ruin. We first came to the large aqueduct of Valens. It is a genuine antique, built of stone intermixed with brick, but destitute of ornament. It is still used to convey water to the upper parts of the town near the Seraglio. For some reason one part of it is much lower than the rest; so that, after a great expense of labour in raising lofty arches in two tiers, the advantage is all thrown away by allowing the water to descend to a lower level. We followed it to its termination, and then visited the mosque of Mahomet the Second, a handsome and symmetrical structure. In descending towards the Golden Horn, we passed a large domed structure in a ruinous state, which may possibly have been a Christian church. A little further on we came to a church which has been converted into a mosque. The door was locked, so we could only see the exterior. In front is a corridor like that of St. Mark's at Venice, floored with marble, with doors and window-cases of verd antique and red marble. The ceiling has been covered with gilt Mosaic work, now *whitewashed,* as is the case I believe also in St. Sophia. Close beside it, in the street, is a magnificent sarcophagus of verd antique, much weather-beaten. On the top and on each side is a Greek cross. It is said to be the tomb of Constantine.

"March 1.—Took a boat and rowed to the Seven Towers, thus coasting along one of the three sides of Constantinople. The ancient walls are standing throughout this distance, and seem to have had little done to them since the time of the Greek empire; they are flanked at intervals with square towers, on many of which are Greek inscriptions, recording their erection or repairs. In many places are fragments of columns and other marbles built into the wall, and in one place is a column standing upright on marble foundations, probably the undisturbed remains of a temple. Further on is a marble monument worthy of the classic ages, though, from its standing on the top of the patchwork-wall, it cannot be of any great antiquity: it consists of a square facade representing three doorways, the side ones arched at the top, and the middle one with a pointed pediment. The panels are filled up with a zigzag ornament. At each side is a sitting lion or griffin. The fortress called the Seven Towers is the only part of the ancient fortifications which is kept up by the Turks. We landed, and followed the walls on the west side of Constantinople. This side is about three miles in length, and is a most picturesque and interesting walk. The wall next the sea is single; but on this side there are three parallel ranges, with a broad ditch outside of all. The inner wall is of considerable height, and is beset with numerous towers. In the middle wall are smaller towers, placed alternately with the others. These walls are for the most part too perfect to deserve the name of ruins, but nothing can be more picturesque than the parallel ranges rising in gradation one above the other; the towers, some square, some octagonal, covered with ivy, and the whole interspersed with numerous trees which have sprung up spontaneously. There are five or six gates on this side of the city; the southernmost, now walled up, was anciently the grand entrance, and called the Golden Gate; at the sides of it are projecting square wings, or *propylea* of marble, and some columns with an architrave are still standing in the middle wall.

"Towards the northern end of the wall is a large ruined building called the Palace of Belisarius; with what justice I do not know, but, from the antique style, I can believe it to be as old as the time of that general. Beyond this the walls descend towards the Golden Horn; they do not however terminate, but turn towards the south-east, parallel with the port, till they join the sea-wall at the Seraglio point, thus completely surrounding Constantinople.

"4th.—I had intended for some days past to procure specimens of the water-birds which abound near Constantinople. The Turks do not allow them to be shot in the vicinity of the town; I therefore took a boat, and went for that purpose to the Sweet Waters at the head of the Golden Horn. The Sultan took it into his head that day, instead of going to the mosque near his palace, to drive three or four miles into the country to go to a mosque at the Sweet Waters; and as it is forbidden to fire a gun within reach of Imperial ears, after devoting the best part of the day in getting to my shooting-ground, I was obliged to return without effecting my object.

"5th.—Resolved to go in quest of my ornithological game to the Seven Towers, and took a boat for the purpose; but was again unlucky, the wind being fresh from the west, and after pulling for some time in the teeth of it, I was obliged to give it up, so I went to have another trial at the Sweet Waters; the coast was clear of Sultans, but almost equally clear of game; however, after rowing backwards and forwards for some time, I at last succeeded in procuring the birds I wished.

"6th.—Made my way through the entangled streets to the Seven Towers, and walked along the shore beyond, for some way, conchologizing and geologizing. About half a mile from the shore are some extensive remains of stone-quarries now deserted, on the east side of which I came upon a singular relic of antiquity:—a large square space, in extent about three acres, enclosed by a massive wall about 40 feet high, and 10 feet thick; it is built of alternate layers of brick and stone, apparently Roman, or of the early part of the Greek empire. On the west side are a succession of semicircular niches, and on the south the remains of a large arch, with two smaller on each side. On the north is the entrance by a small doorway. This singular enclosure is now used as a sheep-fold: it may have been a reservoir for water, but I could find no traces of an aqueduct leading to or from it.

"7th.—Hamilton went two days ago to Therapia, a village on the European side of the Bosphorus about twelve miles from Constantinople, and I joined him to-day. In the afternoon we walked from Therapia towards Buyuk Dere: I here caught the first sight of the horizon of the Black Sea. Near Buyuk Dere is a fine verdant valley leading towards the Forest of Belgrade; the upper part of this valley is crossed by a large aqueduct, which forms a fine object from the Bosphorus, and near its mouth is a celebrated cluster of nine immense plane-trees growing in a circle, united at their bases, and evidently offsets from one common ancestor.

"8th.—We started in a caique to visit the Black Sea. We followed the Asiatic shore, landing occasionally to examine the rocks, which are here of a volcanic nature, and in some cases occur in a columnar form. As we approached the entrance of the Bosphorus, the weather-beaten cluster of rocks, known to the ancients by the name of Symplegades, came in view. We landed at the promontory called Fanaraki, from the lighthouse which stands on it, and walked across to a sandy bay, called Kabakos. Here I found many white and red cornelians, like those of old on the Yorkshire coast. Beyond this bay is the promontory of Koom Bornou, a mass of columnar volcanic rocks. The entrance to the Black Sea is very striking, from the suddenness with which one emerges from the narrow Bosphorus into the open ocean.

"On the Asiatic side the volcanic formation commences at Kavak under the old Genoese castle, and reaches to Koom Bornou, or further. It consists chiefly of conglomerate of angular fragments of trachytic rocks, imbedded in trachytic paste or in tufaceous matrix. These conglomerates viewed on the large scale are distinctly stratified, and from Kavak to Fanaraki have a regular inclination towards the north. They rest upon, or alternate with, trachytic rocks more or less compact, and sometimes pass into phonolite and basalt. In some places, especially near Fanaraki, dikes of the latter substance are seen intersecting the conglomerate beds. The basaltic rocks occasionally assume a columnar form, seen in greatest perfection at Koom Bornou. The prevailing colour of all these rocks is greenish, though when viewed in detail, they exhibit more variety. There is also a species of trachyte of a yellow colour, forming entire hills, especially on the European side, much decomposed, and resembling at a distance loose sand. Near Fil Bornou the conglomerate contains boulders of a brownish trachyte, softer than the greenish trachytic paste which envelopes them; hence they decay with the weather, and form cavities. The reverse, however, is generally the case; the fragments, being hardest, project from the surface. The trachyte and conglomerate contain veins of chalcedony. In one place, near Fil Bornou, are many of these veins, perpendicular and straight, intersecting the conglomerate, and cutting indiscriminately through the paste and the imbedded fragments. On the European side the volcanic formation begins *en masse* on the north of Buyuk Dere, and a similar series of conglomerate beds dipping north extends to Cyaneæ.

"9th.—Crossed over from Therapia to a place called Hunkiar Skellessi, a verdant valley full of trees. From this we ascended a hill, called the Giant's Mountain. It commands a most lovely view, and is clothed with dwarf oak, arbutus and bay, amidst which was a profusion of sweet violets and primroses, the latter of a purple colour. In the afternoon we returned to Pera, highly pleased with the beauty of the Bosphorus. The lower part of the Giant's Mountain is schistose limestone, becoming massive and compact higher up, blue and hard, with a few crinoidal stems and traces of a plicated Terebratula. It is exposed in some large quarries, in one of which there is a remarkable perpendicular dike, 4 or 5 feet thick, of trap, resembling green sandstone. The limestone is quite unaltered at the contact, and some fragments are enveloped in the dike. Lower down is another dike of white quartz rock, crossing the direction of the former one, and another of gritty quartz rock. Above the limestone is argillaceous schist extending to the top of the hill, containing numerous fossils, corals, spirifers, terebratulæ and crinoidal stems.

"13th.—Accompanied Mr. Cumberbatch, the Vice-Consul, to the Forest of Belgrade*. We proceeded to the large aqueduct near Pyrgos, said to be built by Justinian, but probably of a later age; it is more than 100 feet high, and conducts one of the chief supplies of water to Constantinople. The system of aqueducts which supplies the town is very extensive and curious, the water being brought in some instances from a distance of more than twenty miles. In order to ensure a supply in the dry season, many of the valleys in the Forest of Belgrade, where most of the water comes from, are walled across, forming extensive lakes denominated *bendts*. The streams of water are carried either by means of arched aqueducts or by a contrivance called *souterazis*, peculiar to the Turks, who are said to have borrowed it from the Greeks at the capture of Constantinople. Columns, or high piers of stone, are erected at equal distances across the valley, and the water is conducted in pipes up one side of the column and down the other, then underground to the next, and so on; the Turks fancying that this is a very cheap and excellent substitute for the arched aqueduct.

"22nd.—At any other time I might have felt regret at leaving this beautiful and interesting country, but I had been waiting with much impatience for our departure; it was therefore with great pleasure that I at last felt myself afloat and crossing the Sea of Marmora for Moudania. In order to save time, we went this way to Broussa, instead of going round by Nicomedia, which would have occupied three days more. Our vessel was a large caique, with four rowers and a sail. Towards sunset we approached the picturesque promontory of Bozbornou, and landed for a few minutes to examine the rocks. We arrived at Moudania about 9 P.M. It is a small town, but enjoys considerable trade, from being the port of Broussa. We succeeded in finding a spare chamber in a cafinet, where we passed the night.

"23rd.—Moudania to Broussa. Started at 7·20. Followed the sea for a short distance towards the east, and then turned south, ascending some low hills. Soon after leaving Moudania we passed a

* On the north of Constantinople are some extensive deposits of ancient alluvium, analogous to the Diluvium of England. They contain sand, clay, gravel and large boulders, much rolled. The latter are chiefly of quartzose and sandstone rocks. The clay beds grow the oak forest of Belgrade, the schistose rocks being barren and devoid of trees. They probably skirt the base of the Lesser Balkan towards the north-west.

hill on the right, which is said to have a palæocastron or ruin on the top. This may have been the site of Myrlea, which is known to have been near Moudania. The country here is well cultivated, with mulberry-trees, vines and olives. We saw many signs of spring, which presented an agreeable contrast with the country about Constantinople, where all was wintry in appearance when we left. But here the storks had arrived, and were seen stalking about the fields within a few yards of the ploughmen. Many plants were in flower, including the purple globe hyacinth, and the beautiful large anemone, which is the parent of our cultivated varieties. Even in its wild state it exhibits every variety of colour,—crimson, pink and purple.

"In about an hour we reached the top of the range of hills above Moudania. This range, which I have not seen marked in any map, is the continuation of the mountains on the south side of Lake Ascanias and the Gulf of Isnik. It runs parallel to the coast as far as Mohalitch at least, and probably much farther, having on the south a wide vale or plain, the only drainage of which is effected by the Rhyndacus, which passes through a gap in this chain of hills below Mohalitch.

"On descending, we crossed the river which on the maps is called the Nilufer, but we were told that its name is the Delhi-Chay. Great confusion exists in Turkey with respect to the names of rivers. With some few exceptions, they have no proper name of their own, but are called after the towns which stand upon them; hence a river often has half-a-dozen different names at various parts of its course, according to the towns which are nearest each portion respectively.

"After leaving the Delhi-Chay, we crossed some low hills of rich black mould, but uncultivated, and soon after recrossed the Delhi-Chay, here a small rapid stream, flowing towards the north-east, where it receives many streams from Olympus and the surrounding mountains. One of our principal objects was to explore the Rhyndacus upwards from the Lake of Apollonia, where in the maps it enters at the south-east angle. All day we had in view a ridge of mountains, extending west from Olympus as far as we could see, and looking so steep and impenetrable that I almost despaired of effecting the Rhyndacus expedition. We resolved to follow the base of this ridge till we came to the river, and we could then decide whether its farther exploration was practicable.

"We soon passed the baths, and in twenty-five minutes more entered Broussa: it is still a flourishing place, notwithstanding the decline of the silk-trade. At the west end of the town is a singular platform of travertine, formed, no doubt, by the hot springs in ancient times, which have in consequence become stopped up, and have changed their position to the place where we now find them. On this platform, which is more than 100 feet high, stood the ancient citadel of Broussa. On entering the gateway some walls of Hellenic masonry still exist on each side, and ascending a few steps to the left is a splendid view over the town of Broussa. We then visited a Greek church, now a mosque, in which is the tomb of Orchan, a celebrated Turkish Sultan. This church is built in the same style as the old churches at Constantinople and St. Mark at Venice. A large cross of black marble still remains in the wall, affording an unusual instance of Mahometan tolerance. Beyond this is another mosque, the minaret of which stands on a cubical mass of masonry, too solid to be of Turkish workmanship, and I think it must have been the base of an ancient column. Opposite this is a castle built of brick, of Byzantine or early Turkish workmanship. On each side of the gateway are some defaced bas-reliefs. On our return from seeing these antiquities we called on Mr. Tohrab, a merchant to whom we had an introduction.

"24th.—We walked on the side of Mount Olympus, overlooking Broussa; it is clothed with oak and a variety of shrubs. We then called at Mr. Tohrab's office, and discussed with him and his son the various routes which lay before us, and we at last resolved upon going to Kirmasten, a town said to be on the Rhyndacus, above the Lake of Apollonia. We went to see the hot baths, taking a Jew with us as a guide. Jews are very numerous in the Levant, particularly at Broussa; they are descended from those Jews who were expelled from Spain some centuries since, and they still retain the Spanish language. The hot baths are distant about two miles from Broussa. The platform of travertine continues at intervals all the way from thence to the point where it is now daily forming from the hot springs. There are two baths, the largest for men, the other for women, each supplied by a spring of hot water. The temperature at the first is 184° Fahr.; at the 2nd, 182°. We entered the large bath, which is a handsome building paved with marble, and delightfully clean. The stream of hot water that flows away from the baths swarmed with *Melanopsis buccinoidea* (Oliv.), the same species that occurs in the Meles at Smyrna, and many other streams in Asia Minor. It seems, however, to prefer the hot water, for here it existed only in the hot stream, and not in a rivulet of cold water close by. It was living in a temperature of 97° Fahr.

"25th.—Before leaving Broussa we called again on Mr. Tohrab, whose son was educated in England. He is fond of ornithology, and showed me a great black woodpecker, shot on the top of Mount Olympus; also a rose-coloured ouzel, roller, buzzard, otter, &c. Mount Olympus abounds with wild boars, bears and lynxes. We left the beautiful town of Broussa with regret at half-past 2, and retraced the road to Moudania for about two miles, then crossed a bridge over the Nilufer. We crossed several streams flowing to the north into the Nilufer, and at forty minutes past 5 the village of Chetéh was in sight, where we could see the walls of an old fortress. We turned to the left to the village of Tartali, hoping to find a lodging; but as the head man of the village was absent, we failed in our object, and were obliged to proceed. Vast flocks of storks were flying to the village from the fields, perching on the roofs, and clattering their beaks. In about an hour from Tartali, we came to some low hills on the north, trending round and joining the mountains on the south. Ascending to the top, I could just see the Lake of Abulionte or Apollonia, over the opposite hill. Descending a valley, we proceeded by moonlight, and at half-past 7 entered the village of Cassan-Aga, where, after some delay, we found a lodging.

"26th.—Rose at 4, and as soon as it was light discovered that we had slept not only in a stable, but in an apiary. The wall of the house was full of cells about 1 foot in diameter, closed in front, and in each of these was a swarm of bees. Honey is a great ingredient in Turkish cookery, and the abundance of wild flowers on the mountains favours its production. The hives commonly used are long cylindrical baskets closed at each end by a circular board, with a hole to admit the bees. The baskets are plastered with mud or cowdung, and are piled one upon another in stacks of a dozen or more. They may often be seen in retired valleys distant from any habitation, a proof of the honesty of the Turkish peasantry. We left Cassan-Aga at 6, and in half an hour reached the top of a hill, whence is a view of the Lake of Abulionte, a fine sheet of water about two miles wide at the east end, and expanding to nine or ten at the west. Its length is about fifteen miles. It contains several islands, and from the point where we stood we saw the town of Abulionte, entirely covering a small peninsula projecting into the lake. This is the site of the ancient Apollonia, only visited, I believe, by Fontanier, who saw many inscriptions built into the walls. To the right of Abulionte, and more distant, we saw the town of Karagatch. The lake supplies much fish to Broussa and Constantinople, and boats come from thence up the Rhyndacus into the lake.

"We now descended towards the lake, having on the left the same range of rocky hills which had accompanied us from Broussa. They consist of white limestone, and are covered with brushwood, exactly resembling in appearance the limestone mountains of Greece and the Ionian Islands. In order to find the Rhyndacus, we followed the south side of the lake, which is nearly straight, being bounded by the steep range of limestone above mentioned. We entered a plain between the hills and the lake, which is cultivated, and produces mulberries, pear-trees, vines, &c. Its width gradually diminishes from one mile, and the lake at last meets the foot of the hills, while the road winds along their side. Nothing can exceed the inaccuracy of the maps. Abulionte is near the east end of the lake, which is not above two miles wide at this part, and the islands, instead of being in front of it, are farther west opposite Karagatch. Instead of the two large rivers, the Rhyndacus and the Sendgean Sou, which are made to enter the south side of the lake, we could find no stream along this side larger than a mere rivulet. An hour farther the coast retires to the south, forming a marshy bay, with a small island in front. Coasting the bay, we came to the village of Cara-oglanlú, or Black-Boy Village. Half-an-hour farther, we reached the top of a hill, where we were at last greeted with a sight of the Rhyndacus, entering the lake not at the *south-east* angle, but at the *south-west*, at least fifteen miles farther west than it is placed on the maps. We saw it winding through a plain and entering the lake nearly west of the small island before mentioned. It leaves the lake a few miles farther north, near Ulubad, whence it was again seen winding towards Mohalitch, and probably entering the Propontis through a gap which we saw in the hills. From the point where we stood, Ulubad bore north-north-east and Mohalitch north by compass. The walls which fortified Ulubad during the middle ages were distinctly visible. To the right of Mohalitch I could see a small lake, probably that of Dascylitis;—if so, the Nilufer, which flows through that lake, must enter the Rhyndacus below Mohalitch, instead of flowing at once into the Propontis, as it is made to do by Leake and Rennel. In twenty minutes more we were at the top of another hill, which presented us with a view of the upper part of the same plain towards the south-west. This plain extends along the west end of the Lake of Abulionte, and thence to Mohalitch, and a long way to the west.

"To the west-by-south we saw a high and distant mountain, probably Mount Temnus, and south-west ½ west was another, distant ten or twelve miles. We were much gratified at the sight of the river and the smiling and fertile appearance of the country towards the south-west, so easily accessible, yet hitherto, I believe, never visited by any European traveller. We descended into a plain covered with shrubs, and at 20 minutes past 3 reached Kirmasten, a small town of about 700 houses, seated on the long-sought Rhyndacus. It is surrounded with gardens, cornfields and plantations of mulberries, interspersed with fruit-trees, all in blossom. We crossed the Rhyndacus, a rapid river, about ninety yards wide, and went to a khan, where we had just established ourselves, when a messenger came from the Aga, desiring

C.D.M.S. litho. W. West Imp.

Denundation of the RHYNDACUS in tertiary lacustrine limestone near AYATCH-HISSAR. From the NW.

us to wait on him. We found him—a fat and comely old Turk—sitting in state, surrounded by his friends and attendants. He was Seraskier of Erzeroum at the time it was taken by the Russians, for which misfortune he was punished with virtual banishment, by being appointed Aga of Kirmasten. He received us with great politeness, offered us a room in his house, and begged us to dine with him. Some of the Turks present excited our curiosity by speaking of ruins in the neighbourhood, which we resolved to visit to-morrow. The Aga offered us every assistance, and after our pipes and coffee were discussed, we took our leave. Soon after our return to the khan we were agreeably surprised by the arrival of eight or ten dishes from the Aga's kitchen: we found some of them excellent; others, with a little practice, would doubtless have become so.

"27th.—Started in quest of the ruins which we heard of yesterday. We followed the road to Mohalitch for about two miles, then turning to the left, we came to a range of hills, which extend from Kirmasten to this place. Here we found some masses of Roman masonry, apparently the remains of a wall which had surrounded a small low hill. The ground was covered with pottery and fragments, which prove that an ancient town must have existed here. On the lower ground to the west is a ruined building, perhaps Roman, apparently containing a vault, which has fallen in. Above this, on the slope of the hill to the west, is another building, with massive walls of rough stones laid in mortar. It is perhaps this building which has given the place the Turkish name of Hammamlu, but the remains are too imperfect to decide its original object. The walls project from the hill on two sides, and are supported by several buttresses. These are all the existing remains of this ancient town, the site of which is now occupied by cornfields. It is not easy to say what may have been its ancient name, but I am inclined to think it was Poimanenus, which lay in the road from Adrian to Cyzicus. Or it may have been Hiero-Germa, which town is placed by D'Anville at Kirmasten, or, as he writes it, Ghirmasti (D'Anville was acquainted with Ghirmasti from the Turkish geographers, though no European had visited it, nor does it appear on any modern map). At the latter place, however, we could find nothing more ancient than the traces of a Byzantine fortress, commanding the pass of the Rhyndacus in the burial-ground attached to the mosque. Chishull mentions some ruins at a place called Hammamlu, nearer Cyzicus than those which we visited. We ascended the hill on the west of Hammamlu, and had a view of the plain on the west side of Lake Abulionte, extending northwards as far as Mohalitch. To the west was another plain, with a stream flowing past the village of *Cavaklu* (or perhaps Bavaklu), and entering the Rhyndacus two miles above Kirmasten. We could see nothing of the Macestus or Susugherlu river, which enters the Rhyndacus below Mohalitch, though that river could not have been many miles distant on the west.

"We now returned to Kirmasten, and called on the Aga. We had heard yesterday of a sculptured stone in the mosque, with figures of men and horses, and asked permission to see it. The mosque has apparently been a Greek church, but contains nothing of interest. We searched within and without for the sculptured stone, but in vain; and the attendant even took up part of the carpet to show us the pavement, where we saw a tombstone with a Greek cross, but no inscription, and a broken marble slab with the word XAIPETE.

"We left Kirmasten amidst the admiration of a throng of people, who learnt for the first time that there were other persons in the world besides Turks. The men wear the largest turbans I have yet seen, and tie a small string round them to hold together the ample folds.

"We recrossed the Rhyndacus, and proceeded to follow its right bank. We ascended the low range of hills which passes Kirmasten, where it is cut through by the Rhyndacus. It is evident that these hills have once sustained a lake, which has been drained in consequence of this barrier being worn down by the river. Immediately above the gorge at Kirmasten the valley is distinguished by a level platform the deposit of this ancient lake, 50 or 60 feet higher than the present river. This ancient deposit has been extensively denuded by the modern stream, which flows in a lower plain; portions of the platform occupy one or both sides of the valley for many miles up, and where it is denuded, its former surface is marked in many places along the sides of the hills by a level ridge or bank, like the 'parallel roads' of Glenroy in Scotland. The lacustrine deposit consists chiefly of black clay, sometimes of sand and gravel horizontally stratified, but in the cursory examination I was able to give, I could find no organic remains.

"We now followed up the beautiful valley of the Rhyndacus, surrounded with shrubby hills, its fertile bottom partly cultivated, partly wild and neglected. After passing the village of Hadji-keui, the valley becomes more covered with bushes and less cultivated. Among the trees I noticed two or three species of oak, hawthorn, ilex, pear, cherry, wild vine, agnus-castus, walnut, judas-tree, butcher's-broom, alder, willow and hornbeam; these trees gradually disappear, and the road passes through a regular oak forest. Three hours from Kirmasten we crossed a small hill, and beheld a fresh reach of the Rhyndacus valley, extending eastward. A river of some magnitude here joined the Rhyndacus from the southward. On looking up the valley, we beheld the snowy summit of Olympus bearing east-north-east, by compass. At 5 P.M. we reached the village of Kesterlek, the residence of an Aga, who received us, and gave us a chamber in his own house.

"28th.—We were up by daylight, but our friend the Aga had already gone to mosque. No sooner has the first ray of dawn appeared than the Muezzim ascends the minaret, and his loud, clear chant adds to the solemnity of the hour. Soon after leaving Kesterlek we forded the Rhyndacus, here a rapid stream about 2½ feet deep. Ascending the hill, we came to an old castle on a promontory which juts across the valley from the south, and compels the river to form a horse-shoe round it. It is apparently of the middle ages, and consists of an area with a round tower at the west; the walls are massive, and built of rubble and mortar. The river now ceases to flow through a wide and fertile vale, and becomes pent in on each side, with mountains descending to the water's edge, and covered with fir forest. At a little before 8 we passed the village of Karadja-keui, consisting of a few huts built of logs, the stems of trees being placed one upon another, and dovetailed together at the corners. They are roofed with boards, made by splitting the trees. '*Nam primi cuneis scindebant fissile lignum.*' These boards are kept on by stones, like the cottages in Switzerland; so that the whole edifice is constructed without the use of nails or ironwork. During the next two days we passed many villages built in the same manner. A short way beyond Karadja-keui, the Rhyndacus forces its way through a gorge, between picturesque cliffs of grey marble covered with shrubs and fir-trees, after which the road leaves the vicinity of the river, and ascends the mountains on the right. From the steepness and irregularity of the track, our progress for the next three hours was very slow. At half-past 9, in passing along a ridge with a ravine on each side, a view opened down the valley of the Rhyndacus, apparently as far as Kirmasten. At ten minutes to 11 a distant lake was in sight, bearing by compass north-west-by-west. This must have been the Lake of Miletopolis, which we had not seen from Hammamlu, owing to the inferior elevation. We still continued through forests of pine, enlivened by an abundance of yellow crocuses, hyacinths, sweet violets, and a species of wood anemone with a blue flower. At five minutes to 1 P.M. we emerged from these pine-clothed mountains into a wide valley running nearly east and west. Olympus was again in sight to the east. In crossing this valley we passed the village of Chivilu, a mile distant on the right, near which we saw an old castle, apparently of the middle ages, proving that this district has been well defended under the Byzantine empire. Crossing this valley we soon came to another, in which are the villages of Doondár and Bourmá. We again ascended

i 2

the hills, and in forty minutes came to the village of Kayah-jik, which may be translated *Clifton*. We here rejoined the valley of the Rhyndacus, from which we had been separated for some hours. It here receives a tributary from the south, which we crossed soon after leaving Kayah-jik. In a few minutes more we came to a large and neat-looking country-house, surrounded by a large courtyard, 'with stables, outhouses and other appurtenances.' I had no time to admire this unexpected apparition, for our whole party turned unceremoniously into the yard, and commenced to unload the luggage. I found that this was the *Konak*, or residence of the Aga of Adrianós, where we were to pass the night. There is no modern village of Adrianós, that name being applied to the district over which the Aga's jurisdiction extends. It is commonly pronounced *Adernás*, but as the Aga called it Adrianós, which is but the accusative case of the ancient name of Adriani, the latter may be considered the most correct orthography. The Aga is a regular Turkish country squire. He asked many questions about the English and their customs, whether we kept Lent, and whether we believed in the creation of Adam, the general Deluge, &c. He showed us an atlas of Turkish maps printed at Constantinople, tolerably accurate; but in the map of Asia Minor I noticed the same errors as in all European ones. He said that Adrianós was called *Ornos* by the Turkish historians, who describe the conquest of this country from the Greeks. He mentioned a place in the vicinity called Kaya-hammam, where there is a hot spring of inferior temperature to that at Broussa, and the remains of an ancient bath. The Aga showed us an English thermometer which he had purchased at Constantinople; he was unable to read the figures upon it, but had marked with a pencil the extremes of heat and cold which it indicated. The greatest cold had been during the past winter, when it had sunk to 5° Fahr., as indicated by a pencil-mark.

"29th.—The valley of the Rhyndacus expands at Adrianós, where it becomes wide and well cultivated. We went this morning to visit an old Byzantine castle with square towers, standing on an isolated hill of marble, close to the river. On the slope of the hill outside the castle are traces of habitations, and on the north side is a ruined bridge, perhaps of Roman construction. On our return to the north end of the hill east of the Aga's konak, we found some men digging up Roman tiles for building. Still we had not found the site of Adriani, which we knew could not be far distant. The Aga spoke of an *old church*, which we thought might assist our researches. A traveller in Asia Minor should never pass an *old church* without visiting it. If he is told of an *Eski-kalé*, or old castle, it will probably turn out to be an old Byzantine or Turkish fortress; but an *Eski-klissa*, or old church, will commonly prove to be a temple, gymnasium, or other ancient Greek building. Proceeding south from the konak, we turned south-east, and in half an hour came to some ruins which at once proved that we were on the site of Adriani. We first passed four piers of a gateway of Roman workmanship, leaving a wide entrance in the middle and a smaller one on each side; this has probably been the entrance into the city on the west side. Beyond this is a large oblong building composed of marble blocks, squared and closely fitted without mortar. The walls may easily be traced all round, but are most perfect at the west end, where they attain the height of 25 feet. They are about 2 feet thick, and one stone in thickness throughout. The layers or courses vary in height from 1 to 3 feet. At the east end are some small apartments formed by an inner wall, distant 17 feet 6 inches from the extremity. This large oblong building is the *old church* of which we had been told. We could not satisfy ourselves as to its original destination. It has much the appearance of the cella of a temple, but we could find no remains of columns of proportional magnitude. Near the north-west corner of the building is a large portion of an Ionic architrave, richly sculptured. This is the only fragment we saw of sufficiently large proportions to have belonged to a temple the size of the large building. We found many fragments of columns dispersed, but none more than 2 feet in diameter. No other buildings remain at Adriani sufficiently perfect to allow their form to be traced. The site is all cultivated; the stones have been collected into heaps, and are overgrown with briars and bushes. We found many capitals and sculptured fragments of several different orders, proving that a variety of public buildings must once have existed here. Adriani stands in a small plain, communicating on the north-east with the valley of the Rhyndacus. It is at the south base of an isolated hill of marble, which extends as far as the Aga's konak.

"About two miles further is the village of Bayjik, where we saw some inscriptions in the wall of the mosque, and we stopped to copy them, amidst the astonishment of the whole male population. It was now the Beiram, a grand Turkish festival, and all the inhabitants were in their holiday costume. The arrival on such an occasion of a set of strange-looking beings like ourselves was a source of vast delight and amusement, nor was their wonder diminished when they saw us proceed unceremoniously to copy the unknown characters cut in some old stones in the wall of their mosque. In order to divide the spectators, Hamilton copied the inscriptions on one side of the mosque, while I was at work at the other*. We were each surrounded

* See Hamilton, ii. App. pp. 399, 400.

by forty or fifty dashing-looking fellows, dressed in their gayest costume, who pressed round us with intense curiosity, yet with perfect good humour and freedom from rudeness.

"At a fountain in the village is a small bas-relief, but in bad preservation. These marbles had been doubtless brought from the neighbouring ruins of Adriani. We now left Bayjik and ascended some hills or mountains of fine white marble, bare and rocky, with shrubs growing between the crevices. Further on, these hills become covered with oak-trees and a shrubby species of cistus. On our left, at about two miles distant, the deep valley of the Rhyndacus could be traced, but the river was too low down to be visible. Beyond this the snowy range of Olympus covered three or four points of the horizon, for we were now no great distance from it. At last we reached a brow overlooking the valley of the Rhyndacus, and presenting a magnificent prospect. The river was below us at the depth of 800 or 900 feet, flowing through a vast ravine, for it was too narrow and deep to be called a valley. In front, the river entered through a gorge between cliffs of limestone 600 or 700 feet high. This limestone, which is proved by the fossils it contains to be a lacustrine deposit, occurs in strata nearly horizontal, which are exhibited on each side of this gorge.

"A few minutes brought us to the village of Ayatch-Hissar-keui, where we intended to have stopped for the night, but the population seemed panic-struck at our arrival, apparently thinking that we were some government officers come to make conscription, and seemed so unwilling to receive us, that we proceeded to the next village.

"We now descended along a precipitous track to the river, looking at a small fortress called Ayatch-Hissar on our way; it is built on a narrow promontory above the river, which we crossed by a wooden bridge. The proper name of this river is Ghiuk-Sou; but it is called by various names, according to the towns which stand upon it. We now mounted the opposite side of the valley, and after a long and steep ascent, reached the village of Haidar. We found the men all employed in some religious ceremony at a short distance from the village, which was quite deserted; in a few minutes they returned, and seemed much astonished; however, they did not show the same alarm, but conducted us to the Oda or public chamber for strangers. From Haidar there is an extensive prospect to the south, bounded by a distant snowy mountain. This must be the '*Mount Tamonedje*' mentioned by Rennel, as seen from Olympus*. From Haidar it bears south-half-west by compass. I am inclined to think this is Ak-Dag, the mountain on the west of Ghiediz, which we passed some days after.

* Rennel, Geog. Asia Minor, i. p. 246.

"30th.—Leaving Haidar and proceeding south-east, in an hour and a half we again crossed the Rhyndacus, here a rapid stream 15 yards wide, overgrown with plane-trees, alders and willows, matted together with wild vines. On the opposite, or north side of the river are some remarkable cliffs of the white lacustrine limestone in horizontal strata.

"After following the river upwards for a short way, we left it, and turned towards the south-east: ascending some stony hills we reached the village of Harmanjik, a small, wretched place, but it derives some importance from being the residence of an aga, and therefore the chief place of the canton or *Kazay*. These kazays contain ten or twenty villages, and are governed by an aga. Each village is under the superintendence of a *Mochtar*, or sometimes two when the village is large. Harmanjik is perhaps the place called Karmen-Kiahia on the French map of Guilleminot.

"After leaving Harmanjik we crossed some bleak hills to the south-east, with Olympus in sight on the left, and the mountain which we saw from Haidar on the right. We crossed a deep valley, through which flows the stream which joins the Rhyndacus to the north-west of Harmanjik. Ascending the other side we came to Eshén, where we stopped for the night; it is a miserable village of log-huts roofed with split boards, the public oda being the best house in the village. The hills around Eshén are composed of trap and volcanic rocks, with an abundance of iron ore. The ore appears to be very rich, and might be worked to great advantage in a country covered like this with forests, and possessing abundant water-power. At present no advantage is taken of this source of wealth, and the iron-ore of Eshén remains in store for posterity.

"31st.—This morning was miserably cold and wet, and we travelled some way through sleet and rain; our road lay still over hills covered with pine. In two hours and a half we reached the village of Debrent, where we gradually merged into a high plain, at the end of which is an abrupt descent into a valley, in which are several hills of a volcanic appearance, and which we found to consist of igneous rocks. On our left were some fantastic-shaped rocks of white sandstone, and in a few minutes we came to a curious sepulchral monument*, sculptured in the rock. It is a level façade divided into two portions, the upper part of which stands back about three feet behind the lower, and is crowned by a pointed pediment. From it descends a kind of well about 20 feet deep behind the lower part of the facade, which forms a wall of solid rock in front of it, about 2 feet thick. An oblong recess is cut in front, opposite the lower part of this well, with three semicir-

* See Hamilton, i. p. 97.

cular projections on the upper part, but without any other ornament. The only original entrance to the well has been from above, the recess in front not opening into it; but in later times a circular opening has been made in the centre of the recess, about 18 inches in diameter, for the sake of examining the interior, the opening at the top not being accessible; the total width of this monument is 21 feet, and its height about 40 feet. It is cut in a kind of white sandstone containing hexahedral crystals of hornblende. In the surrounding rock are many cavities which resemble sepulchres, but I could find no proofs of their being artificial, and they seem to be caused by the natural washing-away of the softer parts of the rock.

"The curious tomb above described is probably a monument of the ancient Phrygians, as it is similar to others which have been found in various parts of Phrygia. Some of these are described as being a mere façade without any internal cavity for depositing the body, but it is possible in such cases that a cavity may exist in the form of a well, the mouth of which may be inaccessible and invisible from below. We should never have known of the existence of the inner cavity, had it not been for the opening which has been broken through the lower facade. After leaving this curious monument we continued to follow the same valley, surrounded by volcanic hills, till it led us out into that of the Rhyndacus, of which we had lost sight early the day before. It here diminishes much in size and rapidity. We soon crossed the Rhyndacus, and reached the village of Moimoul, chiefly inhabited by blacksmiths. The hill above it consists of rich iron-ore, but I could not learn that it is smelted, and was told that the iron used is brought from a distance.

"A mile further is Taushanlu, a considerable town governed by an aga. In the Armenian burial-ground on the east side of Taushanlu are some ancient marbles, on which we found one or two very imperfect and unimportant Greek monumental inscriptions. These relics render it probable that the fine plain of Taushanlu was the seat of some ancient town, but at present there is no clue either to its name or exact site. From the time of our leaving Broussa we had been traversing a district hitherto unexplored, but we had now emerged into a better-known country. Taushanlu was visited first by Brown, and afterwards by Keppel, in his way from Azani to Broussa. To the south-west of the line which we followed, there still remains a large district wholly unknown, extending south as far as the river Hyllus, and west as far as the high road from Smyrna to Constantinople. This is the district anciently called Mysia Abrettene, watered by the river Macestus, and formerly possessing many ancient towns, as Ancyra, Synaus, &c.

"April 1st.—Leaving Taushanlu, we crossed the plain towards the south and forded the Rhyndacus, here 10 yards wide and 18 inches deep. In 20 minutes more came to the village of Goorchay on the south side of the plain, and began ascending the hills, which assumed the universal clothing of Anatolian mountains, evergreen shrubs, dwarf-oak and pine-trees. I here noticed mistletoe, apparently the English species, growing on the pine. We lost sight of the Rhyndacus some three or four miles distant on our left, and saw no more of it till we reached Azani. We ascended these mountains to a great height, and passed many patches of snow. We then descended near the village of Gozuljah, and entered a formation of soft white limestone with beds of flint, assuming exactly the character of some of the English chalk districts,—smooth, rounded white hills, covered with fir and juniper, and intersected by dry valleys. Further on the fir-trees cease, and the only tree that remains is a prickly species of cypress with small cones.

"About seven hours from Taushanlu we came to a wide and shallow valley opening towards the north-west and containing two villages, Sheklar or Isheklar being one of them. Leaving them to the right, we suddenly came to the brow of a hill, and beheld the fine plain of Azani, stretching many miles to the east and south. We descended to the village of Ooranjik. We here saw many inscribed marbles brought from Azani, but did not stop to copy them. We proceeded to Tchavdour, the village on the site of Azani, about an hour and a half further, the Aga of Ooranjik giving us a note to the Kaya or head man of that village. Crossing the plain, we passed several streams flowing into the Rhyndacus further north. Keppel seems to have mistaken them for that river, as he says he crossed it several times in going from Azani to Ooranjik. In about an hour the beautiful temple of Azani came in view. The sun had set as we entered this ancient city, and the full moon had risen over the eastern hills, harmonizing well with the sombre hues of the Ionic temple and the other ruins.

"This small village, which is built in the midst of Azani, is properly called Chavadour or Tchavdour, not Tjavdere-Hissar as Keppel calls it*. The ancient town of Azani is scarcely mentioned by classic writers, and plays no part in history; yet the extent and beauty of its public buildings prove it to have been a town of great wealth, and the inhabitants have shown a remarkably good taste in the style of their sculpture and monuments. The greater part of the antiquities remain on the spot, a few only having been removed to the neighbouring villages to decorate mosques and fountains.

"2nd.—Spent a delightful day amidst the ruins of Azani. The best-preserved ruins are the temple and the theatre, including the

* Keppel, ii. p. 201, &c.

stadium which extends in front of it. Besides these, are traces of a Doric temple, and the massive piers of two buildings, one on each side of the river, placed on elevated platforms. There are also two bridges, each of five arches, in a perfect state, and still spanning the stream which flowed under them 2000 years ago. Azani was first discovered by Lord St. Asaph nearly twelve years ago; it has since been visited by Dr. Hall, Dr. Millinger, M. Texier, and other travellers, but the only published account is in Keppel's Travels. We first visited the theatre, which stands on the *left* bank of the Rhyndacus, not on the right, as stated by Keppel. It is remarkably perfect, the whole of the marble seats remaining, and most of the proscenium, showing the doors through which the actors had their exits and their entrances, and the rooms into which they retired. The circular part of the theatre is wholly of marble, but the proscenium is built of massive blocks of common stone, in a much ruder style; and I should have been inclined to consider it the work of a different age, were it not common for this disparity to subsist between the proscenium and the cavea of the ancient theatres. The stage is a stone platform. In front it was decorated with columns sustaining sculptured friezes, which in the theatre at Azani are ruined, and are fallen into the area or pit, which at present was a pool of water. The stadium which fronts the theatre is less perfect, most of the seats having been removed. On each side is a ruined building of massive construction, which have contained apartments connected with the games of the stadium. The people of Azani enjoyed a fine view from their theatre over the proscenium. In front was their town, with its beautiful Ionic temple. Beyond, they looked over the fertile plain to the snow-covered mountains of Ak-Dag and Morad-Dag, the ancient Dindymene.

"The ancient cemetery was on the hill behind the theatre. The hill is still strewed with enormous marble sarcophagi. Another kind of monument which prevails is an upright block of marble with a pointed pediment and a hollow niche in front, in which a panelled doorway is carved. Many of these have been removed from the cemetery and placed by the river-side, where they perform the humble office of washing-tubs. A third kind is the square altar, commonly bearing inscriptions and sculptured wreaths. There are many of these in the Turkish cemetery, where they perform their original office at second-hand.

"We next explored the temple, which is one of the very few Ionic temples now remaining*. The east and south sides have fallen, but the other two are nearly perfect, and the taper columns, each of a

* See woodcut at commencement of chapter, p. lxxxviii.

Volcanic Hill. Basaltic Rock

C.D.M.S. litho. W. West Imp

single stone, seem to be balanced by magic. In the wall of the cella are some interesting inscriptions copied in Keppel's work. The only inscription I could see which he has not copied, is a rude scrawl on the north wall of the temple. Under the temple is a large vault occupying the whole area. The fitting of the stones in the arch of this vault is admirable. Its main object seems to have been to elevate the platform of the temple above the plain, but it was probably used also for purposes connected with the services of the temple. It is lighted by several lateral openings or windows. The rising ground on which the temple stands is probably wholly artificial. On the side next the river it is supported by a row of massive arches, like that called the portico of Eumenes at Athens. On the south-east of the Ionic temple are traces of a Doric temple. If I remember right, there are seven columns standing, at irregular intervals, yet in two rows at right angles to each other. They are without capitals, and no other vestiges of this temple are now traceable.

"The arches of the two bridges over the Rhyndacus are beautifully perfect. The roadways over them are gone, and the wheels of the Turkish arabahs have very nearly worn through the stones which form the arches, and in a few years more will effect their destruction. Azani abounds with inscriptions, chiefly sepulchral.

" 3rd.—Proceeding towards the south, we passed over some low hills of the same chalky lacustrine formation before mentioned, and came to a smaller plain. Here the Rhyndacus may be said to rise, for several small streams here unite, and passing through a gorge in the chalky hills, form the stream which flows through Azani. We now ascended a mountain range of schistose rock, covered with shrubs and evergreens. This range seems to unite on the east with the snowy mountain of Morad-Dag, and on the west it is connected with Ak-Dag. It forms the watershed which separates the valley of the Rhyndacus from that of the Hermus. After passing the summit we descended to the latter valley, which, like most others in this country, is filled up to a certain height with a lacustrine formation of chalky limestone. After passing some low hills of this rock covered with vineyards, we descended to the town of Ghiediz, which stands in a singular situation at the bottom of a narrow valley, and in summer must be intolerably hot. The surrounding hills are chiefly of the white limestone formation, but immediately below the town the valley is blocked up by an outburst of basalt, forming a rugged black conical hill. It is divided into two portions by a narrow and deep ravine, through which the stream which Keppel calls the Hermus makes its way, in a current not more than 6 feet wide, with steep cliffs of basalt more than 100 feet high on each side.

"In the afternoon we visited the bridge in which is the inscription relating to the Council of the Mysian Abbaitans. There are two marble statues built into the bridge, the female figure perfect from the neck downwards. We were pursued through the town by a crowd of boys, whom we effectually baffled by climbing to the top of the basaltic rock, an undertaking far too laborious for a Turk, even though a boy, to perform. Ghiediz is nearly in a line between two snowy mountains. That on the west not above seven or eight miles distant, and is named Ak-Dag or the White Mountain. It is erroneously marked Morad-Dag in Keppel's map. It gives rise to the Hyllus, and probably to the Macestus. Keppel, supposing the small stream which flows past Ghiediz to be the Hermus, has applied to this mountain the ancient name of Dindymene, in which the Hermus is said by Strabo to rise. The other mountain bears east-south-east from Ghiediz, and may be twelve miles distant. This is the true Morad-Dag, being commonly called by that name through the country, and as we discovered that the Hermus rises in this mountain, it may therefore be regarded as the ancient Dindymene. It also gives rise to the Thymbres and to the Banas-Chay, which is the most distant branch of the Mæander.

"4th.—Left Ghiediz and proceeded south: in an hour and a half we arrived at the true Hermus, to which the Ghiediz stream is a mere tributary. It is a river about 30 paces wide and 18 inches deep, descending from Morad-Dag in the east. It is here called Morad-Dag-Chay, and then takes the name of Ghiediz-Chay, which it retains to the sea. Keppel supposed that the rivulet at Ghiediz was the Hermus, from taking Ghiediz to be the site of the ancient Cadi. There is no proof that Cadi was on the Hermus, though it was clearly in the vicinity of that river. There is reason to doubt whether Ghiediz is the site of Cadi; the only argument for it is the similarity of name. The antiquities are so few, that it may be doubted whether it stands on an ancient site at all, and the only inscription which exists refers, not to the people of Cadi, but to the Mysian Abbaitans. We were informed of a village named Ghioklare, on the north or north-west side of Morad-Dag, where there are some ancient ruins standing, 8 or 10 feet high, and an *old church* (temple?). Possibly this may be the site of Cadi near the sources of the Hermus.

"Soon after crossing the Hermus we ascended the mountains on the south. This range separates the valley of the Hermus from the plain of Hushak. Soon after passing the highest point we stopped to collect some curious specimens of obsidian, while the baggage and attendants proceeded; in consequence we missed our way, and came to the village of Gunáy. We did not regret this mistake, as it showed us a very curious *coulée* of trachyte, which has flowed down the valley beyond Gunáy. Descending a deep valley on the south of Gunáy, we found our party waiting for us, near a stream which flows westward, and doubtless joins the Hermus.

"Proceeding south we again ascended to as great a height as before, crossing hills covered with cistus and fir-trees. We at last reached a brow from whence is a splendid prospect towards the east, south and west. In front is the immense plain of Hushak, extending south to the Mæander and the snowy range of Mount Cadmus. Some peaks in this direction were evidently at a vast distance, and probably belong to Mount Taurus. On the east is a range of mountains called Bourghas-Dag, running south from Morad-Dag towards the Mæander. On the south-west were some mountains connecting the parallel ranges of Messagis and Tmolus.

"The whole drainage of the plain of Hushak falls into the Mæander, not into the Hermus, as it is made to do in the maps. We now descended from this mountain range and reached the village of Sorkoon, where we stopped, as it was too late to go on to Hushak.

"5th.—Following a valley towards the south, in an hour we arrived at Hushak. This valley and the vicinity is called Chok Koslar, and it is said in Arundell's work, that the antiquities which exist at Hushak were brought from thence. On inquiry, we were told that no antiquities exist at Chok Koslar, and that the marbles, &c. at Hushak were brought from Ahat-keui, a place which we afterwards visited. On our way to Hushak this morning I heard a nightingale, for the first time this year; they abound in this country in spring, and are great favourites with the Turks. Hushak is one of the chief places for the manufacture of Turkey carpets, and we therefore requested to see the process.

"We examined the inscriptions mentioned by Arundell, which occur in the wall of a mosque, but found nothing new to add to his account of them. It does not appear that Hushak stands on or near an ancient site. The few antiquities in the town were all brought, as we were again told, from a place called Ahat-keui. As we had never heard of this place, we resolved to visit it; we therefore left Hushak and followed the road to Sandukli, in a direction nearly east. On the left, forming an acute angle with the path we were on, was the road to Assioum-kara-hissar, and on the right flowed a small stream, which probably joins the Banas-Chay. The plain which we were traversing produces a considerable quantity of opium. A little past 3, we passed the village of Eekee-Serai, or the Two Palaces, but searched for them in vain. In another hour we came to the village of Chorek-keui, standing on a gentle eminence. Here I saw 'two women grinding at the mill.'

They were sitting on the ground face to face, turning a small mill-stone by means of a wooden handle fixed upright in it. This truly primitive operation seemed a very laborious one.

"We had nearly passed through the village when I chanced to look back at the mosque, and thought I perceived some letters on a marble slab in the wall. Turning back to look at it, I found two inscriptions, which I copied amidst the silent admiration of the whole population of the village.

"One of these is interesting, as it records the erection of a statue by the city of Trajanopolis to Trajan, their benefactor and founder. Trajanopolis is one of the cities of Phrygia, of which the site has hitherto been undetermined. We did not learn whence the stones had been brought, but there can be little doubt that they came from Ahat-keui. The ruins at that place have supplied all the country round with marbles for the fountains and mosques; and as those at Hushak have been brought from thence, there is the more probability that these at Chorek-keui, which is eight or nine miles nearer to Ahat-keui, came from that place also. Therefore, till any proof to the contrary can be shown, we may be allowed to call Ahat-keui the site of Trajanopolis.

"On the south side of the village is a large burial-ground, containing many columns. We now entered a road which was said to be one of the great caravan roads from Constantinople to Aleppo; but this can hardly be the case, being too far to the west. It is not impossible that it may lead to Adalia. At 7 we crossed the Banas-Chay, a considerable river, by a wooden bridge. It here flows south into the Mæander, not into the Hermus, as it is marked in most maps. This error was first detected by the Rev. F. Arundell. Ten minutes more brought us to the village of Soosoos, or *Waterless*, which is a misnomer, as it is well supplied with streams. This place is said to be eleven hours from Sandukli. We were told that the Banas-Chay rises eight hours off to the north, in Morad-Dag. This therefore has the best right to be considered the source of the Mæander, as it is far more distant from the sea than Apamea, where it was considered by the ancients to rise.

"6th.—Started from Soosoos, and at the outset found on the wall of the mosque several inscriptions. When we had copied them, we proceeded to Ahat-keui, a village in a retired valley at the west base of the range of Bourghas-Dag. As we approached the village, remains of antiquity were seen on every side, which convinced us our search had not been in vain. The first object we noticed was a piece of massive masonry at the top of a hill. It proved to be the proscenium of a small theatre, whose massive blocks had bid defiance to time. The seats of the theatre have all gone, but some sculptured remains of architraves denote the past glories of the building.

"The hill on which the theatre stands has clearly been the acropolis of a Grecian town. It is a long narrow ridge, steep and difficult of access, except on the east end. It may be in length about one-third of a mile, and from 100 to 200 yards in width. It has been surrounded by a wall of massive blocks of stone, some of them 5 feet by 4, the foundations of which may be traced in many parts, especially on the south side. It has been repaired in later times, perhaps by the Romans, who have not scrupled to mix the marble seats of the theatre and the dentals of an Ionic temple with the massive blocks which they found on the spot. Part, at least, if not the whole, of the hill has been again fortified in the middle ages, as is proved by the traces of a wall of mortar and rubble surmounting the old foundations. On the north side is a large mass of wall of the same age, perhaps part of a fortress. Near this are the foundations of a temple, raised on a vault, but the columns and decorations have all disappeared. No other relics exist within the area of the acropolis, which is now occupied by vineyards. The entrance has been on the east end, which seems to have been decorated by marble propylea, like those at Athens, of which only the foundations exist. A paved road may be traced leading to the propylea. Above them are three semicircular towers, projecting from the walls at the east end of the acropolis. These propylea are probably alluded to in the inscription at Soosoos. In ascending to the theatre are some tombs cut in the rock; one of them contains two chambers, with niches at the sides to hold bodies.

"A few yards east of the theatre I noticed a bas-relief, half-buried in the ground. I descended to the village, and borrowed a pickaxe of an old woman, who, strange to say, did not hide her face or run away. It turned out a beautiful piece of sculpture, representing one of the Titans fighting. The upper part of the figure resembles Hercules, the right arm holding a club, and the left being protected by a lion's skin. The head is wanting, but the rest is tolerably perfect, and the back and right arm are admirably executed. Gladly would I have taken this ponderous mass to England, but was obliged to content myself with sketching it, and it will probably ere long be carried off to serve as a stepping-stone across a stream, or the corner stone of a mud cottage*.

"Between the village and the mill are the foundations of a temple, of which nothing else remains. Near this temple are numerous vestiges of buildings, proving that the city extended far beyond the wall of the acropolis. We afterwards found that this place was visited some

* See Hamilton, i. p. 117, woodcut.

years ago by Seetzen, and considered by him the site of Acmonia. Not having seen his 'Travels,' I do not know his reasons for assigning it so. Acmonia was on the ancient road from Philadelphia to Cotyæum, whereas Ahat-keui is too far to the south to have been on that road. If the inscription of Chorek-keui previously mentioned could be proved to have come from hence, it would show beyond doubt that this is the site of Trajanopolis. It may have been a Grecian town before the time of Trajan, as the pure Hellenic style of the antiquities seems to show, and may have afterwards changed its name in honour of that emperor, who is styled, in the flattering language of the inscription, its *founder* and *benefactor*.

"Leaving Ahat-keui, which is the most eastern point of my travels, we turned towards the south-west, in order to visit Suleimanli, the place described by Arundell as the site of Clanudda. The hills which we passed are composed of schist, rock and marble. On our right, at the distance of three or four miles, we could see the valley of the Banas-Chay, running parallel to our course. We stopped for the night at a village called Segiclare, which is three or four miles from the base of the range Bourghas-Dag, the south termination of which bears south-east by south, about five miles distant. A stream on the south side of the village was the first we had seen since leaving Ahat-keui. Beyond it, on the south, are two large tumuli, one of them about 50 feet high, probably relics of the ancient dynasties of Lydia or Phrygia. It is most likely that the range of Bourghas-Dag formed the boundary of Lydia on the east. Mr. Arundell visited Segiclare, and supposed from its situation that it was the site of Eucarpia. He mentions some inscriptions in the mosque, but did not perceive that one of them clearly proves it to be the site, not of Eucarpia, but of Sebaste, a town which Mr. Arundell supposed to be in this neighbourhood, from finding a stream near here named Sebasli.

"There are no existing ruins at Segiclare; but the frequency of ancient marbles in the village appears to indicate an ancient site, and the inscriptions prove its name to have been Sebaste. We had no time to visit the burial-ground, which is apart from the village, but could see in it many marble columns and other remains, and it very probably contains inscriptions.

"7th.—We left Segiclare, and proceeded west across the plain, with the serrated outline of a group of hills distinctly marked in the distance. These we supposed to be the volcanoes of the Catacecaumene, but afterwards found that, though of igneous origin, they were older in point of date, as well as geographically distinct from the hills of that district. We passed the village of Hadjilar and recrossed the Banas-Chay, flowing south into the Mæander. We again ascended to the surface of this high plain, which is furrowed by numerous valleys of denudation not visible till one is close on their brink. At first the country is well clothed with oak, but further on the trees totally disappear. The lacustrine formation, which forms this plain, closely resembles English chalk, both in its mineral and geographical characters. At the village of Kalinkesé we inquired for the mill mentioned by Mr. Arundell as the site of some ruins which he did not visit, but could gain no information respecting it. We were told that the nearest mills were very distant. We afterwards crossed a deep valley of denudation, with a small river at the bottom running south, and at half-past 3 reached Kobek. Before entering the town, we collated the inscription on a fountain given by Arundell, and corrected some errors in it. Mr. Arundell passed through the town without stopping, but recommends all succeeding travellers to examine carefully the marbles in the town which have been brought from Suleiman, as they may throw light on the ancient name of that town. We were told that only one inscription exists, and were conducted to the burial-ground to see it: it turned out the very thing we wanted, and proved beyond any reasonable doubt that Suleiman is the site, not of Clanudda, but of Blaundus*. This is confirmed by the fact that Mr. Arundell obtained at Suleiman coins of Blaundus, but not of Clanudda, and his only reason for giving it the latter name was that it stands nearly at the distance from Philadelphia which is assigned on the Pentinger Table to Clanudda. It is true that Strabo places Blaundus near the source of the Macestus; but then he calls it a city of Lydia, which proves that it must have been on the south of the mountain range of Ak-Dag, where the Macestus takes its rise. We must therefore suppose that some error exists in the text of Strabo.

"Our guide now led us to see an ancient castle on the east of the town. A few minutes' walk brought us to the brink of a most singular ravine, traversed by the stream which we had crossed before we reached the town. It is a most curious and striking scene, equally interesting to the painter or the geologist. The stream, flowing through countless ages, has excavated the plateau of white limestone to the depth of at least 600 feet. Its course is singularly tortuous, and the impending cliffs of limestone, furrowed into every possible variety of form, produce a scene which can hardly be depicted by pen or pencil. The stratification of the rock is quite horizontal and undisturbed, and this circumstance gives to the broken crags the appearance of castles, towers, buttresses, and ruins of every variety of architecture. In the midst of this vast ravine rises a conical hill, almost insulated by the contortions of the stream, the summit of which is crowned by a real castle, hardly distinguishable from the numerous

* See Hamilton, i. p. 131.

natural imitations around. We passed a narrow isthmus, which unites it to the side of the ravine, and, ascending to the castle, found only some ruined walls of rubble, the work of the middle ages. We now hastened on to Kobek.

"8th.—Left Kobek, which is a miserable-looking town. We proceeded south-west over the same plain, which extends far to the south, and is only terminated by the range of Mount Cadmus. In the neighbourhood of Kobek it is cultivated, and partially enclosed by hedges of wild almond, which was now in blossom. In two hours we descended into another valley of denudation, in which stands the small village of Suleiman, or, more properly, Suleimanlú. Immediately above it, rises the narrow acropolis of Blaundus, the Clanudda of Mr. Arundell. The acropolis stands on a platform between two ravines, which in wet weather are traversed by the branches of a rivulet uniting below, and flowing into the Mæander. At the time of our visit this rivulet was quite dry.

"The description given by Mr. Arundell is very accurate. Many of the houses of the ancient town probably stood on the sides, and at the bottom of the valley; but all the public buildings, except the theatre, stood on the top of the platform, either within the area of the acropolis, or near the entrance. At the narrowest part of the isthmus, which unites the acropolis to the surrounding plain, is the gateway, which seems to have formed the only entrance. It is defended by two square towers or propylea, projecting in front of the entrance, like those at Athens; they are in good preservation, and well built, the courses of stone varying in height. The only ornament which they exhibit is on the uppermost course of stone, which has been adorned with a row of Doric triglyphs. The doorway, though square, is surmounted by an arch built into the wall to sustain the weight. This and other arches in the interior of the propylea prove that they must have been built after the Romans had conquered the country. We could not satisfy ourselves whether the oblong space on the left after entering the acropolis be a stadium or not. It is the only spot that resembles one in the vicinity of these ruins, and yet the space within the acropolis would probably have been too precious to devote to the purposes of a racecourse. In front of the propylea are remains of three marble temples, and within the acropolis is a fourth, but they have all fallen, and present only confused heaps of ruin. The largest is near the centre of the acropolis, and is called the temple of Claudius, by Mr. Arundell. We found the inscriptions on the architrave given in his book, with some other fragments which have formed part of the same; they record the erection or repair of the temple and Doric portico, or row of columns, some of which are still standing in front of it.

"Near this temple are two statues nearly buried, which might well repay the labour of excavation. There is no marble within many miles of Blaundus, and hence the construction of these temples must have been attended with great labour. The walls of the acropolis and other buildings are composed of the white limestone which occurs on the spot. It is owing to the softness and coarse texture of this limestone that so few inscriptions have survived at Blaundus. We could only find one on the spot, and that too imperfect to decipher. We were, therefore, singularly fortunate in having found at Kobek the inscription before mentioned, proving these ruins to occupy the site of Blaundus. In front of the propylea are several sarcophagi, some of which look as if they were unopened. The usual mode of interment, however, at Blaundus was in tombs cut out of the rock. Vast numbers of these exist along the sides of the valley for some distance from the acropolis. They follow the horizontal strata of the limestone, which, from its softness, afforded great facility for these excavations. I entered several of them, which now afford shelter for sheep; they consist of two or three chambers, with troughs or sarcophagi cut in the rock at the sides for the reception of bodies. In one the walls had been stuccoed, and painted with figures of birds, &c.

"The country to the south and south-east of Blaundus, as far as the Mæander, does not appear to have been explored, and may possibly contain other ancient towns as yet undiscovered. The three arches on the neck of the acropolis mentioned by Arundell have clearly belonged to an aqueduct. After leaving Suleimanlú we followed the line of this aqueduct to the north for a mile or more, and then turned towards the north-west. We now made for the isolated cluster of hills which we had seen the day before across the plain from Segiclare. We first crossed the valley through which runs the great road from Hushak to Allah-Shehr; we passed it about a mile to the south of the village of Euré. Passing the village of Karadjá, we entered the cluster of volcanic mountains by such an exceedingly rugged road that we were unable to reach Takmak that night, it being nearly dark when we emerged from this cluster of hills into a plain on the west side. A few scattered tents of Eurukes were the only habitations to be seen, and we were anticipating a couch in one of them, when we discovered a solitary chifflik, belonging to the Aga of Takmak, who lets it during summer to an Euruke herdsman. These nomadic shepherds are generally hospitable and harmless, though, in point of civilization, hardly raised above savages. The tribe among whose tents we now were, retire to the vicinity of Cutayeh in the winter, and in summer spread themselves over the uncultivated parts of the plain of Hushak.

k 2

"9th.—The morning was dark and gloomy, and suited well with the bleak-looking country, covered with masses of grey rock. In rather more than an hour we reached Takmak, a small town, where an Aga resides in a castellated konak. We called upon him, and found him a polished Constantinopolitan, who had only entered on his appointment a few days before. He was delighted to see us, as he already felt the difference in point of gaiety between Takmak and Stamboul. We proceeded from Takmak by a different route from Arundell's, and consequently missed all that highly interesting geological country he describes near Sirghie. We passed over some hills of micaceous schist, covered with stunted oak, producing abundance of large galls, of two kinds, one smooth and glossy, the other hirsute. In about three hours from Takmak are some curious examples of the effect of time and weather on the schistose rock; vast blocks of it are seen projecting above the surface, one of which is exactly similar to those weather-beaten masses in Devonshire, known by the name of *cheesewrings*. On the left, three or four miles distant, was a mountain of considerable height, probably continuous with those at the head of the Cayster valley, which unite the Tmolus and Messogis ranges. On the right, at intervals, we caught glimpses of the vale of the Hermus, seemingly six or seven miles distant. At 2 we reached Aktash, a miserable village, still among the mountains. In the afternoon we ascended the hill east of the village, and were rewarded by a magnificent view of the vale of the Hermus, and the ranges of mountains which enclose it on the north, extending from the distant Morad-Dag to the neighbourhood of Thyatira. In the latter direction we saw a remarkable mountain, resembling a truncated cone, very distant and of great height; it bore north-west by compass, and we never saw it again in our subsequent journey. Towards the west-south-west we saw some snowy ranges, which turned out to be part of the Tmolus range between Sardis and Allah-Shehr. In the bottom of the ravine below Aktash is a strong chalybeate spring.

"10th.—Crossing the hill on the west of Aktash, we descended into a valley, two miles wide, opening on the north into the valley of the Hermus. A small river, descending from the mountains on the south, flows through the vale into the Hermus.

"We were now approaching that volcanic district called by the ancients Catacecaumene, or the *burnt country*. In the midst of the vale, towards the north, we saw two conical hills, whose form at once proved them to be extinct volcanoes. We ascended the west side of the valley; having gained the top, we saw in front a scene of the highest singularity and interest. Two miles distant rose Kara-dewit, the *Black Inkstand*, a lofty cone of ashes and scoriæ, rising at a rapid inclination, and nearly devoid of vegetation. A vast stream of lava, blacker than night, has inundated the verdant plain to the south, and after flowing two or three miles, has at last congealed, terminating with a rounded slope, 20 or 30 feet higher than the alluvial plain*. The lava having flowed completely across the valley, has intercepted the drainage and produced a small marsh. In order to avoid the floods to which the plain is consequently subject, the town of Koola has been built on the extremity of the *coulée* of lava, while the gardens and cultivated ground are on the plain below. A raised causeway connects the town in time of flood with the hills which form the sides of the valley. We entered Koola, which is a tolerably well-built town, and enjoys considerable trade. When we had established ourselves in a khan, we set forth to visit Kara-dewit. A path leads to it across the *coulée* of lava, which has been partly cleared of the loose blocks to form a road. On either side the lava is tossed about like the waves of the most tempestuous sea, presenting an extraordinary scene of desolation. The lava is wholly devoid of vegetation, and though erupted at a time long anterior to history, shows not the slightest tendency to decomposition.

"After a walk of about two miles over the lava, we reached the base of the cone of Kara-dewit, and commenced the ascent. It consists of loose ashes and scoriæ, in which a few straggling plants contrive to vegetate. It is in much the same condition as the Gravenne de Montpezat in the Vivarais, and perhaps about the same height. The crater is not on the summit, but about half-way down on the north side; it is of small size, but very perfect. We remained some time on the summit, admiring the view, which is magnificent. The *coulée* of lava which we had crossed was now seen to surround the mountain on the west, and to have flowed for several miles in a northerly direction. Enormous as is its bulk, it seems all to have resulted from one eruption. The manner in which it has followed the inequalities of the surface is strikingly displayed. In one place it has encircled an older cone of a horseshoe form, which rises from it like an island. On the surface of the *coulée* are many small cones of scoriæ, terminating in acute points.

"Within the circuit of a few miles we could see numerous other volcanic cones. It was very evident that, however ancient might be the volcano on which we stood, all these others were of far greater antiquity. Kara-dewit is a vast mound of naked scoriæ, rising nearly to a point, at an angle of about 30°. The surrounding cones, on the

* If a cauldron of pitch were upset on one end of a green carpet, and a cartload of ashes shot down by the side of it, the product would represent in miniature the scene which we now beheld.

contrary, have been softened down by the hand of time, till their inclination does not exceed 20°. Their summits are rounded, and they have evidently lost much of their original height. Their craters, where they exist, are often nearly obliterated by the falling-in of the sides. Kara-dewit, moreover, is wholly unfit for cultivation, both from the steepness of its sides and the barrenness of the soil, while the other cones are cultivated to their summits with vineyards, which still, as in the time of Strabo, produce an excellent wine. But the most marked distinction between these two periods of volcanic eruption is seen in the rugged and naked condition of the lava at the base of Kara-dewit, compared with that which has flowed from the other volcanoes. The latter is rarely discoverable, its surface being covered with vegetable soil, resulting either from its own decomposition or from ashes thrown out by later eruptions.

" Recent as is the appearance of Kara-dewit, it has certainly been extinct for 3000 years, and perhaps for twice that period. Yet the action of the elements has effected almost nothing towards reducing it to the state of those other cones which are scattered around. The age of the latter must therefore be something enormous, if compared to the limited epochs by which human history is measured. Old as these volcanoes are, they are more recent than the vast lacustrine deposit which once filled the valley of the Hermus. From Kara-dewit we could see platforms of this lacustrine formation, capped with beds of basalt, resulting from the volcanoes scattered around. These platforms are deeply denuded and furrowed by the Hermus and its tributaries, and their investigation would well repay the geologist who had time to devote to this highly interesting region.

" 11th.—Leaving Koola, we ascended the hill west of the valley in which it stands. We now passed numerous cones of ashes, reposing on the marble and schistose rocks which form the substratum of the country. All these hills possess the rounded outline and high state of fertility which denote a great antiquity. In about two hours from Koola we saw on the right a volcano of a later age, black and steep, with a perfect crater on the top, and a *coulée* of the blackest lava flowing from its base towards the Hermus. In another hour we passed a third *modern cone*, surrounded by others of higher antiquity.

" Having arrived at a fountain at the top of a hill, we took our last view of the Catacecaumene. In front was a valley which led us out into that of the Cogamus, a wide and fertile plain, bounded on the south by the snowy range of Tmolus, skirted by a highly picturesque fringe of broken sand-hills. Towards the head of a valley was a town, which I supposed to be Allah-Shehr. As we advanced westward, the magnificent plain of Sardis opened to our view. On the left it was

Cliff of columnar lava.

Hermus.

Schistose Hills.

C.D.M.S. litho.

W. West Imp

LAVA coulée of ADALA.

bounded by the Tmolus range running west till it terminated in the distant Tartali. Mount Sipylus lay like a barrier across the furthest extremity of the plain, and in front we could see the Gygean Lake, the eternal tumulus of Halyattes, and the mazy windings of the Hermus.

"We now descended to a lower level, to the miserable village of Toubailú. I found afterwards that Dr. Cramer makes this the site of the ancient Tabala, but Mr. Arundell seems to prove that Tabala was at the modern Davala above Koola. We soon after reached the bank of the Hermus opposite Adala, and forded it without accident, though it was 160 paces, and 2 feet deep. On reaching Adala we visited the Aga, an enormously fat man, who was sitting in the gateway of his konah, surrounded by splendidly dressed attendants. Adala is a small, wretched town, and nothing but the similarity of names induces the belief that this is the ancient Attalia. The only antiquity in the place is a ruined Turkish tower. We were surprised at finding that Adala stands in a situation similar to Koola, on the extremity of a *coulée* of lava. This discovery was the more unexpected as we had left the Catacecaumene behind us, and no volcanic mountains were now in sight. We lost no time in going to examine this lava stream, which is as black and rugged as that at Koola. It has flowed down the narrow gorge by which the Hermus enters the plain of Sardis, and has proceeded from some volcano higher up, here out of sight. The first effect of this vast stream of lava has been to stop the course of the river, but in the lapse of ages it has cut itself a channel between the lava and micaceous schist, at the side of the valley, and the river now flows at the foot of a cliff of solid lava from 60 to 80 feet high. We followed this cliff for nearly two miles, to a narrow gorge once filled by the lava, which has been worn away by the river on one side and by a torrent on the other, till it becomes very narrow, and in one place not above 3 yards in width. It is here smooth on the upper surface, and the perpendicular cliffs exhibit an imperfect columnar structure. At the furthest point which we reached, the river has completely cut through the bed of lava, shifting its course from the right side of the *coulée* to the left. This is a most interesting spot, bearing the closest resemblance to the similar cases of denudation in Central France, and proving that although the volcanic outburst has been posterior to all the regular formations of the country, yet that it must be sufficiently ancient for the Hermus to have cut a channel 70 or 80 feet deep, between the lava and schistose rock.

"12th.—Left Adala, and proceeded to the westward, steering for the tumulus of Halyattes, which stands conspicuously on an eminence between the Hermus and the Gygean Lake. The plain of Lydia over which we were passing is generally very marshy, but might be easily drained, as there is sufficient fall into the Hermus. In the neighbourhood of Adala the plain is covered with tamarisk, and was melodious with the song of nightingales. This marshy plain suits the aquatic habits of the buffaloes, which were wallowing in the miry pools like hippopotami. We passed some miserable villages of mud-huts inhabited during the summer by Turcomans, who are now almost the sole occupants of the once glorious domain of Gyges and Halyattes. I saw in the course of this day several rare species of birds, which find their proper habitat in this marshy plain. A large species of Plover was frequent. We also saw a flock of Cranes, *Grus cinerea*, a pair of the *Ardea purpurea*, vast flocks of *Larus ridibundus* ?, also *Anas rutila*, &c. &c.

"We now approached the low grassy down which is covered with the tombs of the ancient Lydians. It consists of undulating ground, rising from 100 to 200 feet above the marshy plain, and separating the Hermus from the Gygean Lake, which washes the north foot of these low hills, and in wet weather sends its superfluous water past their eastern front into the Hermus. On ascending the rising ground we first visited its northern brow, which overlooks the Gygean Lake; this fine sheet of water lies in a transverse valley which connects the valley of the Hyllus with that of the Hermus, and isolates the mountain above Mermère; the lake appeared to be about two miles wide and six miles long. A village which we saw at the north-west extremity was probably that of Mermère.

"We now turned south for about a mile, and arrived at that majestic tumulus which is generally admitted to be the one described by Herodotus as the tomb of Halyattes, king of Lydia. The first *coup d'œil* of this extraordinary tumulus, probably the largest known, strikes the beholder with amazement at the persevering toil of thousands who have laboured to raise this magnified ant-hill. It is a monument which required no art or ingenuity for its construction, nothing but brute labour concentrated to a common focus, and as such it carries one back to those primitive ages when political power consisted, not in the perfection of arts and of civilization, but in the accumulation of vast masses of physical force.

"Between fifty and sixty other tumuli are scattered near the tomb of Halyattes, and two of them nearly, if not quite, equal it in size. Their coating of fine green turf, and the low angle at which they rise, will probably preserve them to the end of time. It is curious to contrast the permanency of these rude monuments with the transitory existence of the various dynasties who have held this country in succession. The princes whose bones moulder beneath these mounds succumbed to the Persian yoke, and that in turn was destroyed by Alexander, and they were followed by the Goths, Crusaders and

Turks, who have in turn desolated the beautiful and fertile vale of Lydia. At present the half-savage Turcoman feeds his sheep and camels on the grassy tombs of the Lydian monarchs.

"We had not time to make an accurate measurement of the tomb of Halyattes; but by means of pacing I estimated the direct ascent at 550 feet. It ascends at an angle of about 18°. Its form is not perfectly circular, but obscurely quadrilateral, agreeing with the account of Herodotus. On the summit is a large, nearly globular block of stone, about 10 feet in diameter, with a flattened surface on one side. It appears originally to have stood in the centre upon its flat side, but is now upset, and lies in a small depression on one side. I conceive it to have been a rude representation of a human head, as there is a wide cavity corresponding to the mouth, but the upper part containing the eyes and nose has been broken off. On the south side of the tumulus is a considerable cavity formed by the washing away of the loose materials which compose it. This would be the most eligible point for excavating. It is probable that an arched vault within would be found, similar to the tomb of Agamemnon at Mycene; and the antiquarian would probably be rewarded by finding valuable and interesting relics of ancient Lydia, as there is no reason to suppose that this or any surrounding tumuli have ever been opened. At present, however, the jealousy and suspicion which such an attempt would excite in the minds of the Turks, would render the undertaking highly difficult, if not impossible.

"We would gladly have devoted one or two days to a minute examination of all the surrounding tumuli, but we were obliged to content ourselves with a cursory inspection of this one, and then bent our steps towards Sardis. We first crossed the marshy plain, and then forded the Hermus, not without some difficulty, for it is wide and rapid, and in wet weather impassable. In front was the fine mountain of Tmolus, with patches of snow scattered along its ridge. Along its base was a belt of sand-hills, whose furrowed sides and fir-clothed summits exhibit a most picturesque and varied scene. On one of these, isolated from the rest, once stood the citadel of Sardis, and the proud metropolis of Asia Minor was spread around its sides and base. Nothing remains of the ancient citadel except a few fragments of wall, which still totter on the brink of the mouldering precipices. The sand and gravel washed down has accumulated on the ancient site of the city, and adds to the obscurity which envelopes its desolate remains. Sardis presents little except some unintelligible masses of Roman brickwork, scattered among undulating grassy mounds. Its only resident inhabitants are a miller, and a Greek who keeps a wine-shop.

"The building supposed to have been an ancient church is the best-preserved ruin at Sardis, it being the furthest removed from the crumbling sides of the acropolis hill. It consists of four pairs of square marble piers, sustaining portions of brick arches which have formed a groined roof. It is the work of the Roman empire; but the only proof of its having been a church is its direction, which is exactly east and west. The semicircular projection at the east end proves nothing, for there is a similar one at the west extremity. On the slope of the hill above the mill is another building very similar in construction. The piers consist of blocks of sculptured marble stripped from Grecian edifices, and laid in mortar; the brick arches which the piers support are of Roman workmanship. A little higher are the remains of the stadium. The lower side is formed by an arched passage running the whole length, and supporting the seats on its upper surface.

"The theatre is on the opposite side of the stadium. The wings are standing, and consist of rubble-work faced with squared masonry. No seats remain in this theatre. We ascended round the east side of the acropolis, up some almost perpendicular precipices. I can fully appreciate the prowess of the Persians who captured this almost inaccessible fortress from the Lydians, of which event Herodotus gives an interesting recital. The whole of this hill consists of sand and gravel, which has suffered such destruction from the lapse of time, that the area of the citadel is now reduced to a few yards square, and the greater part of the walls have long since perished. The portions that remain totter on the brink of lofty precipices, and seem ready to follow their predecessors at any moment. These walls seem to belong to the Roman times. They consist of brickwork and blocks of marble derived from ancient buildings, and brought by immense labour to this elevated spot. Among them are many inscriptions, which I find have been given by many other travellers.

"The acropolis of Sardis commands a magnificent view, extending to the north over the rich plain of the Hermus, and to the south over the picturesque and wooded sand-hills which skirt the sides of Tmolus, whose snowy heights appear in the background. We now descended to the temple of Cybele, which exhibits a melancholy proof of the ravages, not of time, but of wanton barbarians, who, from the absence of limestone in the vicinity, resort hither as to a quarry, and consign these marble columns and capitals to the lime-kiln. Of these colossal columns seven were standing in the time of Chandler. There are only two now remaining perfect, the fragments of the remainder being strewed around. The sculpture of the capitals, which are Ionic, is beautiful, and the temple, when perfect, must have been one of the finest. There is an account of this temple by Mr. Cockerell in Leake's 'Asia Minor,' p. 342. Mr. Cockerell considers it to be as old as the time of Crœsus, but I find it difficult to believe that so pure a specimen of Grecian architecture could have been built by the Lydians of that early age.

"13th.—Left Sardis, passing by a large brick building, probably a Roman gymnasium, but supposed by some to be the Gerusia or almshouse for old men founded by Crœsus, and praised by Vitruvius as a remarkable specimen of the stability of a brick building. This, however, I much doubt; and I do not think, with the exception of the temple of Cybele, any of the visible remains at Sardis are older than the Roman times. We now crossed the far-famed Pactolus, a brook about 4 yards wide, and followed the caravan road towards Smyrna. The ancient road must have followed nearly the same line, and we noticed traces of the pavement of it in one or two places. No particular object arrests the attention between Sardis and Durguthli, except three or four small tumuli near the road. In the afternoon we arrived at Durguthli, more commonly called Cassaba, which signifies *Town*. From Cassaba there are two roads to Smyrna: one goes direct by Nimphi and the Pass of Cavaklederi, the other goes by way of Magnesia and Mount Sipylus. Being very desirous of seeing Magnesia before leaving the country, and having already visited Nimphi from Smyrna, I resolved on going by the way of Magnesia, leaving Hamilton to proceed direct to Smyrna by the Nimphi road. I therefore set out for Cassaba about 4 P.M., accompanied by an old one-eyed Greek *suriji*, whose utterance was very indistinct, and as he mixed a large portion of Turkish with his Greek, I had much difficulty in understanding him. We followed the plain for about six miles, and then crossed by a good stone bridge the river which flows down the valley of Nimphi, round the east end of Sipylus, into the Hermus. The excellence of the agriculture in this neighbourhood, the good roads and neat fences excited my surprise, and almost reminded me of England. These proofs of prosperity are owing, no doubt, to the proximity of the large and flourishing city of Magnesia, the governor of which, one of the Cara-osman-oglou family, is an enlightened and superior man, well inclined to Europeans, and ready to profit by their greater civilization. As it was now getting dark, I and my guide turned out of the road to a village on the right named Banishá, wholly inhabited by Greeks. The comparisons which I drew between it and the Turkish villages were certainly not favourable to this Christian community. Instead of the sober cafinets, with their contemplative inmates whiffing chiboques and sipping coffee, this village was as full of wine-shops as an English one is of ale-houses, and like them filled with drunken brawlers and gamblers. Nor was the hospitality of the villagers more remarkable than their temperance, for after applying at several houses for a lodging, I was on the point of pushing on to Magnesia, when a good Samaritan at last appeared who offered me a berth in his house.

"14th.—Started at daybreak, and soon regained the high road from Cassaba to Magnesia, which here runs close to the foot of Mount Sipylus. This mountain, composed of hard grey cretaceous limestone, rises almost like a wall abruptly from the plain to a vast height. I passed several sepulchral excavations which have been formed in the rock, but which exhibit no feature of interest. Numerous springs flow from the base of the cliff, and form a marshy spot, supposed by Chandler to be the Lake Sale mentioned by Pausanias as the site of the ancient town of Sipylus, destroyed by an earthquake. One of these springs, near a cafinet, is very abundant, and at that early hour felt warm to the touch, though its temperature is perhaps not higher than the mean for that latitude. Immediately above this spring in the side of a cliff, about 100 feet above the road, is a curious colossal statue rudely sculptured out of the solid rock; it represents a figure sitting on a throne, contained in a niche. The height from the bottom of the niche to the top of the head may be about 20 feet: it has suffered so much from time that the features are hardly traceable. I climbed up to the base of this singular specimen of sculpture, and made a rough sketch of it. There can be little doubt that this is the ancient statue of Cybele mentioned by Pausanias as the work of Broteas, son of Tantalus (Pausan. 'Lacon.' c. 22). Its form is too regular and artificial to agree with his account of the figure of Niobe. No modern traveller seems to have described it except Chissull, who merely surveyed it from the road below, and considered it to be the figure of Niobe. Chandler and Emerson both passed within 100 yards of it without seeing it. The former fancied that he saw the form of Niobe in the rude outline of some cliff about a mile from Magnesia; but his account of it is very obscure, and I believe no succeeding traveller has verified his conjecture. Emerson involves the matter in confusion by confounding Chissull's description of the statue with Chandler's fanciful account of the lights and shadows on the cliff at Magnesia*. Whether or not the latter is right in his identification of Niobe, it is, I think, pretty clear that the statue here described is not that nymph. The rudeness of the workmanship denotes an extreme antiquity; and, if it be not the statue of Cybele which Pausanias mentions, we may suppose that it represents some deity presiding over the spring below, or perhaps some hero whose tomb may be excavated in the cliff behind †.

* Letters from the Egean, vol. i. p. 225.

† Archæologia, vol. xxx. p. 524. Collected Papers.

About four miles further is Magnesia, a large and picturesque town, which has been so frequently described that I need say no more concerning it. (*Magnesia is sometimes written* MANISA, *but it is more correct to retain the ancient orthography. The modern Greeks spell it* Μαγνεσια, *but the* γ *being mute and the penultimate accented, they pronounce it as an Englishman would* MANISEEAH.) Proceeding to the Menzil-khan I engaged a fresh suriji, and was soon on the road to Smyrna, which runs for about a mile in the direction of Menimen, and then turns to the left up a ravine in the side of Mount Sipylus. Bearing in mind the assertion of Chissull that his compass was affected by the rock at Magnesia, I was on the look-out for magnetic iron-ore, but saw nothing of the kind, and would strongly recommend that Chissull's experiments should be repeated before the parentage of the magnet be finally awarded to this city. All this part of Sipylus appears to consist of a compact grey limestone with occasional beds of schist, the whole being, I have no doubt, a cretaceous formation*.

"After a long and very steep ascent I reached the summit of the ridge, whence a magnificent view opens in front, embracing the vale of Bournabat and the city and bay of Smyrna; half an hour further I halted for a short time near the village of Djaki-keui. I soon after crossed two small streams running east into the valley of Nimphi, and then, descending near the village of Had-jilar, I entered the rich groves of olives, pomegranates and figs, which were now in the height of vernal beauty. I soon reached my old quarters at Madame Marracini's; but when that lady heard that I had come from Magnesia, which was then under the ban of plague, she looked askance, and seemed half-disposed to pronounce me *compromis*. Having, however, assured her that I had made no stay in that suspected city, she at length opened her door and admitted me into her hospitable mansion, where Hamilton had arrived a few hours before.

"My Asiatic travels were now brought to a close, and I was impatient to start for Europe. I therefore, on the morning of the 23rd of April, left Hamilton to the further prosecution of his travels, and embarked in the 'Levant' steamer for Athens. We followed the same track by which I had come to Smyrna, passing between Scio and Carabournou. The next morning we cast anchor in the harbour of Syra, but were unable to go ashore owing to the quarantine. In the evening we proceeded on our voyage, and next morning entered the harbour of Piræus. I had been told that the quarantine at Piræus would be only seven days; but owing to a rumour of the plague being at Scio, it had been increased to ten days. Capt. Ford of the 'Levant' kindly allowed me to remain on board his vessel till the 26th, when

* See Collected Papers, pp. 15, 16, and Geol. Trans. vol. v. p. 393.

he sailed for Smyrna, and I, along with the other passengers, was transferred to a small fishing-boat anchored in the harbour, and used as a temporary lazaretto. I had the best cabin, 5 feet by 4 and 3 in height, with a trap-door in the top for ingress. The hold was occupied by the other passengers, consisting of four Greeks, a young Turk, and a Corfuote whom I had engaged as servant and interpreter, besides the owner of the boat and his two boys, and a crusty old *guardiano* who had the charge of us. In this narrow domicile we remained a week. I generally began the day with a bug-hunt; I then took the old guardiano in the boat, and used to pull to one of the square piers built by Themistocles at the entrance of the harbour; there I landed, and sent the old man to catch echini and limpets for *his* breakfast while I took a swim in the crystal wave, and often brought home a rich harvest of conchology. I had breakfast and dinner in a little wooden hut erected on shore, and called a *parlatorio*. The formalities of the quarantine system are very amusing to a novice in them. No sooner did I step ashore than the filthy bare-legged *guardianos* retired to a respectful distance, keeping me at bay like a wild beast with their sticks, and forming a lane for me to pass to the *parlatorio*. I had more need of these precautions than they; for, had I come in contact with their filthy and populous persons, I should have been the chief sufferer. During the day I was writing my notes and fishing for shells over the side of my floating prison, which I drew up from the bottom of the clear water by means of a split stick. On the 3rd of May I was released from quarantine, and spent the day in Athens. I called on my old friend Pittakis, and accompanied him to the Acropolis. The visit of the king of Bavaria, a great antiquarian, during the winter, had given a new impulse to the operations at the Acropolis, and I found the temple of Victory, without wings, was now nearly restored. Great excavations had been made at the east end of the Parthenon, and numerous interesting fragments had been found. Many large unfluted drums of pillars had been dug up similar to those built into the north wall of the Acropolis, and probably belonging to the old Hecatompedon, which preceded the Parthenon of Pericles. Several beautiful portions of the frieze of the latter building had also been found. The propylea are also to be cleared out from the rubbish, and the steps leading up to them restored to their original state.

" May 4th.—Took my gun and strolled over the Phaleric hill and round the shore of the Phaleric bay as far as its eastern side. I aimed at the promotion of ornithology, geology and antiquities, but did not make any very important additions during my walk. In the evening I embarked in one of the common passage-boats for Pidavro, the ancient Epidaurus, on the opposite coast of the Saronic Gulf. Next morning we left Ægina on the left, and gradually neared the coast of Argolis, a steep iron-bound shore, descending abruptly to the water, and clothed with shrubs and thickets, through which Theseus once fought his way from Trœzene to Athens. In the afternoon we landed at Pidavro, a small village on the west side of a bay, the opposite side of which is formed by the rocky peninsula, on which stood the ancient Epidaurus. Nothing could exceed the verdure and beauty of this spot. I walked round the head of the bay to visit the ancient city of Epidaurus. The walls are in a tolerably perfect state. The area within is occupied by corn-fields, and the more rugged parts are a perfect thicket of shrubs, which conceal any remains of antiquity that may exist.

"I now set off for Napoli di Romania. That night I reached Ligouzio, a village once of some importance. The valley from Pidavro to this place, and hence to Napoli, has in all ages been the scene of active operations, as it affords almost the only pass from the Argolic to the Saronic Gulf, across the rocky mountains of Argolis.

"6th.—In my way to Napoli I made a detour to visit Hiero, the site of the ancient grove of Æsculapius, celebrated for its medicinal springs. No ancient buildings are standing on the ancient site, but the foundations of several baths and temples may be traced; also the stadium, theatre, and a large cistern coated with thick stucco, like that of the 'Piscina mirabilis' at Misenum. It has five cells on each side, formed by transverse walls jutting out from the sides, probably to support the roof. The aqueduct leading to this cistern still remains; it brought water from a spring, which is still flowing, but does not seem to possess at present any mineral property. The theatre, which is excavated in the side of a hill, is in a very perfect state, nearly all the seats remaining. The wall of the proscenium is removed. The rows of seats are about fifty-six in number. Behind the seats is a space for the feet of the persons sitting in the row above.

"The site of this ancient town is now covered with bushes growing among the heaps of sculptured marbles and other fragments. In searching about I found some inscriptions, but they bore marks of having been copied before. These are no doubt votive inscriptions raised in gratitude for cures effected by a residence at the Grove of Æsculapius. I now proceeded on my journey to Napoli, along the same valley which leads from Pidavro to Ligouzio, flanked with ruined forts on either side, some of which are old Hellenic, and emerged from the valley into the plain of Argos. On the left was a precipitous hill, on which stands the redoubtable fortress of the Palamede, called the Gibraltar of Greece. On a point of land below this castle stands the town of Napoli. In the afternoon I went to visit the celebrated walls of Tiryus, about two miles from Napoli. The site of Tiryus appears to have been originally a small rocky hillock, rising in the midst of the level plain of Argos. The apex of the hillock seems to have been cut off, and the materials used to construct the wall which encloses it, which being filled up solid within, a level platform with perpendicular sides was formed, whose upper surface is between 40 and 50 feet above the plain. The hill thus modified by art, became a miniature resemblance of that grand natural fortress, the Acropolis of Athens. Tiryus is a rare, perhaps unique instance of a Pelasgic fortress built on so slight an eminence in the midst of a plain. The fortress is of an oblong shape, divided into two parts by a transverse wall; the western wall of these two divisions is lower by several feet than the other. The whole area is enclosed by a wall 30 to 40 feet high, by far the greater part of which is standing uninjured; the wall is of the rudest kind of Cyclopean architecture, formed of irregular hewn blocks, and the interstices filled up with smaller ones: those for the corners, though not fitted with the same nicety as in the more perfect styles of Cyclopean architecture, are evidently squared by art. On the north side a square tower, and on the south two semicircular ones project beyond the main line of the walls. It is commonly supposed that Tiryus has been unoccupied since its conquest by the Argines, about 500 years B.C.; yet I think it must have been inhabited by at least one family in later times, for a small excavation lately made by the king of Greece, disclosed the foundations and plaster floor of a house, which had more the character of Roman workmanship than of anything more ancient.

"The entrance to Tiryus is by an ascent at the east end, which seems to have been provided with steps; it is sufficiently wide to admit chariots. In the walls on each side are those extraordinary vaulted passages, the use of which it is so difficult to determine. They are three in number, that on the north side being the longest and most perfect. It is contained in the thickness of the wall, which must be here more than 20 feet. The top is not arched, but vaulted by the progressive advance of each layer of stone, like the vault in the so-called tomb of Tantalus near Smyrna. This passage is walled across at the end, and I therefore suppose it and those on the other side to have been merely chambers to contain the persons who guarded the entrance. In one of the semicircular bastions or towers on the south is a small doorway, vaulted like the passages above mentioned, probably a postern gate communicating with the interior of the fortress.

"7th.—Left Napoli for Mycene. I sent my attendant and baggage direct to Carvathi, the village near Mycene, while I made a detour to visit Argos, about five miles from Napoli. I had a letter to Mr. Riggs, an American missionary, who has settled at Argos, and established a school. There are several of these schools in Greece, at Athens, Syra, &c. Their object is merely to diffuse useful education, without interfering with the religious tenets of the scholars.

"The modern town of Argos stands at the foot of a steep rocky hill, whose summit is crowned by a Venetian castle on the site of the ancient citadel. The theatre is at the foot of this hill, where the central part of the curve yet remains with all the seats cut in the solid rock, but the two wings which completed the semicircle are nearly annihilated. The seats cut in the rock have been lately cleared out by the Government, and they are found to be in a perfect state of preservation; the rows of seats are about seventy in number. This theatre has recently been used for public meetings by the Greek Government. In front is a large brick building, apparently of Roman workmanship, but its destination is not easily decided. The only other relic of antiquity which I noticed at Argos is a square chamber, partly cut in the rock and partly built of brick. In the centre of the inner side is a semicircular niche, and a small opening or passage leads to it from without. Some travellers think that this is a temple or oracle, and that a statue placed in the semicircular niche was made to utter prophecies by the help of a priest, who entered the secret passage behind. It seems, however, more probable that this was a bath, and that the supposed secret entrance was merely an aqueduct to admit water.

"I now left Argos for Carvathi, where I had sent my baggage, and being told that Carvathi was near the road, and that I could not fail to see it, I went on for some miles, till the road quitted the plain of Argos and entered a narrow valley. I now suspected I had missed my way, but continued on, in hopes of meeting some one to direct me to Carvathi. I was now on the high road from Napoli to Corinth, anciently called *Tretum*, and was rather surprised when emerging from the valley at perceiving the walls of Acrocorinthos a few miles distant. At last I reached Courtessa, where there is a guard-house and *café*, and I here learnt that I had left Carvathi several miles in the rear. I lost no time in returning, and was directed by a shorter way across the mountains. Meantime my servant and baggage, finding I had gone forward, followed me, and he reached Courtessa about the time I got back to Carvathi. As it was now getting late, I started for Mycenæ, about a mile north of Carvathi, and meeting a shepherd, he undertook to guide me to Agamemnon's tomb, as the spot is commonly called by the country people. After ascending a hill I came to the Necropolis of Mycenæ. There are several tumuli of earth, which cover large circular vaulted chambers; the largest has been called the Treasury of Atrevo, and by others the Tomb of Agamemnon, but erroneously, as he was buried, according to Pausanias, within the walls of Mycenæ. It is perhaps the tomb of Ægistheus or of Clytemnestra, who were buried without the walls. The entrance to this subterranean chamber is by a square doorway, over which is a flat slab of stone, said to be one of the largest hewn stones in existence. Over the door are some small holes for the attachment of bronze ornaments. Above this is an aperture to admit light to the interior, in shape like a Gothic arch, yet formed only by the successive advance of the horizontal layers of stone, the principle of the arch being unknown at the time of its construction. Within is a vast chamber, with a domed roof, capped by a large circular keystone. My guide lighted a fire, to show the construction of this chamber; the courses of masonry are beautifully regular, and the construction of the edifice so mathematically accurate, that it has stood uninjured through at least thirty centuries. A doorway on one side leads to a lateral chamber of smaller dimensions, not built up like this one, but cut out of the native rock.

"Further west is another of these hollow tumuli, which my guide informed me had been opened about twenty-five years since by Veli Pasha, a son of the famous Ali. He found here, according to my informant, many statues, sculptures and inscriptions, which he carried off to Janina. I have also heard from other sources that the walls were lined with plates of bronze, which were also carried away by him and melted down. These antiquities must have been of the highest interest, from their undoubted antiquity, and their connexion with the romantic history of Troy and Mycenæ. A third tumulus remains, the roof of which has fallen in, and the antique relics, whatever they be that were first deposited here, probably still remain beneath the ruins. From these tumuli it is a few hundred yards to Mycenæ, which occupies a rounded hill nearly surrounded by deep ravines. Lofty mountains rise at the back, while in front is a view of the plain and Gulf of Argos. The form of this ancient town is nearly circular, and its circuit cannot be much less than a mile; it has been surrounded by a massive wall, the greater part of which is standing, and in some places must be 50 feet high. Within this several walls of less altitude form successive lines of fortification, till a small area or citadel remains at the top of the hill, to which the inhabitants could retire when hard pressed by their enemies.

"On the side next the tumuli above mentioned is the celebrated Gate of the Lions, so called from a well-sculptured bas-relief representing two of those animals in a rampant attitude, with a pillar between them. This sculpture derives great interest from being mentioned by Pausanias. Many assert that this stone is of basalt or porphyry, whence some have fancied that it was brought over from Egypt by the early colonists of the country. This was a point which I was particularly anxious to verify; I therefore climbed up to the sculptured block, when a cursory glance sufficed to prove that the supposed Egyptian basalt was *Grecian limestone,* rather more compact, and of a browner colour than usual, yet probably quarried within ten miles of the spot. The walls on each side of the gateway, and the salient angles in other parts, are formed of blocks of limestone-conglomerate, squared and fitted with the greatest nicety. The intermediate parts of the walls consist of rough unhewn masses of common limestone, as rude and shapeless as those of Tiryns: it is therefore evident that the different styles of Cyclopean architecture depended more on circumstances than on their relative antiquity. Similarity of style does not necessarily imply either contemporaneity of age or relationship between the builders, and I have myself seen the Turks building a sea-wall on the Bosphorus of polygonal blocks fitted with as much nicety as in any Cyclopean wall in Greece.

"After examining this gateway I made the circuit of the walls, which disclose some interesting feature at every step, and impress the mind with wonder at these mighty relics of the heroic ages of Greece. If we consider the important share which Mycenæ took in the Trojan war, its subsequent destruction by the Argives, the accurate description given of its ruins by that polished antiquary Pausanias, and the perfect preservation of those remains to our own times, we may regard Mycenæ as little, if at all, inferior to Athens in historic interest. On the north side of the town I found a small postern gateway, equal in point of workmanship to the large one on the south, but in place of the sculptured lions there is only a plain triangular slab. The holes for the bolts and hinges of the gate still remain, and lines caused by its friction may be seen in the doorposts and lintel. Near it is another small entrance, pointed like that at Tiryns. A small aqueduct, cut in the rock, runs round the outside of the wall on the west at too low a level to supply the city; it descends towards the tumuli, and is probably the work of a later age. Within the circuit of the walls is much accumulated rubbish; from the occurrence in it of mortar and pottery, I am inclined to suspect that Mycenæ was partially reoccupied in the Roman times. On the south side, near the top of the wall, is a flat terrace of stucco with a parapet. Coins are frequently found among the rubbish by the shepherds, and I succeeded in procuring a few from them. It was nearly dark when I completed my hasty survey of this interesting spot; and I went to the small coffee-shop in the plain below Carvathi to pass the night. Sleep was impossible from the formidable attack of tiny foes; and, anxious to escape from them, I started before daylight on the 8th in quest of my lost baggage and attendant. At Courtessa I found my attendant anxiously awaiting me, and I was not less glad at rejoining my baggage.

"Leaving Nemea I ascended the other side of the valley, and descended to the populous village of Agios Georgios, or St. George. On the left or south side of this valley is a singular cliff of horizontally stratified limestone, against the almost perpendicular face of which the monastery of the Panagia is attached like a swallow's-nest.

"On leaving Agios Georgios proceeded across the plain to the westward, and up a branch valley in which stood the ancient city of Phlius. At noon stopped to rest at a khan near a rivulet, a quarter of a mile from the village of Boutchika. This exactly corresponds with the place indicated by Sir W. Gell as the site of Phlius, and he makes mention of city-walls, foundations of temples, and other remains: during the two hours I remained here I searched in every direction, and made every inquiry of the inhabitants without gaining the slightest clue to the ruins of Phlius; I therefore cannot but think that Sir W. Gell was here run away with by antiquarian enthusiasm, like M. Texier, when he recognized the ruins of Sipylus, the capital of the kingdom of Tantalus, in some stone walls of a sheep-fold near Smyrna.

"In despair of finding the ruins of Phlius, I proceeded on my journey. After a very long, rugged and steep ascent, I reached a high and desolate country covered with scattered bushes and herbage. I was now in Arcadia, that land of poetical associations, to whose name we attach the idea of verdant groves, and all the beauties of pastoral life. Yet what is the reality?—a barren, bleak country, incapable of cultivation, and therefore necessarily pastoral, yet not the more sentimental on that account. Shepherds are to be seen in plenty, but they are as unpoetical as an Esquimaux. Clad in sheepskins, with a large knife stuck in their girdle, their air is ferocious and savage; and though the crook is still retained, yet the oaten pipe is no longer to be seen. I passed several of these modern Damons, surrounded by vast flocks of sheep and goats, and guarded by fierce wolf-like dogs.

"After passing a large mound of rough stones, supposed by Sir W. Gell to be the tomb of Æpytus, mentioned by Pausanias, and one or two other tumuli, I crossed a small valley with a stream running south, called by Sir W. Gell the 'Trano-potamo.' That traveller mentions the ruins of an ancient town in this valley, supposed to be Alea, but I saw nothing of them. Ascending another height, I beheld in front the Stymphalean Lake, lying at the bottom of a deep crater-like valley. Like many other lakes in Greece, it has no visible exit for its waters, being surrounded on all sides by mountains, but the surplus water makes its escape through a subterranean passage or *catavothron,* as they are called. Descending to the lake, and passing along its south side, I passed near the catavothron: a small river here flows from the lake for a short distance, and terminates against the face of a cliff, beneath which it is slowly engulfed. The lake, which is shoal and marshy, covers about half the bottom of the valley. At the western extremity of the plain, in a secluded dell, stands Leucæ, or Lefké, a considerable village, to which we pushed forward: it was dusk when we got to the village, which presented a scene more truly Arcadian than anything I had yet seen. Large herds of cows, sheep, goats, pigs and asses were marching homewards from their distant pastures on the hills; for in this unenclosed country, where wolves and jackals abound, it is always necessary to fold the herds at night. Each little group of quadrupeds was guided by a young shepherd with his wooden crook, or by a shepherdess plying her busy distaff as she walked. On arriving at the village, I called on the *demarch* or governor, who kindly lodged me in his house.

"9th.—At 4 A.M. my host and I started for Stymphale, which is on the north side of the lake, about four miles from Leucæ. Beautiful wreaths of clouds were rolling up the sides of the surrounding mountains, which are truly Alpine, covered with pine-forests, and their higher parts with patches of snow. On leaving Leucæ we passed some remains of ancient walls in Cyclopean style, surrounding an elevation on the left, and indicating an ancient site. No town is, I believe, mentioned by ancient authors as occurring here; but it is probable that the name of Leucæ has been handed down from classical times. About an hour from Leucæ, passing along the north side of the lake, we came to a small catavothron, which aids the larger one on the other side in discharging the waters of the lake. As it is an important object to gain as much land as possible from the lake, the Government last year employed a number of troops in cutting a channel to this catavothron, and enlarging the exit for the water, but without much apparent success. The lake here approaches the base of a rocky ridge on the left, on and around which stood the city of Stymphale. The remains of that place, though numerous, are of small interest, consisting of fragments of walls in the polygonal style, and steps, niches and small chambers cut in the rock. On turning round the east end of this ridge of rock, I crossed a cultivated plain bearing many signs of former habitations, and in ten minutes reached an old church and tower, built of antique fragments. Pointed arches and capitals, carved in a kind of Gothico-Byzantine style, indicated a barbarous age, and deprived these ruins of interest. The village of Chionia is near this spot, but the fog was so thick I could not see it. In returning I crossed over the rocky ridge of Stymphale, instead of going round it. The road in descending to the lake has been in ancient times cut with much labour in the face of the precipitous slope.

"Leaving Leucæ, I ascended a valley towards the north-west, and in an hour and a half reached a ridge which divides the land-locked basin of Stymphale from the similar one of Phonea. The latter is a complete crater, surrounded on all sides by lofty mountains, and, were it not for a catavothron, the waters would here form a lake many hundred feet deep before they could escape over the lowest lip of the crater. The bottom of this valley is nearly flat, and traversed by a meandering stream, which suddenly disappears in the catavothron. This plain has been subject to some singular adventures in consequence of the peculiarity of its situation. About the commencement of the Greek revolution, the catavothron, whether from natural or from human agency is not clear, became partially stopped up, and the water began to flood the richly cultivated plain. The inhabitants of the surrounding villages were in dismay. The water continued to rise, and in a short time a large tract of fertile land became a lake about 50 feet deep. This continued till about three years ago, when an earthquake, or some other cause, reopened the subterraneous channels, and the lake was drained as rapidly as it had been formed. The delighted peasants reoccupied their long-lost fields, which at the time of my visit were covered with luxuriant crops. A horizontal line of gravel still skirted the foot of the hills, at about 50 feet above the plain, and marked the height to which the waters had risen. Parallel to it, about 50 or 60 feet higher, was another shingle beach, of much older date, and overgrown with bushes, yet still very discernible, proving that the plain of Phonea has been subject to more than one of these immersions. Indeed Strabo relates that such had been the case, and that the sudden escape of the pent-up water by the opening of the catavothron had caused a flood of the Alpheus, into which this subterranean river flows. Descending to the village of Mosha, I followed the eastern margin of the plain towards the north for about three miles, and then turning westward, crossed the plain and began to ascend the western side of the valley. An exceedingly difficult ascent of two hours brought me to the top of the pass at the head of the valley of Peristera. Here, as in most of the passes in Greece, was a guard of about a dozen armed peasants encamped in a hut of pine branches. After giving them a drachma or two to buy *καπνό*, or tobacco, I descended into the valley in front. All the mountains which I passed today are thickly covered with pine-trees, the aboriginal clothing of the country, which in the more accessible parts of Greece has been long since destroyed.

"The valley I was now descending is traversed by a river which flows into the Corinthian Gulf at Acrata. This valley is only a vast ravine, whose sides converge at the bottom, and leave scarcely more than space enough for the torrent. The sides of the mountains are occupied with vineyards, and but rarely admit of the growth of corn; yet this secluded valley presents the rare phænomenon in Greece of a redundant population. Its steep sides are crowded with villages, many of whose inhabitants migrate during summer in quest of work, and return to their homes in winter. In most of these villages are great ugly *pyrgoi*, or country-houses, resembling manufactories, the former abodes of Turkish landed proprietors, who have long since fled the country. I passed the night at Peristera, a large village on the steep slope of a mountain.

"10th.—From Peristera there are two roads to the Monastery of Megaspeleon, whither I was bound. The easiest way is to descend the valley, but the other, which I followed, is considerably shorter, though more difficult. I ascended the mountain immediately above Peristera, by an exceedingly steep zigzag path. On reaching the summit a wild and alpine scene presented itself. To the north the situation of the Gulf of Corinth was marked by a line of separation in the clustered mountains, but its waters lay at too great a depth to be visible. The path passed at the brink of a precipice, at least 1000 feet deep; it was here covered with snow frozen over, and any one who attempted to cross it would probably have paid a rapid visit to the valley below. We were therefore obliged to make a detour round the upper side of the snow, an operation by no means easy, on account of the steep and rugged rocks. Having cleared this patch of snow and another larger one further on, I arrived at the highest point of this mountain track. I was now at the head of a ravine which runs northward, and probably joins the valley which descends from Peristera. Descending this ravine for two or three miles, and then ascending its west side, you cross over to the valley of Megaspeleon; after following a steep and winding descent, I arrived at that celebrated monastery. It stands on the east side of one of those magnificent valleys which abound in this part of Greece, whose fir-clothed sides, rising from the bed of a torrent to an elevation of many thousand feet, present every variety of romantic scenery. At the upper part of the steep side of the valley is a perpendicular cliff 200 or 300 feet high, in the base of which is a large cavern, of great width and height, but of no great depth. This almost inaccessible cavern was selected in remote ages as a retreat from religious persecution; it now contains the largest and most richly endowed monastery in Greece. Upwards of 300 monks are maintained in this establishment, which is denominated Megaspeleon, or the *Great* Cavern. The front of it is walled in a line with the face of the cliff, and it resembles a great manufactory eight stories high. In front the ground slopes rapidly to the bottom of the valley, and is adorned with gardens, vineyards and water-mills. I stayed two hours at Megaspeleon, and found the monks very hospitable and glad to see strangers. They showed me all the *lions* of their den, consisting of some large casks full of wine, the bakehouse, refectory and church, which contains much valuable plate and many relics, including a Madonna and Child, modelled in wax by St. Luke, and dug up in this cave, which occasioned its selection as the site of a monastery: so runs the legend.

"I called on the Hegoumenos, or Superior, and found him a venerable-looking old man, and well informed on Grecian politics. He is very friendly to the English, and showed me numerous cards with the names of his visitors, among whom I noticed Sir John and Lady Franklin. After leaving Megaspeleon, I descended to the bottom of the valley and crossed to the opposite side. After crossing one or two hills, I was rewarded with a magnificent view of the Gulf of Corinth, with a beautifully picturesque foreground, formed of the nearly isolated crag, on which anciently stood the city of Bura. It was destroyed by an earthquake, and a few vestiges of it now remain. After a long-continued descent we at last reached the alluvial plain which skirts the coast. After crossing the Selinus, a small but rapid river, I reached Vostizza an hour after dark, and was lodged in the house which I had occupied on my outward journey in October 1835.

"11th.—Proceeded from Vostizza to Patros by the same route which I followed last autumn. The country had undergone an agreeable change in the interim. In October last the fields had been parched by the summer heats and by a drought of seven months' duration; not a green leaf was to be seen. Now, on the contrary, all nature was revelling in the spring, the ground adorned with flowers and the bushes animated by numerous birds, some of them almost unknown in other parts of Europe. In the afternoon I reached Patros, where I was detained four days by contrary winds, which prevented my starting for Zante. I amused myself with rambling in the neighbourhood, and shooting and preparing ornithological specimens.

"14th.—Great excitement was caused today by a visit from King Otho, who spent a few hours here on his way from Athens to Corfu. With a view to gain popularity, he appeared in the Greek, or, more properly, the Albanian costume. After parading through the town and bowing to his subjects from a window, he re-embarked and proceeded to Corfu.

"On the evening of the same day, about 6 o'clock, a pretty smart shock of an earthquake occurred. I was sitting quietly in my room, when a violent vibration seized the house, made the timbers crack, and detached small fragments from the ceiling. It lasted three or four seconds, during which I lost no time in rushing into the street, where I found everybody else in the town doing the same thing. The earthquake, however, was now over, and no damage done, so the people slunk back to their houses, and the priests, as is customary on such occasions, restored the harmony of their nerves by ringing the church bells. I learned afterwards that this earthquake was felt all over Greece and the Ionian Islands, and that its effects had been limited to shaking down a few cottages, which, considering the manner in which they are built in this country, implies no excessive degree of violence in the shock.

"I had long wished to feel an earthquake, and was now gratified. The sensation during the few seconds that it lasted may be compared to the vibration felt on board a steamer when in motion, but was considerably more violent. The Greeks were not slow to draw an ominous connexion between this earthquake and the King's visit, and to remember that on a former occasion, a visit which he paid to Corfu had been in like manner attended with an earthquake.

"The 15th, the day following, was signalized by a great solar eclipse, which in the days of Thucydides would no doubt have been regarded as an additional omen. On the evening of the 15th the wind was sufficiently favourable to enable me to set sail for Zante in a small fishing-vessel. It was nearly dusk the following evening before we anchored in the harbour of Zante. An officer from the Lazaretto hailed the captain of the boat from the shore. '*Πράγματα ἔχεις?*' says he: 'Have you a cargo on board?' '*Δέν ἔχω πράγματα,—ἔχω ἕνα Μιλόρδον:*' 'I have no cargo,' replied the captain, 'but I have a Milord.' With this exalted title was my arrival announced to the citizens of Zante, and in a stentorian voice that made the whole harbour re-echo. It was too late to undergo all the ceremonies at the Lazaretto, so I passed a second night on the planks of my yacht.

"17th.—Having undergone a long cross-examination through the gratings of the Lazaretto, the authorities were at last satisfied of my state of health; the officers then gave me '*pratique*' by shaking my hand, and I was released. I remained at Zante eighteen days, as no opportunity occurred for leaving it. I employed myself in exploring the island, and in collecting birds, fish, fossils and other specimens. Zante presents a much greater surface of level and fertile land than the other Ionian Islands. A mountainous ridge runs along the western coast, and some detached hills occur on the eastern side of the island. Between these is an extensive vale, highly cultivated, and abounding with olive-trees, corn, and especially currant-grapes, which form the staple produce of the island.

"Although within a few hours' sail of Greece, the Zantiotes are much more civilized and less oriental than the other Ionian islanders. Their costume is more nearly that of Italy than of Greece; straw-hats and trowsers have superseded the *fez* and *fustanella,* and the practice of carrying arms, so common in the other islands, is here nearly, if not quite obsolete. I was much struck by the civility and good-humour of the people, as proved by the friendly salutations with which I was always greeted in my solitary rambles.

"The town of Zante stands in a semicircular bay; it is mostly well paved, and there are numerous small churches like the Italian churches, both in the town and in the surrounding country. At the back of the town rises a high insulated hill, surmounted by a large Venetian fortress, occupied by the English garrison. The view from hence over the interior of the island is extensive and beautiful, while to the north are seen the rocky islands of Cephalonia and Ithaca, and to the east the coast of the Morea with the fortress of Chiarenza, whence our Dukes of Clarence derive their title. On this hill anciently stood the city of Zacynthus, but nothing remains except some fragments of pottery, which may be found in the sides of the ravines in ascending to the citadel. *Sic transit gloria mundi.* Zacynthus has shared the fate of many of her betters. On the south side of the Bay of Zante is a lofty hill with a celebrated convent on the top. I set out twice with the intention of visiting it, but found so many rarities in botany, ornithology and geology to detain me by the way, that my pilgrimage was never performed.

"During my stay at Zante I engaged a small boat with two men to take me round the island, an excursion which occupied three days. The first day was perfectly calm, and we were only able to proceed by rowing. I sat looking over the side of the boat, admiring the submarine prospect: the branching sea-weeds, sponges, shells and star-fish, and the shoals of fish sporting among the rocks, all of which were distinctly visible in six or eight fathoms of water, while the surface was so smooth and transparent that the eye looked *through* the glassy wave, as if it had no existence. We fell in with some sponge-fishers, who do not here, as in the Archipelago, obtain their prey by diving, but by drawing them up from the bottom by means of a hook attached to a long pole upwards of 30 feet in length. The sponges, when first taken, resemble india-rubber, or black leather, and are prepared for sale by maceration and stamping them with the feet. Towards noon our boat stopped to fish, and in a short time we hooked up a variety of beautiful species, many of which were new to me. I prepared a skin of each of these for a specimen, and the men prepared the rest for dinner. With this intent we landed on a smooth beach, in a beautiful little cave only accessible from the sea. The walls of this cave were covered with pendant ferns, forming a natural tapestry. Near this cave is a remarkable spring of milky water with a strong smell of sulphuretted hydrogen. It rises in the sea at the foot of a limestone cliff, and is doubtless connected with some volcanic phænomena below. We stopped for the night at a small port near the north extremity of the island, which we rounded early the next morning, and proceeded along its western coast. This side of Zante is very different in aspect from the low coast on the east. It is occupied by perpendicular limestone cliffs 600 or 800 feet in height, and descending with the same abruptness under water. There is only one small anchorage for boats along the whole of this iron-bound coast.

"The scenery is here magnificent: above are lofty crags and pinnacles of rock with scattered pine-trees growing on their inaccessible summits, and below are vast time-worn caverns echoing to the ceaseless roar of the waves. Numerous brown hawks, of a species which I take to be the lanner, inhabit these caves; also rock-pigeons and alpine swifts. We again applied ourselves to fishing, and though not more than 200 yards from the shore, it required lines of from 30 to 50 fathoms to reach the bottom. The sea was here of the most lovely ultramarine, and pure as crystal. From the difference of situation, the fish which we caught were mostly of different species from those which we had taken yesterday. As soon as drawn up, they burst from the expansion of air in their air-bladders. In the evening we stopped at the anchorage above-mentioned,—a little secluded nook which would only admit a few small boats: we landed to cook our fish, and then went to sleep in the boat. Starting before daylight next morning we proceeded to the southward, within a short distance of the cliffs, which still preserve the same sublime aspect. Soon after sunrise some human voices broke the solitude, and after some search I at last discovered two human beings descending the almost perpendicular cliffs many hundred feet above my head. Each had a long rod in his hand and a basket on his back. I now found that the inhabitants of this mountainous part of Zante are accustomed to descend these frightful precipices, to fish on the rocks at the bottom. One of these poor fellows whom we took into the boat, told me that fatal accidents are the frequent result of this dangerous trade. After rowing for many hours we at last reached the south point of the island, and a fresh breeze springing up, carried us back to the town of Zante on the evening of the third day.

"Another excursion which I made was to the famous *pitch-well*, which is nine or ten miles from the town to the south-west. This source of bitumen rises in a marshy flat at the bottom of a small bay, abounding with luxuriant vineyards, but not inhabited on account of the malaria in autumn. The pitch oozes from the bottom of a small well walled with stone, about 4 feet deep, and filled with excellent water. It is collected for covering the bottoms of boats and other purposes. Mr. Walsh states, in his work on Constantinople, that bubbles of gas rise from the bitumen, but I observed nothing of the kind. Mr. Hawkins, in 'Walpole's Travels in the East,' gives a good account of this pitch-well. The bitumen is doubtless derived from some deep-seated volcanic source. In this and other excursions which I made in the island, I took my gun and collected many rare species of birds. The preparing these, and the various fish which I procured in the market, filled up the remainder of my time.

"On the evening of the 3rd of June, I left Zante by the 'Falmouth' steamer."

Note.—In the termination of many of the names of places mentioned in the preceding Journal, in Turkish,—

A large town is called . . . . Memleket.
A small town . . . . Cassabá.
A village . . . . Keui.

In the woodcut of Cyclopean walls at Cranii, at the end of last chapter, and described at p. lxxii, the size of the large stone in the middle figure is 10 feet 9 inches long by 4 feet high, and 1 foot 6 inches in thickness. The stone with a dotted line in the lower figure had fallen down, but is replaced in the sketch, and that on which it rests is the only instance of a curved line noticed.

LAVA COULEE AT KARA-DEWIT.

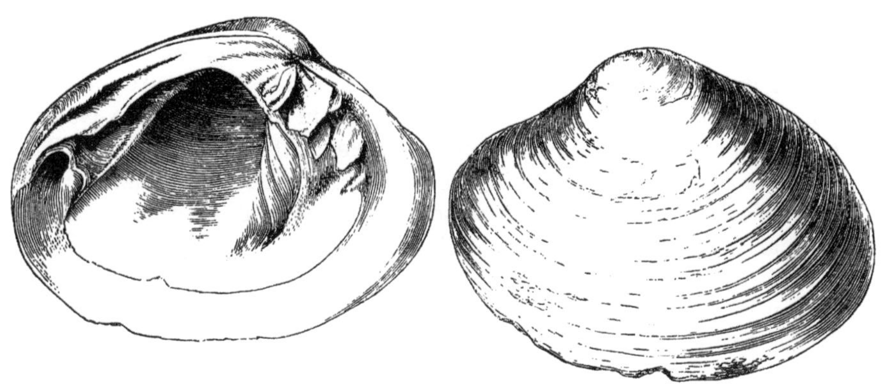

UNIO ANTIQUIOR, STRICKLAND.
Nat. size.

CHAPTER VI.—1836 TO 1842.

DEPARTURE FROM ZANTE AND RETURN TO EUROPE.—RESIDENCE AT CRACOMBE.—RESULTS OF TRAVELS IN ASIA MINOR.—HOME GEOLOGY.—TRANSPORTED DRIFT.—UNIO ANTIQUIOR.—CORRESPONDENCE WITH SIR RODERICK MURCHISON.—JOINT PAPER.—GEOLOGY OF CHELTENHAM.—ORNITHOLOGICAL PURSUITS.—BRITISH ASSOCIATION AT GLASGOW.—VITALITY OF SEEDS.—ZOOLOGICAL NOMENCLATURE.

THE last chapter terminated the Journal kept in the East, and we left Strickland on board the 'Falmouth' steamer about to sail from Zante. His return to Europe was performed as rapidly as possible. Intelligence from home made it necessary that he should relinquish a second excursion with Mr. Hamilton to the interior of Asia Minor and to Armenia; and a letter to his father, written when he arrived at Zante, will show the rapid journey he accomplished after leaving his companion.

"Zante, May 18, 1836.

"MY DEAR FATHER,—On arriving here two days ago, I was glad to find two letters for me at the post-office, as I was very anxious to hear some tidings of you and the rest of the party. . . . . . You will perceive by the date of this letter that I have made slower progress than I originally calculated upon, but I can only say that I have done my best, and never wasted time voluntarily. There is no part of the world where there are more impediments to locomotion than in the Mediterranean, in the shape of quarantine, bad roads, contrary winds, &c. I have written to you two letters addressed to Naples; if you have not received them, perhaps you can have them forwarded from Naples to Rome, unless you like to come to Naples and meet me there. The first was from Constantinople, the second from Athens, where I was compelled to remain eight days cooped up in a small boat anchored in the harbour, and found the operation much less agreeable than I expected. However, I had plenty to do, and employed myself in writing up my journal, &c. On the 3rd of May I was released, and spent the day at Athens. Next day I embarked for Epidaurus, on the opposite side of the Gulf, and on the 6th reached Napoli de Romania. Here I visited the interesting remains of Tiryns, Argos and Mycenæ, with the latter of which I was particularly interested, on account of the immense antiquity and rude simplicity of the remains, which continue in the same state as they are described by Pausanias. On the 8th I reached Leucæ, a village near the Lake of Zaraca, formerly the Stymphalean Lake. This lake is in the midst of the mountains of Arcadia, without any visible outlet, the waters escaping through a subterranean passage. I never saw a more miserable country than Arcadia, beautiful indeed as far as the scenery goes, but cold, barren and unprofitable. Shepherds there were in abundance, but they were rude, clownish and ragged, and far removed from the *beau ideal* of the Damons and Corydons of antiquity. On the 9th I crossed some more mountains by a very bad road, which caused my progress to be slow, and reached a village named Peristera. On the 10th I descended to Vostizza, on the Gulf of Corinth, passing on the way the celebrated monastery of Megaspeleon. This is one of the lions of Greece, on account of its singular situation, built in a large cavern beneath a perpendicular cliff 300 or 400 feet high. The scenery around is exceedingly picturesque. On the 11th I reached Patras, coming from Vostizza by the same road which I passed last year in my way out. Thus, with the exception of the last day, I returned through Greece by a different route from that by which I passed before. I forgot to mention that I had with me a travelling servant, who wanted to go from Smyrna to Corfu, and agreed to accompany me if I would give him his passage. By his assistance I made my way through Greece with ease and safety. I engaged some horses at Napoli, and came with them as far as Patras. I was detained three days at the latter place by contrary winds, and on the 15th sailed for Zante, which I reached next day. I am now looking forward with impatience to our meeting, and shall therefore use every means to reach Naples without loss of time. From all that I can learn, there is nothing to choose in point of *time* between the route by way of Otranto or by Malta. Either way there is a quarantine of seven days to perform. At Malta there are always plenty of opportunities of reaching Naples. If I can, I will visit Etna on the way, but I fear I cannot spare time for Stromboli. In consequence of the unavoidable delays above mentioned, I shall not be able to reach Naples before the 10th or 12th of June. I would have written from hence to my dear mother, but do not exactly know where to direct to her, as I presume she is at present on the continent. Perhaps, therefore, you may be able to forward this letter, or the substance of it, to her. I have not told you anything about Zante. It is a beautiful island, and at this season is a perfect garden. It is much less rocky and mountainous than the other Ionian islands, and is consequently in a high state of cultivation and prosperity. I almost seem to have reached home again, now that I am again among people with hats on their heads.

"Adieu, my dear father, may we soon meet; mean time believe me

"Yours most affectionately,

"HUGH E. STRICKLAND."

During his absence a gloom had been cast over the family by the death of his brother Algernon. The yellow fever broke out on board his ship, stationed at Jamaica, and Algernon sunk under it, in the hospital of Port Royal, at the early age of nineteen. The return of Strickland was naturally wished for, and arrangements were made that he should, if possible, meet his parents in the South of France or Italy, whither they proposed to make a short excursion.

From Zante the route by Malta was chosen, but quarantine again caused a delay. That time was employed in watching the occupations of the fisherman and in examining the marine productions from his little boat; and it is to the Maltese quarantine that we owe the observations on the mode of progression of the genus Lima*. From Malta he sailed to Naples, where, finding no letters from his family to direct him, he resolved to wait for a few days, and to ascend

* "On the Mode of Progression observed in the Genus Lima," Mag. Nat. Hist. 2nd Ser. i. p. 23. Collected Papers.

Vesuvius. On the way to the summit of the mountain he slept at the 'Hermitage,' to be ready for an early start. Mr. Strickland had in the meanwhile crossed the continent to Naples, in the hope of meeting with his son, and found a letter from him there, announcing that he would leave Malta as soon as quarantine would permit. While waiting for his arrival, his father also made an excursion to Vesuvius, with a small party who had chosen the same day. On reaching the 'Hermitage,' and opening the travellers' book, the last and newly written name on the list was "*H. E. Strickland.*" It could only be his son: he was in the adjoining room, and thus in a place the most unexpected was the long-looked-for meeting accomplished, after the parties had been unconsciously near each other for several days. Mr. Strickland had left a letter in the post-office for his son. This had never been delivered, or acknowledged when asked for, and the movements of the travellers were unknown to each other.

From Naples, now accompanied by his father, he travelled through Italy, and early in August joined his mother and sisters at Bex in the Pays de Vaud. From thence the journey homewards was commenced, but further trials still awaited him: at Brussels his youngest sister was seized by fever, which in a few days terminated fatally. Shortly after the sorrowing family returned to England.

Again residing at Cracombe, and resting from the fatigues and anxieties incident to his long travels and the distress of his family, his mind gradually turned itself to the interesting objects around him, and his attention became chiefly directed to geology. This science was every day becoming more important, as well as more fashionable. Worcestershire and Gloucestershire were now intersected by railways, and their cuttings and excavations had opened a wide field for observation. Sir Roderick Murchison was completing his 'Silurian System,' and had already entered into that contest with Buckland and Sedgwick, which was so often and long fought in the meetings of Section G.; and Strickland's residence being within one great area of the disputed ground, and Cambria lying at no great distance, constant appeals were made to him by Sir Roderick to verify the Sections, and the subject was eagerly taken up. Had his previous studies not prepared him for this, the interest would still have been irresistible; but for some time previous to the Eastern expedition geology was foremost in his mind, and finding so many additional avenues opened to him on his return, the occupation became still more engrossing.

Before, however, entering into the *Home Geology*, let us look at the discoveries made in conjunction with Mr. Hamilton and during his route in returning to Europe alone; this we shall do, only very shortly referring the reader to the papers themselves, and the notes appended to them, published in another part of this volume.

At the period when Strickland visited Asia Minor with Mr. Hamilton, that country was little known, and in fact the results of this expedition, communicated in the published observations of these travellers, formed very important additions to our knowledge. He was now therefore busily engaged in working up his Notes and Sections; and the papers as finished were read to the Geological Society, and published in their Proceedings or Transactions. During his absence an occasional letter was read from him at the meetings of the Society, and one of these, dated 'Athens, October 1835,' contained his paper 'On Currents of Sea-water running into the Land in Cephalonia' *; but now, the matured results were from time to time published. Those relating to geology were drawn up and read chiefly in 1836–37. 'A General Sketch of the Geology of the Western Parts of Asia Minor;' Geologies 'Of Thracian Bosphorus,' 'Neighbour-

* This extraordinary phænomenon occurs about a mile north of Argostoli, at the extremity of a promontory composed of hard, white secondary limestone. The streams of water have been noticed for many years rushing in between the rugged masses of rock, and Mr. Stephens of Argostoli has been enabled to obtain a sufficient supply to turn a mill. After passing the wheel, the current flows for six or seven yards, and is then partly absorbed in swallow-holes, and partly disappears under the rocks. See Abstract, Proceed. Geol. Soc. ii. p. 220.

hood of Smyrna,' 'Island of Zante;' and, conjointly with Mr. Hamilton after his return, two papers, one on the 'Geology of the Western Part of Asia Minor,' the other 'On a Tertiary Deposit near Lixouri in Cephalonia.' The latter very interesting paper was read before the Geological Society in May 1837, but was not published until ten years afterwards, on account of a series of the fossils having been sent to M. Deshayes for identification, but who was at the time in Algeria. The delay, however, was favourable, as it obtained later opinions on the fossils both from Deshayes and Edward Forbes *.

During these travels geology was not attended to alone, to the exclusion of other branches; a general collection, more or less numerous in its departments, was also made. Many insects were brought home, a large collection of land shells was made, and attention was given to the Mammalia and Birds. A list of the latter only was published, amounting in number to 129 species †. The greatest number of the specimens were obtained near Smyrna in winter, and we have no doubt that a rich harvest would yet reward any traveller who would devote himself to the study of the zoology of that country. Some interesting facts are mentioned in relation to migration. The Song-Thrush was a winter visitant to the vicinity of Smyrna, and with it some of the birds which are only seen in Britain during the winter, such as *Turdus pilaris* and *iliacus*, and *Fringilla montifringilla*. It seems to be also a winter retreat for many of the Sylviadæ, which visit us in summer only; and from among these one was looked upon as new, and is described as *Sylvia brevirostris*, Strickl., 1836. The specimens are now in the ornithological collection with Strickland's own labels attached. The label of *S. brevirostris*, No. 34 of the list, has, at a later date, written below the first name, "*Phylloscopus rufus*?," and on the back, "Same as *rufa* of England?" 2 rem.=8. On comparing the Smyrna specimen with *S. rufa*, killed in this country, no great

* See 'Collected Papers,' p. 70, &c.

† See Proceed. Zool. Soc. 1836, p. 97. Collected Papers, p. 223.

J Wolf, lith

M & N. Hanhart.

EMBERIZA CINEREA, *Strickland.*
(October, 1836)

variation in size is observed, but in colour there is no tint of yellow on the under parts, and the under wing-covers are also much paler. The proportion of the quills in the Smyrna bird are 2 rem. = 8, in British specimens 2 rem. = 6. Again, on comparing it with species from India, it is very closely allied, if not identical with, *Phylloscopus tristis*, Blyth, which is " devoid of any greenish or yellowish tinge on the plumage, except on the fore-part of the wing underneath*," and the proportion of the quills in our specimens received from Mr. Blyth is 2 rem. = 8. If our opinion, therefore, is correct, *Ph. tristis*, Blyth, 1844, will become a synonym of *Ph. brevirostris*, Strickl., discovered near Smyrna in 1836. *Sylvia olivetorum*, Strickl., was also discovered during this expedition, the specimen brought home serving as the type for the figure in Gould's 'Birds of Europe,' &c.

The other new species described, No. 56, *Emberiza cinerea*, Strickl., is a beautiful Bunting, and the specimen is still unique, no others having yet been obtained. The plumage is perfect, and is that of an adult. The remark of Prince Bonaparte, " pourrait bien être un jeune Hypocentor†," cannot therefore apply to it. The Prince had not seen this specimen for several years.

Home Geology, as we observed, had exceeding interest, and the almost constant correspondence with Sir Roderick Murchison in regard to the distribution and position of the Sandstones of Gloucestershire and Warwickshire, forced him to enter minutely into an extensive examination of the whole area. Sir Roderick was entirely at variance with Dr. Buckland as to the supposed analogy, or rather, identity of the great Red Sandstone of Warwick with the Keuper. He looked upon the Gloucestershire marls, with subordinate bands of sandstone, to be *Keuper*, having examined the formation in detail in Swabia, Alsace, &c., where its best types exist‡, and considered the sandstones of Ombersley, Bromsgrove and Warwick, not the *Keuper*, as Buckland conceived,

* Ann. Nat. Hist. xiii. p. 178. † Rev. Zool. 1857, p. 161.

‡ See Letter, 24th April 1837.

but a portion of the *Bunter Sandstein* *. It was to work out these details that he was now so anxious; and Strickland being close to the localities, had every opportunity of verifying the supposed points of variance. By-and-by, becoming the teacher rather than the pupil, Sir Roderick proposed that they should write together a joint memoir for the Geological Society. This was the origin of the paper 'On the Upper Formations of the New Red Sandstone System in Gloucestershire, Worcestershire and Warwickshire,' remarkable and valuable as being the first to mention foot-prints in the English Keuper.

Mr. Jabez Allies had also written to Sir Roderick that he had found marine shells in the gravel deposits at Kempsey near Worcester. The subject of transported gravel or drift had been one of the earliest to attract Strickland's attention, and, when not yet a member of the Geological Society, he had communicated papers on this subject, and made discoveries of fossil remains in the fluviatile drift of the Avon, that are now turning out to be of much importance. Just before starting for Asia Minor, he had drawn up a detailed paper on the various kinds of drifted materials of the Severn Valley, which was sent in to the Geological Society in 1835, but, probably from the absence of the author, it was never read On his return to England that paper was submitted to Sir R. Murchison, who had already studied the subject, requesting his opinion:—

"DEAR SIR,—I send you a paper 'On the Superficial Gravel of Worcestershire,' &c., which I wrote just before leaving England last year. It is in part a recasting of a former communication on the Gravel with Shells of the Avon Valley. I should be obliged if you can look it over at your leisure, as it is a good deal connected with your own researches on the same subject. You will thus see what my views were before the publication of your remarks on the *Northern drift*, a paper which has thrown a new light on the subject, by supplying a cause (viz. a marine current) for a result which it was otherwise very

* See Murchison's and Strickland's joint paper, Trans. Geol. Soc. v. p. 347; and Collected Papers, p. 111; particularly Professor Ramsay's notes to the last, pp. 125, 127.

difficult to account for. I have now added a postscript to my paper, in which I have shown how your views may be reconciled with my own. You will see that in one point I differ from you, viz. in supposing the fluviatile gravel of the Avon Valley to be posterior to the general northern drift which is scattered over the country.

"I have sent you these papers because they are closely connected with one branch of your present inquiries. I propose to work the whole paper over again, according to the new views which have arisen from your paper on the Northern Drift, and present it to the Society in a better form than it now exhibits. Before doing this, however, I should be glad of your opinion on it, more particularly on the point where we differ, viz. the relative ages of the fluviatile and the marine drift."

Sir Roderick wrote to him very encouragingly about it, and afterwards used the details for his Silurian System*.' In this paper the position of the Avon beds then exposed is pointed out, and forms an important record where the localities are now covered up and rendered inaccessible. Some shells considered new are described, among them the *Unio antiquior*, Strickl., represented in the woodcut heading this chapter, p. clvii, the drawing of which is made from the original typical specimens in Strickland's collection. The fossil under this name is now generally considered the same as the recent *U. littoralis*; but on comparing the fossils in the Strickland collection with recent specimens of *U. littoralis* from France and Spain, the latter is always of a more oblong shape, the shell not so strong, and the form constant; and notwithstanding the great variation we know takes place among the Unionidæ, we have retained the distinct name. This is further confirmed by comparing a fossil Unio from Clacton, obligingly sent to us by Mr. Morris. This is undoubtedly *U. littoralis*; it possesses all the characters of the recent shell, and presents the same differences which we have mentioned when placed beside the *U. antiquior* of the Strickland collection.

The subject of Drift was still further continued; a memoir taking a more extended range was read before the meeting

* Silurian System, p. 555.

of the British Association at Liverpool, and geologists cannot attend too much to the points respecting which it is most desirable to obtain information; an enumeration of these concludes the paper*.

The correspondence with Mr. J. Allies alluded to, and the discovery of mammalian remains in the Cropthorne drift and in the valley of the Avon, and the association in the latter deposits of recent land and freshwater shells in immediate contact with them, were viewed by Strickland as points of very great geological interest; and their importance has been since recognized† in the researches of Dr. Falconer, who looks upon the Elephant-remains of the two deposits as belonging to distinct species. Dr. Falconer conceives that there exists in the valley of the Severn an old *fluviatile* pliocene deposit of great extent, containing the usual association of pliocene mammalia,—*Elephas antiquus*, *Hippopotamus major*, Bison, Deer, &c., with freshwater shells; and also *overlying* beds of gravel which are described as estuarine, and drift gravel containing a few marine shells, Cardium and Turritella, and the mammalian remains of the glacial period—*Elephas primigenius*, *Rhinoceros tichorhinus*, Rein-Deer, Horse, &c.‡ This is the present state of the question of these deposits and their contents, to which attention was first directed by Mr. J. Allies, and the papers of Strickland already referred to.

We now see Strickland taking an active and a public part in science. Sir R. Murchison, who had gained for himself a European name, had been exceedingly kind to him from his earliest acquaintance. He had maintained a confidential correspondence, he had actively interested himself in endeavouring to obtain a fellowship at Oxford, previous to the Asia Minor expedition, and subsequently thought of Strickland as a fit person on whom to bestow the charge of the Geological Society; and although Mr. Hamilton was ultimately chosen, the claims of Strickland stood very high, and the propriety

* See "Memoirs on Drift," Collected Papers, pp. 90 and 105.

† Collected Papers, p. 96.

‡ See Dr. Falconer's views, Proc. Geol. Soc. 1856–57, p. 5, &c.

of the proposed choice was universally acknowledged; he consulted him largely during the progress of the 'Silurian System,' and the proof-sheets passed through his hands, as afterwards did a part of those of 'Siluria' before they were sent to press. Mr. Hamilton made a proposal to him to accompany Capt. Blackwood, who was appointed to survey New Zealand, as naturalist.

All these were the natural steps towards scientific promotion, when another instance of confidence in Strickland's geological acquirements took place. Just as Sir R. Murchison was about to set out on his second expedition to Russia, he was applied to by Mr. Davies, a bookseller in Cheltenham, to republish his work on the geology of that neighbourhood, and he writes to Strickland March 15th, 1841:—

"I hope I may have time and opportunity to attend to your wishes at St. Petersburgh; in the meantime I beg to ask you if you will undertake to prepare the second edition of the little work on Cheltenham which I published some years ago. Mr. Davies, the bookseller to whom I gave it, is desirous of having a reprint with additions, and I have told him that you could double the concern with enrichments, and therefore I should ask you to do it, and put your name to the same.

"You can take your own account of the railway-cuttings for the chief extracts, and you can extract a few leaves from my big book on any subject which you think may best stir up the *Cheltenhamians*; in short, do as you like about it."

This work afterwards appeared as "Augmented and revised by James Buckman, F.G.S. and H. E. Strickland, M.A., F.G.S.;" and whatever additions and discoveries may have been made since its publication, the 'Geology of Cheltenham' is indispensable to any visitor who has the slightest wish for an acquaintance with the formation of the surrounding country*.

In the summer of 1837, accompanied by his father, he made a tour extending to the North of Scotland and the

* See also 'Memoirs of Geol. Survey of Great Britain,' Geology of the Country around Cheltenham, by Edward Hull, Assistant Geologist, 1857, which has drawn largely from Strickland's papers, and the volume edited by Buckman and Strickland. The 'Cotteswold Hills Handbook,' by John Lycett, will also be a good help to geologists visiting Cheltenham.

Orkney Islands. From home he travelled through the principal manufacturing towns to Lancaster, the Lake district and Carlisle, and entering Scotland by the Longtown and Langholm line of road, visited Abbotsford and Melrose on his way to Edinburgh, which he left as commencing the main part of his Scotch excursion on the 19th of July. The usual note-book is kept, illustrated by some very good pen-and-ink sketches of the more remarkable places visited. He travelled by Inverness and Strathpeffer (remaining a day with Sir George Mackenzie) to Dingwall, and along the Cromarty firth, where he records a "call on Mr. Miller, a self-taught genius, who shows us his fossil fish, and gives me one of his works; then go along the shore to South Sutor, and collect many specimens of fish," &c.

Visits Dunrobin Castle and Brora, the latter of much interest after his work among the oolites at home. Thence to Wick, where the Leith and Orkney steamer was taken up to Kirkwall; examined the antiquities, &c., and crossed to Stromness; sketched the Druidical stones of Stennis, and saw the slate-quarries containing fossil fish, and the Hill of Hoy.

"Went to the 'dwarfie stone,' a vast mass of stone, in which a small central chamber and two lateral ones have been excavated. I believe it to have been a tomb, as it is not unlike some Græco-Roman ones at Ephesus and elsewhere; but by what people excavated it is difficult to conjecture. In front is a square block, which has once been rolled away from the aperture. It would be desirable to turn this stone over, as there may be an inscription on its lower face*."

Landed from Orkney in Sutherlandshire, the scenery of which was much enjoyed, particularly the remarkable forms of some of the magnificent mountains of Assynt†. Visited the Isle of Skye, and examined and sketched the singular and rugged "Torrs." At Portree everything was extremely primitive; that of the jail particularly so; perhaps in 1837

* "Others (caves) are the retreats which the primitive confessors of Scotland excavated or enlarged for their oratories or cells. Possibly the singular dwarfie stone of Hoy, in Orkney, owes its origin to a similar source."—Dr. Wilson's *Prehistoric Ann. of Scotland*, p. 89, 1851.

† See woodcut at end of chapter.

not very creditable to the authorities, but it was before the "Prison Board" days, and had the "Chapter on Prison Discipline" he gave been quoted, some objections to that board might have been silenced:—

"Portree is the chief town of Skye, which is in the county of Inverness. It therefore contains a prison, in which persons are confined for petty offences, or till they can be sent to Inverness for trial. This prison is a small house, with magistrates' office above, and four cells below, in each of which is a grated opening, without either glass or shutters, looking into the street, and enabling the prisoners to gossip with their friends. No persons besides the prisoners sleep in the house, nor was there till within this fortnight any night-watch. The prisoners are allowed sixpence a day to keep themselves, and their provisions are sold to them by the inhabitants through the grating of their cell. A few weeks ago two sheep-stealers were supplied by their friends with files, and succeeded in cutting through the bars of their cell and making their escape. Now as the jail was built by the late Lord Macdonald, the county economically refuses to *repair* it, and accordingly during our stay at Portree, the wrenched and broken bars of the window remained *in statu quo*.

"In the meantime a sailor was brought to the jail charged with stealing a large sum from his employer. He was no sooner left to himself than he ground off his handcuffs by rubbing them against the stone walls of his cell. A second and third pair were put on him with the same result, so the authorities, grudging the useless waste of iron, allowed him to remain unshackled. They, however, appointed a night-watch over the jail, but left it unguarded by day; and this day at noon, the 21st of August, 1837, this audacious culprit burrowed through a thin partition wall into the cell from which the bars had before been removed, and made his escape by the same aperture as the sheep-stealers. Such is the state of penal affairs in Skye."

He finished this tour with the Falls of Glomak and a passage through the "Great Glen":—

"Went from Invergarry to the solitary and small inn of Letterfinley. Our road was along that remarkable valley called the *Great Glen*, strikingly distinguished from every other glen in the Highlands. Most of them run from east to west, and are irregular in form, with salient and retiring ridges and ravines along their sides: they seem to have been brought into their present form in great measure by the slow denudation caused by the torrents which flow along their bottoms and descend their sides. The Great Glen, on the contrary, runs through the whole country from south-west to north-east; its average width at the bottom is about a mile. Its course is perfectly straight, and there are none of those alternate projections from either side which distinguish the smaller glens of Scotland. The mountains on each side seem shorn of all their inequalities, and slope down with perfect regularity till they meet the flat gravelly bottom of the valley. On looking along the glen the appearance is precisely that of an artificial excavation, or open cutting for a railway or canal on a most gigantic scale. Such a remarkable difference of character in this glen appears to indicate some peculiarity in its origin. I am inclined to think that the straightness and smoothness of its sides is owing to the excavating force of a vast current of water, different in kind and degree from those petty rivulets which are still at work enlarging the other Highland glens. These rivulets are easily diverted from their direct course by local obstacles, and hence arise the sinuosity of their current, and the irregular form of the ravines through which they flow. But when a powerful and rapid body of water flows through a narrow channel, any obstacles which exist, instead of diverting the current, are themselves removed, and a straight smooth-sided channel is ultimately formed. Such appears to be the origin of the Great Glen of Scotland. Nor is it difficult to find a current adequate to the effect. The whole coast of Scotland is surrounded by raised beaches and platforms of gravel overhanging the sea at heights of from 10 to 100 feet. The country must therefore have undergone at least that amount of elevation at a very recent period. The Great Glen is the only valley which runs through the Highlands from sea to sea at the low level of 100 feet above high-water mark, and the highest point in its bottom is precisely level with the raised beaches near Inverness. It is therefore certain that the sea flowed through it at the time when those beaches were formed, and that the North Highlands formed an island separated by a narrow strait from the South Highlands. Now what would be the result if North Britain were again lowered 150 feet, and the sea admitted into the Great Glen? That vast tide which now daily creeps up the west coast of Britain in quest of an outlet, and at last rushes, at the rate of ten miles an hour, through the Pentland Firth, would then find an exit through the Great Glen. This channel not being one-sixth the width of the Pentland Firth, the velocity of the tidal current would probably be much greater than ten miles an hour; but even in a tide of this speed we possess an agent which would soon level down all obstructions, and give to the Great Glen that smooth and straight outline which so remarkably distinguishes it from any other valley in Scotland."

It was during this tour that the material for his paper on 'Some remarkable Dikes of Calcareous Grit at Ethie in Ross-shire' was collected*.

We have hitherto mentioned geology as the chief among Strickland's scientific pursuits; but in the expedition to Asia Minor we saw that ornithology had also occupied a portion of his attention, and had been rewarded by discoveries; indeed, at a very early period it had been entered into, and had it not been for the temptations of the peculiar geologic district in which his residence was placed, that study would have perhaps engrossed more of his time. The commencement of his ornithological collection was stuffing such British birds as he could procure, and placing them in glass cases, which he both made and glazed himself. The basis of the foreign collection at one time belonged to his cousins, Nathaniel and Arthur Strickland, some of which were at various times sold to him, or exchanged for books, and the whole ultimately came into his possession. His brother Algernon, whose untimely decease was so much felt, had also a strong leaning to, and inclination for, the natural sciences, and during his service abroad collections in different departments were sent home. The last time he left Portsmouth he writes to his brother from the 'Undaunted':—

"MY DEAR BROTHER,—I write to you again (as you are the chief naturalist of our Society), to inform you that I went on shore, and took two boxes to the Custom-house, where they were overhauled, and paid 1*s*. 6*d*. for an entry of them. The smaller box contains only shells and bird-skins, which are in very fair condition, though I have had them in my hands for a long time."

It was to this branch of natural history that my acquaintance with Strickland is indebted, an acquaintance and intimacy which continued as long as he was spared, with great pleasure and satisfaction to both. Ornithology had made us correspondents for some time; but our first interview

* See 'Collected Papers, p. 150.

with each other was at Glasgow during the Meeting of the British Association there in 1840, when Strickland read his paper 'On the true Method of discovering the Natural System in Zoology and Botany*.' In this paper the systems of binarians and quinarians, &c. are combated, or, in fact, all those which follow any symmetrical or regular figure. The system which was proposed was afterwards carried into effect in the arrangement of his own collection, and is that pursued in the MS. of the 'Ornithological Synonyms,' which we have thought it right to continue in the publication of that work. After the Meeting he returned with me to Jardine Hall, where he examined the ornithological collection, and various plans were formed for the future advancement of that branch of natural history.

It was at the Glasgow Meeting that the important subject of "Vitality of Seeds" was taken up and proposed. Stories had often been told of wheat from mummy-cases and Egyptian catacombs having grown and yielded remarkable increase after thousands of years' entombment; these even now are circulated once a year at least by the popular newspapers, and being taken as *facts*, theories are built up upon them. But the subject is of far more importance than it appears at first sight, or as it is generally treated—a mere curiosity. It may explain the existing anomalies, as they have been called, in the distribution of plants, or it may show us that long vitality being merely a popular error, or a myth of the credulous, some other cause must be looked to, and some other method contrived to explain the extraordinary diffusion of some species, and the extraordinary restriction of others. At the same time we shall have to consider if any artificial process that we can suggest or perform will be similar in all its protecting and preserving effects to the natural burying of seeds, or the submergence of lands. At Glasgow a Committee was appointed, consisting of "Mr. Hugh Strickland, Mr. Babington, and Professor Lindley, to

* See Reports of Brit. Assoc. 1840, p. 128; Annals of Nat. Hist. vi. p. 184; and Collected Papers,' p. 408.

institute a series of experiments with a view to determine the longest period during which the seeds of plants can retain their vegetative powers." But his views on the subject, and the objects of the Committee generally, will be best understood from a letter written to Dr. Daubeny, requesting his cooperation:—

"Cracombe House, Evesham, Worcestershire,
January 6, 1841.

"DEAR SIR,—At the late meeting of the British Association at Glasgow, a Committee was formed to make experiments on the duration of vegetative powers in seeds, with a grant of £10 for the purpose. The Committee consists of Dr. Lindley, Mr. Babington, and myself, but as it is an inquiry which requires the cooperation of a larger number of persons, we are desirous of adding a few more botanists to the Committee. I have therefore written to ask whether you would be willing to join in these researches; and that you may the better be informed on the subject, I enclose you a copy of a rough outline of the experiments which it is proposed to make.

"The recommendation having originated with myself, my name was placed on the Committee, but as I do not profess to be a botanist, I am desirous of seeing the subject placed in abler hands, and undertaken by persons who have the requisite leisure, knowledge and horticultural appliances at command. I have always been very sceptical as to the various accounts which have been published of the vegetation of seeds after the lapse of thousands of years, and it was in the hope of eliciting some facts which might tend to refute or corroborate these statements that I set the inquiry on foot.

"I think that a portion of the £10 might be advantageously spent in printing for the use of botanists a series of inquiries and suggestions on the subject, similar to the outline which I now enclose, and to which I should be greatly obliged if you can add any further suggestions.

"If you will allow us to add your name to this Committee, we shall feel greatly indebted to you for any assistance which you may give us in this undertaking.

"Mr. Babington has, I believe, secured the services of Prof. Henslow by adding him to the Committee.

"The plan being as yet in an immature state, any hints which may occur to you on the subject will be very valuable.

"Believe me, dear Sir,
"Very truly yours,
"HUGH E. STRICKLAND."

"*Prof. Daubeny, &c. &c.*"

The Committee immediately published "Suggestions for Experiments on the Conservation of Vegetative Power in Seeds," and worked on, giving the results of their labours* each year in a report to the British Association. A general summary of the facts ascertained up to that time was drawn up in 1848 for the meeting at Swansea, and in the excellent presidential address of Dr. Daubeny at Cheltenham, the results are brought down to 1856:—

"An inquiry has accordingly been carried on for the last fifteen years under the auspices of, and with the aid of funds supplied by, this Association, the results of which, it is but fair to say, by no means corroborate the reports that had been from time to time given us with respect to the extreme longevity of certain seeds, exemplified, as it was said, in the case of the mummy-wheat and other somewhat dubious instances; inasmuch as they tend to show that none of the seeds which were tested, although they had been placed under the most favourable artificial conditions that could be devised, vegetated beyond a period of forty-nine years; that only twenty out of 288 species did so after twenty years; whilst by far the larger number had lost their germinating power in the course of ten.

"These results, indeed, being merely negative, ought not to outweigh such positive statements on the contrary side as come before us recommended by respectable authority, such, for instance, as that respecting a Nelumbium seed, which germinated after having been preserved in Sir Hans Sloane's Herbarium for 150 years; still, however, they throw suspicion as to the existence in seeds of that capacity of preserving their vitality almost indefinitely, which alone would warrant us in calling to our aid this principle in explaining the wide geographical range which certain species of plants affect†."

The reform of the *nomenclature* of zoology, as well as that of classification and system, was also a subject to which he had devoted a good deal of consideration. The difficulty and inconvenience attending some systems lately invented, and the confusion involved by a nomenclature made without any principles, and often with the miserable haste of being first, were great drawbacks to the systematic progression of natural science, and Strickland resolved to bring in his "Reform Bill," as Sir R. Murchison styled it.

* Reports of Brit. Assoc. 1841, p. 50.

† Address, p. 17.

In July 1841 he writes to me—

"I have some thoughts of moving in the Zoological Section at Plymouth for the appointment of a Committee to prepare a set of regulations with the view of establishing a permanent system of *zoological nomenclature.* I should be glad to have your opinion on the subject. The plan which I propose will not interfere with *zoological classification,* in which every one must in the present state of the science be left to please himself."

Later in the year his "*plan*" was printed in a small form and sent to various naturalists at home and abroad for their advice and opinions, and it cannot, we trust, be considered as any breach of confidence if we transcribe a few extracts from the replies that were obtained. Professor Owen writes—

"I have read with interest your concisely and clearly-penned plan, &c., and if the British Association had power to enforce the code, I should be glad to see the one so ably drawn up by you enforced by the authority of the Association. The nomenclature of botanists owes its influence to the individuals who have, on their own responsibility, laid down its rules. May not zoologists yield to similar guidance?"

Mr. Darwin, who afterwards took a prominent part in the discussions, and gave great assistance, writes—

"I have read carefully your laws and suggestions, and have been able to make only one or two unimportant notes. As far as my judgment goes, the laws appear well digested and clearly written."

Sir John Richardson replies to his request—

"I have some hesitation in joining the Committee on 'Nomenclature,' because my other avocations will prevent me from being an efficient member, though I should be glad to be able to aid in your very important task. The main difficulty will be in gaining the hearty assent of other European naturalists to the proposed plan, and nothing short of that will be sufficient, however beneficial the proposed rules may be."

The Rev. Leonard Jenyns wrote a long and able letter in connexion with the printed plan sent to him, entering fully into the subject, criticising and adding valuable hints, many of which were afterwards adopted by the Committee; he concludes this friendly letter—

"I consider your plan, as a whole, extremely good, and likely to convey a great benefit to science; most of the rules must, I should think, appear so reasonable to all persons, who will be bound by any fixed principles at all in such a case, that they can hardly be overthrown by any criticism."

The replies from foreign naturalists had not yet been received, but the opinions of such men as we have quoted, representing high authorities in zoology, could not be taken as nothing, and the subject was introduced at the Plymouth meeting in 1841, where it was discussed, its general necessity acknowledged, and after some opposition it was moved that a Committee be appointed "to draw up a series of rules with a view of establishing the nomenclature of zoology on a uniform and permanent basis, and be requested to present their report to the next meeting of the Association." A small grant was at the same time given, to print and circulate the proposed rules. Meetings of this Committee were held in London, and new members were from time to time added to it. The clauses of Strickland's plan were gone over *seriatim*, copies of the revised rules were printed for the use of the Committee, and they were again widely circulated among both the home naturalists and those of the Continent and America, and thus every care was taken, and exertion made, to render the case as clear and strong as possible.

In the meantime replies from foreign naturalists came in; and to continue the history of "Zoological Nomenclature," we think it right to copy some of those as exhibiting the opinions of men of science abroad; and we also give Strickland's replies, as showing the clear manner in which he dealt with the subject.

The Prince of Canino writes from Florence, Jan. 20, 1842—

"Your PREFACE is perfect, only it must not be exclusively English, but directed to the zoological world at large. I should insist still more on the distinction of *past* and *future*. We may tolerate a great deal that exists (at least till fixity is obtained), but will be severe for the future, and the punishment of whatever eminent man that will not submit to our laws *when duly sanctioned*, will be to have his names disregarded. LIMITATION is necessary: we have nothing to do with common names. PRIORITY our only guide, with the unavoidable exceptions as *few as possible.* In one of my writings I have denied to a naturalist the right of withdrawing a name of his own, on the same ground that a banker cannot withdraw *une lettre de change* (bill, &c.). NOT TO EXTEND TO AUTHORS OLDER THAN LINNÆUS has always been my maxim: that compliment is due to him even were he not the founder of binomial [nomenclature]. I recollect some twenty years ago making a scientific meeting start at my saying that we ought to treat those that went back beyond Linnæus, and against that great man, as good Christians do the instances of *pre-Adamites*! I even give the preference to Linné (*but to him alone*) when he has made a mistake. and call with his classical name the Swedish species, rather than the Greek or Roman, when distinct. As to the genera, there can be NO DOUBT. Brisson, though often apparently binomial, is not to be taken as such; still, a naturalist now, who, wanting a name for a genus would not take up Brisson's or even Ray's if possible, ought to be *whipped.* The case you bring of *Perdix rufa* instead of *rubra,* although perfectly correct in theory, I should not carry into practice in this *particular case*; and therefore I should add still, when an old author has well discriminated good species confounded or included (even not explicitly, for Linné did not confound the two) posteriorly, then, as in the case of *P. rubra,* Briss., the name must be retained. As some say, every rule has its exceptions. But I should submit in this case, as in that of Gmelin which follows, to the decision of the majority. In my opinion Gmelin deserves no regard at all, for he kept back science; and if I was to make an exception, it would be for Latham's species and for Illiger's genera; but no exception at all is preferable, although your reasons are most excellent. A point on which I totally differ from you (*and for that I shall fight*!) is about names of Orders. I admit that generic and specific names alone are to have a positive law; but I maintain (and I could give you many reasons of *all kinds*) that GLIRES, PECORA and ANSERES are better than RODENTIA, RUMINANTIA and NATATORES, although I should never keep NANTES (observe—it was AMPHIBIA NANTES, not PISCES NANTES) instead of Chondropterygia (which, by-the-by, I have also sent to the devil). As to the family in IDÆ and subfamily in *inæ, ina,* or *ini* (*inæ* of course for Birds), there can be no doubt. If an improvement was called forth for the *Order* and *Class,* it would be to make them a collective name (class *Ornithia* instead of *Aves*; *Monadelphia Decandria* in botany, and why not in zoology? *take that hint*!).

"That a name (generic) must be retained for a portion of its own subdivisions, there is no doubt; but a great deal of *tact* is necessary to see what portions, and on that many laws could be made. (I don't care if the code is complicated; he who wishes to forge names must have studied as physicians, lawyers, &c. do.) The case you quoted of Swainson's *Asthenurus* is hard, but just: your § 6 unexceptionable. The 7th I should modify by changing your word *should* [be cancelled *in toto*] into *might* be cancelled, &c. If I wish to abandon my right, and be merciful to a sinner, why prevent me from so doing? besides, less confusion arises from restricting an old name than from creating a new. (I hate new names to such an extent!) I have done it myself in several instances, Heliornis and Podoa, &c. I object equally to your other consequences, and specially in the case of *Pastor*. To be consequent in your false footstep (excuse the expression to my naming peculiarity), you ought to think yourself authorized to change a name for the least transposition of species (as so many fools do!). Your § 9 comes quite into my notions; but it is no great CONCESSION; pray concede a little more, in order to *avoid multiplication of names* (*my hobby* !). § 10 *agreed,* but only perfectly similar names; for instance, STELLERIA, nob., a Duck, has stupidly been changed on account of *Stellaria,* Linn., a plant, owing to the English *a* being the French *e* ! ! ! ; a letter, as in this case, is more important than three or four (etymology and euphony must be consulted more than orthography itself !).

"§ 11. A downright false name must of course be changed, such as *Fringilla hispaniolensis* ! ! (St. Domingo, while he meant Spain, and not even there !) ; *P. hyacinthinus,* meaning red instead of blue !, &c. Caprimulgus and *Paradisea apoda,* however, I consider excellent ! ; they remind us of prejudices, &c., as Whitehall does of the beheading of Charles the First ! I shall defend these. I have also my doubts about *dubia, hybrida,* &c. As to the erroneous geographical names I agree with you, but do not take *helveticus,* as some have done, for a geographical name.

"§ 12. Agreed, in your precise terms. But he who prints a stolen name is a great rogue, as well as he who, knowing of one unclearly defined, gets it to himself and changes, is to be blamed. § 13 is too clear. § 14. I shall resist any attempt at giving credit and currency to Gmelin's edition of Linnæus, which ought to be burnt. No injury (did I print somewhere?) can be made worse than to attribute Gmelin's names to Linné; besides, even if you grant that Gmelin's names should have the preference over some older, why declare that a man is not now to look in other books? Boddaert and Scopoli do not deserve that contempt, nor Gmelin that honour. Expediency, however, is, I confess, on your side! § 15. Latin orthography is excellent, and your country naturalists want it badly. Some barbarous names, however, I should adopt no way at all. As to the genitive, I never double the *i* when the word finishes by a consonant, and only do it when by a vowel; the rule is clear, and let it be adopted :—STRICKLANDI and not *Stricklandii, urvillii* and not *urvilli,* the name being *Stricklandus* and *Urvillius*.—Part II. A. Adopted with all its subdivisions, and very wise and temperate observations, especially *h, k,* (see Illiger, Rafinesque, whose laws I hope you have consulted, &c.). Great taste is required in selecting names, but *non omnibus datur habere nasum* !—B. I do not agree that in any case the families should end in *adæ* and *anæ*, rhyme being as useful as anything else. Besides, you English are not justified in saying *Alcanæ* instead of *Alcinæ*. I should therefore reject your exceptions (even of *Laniadæ*) into the rule (*Lanidæ*). Of course a great deal of judgment is necessary to select the genus from which to call the family. I have often been perplexed, but never so much as for my family *Menuridæ*, containing *Menurinæ, Malurinæ* and *Troglodytinæ* (no more *Thryothorinæ*, and much less *Anorthurinæ*), for which I can prove the name must be *Menura*, going as well from form as from language. But you do not come to that conclusion at first sight.—C, the last, is the best rule of all, though I have only lately adopted it; and it is in this especially that our brothers the botanists (who must form a conspicuous part of the Committee) will be obliged to us! It is a long time since I decided that all species being *equal*, they ought to be written all alike (Linnæus only wrote them with a capital when they had been genera, or might be made so henceforth). As some were to be written with a capital, as proper names, I concluded that all ought to be so, and you will see that plan executed in several of my works, *ex. gr.* as the tables of my Iconography. It was a step, though bad, that led to my present practice, in which I rejoice to see that you have almost preceded me. This trifling circumstance would make your code valuable, as it is on a thousand other grounds. It is the most rational now to write with small letters."

M. Agassiz replied from Neuchâtel:—

"MON CHER MONSIEUR,—Le sujet sur lequel vous désirez connaître mon opinion, est un de ceux qui m'a le plus occupé. Rien ne serait maintenant plus important aux vrais progrès de la science, qu'une législature zoologique invariable; le besoin de règles fixes se fait chaque jour sentir d'avantage, à raison de l'incertitude qu'offre maintenant la nomenclature et de l'épouvantable confusion qu'on rencontre partout dans la synonymie, confusion qui va chaque jour en croissant. Comme vous, Monsieur, j'ai tenté plus d'une fois de tracer les bases d'une nomenclature invariable; mais les difficultés qu'offre une pareille entreprise m'y ont toujours fait renoncer. Je dirai plus:

n 2

cette question a des côtés si divers, qu'on se surprend soi-même à avoir sur divers points une opinion différente en diverses circonstances. Il n'est pas un des principes que vous posez, dans votre brochure, dont je n'aie tour-à-tour envisagé l'adoption comme indispensable, puis repoussé l'application comme impraticable. Cette incertitude ne provient point d'une indécision dans les principes, mais résulte toute entière des antécédens de la science elle-même, et de l'inégalité dans le développement de nos connaissances de différentes classes du règne animal. Tel principe auquel on est conduit sans hésitation par l'étude des ouvrages ornithologiques paraît funeste dès qu'il s'agit de la classe des Poissons, de celle des Mollusques, de celle des Echinodermes, ou de telle autre. Et cependant, pour réformer la nomenclature dans toutes les branches de la zoologie, il importerait d'établir des règles également applicables à toutes les classes du règne animal. Une première question, que vous n'avez pas traitée, se présente tout naturellement, et il est indispensable de la traiter d'abord.

"En attachant une autorité à un nom quelconque, cette autorité n'implique-t-elle que le nom spécifique, ou le nom spécifique dans son association avec le nom générique? ou enfin, conviendrait-il peut-être d'inscrire l'autorité à double, une fois à la suite du nom du genre et une fois à la suite du nom spécifique? Voici l'opinion à laquelle je me suis arrêté en dernier lieu. Depuis que les objets de la nature ont commencé à être décrits et envisagés sous un point de vue scientifique, il n'a existé que trois manières de les désigner: ou bien on s'est servi d'un nom propre emprunté au langage ordinaire et variable selon les différentes langues, puis précisé dans le domaine de la science; ou bien on a emprunté un nom propre au latin ou au grec, et on y a annexé une circonlocution explicative; ou enfin on s'est servi d'une dénomination bivocale qui a été généralement adoptée depuis Linné, mais que plusieurs auteurs anciens, et en particulier Rondelet, avait présentie et même mise en pratique dans beaucoup de cas. Ces noms propres, ces noms paraphrasés, ces noms doubles ont toujours dû désigner dans leur ensemble une espèce déterminée; cela est bien évident pour les deux premiers modes d'appellation; c'est ainsi qu'on citera toujours le nom d'Elephas, déjà employé par Aristote, pour désigner cet animal, abstraction faite de toute idée systématique. Il en est de même des dénominations paraphrasées: le *Lepusculus cuniculus terram fodiens* de Klein désigne le *Lepus cuniculus*, Linn., comme on le désignerait peut-être dans d'autres circonstances par un simple nom propre (Lapin, Rabbit). Si j'ai bien saisi l'esprit des loix prescrites par l'immortel réformateur des sciences naturelles du 18e siècle, il faut envisager également ses dénominations bivocales comme l'équivalent d'un nom propre, et dès lors il n'est pas douteux qu'il ne faille entendre l'autorité attachée au nom spécifique comme se liant aussi intimément au nom générique. Il faudra donc nécessairement écrire *Falco peregrinus*, Linn., *Diodon peregrinus*, Spix, et non point *Diodon peregrinus*, Linn.; car évidemment Linné n'a pas pensé que cette espèce se distinguât génériquement des autres oiseaux de proie diurnes, sans quoi il aurait établi un genre au détriment des Falco, pour l'y placer. En conservant l'autorité d'un ancien auteur pour un nom spécifique que l'on associe à un genre nouveau, on s'exposé à un nouvel élément de confusion inextricable dans la synonymie, sans atteindre pour cela son but, qui est de rendre justice à ses prédécesseurs. Si la nomenclature constituait à elle seule la science, il y aurait évidemment dépréciation des mérites d'un auteur à évincer de nos registres systématiques les noms qu'il a donnés. Mais comme la nomenclature n'est qu'un échafaudage formulaire, il importe peu que ce soit le nom d'un auteur ancien ou celui d'un auteur moderne qui prévale en définitive, pourvu que la nomenclature soit toujours en rapport réel avec l'état de la science à une époque donnée. C'est maintenant une exigence indispensable de la science, de joindre une bonne synonymie au nom systématique de toutes les espèces, et cette synonymie ne devra pas seulement comprendre l'indication des noms divers qui ont été donnés successivement à une espèce, mais encore indiquer toutes les sources où l'on trouve des renseignemens importans qui la concernent: que ces renseignemens aient été publiés en latin, en grec, en allemand, en français, ou en anglais, peu importe. Les faits rapportés par Aristote, par Rondelet, par Ray, par Lyonnet, par DeGeer, et par tant d'autres auteurs devront toujours être mentionnés dans l'histoire d'une espèce: que cette espèce soit étiquetée dans nos collections du nom de ces auteurs ou d'un nom moderne, que ce nom figure en tête d'un article dans un ouvrage systématique ou dans la synonymie, peu importe. Ce sont là deux domains distincts de la science, quoiqu'ils s'entr'aident mutuellement; et la nomenclature doit toujours rester l'expression des rapports admis entre les espèces et les genres dans leurs diverses délimitations.

"Comme il n'y a pas long-temps que l'on attache en zoologie une importance réelle à l'étude des genres, il reste encore trop de vague dans leur délimitation pour qu'on puisse espérer de voir des principes quelconques fixer l'opinion sur ce sujet. Tel auteur croira pouvoir réunir toutes les espèces d'oiseaux connues maintenant dans quelques centaines de genres; tel autre auteur en admettra quelques milles; et il en est de même de toutes les classes du règne animal. Entre des manières de voir si opposées il n'y a pas maintenant de transaction possible, et il est indispensable de consacrer pour chaque auteur consciencieux le droit d'envisager ces choses comme il les comprend.

Aussi l'idée, juste d'ailleurs, de conserver l'autorité d'un auteur à la suite de toutes les espèces qu'il a établies, à quelque genre qu'on les rapporte, ne saurait être mise en pratique maintenant; la chose n'est possible que pour le nom spécifique uni au nom générique auquel il a d'abord été rapporté. Peut-être un jour en viendra-t-on à employer l'autorité à double, et à dire par exemple, *Diodon* (Spix) *peregrinus*, Linn., au lieu de *Falco peregrinus*, Linn., ou de *Diodon peregrinus*, Spix. Mais la chose ne sera praticable que lorsqu'il règnera plus d'accord entre les zoologistes sur l'importance et la délimitation des genres. L'introduction actuelle de cette méthode entraînerait à de graves inconvéniens, parmi lesquels la nécessité de changer une foule de noms ne serait pas la moindre. Je suis dès lors convaincu que ce qu'il y a de mieux à faire est de maintenir la double appellation usitée jusqu'à présent, en envisageant l'autorité comme nécessairement liée aux deux noms réunis. Mais je vais plus loin. J'ai acquis la conviction qu'il serait également préjudiciable aux vrais progrès de la science de changer quoi que ce soit aux noms existans, par qui qu'ils aient été proposés, et en quelque temps qu'ils l'aient été. Une assertion aussi générale a besoin d'être expliquée. Je ne veux point dire par là qu'il soit préjudiciable à la science d'établir de nouveaux genres là où une étude plus approfondie des objets nous y conduit naturellement; j'ai moi-même établi un trop grand nombre de genres dans la classe des poissons et dans celle des échinodermes, pour ne pas sentir l'importance de pareilles innovations lorsque le temps et nos connaissances les rendent nécessaires. Je pense qu'il doit en être de même des coupes plus étendues que l'on désigne sous les noms de sections, de tribus, de familles, d'ordres, &c., &c.; j'entends seulement que ces innovations s'opèrent sans mutilation quelconque de la nomenclature de nos devanciers, et que sous prétexte de perfectionner la nomenclature on ne le surcharge pas de noms nouveaux uniquement destinés à en remplacer d'autres, souvent meilleurs, et le tout parce qu'il aura plû à tel ou tel auteur de n'admettre dans ses cadres que des noms façonnés d'une certaine manière.

"Après avoir long-temps partagé l'opinion de la plupart des naturalistes, que le même nom générique ne saurait être admis simultanément dans différentes classes, ni même à la fois dans le règne animal et dans le règne végétal, et que la priorité devait décider d'une manière absolue lequel des noms devait être conservé, et lequel devait être remplacé par un nom nouveau, je me suis convaincu, après avoir examiné sous ce point de vue la nomenclature de tout le règne animal, qu'une semblable pratique jetterait la plus grande confusion dans la science, sans remédier à l'inconvénient que l'on voit aux doubles emplois d'un même nom dans différentes classes. Et d'abord, pour être changés, les noms anciens n'ont point disparu des ouvrages où ils se trouvent et on nous les rencontrons chaque fois que nous devons les consulter. Ils devront d'ailleurs toujours être reproduits dans la synonymie, quelque peine que l'on se donne à leur en substituer d'autres. Un relevé que j'ai fait de tous les noms de genres du règne animal m'a fait connaître plus de 700 genres dont les noms se répètent dans différentes classes du règne animal, sans compter les doubles emplois avec le règne végétal. Est-il quelqu'un qui puisse raisonnablement penser que changer ces 700 noms de genres, et par cela-même le nom des milliers d'espèces qu'ils renferment, uniquement en vue d'éviter la répétition du même nom générique dans différentes classes, compense l'inconvénient d'introduire des milliers de noms nouveaux que rien ne rappelle dans les ouvrages où se trouvent peut-être consignés les faits les plus intéressans de l'histoire de ces animaux?

"La stricte observation de la loi de priorité entraîne à des conséquences tout aussi fâcheuses; car pour réintégrer dans ce qu'on appelle leur droit, certains noms génériques tombés en désuétude, il faudrait souvent faire disparaître de la nomenclature des noms maintenant généralement adoptés. Par exemple, Klein, qui fait autorité pour la classe des Echinodermes, a établi en 1734 * un genre sous le nom de *Cassis*, pour désigner les fossiles dont Lamarck a fait en 1816 son genre *Ananchytes*, qui est maintenant généralement adopté, tandis que le nom de *Cassis* de Klein n'est mentionné par aucun auteur moderne. D'un autre côté, Bruguière a établi un genre *Cassis*, adopté par tous les naturalistes, et qui compte de nombreuses espèces. Faudra-t-il maintenant, pour être conséquent à la loi de priorité, réintégrer le genre *Cassis* de Klein pour désigner les *Ananchytes* de Lamarck, supprimer le nom d'*Ananchytes* et le remplacer par celui d'*Echinocorys*, que Breynius avait proposé dès 1732 pour le même groupe que Lamarck a appelé *Ananchytes* en 1816, et supprimer enfin le genre *Cassis* de Bruguière pour lui donner un nouveau nom, ainsi qu'à toutes les espèces de coquilles qu'il renferme? Il est évident qu'en le faisant on se conformerait strictement aux préceptes qui prévalent maintenant; mais ce serait incontestablement au détriment de la science †. Que diraient, en particulier, les auteurs de tous les noms qui font double emploi, si, connaissant maintenant tous les genres du règne animal dont les noms se répètent dans différentes classes, j'allais de gaieté de cœur leur en substituer d'autres, suivis de ce *mihi* qui fait l'unique ambition de tant d'auteurs? J'y serais ce-

* Before date of 'Syst. Nat.'—H. E. S.

† Which is most detrimental to science: the want of fixed principles, or the occasional inconvenience which the enforcing of those principles produces in particular cases?—H. E. S.

pendant pleinement autorisé d'après les principes que tous les zoologistes invoquent de nos jours; et cependant de quelle réprobation un pareil fait ne serait-il pas justement frappé? Et moi-même parcourant telle ou telle collection étiquetée d'après ces lois rigoureuses, ne serais-je pas surpris de voir mon nom figurer comme autorité à la suite d'une foule de genres que je n'aurais point institués, et à côté d'espèces que je pourrais n'avoir jamais vues antérieurement? Le registre dont je viens de parler, et que je publie dans ce moment sous le titre de *Nomenclator Zoologicus,* montrera mieux que toute espèce de raisonnement les inconvéniens auxquels exposeraient les effets rétroactifs que l'on voudrait donner à de pareilles lois. Aussi longtemps que l'on n'a pas possédé de registre général des noms employés en zoologie, on ne saurait reprocher à un entomologiste ou à un conchyliologiste, patient observateur, de n'avoir pas connu les travaux des ornithologistes et des ichthyologistes. La prétention de faire table rase du passé, pour élever un édifice nouveau, uniforme dans toutes ses parties, me paraît aussi insensée que le serait un historien qui, pour faciliter l'étude de la chronologie, n'admettrait dans la succession des souverains de telle ou telle dynastie que des noms de son choix, et substituerait à l'histoire réelle une histoire factice, en faisant rentrer la première dans les synonymes de la seconde. Je l'ai dit ailleurs, et je le répète ici, il ne résulte pas plus de confusion dans la nomenclature de la zoologie de la répétition de quelques noms dans différentes classes, que de la répétition des mêmes noms dans l'histoire de différens pays ou dans la géographie de différentes contrées. Que diraient les anglais, si un français avait la prétention de débaptiser dans une histoire de l'Angleterre tous ceux de leurs souverains qui ont porté des noms qu'auraient portés avant eux des souverains français? Que diraient les américains, si un géographe d'Europe avait la prétention de donner des noms nouveaux à toutes les localités qui dans les Etats-Unis portent des noms déjà employés pour des localités de nos continens? A quoi en serions-nous nous-mêmes si nous devions changer les noms de tous nos amis qui s'appellent d'un nom qui ne leur va pas, et si nous voulions appeler Grands tous les Petits qui sont grands, et Petits tous les Grands petits, &c. &c. &c.

"Le respect pour le passé doit être la loi souveraine de toute institution qu'on veut faire prospérer; et si la science, au lieu de viser à se consolider en fortifiant son organisation intérieure, tend continuellement à renouveler seulement sa forme, elle court risque de tomber dans l'anarchie. Ces principes ne mettront pas un frein à l'exagération, ils ne retiendront point les novateurs bouillans; mais il serait tout aussi fâcheux que ces tentatives de réforme ne pressassent pas continuellement ceux qui seraient disposés à s'endormir que de les voir envahir sans frein l'enceinte sacrée de la science. De vrais progrès ne sont possibles qu'en luttant continuellement contre les entraves qui gênent la marche de l'esprit humain, de quelque nature que soient ces entraves.

"Si, après ces considérations générales, j'en viens encore à quelques remarques spéciales, c'est moins pour proposer des règles précises que pour vous présenter quelques réflexions sur celles que vous proposez.

"Dans les cas que vous citez, page 5, du *Perdix rubra,* Briss., j'écrirais d'après mes principes réellement plutôt *Perdix rubra,* Briss., que *Perdix rufa,* avec une nouvelle autorité. Je ne crois pas que l'époque de l'adoption de telle ou telle nomenclature puisse être fixée pour aucune classe, et qu'il puisse y avoir prescription pour les noms des auteurs même les plus anciens, lorsqu'ils sont bons; seulement je ne voudrais pas les réintégrer forcément lorsque le temps en a sanctionné d'autres. Un semblable rigorisme aurait toutes les conséquences fâcheuses du principe politique de la légitimité poussé à rigueur dans tous les cas, contrairement à l'ordre de choses régnant de fait dans la société.

"Il me semble à peine qu'il puisse y avoir un avantage réel à donner une terminaison uniforme à toutes les familles; il n'y a pas deux naturalistes qui les délimitent de la même manière; et si en Angleterre l'usage de terminer les familles et les sous-familles en *idi, idæ, ida* et *ini, inæ, ina,* prévaut généralement, il en est à peine question chez quelques auteurs allemands et chez un petit nombre de francais. Puis, vous voyez à chaque instant une division changer de rang aux yeux des naturalistes, et d'un *ini* qu'elle portait pour finale se revêtir d'un *idi,* puis les divisions se multiplier ou diminuer en nombre; ce qui rendra nécessairement un jour cette nomenclature des familles très-compliquée, lorsqu'elle aura fixé plus généralement l'attention. Pour ma part, je ne repousserais pas l'emploi des noms radicaux, quoique je désire voir les noms patronimiques prendre de l'extension.

"Le rigorisme avec lequel vous scindez la difficulté au sujet du démembrement des genres et de la conservation du nom primitif, pour les espèces types, est très-louable, et dénote chez vous un sentiment d'équité qui vous fait honneur, et dont, pour ma part, je ne dévierai jamais dans le cercle étroit de mes travaux. Cependant vous n'empêcherez pas les esprits égoïstes, pauvres, et surtout les intrigans (qui sont peut-être plus à redouter dans la science qu'ailleurs) de présenter les choses sous un jour qui excuse momentanément les petits moyens qu'ils emploient pour placer leur *mihi* à la suite de quelques noms de plus. Eh bien, si les divisions qu'ils proposent sont acceptables, il vaut, je crois, en définitive mieux les admettre que de les changer de nouveau. L'histoire de la science stigmatisera tôt ou tard ces pratiques mesquines.

"Dans un moment où il règne une tendance si marquée à l'établissement de nombreux genres circonscrits dans des limites très-étroites, il importerait de chercher à donner une acception fixe à tous ces noms de genres anciens qui sont restés dans l'oubli, et qu'on pourrait en grande partie réintégrer sans altérer ceux qui sont déjà généralement admis. Dans le cas où des noms qui font double emploi maintenant pourraient être affectés à des genres nouveaux dans le voisinage de ceux que l'on doit conserver, j'encouragerais cette pratique plutôt que de la préscrire comme vous le faites; par exemple, page 7.

"Il y a plusieurs de vos préceptes, pages 8, 9 et suivantes, qui me paraissent excellens comme directions pour l'avenir, mais qui auraient, je crois, un effet fâcheux si on leur accordait un effet rétroactif. En dernier ressort, la meilleure de toutes les règles sera toujours l'exemple, et je pense que les auteurs qui se sentent la mission d'influer sur la marche de la science, en s'observant dans le choix qu'ils font des noms qu'ils proposent et dans les modifications qu'ils introduisent dans la nomenclature déjà existante, exerceront une très-salutaire influence sur l'adoption des mesures les plus propres à purger la nomenclature des incorrections qu'elle présente. Et si nous voulons mettre un frein efficace à des changemens sans fin, adoptons pour devise: *Faire bien à l'avenir, mais respecter ce que nos prédécesseurs ont fait, alors même que nous aurions des améliorations à apporter dans leur nomenclature.* J'ai déjà énoncé quelques-uns de ces préceptes dans la préface de mon *Nomenclator Zoologicus,* que vous aurez sans doute parcouru; la réception de votre lettre et de votre programme m'a fait faire de nouvelles réflexions sur ce vaste sujet; je vous les aurais bien volontiers exposées plus en détail que je n'ai pu le faire ici, si mes occupations ne m'avaient pas forcé de renvoyer d'un jour à l'autre ma réponse, et si maintenant, pressé par le temps, je ne devais terminer mes observations pour qu'elles vous parviennent pour la réunion de Manchester. Je regrette infiniment de ne pas pouvoir assister aux intéressantes discussions que ces questions soulèveront sans doute dans la Section de Zoologie; si vous voulez bien à cette occasion communiquer à la Société ma manière de voir sur ce sujet, ce sera me rappeler au souvenir précieux des membres de la Société, et leur prouver qu'en toutes circonstances je prends un vif intérêt à leurs travaux, même lorsque je ne puis les partager personnellement.

"L. AGASSIZ."

"Neuchâtel, le 18 Juin 1842."

Strickland replied to M. Agassiz as follows:—

"Cracombe House, Evesham, Worcestershire,
July 8th, 1842.

"MY DEAR M. AGASSIZ,—I beg to thank you sincerely for the valuable and important observations which you were so good as to address to me during the Meeting at Manchester, and I take the earliest opportunity of informing you of the progress of the nomenclature question. The Committee originally appointed for this purpose, having obtained a large mass of correspondence on the subject from the various geologists to whom the printed pamphlet had been sent, held several meetings in London in April and May last, and gave the whole measure a very careful consideration. They also endeavoured during the Manchester meeting to collect a variety of opinions on the subject, and having thus, as they conceive, viewed the subject in all its bearings, they finally agreed upon their report, which was then read to the meeting, and will be printed in the next volume of the 'Transactions' of the Association. A number of extra copies will also be printed for private distribution, and I will take care to send you one by an early opportunity. You will there find that many modifications have been introduced into the original plan, with a view of increasing the practical efficiency of the measure, and of diminishing the difficulties which are in some degree inseparable from the subject. The Committee have chiefly aimed at forming such a plan as will meet the views of the greatest number of geologists, feeling it to be impossible to please all.

"On the subject of the authority to be cited after a binomial name, in which the *genus* dates from one author, and the *species* from another, the Committee had already introduced a regulation since the printing of the original outline sent to you. Feeling, as you do, the CONFUSION that would arise from writing *Diodon peregrinus,* Linn. (that binomial phrase not being used by Linnæus); and also being sensible of the INJUSTICE of the expression *Diodon peregrinus,* Spix (which implies that Spix discovered the species, and presents no distinction from those species which were really first described by Spix), they propose to remove this difficulty by adopting the *distinctive mark* (*sp.*), implying that the authority refers to the *specific name only* in such cases. Thus they would write *Diodon peregrinus,* Linn. (sp.); but when the authority refers to both genus and species, they omit this mark, as *Ostrea edulis,* Linn. If this *distinctive mark,* or some other which might appear preferable, is generally adopted, it would meet the difficulty and prevent confusion, and render justice to original describers!

"The Committee illustrate this by the *Muscicapa crinita,* Linn., which was first referred by *Swainson* to the genus *Tyrannus* of Vieillot. They say, 'Of the three persons concerned with the construction of a binomial title, as in the case before us, we conceive that the author *who first* describes and names a species which forms the groundwork of later generalizations, possesses a higher claim to have his name recorded, than he who afterwards defines a genus which is found to embrace that species, or who may be the mere accidental means of bringing the generic and specific name into contact. By giving the authority for the *specific* name in preference to all others, the inquirer is referred *directly* to the original description, habitat, &c. of the species, and is at the same time reminded of the date of its discovery; while genera, being less numerous than species, may be carried in the memory, or referred to in systematic works, without the necessity of perpetually quoting their authorities.' Instead, therefore, of writing either *Tyrannus crinitus*, Linn., *Tyrannus crinitus*, Vieill., or *Tyrannus crinitus*, Swains., the first of which is *incorrect*, and the last two *unjust*, they propose to write *Tyrannus crinitus*, Linn. (sp.), showing that Linnæus was concerned with the *specific name only*. The Committee hope that geologists will in future be willing either to adopt the expression (sp.) in such cases, or to point out some other distinctive mark which may be preferable to it.

"There is no proposition in the Report of the Committee which has cost more careful and anxious consideration than that which relates to changing generic names when they have before been used in other departments of natural history. After a diligent examination of the various opinions on this subject to be found in books and in the letters of their correspondents, they have decided that it is more for the interests of science to change such names than to retain them in '*double emploi*.' In this opinion I regret to find you do not concur. But we should remember that it is distinctly enforced by the great Linnæus himself (Philos. Botan. 214, 217, 230), and has been ever since adopted, as you yourself admit, by 'la plupart des naturalistes.'

"2ndly. I find, on looking over the correspondence from eighteen zoologists now in my possession, who have sent me remarks on the subject of nomenclature, that fourteen make no allusion to the point in question, thus implying that they are content with the rule 10, as laid down in the printed plan: one (the Prince of Canino) distinctly states that he *agrees* to it; one (O. Westwood) suggests that *subgeneric* names in botany may be retained in *double emploi* for subgenera in zoology; one (E. H. Bunbury, Sec. Geol. Soc.) says, that 'though very *desirable* that no two genera should have the same name, yet where they belong to classes altogether different, and have been long *established,* he *doubts the policy* of changing them;' and *one only* (yourself) is decidedly of opinion that they should be retained. Therefore it appears that the majority of naturalists are decidedly in favour of excluding duplicate generic names.

"3rdly. Although it appears at first sight a severe measure to set aside 700 generic names (as stated in your letter), yet it must be remembered that the total number of generic names of animals is said by yourself to amount to 17,000, so that only a proportion of 4 *per centum* would require to be changed; and if zoologists would submit to this slight inconvenience *now,* and always consult your 'Nomenclator Zoologicus' *in future,* before giving a name to a genus, the science would be permanently purged of these objectionable names. The names which Gray has altered in his 'Genera of Birds' in compliance with this rule are about 58 out of 1119, and these I shall gladly adopt in future, in the hope that ornithologists will hereafter agree on a uniform nomenclature.

"4thly. The evils of creating new names in place of old ones adopted elsewhere, are much diminished by the fact that synonyms *already exist* in numerous cases, and it is only necessary to take the *oldest synonym,* and not to form a new word (e. g. *Cichla,* Wagl., 1827, being preoccupied for fishes, Mr. Gray adopts the synonym *Donacobius,* Swains., 1831).

"5thly. We do not wish to enforce this rule, except when the earlier name is *still used in the science.* When a duplicate name has been reduced to a synonym by the operation of our other laws, we may tolerate its adoption in another part of the system (e. g. *Lophorhynchus,* Vieill., 1816, is a mere synonym of *Cariama,* Brisson, 1760, and he therefore does not object to Swainson applying the word *Lophorhynchus* in 1837 to a genus of Pigeons, though it would have been more judicious if he had not done so).

"6thly. We recommend that when it is necessary to form a new word in place of an old *double emploi,* the new word should have an *analogy* to the old, as *Plectorhamphus* in place of *Plectorhynchus.*

"7thly. We recommend, that when the original author of a duplicate generic name is still living, the person who may discover the oversight, should, as an act of *courtesy,* communicate it to the original author, and leave it to him to propose a new name, so that *his authority* may still be quoted for the new name.

"8thly. The cases which you cite of *Cassis* and *Echinocorys,* do not apply to our rules, because the works of Klein and Breynius were published before the 12th edition of the 'Systema Naturæ,' which we propose to take as the limit of the law of priority. Therefore, according to our views, the names *Ananchytes*, Lamarck, and *Cassis*, Bruguière, should be retained as they are now used.

"9thly. Your analogy from the existence of *doubles emplois* in geography and history will not apply, because in these cases we prevent confusion by the use of additional phrases; thus we speak of *Friburg en Suisse, Friburg en Brisgau, Frankfort-en-Maine, Frankfort over Oden, Boston in England, Boston in United States*, or, for brevity, *Boston, U.S.* So in history we say, *Louis le Gros, Louis XIV., Louis XVIII., Charles the First of England, Charles the First of Sweden*, &c. &c. But in zoology, unless we are prepared to give up our *bivocal* system of nomenclature, and to adopt a *trivocal* or 4-*vocal* method, so as to denominate the species of animals, e.g. *Temnurus albicollis Avis Trogonides, Temnurus leucopterus Avis Corvides*, and the genera *Brachypus Reptile Chalcides, Brachypus Insectum Coleopterum Curculionides, Brachypus Avis Insessor Turdides*, &c.,—unless, I say, we adopt this cumbrous, and I may say impracticable method, I do not see how we are to retain the use of duplicate generic names without producing serious confusion.

"10thly. If we were to permit the unlimited use of duplicate and triplicate generic names, one evil consequence would be, that indolent persons, when they want a name for a new genus, would not take the trouble to form a new word, but would take the first name of an animal or a plant which offered itself, and would adopt it for their new genus, thus making the evil worse than it is.

"11thly. Another evil consequence would be, that numerous generic names, which have long since been discarded from the science in consequence of their *double emploi*, would again assert their claim, and would supersede many words now in use. Thus, the name *Arenaria* was given by Brisson to the bird called Turnstone (*Tournepierre*), but as the word *Arenaria* is used in botany, the name *Strepsilas*, of Illiger, has been universally used in modern times for this bird. Now, if we permitted the use of words in *double emploi*, it would be necessary to discard the name *Strepsilas*, and resume the obsolete term *Arenaria* of Brisson, on the ground of being prior to *Strepsilas*. The question is simply this:—Which is most detrimental to science: the want of fixed principles, or the occasional inconvenience which the enforcing these principles produces in particular cases? We are willing to hope that it is not yet too late (though it may be too late twenty years hence) to apply a fixed principle in the case before us. I will therefore conclude this long discussion (which the importance of the subject has induced me to carry into great detail) with quoting a few words from the Report which the Committee have now agreed to and signed *.

* See Collected Papers, p. 375;—Report, pt. 1, § 10.

"Trusting to your candour to give an impartial consideration to these remarks, and hoping that both you and other zoologists will, for the sake of uniformity, be willing to adopt this rule in future, I now pass on to another subject.

"Respecting names of families and subfamilies, we do not recommend any change of *gender*, as *idi, idæ, ida,* but the uniform termination *idæ,* which was by the ancients used for the members of families, both male and female. The advantage of this method is, that instead of loading the memory with *new* radical words for families, it adopts the name of some already well-known genus, and by the termination shows the rank which such or such an author assigns to it. If an author, A, sees reason to reduce the *family* of the author B to a *subfamily,* he changes the termination from *idæ* to *inæ,* and thus at once shows the *rank* which he assigns to it; just as 'Your Majesty,' 'Your Highness,' 'Your Excellency,' &c. denote the *rank* of an individual. In conclusion, the Committee do not expect to enforce these laws except by means of the *volition* of such zoologists as may approve of them. Of course they cannot prevent authors from departing from these rules as far as they may think proper; and the Committee do not therefore expect an *immense* acquiescence in them. But they hope, for the sake of arriving at uniformity of practice, persons will consent in some cases to follow these laws, even when differing from their own private opinions. If the *majority of naturalists* will only adhere to the *majority of these rules,* an approach to uniformity will be attained, and a very great amelioration in that which you so well denominate 'l'épouvantable confusion de la synonymie,' will be effected.

"The Zoological Section at Manchester could not devote much time to the public discussion of the nomenclature question, but the Committee derived considerable assistance from many of the members present, and when they had agreed on their Report, it was read to the public in the Section. I hope in a very few weeks to have some printed copies ready, and will take the earliest opportunity to send you one.

"The Meeting at Manchester was a very successful one; many important communications were read, and everything passed off very pleasantly. I was sorry to find that the state of our funds prevented the Association from giving a grant of money in aid of your publications this year, but I hope that next year we shall be able to do so. Your 'Nomenclator Zoologicus' and 'Bibliothèque Zoologique' were exhibited both in the Geological and Zoological Sections, and excited general approbation.

"I have received from Dr. Buckland your 'Nomenclator Zool. Aves' to the end of page 76, but have not yet received from Mr.

Murchison the concluding pages which you say you had sent him. I have looked over this portion with some care, and am happy to bear testimony to its general accuracy. I have but very few names of *genera* to give you by way of *addition*; indeed, it is so short a time since Gray's work was published, that scarcely any additions have been made to the science. Have you seen the *Appendix* to the 'Genera of Birds' which Gray has published? He there includes all Kaup's genera, and also the new names proposed by Hodgson, which I see you have also entered into your list. I enclose you some remarks of mine on Gray's work, which you may not have seen.

"I have sent you several remarks on the *etymologies* of the names, some of which, as given in your work, seem to be inaccurate; in fact, unless a person is well acquainted with the *characters* of a genus, he cannot always pronounce with certainty as to the allusions contained in the etymology of the name, and therefore, as you have been less occupied with ornithology than with other branches of zoology, it is to be expected that you should sometimes be misled; and the *orthography* of generic names is so imperfect as often to cause errors in *etymology*. I should suppose that ten or fifteen *per centum* of the names given in your list have been erroneously written by their authors, and it would have been well to have corrected all such *cacographies*, and to have referred from the *wrong* name to the *right* one, thus:—*Crataionyx* (vid. *Cratæonyx*), *Crypsirina* (vid. *Crypsirinæ*), *Cryptonix* (vid. *Cryptonyx*), &c. I have myself corrected some of these wrong spellings in my review of Gray's work, and I see that you have inserted these alterations, *on my authority*, in your list. I beg, however, to disclaim all wish or intention of having *my name* attached to these corrections, and of thus robbing the original author of his credit. The authority for the genus *Microura* still belongs to Gould, and that of *Oidemia* to Fleming, although I may have amended the orthography to *Micrura* and *Œdemia*. I shall therefore always write *Œdemia*, Flem., *Siurus*, Swains., &c. I have no wish that my name should be attached to any words except such as I have myself constructed *de novo*, such as *Sphenœacus*, &c. I should be much obliged if you would state this in the Appendix to the 'Aves;' for though I think it very desirable to render the orthography of zoological words fixed and consistent with classical rules, yet I agree with you that it is highly culpable to rob other authors of their credit, and to substitute '*ce mihi* qui fait l'unique ambition de tant d'auteurs' (except in the one case of *double emploi*, for which I have given the reasons above).

"Some of the words in your list appear not to have been used as Latin by their authors. Such words as *Barbacon*, *Corbicaluo*, &c., were, I believe, never used by Levaillant, except as *French* words, and might therefore have been omitted in their list. Some of Lesson's words are also *French,* and Azara's are *Spanish,* as *Batara,* &c.

"I am always sincerely,
"Your most devoted,
"HUGH E. STRICKLAND."

Previous to the Meeting at Manchester, the Council of the British Association, at one of their preparatory Meetings in London (Feb. 11, 1842), "Resolved,—That (with a view of securing early attention to the following important subject) a Committee, consisting of Mr. C. Darwin, Professor Henslow, Rev. L. Jenyns, Mr. W. Ogilby, Mr. J. Phillips, Dr. Richardson, Mr. H. E. Strickland (reporter), Mr. J. O. Westwood, be appointed to consider the rules by which the Nomenclature of Zoology may be established on a uniform and permanent basis; the report to be presented to the Zoological Section, and submitted to its Committee at the Manchester Meeting." To this Committee its Members added the names of W. J. Broderip, Professor Owen, W. E. Shuckard, G. R. Waterhouse, and the late W. Yarrell. It was to the charge of those gentlemen that the whole plan was now entrusted; it was by them that it was brought to maturity and presented at Manchester; and it was under their sanction and signatures, with the exception of Mr. Ogilby only, that it was afterwards completed and given to the public.

At Manchester unfortunately there was a small number of the Nomenclature Committee present, and on submitting the subject to the Sectional Committee of the Association, it was evident, that to enter upon so extensive a question, when to so many of the Members it was in a great measure new, would only lead to a protracted discussion, which could not be terminated within the very short time at their disposal. The Sectional Committee therefore resolved to leave the matter entirely in the hands of the gentlemen appointed to prepare the report, and requested them to present it in a finished state to the Section. The following day this was accordingly done; but after being read and explained, as far as time and circumstances would allow, it encountered an opposition that was scarcely expected, couched in a spirit of prejudice, and almost jealous animosity, which was discreditable to the discontents, no matter what their opinions might be. But in all this, the opposition never assumed a definite form; and it is remarkable that among all the correspondence, and in all the discussions, we have scarcely a dissentient voice on the general question, and that the objections and criticisms lay almost entirely in the impropriety of making such radical changes as those proposed appeared to be, and in the difficulty of getting the "plan" worked out and adopted. Some modern inventors of names felt sorely the criticism of their views and compositions, which many of the clauses exposed; and although no reference was made individually, or possibly could have been allowed in a report of the kind, and sanctioned by such authority, yet over-sensitive minds took many of the clauses as aimed at themselves, and hence the almost acrimony of some of the observations in the Manchester discussion. But these very circumstances caused their fall, and prevented any distinct motion being made for either censure or delay; and the report, after being well thrashed, was left in the hands of the Committee,—was printed in the Reports of the British Association for 1842 as finally approved by them, and "was so distributed to the academies and natural-history societies throughout the world*."

It must have been a pleasing feeling to Strickland, after such a discussion, to have expressions of satisfaction and congratulations coming from his friends; but it must have been still more gratifying to him to have the sanction of zoologists abroad awarded to him, of foreign societies and foreign reviewers. To the latter, all the license of anonymous writing was open; and although objections were stated, the main bearing of these papers was entirely favourable.

Such is the history of this subject in modern days. The

* Sundevall, Zool. Framst. 1840–42.

report of the Committee now forms the basis for *Zoological Nomenclature*, and its axioms are daily becoming more widely and decidedly acted upon.

SUGAR-LOAF MOUNTAIN, ASSYNT.

WAINLODE CLIFF.

CHAPTER VII.—1842 TO 1846.

SURVEYING EXPEDITION OF H.M.S. 'BEACON.'—HALCYON SMYRNENSIS.—FOSSIL ENTOMOLOGY: REV. P. B. BRODIE. — NOMENCLATURE OF COLOURS.—REPORT ON ORNITHOLOGY.—ORNITHOLOGICAL COLLECTOR.—RAY SOCIETY.—LORENZ OKEN.—AGASSIZ' BIBLIOGRAPHIA.

THE Surveying Expedition of H.M.S. 'Beacon,' under the direction of Capt. Graves, greatly interested Strickland. His friend Edward Forbes had just sailed with it, and expected to travel over much of his old ground in Asia Minor, and a good deal of correspondence had passed between them connected with natural history, antiquities, &c. Strickland, during his expedition, had been extremely anxious to rediscover there the true *Alcedo smyrnensis*, unknown since Sherard's time, but had been unsuccessful from his visit to the country having been made in winter, when the bird was most probably migratory; and one of his letters, which has been recovered by the kindness of Mrs. E. Forbes, also otherwise interesting, calls attention to this bird:—

"Cracombe House, Evesham, April 20th, 1842.

"MY DEAR FORBES,—I was much gratified at the receipt of your letter, giving as it does so satisfactory an account of your proceedings. I only wish I was along with you; but as that cannot be at present, I must content myself with learning the results of your researches as soon as you may be able to communicate them to your English friends. As a few additional suggestions of points of inquiry have occurred to me, I will proceed to give them. First then, did you see any *Lions* in Lycia? You will doubtless have seen in Fellows's last book that there is a probability of a small breed of lions (called *aslan* by the Turks) being found in the mountains of Lycia. I had much talk with Fellows about them at Plymouth last year, and he promised to make further inquiries should he go there again, so that I hope this *exceedingly* interesting question will now be settled. Its bearings on geology, zoology, and ancient history are very important. The *Nemean Lion* will perhaps become an historical personage! With respect to ornithology, I well know the time and trouble required to collect specimens, and therefore should be sorry if your valuable time were to be consumed in this occupation. But, as you justly remark, *observations* are more important than *collecting*, and therefore, if you can only collect *facts* as to the distribution of species, it will be *nearly* as valuable to science as if you sent home boxes full of skins. To do this, it is only necessary when you get hold of a bird which you do not know as a common British species, to cut off the *head, one wing* and *one leg*, take out the brains and muscular parts and dry them (*i. e.* the head, wing, and leg) thoroughly, then wrap them altogether in one paper, and write on it the locality and date. In this way you may, with very little trouble, collect data for an extended list of the birds of those countries, as the parts above mentioned are quite sufficient for identifying the species. It is principally the SUMMER birds that will contain novelties, as they emigrate from Syria and Egypt. The birds of Crete will be *particularly* interesting; first, because it is quite unexplored; and secondly, because, being the extreme south-east point of Europe, it will doubtless contain exotic forms new to the European fauna. I expect you will find *three* species of *Turtledove*, one of them the British one, and two others not yet recorded as European. Also look out for *Shrikes, Finches*, and the smaller *Warblers*, &c. I saw lately at Oxford a collection of unpublished drawings made by Dr. Sibthorpe forty years ago in Greece and Cyprus, and among the birds are many not recorded as European.

"The large Kingfisher I mentioned is described as having the 'head, neck, and under parts chestnut, throat white, wings and tail dull green.' It was 'shot by Consul Sherard in a river of Smyrna' more than 100 years ago. When I was at Smyrna I could hear nothing of it, though I procured a large *black and white* Kingfisher, which is perhaps the one you mention as having procured at Macri.

"I would further suggest to you (though I dare say you have not neglected it) the desirableness of noting down the vernacular names used by the modern Greeks for all classes of animals. This is a most important point for elucidating the works of Aristotle, Pliny, &c.

"I should be glad to learn the localities and geological position of the *whetstones* used by cutlers and carpenters in this country, called *Turkey stones,* but from what part of Turkey they come I never could make out. I believe also that the *hones,* used for setting razors with, come either from Crete or Cyprus, but I am not certain. The hones seem to be fine slate, and contain no lime, but the *Turkey stones* effervesce strongly with acids. Look out for *fossil fish* in the cretaceous rocks of Crete; they have long been known in the limestone of Lebanon, and I think in some parts of Taurus.

"I wish I had you to help me just now: I have undertaken the difficult task of preparing a code of laws for the regulation of zoological nomenclature. I enclose you a copy of the original outline of the plan. It is now in the hands of a committee of the British Association, consisting of Leonard Jenyns, Ogilby, Henslow, Yarrell, Westwood, and one or two others. It will undergo several modifications and additions, and will then be submitted for approval to the Association, when they meet at Manchester in June; and though I anticipate some dissenters from it, yet I hope the measure will be so far adopted by zoologists as to diminish in some degree the evils of anarchy, which are now so rife. I fear there will not be time to hear from you before the 26th of June next, when the meeting takes place, but if you *could* give me any suggestions respecting this plan and its details they would be most valuable.

"Hamilton's journal of his and my travels is on the eve of publication. It is waiting for Arrowsmith's new map of Asia Minor, which is not quite ready.

"My father joins with me in wishing all success to your researches. Pray remember me to Graves, and believe me, dear Forbes,

"Sincerely yours,

"*E. Forbes, Esq., &c.*" "HUGH E. STRICKLAND."

The Kingfisher alluded to in the preceding letter as procured at Macri, turned out to be the lost species, and was the foundation of his short paper read at Section D. of the Manchester meeting, afterwards published in the 'Annals of Natural History *.' By some it has been thought that the Smyrna bird was distinct from those of Continental India, &c., and Mr. G. R. Gray, in his 'Catalogue of Fissirostres' in the British Museum, and also in his 'Genera of Birds,' has

* Vol. ix. p. 441. See also 'Collected Papers,' p. 316.

kept that of the Continent under the name of *H. fusca*, Bodd., the type of which is the figure in the 'Planches Enluminées,' 894; but after comparing the Smyrna specimen, no tangible difference can be perceived between it and specimens from Continental India, Nipal, or Ceylon. These cannot be separated as species.

The Rev. P. B. Brodie was now working on the insects of the Lias, to which Strickland had already paid some attention, and had described a very fine and perfect specimen of a fossil Dragon-fly from the Lias of Warwickshire*. He was

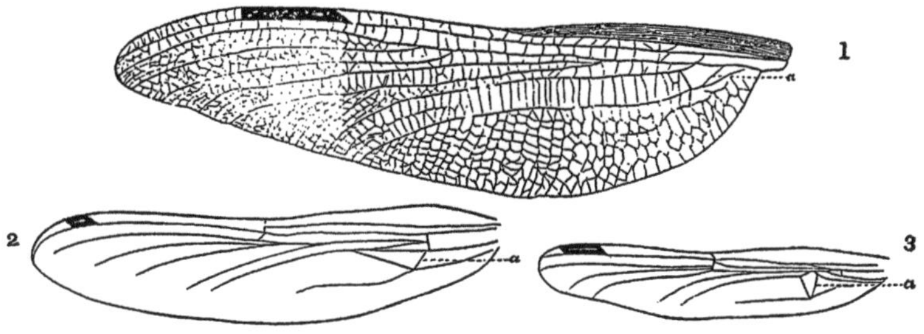

1. *Æshna liassina*, Strickl. 2. *Æ. grandis*. 3. *Libellula depressa*.

thus naturally consulted about the possibility of bringing out a work devoted to this branch of palæontology, not confining it exclusively to the lias beds; and recent discoveries had so much increased the remains of those fragile beings, which not long since could only count a few individuals, that the undertaking seemed most desirable; every assistance was therefore afforded, both by advice and by the more substantial help of subscribers. Strickland's locality was also most favourable for research. Wainlode was in his immediate vicinity, so also Combe Hill. His father had opened a quarry on the Apperley estate, in which the insect-bed was exposed; and there were several other places not far distant where it also could be easily examined. The following letter refers generally to this subject, and to the examination of some of the localities in Warwickshire, where other interesting fossils also occurred. The crustacean referred to is from the

* Mag. Nat. Hist. 2nd Ser. vol. iv. p. 302. See also 'Collected Papers,' p. 152.

"Insect-bed" of the Lower Lias from Temple Grafton in Warwickshire, and Mr. Hartland's collection is now inaccessible. It appears to be the *Eryon barrovensis*, M'Coy, which is from the Lias of Barrow-on-Soar in Ireland *. Strickland made a drawing at the time, and the accompanying figure of a crustacean yet indistinctly known in this country, has been cut from that. The Rev. W. S. Symonds informs us, that Miss Holland of Worcester possesses a fine specimen of this species, or one nearly allied to it, from the Upper Lias of Dumbleton †.

"Cracombe House, Oct. 18, 1843.

"MY DEAR BRODIE,—I have not been able to revisit the Bitford country since you were here, but hope soon to do so. I have been rather unlucky about the fine Crustacean which we saw, as it has found its way into the possession of Mr. Hartland, the banker at Evesham, who is so much in love with it that he is unwilling to part with either of the halves, so I must content myself with borrowing it to make a drawing of it :—

"The following are the sections I took when with you :—

I. *'Nook' Quarry, Bickmarsh.*

		ft.	in.
	Clay	14	0
1.	Top rock	0	8
	Clay	3	1
2.	Thick rock	0	7
	Clay	2	3
3.	Gravestone bed	0	4
	Clay	1	2
4.	Bottom flooring bed	0	4
	Clay	3	7
5.	White bed	0	4
6.	Second white bed	0	4
	Limestone and dirt	1	5
7.	Blocks	0	9
8.	Headstone	0	6
	Clay	4	8
9.	Firestone		
16	Total	34	0

II. *Mr. Harwood's Quarry at Temple Grafton.*

		ft.	in.
	Clay	4	0
1.	Top rock	0	6
	Clay	3	8
2.	Thick rock	0	6
	Clay	3	0
3.	Gravestone bed	0	5
	Clay	1	0
	Middle bed (sometimes wanting)	0	2
	Clay	0	4
4.	Bottom paving rock	0	4
	Clay	5	0
5.	White bed	0	3
	Clay	1	6
	Potstone	0	5–10
	Dirt	0	9
7.	Blocks	0	8
	Clay	0	5
8.	Headstone	0	7
	Stone	0	2
	Bottom rock	0	3
	Clay	5	6
9.	Firestone	0	7
22	Total	96	0

* Ann. Nat. Hist. iv., 2nd Ser., p. 172.

† Miss Holland's specimen has been examined by Mr. Salter, and identified with *Eryon barrovensis*, M'Coy.

"The same numbers refer to the corresponding beds in both sections, and there seems no reason to doubt of their identity. You will remember that the 'Firestone' is the hard marble—like the stone of which we each carried away a polished fragment, and which I compared at the time both to the 'Cypris' bed of Gloucestershire and to the landscape-stone of Bristol. I had no sooner got home, and looked a little closer into my polished specimen, than I found that the broken surface presented numerous specimens of Cypris, thus confirming the above view.

"I lately re-examined and measured the beds at the junction of marl and lias at Dunhamstead, near the Droitwich station of the Birmingham and Gloucester Railway. The insectiferous beds are unfortunately concealed, but I recognized the Cypris-bed with Cyclas, the Pecten-bed, Bone-bed, &c., nearly the same as at Wainlode. I was at Apperley for two or three days last week; my father has opened the quarry on the hill between there and Wainlode, and raised about two cartloads of stone, partly the insect-bed, and partly the superjacent bed with Ostreæ, &c. I broke up a good quantity of the former, but without any remarkable success. However, I laid aside a heap of the best slabs, and gave directions for it to remain till I go there again, when I will get you to help me to break it up; or, if you like to go there at any time and work at it, you are very welcome.

"I should like to see a section of the different beds of Lias near Bristol, with measurements, more especially showing the position of the *Landscape-stone* with respect to the *insect-bed*, and should then be better able to decide on the true equivalents in this neighbourhood of the former. Respecting the publication of the Lias insects, the following plan has occurred to me. De la Beche contemplates publishing, in connexion with the Ordnance Survey, a complete series of all British fossils (each species on a separate plate), the impressions to be supplied to any amount, for the illustration of works on geology. Now I think that if proper application was made to Sir H. De la Beche and to Phillips, they might perhaps be induced to figure in this way a selection of the best and most illustrative specimens, which would save you all the cost of making engravings, and ensure its being done well. In that case I should say that either Curtis, Westwood or Denny could make these engravings of fossil insects better than any other artists I know, they being exclusively *entomological* artists. If this could be effected, you would only have the expense of printing the letterpress, and could easily obtain sufficient subscribers to cover that: you can think over this proposition, and if you wish it, I shall be happy to write to Phillips on the subject; or you can, if you like, get Sedgwick to make the application. It is really a pity that these numerous and interesting fossil insects, presenting a perfectly new feature in the palæontology of the Lias, should remain longer unpublished.

"I will not forget the cast of the Mastodon's tooth, nor the section of the *Avicula longicostata* quarry.

"Yours very truly,

"*Rev. P. B. Brodie.*" "HUGH E. STRICKLAND."

The plan of publication proposed in the foregoing letter could not have been accomplished without too much delay; after some correspondence it was abandoned, and it was resolved to publish the work independently. The MS. was submitted to Strickland, who wrote, after perusing it:—

"Cracombe, July 13, 1844.

"MY DEAR BRODIE,—I have perused your MS., which appears to me to give a very clear account of the localities, and to require nothing but a few slight verbal alterations which I have marked in pencil, and which you can adopt or not as you think proper. Of course you will not begin to print till the plates are nearly ready, as the printing will not take long, and otherwise you will not be able to refer correctly to the plates and figures.

"I think Buckland and Conybeare referred to the whole of Aust Cliff, when they said it was 60 feet high, &c.; but they probably wrote only from guess or memory in this particular instance.

"Westwood's conclusions on the Wealden insects are very interesting, and I shall be curious to hear what he has to say about the Lias ones. I hope you took care so to mark the latter as to prevent their being mixed or confounded with the Wealden, otherwise the mere *mislaying* of a specimen might spoil a geological theory.

"I have seen the counterpart of Gibbs's *second* insect which you mention: it is very different from his former one, and, I imagine, not a Libellula at all. I forgot to tell you that I saw Mr. Hope in London, and he said he would take *ten copies* of your book! Also the Rev. W. S. Symonds of Offenham, a very zealous geologist of *two months'* standing, will subscribe for your book. He was infected with geology by his parishioner Gibbs. I am going from home on Monday for a fortnight or three weeks.

"Yours very truly,

"H. E. STRICKLAND."

Strickland's correspondence with myself now became very frequent, chiefly upon ornithological subjects, but by no means exclusively so. Nomenclature of colours had attracted his attention, as it had my own, on account of the vague terms applied, and the equally random way in which modifications of shades were expressed, which gave rise to great difficulty, and to a want of uniformity in describing zoological or other objects. It is true we had Symes's edition of Werner's 'Nomenclature,' a very good guide; but its terminology was often extremely awkward, and it referred to objects for examples which were either unknown, or inaccessible to nine-tenths of its readers. He alludes to this subject in a letter:—

"The object should be to form a good, simple, standard nomenclature, adapted to the practical purposes of *descriptive natural history*, though it would probably not be sufficiently comprehensive for the use of *artists*, who might form another and more full table of colours for their own purposes*."

He after this drew up an extended plan as a basis to work from; it was a little complicated to express or write from, but this could have been easily modified had other occupations not prevented its completion.

When attending the Meeting of the British Association at Cork, he writes to me:—

"I made a most delightful tour through the West of Ireland, and terminated at Cork, where the Meeting, though not a large one, was in all other respects as satisfactory as fine weather, good fellowship, and abundance of scientific communications could make it."

To the list of communications, he contributed to Section D. two ornithological papers; one, the "Description of a Chart of the Natural Affinities of the Insessorial order of Birds," the other "On the Structure and Affinities of *Upupa*, Linn., and *Irrisor*, Less."†

The "Description of the Chart," &c.‡, was in further illustration of his paper "On the true Method of discovering the Natural System in Zoology and Botany," read at the Glasgow

* Letter, Cracombe, April 19th, 1843.

† See Reports of Brit. Assoc. 1843, Sections, p. 69; Ann. Nat. Hist. xii. p. 238. Collected Papers.

‡ See Reports of Brit. Assoc. 1843, Sections, p. 70.

Meeting. It exhibited the affinities of allied organic forms by a process analogous in some respects to map-making. The connexion of the families and genera of the Insessores was chiefly dwelt upon, but a complete map of the whole class Aves was hung up in the Section-room. Upon this the names of the genera were placed, each within an oval cartouche; and their affinities with others were indicated by the length of connecting lines corresponding with a graduated scale. The forms which were supposed to join each family and larger division were also pointed out by lines, so that the value and position of all could be at once comprehended. This will be best understood by the subjoined diagram of the value of affinities, and by a portion of the chart which has been reduced from the original. The latter, however, is given only to explain his manner of treating the subject; some of the affinities and their positions would have been changed in a matured copy; and it was intended also to have used colour as an accessory, which would have rendered the whole disposition of the map more distinct.

The affinities of the genera are indicated by the connecting lines, which are longer or shorter, according to the remoteness or proximity of the affinity, as shown in the following scale:—

Degrees of Affinity.

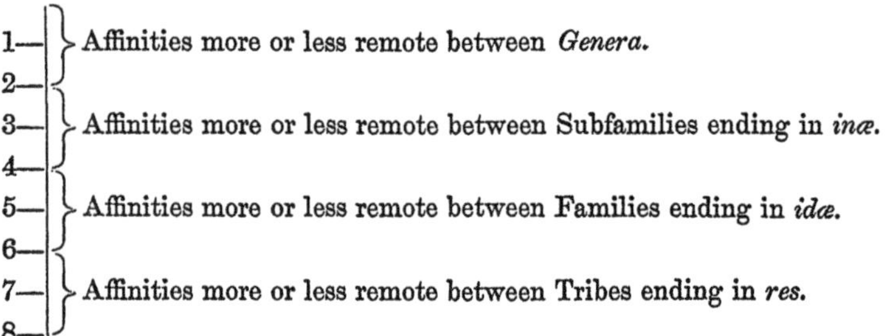

At the same Meeting the Committee of Section D. requested him to draw up a report "On the Recent Progress and Present State of Ornithology." In describing the Cork Meeting to me, which I was unable to attend, he writes:—

"The Committee of Section D, understanding that Mr. Selby had declined writing a Report on Ornithology, requested me to undertake

Part of the Chart of the Natural Affinities of the Class of Birds.
1843.

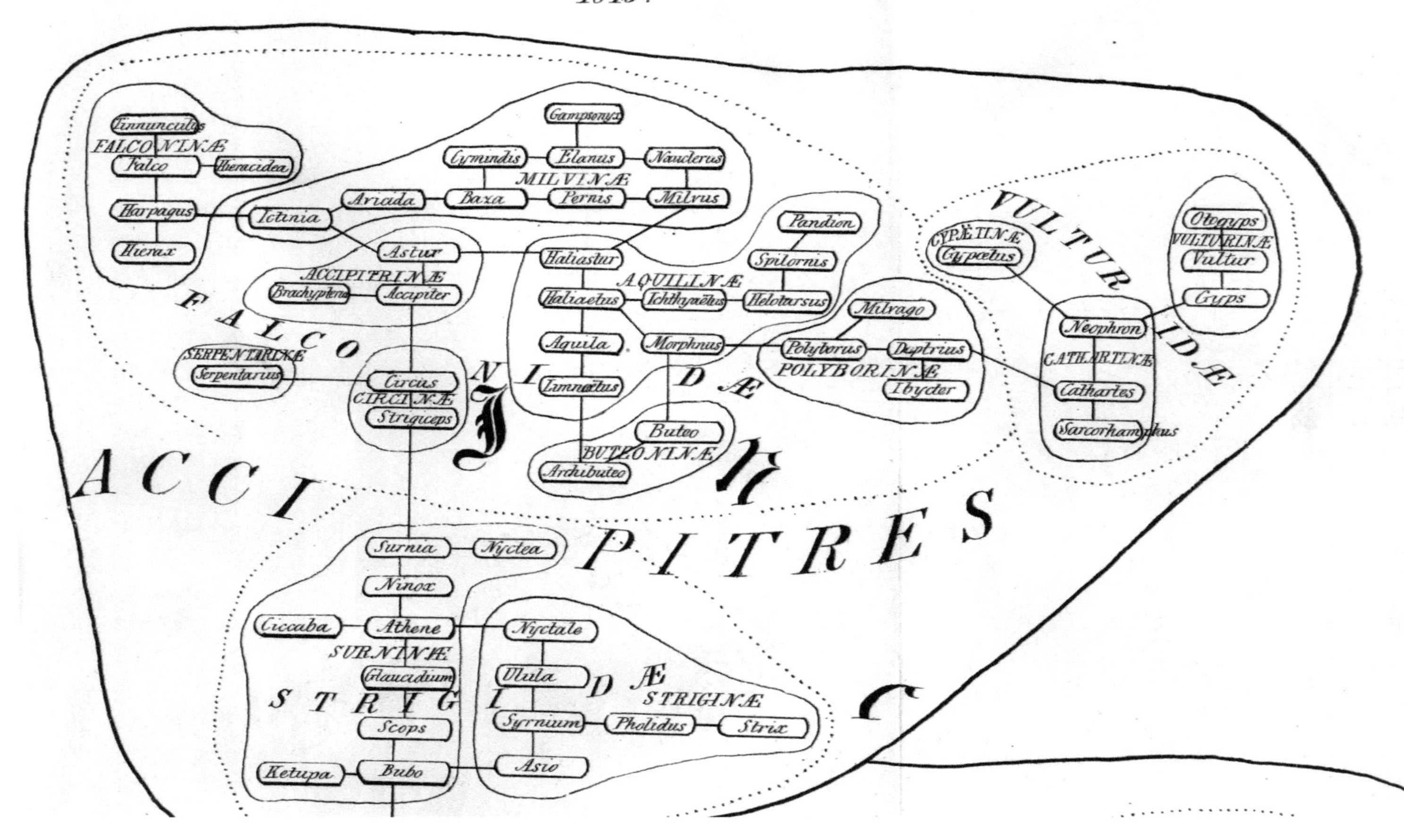

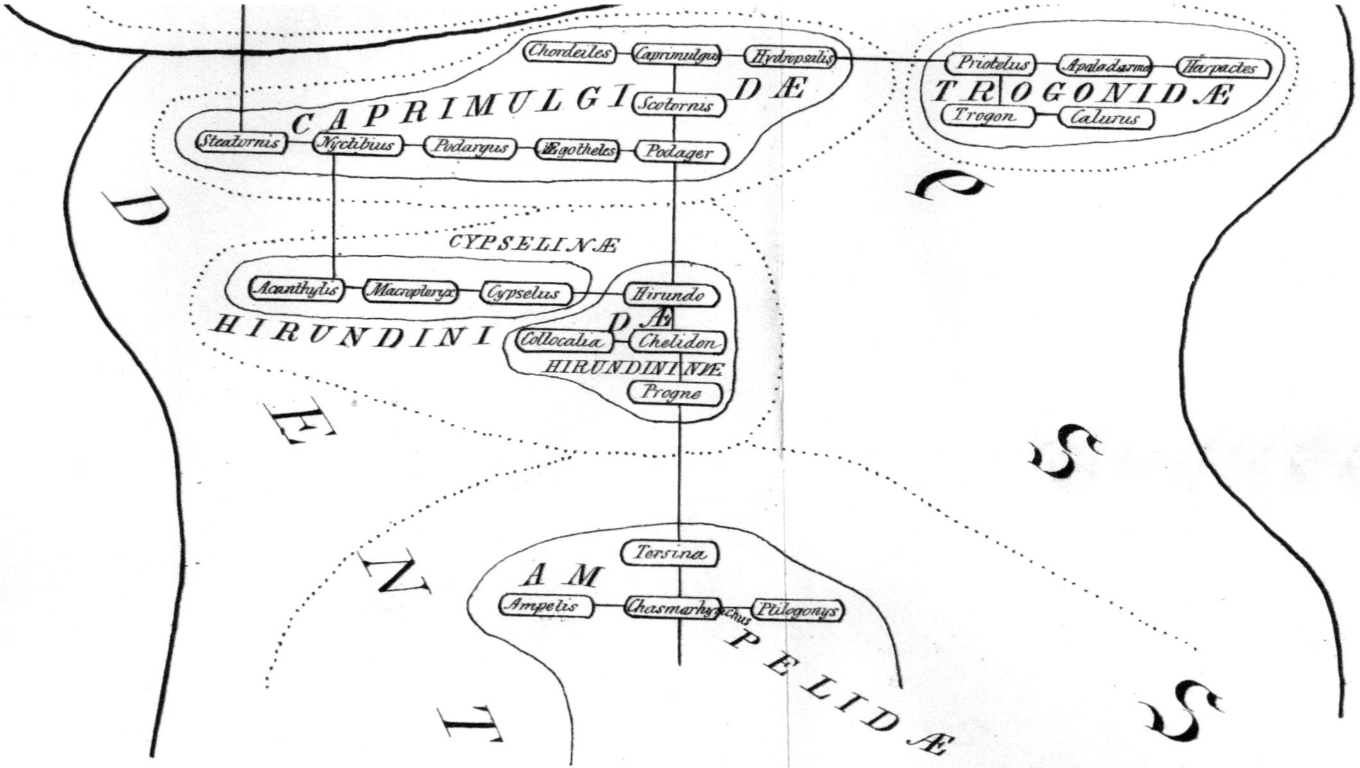

CAPRIMULGIDÆ
Chordeiles
Caprimulgus
Hydropsalis
Scotornis
Steatornis
Nyctibius
Podargus
Ægotheles
Podager

TROGONIDÆ
Priotelus
Apaladarma
Harpactes
Trogon
Calurus

HIRUNDINIDÆ
CYPSELINÆ
Acanthylis
Macropteryx
Cypselus
HIRUNDININÆ
Hirundo
Collocalia
Chelidon
Progne

AMPELIDÆ
Tersina
Ampelis
Chasmarhynchus
Ptilogonys

DENT.

SS.

C.D.M.S. litho

W. West Imp.

it instead. I promised to try what I could do, but I fear it will be a very tough job. The great difficulty will be to have access to a sufficient number of works on the subject, living as I do in the country, where I can get none but what I either buy or borrow. I must of course make some stay in London for this especial purpose; but when there, I generally find too many sources of interruption to do much steady work. However, I will do what I can, and if one year be not sufficient, I will take two for it*."

Notwithstanding these difficulties, the report was read next year at the Meeting in York, and was published in the Reports for the year 1843†. Considering the time allowed, and the deficiency then of ornithological works in this country, it was very ably and completely drawn up, and included Fossil Ornithology, in which the existing knowledge of the Dodo was rapidly run over, and was probably the first hint to examine that curious subject more fully. The Report is concluded by a list of the desiderata in Ornithology, in which some important observations occur on classification and arrangement, and particularly on the unnecessary multiplication of genera, which still continues to an alarming extent. It is much to be regretted that a report so well arranged should not have been annually continued. The main labour was completed, omissions could easily have been supplied, and the works and discoveries of each by-gone year added and filled in by supplements; but the fifteen years that have elapsed since its publication have been so extended and varied in their information and progress, that the labour now to bring up the subject would fully, if not more than equal that of the first.

His ornithological collection was now rapidly increasing, which the publication of this "Report" materially assisted; collections from correspondents abroad came in, exchange and occasional purchases contributed their share, and the project of sending a collector to some unexplored land was seriously contemplated. Although this scheme ultimately

* See letter, 'Cracombe, 23rd September, 1843.'

† See 'Collected Papers,' p. 247.

fell aside, its notice is here introduced, both to show the increasing interest he took in ornithology, and because the plan of the expedition will be useful to others who may contemplate anything similar. In October 1843 he wrote to me—

"Cracombe House, Evesham,
18th October, 1843.

"MY DEAR SIR WILLIAM,—I have lately been reading Stephens' 'Incidents of Travel in Yucatan,' in which work there is a brief and unsatisfactory notice on the birds of that hitherto unexplored country, which seems to offer a most tempting field to the ornithologist. Two species only from that peninsula have yet been described, viz. the magnificent *Meleagris ocellata* from Honduras, and the *Momotus* or *Crypticus superciliaris* from Campeachy. It is, however, evident that this country possesses a peculiar fauna, intermediate between that of Mexico and of Columbia; and though only five specimens of *Meleagris ocellata* have been brought to Europe (not one of which I believe is in Britain), yet Mr. Stephens assures us that this bird is abundant throughout Yucatan. Now seeing that hundreds of ships sail from our ports every year for Honduras for mahogany, besides others to Campeachy for logwood, it seems extraordinary that we should be almost wholly ignorant of the zoology of that country, and I wish to consult you as to the practicability of raising a subscription to send some person thither for the express purpose of making a zoological collection. This plan has, as you are aware, been adopted with great success by the botanists, and I do not see why ornithologists should not do the same. I should suppose that some active and intelligent birdstuffer or collector might be found, who, for about £100, would go out in a mahogany ship to Honduras, and remain in that country a sufficient time to form a good collection of natural history to be divided equally among the subscribers. I should be glad to contribute £10 for this purpose, and I suppose there would be no difficulty in getting nine other subscribers of £10 each. If the said collector gave satisfaction, he might be continued in that country at the same salary for one or more additional years, during which time he might proceed through Guatemala (where he would get *Trogon pavoninus*, &c.) to Columbia, the zoology of which is highly interesting. Each cargo of specimens as it arrived should be first carefully catalogued, and the new species in it made public, and the whole then divided equally. If a collector could be found to act honestly and to exert himself, I have no doubt the subscribers would be well repaid, as a single specimen of *Meleagris ocellata* would, in the present state of the market, be well worth £10, without reckoning other birds which would be obtained. The great difficulty would be to find a trustworthy and suitable person for the purpose, but as the botanists have succeeded in this, we may also. Unless you know of any such person, I should say that it would be well to consult Gould on the subject, as he is well acquainted with the practical part of these arrangements, and would perhaps become one of the subscribers. He has had an excellent collector in Australia named Gilbert. I do not know if he is now in England, but if so, and is willing to go to Honduras, we could not have a better man. I have as yet mentioned this subject to no one, but shall be glad to hear your opinion respecting it, with any improvements on it which you may suggest. I would not recommend a larger number than ten subscribers, otherwise the *dividend* would be too small. Probably Lord Derby, Messrs. Eyton, Selby, Barwick Baker, Gould and others would join, and some of the public museums or natural-history societies would make up the number."

A very zealous collector was soon after recommended, and instructions were drawn up for his guidance; Strickland had the young man at Cracombe for two days, and was much pleased with him; our share list was filled up, with Lord Derby at its head, when the whole matter was unfortunately broken off, and Dyson went out under other arrangements, with which we had nothing to do.

Another project was started about this period, in which Strickland took very great interest; indeed, the editing of one of the publications* brought out under its auspices occupied so much of his time as to interfere with the commencement of his great and laborious work, 'Ornithological Synonyms.' We allude to the Ray Society; and as that association was the first of the kind applied to the publication of zoological and botanical works, a short outline of its history, and the part he took in it, may not be unacceptable. It was, moreover, first thought of by Strickland himself, and the plan was afterwards developed by the activity of a fellow-labourer and a much-valued friend. It was also intimately connected with the attempts made at that time to diffuse more widely the principles of natural science, and the correspondence

* Bibliographia Zoologiæ, Agassiz.

which its progression called forth contains valuable opinions and observations.

It is well known that the publication of scientific works of value and importance is often impeded or altogether prevented by the great expenses incident to such works, and the small returns that can be expected from their sales. Strickland thought that the encouragement of the publication of such works came within the legitimate objects of the British Association, and although assistance was at times given, that a more systematic and extensive mode of proceeding might be adopted. In pursuance of those ideas, he drew up a series of resolutions, in which the general principle was to be confirmed, and power given to the Sectional Committees (subject, of course, to the General Committee) to vote money for the assistance of such works as were decidedly deserving of support. This idea was mentioned to the Council in 1842, and was brought before the Committee of Section D. at their Meeting at Cork in 1843, where it was partially discussed, but did not go farther; it was thought that the General Committee, under whose control and sanction it must stand, would scarcely agree, as it might open a door to various abuses, and that money assistance was already occasionally voted for such purposes. The proposal of a society for publishing works on natural history then suggested itself, and in August 1843 Strickland drew up the sketch of a series of rules for that purpose, which, as we shall presently see, embodied most of the principles ultimately agreed upon. This was the real origin of the Ray Society. To Dr. Johnston of Berwick devolved the merit of afterwards bringing the plan into working order.

On the 25th of November, 1843, Dr. Johnston wrote to me as follows:—

"DEAR SIR WILLIAM,—I have been dreaming of late of a '*Ray Club*,' and you shall know my dream,—a little shadowy, no doubt, for I have not information sufficient to draw the outline more distinctly. You know very well that there are many clubs established for the printing of books of antiquaries, reverends and doctors. I am a member of a Sydenham Club, and for the sum of £1 we get annually three handsome 8vo volumes. Now the objects of the '*Ray Club*' would be to print, or to assist in the publication of works in Natural History, illustrative principally of the Natural History of Great Britain. The more I think of it, the more I am satisfied that such a Club would do more good than all our present societies put together, always, of course, abating the Berwickshire Club. I think if we could get 500 subscribers at a guinea, a great deal might be done. Could you not move in such a matter? My idea is that the Club should print *only* for subscribers, and the reason is the expense and complexity in our management; that *publication* would introduce the accounts of booksellers; their discount, extra charges, advertisements, &c. would amount to far more than any profit. We would require a good steady Secretary, a man of business resident in London or Edinburgh, who should have a small salary: I thought of Dr. Greville. A Council would be easily named, and the communication between them is now easy and cheap. The objects I would have in view might be thus classified :—

"1st. The printing of original works approved of by the Council, *e. g.* Alder and Hancock's 'Mollusca Nudibranchiata,' Denny's 'Intestinal Worms,' Parnell's 'Jamaica Fishes,' Sir William Jardine's 'Salmonidæ,'' letterpress, &c.

"2nd. A uniform edition of the works of our older authors who treat of the Natural History of Britain, edited by competent individuals: this head would admit of several divisions. There are in the British Museum several curious MSS. deserving of printing, as illustrative of the history of our Natural History. The works of Ray, Lister, Sibbald, Llhwyd, the essays in Transactions collected together, and so on. To begin with this, I thought of a new edition of Pulteney's 'Sketches of Botany,' by Dawson Turner.

"3rd. A complete and systematic Natural History of Great Britain. This might be effected by a judicious distribution of the labour.

"4th. A 'Systema Naturæ.'

"5th. A scientific catalogue of all works ever published in Natural History.

"6th. A Dictionary or Nomenclator Hist. Naturalis.

"Such are a few things that occur to me. We might have two volumes a year, one Zoological, one Botanical; and by economical printing 500 subscribers would effect this. The Club would, of course, be opposed by some, and perhaps is impracticable: I don't know. We get on with our Medical Club quite easily: I am willing to work an oar, if my situation did not make it unadvisable.

"Yours, GEORGE JOHNSTON."

The above letter was immediately replied to, entering into the Doctor's views; but, as the original cannot be recovered, the exact terms of doing so cannot now be given. Dr. Johnston, however, answered it by writing,—

"Your letter was so encouraging that I immediately followed your advice, and had the enclosed Prospectus printed. I send you 6*d.* worth of them, but to-day I have posted besides fifty-two to naturalists *resident in Scotland,* so that the intention as to a Ray Club will in a few days be scattered over the land. Letters have gone even to Orkney and Shetland. I think it must succeed, for such a thing is really wanted . . . . I will write soon to report progress; *at present* there are eight members, of whom one is a lady. I have written to W. Thompson for the names and addresses of the Irish naturalists."

This was the second origin and *birth* of the Ray Society, which has now been carried on successfully for fourteen years, and has brought out a valuable series of works that could not otherwise have been published. Most of the leading zoologists and botanists of the time—many of whom, as well as our two leaders, have since passed from among us,—took much interest in the scheme, and wrote at considerable length either to myself or Dr. Johnston, and we shall record some of their opinions as a small testimony of the value in which we hold them; but before doing so, we shall go back to the original conception sketched at Cork by Strickland. Dr. Johnston's Prospectus had been enclosed to him, and counsel asked. The following was his reply, from which it will be seen how nearly his plan agreed with the rules ultimately drawn up and confirmed:—

"Cracombe House, Dec. 19th, 1843.

"MY DEAR SIR WILLIAM,—The proposition of a Society for publishing works on Natural History was the subject of conversation among several members of Section D. at least, but nothing definite came of it at that time. In the hurry of the moment I drew up a rough plan, which I showed to several of my friends there, who approved of it; but there was no time to discuss it in detail. The following is a copy of the project which I then drew up:—

"'Cork, August 1843.

"'It is proposed to establish a Society, to be called the MONTAGU SOCIETY, for the purpose of publishing original *illustrated* Memoirs on Recent or Fossil Zoology, and Comparative Anatomy.

"'The subscription to be £1 per annum.

"'The number of the Subscribers to be unlimited.

"'Each Subscriber to receive annually a copy of all the publications of the Society, which are also to be sold, without restriction, at a reasonable price to the public.

"'The Society not to accumulate capital, but to publish annually a greater or less amount of matter, according to the state of its funds. The illustrations to be confined to such objects as are either new to science, or have been inaccurately or incompletely figured previously. The Society may (if thought advisable) present any of their plates or letterpress to the Transactions of scientific societies, retaining a sufficient number of copies for distribution among their own members. Memoirs which have been already read to other Societies, may be published by this Society, provided such memoirs have not been previously printed.'

"Such was the outline of my proposed plan, the idea of which originated in the evidence which came before us at Cork of the number of valuable researches which seem likely to remain in MS., from the great cost of making engravings. There were, Forbes's 'Dredgings in the Mediterranean,' Denny's 'Entozoa,' Alder and Hancock's 'Nudibranchiata,' Dr. Carpenter's 'Microscopic Researches on Shells,' and a host of other valuable communications, all appealing to the Association for help, which the Association was obliged, from want of funds, to refuse. And it further appeared, that most of the great scientific societies were equally unable to publish memoirs which required any great outlay of plates for their illustration. A Society specially devoted to this object seemed to offer the only chance, and I rejoice therefore to see that such a project is likely to be at last realized.

"On comparing Dr. Johnston's plan with my own, I think there are some points in which both might be improved. Ray is, I admit, a more appropriate name than Montagu to place at the head of it. I think, however, that it would be better to confine its objects to ZOOLOGY, and to let the botanists have a club of their own if they wish it.

"The two subjects are now become so distinct as to be rarely combined in the same person, and it is therefore hard to make me buy 10*s.* 6*d.* worth of botany out of my guinea, which I should never look at. Besides, there is plenty of zoological matter now cut and dried to employ the Society for years to come, so they will have no lack of materials, and I believe they would obtain quite sufficient subscribers among British and foreign *zoologists* to make it answer without including *botanists.*

"I think the Society should confine itself *chiefly* to the publication of *new* memoirs which require *illustration*. Memoirs *without plates* are readily published by scientific societies or by periodicals, such as the 'Annals of Nat. Hist.,' and I think it would be a pity to trench upon ground which is already well occupied.

"I would not, except in rare cases, recommend the publication of old MS. tracts on zoology. The proposed Society is not a *literary* or *antiquarian*, but a *scientific* Society, and in a science so eminently progressive as zoology, we want the *new* and not the *old* researches to be published. I, however, would make an exception in favour of any original unpublished researches of sufficient merit, made *since the time of Linnæus* (say since 1750), but would admit scarcely any beyond that date, such being mere *literary curiosities*, and not *scientific data*.

"It seems to me, that there is rather too much stress laid in the Prospectus upon *British* Natural History. Now this is of all branches of zoology the most *popular*, and therefore the best able to swim without corks. The real interests of science, and not the amusement of the public, form the object at which the Society must aim, if it would stand high in the eyes of the scientific world. Now it is *exotic* and not *British* zoology which is most in arrear at present, and most in want of artificial aid. I except of course such departments of the British fauna as are *not* popular, and are consequently neglected, such as Entozoa, Arachnidæ, Crustacea, &c. These of course would well deserve the aid of the Society; not, however, because they are *British*, but because they are *neglected*. Another objection to confining the Society too much to *British* zoology is, that foreigners would take less interest in it, whereas if it is made more cosmopolitan, a large number of foreign zoologists would subscribe to it, and it would thus acquire both more *circulation* and more *influence*.

"The reprinting of *scarce but valuable* zoological tracts, whether originally published separately, or in Transactions, &c. (especially the *foreign ones*), is a perfectly legitimate and very desirable object, though I think it should be kept subordinate to the *primary* object of the Society, that of PUBLISHING ORIGINAL MEMOIRS WHICH WITHOUT SUCH AID WOULD REMAIN UNPUBLISHED.

"A 'Systema Naturæ' would, I fear, be too great an undertaking for such a Society as this. Nothing short of a Government could carry out such a project at present, and as a 'Systema Naturæ' is probably the last thing that Governments dream of, it will, I fear, never be realized. However, it is very right that the Ray Club should keep such an object in view, as every contribution towards it is so much gained for science.

"A descriptive catalogue of zoological works is a very desirable undertaking for the Society. I have made a very extensive catalogue of such works for my own use, which I would readily communicate to the Society, if desired.

"I see several objections to Rule VI., 'That the publications of the Society shall be confined to members only, except the Council shall otherwise determine.' I think it would be far better to say, that a certain number of copies of every publication should be printed, in addition to those required for the subscribers, and that these surplus copies shall be sold to the public at a small advance on the price at which the members obtain them. By so doing the Society will,—1st, obtain a considerable surplus to its funds in addition to the subscriptions of its members; 2ndly, they would forward the true interests of science, by more widely diffusing their publications; 3rdly, they would give greater satisfaction to the authors of original memoirs, who naturally wish them to be made known as extensively as possible; and 4thly, they would avoid the imputation of selfishness or exclusiveness in keeping their stores of information to themselves.

"I conclude that when the Ray Club is matured, its head-quarters will be established in London, that being by far the best locality for conducting its business on the extensive plan which I hope to see it assume. Next to London, of course Edinburgh is the best place.

"I have now, as you requested, stated freely and candidly my opinion on various points connected with the scheme. Considering it, as I do, a very important and very valuable undertaking, it is the more desirable that its details should be well considered, so as to make a good start in the first instance. I can only say, that I shall be happy to subscribe to it and give it every aid in my power.

"Yours, H. E. STRICKLAND.

"P.S. I should suggest that the subscription to the Ray Club should be £1, and not £1 1*s*. *Guineas* have long been extinct, and accounts are much more easily kept in *pounds*. Besides, foreigners, of whom I hope some hundreds will join the Ray Club, can understand a 'livre sterling' (25 francs), but they would be much puzzled how to convert a guinea into foreign money. (In France it would be 26 francs 4 sous, a very inconvenient sum to transmit.)"

This letter was sent to Dr. Johnston, who immediately replied to it, entering fully into the scheme, and acknowledging that he had heard of the Cork proposal previous to his own dream. We give the letter entire: it shows well the views

Dr. Johnston took, and the interest exhibited in it was kept up unflagging to the last*:—

"Dear Sir,—I have just read your letter to Sir W. Jardine, and in it there are several suggestions which, if the Ray Club can be instituted, ought to be adopted. It will amuse you to be told, that the same post that brought me your letter brought one also from Dr. Neill, a very kind friend of mine, who has *no doubt* the Club would succeed if it confined itself to the reprinting of the older authors of note and rarity amongst us, and of unpublished MSS. and letters. So great is the diversity of opinion amongst even the friends of the scheme, which I believe *now* it is not probable can be carried into effect! Should the present proposal fail, I shall be glad to join you or others to do my best in carrying into effect your more limited scheme.

"In reference to your letter I would say, that to limit ourselves to zoology would be so to narrow the field that we could not muster enough of members. This is my impression, but I may be wrong. It was this impression that led me to give to the scheme so great an extent and variety, holding out an inducement to greater numbers to join, and leaving the field open to the Council, who could then, untrammelled, work out any *head* that was deemed by it most conducive to the interests of the science among us. I fear now that more have been deterred from joining by this extent than have been gained, and on a revisal I would certainly limit the objects more; and, as a specimen, I should also now append a list of works to be undertaken by the Club. What I would exclude entirely is the 'Systema Naturæ' and the 'Nat. Hist. of Great Britain.'

"It is just to you to say, that I had heard something of the proposal made at Cork, although I was not aware it had gone to the length you mention.

"In regard to confining its publications to members, except in particular cases, this was done to keep ourselves free from booksellers, and to save expense. If it could be shown that no expense nor complexity in accounts would be introduced by publishing, in the proper sense of the word, then I would agree to the publication. I attach little importance to where the management is placed. What we want is a zealous, accurate, business man for a Secretary, and a good common-sense Council. It is most probable that the Council would print in Edinburgh, on account of the greater cheapness of printing there, because they of course would only print on estimates;

* The Doctor sometimes forgot to date his letters, but this was written at the end of December 1843.

but the members of the Council will be scattered throughout the country. I anticipate no difficulty in management; all I foresee of difficulty is the want of members. You may have observed, that in the Prospectus the Council have liberty given to *pay* the Secretary. This was done because *I know* that in similar Clubs *honorary* payments are made to the Secretary, and these are always higher than the salary would have been. To prevent this, I thought of a salary, a few pounds; but the duties may be done gratuitously. Perhaps it may be expedient to wait a few weeks to see how far the proposed Club may be workable; if a sufficient number come forward, then we should circulate a new Prospectus; if not, I will gladly assist you in any more feasible plan, by doing any *fagging* work you may assign me. But let us always keep in mind, that we must yield up our own wishes sometimes in part to get them carried out. I do sincerely wish that naturalists would really cooperate in some such Club, which I am satisfied is *necessary* to the promotion and diffusion of natural science. The cry that has been raised against us for combining against the Linnean Society, or any other, is foolish and absurd, and I fear those who raise it know that it has no foundation.

"It would be presumptuous in me to nominate a Council, but the following has been suggested, principally by Mr. Thompson of Belfast (a very kind friend of mine), who, however, put in three names of gentlemen who refuse to join us:—

1. Alder, Joshua, Newcastle-on-Tyne.
2. Balfour, Professor, Glasgow.
3. Babington, C. C., Cambridge.
4. Ball, Robert, Dublin.
5. Bowerbank, J. S., London.
6. Daubeny, Professor, Oxford.
7. Forbes, Edw., Professor, London.
8. Haliday, A. H., Ireland.
9. Jardine, Sir William.
10. Johnston, Dr. George, Berwick.
11. Lankester, Dr., London.
12. Strickland, Hugh E.
13. Thompson, William, Belfast.
14. Yarrell, William, London.

"The Council should continue two years, when one-third ought to retire. With your permission, I shall be glad to have the honour of telling you by-and-by how matters stand.

"I am, dear Sir, most respectfully yours, &c.,

"George Johnston."

There were great variations of opinion among our London supporters; Mr. Bowerbank and Edward Forbes were of the greatest use in explaining and advising. To Strickland the latter writes:—

"Dear Strickland,—I agree with every word you write about the Ray Club, the plan of which I fully approve of, and do my best to promote. At first it was very coldly received here, but now there is a strong feeling in its favour. Robert Brown has put his name down, with an objection, however, against the clause about original works. So has Thomas Bell. Bowerbank is writing enthusiastically to get subscribers. Owen, I am told, is strongly opposed. Gray was, but has since joined. Bowerbank has lists, &c., and a correspondence between him and you might be useful. "E. FORBES."

To Dr. Johnston's letters Forbes replied more fully, and advises on many points:—

"London, Christmas-day.

"DEAR DR. JOHNSTON,—I have delayed a few days in answering your letter, in order the better to consider of your proposed Club. I think the notion of a 'Ray Club,' with such intentions as those proposed in the Prospectus, a capital one. There has been a feeling of the necessity of such a Club for some time, and sundry proposals were partially made by Strickland for zoology, and by Daniel Sharpe for geology, but never completed. If we can get men enough, or rather guineas enough, together, I doubt not it will be of the greatest service to natural history, for the scientific societies have mostly run themselves into pecuniary difficulties, and none of them take up the special object in view. So you say London is not the best place to look for help. The men here feel half-jealous at things coming from the country, or else have so little public spirit and sense of the great truth-mission for which naturalists are sent into the world, that they coldly turn the shoulder on all proposals having a philanthropic object. There are, however, honourable exceptions, and doubtless they will join the ranks.

"Permit me to make a few comments on the provisional Prospectus, taking the several articles successively:—I. As good as can be. II. I fear a guinea is too little to effect the object. It is true that the Shakespearian Society and others have succeeded with a guinea, but the body of naturalists is small compared with that of lovers of literature. As an after-consideration, I should recommend printing in Edinburgh, being so much cheaper than in England. III. Where will the money come from to pay the Secretary? It must be a labour of love: there is no remuneration to naturalists in this world. IV., V., VI., VII. All right. VIII. It would be desirable in an after Prospectus to give *examples* of each of the classes of works to be printed. *Class* 1. of your Prospectus I should be inclined to confine to such works or papers as regard *coloured* illustrations. *Class* 2. is too extensive ever to be done. All the other objects should take precedence of it. Besides, the publishers are more likely to do it well than we are. *Class* 3. Excellent. As an instance of this I would suggest, in the beginning, the collection and publication of Robert Brown's botanical tracts,—if he would allow us. *Class* 4. Very good. Leach's Molluscan manuscript might be printed in this Class; it is in Bell's possession. *Class* 5. This is going on spontaneously. Your own works, Yarrell's, &c., for instance. Perhaps a Latin synopsis of British An[nelides] and Ver[mes] might be done. *Class* 6. A bold proposal, which most people seem to oppose. I think with you it might be done. *Class* 7. Add to 'printed books' Memoirs in Transactions and Journals, and no better work could be executed than a 'Bibliographia Historiæ Naturalis.' I would suggest as an 8th *Class*, an 'Iconographia Linnæana,' in which good figures should be given of all the original specimens in the collection of Linnæus. It might be extended to a publication of figures (authentic) of species described by the older authors, but never engraved.

"Yesterday I dined at Bowerbank's, who enters warmly into the idea. Thomas Bell thinks well of it. Dr. Lankester has desired me to send you his name as a subscriber, along with my own.

"Ed. Forbes.

"I should like to hear from you farther, whether any provisional Council has been formed, and what are the prospects of success in the North, before giving an opinion in the Council."

In Ireland William Thompson took the charge, and he writes:—

"Donegal Square, Belfast, Feb. 17, 1844.

"My dear Strickland,—About the Ray Club (Society!) we are, I hope, quite of the same mind. The grand object of the Society should be the publication of works which cannot otherwise appear, either through the medium of Societies' Transactions, or by being undertaken by publishers. The works which first occur to me are Alder and Hancock's 'Mollusca Nudibranchiata of the British Seas,' and Dr. Johnston's 'British Annelides.' The illustration of the British fauna and flora should be the primary object; perhaps next, that of other countries by British authors. Dr. Parnell, for instance, has materials for a work on the Fishes of Jamaica, which no publisher will undertake; no other person could treat of this subject so well. . . . . The last thing, in my opinion, that the Society should think of, is the reprinting of old, or rare tracts, or the translation of them. Our object should surely be to move *onward*, and not *backward*; at the same time we must, in our writings, render full justice to all who have gone before us. If it be converted into an Antiquarian Natural History Society, I shall have nothing to do with it. We are agreed about a second edition of a Prospectus. I wrote Johnston some weeks ago that I would not circulate any of the original Prospectuses, lest injury might be done by them.

"I am, yours,

"W. Thompson."

And thus it was that the Ray Society went on: the correspondence during the first year was very large; no one who has not had it before them can be aware of its extent. Many hundred persons were written to by Strickland, Dr. Johnston and myself, and with that of our supporters, which was very frequently sent to us, it forms altogether a curious collection. Many of the writers being on terms of intimacy, the letters are sometimes written with great freedom, just as the impulse prompted, mentioning persons and things by their proper names; but it would be unjust to the writers, and serve no useful purpose now, to copy these, and when the present generation has passed away they will not be understood.

Many who at first were our opponents in the scheme were converted before our first publication; some joined us at a later period, and very few, if any, have kept up their ideas of the inutility or impracticability of such a Society. The geologists did not cordially join us, because geology was not admitted as one of the Ray branches; they scarcely began yet to perceive how much recent zoology and botany bore upon their studies. It is a curious fact that few early geologists knew much of recent forms, and it was only when palæontology rose into importance that the necessity for their study and comparison became apparent and really felt. No geologist can now work without a knowledge of recent forms, and, though not so essential, the inverse of the proposition must also be borne in mind. But the value of Strickland's plan was proved in another way: geologists immediately started a Society of their own, conducted upon the same principles, and managed nearly by the same rules, and we rejoice to say successfully. The Palæontological Society has furnished to its subscribers a series of volumes, beautifully and elaborately illustrated, and otherwise of the greatest interest. Geographers followed, and the name of Hakluyt is recorded by another Society, which has printed a series of voyages and travels of olden time that were before nearly, if not quite inaccessible.

The circumstances of all societies or combinations undulate —undergo waves of success or adversity. To satisfy 1000 or 1500 minds, to give each what he thinks a sufficient amount of paper, print and picture for his money, and bring out everything in accordance with his particular views, is not an easy task. The Ray Society, however, has gone on very pleasantly; the publication of one work only, at an advanced period of its existence, threatened a little disruption; that was, the 'Elements of Physio-philosophy,' by Lorenz Oken: it was published perhaps without sufficient consideration of the "mind" in this country, and a good deal of correspondence ensued*.

The President was written to by several of the members, remonstrating, and condemning the publication of the book; and Professor Bell, among others, consulted with Strickland and myself. His reply is very characteristic, and I believe agreed with the opinions of a great proportion of the subscribers:—

"Oxford, Nov. 13th, 1847.

"I have been considering, &c. the matter of 'Oken,' and I really think that the wisest thing the Council can do is to say nothing about it. It will be very hurtful to the Society if the body of the members discover that there is anything like a split in the Council, and I am not aware that any important or influential members who are not of the Council have remonstrated or protested against the character of the work. It would undoubtedly have been more prudent in the Council not to have published it, and I for one was quite unacquainted with its nature till the volume was published; but now that the thing is done, the less said about it the better, and succeeding volumes will soon efface the recollection of this one injudicious step.

"But though I regret that the Ray Society have published the work, yet I am glad in the abstract that an English translation of

* The Members of the Council at a distance were not consulted; German mysticism and German authors were a little in fashion, and Oken came out.

'Oken' has been published. It is really a very interesting work. Unquestionably the greater part of it is *nonsense*, but the work is interesting; first, *psychologically*, to observe the kind of nonsense which a powerful mind produces when it attempts to grapple with subjects which are beyond the reach of the human faculties, and when it deserts the Baconian, or inductive method of reasoning. It is a kind of nonsense very different from the *trash* which inferior minds produce abundantly in these days.

"And secondly, the work is interesting *ethnologically*; for the fact that the original work has gone through three editions, and been looked upon for thirty years as an oracle of wisdom, shows how very different Germans are from Englishmen. Nor must we forget, that amidst all the nonsense and dogmatism there are certain original ideas on morphology and physiology, which are considered by eminent British zoologists and botanists to be both true and valuable.

"I presume the chief objections felt against the work are the seeming irreverence in dealing with the name of the Deity. But I have no doubt that in Germany it passes for a very religious work, inasmuch as it speaks of *God* by name, instead of using the vague and unmeaning term *Nature*, which most of the Continental philosophers now prefer. Some of Oken's views of the Deity are sublime and commendable, others are simply unintelligible; but none, as far as I see, are *purposely* irreverent or profane. I do not, therefore, consider the work as likely to be really injurious either to religion or morals; and though I would rather have seen it published by an individual than by the Ray Society, yet I think that any step which the Council can *now* take respecting it will only cause the members generally to criticise their proceedings, and make matters worse.

"I am, &c.,
"H. E. STRICKLAND."

Strickland's interest in the affairs of the Ray Society continued unabated throughout. In one of its early volumes he translated and superintended the publication of the 'Report on the State of Zoology in Europe,' by Bonaparte, then Prince of Canino *, and it was afterwards indebted to him for one of its most important publications, in which many of

* "Observations on the State of Zoology in Europe, as regards the Vertebrata," read at the third meeting of the Italian Congress of Science. Florence, 1841. By Charles Lucien Bonaparte, Prince of Canino and Musignano. Translated for the Ray Society by H. E. Strickland, M.A., F.G.S.; and published in the volumes for 1841-1842, Reports on the Progress of Zoology and Botany.

its members took a more than usual interest, the 'Bibliographia Zoologiæ et Geologiæ.' This work, a catalogue of all the publications relating to zoology and geology, was originally begun by M. Agassiz as a compilation for his own private use; it gradually increased, and from a limited Index, swelled into bulk and importance, which rendered its publication an object. M. Agassiz had frequently consulted him about this work, which agreed well with Strickland's views, and the institution of the Ray Society presented an opening which it occurred to him might be made useful to the Swiss naturalist, and at the same time advantageous and honourable to the Society. It was proposed accordingly to the Council, and ultimately the whole arrangement of the transaction was given up and entrusted to him.

When the publication of his MS. was proposed to M. Agassiz, he expressed himself with gratitude, and with feelings of much satisfaction, at the opinion entertained by the Ray Society of his labours. But the price demanded, £400, was beyond what their income could then afford, and the transaction was for a while postponed. Strickland explained the financial position of the Society, and made a more limited offer; and some time afterwards the following letter was received:—

"Mon cher Monsieur,—Je viens de recevoir votre lettre concernant ma 'Bibliographie,' et je m'empresse d'y répondre. J'ai mûrement réfléchi à tout ce que vous me proposez, et je dois par quelques détails justifier la restriction que je fais à l'acceptation de vos propositions. Sans fortune et sans ressources, provenant de la place que j'occupe, et qui me donne à peine de quoi vivre, j'ai dû faire des efforts inouïs pour gagner par mon travail de quoi suffire aux recherches que j'ai faites dans le domaine de la science et aux sacrifices que plusieurs de mes travaux ont exigé de moi. Aujourd'hui je sens que ce fardeau dépasse mes forces; mais modéré dans mes désirs je n'ai d'autre ambition que celle de pouvoir continuer à me rendre utile aux sciences que je cultive. C'est ce qui m'avoit engagé à vous demander pour la cession de mes matériaux bibliographiques purement et simplement le rembours de ce qu'ils m'ont coûté jusqu'ici. Depuis que je vous ai fait cet offre, un nouvel incident est venu modifier, non pas mes dispositions, mais ma position. Grâces à la recommandation de M. de Humboldt, Sa Majesté le Roi de Prusse m'a confié l'honorable mission d'aller explorer pendant un an et demi ou deux ans l'Amérique du Nord. Je dois partir dès que j'aurai pu mettre ordre à mes affaires. Les ressources qui sont mises à ma disposition sont suffisantes pour rendre ce voyage utile aux progrès des sciences, si j'apporte à le faire la plus stricte économie. Mais d'un autre côté, je laisserai en Europe ma femme et mes enfans dans une position peu aisée et sans ressources si j'avois le malheur de succomber dans cette entreprise. Ce sont ces considérations qui m'ont fait réfléchir un moment avant de me prononcer sur votre proposition définitive.

"J'accepte les £300 que vous m'offrez pour mes matériaux, mais pour le cas seulement où je périrois dans mon voyage. Dans l'intervalle nous pourrions discuter à loisir les modifications que vous proposez d'y apporter; comme vous j'ai senti qu'il y a des limites plus précises à déterminer dans la circonscription du domaine à embrasser et un plan à mûrir sur la forme à donner à l'ensemble et aux registres spéciaux des différentes rubriques. Un retard dans l'impression loin de nuire à l'ouvrage, ne sauroit que lui être favorable. Sur environ quarante personnes qui m'ont promis des documens de toutes les parties de l'Europe, il n'y en a que sept ou huit qui me les aient réellement déjà transmis. A je mis certain d'en recevoir encore de très-considérables de plusieurs de mes amis qui me les ont promis positivement, entr'autres de MM. Zeuschner et Nordmann pour la Pologne et la Russie, de M. Märklin pour la Suède, de M. Eschricht pour le Danemark, de M. Wagner pour l'Allemagne, de M. Gené pour l'Italie, de MM. Milne-Edwards et Gervais pour la France, de M. Boué pour les sciences géologiques et leurs aboutissants. De mon côté, je ne resterai pas inactif pendant mon séjour aux Etats-Unis, et tous ces documens réunis constitueront un corps d'ouvrage aussi complet qu'on peut le former maintenant. Pour cela je continuerai à occuper un de mes copistes du dépouillement et de l'arrangement méthodique de tout ce qui me sera adressé pendant mon absence et de tout ce que je pourrai lui transmettre moi-même. En revanche et comme compensation de ce nouveau travail, si je reviens sain et sauf d'Amérique, vous me rembourserez £400 au lieu de £300; mais alors je me considérerais aussi comme engagé à mettre en ordre tout ce que j'aurais pu recueillir ou faire recueillir d'ici là, et au moment de mon départ je vous expédierais mon exemplaire avec toutes les additions manuscrites qu'il renferme pour vous éviter de refaire ce que j'ai déjà fait.

"Je crois cette proposition additionnelle tout à fait dans l'intérêt de l'ouvrage et de la société Ray, dont les ressources s'augmenteront sûrement d'année en année. Si cette nouvelle close seroit cependant souffrir des difficultés je ne voudrois pas qu'elle fit revenir la Société de son offre, et pressé par les circonstances, j'accepterois purement et simplement ce que vous m'avez proposé, et je vous expédierois sans délai mon manuscrit. J'ose croire cependant que vous apprécierez la convenance de la modification que je vous propose et que vous pourrez la faire accepter à la Société. Si je puis je tâcherai de me mettre en route vers la fin de Juin ; mais j'ai encore tellement à faire que je commence a désespérer d'en venir à bout.

"Adieu, mon cher Monsieur, croyez à mon sincère attachement.

"Votre dévoué,

"Neuchâtel le 23 Avril, 1845." "L. Agassiz."

The preceding letter was replied to as under, and the terms were accepted by M. Agassiz:—

"The Lodge, Tewkesbury,
May 23rd, 1845.

"My dear M. Agassiz,—I have laid your letter to me of April 23rd before the Council of the Ray Society, and they have consented to make the following modification in their former proposition. The Council are strongly impressed with the value of your bibliographical labours, and as lovers of science they are grateful for the sacrifices of time and money which you have made in effecting this object. Still the prosperity of the Ray Society requires that much economy be practised in the administration of its funds, otherwise it may, in its present incipient state, fall into difficulties which may end in ruin. Should the Society, however, as they trust it will, continue to prosper and to acquire new members, the Council will be able to act more liberally than they can venture to do at present. They now, therefore, propose, as formerly, to give you for the materials of the 'Bibliography' £300, in sums of £50 *per annum, and* £100 *more if the Society reaches* 1000 *members before the whole* £300 *is paid.* As the Society already consists of nearly 700 members, I have no doubt that it will soon exceed 1000, if its friends are sufficiently active in making known its objects and merits. I have, therefore, no fear but that if you accept this proposition, you will be entitled to receive the £400, and the sums which the Society will have to pay will thus be, in some degree, proportionate to its power of paying them. In return for the above sum, the Society will expect to receive the whole of the materials which you already possess, and all those which you or your correspondents may contribute till the work is completed.

"If you consent to this proposal, the Council are desirous that the printing of the work should commence *if possible* in the course of the autumn, so that the first volume may be published during the present year. Nevertheless, rather than the perfection of the work should be diminished, I have no doubt that they will agree to such a delay as is necessary to collect the contributions of your numerous correspondents. Some time may, however, be gained if your correspondents would send you *at once* all the titles of *journals and periodical works,* as we could print them first, and possibly fill a volume with them, before we commence printing the *alphabetical* series of titles.

"Be so good as send me your answer to the above proposal as soon as possible. I hope you will be able to come to the Cambridge meeting of the British Association, which begins on June 19th.

"I send you some additional genera of birds taken from all the published portion of Gray's large work. I also send an outline of the families of Birds according to *my* views. They nearly agree with Gray's system.

"Yours, &c.,

"H. E. STRICKLAND."

The price paid for this work was a very large one, as given by a Society, in addition to the very heavy printing expense which its character entailed; but being spread over a number of years, and considering the position the publication of such a work would give to the Society, Strickland judged rightly that the money would be advantageously spent. The time and labour and outlay of the compiler could not have been repaid by it, for copiers were employed, and what Agassiz termed his "matériaux" were printed. This was just the work as far as prepared, set-up, and printed on large sheets with broad margins for additions; these proved of immense value and saving during the progress of the volumes, and several copies being furnished with the MS., the spare ones were cut in pieces, and transmitted to the various authors for correction, thus ensuring, as far as possible, a perfect text. The editing of these volumes occupied a large portion of Strickland's time and attention, even to his last days; for although the labours of Agassiz formed an excellent skeleton, innumerable additions had to be made, and Agassiz having employed different individuals and different countrymen to make his copy, the style was far from uniform, and required unceasing attention; but the labour attending the work is best explained in his own words. Dr. Johnston had got hold of a copy of the uncorrected "*matériaux**," and not aware of the charge Strickland had undertaken, and ever-watchful, he was alarmed for the credit of the Society, and wrote to the Secretary regarding it. The following is the reply:—

"4 Beaumont Street, Oxford, Oct. 9, 1847.

"MY DEAR SIR,—Dr. Lankester has just forwarded to me a letter from yourself, by which it appears that Mr. Gray has sent you for inspection his copy of Agassiz's '*matériaux*.' The remarks which you make upon that work are perfectly just; but you should remember that it was merely the *rough draft* of his compilation, of which a few copies were printed five or six years ago for private circulation, in order to procure additions and corrections from his friends. The work, as delivered by him to the Ray Society, is in a very different state, most of the errors being corrected, and innumerable additions having been made to it. Still it is very far from perfect, and I am labouring as editor to improve it as far as possible. The Council allowed me to expend £5 in employing a clerk, at 6*d*. an hour, to copy out the titles of memoirs which had been omitted by Agassiz, and several thousand titles have thus been inserted, besides a great number which I have added myself. I have also had to cut up and re-arrange all Agassiz's matter, in order to introduce these fresh materials. The English titles have been as far as possible substituted for their foreign translations, and the references to the foreign works, where those translations exist, have been appended. In short, I have given up to this gratuitous labour an amount of time which has interfered much with my other pursuits, and which the utility of the work, when done, has alone justified.

"The printing is going on steadily; about 240 pages are completed, and the first volume will, I hope, be ready by January. The *first part* of the work consists of *periodicals*, arranged *geographically*, according to the place of publication; it only occupies eighty-five pages. The *second* part contains the names of authors arranged alphabetically, and a list of the works and memoirs of each. The first volume, of at least 600 pages, will only come down to the end of letter B, which will give you some idea of the vast extent of the work.

"Were I able to devote my whole time to it, I could no doubt render the work far more complete and perfect than I can now do, especially by classifying the titles according to the periodicals from

* We perceive that some copies of these "*uncorrected matériaux*" have got into the catalogues of old and second-hand books as *Agassiz's 'Bibliographia,'* and are now on sale. The above account of them will be a warning to persons proposing to purchase.

which they are taken, in the way you recommend. As it is, the list of each author's works is necessarily arranged somewhat promiscuously; for I find that, to cut up each separate *title,* and arrange it according to *date,* to *subject,* or to *periodical,* would involve an immense addition to my labour, both manual and mental.

"I have hitherto, as far as possible, sent either the original MS., or a proof, to the authors of papers, for their corrections and additions, before going to press. I shall, however, be very glad of farther aid in this respect. Several important British periodicals still require to be gone over, and the titles of all zoological or geological papers of any scientific value extracted. I enclose you the sheet of British periodicals, and shall be much obliged if you can procure me the titles of the papers (zoological and geological) in any of those marked 1. These I will incorporate in the text, as far as the alphabetical order allows; and all which are *too late* can come into a supplement.

"I quite agree with you that many of Agassiz's titles are *redundant.* Some of these I have struck out, but have been cautious of doing so to any extent, where I was not acquainted with the works themselves, as a work may appear *from its title* to be purely medical, or otherwise irrelevant to zoology or geology, and yet may contain facts of importance bearing on those sciences, which induced Agassiz to insert it in his list.

"If the Ray Society's purse were in a plethoric state, I would certainly recommend them to employ some scientific and industrious man, at a salary, to work exclusively at the 'Bibliographia,' and render it as perfect as possible. But as the case stands, I cannot be expected to give more time to it than I am now doing, and the work must consequently be somewhat imperfect, though it will unquestionably be far more complete than anything of the kind ever yet attempted in the department of natural history.

"Believe me, my dear Sir,
"Yours very truly,

"*G. Johnston, Esq., M.D.*" "H. E. STRICKLAND."

To this long explanation the Doctor replies:—

"Berwick-on-Tweed, Oct. 23rd, 1847.

"MY DEAR SIR,—When I wrote to Dr. Lankester concerning Agassiz's 'Bibliographia,' I was aware that you had undertaken the post of editor, but I could not be aware of the great pains and labour you have bestowed on the work; nor was it a *reasonable imagination* that would have gone the length to believe any one would have done so much for what was not his own. You can truly say of the book, 'cujus magna pars fui;' and my letter to Dr. L., and this intrusion on your time, would have been spared had my belief in your friendship to Agassiz and your love of labour been more lively. I am really surprised you should have done so much for it; and we Raians cannot thank you too warmly."

The volumes of Agassiz's compilation are now published, and no one but those engaged can know the labour and accuracy required for such an undertaking. There may be, and we know that there are, faults in the plan and in the detail; but it will be long before such a complete and useful work will be again drawn up, and for which, whenever it is done, Agassiz and Strickland's 'Bibliographia' must form the base. It cannot be too much regretted that the Ray Society did not think it right to fulfil its pledge of bringing out Supplements*. Strickland left the working staff ready, and in fact proceeding, and every day lost will render the continuation more arduous and expensive.

* Minutes of Ray Society, Feb. 29, 1848.—"5th. That it be announced in the first volume of the 'Bibliographia,' that it is the intention of the Council of the Ray Society to publish Supplements from time to time after the work has been finished."

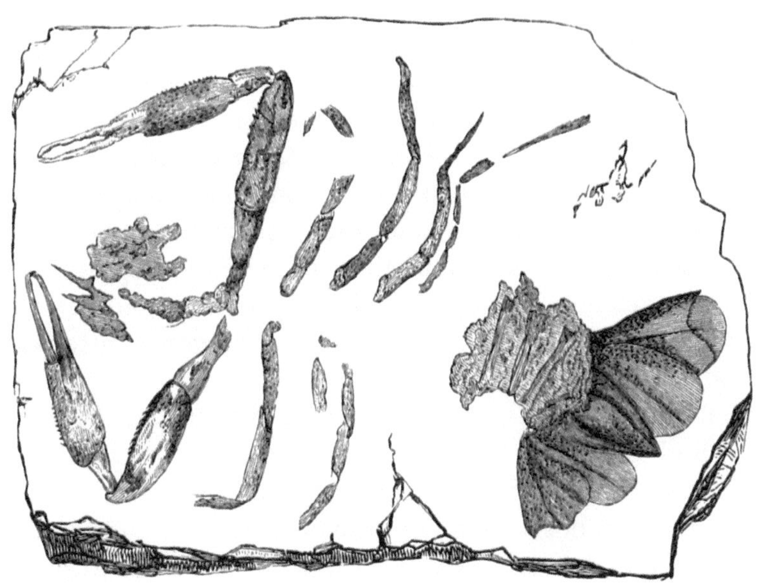

ERYON BARROVENSIS, M'COY?

1. FUSUS FORBESI, STRICKLAND. 2. NASSA PLIOCENA, STRICKLAND.

(Nat. size.)

CHAPTER VIII.—1846 TO 1853.

MARRIAGE. — TOUR IN SWEDEN. — RESIDENCE IN OXFORD. — BRITISH ASSOCIATION AT OXFORD.—DODO AND ITS KINDRED.—PROFESSOR REINHARDT. — ANASTATIC PRINTING. — RESIDENCE AT APPERLEY GREEN.—READER OF GEOLOGY AT OXFORD.—OXFORD STUDIES AND GEOLOGICAL LECTURES.—FIELD NATURALIST CLUBS.—CRUISE OF THE 'ADA.'—MEETING OF BRITISH ASSOCIATION AT HULL.

ON the 23rd of July, 1845, Hugh Strickland was married at Jardine Hall to Catherine D. M. Jardine, second daughter of Sir W. Jardine, Bart., and on the same afternoon set out upon a tour to Sweden. This was rather an unusual excursion for a marriage jaunt, but Strickland had long wished to visit some parts of Northern Europe; it proved agreeable to both parties, and the memorials of a time happily spent amidst scenes not usually visited are preserved, and some of the results are recorded, in papers and works afterwards published.

The route chosen was by Carlisle, Newcastle, York and Hull, and from the latter seaport a passage was taken to Rotterdam. Many interesting observations were made by the way which bore upon his previous studies, and were frequently alluded to afterwards; we shall therefore follow the party. At York Mr. Allis was visited: he possessed a large collection of living ferns, and their forms and manner of growth were very interesting to Strickland as a geologist:—

"The genera being based on the fructification, often embrace species in which the forms of leaf and modes of growth are very various. Hence no fern, whether recent or fossil, can be correctly referred to its genus unless the fructification be shown, and the genera now used for fossil ferns are therefore wholly conventional. It is very interesting to see how the ferns *imitate* other plants, as well as each other, in the forms of their leaves, and their mode of growth and of propagation. One species has a smooth glossy leaf like that of Vaccinium, and if found fossil would be pronounced Dicotyledonous. Some grow like palm-trees; some throw branched roots into the ground, as that which gave rise to the 'Vegetable Lamb' of Tartary, which Mr. Allis has growing; and one creeps on the ground much like Stags-horn moss (Lycopodium), and I suspect will furnish a better clue than that plant to the Lepidodendrons of the Coal-measures."

After a long and very stormy passage from Hull, Rotterdam is reached:—

"We examined the various objects of interest in this very picturesque and remarkable town, which combines the busy docks of Hull, the shady boulevards of Paris, and the animated canals of Venice with the brilliantly painted houses of Holland. In no town does one see such large and beautiful ships, built for distant voyages to the East Indies, brought into the midst of crowded streets, and combined into beautiful groups with the rows of lofty trees which are planted along the wharfs. And in no country but Holland does one see so universal an appearance of comfort and prosperity, every house seeming as though fresh from the painter's hands, and every town looking fresh scoured and decorated, as though in immediate expectation of a visit from royalty."

At the Hague, the painting by Roland Savery of "Orpheus playing to the Birds and Beasts" was a great attraction, as containing a representation of the Dodo:—

"This figure, which is about three inches high, is in precisely the same attitude as the large painting by John Savery of the Dodo at Oxford, and has no doubt been taken from the same original design. All the other animals in the picture are very correctly designed, without any tendency to exaggeration, and there is therefore no doubt that the artist intended to represent the bird correctly."

From the Hague they travel by railway to Leyden:—

"One of the cleanest towns in the world. The greater part of the traffic is carried on by the canals which intersect it, so that the streets have a comparative sinecure, and like those of Venice, when once cleaned, do not soon become dirtied again. We soon visited the Museum of Natural History, where I was unfortunate in not meeting M. Temminck, who had just departed to his country-house. The extent of this museum is immense; the collection of skeletons and stuffed specimens of Mammalia and Birds is one of the finest in the world, and it is also very rich in Fish, Reptiles, Insects and Shells. The general arrangement is also good; but in the department of Birds, the ticketing of the specimens is not so complete or modernized as I should have expected. I called on M. Van der Hoeven, a zoologist of much eminence, who edits a valuable Magazine of Natural History.

"Passed a couple of hours at the Museum, looking over the collection of birds in detail. There are here a vast number of species of extreme rarity, including many of the original specimens figured by Levaillant and Temminck. But though the materials are of the highest value, there is yet much to be done in the arrangement to bring it up to the present state of science."

At Haarlem, Professor Van Breda, an eminent palæontologist, was visited, and under his guidance they examined the Teylerian Museum there:—

"There is here a very splendid geological collection, including a fine series of Solenhofen fossils, such as Fish, Insects, Crustacea, Fuci, Ammonites with various species of Aptychus in them, Sepia, &c. It certainly seems from these specimens that the Aptychus belongs to the Ammonite: first, because certain species of Aptychus are peculiar to certain species of Ammonites; secondly, there is never more than one Aptychus in an Ammonite; thirdly, the Aptychus is always symmetrically placed, or nearly so; fourthly, the Aptychus is always proportionate in size to that of the Ammonite. I was shown one very young Ammonite with a small Aptychus in it, in exact proportion. These specimens re-established my faith in the connexion of Aptychus and Ammonites, which had been somewhat shaken by the specimen which Charlesworth showed me at York on the 28th ult. There are here many specimens similar to the latter, both from Solenhofen and the Lias of Boll; and though they have some similarity to the *corneous* Aptychi, yet they are certainly allied to the Belemnites and the Sepiæ*. In this Museum are also a fine collection of Crocodilian remains, Am-

* See 'Collected Papers,' p. 181.

monites and Belemnites, from the Lias; Mosasaurus and Turtles from Maestricht; a beautiful new species of Pterodactylus with a long tail from Solenhofen; and a new Rodent and other fossils from the tertiary beds of Œningen. In the same institution are a series of philosophical instruments, and an immense loadstone from Siberia, supporting a weight of 270 lbs. Also a very choice collection of paintings by living Dutch artists of great merit.

"There is also a capital library of scientific works, voyages and travels, especially rich in the Transactions of foreign scientific societies, many of which are almost unknown in Britain. It was pleasing to see so well-managed and efficient an institution as the Teylerian Museum."

At Utrecht the Museum in the University is of moderate extent, but the superintendent, Professor Van Lidth de Jeude, possesses a very fine collection:—

"He is an elderly man, and by a process of gradual accumulation has collected the largest general museum of zoology which I ever saw in the possession of a private individual. It includes a very valuable collection of osteology, as well as of Mammalia, Birds, Reptiles, Fish, Shells, Insects and Zoophytes, all in the finest state of preservation, and well arranged in a large building constructed on purpose. The Professor assured me that his collection had cost him 200,000 florins (£16,666), and he was anxious to sell it to some university or public institution, in order to ensure its preservation in its present state. He is a man of considerable scientific attainments, but has studied the lower animals rather than the Vertebrata, and his collection of birds, though a very valuable one, is deficient in its arrangement and nomenclature*. Indeed, he assured me that Temminck is the only scientific ornithologist in Holland, which I am disposed to believe from seeing the general want of arrangement in this department of the public museums."

Arrived at Bremen, they called for Dr. Hartlaub:—

"A highly scientific ornithologist, and most obliging person, with whom I formerly had some correspondence. He accompanied us to the Museum, under which name is comprised a public reading-room, an excellent library, and a very admirably arranged collection of natural history. The Birds have been entirely arranged by Dr. Hartlaub, and include many specimens of rarity, the whole in excellent condition. It is really refreshing, after seeing the masses of antiquated confusion which commonly go by the name of museums, to find a collection in

* This collection is advertised for public sale on 6th April next.

which every specimen is put in its right place, and marked with its right name, according to the latest and most approved zoological principles.

"In the afternoon Dr. Hartlaub called upon us, and accompanied us to his mother, a very agreeable and cultivated lady, who, though she has never been in England, speaks our language very fluently, and is well acquainted with our best authors. She lives in a handsome house on what was once the ramparts of the town, now pleasantly laid out with walks, shrubberies and pieces of water. It is one of the most pleasing evidences of the peaceable state of Europe, that in almost every town we have come to, we have found the ramparts and fortifications converted into pleasant boulevards and public gardens for the amusement of the people. The library contains many rare and scientific works. Dr. Hartlaub has compiled a very complete and admirable *catalogue raisonnée* of ornithological works and memoirs, which he has offered to add to the 'Bibliographia Zoologiæ' of Agassiz, whenever that goes to press."

At Hamburg:—

"The ramparts round the town are now converted into beautiful walks and pleasure-grounds, laid out in shrubberies and flower-beds, entirely unenclosed, yet injured by no one. It is certainly a remarkable phænomenon to see in these continental towns, flower-beds and grass-plots in the midst of the busy crowd, yet untrodden and respected by all. As far as it goes, this is a pleasing feature in the people, were it not, at the same time, an indication, among many others which we see daily, of a habit of implicit and unreasoning obedience to established rules, which keeps the people in an infantine condition, and too often tempts the governments to tyrannize. In these countries the *State* does everything, the *people* nothing; all those public affairs which in England are transacted by the magistracy, by turnpike trustees, commissioners and parish vestries, are here centralized in the Government, and the consequence is, that the country is almost wholly deserted by the upper classes, who spend their time in the frivolous amusements of the great towns, or in filling the official appointments of the State. One might have expected in the free towns of Bremen and Hamburg to find indications of a more independent spirit in the people, but these microscopic republics seem completely to fall into the general continental system, and to imitate their superiors in all the paraphernalia of absolutism, such as custom-houses, passport-offices, soldiers drumming from morning to night, liveried officials of all degrees, an incomprehensible coinage, locking of the town gates at night, '*warnungs*' and '*bekannt-machungs*' written up on all sides."

Kiel was at last reached, and after M. Boie and the Museum there were visited, they embarked in the 'Löwen' steamer for Copenhagen: remarks on the Museum, and on the remains of the Dodo which Professor Reinhardt, jun., had examined, will be found further on in this chapter. The Museum of Antiquities in the Christiansburgh Slot, or winter palace of the king, is one of the most interesting sights, on account of its Scandinavian and Danish specimens, many of which closely resemble those found in the British Barrows and in Ireland. Thorwaldsen's works and models were found worthy of a visit, arranged in a Pompeian house paved with mosaic; the ceiling blue, with golden stars, like one in Adrian's villa.

Leaving Copenhagen for Malmo on the way to Lund, the Museum in the latter town was examined with Mr. Kinberg; it was unfortunately Professor Nilsson's turn to be Rector or chief of the University, and the Crown Prince was at this time a visitor to Lund to celebrate the 700th anniversary of the cathedral. This rather interfered with scientific intercourse, and an invitation to the dinner given in honour of the Prince, over which Nilsson presided, and to the ball afterwards, had to be taken instead. The journey was pursued by Stralsund, and onwards to Berlin. The country from Stralsund is one wide extent of sandy level, chiefly cultivated with corn, unenclosed, and with very few trees:—

"Over all this district an abundance of rounded granite boulders, averaging from a foot to one or two yards in diameter, are scattered. Some of them lie on the surface, others are buried at various depths in the sand, but they are accompanied only by scattered pebbles of smaller size, and rarely or never by beds of regular *gravel.* From this, as well as from other considerations, I am more and more convinced that these boulders could only have been transported by *floating ice.* A current strong enough to have carried these blocks would have washed away all the sand, and only left accumulations of coarse boulders; and the idea of subaërial glaciers propelling or carrying blocks over these vast plains, hundreds of miles from their parent mountains, is utterly untenable. There therefore only remains the *floating iceberg,* and this we know can and does carry such materials at present into the North Atlantic. The boulders are often assembled into groups, as if from the grounding and melting of one particular iceberg, or from the drifting of them by currents into particular localities. These granite boulders are cut up and applied to all kinds of purposes. In front of the Museum at Berlin is a magnificent granite basin, probably the largest in the world, 22 feet in diameter, hewn out of one of these enormous transported masses."

The Zoological Museum at Berlin is under the charge of Professor Lichtenstein, who obligingly showed them over the vast collection. A series of the Urus of Lithuania was of particular interest. The ornithological collection contained then (1845) above 10,000 specimens; but there are many labelled with Lichtenstein's manuscript unpublished names only, which renders it difficult, without study, to say whether a species is undescribed or not. In the Royal Museum of Fine Arts there is a very large collection of pictures, and among them another representation of the Dodo.

"The pictures are admirably arranged in chronological order, beginning with the stiff formal figures of saints, &c., with gilded backgrounds, and glories on their heads, down to the latest periods of the Flemish and Italian schools. I was much pleased by finding a picture bearing the name of '*Roelandt Savery*, 1626,' containing a figure of the Dodo, exactly like the one by the same artist at the Hague. It represents Adam and Eve in Paradise, and is crowded with animals of all kinds most accurately depicted. Among them is a figure of the European Bison, exactly like the stuffed one in the Museum, proving either that the artist had seen these animals alive, or had access to accurate figures of them. The figure of the Dodo is in the usual attitude in which that bird is represented, but the beak is less hooked, and more like what we now know to be its real form. This then is the third picture by Savery in which the Dodo is represented, and the fourth oil-painting (including the one in the British Museum) taken in all probability from the living bird."

Lichtenstein has an excellent zoological library, which contains many rare tracts. He has also a complete set of all the editions of the 'Systema Naturæ' of Linnæus, including the first edition of 1735, of twelve large folio pages.

The travellers next visited Leipzig and Dresden. At the latter, a place of great interest, some days were spent, and an excursion to the frontiers of Saxon Switzerland and the

Bastei was made. After a journey of five days, Frankfort is reached, and Dr. Rüppell conducts them over the Senckenberg; at Liège, M. de Selys-Longchamps is visited at his country-seat; Brussels is the next stage; next Ghent, Bruges and Ostend, and thence to Dover, where

"The porters extorted 3*s*. 6*d*. for taking our goods to the Custom House. Sir Robert Peel pocketed £1. 1*s*. 11*d*. for duties on a few books, glasses and China cups which we imported, and the railway people charged us as much for going eighty miles to London as would have taken us 300 miles on a German railway. So we gradually became aware of the fact that we were once more in Old England, with its taxes, its high charges, its comforts, and its FREEDOM."

Strickland's family had removed a short time previous to his marriage from Cracombe House to 'The Lodge,' near Tewkesbury, in the vicinity of Apperley Court, the residence of his aunt, Miss Julia E. Strickland, whose increasing years rendered the proximity of some relative agreeable, as well as necessary, on account of the management of her fine property on the banks of the Severn. Strickland and his wife, after returning from their northern tour and visiting Jardine Hall, took up their abode at the Lodge until some independent residence could be chosen, and here, when comparatively settled, the old pursuits were again resumed.

It was wished that some residence near his own family could have been found, but none that was suitable could then be obtained, and his old associations towards Oxford, with the advantages of easy and unrestrained access to her libraries prevailed, and it was resolved to settle, for a time at least, in that city.

On 31st January 1846, he writes of his plans:—

"I have for some time past delayed writing to you, until I should be able to give you some definite account of our plans. We now propose to start on Tuesday next for Oxford, and to establish ourselves at first in a lodging, until we can find time to look about us, and make inquiries about houses. I need not say how happy we both feel to get settled in a house of our own, be it ever so humble a one."

A small house was soon afterwards found, and he was able

to unpack and arrange the ornithological collections which had been for some time locked up in boxes. It should have been mentioned, that previous to removing from Cracombe to the Lodge, the geological collection had been all packed up. In a letter to Mr. Thompson at this time he wrote:—

"My father is just on the point of removing from this place (Cracombe) to a house near Tewkesbury, and I shall take the opportunity of the general move to pack up all my collections, including many thousand specimens of fossils and shells, which I shall lay aside for the present and stick to birds, taking all my collection of the latter with me to Oxford."

It was impossible to carry about a large and heavy collection of fossils while his residence continued liable to change, and they were not unpacked again until his removal to Apperley Green; not that geology was by any means placed in a secondary position, for the vicinity of Oxford was nearly as seductive as the Vale of the Severn, and the several railways that were then commencing, opened up by their cuttings some remarkable sections. Many a geological walk have we had at this time through a country entirely different from anything before seen, and many a lesson has a sometimes restive pupil received. Large collections were formed during his residence here, which were afterwards worked into their places in the cabinets at Apperley Green. But ornithology now occupied the greater share of his time; and the libraries were taken advantage of to examine their stores, in reference to the progress of the 'Ornithological Synonyms,' now very far advanced. A press in the Radcliffe Library was given up for his MS., and the works immediately used for consultation; he had free access to all the treasures therein contained, and some hours daily were spent there at work on the 'Synonyms.' Correspondents materially increased; of whom one of the most prolix, at the same time valuable, was Edward Blyth, zealously employed in working out the Zoology of India for the Museum at Calcutta, sparing neither in his information nor in his queries; every letter gave rise to some points, and they often required much time and research to reply to. Scarcely a letter to myself passed that was not at least half ornithological, either queries, answers, or news; in fact, nothing escaped; and beyond the mere scientific range, literary or antiquarian research occasionally intervened.

The 'Ornithological Synonyms' alluded to was a very laborious undertaking; they were commenced at an early period, as well as a 'Synonymy of Reptiles'; and a series of uniform skeleton papers were printed, which were filled up as each author was consulted. The object of the work was to facilitate the tracing out of any species. The multiplicity of names given to each rendered this a labour always requiring much time, and frequently much research and care; but by turning up a species in the 'Synonyms' the whole descriptive and iconal history could be at once seen, and it only required short specific characters to have made it a complete synopsis. The task of editing the 'Bibliographia' of Agassiz retarded the 'Synonyms'; had it not been so, some portion of them would have been published during his lifetime, and it was his intention to have commenced this undertaking immediately on the completion of the former work. For the extent of labour devoted to it we can best refer to the volume which has been published from his MS., containing the complete synonymy of the Rapaces up to 1855. For this volume between 200 and 300 works were examined by him, and the rest of the MS. is worked up with equal care*.

His general ornithological correspondence had for some time so much increased, and was now become so extensive, that the insertion of it here would extend the volume far beyond its intended bounds; it nevertheless embraces so wide a range that a selection from it would be useful, and perhaps not unacceptable to present working ornithologists, Strickland's replies to the more important letters being mostly preserved.

* 'Ornithological Synonyms, by the late Hugh Edwin Strickland.' Edited by Mrs. Hugh Strickland and Sir W. Jardine, Bart. Vol. i. Accipitres. London: Van Voorst, 1855. In this volume 520 separate works are quoted from.

There were three subjects at this time which consumed a great deal of time, and required much of his attention,—the editing of the 'Bibliographia' for the Ray Society; the 'History of the Dodo,' upon which he and Dr. Melville were now seriously at work; and the preparations for the meeting of the British Association at Oxford. In the preceding chapter we have seen that the first of these already occupied much time in the necessary correspondence. The first volume had not yet been published, and the sheets required constant care and attention while passing through the press.

The desire to render Oxford worthy of the great scientific Association that was about to meet a second time within its halls, was universal among all those connected with her.

The Meeting proved an eminently successful one, and the number of foreigners who attended and contributed to the work of the Sections was much greater than usual. It was imbued also with a more social character, and there was a freer interchange of conversation and opinion among the Members. Ladies took a more active part, went about more independently, and threw off much of the disagreeable restraint which had formerly prevented them from visiting the Sections alone. They took a real interest in the work, and for the first time they attended the ordinaries. Strickland worked hard at everything that could assist the Meeting; but this, as well as his duty as Chairman of Section D, prevented him from reading papers at the Sections. He contributed, however, an evening lecture upon the Dodo, which was numerously attended and well received. Professor Milne Edwards had brought over the Parisian bones with him. The Andersonian Museum in Glasgow liberally sent the bones of the Solitaire belonging to their collection. The British Museum sent her foot; and Oxford, as the original holder and destroyer of that valuable specimen, from which almost our only figures exist, contributed what remained of her treasures; so that almost every authority was present which could render the lecture perfect and attractive*. In the concluding words of the lecturer,—

"Kindred spirits from all countries brought their minds and the objects which engage them into comparison and contact. By such a process it is that we now behold on the table the relics of extinct birds entrusted to us by the curators of five different public establishments. The whole subject is thus placed before us at once; tedious journeys and years of delay are spared to the scientific student, while a kindly spirit is engendered among these fellow-labourers, which will diffuse itself into all departments of life."

It was at the close of this lecture that Dr. Buckland first indicated symptoms of that dire malady which soon after caused his removal from public life and usefulness. In rising to congratulate the reader, he commenced in his usual happy way, but soon rambled over every conceivable subject, from the first missionary to the potato disease and Penny Magazine. The general company did not suspect anything wrong, and it was not until some time afterwards, when the melancholy position of the once great mind became certain, that the Dodo scene of this Meeting was recalled to our minds.

It was altogether a very pleasant week. Large private dinner-parties had been avoided; they were very inconvenient; and the ladies could not attend both to the comfort and detail necessary to receive numerous guests, and at the same time partake of the mental recreation provided by the *savants*. Breakfasts, generally very limited in time, gathered the Members together; but the chief *réunion* was in the evening, after the special labour of the day was over, and the pleasant evenings spent in the small dining-room in Beaumont-street will be well remembered. At first the assemblage was small and shy, brought together by invitation; but as the week wore on, it became a settled point, and rap after rap announced some expected guest, oftentimes accompanied by a friend picked up on the way. Work

* Casts of the Copenhagen head had been procured, but they could not be got out of the London Custom-house in time to be exhibited.

was over, and the school was let loose; information came and went, and opinions, which would not come out in the Sections, were now freely exchanged. The Prince of Canino was a sure guest; he was always active and employed. In the Section he had a corner to himself, with a small table, on which lay a confusion of his works and catalogues, which he was constantly adding to, or annotating upon, as information occurred; he now lost no opportunity of asking questions, and was particularly sceptical upon the fact of the *Sterna velox* occurring in Ireland, and of the Irish Atherine being identical with that of the Mediterranean. Van der Hoeven was often present; Nilsson from Lund, Puggard a young Swedish geologist, Gould, Dr. Latham, William Thompson, Edward Forbes, &c.

Nilsson, an intelligent and gentlemanly old man, full of information, was just completing his Scandinavian Fauna, and felt great interest in many of our species. He was exceedingly anxious to see our red grouse alive, as he thought his northern *T. subalpinus* distinct, from its more arboreal habits, frequenting willow-beds, as one of its names, "*saliceti*," implies, and perching freely upon trees*. This becomes a very interesting question in the distribution of species, and can only be rightly solved by one conversant with the habits of each. Some of our yachting gentlemen, who spend their summers in the North, might gain for us much information; and we trust Mr. Wolley, who has lately spent much time in Sweden and Denmark, and who knows the habits of our own and of the northern birds so well, will on his return place all the facts before British ornithologists. Mr. Gould also lately made an excursion into those countries, saw the birds alive, and brought over specimens; but he has not yet given or published his opinion on the identity of the two birds.

The red grouse has always been considered as a species confined to the British Isles, and it is wondered why it should not cross to the mainlands of the Continent south-

* This is *Tetrao albus,* Gmel., *Lagopus saliceti,* Less. and Swains., *Lagopus subalpinus,* Nilss.

ward. But the causes of animal distribution have not been much looked into; this is a subject now only beginning to attract attention. The Red-grouse, in its entirely brown plumage, is undoubtedly not found out of the British Islands, and is there the *Lagopus scoticus* of ornithologists; but this state is the abnormal plumage of the bird, and Great Britain is the limit of its range, as it almost entirely ceases to be found before it reaches the more southern parts. The winters in Scotland (the grouse-land in the North also often bordering on the sea) are not in their character so severe and dry as to bring on the northern change of plumage, upon the same principle that a southern climate does not affect the colour of the fur of the Ermine, or the winter in Ireland that of the Alpine Hare; but we trace by indications in the rougher covering of the legs, the more hoary under-parts and flanks, and an occasional white quill, that a very little modification of climate would supply the complete Arctic plumage. We look, then, upon this bird as being essentially a northern species, disappearing southward; and if we are right in this, its geographical distribution will cover a very wide extent. The remains of the animals found in the "Tourbières" of Scania also gave Nilsson great interest as compared with ours. The Bear of these bogs is identical with the remains of that found in the British mosses, and he thought them the same with the Cave-bear, and large varieties only, of the present European species, an opinion which others are also gradually falling into.

The remains and records of the Dodo being preserved in Oxford, and the mystery attached to them continuing unravelled, were sufficient inducements to make him take advantage of being present at head-quarters, and endeavour to investigate the subject more fully than it had yet been done. Various papers had been already written in this country by Duncan, Broderip and others, and pictures, with figures in them, representing the Dodo, had been discovered, besides the one in the British Museum. He had examined bones in other collections, which proved that the remains of more than one specimen had been collected, and the disparity of some of the bones hinted the probability that more than one species of this singular group might have existed; in fact, the more he examined, the more he found wanting, and he saw plainly that to work out the materials a rigid historical and antiquarian research, as well as an anatomical investigation, must be instituted. Strickland felt himself quite competent for the historical and zoological parts, but being diffident in his experience and the extent of his knowledge of comparative anatomy, Dr. Melville *, who had studied in the Edinburgh school of comparative anatomy, and who was at this time assisting Dr. Acland, was associated with him in working out the nice osteological details, which were necessary to prove the position of the bird in its own class. The pains and research with which the whole are carried out, and the reasonings deduced from minute modifications of structure, are models of perseverance in the authors, and point out how very essential varied information is for the elucidation of natural history, and that, far from being a science requiring little knowledge, it is perhaps the one where the greatest amount of training and the greatest versatility of mind have to be called in. The 'Dodo and its Kindred' will stand in every way a fine example of patient research and combination of talent†.

It is a pity now that any misunderstanding should exist as to the manner in which Strickland viewed Professor Reinhardt, jun., as the first to point out the true affinities of the Dodo. He visited Copenhagen in 1845, and was conducted over the Museum there by Reinhardt's father, his son being then abroad. The bones of the Dodo naturally attracted attention, as the pictures of Roland Savery had done in other places, and the memorandum in his journal, entered at the

* Now Professor of Zoology at Queen's College, Galway.

† "It is a striking example of refined investigation, an elegant specimen of the elaborate manner in which research is often carried forward at the present day; and the costume of the work is in full keeping with its contents. The lithographic illustrations are beautiful, and the diagrams of the bones must, we should suppose, satisfy the most fastidious anatomist."—Silliman's American Journal, 1849, p. 67.

time, communicates all that he then learned upon the subject:—

"We went to the Museum of Natural History in the Strongade, where Professor Reinhardt, sen. showed me the Osteological Collection, which includes many good skeletons, especially of Cetacea and of Birds. Among the latter, the greatest treasure is a Dodo's skull, stripped of its integuments, in excellent preservation, and apparently just the size of the one at Oxford. It once belonged to an ancient 'Raritäten-kammer,' called the Gottorf Museum, the contents of which were transferred a few years ago to the present collection; and among its miscellaneous contents the Dodo's head was recently discovered. This head is mentioned by Olearius in a Catalogue of the Gottorf Museum, published in 1666. Professor Reinhardt, jun., who is now on a voyage round the world, has prepared a memoir on it, with figures, but has not yet published it. He is of opinion that the bird is not *Struthious,* but intermediate between the *Rasores* and *Columbidæ,* and certainly on comparing it hastily with the skulls of the Ostrich and Cassowary which lay by it, very great differences were apparent."

This information was communicated in conversation, and the affinity once hinted at was afterwards examined. Professor Reinhardt, jun., was absent, and he afterwards writes:—

"I never have published my opinion in any paper; it is only in letters three or four years ago that I have communicated it to several of my correspondents*."

Strickland's knowledge of Professor Reinhardt, jun.'s views was entirely accidental, and in fact communicated to him incorrectly, for the latter acknowledged no alliance with the gallinaceous birds. These views, as obtained at Copenhagen, were mentioned in the discussions at the Oxford Meeting; but they would not be given as *authority,*—a second-hand communication would not entitle them to that. He was therefore not acquainted with the Professor's real opinions, and could not be expected to know what had been written to correspondents whose intercourse with Britain was by no means frequent. The acknowledgement to Professor Reinhardt, jun., in different parts of the volume appears, under

* Letter to Strickland, 7th February 1848.

these circumstances, most ample; his opinion of the affinities of the Dodo is unreservedly spoken of*; and in the numerous reviews which were made of this work, notice was often taken of the position it should hold. In the notice given in the 'Athenæum,' written by a friend who knew Strickland's views, it is said,—"Professor Reinhardt first suggested that it might be a large form of the Columbidæ." Professor Owen wrote:—

"The idea entertained by Dr. Reinhardt of its affinity to the Columbidæ, was SUPPORTED by new arguments adduced by Mr. Strickland in his elaborate and interesting communications and Lecture before the British Association at Oxford†."

While the notice of the book in the 'British and Foreign Medico-Chirurgical Review‡,' written by Dr. Carpenter, and *revised on this very point by Strickland himself*, leaves the matter without any doubt:—

"The idea of referring the Dodo to the neighbourhood of the Pigeons originated with Professor J. T. Reinhardt of Copenhagen, the discoverer of the imperfect cranium which lay hid in the Gottorf Museum, who had pointed out to many Swedish and Danish naturalists, previously to leaving Denmark in 1845 on a voyage round the world, 'the striking affinity which exists between this extinct bird and the Pigeons, especially the Trerons.' This view was adopted by Mr. Strickland, who succeeded in bringing round several eminent naturalists to his opinion, and at the Meeting of the British Association at Oxford in 1847, he formally laid the subject before the zoologists and comparative anatomists there assembled."

* "The first idea of referring the Dodo to the neighbourhood of the Pigeons originated with Professor J. T. Reinhardt of Copenhagen, the discoverer of the cranium in the Gottorf Museum. When I was at Copenhagen in 1845, Professor Reinhardt was then absent on a voyage round the world; but I was orally informed that he considered the Dodo to be intermediate between the Pigeons and the gallinaceous birds. On subsequently examining the remains which we possess in Britain, I soon saw reasons for classing this bird even nearer to the Pigeons than I understood it to be placed by Professor Reinhardt. This gentleman, however, has lately visited London on his return from his distant voyage, and has informed me that, before he left Denmark in 1845, he had pointed out in his letters to several Swedish and Danish zoologists, 'the striking affinity which exists between this extinct bird and the Pigeons, especially the Trerons;'" and he adds in the Postscript, "I have already acknowledged the originality of his idea as to the affinity between the Dodo and the *Columbidæ*."—*Dodo and its Kindred*, pp. 40, 65.

† Trans. Zool. Soc. iii. p. 345.

‡ British and Foreign Medico-Chirurgical Review, 1848, p. 493.

In fact, Professor Reinhardt, jun., could never have seen the numerous reviews and notices of this book, and thought for a moment that Strickland wished to appropriate his labours; and we cannot trace throughout the whole work, or in anything that has passed or been written, the slightest wish to take advantage of the oral information received at Copenhagen, or to detract from the Professor's merit in having discovered the affinities rightly. He arrived at his opinion from the structure of the skull. Strickland and Dr. Melville confirmed and supported the correctness of that opinion, from the structure of the tarsal bones. We can state moreover confidently, that if there was a leaning anywhere in all such discussions as the above, it was to give credit where due, and never to claim for himself the merit of any discovery which was not certainly his own.

The preparation of the plates to illustrate the 'Dodo and its Kindred,' turned his attention to the easiest way by which accurate copies from any old engraving could be obtained, and also to any system which would lessen the expense attending the publication of figures illustrating natural history. Mr. De la Motte, at this time established in Oxford, had introduced the anastatic process of printing in several of his illustrated works, and the facility and exactness with which copies from various subjects could be made, induced Strickland to attempt the copying of some old plates representing the scenery of the islands where the Dodo was known to have formerly existed. These attempts were so successful as to give rise to the letters printed in the 'Athenæum,' in the beginning of 1848*, and afterwards to the little work by De la Motte, entitled 'On the various applications of Anastatic Printing and Papyrography,' in which the process was explained, and specimens of the various styles of execution were given, the ornithological examples being drawn by Mrs. H. Strickland.

At this time we contemplated and began our work, 'Con-

* Athenæum, 1848, pp. 172, 276.

tributions to Ornithology,' intended for exchange among our correspondents; and as economy in the production of the plates was of great importance, our attention was turned to anastatic printing. Strickland had made some sketches on paper with common lithographic chalk, which, when subjected to the anastatic process, were found to print freely; and as drawings could thus be sent by post from any locality, an expedition or tour could be illustrated by sketches actually made on the spot, and at an expense far below that of line-engraving or lithography. This Strickland termed *Papyrography*; and in one of the early parts of the 'Contributions,' his paper on 'Papyrography and its Applicability to Ornithology' was printed; and in the same Number, and in many of the succeeding ones, the plates were mostly executed by this process,—drawn upon paper, and transmitted by post to Mr. De la Motte, by whom they were treated and printed.

The decease of Miss Strickland at Apperley Court in 1849, caused his father to remove from his temporary residence, "The Lodge," to the house and property of which he now became the possessor; and a house and grounds in the vicinity, closely adjoining to the Apperley property, being at this time vacant, a short lease of it was taken, and Strickland removed to "Apperley Green," with the intention, if suitable, of making it his permanent residence. From the vicinity of the Green to Gloucester, Tewkesbury and Cheltenham, railway communication in every direction was easy, and there was now room for the collections which had been so rapidly increasing. The birds were all arranged in cabinets; and the old geological collection, which was packed up before leaving Cracombe, and which had since remained in that condition stowed away in a large ware-room, was brought home, and gradually unpacked and arranged in cabinets made upon a uniform plan, and added to as each successive series of drawers was filled. These matters, with the general correspondence incident to natural history and geology, now fully occupied him, interrupted occasionally by a visit to

London or Oxford, or to the Meetings of the British Association, which were regularly attended; it was at this time also that circumstances rendered our own residence in Cheltenham necessary for some months, and scarcely a week passed without an excursion, which his intimate knowledge of the country and its formations rendered most instructive, independently of the mutual interest which was taken in science generally. These walks have invested that county with peculiar associations, and will be long remembered.

Some terms had elapsed since lectures were given by Dr. Buckland; his recovery was now most improbable, and the blow that had silenced the eloquence and wit which had spread a charm of delight over the driest paths of physical science, had come so suddenly that the sufferer was unable either to resign the offices which he held, or to make temporary arrangements for the performance of their duties, and, under these circumstances, it was felt by the Heads of the University of Oxford that some provision should be made for continuing the duties of his professorial chair. It was proposed to separate the subjects of geology and mineralogy, which Dr. Buckland had held conjointly; and a young Oxonian, who had made the latter his particular study, and was fully qualified, would willingly undertake the mineralogical department; but there was a greater difficulty in finding one connected with the University who would at the same time prove a worthy successor to Dr. Buckland, whose high scientific reputation had long since rendered Oxford and its vicinity classic ground to the geologist. Dr. Daubeny had been consulted by the authorities, and, always the friend of Strickland, corresponded with him in regard to this matter. Some eyes were turned to the Continent for a successor; but the Doctor writes:—

"Having two such good men as Lyell and yourself, there can be no reason for looking *abroad* for a successor to the Chair of Geology, and I do not think the University would, or ought to pass over the person they regard as most eminent, on political grounds."

Sir Charles Lyell, from his long scientific career and great experience, would undoubtedly have stood before Strickland in an offer to fill this chair; but it was understood that Sir Charles would decline the appointment, and Dr. Daubeny had been requested to ascertain privately if Strickland would accept the Readership. The correspondence which ensued was creditable to Dr. Daubeny as a friend, and at the same time as one who kept the interests of the University steadily in view; and we copy Strickland's letter, expressing his readiness conditionally to accept the office, as characteristic of his disposition,—willing to be of use, but at the same time diffident of his own abilities:—

"Apperley Green, 1850.

"MY DEAR DAUBENY,—I feel greatly obliged by the kind interest you take in proposing me as a Deputy, or possible successor to Dr. Buckland; but I really feel much hesitation in the matter on several grounds. I may, however, by way of simplifying the question, say at once that I should be very glad to be released from all connexion with the *Mineralogical* Chair; first, because I have never given that amount of attention to mineralogy which would justify me in accepting such an office; and secondly, because it would consume more of my time than I could spare from other objects and duties.

"In regard to my giving lectures on geology during Dr. Buckland's illness, I also feel considerable misgivings; for, although some ten years ago I took great interest in that study, yet of late years, partly from want of opportunity, and partly from giving most of my attention to recent zoology, I have been a less active worker in the field of geology than formerly. I can therefore conscientiously say, that if any other Member of the University can be found who is in any way competent to such an undertaking, I would most gladly make way for him, as I am fully occupied and very comfortably settled where I now am. If, however, the case is really urgent, and no other suitable person can be found, I should feel it a duty I owe to the University to come forward and do my best.

"I can hardly say, however, whether it would be possible for me to prepare an entirely new course of lectures in the short interval between now and Easter Term, even if I set to work immediately, on the expectation of the appointment being made after Easter. To give lectures which would do credit to the appointment, much study and preparation, both of matter and illustration, would be requisite; and until I begin to collect materials for the lectures, I really cannot pledge myself to be ready to deliver a course in Easter Term.

"Would there be any objection to my postponing the course till Michaelmas Term, in the event of not feeling prepared to commence it in Easter Term?"

Arrangements were soon afterwards made; Mr. Storey Maskelyne, M.A., was placed in the Mineralogical Chair, and Strickland was installed as Deputy Reader in Geology, an appointment which acknowledged his ability and reputation, gave him undoubted right of precedence over many, and enabled him to take his place alongside of those whose hónours had been already worked for and won.

The studies of Oxford had been hitherto conducted upon that plan of devoting all time to the classics or mathematics, to metaphysical and abstract sciences, to the exclusion of the physical. All those relating to the properties of matter, or that had a bearing upon tangible objects, were repudiated as almost unworthy, or beneath the notice of the members of that great repository of learning. Nevertheless, under that system she produced many very eminent men, and among those of the later age Dr. Buckland; he made geology popular in this country, and carried with him not only the men of science, but also the general public. He worked at it with his whole heart, entered its deepest mysteries, and explained away many of those points which were beginning to render it unpopular by their supposed discordance with the records of the Bible; and he was remarkably gifted with the power of rendering the drier details agreeable and easily understood; yet he could not change the University prejudices, and could only induce a few men to join his pursuits and attend his lectures.

The courses given were very short, consisting of about twelve or fifteen lectures each; and thus, at first sight, the appointment did not appear to require a great deal of labour. But, to succeed to the reputation of Buckland was sufficiently arduous, while, to a conscientious mind, thinking that a trust was imposed which should be faithfully executed, and to one whose lectures were yet unwritten, the preparation for the first session consumed the greater part of the intervening time. The address, 'On the Studies of the University of Oxford,' forming one of his opening lectures, will best show the views he took of mingling abstract and purely scientific information with what was practical*. No doubt Oxford was too restricted and too exclusive, but the curriculum was a fine school and training for the mind, and the men at the completion of their terms were fit for anything, whether scientific, political, or commercial. The modern system has run into the opposite extreme; the Greek and Latin of the old colleges lose their importance. "*Education*" has become a political war-cry, and it is fashionably, but unthinkingly, heard of in the gossip of almost every dinner or tea party. Many of the new schools profess an extent of information beyond their powers or means of teaching, and the men are turned out highly crammed with a supposed knowledge of "something of everything."

From the shortness of the courses it was scarcely possible in one term to run over all geology, indeed it was never attempted; different departments were taken up, and a short survey of some formation, or the action of some leading power, was given. Thus the first course of twelve lectures, delivered at Michaelmas Term 1850, after three lectures 'On the Chronology of the ancient Earth,' as introductory to the whole general subject, was devoted to the operations resulting from the action of fire and water, the two chief agents, which have had almost an equal share in bringing the earth to its present state. The second course at Easter Term 1851, from the more simple aqueous action, continued and treated of the second great agent, by a description of 'Volcanic Phænomena, the Elevation of Mountains, and other Disturbances affecting the Earth's Crust; Principles of Geology; Earlier Stratified Deposits.' The succession of the Secondary and Tertiary formations were in their turn lectured upon; and the last course he was spared to deliver, at the Lent Term of 1853, carried on the subject to the 'History of Tertiary and recent Epochs.'

* See 'Collected Papers,' p. 207.

The number of pupils who attended his first course was only eight, and they increased in the last to eighteen,—small encouragement for an ardent Professor, but strongly illustrating the system of the University, and confirming the remarks we have already made upon it. The excursions, however, which were taken to interesting localities during the term, were attended not only by the class, but by others, who, attracted as amateurs, showed that geological impressions were slowly dawning.

With the exception of the "Dynamic" course, which was written out at length, and had been very carefully studied, he lectured chiefly from notes, and illustrated the lecture by diagrams and sections, and by specimens from the large collection his predecessor had so perseveringly brought together, and to see which had so much "delighted" him on his first entrance to his college*.

Strickland was ever ready to communicate his information, and he was frequently requested to lecture for different institutions. The Natural History Society of Worcester often received his advice, and he drew up for it a plan for the arrangement of the Museum, possessing a very considerable collection, among which were some fine specimens of fossils from the Severn Valley. He gave several lectures in the Society's rooms, chiefly such as afforded local interest, as, 'On the Geology of the District between Worcester and London,' &c., but also on other subjects, 'On the Great Exhibition—Coal and Iron,' and we find the notes of one 'On the Caves of Carniola.' He was President of the Tewkesbury Mechanics' Institute, where he also lectured on similar subjects, 'Geology of the Neighbourhood of Tewkesbury,' &c.

The formation of clubs for the investigation of local natural history also began to take up new ground. These are of much importance. The preservation of the condition of the present physical characters of our country will be far more dependent on them than at first appears. The last fifty years have made a great change in the surface of the

* Page xxii.

country; population has increased; so has agricultural improvement, plantations, drainage, enclosure of waste lands, in short, artificial works of every kind. These have often completely altered the nature and aspect of the country, and in consequence the productions, both animal and vegetable. In parts of the north of Scotland, another cause, that great rage and fashion for "sporting," as it is termed, has influenced the distribution of the higher orders; the wild animals and birds have been reduced in numbers, as "vermin," sometimes almost extirpated, and many will in a few years stand side by side in history with the Bear and the Wolf. It will be to these clubs that we shall be indebted for a record of what in their days did exist; and in the still untouched mountains and valleys we may have the discovery of insects and plants not known to our geographic range; and when the country shall have been mapped on the large scale by the Government Surveyors, there is nothing that should prevent an active club to fill up in a few years a list of the productions within their beat, and so lead on to a complete and accurate Fauna and Flora of our own time and age; and generations succeeding would be able not only to mark the changes of the productions, but to judge and reason upon the effects which these now so-called improvements have produced on the climate and soil, and the fertility and increase of the latter. These clubs have yet to write the natural history of Great Britain.

In Gloucestershire and Worcestershire natural-history clubs took their rise from the example of a Border association, due to the activity and restless inquiry of our already mentioned friend, the late Dr. Johnston of Berwick; under his management the 'Berwickshire Naturalists' Club' was founded in 1831, having Mr. Selby and Dr. Baird among its first members: circumstances made our family at the time resident in Roxburghshire, the neighbouring county, and we had the honour also of becoming an early member. Its rules were most simple; its objects, to investigate the natural history of the county. The fourth volume of its 'Transactions is now in progress; and though its founder and some old members have passed away, and others have removed to a distance, it still goes on with energy, and finds ample employment. It is curious that the example of this Club has only extended southward. The Tyneside Club is now in its twelfth year. Sir Thomas Tancred, the son-in-law of Mr. Selby, was a frequent visitor to the excursions of the Berwickshire Club, and afterwards became a member. He thought highly of its objects, and when settled for a time in the vicinity of Cirencester, where he took much interest in the progress of the Agricultural College then just forming, he thought that that neighbourhood could well afford to support a club, and Strickland, among others, was one of those applied to. The Cotteswold Club was thus formed on the 7th of July, 1846, at the Black Horse, Birdlip: there Dr. Daubeny and Strickland, the Professors of the Agricultural College, and many more, were enrolled, with Sir Thomas Tancred as their Secretary, and Brodie, Wright, Lycett and Woodward became early members. This Club still goes on successfully; and from it again has sprung the Woolhope and Malvern Field Clubs, all publishing their Transactions, and forming a strong exploratory phalanx, as well as a most excellent training school for young naturalists.

The meetings were often attended, and sometimes the three clubs joined and had a great "field day," and Strickland's position and great local knowledge allowed him occasionally to give open-air lectures on the geology around. We find one recorded as delivered from the Painswick Beacon, another from Stinchcomb Hill, and his last was addressed to the members of the three Clubs, for the day united, and assembled on the summit of the Ragged Stone Hill, near Eastnor. These addresses were always attentively listened to and most gratefully received; this last was peculiarly interesting, from the remarkable characters of the surrounding country and from the success of that day's ramble. In the Valley of the White-leaved Oak, Strickland had in the morning rediscovered the small trilobite, the *Agnostus pisiformis*, a fossil known only in the Alum schists of Sweden, and new to the Malvern district; it holds its position there in the black shale, but in a bed of such extreme thinness, that it is often sought for in vain. On a previous day, the late Mackay Scobie found in the Upper Ludlow rock at Hagley Dome, the remains of a remarkable crustacean, *Pterygotus problematicus* *. This specimen was at the time unique for the district, and formed the subject of the short paper and figure by J. W. Salter, and is now in the Strickland collection. Some time afterwards, another fragment of Pterygotus, one of the jaw-feet, was found by Mr. Burrow † below Eastnor Obelisk, in the Caradoc conglomerate of Professor Phillips ‡: one half of this specimen is in the Strickland collection, the other is in the Malvern Museum, whence it has been lent to us for the figure introduced at the end of this chapter.

These public vocations and private scientific pursuits, with the exercise of his duties as a country gentleman and magistrate, which also began gradually to fall upon him, fully employed his time; and the "Green," of which a new lease had been taken, might now be considered as his fixed residence. The routine of home was at the same time varied by excursions, either before or after the meetings of the British Association. The coast of Wales was several times visited as a sea-bathing locality, where health and science could be alike improved, and expeditions were now and then taken in company with his dredging friends.

One of the last of these, of which a short journal was written in letters home, was a cruise to the Isle of Man and Belfast Lough with his friend T. C. Eyton, Esq., of Eyton Hall, Shropshire, in the yacht 'Ada,' and of which a short account was afterwards published by the latter §.

On the 10th of May, 1853, the 'Ada' sailed from Douglas Harbour in the Isle of Man, with a crew of five men. In Ramsay Bay some of his previous observations in regard to

* Quart. Journ. Geol. Soc. vol. viii. pl. 21. p. 386.

† Hon. Sec. Malvern Nat. Hist. Field Club. ‡ Lower May Hill bed.

§ Hunt's Yachting Magazine, vol. iii. p. 233.

the tertiary beds there were re-examined and verified, and additional specimens of the shells formerly described as new were procured, and from these the woodcut commencing this chapter has been drawn*.

"Eyton and I went ashore and walked some way along to the north of Ramsay. This gave me an opportunity of re-examining the curious cliffs of Pliocene sand and gravel which I described in the 'Proceedings of the Geological Society' in 1842 or 1843. We found several more specimens of shells in this gravel, and two or three additional species to those which I formerly procured. I also ascertained that the singular stalactite-like concretions of sand there mentioned, all lie *horizontally* in the strata in a direction nearly north and south, with their *points* toward the south, showing that they are due to the action of currents from the north at the time of the deposition of the strata. This was a point which I had not an opportunity of deciding on my former visit."

On the 13th the party anchored in Belfast Harbour, visited the Museum of that town with Mr. Patterson and Mr. Hyndman, and the memorandum made of the condition of the collection left to that institution by the late William Thompson will be read with interest:—

"In the evening we accompanied Mr. Patterson and Hyndman to the Museum of the Natural History Society, a very well arranged and creditable institution in the College Square. An additional room has just been erected by subscription, to commemorate the scientific labours of poor Thompson, the most eminent naturalist Ireland has produced. All his cabinets and natural-history collections will be here placed, and kept distinct from the general collection. Mr. Hyndman showed us over the drawers of Thompson's British shell cabinet, which he had partly arranged at the time of his death, by sticking the shells on to cards, and writing the localities at the foot. In one instance I noticed as many as sixteën localities assigned to one species, the individuals being referred to by numerals. The rest of this collection is roughly sorted in the drawers, the shells being still separate from their labels; but Mr. Hyndman is about to get them all stuck down, and Thompson's original labels gummed down to the cards. They can then be examined without fear of derangement."

---

* Ann. Nat. Hist. vol. xi. p. 507; Proc. Geol. Soc. vol. iv. p. 8; Collect. Papers, p. 170; Forbes, Malac. Monen. p. 61. pl. 3.

We trust the intentions of the donor and present trustees will be carried out. As a collection, it is of great interest generally, but to Ireland it is invaluable. Other naturalists we hope are in training, but it will be long indeed before another will be found possessed of the same knowledge of his country's productions, combined with so much patient research and rigorous accuracy.

This cruise was very productive; above a hundred species of recent Mollusca alone were dredged, besides Crustacea, Echinoderms, &c.

On the 7th of September of the same year, Strickland proceeded to the Meeting of the British Association at Hull. He was alone. We met there, and lodged in the same house. The Meeting, though much below the average in amount of numbers, was an interesting one. Professor Sedgwick presided over the Section of Geology, and there read his paper 'On the Classification of the Palæozoic Rocks of Britain,'—a long and eloquent dissertation, in which he worked up all the points and evidences in his 'System and Nomenclature of the Older Rocks,' fighting his own battle with spirit to the last. During the discussion Strickland occupied the Chair as one of the Vice-Presidents, and with much tact smoothed down the excitement which might have arisen among the opponents of Professor Sedgwick and the defenders of Sir R. Murchison's views, who was then absent in Germany.

To the Geological Section Strickland read a paper 'On Pseudomorphous Crystals in New Red Sandstone*;' and to the Zoological Section he contributed a notice 'On the Mode of Growth of *Halichondria suberea*†,' and another 'On the Partridges of the Great Water-shed of India‡,' incidental to his placing on the table the last published part of Gould's 'Birds of Asia,' which contained figures of the large Tetraogalli.

* Reports of Brit. Assoc. 1853, Sections, p. 60; Quart. Journ. Geol. Soc. ix. p. 5; Collected Papers, p. 204.

† Reports of Brit. Assoc. 1853, Sections, p. 71.

‡ Reports of Brit. Assoc. 1853, Sections, p. 72.

On the Saturday we joined an excursion, under the charge of Professor Phillips, to Flamborough Head, and after having seen its caves, Strickland and myself, accompanied by Dr. Dickenson of Liverpool, M'Andrews, Bowerbank and Dr. Lankester, started to walk along the cliffs. This was Strickland's early county; all was familiar to him, and the remarkable sections well known, were in their turn pointed out. The day was fine, the walk much enjoyed, and we are sure will be long remembered by our companions. At the termination of the cliffs, some of the party went on to Scarborough. We descended to the shore, where I was introduced to Speeton clay. There we lingered until dusk, reaching the station in time to return to Hull by the last train.

Next week the business of the Sections was resumed, and the Meeting terminated in the usual way. On the morning of the 13th we parted, the one intending to wander round the coast to Newcastle, Strickland to examine the Clareborough cuttings near Retford.

PTERYGOTUS PROBLEMATICUS.

Jaw Foot.

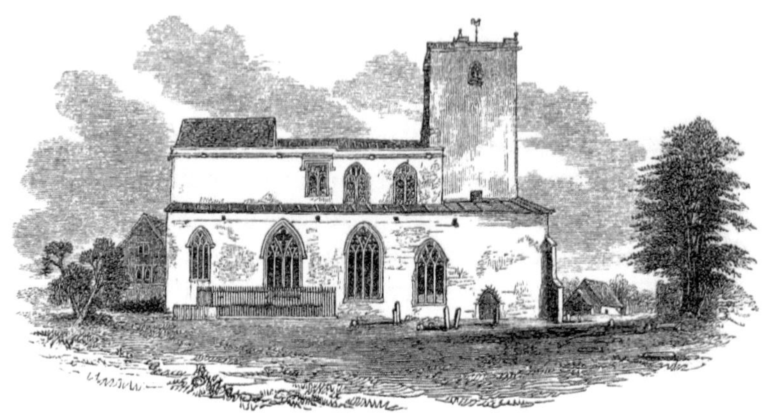

DEERHURST CHURCH.

CHAPTER IX.—1853.

CONCLUSION.

In that beautiful part of the Vale of the Severn which lies between Tewkesbury and Gloucester, there stands a venerable and time-worn church. On its northern side a modern family tomb has been enclosed, and in its front an ancient window has been lately renewed. In the tomb are laid the remains of one, to whom the kind affections of friends have repaired the window as their memorial. The latter is yet conspicuous in the grey and moss-covered walls of the structure, and the visitor or traveller passing by the churchyard foot-path sometimes inquires its purpose.

When necessarily resident for a short period in Gloucestershire, after Strickland had removed to Apperley Green, it was a favourite excursion to examine some one of the highly interesting cuttings which abound in most of the lines of railway in that neighbourhood; and for this purpose, having reached the station from which our day's work was to commence, we obtained permission from the station-master, and at the same time a note of the hours at which the different trains in both directions passed. In his last excursion he

MEMORIAL WINDOW,

CHURCH.

was alone. At the close of the Hull Meeting he was anxious to examine in his way home the cuttings in the Manchester, Sheffield, and Lincolnshire Railway near Retford, in reference to which some discussion had taken place during the sittings of the Geological Section; and for this purpose he got out at Retford, intending to see what he wished, and to be taken up again at Gainsborough, where his luggage was directed to meet him. The chief point of interest was a short way from Retford, at the deep cutting near Clareborough, where the railway enters the tunnel bearing the same name. The station-master at Retford was absent, and a pass could not be obtained to travel upon the line; he therefore walked along the banks, descended at the cutting, and was seen examining its sides. He had afterwards gone upon the line, at a part near Gringley Balk Bridge, where there is a pretty sharp curve. Here two trains were coming up in opposite directions; the one a coal-train, which was seen, and to avoid this he stepped upon the other side, on which an unseen passenger train was approaching round the curve, at a speed of twenty-five or twenty-six miles an hour, and at a distance from him of less than a hundred yards. The driver of this train reversed his engine, put on breaks, and whistled loud and very alarmingly; "but he took no notice; his attention seemed to be on the coal-train*." Death must have been instantaneous; there could be no pain, not even a momentary feeling of alarm or fear. The double noise gave no warning of the engine behind, and he was struck, unconscious of danger, and not even turning from his watch of the other receding train.

These are the simple facts of this deeply-felt dispensation; and we have been particular in giving the short and melancholy detail, that it may stand as the correct and true narrative made from personal observation at the time, and that the exaggerated statements which find their way into the newspapers of the day may not be retailed when passing strangers or posterity call for the history of the Deerhurst

* Evidence given by the driver at the inquest.

window, and ask the attendant to relate the story of his death.

Another window is erected to his memory in Watermoor Church, Cirencester, by Professor Buckman and other friends. His name is recorded to science by the names of various species, and a genus, STRICKLANDIA, has been dedicated to him by Professor Buckman. This is a fossil plant, of which a single specimen only has yet been found in the Stonesfield slate at Sevenhampton, near Cheltenham. That specimen is now in Professor Buckman's collection, who kindly entrusted it to us for the woodcut which we have placed on the title-page of this volume.

Of the collections which Strickland had formed, a series of specimens and skeletons illustrating the comparative anatomy of birds, has been presented to the Museum at Worcester. And the entire ornithological collection has been in like manner presented to the new Museum of the University of Oxford, under the condition of being always open for scientific consultation. It was thought that this distribution was most in accordance with what he might have himself proposed, while at the same time it prevented the dispersion of the collections, and offered suitable memorials of his attachment to the institutions themselves, as well as of the interest he took in that department of zoology. His manuscripts and correspondence, with the geological and palæontological specimens, have been retained by Mrs. H. Strickland.

In the preceding pages of this volume we have endeavoured to fulfil a promise made to his parents, to collect together whatever was most worthy of preservation among the works and correspondence of the short, but very active scientific career of their son. This has been confined, however, mainly to those subjects with which he was most intimately conversant, and such as were more immediately in connexion with his profession, if it may be so called, of *physical science*. His acquirements in general science and literature were at the same time of a high class, and very varied, as may be judged of by the complete list of his collected writings which we have appended to the "Second Part." His college education fitted him to take part in any department that inclination, necessity, or the welfare of his fellow-men might suggest; and although he did not put his name to essays, except they were upon subjects he had completely studied, he was a very frequent contributor to the periodical press of the day, upon occasions where general or private rights were assailed, or individual oppression was attempted to be exercised.

His classical studies were kept up necessarily during his Eastern travel, and they were applied at home on various occasions. He took considerable interest in a paper 'On the Zoology of Homer and Hesiod,' by M. Groshans*, republished in English by Mr. B. Macdonald, of Rammerscales, and was consulted during the translation. Aristotle's 'Historia Animalium' was a favourite book, and notes were made upon various passages. A paper 'On Accent and Quantity, and the practicability of introducing Ancient Metres into Modern Languages,' was prepared, and sent to the 'Classical Museum;' but the cessation of that work at the time prevented its publication†; and in connexion with ancient art, we have an account of 'An ancient Colossal Statue near Magnesia,' printed in the 'Archæologia'‡.

In 1846 or 1847, English spelling reform, or an attempt to accommodate our spelling to the sound, was agitated by Mr. John Ellis, and about the same time various phonetic speculations arose. There was a 'Phonetic News,' a 'Phonetic Journal,' and a Phonetic Society. Many able men were interested in this subject, and took part in it, and among others, Dr. Latham. Strickland also took it up, and several letters, with the initials 'H. E. S.,' were printed by him in the 'Athenæum.'

* Geo. Phil. Fred. Groshans, Prodromus Faunæ Homeri et Hesiodi, 1839. From the 'Tijdschrift voor Nat. Gesch. en Physiol.' † See 'Collected Papers,' p. 437.
‡ Vol. xxx. p. 254; and 'Collected Papers,' p. 434.

There is no written language in which the discrepancies between the signs of the alphabet and the sounds of the tongue have attained a greater amount than our own. Every foreigner commencing to speak English will confirm this. Several persons had now started to endeavour to remove these anomalies, and to invent a series of signs, which would exhibit the sounds of the English language, and enable us to spell every word as it should be pronounced. No less than between fifty and sixty distinct characters had been contrived as indispensable; the parties had gone so far as to have types cast representing them; and he thought an undertaking of such magnitude, and apparently conceived with honesty of purpose, was, to say the least of it, deserving of a candid examination. From this arose a correspondence with his friend Dr. Latham, the letters in the 'Athenæum' alluded to, and the preparation of a paper on phonetic alphabets, which was read before the Ashmolean Society at Oxford*.

The miscellaneous subjects which from time to time attracted his attention, were just those which any public incident or event suggested. Thus a letter to 'The Times,' "Long Hours and Long Walks in the Factories," was written after one of Lord Ashley's noble appeals against the overworking of women and children, and suggested mechanical contrivances to relieve them by lessening the labour. Another letter was printed in the 'Morning Chronicle' for 1849, signed "Sigma," in reference to the plan for female emigration proposed by the Honourable Sidney Herbert; and when Lord John Russell proposed the transportation of the Hill-Coolies of India to the sugar-plantations of America, his letter, "Sugar and Slavery," recommended the reversal of the proposal, to cultivate sugar on the Indian hills and equalize the duty. When the colonnade of the Quadrant in Regent-street, "the only scrap of poetry that enlivens the prosaic monotony of London streets," was about to be effaced, he remonstrated, suggesting light to the shops by glazing the ceiling; and an average state of street morality to the Qua-

* Proceedings, ii. p. 215.

drant by the presence of an extra beadle or two, "armed not only with baton and cocked-hat, but with a little extra executive power if necessary." He was one of the objectors to the removal of the Crystal Palace, and wrote under "Waste not, want not," to the 'Morning Chronicle' in 1852. He would have had it a Museum of Art, taken in the widest sense of the term, to relieve the British Museum and National Gallery, and to bring together all the scattered riches possessed by the nation, but unseen and unused by the people. The great mistake in that fine structure was in the manner of its erection; it would not last, and was not of that substantial character necessary for such a purpose in our climate; but the remedy was pointed out, and a comparatively little cost would have adapted it to receive the collections, and "the present generation would have the enjoyment of the magnificent spectacle which its interior would speedily present; whereas, by constructing a National Museum of Art with bricks and mortar, in the old-fashioned way, we should see nothing of it all our lives but a forest of scaffolding."

He was frequently also drawn upon for a review, both by author and publisher. One of the earliest appeared as 'Observations on Sharon Turner's History of the Earth,' published in the 'Analyst.' The others found their way to the journals and magazines, and were generally written with a clear independence, pointing out both faults and merits.

He had a turn for mechanics, which he employed in

planning his cases and cabinets, and the various appliances which a large collection required. He had attempted wood-cutting, and the accompanying cut will show one of his early attempts at this nice and clean employment. He was a very fair artist, and many of his sketches, taken while abroad and on his different excursions, were spirited and characteristic. He had also tried lithography; but all these departments of art he put in charge of his wife, who executed most of the illustrations of those works which were published after his marriage.

But Oxford was his darling; from his first entrance to her halls he was impressed with her advantages; but he also saw her imperfections, and no opportunity was afterwards let slip to improve or reform, and to exalt her among Universities. In his early career his hints were often listened to, and his recommendations, if not at once adopted, were treated with consideration. He took a deep interest in the success of the Ashmolean Society, and it was there, during his Presidency, that he advocated strongly improvements for the sanitary condition of the city, better drainage, and the expulsion of intramural burying-grounds. When the Radcliffe Library wished to reduce the annual sum expended upon books from £500 to £200, it was he who drew up the petition to the Trustees*. In later days, when a Member and installed among her teachers, he assisted in her councils, took charge of the reform proposed in the department of physical science,

* We find the following table in a note to the draft of the petition. The state of other libraries in this respect might be advantageously looked into.

"Numerical statement to show to what extent the science of Zoology is represented in the public libraries of Oxford:—

"Total of known publications on Zoology according to a carefully-compiled catalogue	2419
Of these there are in the *Bodleian*	478
" " *Radcliffe*	954
Works in the *Bodleian* and not in the *Radcliffe*	202
Works in the *Radcliffe* and not in the *Bodleian*	678
Works which occur in both libraries	276
Zoological works which occur in neither library	1263

"Jan. 10th, 1845."

and drew up the 'Explanation for the School of Natural Science,' in the new examination statutes. As Deputy-Reader in Geology, the Oxford University Commission applied to him for answers to their inquiries, and he sent in a very long and clear statement, embracing the details of many proposed reforms in the general constitution of the University, as well as suggestions for improvement in teaching the classes of his own particular "school." He was one of the Committee for the erection of the new Museum; and in the selection of a site, and in its progress while he lived, he was an active and judicious assistant. Architects never understand a Museum; their thoughts are of the walls and outward appearance, or interior decoration; and while Strickland's classical taste and knowledge of architectural art would have guarded against the introduction of anything injurious to the appearance of the building, his practical acquaintance with what was necessary to preserve and display the objects intended to be contained within, would have been of the utmost importance. Truly Oxford had to deplore his loss.

It would be contrary to the objects of this Memoir to write of his private virtues, or enter upon the griefs and sorrows of near relations and friends. The former are rightly appreciated by those who were dear to him. The event of his death, and the circumstances attending it, can never be forgotten by them. Let us hope that they will be soothed by time, and a dependence upon God.

T.H. MAGUIRE. LITH. DE LA MOTTE PHOTO.

HUGH EDWIN STRICKLAND, M.A.

F.R.S F.G.S. F.G.S.A. &c.

1853.

M & N. HANHART, IMP.

SELECTION

FROM THE

SCIENTIFIC WRITINGS.

CONTENTS.

PART II.

MEMOIRS ON THE GEOLOGY OF ASIA MINOR.

MEMOIRS ON THE GEOLOGY OF GREAT BRITAIN AND IRELAND.

Page

ORNITHOLOGICAL MEMOIRS.

NOMENCLATURE AND CLASSIFICATION.

MISCELLANEOUS PAPERS.

LIST OF PLATES.—PART II.

WOODCUTS.

LIST

OF

THE PUBLISHED WRITINGS

OF

HUGH EDWIN STRICKLAND, M.A

CHRONOLOGICALLY ARRANGED.

Besides the Papers and Memoirs which have been reprinted in this volume, there are other communications, which we have not judged it necessary now to republish. These consist of—Abstracts of papers read at the meetings of Societies, and supplied by him for their 'Proceedings';—short Scientific Notices upon various subjects, generally printed in the periodicals under the head of "Intelligence" or "Miscellanea";—Letters upon subjects both scientific and general, written for, and printed in, the principal newspapers and the weekly literary journals;—and Reviews and Notices of Books, which appeared anonymously, and were not intended for after reprint or collection.

He had himself printed a list of his 'Principal Scientific Writings'; and in addition we have now inserted all those of which we are aware, with a reference to the works themselves, or the periodicals in which they appeared, so that at any time they may be referred to, or consulted.

The papers which we have reprinted have a short explanatory note appended to them; and to some of the titles in the following list we have added any memoranda which would explain or elucidate the circumstances or intentions under which they were written.

In the 'Collected Papers,' p. 152, the woodcut illustrating *Æshna liassina* was omitted, but a copy from that in the original paper has been introduced into the Memoir, p. cxcix, where the subject is alluded to. In the 'Notes on a Section of Leckhampton Hill,' Part II. p. 189, the Section has in like manner been unfortunately omitted; for the sake of reference it is introduced here.

In the paper on the 'Thracian Bosphorus,' p. 8, allusion is made to a fragment of bone—"the only fragment of a fossil Vertebrate which I have found in Turkey." This Dr. Baird was so kind as to request Mr. Quekett to examine, and he reports :—"As far as the transparency of the section would allow me to see, the structure is *fish-bone*, and sections of a Gar-fish of large size somewhat resemble it."

CHRONOLOGICAL LIST OF WRITINGS.

1827.

1. Plan for uniting a Wind-Gauge with a Weather-cock. By Boreas. Dated "North Pole, Feb. 1827." Mechanics' Mag. vii. p. 264.
   This is the letter referred to in his Journal, Memoir, p. xii, and shows his early taste for Mechanics. It appeared in the Number for April 28th, 1827.

1833.

2. Notice concerning the Red Viper (*Coluber chersea*, Linn.). Mag. Nat. Hist. vol. vi. p. 399, vol. vii. p. 76.
3. On the Red Marl and Lias of Worcestershire. Proc. Geol. Soc. vol. ii. p. 5. Geol. Trans. 2 ser. vol. v. p. 260.
   This Paper was not printed; abstracts only were printed in the 'Proceedings' and in the 'Minutes' of the Society.

1834.

4. Observations on Classification, in reference to the Essays of Messrs. Jenyns, Newman and Blyth. Mag. Nat. Hist. vol. vii. p. 62. Coll. Papers, p. 398.
5. List of the more rare species of Shells which were collected, in August 1833, at Aberdovey in Merionethshire. Mag. Nat. Hist. vol. vii. p. 159.
6. On the Luminosity of Glow-worms' Eggs. Mag. Nat. Hist. vol. vii. p. 252.
7. On *Vespa britannica*. Mag. Nat. Hist. vol. vii. p. 264.
8. List of Land and Freshwater species of Shells which have been found in the neighbourhood of Henley-on-Thames. Mag. Nat. Hist. vol. vii. p. 494.
9. Observations on Sharon Turner's Sacred History of the Earth. Analyst, vol. i. p. 319. Anon. Referred to in Memoir, p. cclxiii.
10. On the occurrence of Freshwater Shells of existing species beneath the Gravel near Cropthorne, Worcestershire. Proc. Geol. Soc. vol. ii. p. 95, 111. An abstract only. Referred to, Memoir, p. clxvi.

1835.

11. On the Arbitrary Alteration of Established Terms in Natural History. Mag. Nat. Hist. vol. viii. p. 36. Coll. Papers, p. 366.
12. Memoir on the Geology of the Vale of Evesham. Analyst, vol. ii. p. 1. Coll. Papers, p. 79.
13. On the Nomenclature of Birds. Analyst, vol. ii. p. 317.
    This Paper is mostly a repetition of the arguments used in Nos. 11 and 19. The first paragraph is printed as a note to No. 19. Coll. Papers, p. 374.
14. On Currents of Sea-water flowing into the Land near Argostoli in Cephalonia. Abstract only. Proc. Geol. Soc. vol. ii. p. 220. See Memoir, p. clxi.

1836.

15. List of Birds noticed or obtained in Asia Minor in the Winter of 1835 and Spring of 1836. Proc. Zool. Soc. pt. iv. p. 97. Coll. Papers, p. 223. Noticed in Memoir, with remarks on the new species discovered, and a figure of *Emberiza cinerea,* Strickl., p. clxii.
16. A General Sketch of the Geology of the Western Part of Asia Minor. Proc. Geol. Soc. vol. ii. p. 423. An abstract only. The materials are extended in the joint paper by Hamilton and Strickland, Coll. Papers, p. 26.
17. On the Geology of the Thracian Bosphorus. Proc. Geol. Soc. vol. ii. p. 437. Geol. Trans. 2 ser. vol. v. p. 385. Coll. Papers, p. 1.

1837.

18. On the Mode of Progression observed in the genus Lima, Brug. Mag. Nat. Hist. 2 ser. vol. i. p. 23. Coll. Papers, p. 423.
19. On the Inexpediency of altering Established Terms in Natural History. Mag. Nat. Hist. 2 ser. vol. i. p. 127. Coll. Papers, p. 370.
20. Rules for Zoological Nomenclature. Mag. Nat. Hist. 2 ser. vol. i. p. 173.
    Not republished; being the first sketch of the subject fully treated of in the Report to the British Association.
21. On the Errors which may arise in computing the Relative Antiquity of Deposits from the Characters of their imbedded Fossils. Mag. Nat. Hist. 2 ser. vol. i. p. 234. Coll. Papers, p. 145.
22. On the Geology of the Neighbourhood of Smyrna. Proc. Geol. Soc. vol. ii. p. 538. Geol. Trans. 2 ser. vol. v. p. 393. Coll. Papers, p. 9.
23. On the Geology of the Island of Zante. Proc. Geol. Soc. vol. ii. p. 572. Geol. Trans. 2 ser. vol. v. p. 403. Coll. Papers, p. 19.
24. [— and W. J. Hamilton.] Account of a Tertiary Deposit near Lixouri, in the Island of Cephalonia. Proc. Geol. Soc. vol. ii. p. 545. Quart. Journ. Geol. Soc. vol. iii. p. 106. Coll. Papers, p. 70.

25. [— and R. I. Murchison.] On the Upper Formations of the New Red System in Gloucestershire, Worcestershire, and Warwickshire. Proc. Geol. Soc. vol. ii. p. 563. Geol. Trans. 2 ser. vol. v. p. 331. Coll. Papers, p. 111.
26. Anniversary Address of the Council to the Worcestershire Natural History Society. 8vo. Worcester, 1837.
27. On the Nature and Origin of the various kinds of Transported Gravel occurring in England. Rep. Brit. Assoc. vol. vi. Sections, p. 61. Coll. Papers, p. 105.

1838.

28. Remarks on 'Viator's' proposed New Name for the Infusorian genus Proteus. Mag. Nat. Hist. 2 ser. vol. ii. p. 165.
    Printed as a note to the Report on Zoological Nomenclature. Coll. Papers, p. 385.
29. Reply to Mr. Ogilby's Observations on Zoological Nomenclature. Mag. Nat. Hist. 2 ser. vol. ii. p. 198.
30. Remarks on Mr. Ogilby's Farther Observations on Rules for Nomenclature. Mag. Nat. Hist. 2 ser. vol. ii. pp. 326, 555.
    The two last Papers, not republished, refer generally to Zoological Nomenclature. See its history, Memoir, p. clxxiv.
31. On the Naturalization of *Dreissena polymorpha* in Great Britain. Mag. Nat. Hist. 2 ser. vol. ii. p. 361. Coll. Papers, p. 426.
32. On some remarkable Dikes of Calcareous Grit at Ethie in Ross-shire. Proc. Geol. Soc. vol. ii. p. 654. Geol. Trans. 2 ser. vol. v. p. 599. Coll. Papers, p. 150.

1839.

33. Queries respecting the Gravel in the neighbourhood of Birmingham. Rep. Brit. Assoc. vol. viii. Sections, p. 71. Printed at conclusion of paper 'On Nature and Origin of Transported Gravel.' Coll. Papers, p. 110.
34. Review of 'Muscicapidæ, or Flycatchers,' by W. Swainson. Mag. Nat. Hist. 2 ser. vol. ii. pp. 389, 499. Anon.

1840.

35. Memoir descriptive of a series of Coloured Sections of the Cuttings on the Birmingham and Gloucester Railway. Proc. Geol. Soc. vol. iii. p. 314. Ann. Nat. Hist. vol. vii. p. 69. Geol. Trans. 2 ser. vol. vi. p. 545. Coll. Papers, p. 132.
36. Observations upon the Affinities and Analogies of Organized Beings. Mag. Nat. Hist. 2 ser. vol. iv. p. 219. Coll. Papers, p. 401.
37. On the occurrence of a Fossil Dragon Fly in the Lias of Warwickshire. Mag. Nat. Hist. 2 ser. vol. iv. p. 301. Coll. Papers, p. 152. Memoir, p. cxcix, where the woodcut is introduced.

38. On the True Method of discovering the Natural System in Zoology and Botany. Rep. Brit. Assoc. vol. ix. Sections, p. 128. Ann. Nat. Hist. vol. vi. p. 184. Coll. Papers, p. 408.

1841.

39. Commentary on Mr. G. R. Gray's 'Genera of Birds.' Ann. Nat. Hist. vol. vi. p. 410, vol. vii. pp. 26, 159.
40. Review of Pye Smith's 'Relation between the Holy Scriptures and Geological Science,' and of Sidney Gibson's 'Certainties of Geology.' Ann. Nat. Hist. vol. vii. p. 429. Anon.
41. [— and W. J. Hamilton.] On the Geology of the Western Part of Asia Minor. Geol. Trans. 2 ser. vol. vi. p. 1. Coll. Papers, p. 29.
42. Descriptions of Cuttings across the Ridge of Bromsgrove Lickey, on the Line of the Birmingham and Gloucester Railway. Abstract. Proc. Geol. Soc. vol. iii. p. 446. Coll. Papers, p. 143.
43. On the occurrence of the "Bristol Bone-Bed" in the Lower Lias near Tewkesbury. Proc. Geol. Soc. vol. iii. pp. 585, 732. Ann. Nat. Hist. vol. x. p. 147, vol. xi. p. 502. Coll. Papers, p. 154.
44. Report of the Committee for making Experiments on the Preservation of Vegetative Powers in Seeds. Rep. Brit. Assoc. vol. x. p. 50, vol. xi. p. 34, vol. xii. p. 105, vol. xiii. p. 94, vol. xiv. p. 337, vol. xv. p. 20, vol. xvi. p. 145, vol. xvii. p. 31, vol. xviii. p. 78, vol. xix. p. 160, vol. xx. p. 53, vol. xxi. p. 177, vol. xxii. p. 67. Ann. Nat. Hist. vol. viii. p. 77. Memoir, p. clxxii.
45. On the Genus Cardinia, Agassiz, as characteristic of the Lias Formation. Abstract. Rep. Brit. Assoc. vol. x. Sections, p. 65. Ann. Nat. Hist. vol. xiv. p. 100. Coll. Papers, p. 173.

    In reference to the passage in this Paper, p. 175, regarding Agassiz's opinion of the extent of the distribution of Cardinia: in his German edition of Sowerby's Mineral Conchology, he places the shells referred to Unio by Sowerby, along with them. See Paper by Capt. Portlock, R.E., Ann. Nat. Hist. vol. xv. p. 343.
46. Letter signed "A Briton" against opening the Ports to Slave-grown Sugar. Worcestershire Chronicle Newspaper, June 2, 1841.

1842.

47. Review of Mr. G. R. Gray's 'Genera of Birds,' second edition. Ann. Nat. Hist. vol. viii. pp. 367, 544.
48. On the occurrence of *Mergulus alle* at Ipswich. Ann. Nat. Hist. vol. viii. p. 394.
49. On some New Genera of Birds. Proc. Zool. Soc. pt. ix. p. 27. Ann. Nat. Hist. vol. viii. p. 520. Coll. Papers, p. 229.
50. Review of Gould's 'Birds of Australia.' Ann. Nat. Hist. vol. ix. p. 337. Anon.

51. On the occurrence of a flock of *Sterna arctica* in the Midland Counties. Ann. Nat. Hist. vol. ix. pp. 352, 518.
52. Review of Bonaparte's 'Iconografia della Fauna Italica.' Ann. Nat. Hist. vol. x. p. 127. Anon.
53. On some remarkable Concretions in the Tertiary Beds of the Isle of Man. Proc. Geol. Soc. vol. iv. p. 8. Ann. Nat. Hist. vol. xi. p. 507. Coll. Papers, p. 170. Memoir, p. cclv.
54. On certain Impressions on the surface of the Lias Bone-Bed in Gloucestershire. Proc. Geol. Soc. vol. iv. p. 16. Ann. Nat. Hist. vol. xi. p. 511. Coll. Papers, p. 167.
55. Report of a Committee appointed to consider of the Rules by which the Nomenclature of Zoology may be established on a uniform and permanent basis. Rep. Brit. Assoc. vol. xi. p. 105, vol. xii. p. 119. Ann. Nat. Hist. vol. xi. p. 259. Phil. Mag. ser. 3. vol. xxiii. p. 108. Coll. Papers, p. 375.
56. On the occurrence of *Halcyon smyrnensis* in Asia Minor. Rep. Brit. Assoc. vol. xi. p. 70. Ann. Nat. Hist. vol. ix. p. 441. Coll. Papers, p. 316.
57. On the ancient Colossal Statue near Magnesia. Archæologia, vol. xxx. p. 524. Coll. Papers, p. 434.

1843.

58. Remarks on a Collection of Drawings of Australian Birds, the property of the Earl of Derby. Ann. Nat. Hist. vol. xi. p. 333.
59. List of the Birds in the Chinese Collection at Hyde Park Corner. Proc. Zool. Soc. pt. x. p. 166. Ann. Nat. Hist. vol. xii. p. 220.
60. On the Structure and Affinities of Upupa, Linn., and Irrisor, Less. Ann. Nat. Hist. vol. xii. p. 238. Rep. Brit. Assoc. vol. xii. Sections, p. 70. Coll. Papers, p. 418.
61. Notes on Catalogues of the Birds of Corfu and of Crete by H. M. Drummond. Ann. Nat. Hist. vol. xii. pp. 412, 423.
62. Description of a Chart of the Natural Affinities of the Insessorial Order of Birds. Rep. Brit. Assoc. vol. xii. Sections, p. 69. Memoir, p. cciii.

1844.

63. Notes on Mr. Blyth's List of Birds from the vicinity of Calcutta. Ann. Nat. Hist. vol. xiii. p. 32, 204, vol. xiv. p. 54, 114.
64. Review of the Catalogues of Mammalia and of Birds in the British Museum. Ann. Nat. Hist. vol. xiii. p. 380. Anon.
65. Descriptions of several new or imperfectly defined Genera and Species of Birds. Ann. Nat. Hist. vol. xiii. p. 409. Coll. Papers, p. 235.
Figures of several of the birds described, drawn upon stone by

Strickland, accompanied the original paper; but the stones havin been destroyed, these could not now be reprinted.

66. On the evidence of the former existence of Struthious Birds, distinc from the Dodo, in the Island near Mauritius. Proc. Zool. Soc. pt. xii. p. 77. Ann. Nat. Hist. vol. xiv. p. 324.
67. On certain Calcareo-corneous bodies found in the outer chambers of Ammonites. Quart. Journ. Geol. Soc. vol. i. p. 232. Coll. Papers, p. 181.
68. On *Thalassidroma melitensis,* Schembri, a supposed new species of Stormy Petrel. Ann. Nat. Hist. vol. xiv. p. 348.
69. Review of Malherbe's 'Faune Ornithologique de la Sicile.' Ann. Nat. Hist. vol. xiv. p. 431. Anon.
70. Descriptions of some new species of Birds brought by Mr. L. Fraser from Western Africa. Proc. Zool. Soc. pt. xii. p. 99. Ann. Nat. Hist. vol. xv. p. 125.
71. Report on the recent progress and present state of Ornithology. Rep. Brit. Assoc. vol. xiii. p. 170. Coll. Papers, p. 247.
72. On an anomalous Structure in the Paddle of a species of Ichthyosaurus. Abstract. Rep. Brit. Assoc. vol. xiii. Sections, p. 51.

1845.

73. On *Cyanocitta,* a proposed new genus of *Garrulinæ,* and on *C. superciliosa,* a new species of Blue Jay, hitherto confounded with *C. ultramarina,* Bonap. Ann. Nat. Hist. vol. xv. p. 260, 342.
74. Review of Jerdon's 'Illustrations of Indian Ornithology.' Ann. Nat. Hist. vol. xv. p. 274. Anon.
75. Review of Kaup's 'Classification der Saügethiere und der Vögel.' Ann. Nat. Hist. vol. xv. p. 422. Anon.
76. On two species of Microscopic Shells found in the Lias. Quart. Journ. Geol. Soc. vol. ii. p. 30. Coll. Papers, p. 186.
77. [— and J. Buckman.] Outline of the Geology of the Neighbourhood of Cheltenham by Sir R. I. Murchison; new edition, augmented and revised. 8vo. Cheltenham, 1845. Memoir, p. clxvii.
78. On the Results of recent researches into the Fossil Insects of the Secondary Formations of Britain. Rep. Brit. Assoc. vol. xiv. Sections, p. 58. Coll. Papers, p. 187.

1846.

79. On the Structural Relations of Organized Beings, read before the Ashmolean Society. Phil. Mag. ser. 3. vol. xxviii. p. 354. Coll. Papers, p. 348.
80. On the Use of the word Homology in Comparative Anatomy. Phil. Mag. 3 ser. vol. xxix. p. 35.
81. On the Satellitary Nature of Shooting Stars and Aërolites. Phil. Mag. ser. 3. vol. xxix. p. 1. Coll. Papers, p. 429.

82. Notes on Capt. W. J. Begbie's 'Observations on the Natural History of the Malayan Peninsula.' Ann. Nat. Hist. vol. xvii. p. 395.
83. Translation of and Notes on 'The Birds of Calcutta, collected and described by C. J. Sundevall.' Ann. Nat. Hist. vol. xviii. pp. 102, 168, 251, 303, 397, 454; vol. xix. pp. 87, 164, 232.
84. Exhibition of a new species of Corvus, discovered by Capt. H. D. Drummond. Proc. Zool. Soc. 1846, p. 43.
85. Notes on certain species of Birds from Malacca. Proc. Zool. Soc. 1846, pt. xiv. p. 99. Ann. Nat. Hist. vol. xix. p. 129.

1847.

86. Notes on 'Drafts for a Fauna Indica,' by E. Blyth. Ann. Nat. Hist. vol. xix. pp. 41, 98, 179.
87. Letter signed H. E. S., on Phonotypics. Athenæum, Jan. 9, 1847.

1848.

88. Letters on Anastatic Printing. Athenæum, Feb. 12, 1848, p. 172; March 11, 1848, p. 276.
89. On the present state of Knowledge of the Geology of Asia Minor. Phil. Mag. ser. 3. vol. xxxii. p. 137. Coll. Papers, p. 26.
90. Review of Dr. Daubeny's 'Description of active and extinct Volcanos, of Earthquakes, and of Thermal Springs.' Phil. Mag. ser. 3. vol. xxxii. pp. 216, 296. Anon.
91. [— and A. G. Melville, M.D.] The Dodo and its kindred, or the History, Affinities, and Osteology of the Dodo, Solitaire, and other extinct Birds of the Islands Mauritius, Rodriguez and Bourbon. 4to. London, 1848.
92. On Papyrography and its applicability to the illustration of Ornithology. Jardine's Contributions, No. 2. vol. i. p. 18.
93. Illustrations of Ornithology. Jardine's Contributions, 1848, i. pp. 23, 60.
94. Bibliographia Zoologiæ et Geologiæ. A general Catalogue of Books, Tracts, and Memoirs, on Zoology and Geology. By Louis Agassiz. Corrected, enlarged and edited by H. E. Strickland. Printed for the Ray Society. 8vo. London, 1848–1854.
95. On Anastatic Printing. Proc. Ashmolean Soc. vol. ii. p. 184.
96. On the Geology of the Oxford and Rugby Line. Proc. Ashmolean Soc. vol. ii. p. 192. Coll. Papers, p. 184.
97. On the Improvement of the Drainage of the Isis Valley. Proc. Ashmolean Soc. vol. ii. pp. 190, 200.
98. On Phonetic Alphabets, with suggestions for their improvement. Proc. Ashmolean Soc. vol. ii. p. 215.

1849.

99. Supplementary Notices regarding the Dodo and its kindred. Ann. Nat. Hist. ser. 2. vol. iii. pp. 136, 259, vol. iv. p. 335, vol. vi. p. 290. Rep. Brit. Assoc. 1849, p. 81.
100. On the Habits of a living specimen of *Nanina vitrinoides* (Desh.). Proc. Zool. Soc. pt. xvi. p. 142, Moll. pl. ii. figs. 1, 2, 3. Ann. Nat. Hist. ser. 2. vol. iv. p. 379.
101. Illustrations of Ornithology. Jardine's Contributions to Ornithology, 1849, ii. pp. 17, 33, 60, 91.
102. On the various applications of Anastatic Printing and Papyrography, with illustrative examples by P. H. Delamotte. 8vo. London, 1849.
103. On Vegetable Remains in the Keuper Sandstone of Longdon, Worcestershire. Rep. Brit. Assoc. 1849, p. 66.
104. Letter signed H. E. S., on Modern Hexameters. Athenæum, June 23, 1849, p. 644.
105. Letter signed " Sigma," on Female Emigration. Morning Chronicle, Dec. 31, 1849.

1850.

106. A few Dodo Queries. Notes and Queries, vol. i. p. 261.
107. Notes on a Section of Leckhampton Hill. Quart. Journ. Geol. Soc. vol. vi. p. 249. Coll. Papers, p. 189.
108. On the occurrence of *Charadrius virginiacus*, Borkh., at Malta. Ann. Nat. Hist. ser. 2. vol. v. p. 40. Coll. Papers, p. 319.
109. Kordofan Birds, procured by Mr. J. Petherick, with Notes by H. E. Strickland. Proc. Zool. Soc. 1850, p. 214. Coll. Papers, p. 357.
110. On Accent and Quantity, and on the practicability of introducing Ancient Metres into Modern Languages, conformably to the laws of Prosody. Coll. Papers, p. 437.
111. Illustrations of Ornithology. Jardine's Contributions to Ornithology, 1850, pp. 47, 120, 147.
112. *Vidua paradisea*. Jardine's Contributions to Ornithology, 1850, pp. 88, 149.

1851.

113. On the Elevatory Forces which raised the Malvern Hills. Phil. Mag. November 1851, ser. 4. vol. ii. p. 359. Coll. Papers, p. 192.
114. Ornithological Notes. Jardine's Contributions to Ornithology, 1851, pp. 15, 103.
115. Review of Prince L. C. Bonaparte: Schlegel, Monographie des Loxiens. Jardine's Contributions to Ornithology, 1851, p. 27.
116. Notes on some Birds from the River Gaboon, in West Africa. Jardine's Contributions to Ornithology, 1851, pp. 131, 161.

1852.

117. On Geology in relation to the Studies of the University of Oxford. 1852. Coll. Papers, p. 207.
118. On a Protruded Mass of Upper Ludlow Rock at Hagley Park in Herefordshire. June 2, 1852. Quart. Journ. Geol. Soc. vol. viii. p. 381. Coll. Papers, p. 199.
119. On the Distribution and Organic Contents of the "Ludlow Bone-Bed" in the districts of Woolhope and May Hill; with a Note on the Seed-like bodies found in it by J. Hooker, M.D. Dec. 1, 1850. Quart. Journ. Geol. Soc. vol. ix. p. 8. Coll. Papers, p. 161.
120. Notice of recent discoveries of the Foot-prints of Extinct Animals in Ancient Formations. Proc. Ashmolean Soc. 1852, p. 321.
121. Ornithological Notes. Jardine's Contributions to Ornithology, 1852, pp. 28, 41, 127, 161.
122. [— and P. L. Sclater.] List of a Collection of Birds from the Damara Country. Jardine's Contributions to Ornithology, 1852, p 141.

1853.

123. On Pseudomorphous Crystals of Chloride of Sodium in Keuper Sandstone. Rep. Brit. Assoc. 1853, p. 61. Quart. Journ. Geol. Soc. vol. ix. p. 5. Coll. Papers, p. 204.
124. On the Partridges of the Great Watershed of India. Rep. Brit. Assoc. 1853, p. 71.
125. On the Mode of Growth of *Halichondria suberea*. Rep. Brit. Assoc. 1853, p. 72.

SCIENTIFIC WRITINGS

OF THE LATE

HUGH EDWIN STRICKLAND.

MEMOIRS ON THE GEOLOGY OF ASIA MINOR.

1. ON THE GEOLOGY OF THE THRACIAN BOSPHORUS.

[*Read November* 30, 1836.]

[This Memoir, with the two which follow, were drawn up soon after Mr. Strickland's return from Asia Minor, and abstracts of them were published in the 'Proceedings;' the complete papers were printed in the Transactions of the Geological Society for 1836–37.—Ed.]

During a short residence at Constantinople, in March 1836, I made some researches into the geology of its vicinity. A very cursory examination sufficed to show that the district was unexplored, and that the brief geological remarks which exist in the works of other travellers are very unsatisfactory. I therefore communicate my observations to the Society, in the hope that, though slight, they may serve as a guide to those geologists who may hereafter be induced to explore this interesting district * †.

* These observations were made in company with Mr. Hamilton; but that gentleman being still on his travels, I have been deprived of his direct assistance in drawing up this paper.

† On June 5, 1837, M. de Verneuil communicated to the Société Géologique de France a memoir on the Geology of the Bosphorus, which completely confirms the present one, though both papers were published without a knowledge of the other. —H. E. S. ms.

The formations in the neighbourhood of the Bosphorus may be classed as follows:—

1. Silurian schist and limestone.
2. Igneous rocks.
3. Tertiary limestone.
4. Ancient alluvium.

1. *Silurian Schist and Limestone.*

This formation occupies both sides of the Bosphorus for about three-quarters of its length, and extends thence towards the west-north-west and east-south-east, to an unascertained distance in Europe and Asia. It may probably form a great part of the Balcan range, of which the hills above the Bosphorus are a continuation.

Section of west coast of the Bosphorus from the Sea of Marmora to the Euxine.

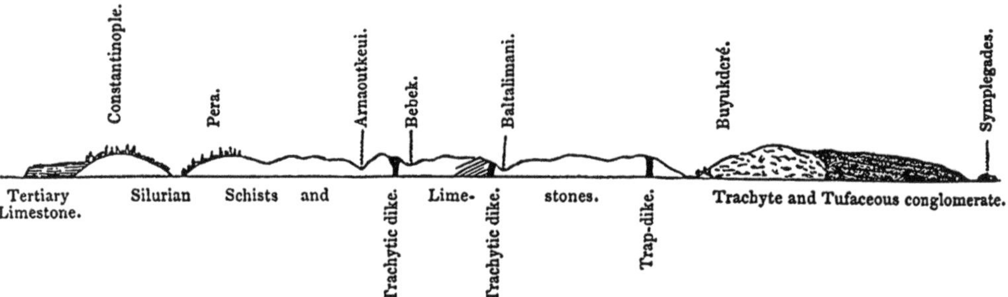

This formation may be described as a mass of argillaceous schist, compact brown sandstone, and compact dark blue limestone, passing into each other by insensible gradations. The argillaceous schist predominates greatly over the other two rocks, and frequently exhibits an oblique or slaty cleavage. The whole formation bears a close analogy to the lower Silurian and upper Cambrian series, exhibited in Wales and Devonshire. From its resemblance to the type of Northern Europe, Andréossy* and an American traveller† referred it to the transition series on mineralogical grounds alone; and this opinion is now confirmed by the discovery of organic remains, analogous to those of the Silurian rocks of Britain. This fact is interesting, from the extreme rarity of fossils in deposits of this early age in the southern parts of Europe.

The stratification of these rocks is in general much disturbed and contorted; and in other cases it is very obscure and difficult to detect.

* Essai sur le Bosphore. 8vo, Paris, 1818.

† Sketches of Turkey, by an American. 8vo, New York, 1833.

Hence it would require a closer examination than I was able to give, to establish the prevailing *strike* of the formation.

Organic remains appear to be extremely local, and were noticed in only two localities. The first of these is in the ravine above Arnaout-keui, a village about four miles from Pera, on the European side. The rock is an argillo-calcareous schist, splitting at an angle oblique to the stratification. Impressions of various Brachiopoda and joints of Crinoideæ occur in it, lying in the planes of stratification. The substance of the shells has perished, leaving a brown dust; and the casts are in general much distorted by the compression and consolidation of the strata.

These remains were first noticed by M. Fontanier*, who mistook the shells for Pectines, and thence inferred that this was a secondary formation. Had he examined them with more care, he would probably have agreed with M. Andréossy, who, without the assistance of organic remains, referred these rocks, as already stated, to the transition era.

The only other locality where organic remains were seen, is on the hill called the Giant's Mountain, on the Asiatic side of the Bosphorus, about fifteen miles from Constantinople. It is singular that of three authors (Andréossy, Fontanier, and the author of 'Sketches of Turkey') who have noticed with more or less exactness the geology of this hill, not one has remarked the numerous and interesting fossils which occur there†.

The lower part of the hill consists of compact limestone, thickly bedded, of a dark grey colour, and slightly crystalline. It is exposed in extensive quarries, where it is traversed by some remarkable trap-dikes, to be mentioned hereafter. Organic remains are rare in the limestone; but I detected a few impressions of Terebratulæ and Crinoidal stems, exhibiting the same crystalline fracture which distinguishes them in the carboniferous and other limestones of England.

Above the limestone is an argillaceous schist with an oblique cleavage, and it extends to the summit of the hill. Impressions of shells, similar to those of Arnaout-keui, occur in it in many places, and in tolerable abundance. I have found only one specimen which can be referred to the family of Trilobites. It is a small fragment, but it so accurately exhibits the structure of the eye of an Asaphus, that I have no hesitation in referring it to that genus.

The following is a list of the fossils of this formation, as far as the imperfection of the specimens admits of identification. I am indebted

* Voyages en Orient. 8vo, Paris, 1829.

† Add to these, De Verneuil in Bull. Soc. Géol. France, vol. viii. p. 268.—H. E. S. MS.

to Mr. Murchison and Mr. James De Carl Sowerby for assistance in determining these fossils.

Brachiopoda.

1. *Spirifer.*—Allied to *Spirifer speciosus,* Bronn, Leth. Geog. ii. 15.
2. *Spirifer.*—An undetermined species.
3. *Producta (Leptæna) sericea.*
4. *Producta (Leptæna) euglypha?*, Dalm.
5. *Producta (Leptæna).*—New species.
6. *Terebratula affinis,* Min. Con.—In the Upper Silurian series of Great Britain.
7. *Terebratula unguis?*—In the Caradoc sandstone of Great Britain.
8. *Atrypa.*—An undetermined species.
9. *Orthis.*—New species.
10. *Orthis.*—New species.

Crustacea.

11. *Asaphus.*—Eye like that in Buckland's Bridgewater Treatise, pl. 45. fig. 10'.

Crinoidea.

12. Detached joints of undetermined genera.

Polyparia.

13. *Cyathophyllum,* Goldf.
14. *Favosites,* Lam.—Similar to one in the Cambrian rocks.

Should any geologist ever devote that share of time and attention to this country which it well deserves, he would probably find several other localities for these Silurian fossils, and perhaps be rewarded by the discovery of many Trilobites and other organic remains, which I did not meet with.

The compact limestone of this formation is quarried in several places near the Bosphorus, especially at Baltalimani, where it dips to the westward. In the lower part the strata are thick, but they become thinner upwards, and, after assuming a concretionary structure, pass into argillaceous schist. The limestone exhibits some very obscure traces of organic remains.

The sandstones are best exhibited in the cliff between Scutari and Mondabornou. They are brown, commonly compact, sometimes schistose, and the stratification is distinct, being much broken and distorted.

The evidence of the organic remains justifies me in referring the formation to the Silurian system of Britain. The occurrence in this remote locality of rocks which approach so closely to the type of north-western Europe is somewhat remarkable; and the more so, as nothing of the kind occurs, as far as I know, in the more southern parts of Asia Minor, or in any portion of the Mediterranean basin. Many geologists are of opinion, that the transition and secondary rocks of Northern Europe belong to a different type from those of the Mediterranean basin; and that the characters, both mineral and zoological, of these two parallel series present scarcely any common points of comparison. A line, nearly coinciding with the Pyrenees and Alps, appears to divide these two great basins, if we may so term them. Further investigations may perhaps show, that this line admits of being continued from the Alps through European Turkey to the range of the Mysian and Bithynian Olympus; and this may explain the appearance at Constantinople of rocks belonging to the type of Northern Europe, which are wanting further south in Asia Minor.

The transition formations which we have been considering, unite on the north to a mass of igneous rocks, and on the south-west to tertiary deposits. At present, I have no clue to the relative age of the two latter, but as the igneous rocks are in more immediate relation with the Silurian group, I will notice them next.

2. *Igneous Rocks.*

The igneous rocks of the Bosphorus may be described, in the aggregate, as consisting of trachyte and trachytic conglomerate. On the Asiatic side they commence, *en masse*, at Kavak under the old Genoese castle, and extend thence to Yoom-bornou on the Black Sea, or perhaps further. They consist chiefly of angular trachytic fragments, imbedded in a tufaceous paste. The conglomerates, viewed on the large scale, are distinctly stratified; and from Kavak to Anadoli-fanar they have a regular inclination towards the north. They rest upon, or alternate with, trachytic rocks, more or less compact, and occasionally pass into phonolite and basalt. In some places, especially near Anadoli-fanar, dikes of the latter substance intersect the conglomerate beds. The basalt occasionally assumes a columnar form, which is seen in the greatest perfection at Yoom-bornou. The prevailing colour of these rocks is greenish, owing to the presence of copper, a circumstance which gave the name of Cyaneæ to the weather-beaten rocks of the Symplegades.

Near Filbornou the conglomerate contains boulders of a brownish trachyte, softer than the greenish trachytic paste which envelopes them; hence they decompose with the weather, and give a honey-combed appearance to the rock. The contrary, however, is more often the case, when the imbedded fragments, being harder than their matrix, project above the general surface of the rocks.

These igneous formations contain many veins of red and white cornelian, and varieties of chalcedony. In one place near Filbornou the veins are perpendicular and straight, intersecting the conglomerate, and cutting indiscriminately through the tufaceous paste and the imbedded fragments.—See the accompanying woodcut.

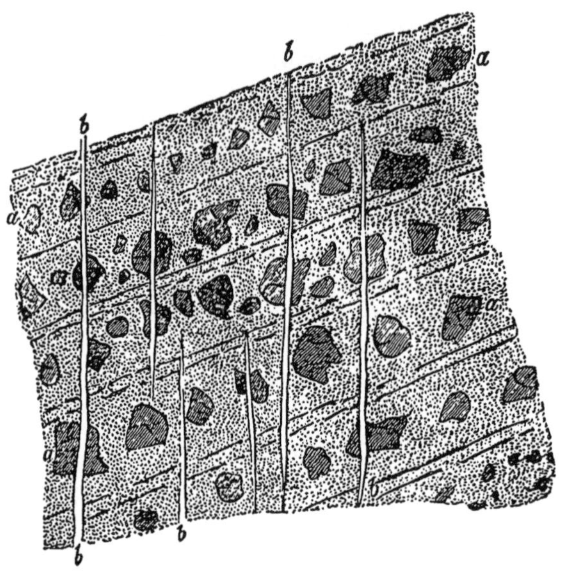

a, a, Angular trachytic fragments imbedded in a tufaceous paste.
b, b, Veins of chalcedony traversing both the fragments and the paste.

The deposition of these conglomerates appears to have been effected with the assistance of, if not solely by, water. The fragments imbedded in them, though commonly angular, are sometimes rounded, and are contained in finely laminated strata of volcanic sand. An instance of this, on the north side of Anadoli-fanar, is represented in the following woodcut.

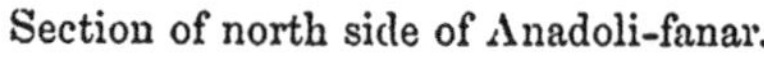
Section of north side of Anadoli-fanar.

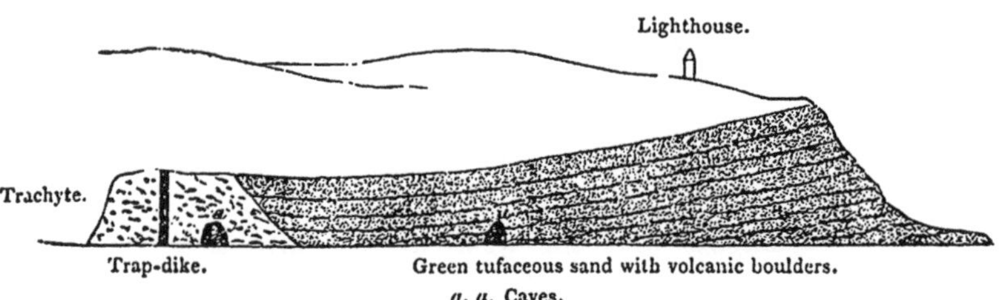

a, a, Caves.

On the European side of the Bosphorus, the igneous rocks commence on the north of Buyukderé. Here a species of trachyte of a yellow colour forms some considerable hills. It is much decomposed, and resembles loose sand when viewed at a distance. Further north is a series of conglomerate beds, similar to those of the Asiatic side, and extending to the Black Sea*.

These are the general features of the mass of igneous rocks which abuts against the northern flank of the Silurian schists above described. The works of MM. Andréossy and Fontanier contain some information on their mineral contents.

A variety of trachytic and trap-dikes occur in the Silurian schists, and are doubtless connected with the general mass of igneous formations. At Baltalimani and on the hills above Bebek, a ferruginous decomposing trachyte appears in the midst of the schistose rocks. At Kiretch-bornou is a dike of fine-grained greenish trap, about 15 feet in thickness, cutting through strata of calcareous and argillaceous schist. A similar dike occurs in the limestone quarries at the base of the Giant's Mountain. It is perpendicular, 4 or 5 feet thick, and from the removal of the limestone on each side, it resembles a wall running across the quarry. Some fragments of limestone occur in the midst of the trap, but neither they nor the rock on each side of the dike, show any signs of alteration.

Lower down in the quarry, near the level of the Bosphorus, are some masses of greenish and white quartz rock, which have more the appearance of an altered sandstone than of an igneous injected rock.

3. *Tertiary Formation.*

This formation commences immediately on the west of Constantinople, and extends along the north coast of the Sea of Marmora for many miles, its western limit being at present undetermined. On the west side of Constantinople the formation ranges inland for about three miles, till it meets the transition series, but the nature of the junction is not exposed by any section which I could meet with.

The beds are best exhibited in the quarries at Baloukli and Makrikeui, where they consist of soft white shelly limestones and marls, resting on sand without fossils. The stratification is in general horizontal, a character common to nearly all the tertiary formations which I saw in Turkey.

That portion of this deposit which occurs near Constantinople appears to have been accumulated in an estuary. The most abundant shell is an oval, subtriangular bivalve, apparently a Mactra, which composes almost the entire substance of some of the beds; there is also a ribbed bivalve resembling a Cardium. These are apparently the only fossils

* M. Texier, in Comptes Rendus, vol. iv. p. 465, says the trachytic rocks extend inland as far as Belgrade (where I saw nothing of them) and Kila.—H. E. S. MS.

which belonged to marine genera, all the rest having been natives either of the land or of fresh water.

Potamides, 1 species.
Neritina, 1 species, the delicate black stripes remaining.
Cyclostoma, 1 species.
Helix, a species like *H. pomatia*.
——, a species resembling *H. rufescens*.
Bulinus, resembling *B. montanus*.
——, similar to *B. acutus*.
Clausilia, 1 species.
Pupa, 1 species.
Planorbis, like *P. alba*.
Cypris, 1 species.

The terrestrial shells are remarkable for their near approach to existing forms. They are less generally diffused than the marine and freshwater species, but are abundant in the quarries at Makri-keui.

A single specimen of a bone occurred in this formation, but it is too imperfect for me to decide whether it belongs to a mammifer or to a reptile.

I found no traces of a tertiary deposit along the banks of the Bosphorus, whence it may be inferred that this channel has been opened at a comparatively late period. Tertiary beds are said to exist along the shores of the Euxine; and lignite has been worked for fuel at Domouzderi, a few miles west of the Bosphorus.

4. *Ancient Alluvium.*

A few miles north of Constantinople are some extensive deposits of clay, sand and gravel, reposing on the Silurian schist. The latter formation is barren and devoid of trees, but the alluvial clay favours the growth of the oak and causes the forest of Belgrade. These deposits are of great thickness, and consist chiefly of ferruginous clay, with pebbles and boulders of quartzose and sandstone rocks interspersed. The general want of stratification and the rounded form of the boulders denote an aqueous action of considerable violence. These accumulations of detritus appear to skirt the southern side of the Lesser Balcan range towards the north-west.

Taken as a whole, the deposits bear a close analogy to those which in England have been called diluvium; and they appear to have resulted from causes acting on a greater scale than any of those, which produced the local alluvia, noticed by me in Asia Minor.

2. ON THE GEOLOGY OF THE NEIGHBOURHOOD OF SMYRNA.

[*Read April* 5, 1837.]

The vicinity of Smyrna presents geological phænomena of considerable interest, and of a more complicated kind than is usual in Asia Minor. A residence there during the winter of 1835–36 enabled Mr. Hamilton and myself to investigate this district with greater attention than we could bestow upon the other parts of Asia Minor which we visited; and its geology may, therefore, deserve to be laid before the Society in a separate communication.

The details of the paper can be understood only by reference to the accompanying map, Pl. I. (Pl. xxxii. Geol. Trans. 2nd ser. v. p. 393), and by a general description of the geographical features of the district.

The Bay of Smyrna and the alluvial plain at its upper end lie between two parallel ridges of high land, ranging from east to west. That on the north commences with the lofty hill of Sipylus, rising abruptly from the plain of Sardis; and this ridge, which separates the waters of the Hermus from the Bay of Smyrna, terminates at Menimen near the estuary of that river, the western part being known by the name of Cordileon. Its total length is about thirty-two miles.

The ridge on the south side of Smyrna Bay, begins with Mount Tartali, the ancient Mastusia, which is the western termination of the Tmolus range. Descending towards the west, it passes into the table-land which overhangs Smyrna on the south. Further west the country again rises, and forms the group of conical mountains, anciently called Corax, which terminates at the Isthmus of Vourla. The distance from Mount Tartali to Vourla is about twenty-four miles.

These two lofty ranges are united at the Pass of Cavakluderé by a transverse ridge, which separates the Vale of Nimphi from that of Smyrna, and completely isolates the latter.

The rocks in the vicinity of Smyrna belong to the following formations:—

1. Micaceous schist and marble.
2. Hippurite limestone and greenish schist.
3. Tertiary lacustrine limestone and marl.
4. Trachytic rocks.

The general features of these and the other formations of Asia Minor were explained in a paper laid before the Society on November 2, 1836*.

In the Smyrna district, the mountains of the hippurite limestone formation apparently once contributed to form the boundary of a great

* See p. 29. (Proc. Geol. Soc. ii. p. 423.)

lake, in which calcareous matter was deposited to the depth of many hundred feet. Two vast eruptions of igneous matter have since taken place, one on the north, the other on the south of the present Bay of Smyrna. These convulsions produced great changes in the form of the surface and in the arrangement of pre-existing rocks, and the study of this district is thus rendered both interesting and difficult.

As the phænomena exhibited on the north of the Bay of Smyrna are not visibly connected with those on the south side, it will be best to consider each portion of the district separately. I will begin with the southern side, and take the formations in the order of age.

1. The micaceous schist and saccharine marble, which compose the ridge of Mount Tmolus, terminate near the village of Trianda, and form part of the eastern boundary of the lacustrine formation. The micaceous schists of Asia Minor being fully described in the memoir before alluded to, it is needless to repeat their characters in this paper.

2. The compact grey limestone with Hippurites is, in this district, associated with an abundance of compact sandstone and schist, of black, greenish, or cream-coloured hues. The whole formation has been so greatly disturbed, that it is difficult to determine whether the sandstone and schists were deposited above or beneath the limestone; but their union with it is too intimate to admit of their being regarded as a distinct formation. They seem analogous to the greenish sandstones of the Morea and the *macignos* of Tuscany and Trieste, all of which have been referred to the cretaceous system.

The western boundary of the hippurite limestone of Mount Tartali is at the village of Cucklujah, whence it extends eastward for perhaps fifteen or twenty miles, along the northern declivity of Mount Tmolus. Further details respecting it at this locality will be found in the memoir alluded to.

The range of Mount Corax consists almost entirely of sandstones and shales, but at one point, about 3½ miles west of Smyrna, a small mass of grey limestone is exposed, and serves to identify these rocks with those of Mount Tartali.

Around the village of Baltchikeui the shales are of a pale cream colour, soft and friable, and contain a considerable admixture of sand, which is sometimes so coarse as to become a conglomerate of fragments of quartz, a quarter of an inch in diameter.

About 1½ mile beyond Baltchikeui is a thermal spring, on the site of a temple to Apollo mentioned by Strabo. In the ravine immediately above it, the beds of shale are dark-coloured and compact, dipping thirty or forty degrees to the northward*.

* The author of 'Sketches of Turkey' calls them limestone rocks, but on the application of acid not the slightest effervescence takes place.

3. Tertiary lacustrine limestone.

This deposit forms an extensive table-land, ranging southward from Smyrna for about fifteen miles. On the east it abuts unconformably against the hippurite limestone of Tartali and the micaceous schist of Tmolus, and on the west it is bordered by the range of Mount Corax. The southern boundary is near Trianda, where the formation rises above the alluvial plain of Tourbalí; but the original barriers of the lake on that side are not now apparent. This area is occupied principally by white or yellowish limestone, varying in texture from lithographic stone to soft chalk. The more compact beds closely resemble the secondary limestones of the Ionian Islands, and like them contain nodules and layers of black flint and quartz resinite. From this resemblance we were at first inclined to refer the compact limestone near Smyrna to the secondary series, but after a close search Mr. Hamilton and myself succeeded in finding lacustrine shells, p. 18 (Geol. Trans. 2nd ser. v. p. 402), which indicate a tertiary epoch. The flints and quartz resinite exhibit an analogy to those in the lacustrine deposits of Auvergne and of the Cantal, as described by Mr. Lyell and Mr. Murchison*.

White and greenish marls are often interstratified with the limestone, and are most abundant in the central parts of the area between Boudjáh and Sedikeui. About two miles north-east of the latter place the marls contain traces of vegetable remains, and are accompanied by extensive beds of gravel, which both alternate with and overlie them. These gravel beds much resemble the ancient drift or alluvium of England, but are merely a local portion of the lacustrine deposit. They contain rolled pebbles of nummulitic limestone and schist, derived from the surrounding mountains, also fragments of red trachyte similar to that which occurs, *en masse*, near Smyrna.

As is usual with the lacustrine basins of Asia Minor, the beds of limestone and marl pass into conglomerate as they approach the foot of the surrounding mountains. The southern boundary of this formation near Trianda consists almost wholly of gravel imbedded in yellow clay.

The pebbles of trachyte, in the central parts of the formation, appear to prove that, at this point, the lacustrine deposits were continued after the eruption of trachyte, which overlies them near Smyrna. The bed of the lake being probably lowest in the centre, water might lodge there, after the main body of the lake had been drained by the igneous eruption; and streams flowing from all sides would carry to this point the pebbles which, during the quiescent state of the lake, formed a shingle beach around its margin. In course of time the denudation of the outlet would complete the work of desiccation, and the whole drainage, of what is now the Plain of Sedikeui, would reach the sea by the ravine at

* Annales des Sciences Naturelles, 1828.

present traversed by the Meles. That all these operations have taken place, will, I think, be evident to any one who will make a careful survey of the locality.

The stratification of the lacustrine series is, in general, nearly horizontal; some local exceptions, however, exist in the vicinity of the igneous rocks.

4. Trachytic rocks.

The greater part of the lacustrine series had been deposited, when the basin of Smyrna became the seat of volcanic action; and an enormous mass of trachytic matter was poured forth, and spread over the bed of the lake. The eruption appears to have ceased as suddenly as it began, for there are no alternations of trachytic and lacustrine rocks,—no injection of dikes, or other phænomena, which indicate a long continuance of igneous action.

These rocks occupy a surface of about five miles by two, and a reddish-brown, porphyritic variety prevails throughout nearly the whole area. A remarkable feature in the trachyte is the close approach which it occasionally makes to a stratified structure. It splits into slabs from a foot to an inch in thickness, and their cross section exhibits stripes of various colours, parallel to the planes of cleavage. These strata (if they may be so called) are in some places accumulated to the thickness of 100 feet, and are traceable laterally for as many yards. This is especially the case on the hills about two miles west of Smyrna. The crystalline texture of this *stratified trachyte* decidedly indicates igneous fusion, and the explanation of this peculiar structure is, therefore, by no means easy. It may, however, be conjectured, that when the fluid trachyte expanded horizontally from the point of eruption, any variations of substance or texture would be drawn out and extended in planes parallel to the horizon, and when cooled would cause coloured stripes and fissures in the rock, in the direction of those planes. It must, however, be observed, that the apparent strata are now far removed from the horizontal position, supposing them to have once occupied it.

These trachytic rocks are, in general, homogeneous, but in some places they contain numerous angular blocks and fragments of black porphyritic trachyte, much harder than the reddish paste in which they are imbedded. This species of conglomerate is well developed, about two miles west of Smyrna, in a ravine near the sea.

On the north side the trachytic rocks descend to the shore, but along their southern border they overlie the lacustrine limestone and marl of the Plain of Sedikeui. The boundary of the two formations is well defined, the trachyte terminating abruptly, with a steep slope, like the margin of a modern *lava coulée*. A stratum of pumiceous tufa is commonly interposed between the trachyte and the lacustrine strata. The uppermost bed of the latter is a conglomerate of rounded pebbles of quartz, hippurite limestone, and schist, but *not the least trace of any igneous rock is discoverable in it*. Beneath this bed are calcareous marls, containing Planorbes and other freshwater shells, and passing downwards into white limestone.

The absence of igneous products in the conglomerate proves the sudden eruption of the trachyte which reposes on it; yet it is possible that this bed of transported pebbles may be connected with that event. Earthquakes preceding the eruption may have caused currents in the lake, which would wash these rolled pebbles from its gravelly margin into its central parts, and terminate that long period of repose, during which the subjacent marls and limestones were deposited.

The highest point attained by the trachyte is about three miles southwest from Smyrna, near the valley which separates these rocks from the Corax range. The view from this eminence proves how much the picturesque and botanical features of a country depend on its geology. On the south is the vast, undulating plain of lacustrine limestone and marl, the greater part of which is cultivated and interspersed with villages. On the west, the decomposing sandstone and shale of Corax form a cluster of conical mountains thickly covered by evergreen shrubs. In the east, the view is terminated by Mount Tartali, where the compact grey hippurite limestone forms crags and precipices, with fir and other trees sparingly scattered among them. Lastly, the trachytic rocks in the foreground produce neither trees nor shrubs, but present innumerable masses of naked rock, jutting through a fine green turf, adapted only for pasturage.

Having now given a general sketch of the sedimentary and volcanic rocks which occur near Smyrna, I will introduce a more detailed account of Mount Pagus, where several of them are brought to view within a very small space. Pl. I. (Pl. xxxii., Sect. Nos. 1 and 2, Geol. Trans. 2nd ser. v.)

The hill anciently called Pagus rises immediately to the south of the modern town of Smyrna. It forms the north-eastern extremity of the mass of trachytic rocks above described; and a deep ravine separates it on the east and south from the *plateau* of lacustrine deposits, whose drainage produces the classic river Meles, which escapes into the sea through this ravine.

The height of Mount Pagus may be estimated at about 500 feet. It is principally composed of trachyte and trachytic conglomerate, which on the east and south extend from the summit to the bed of the Meles; but along the northern declivity they terminate about half-way down, and repose on strata of the lacustrine series. The uppermost of these freshwater beds is here, as elsewhere, a conglomerate of grey limestone, schist and quartz, in rolled fragments. It passes downwards into beds of sand which may be traced by means of ravines and sand-pits for a depth of more than 100 feet, till the houses of Smyrna preclude further search. At the lowest visible point, the sand contains abundant impressions of leaves; also, though more rarely, imperfect shells of Helix and Unio, which identify it with the lacustrine formation.

It is remarkable that these beds of sand and conglomerate dip about 20° *towards* the trachyte of Mount Pagus, which having been erupted subsequently to their deposition, might be expected to have heaved them in an opposite direction. They are clearly not *overturned,* because the conglomerate of limestone pebbles is here in its regular position, on the top of the other lacustrine strata.

At the north-eastern side of the hill, the sand is replaced by beds of marl and white limestone containing Planorbes; and above them is the never-failing conglomerate, extending under the trachyte. Pl. I. (Pl. xxxii., Sect. No. 1, Geol. Trans. 2nd ser. v.)

A few yards lower down, at the foot of the hill near the Caravan Bridge, we come unexpectedly upon a rugged mass of rocks belonging to the hippurite limestone formation. Its whole area does not much exceed an acre, and we must suppose that it has been brought up by some convulsion, connected with the outbursts of trachyte in the vicinity. It consists of friable, marly, cream-coloured shale, similar to that at Baltchikeui, and it contains veins of crystalline carbonate of lime, and some masses of a hard, yellowish, calcareo-siliceous stone. The stratification is very obscure, but as far as it could be determined, it dips towards Mount Pagus, conformably with the lacustrine marls, a few yards higher up. These friable schists are extensively quarried, and when broken small are used in Smyrna for garden walks and terraced roofs. On the east side of the schist are some irregular masses of compact grey limestone, similar to that which, in Mount Tartali, contains Hippurites; and they identify the whole of these dislocated rocks with that formation.

About half a mile north-east of this point, is a small, isolated hillock of trachyte, the elevation of which may have caused the otherwise anomalous dip of the lacustrine strata, on the north and north-east sides of Mount Pagus. Pl. I. Sect. 1.

The right bank of the Meles, opposite Mount Pagus, presents an escarpment of the lacustrine formation. Pl. I. Sect. 2. The trachytic rocks on this side are of small extent, but rise, in one place, about half-way up the south side of the ravine, and appear to abut against the edges of the lacustrine strata, which form the upper portion of the hill. The highest stratum exposed is a compact white limestone, containing, though rarely, a small species of Paludina. A singular substance is also found here, being an aggregate of coarse crystals of carbonate of lime, but too loosely united to constitute saccharine marble*.

Below the limestone containing Paludinæ, strata of marl with calcareous nodules are exposed, till they are obscured by the trachyte below. The only alteration apparent in the lacustrine strata, from their contact with the igneous rock, consists in a greenish colour imparted to the marl, limestone, and accompanying flint or resinite, and not observable at a distance from it.

In descending from this point to the ford through the Meles, a singular mass of conglomerate is found in the midst of the trachyte. It consists of rolled quartz pebbles, imbedded partly in a red and partly in a bright green paste. This mass, which is but a few feet in extent, seems to have been derived from the subjacent lacustrine beds, and caught up by the trachyte. Near this point, the latter rock contains some veins of white chalcedony.

We now proceed to the geology of the north side of Smyrna Bay, where phænomena are exhibited similar to those which are described above.

The grey limestone of Mount Sipylus seems to have formed part of the northern margin of that ancient lake, which was bounded on the east by Mount Tartali, and on the west by Corax. Lacustrine strata still flank the south side of Mount Sipylus; and they probably once extended across the alluvial plain from Bournabat to Cucklujah. Whether that extensive vale has been formed by denudation or by subsidence is uncertain, but its origin is probably connected with the igneous action which has convulsed the district.

A vast eruption of trachytic matter, apparently contemporary with that near Smyrna, has on this side also broken up and overlaid the lacustrine deposits. It forms the western half of the Sipylus range, called Cordileon, a mountain whose height may be estimated at about 2000 feet above the sea.

§ 1. The eastern part of Sipylus consists of compact grey limestone, rising precipitously from the plain of the Hermus. Further westward, the limestone is accompanied by black and greenish shales, resembling those of Tartali, before mentioned. The boundary between this formation and the trachytic rocks coincides in general with the ravine which descends from the lake of Kizghioul to Bournabat, but considerable intermixture of these two rocks occurs along the line of junction.

§ 2. The lacustrine beds repose against the southern slope of Mount

* Strata of the same substance are seen in ascending by the easternmost road from Smyrna to Boudjah.

Sipylus, from Djaki-keui, in the Vale of Nimphi, to Bournabat, where they are covered up by the trachytic rocks. The most interesting sections of them are on the west side of the ravine, about half a mile north of Bournabat; and the right bank of the torrent here exhibits the phænomena shown in Section 4. A small outburst of red trachyte has heaved up strata of white and greenish marl, containing concretionary masses of brown, hard, crystalline limestone. Some of these masses are thickly perforated with sinuous tubes, and resemble travertin formed around reeds; yet it is doubtful whether the tubes originated in any organic body. Above these strata is a succession of whitish calcareous marls, extending about 200 yards to the northward, and dipping 35° N. In one place, a thin bed of brown clay is interposed, and contains fragments of shells and vegetables.

The upper part of these marly beds affords a rich mine to the student in fossil botany; some of the strata being crowded with leaves and other portions of plants, perfectly preserved. The substance of the leaves commonly remains of a ferruginous colour, and contrasts beautifully with the pale cream-coloured matrix.

The leaves belong to about twelve species of trees; and some of them appear referable to the genera *Laurus, Nerium, Olea, Salix, Quercus,* and *Tamarix,* all which still flourish in Asia Minor; but it is yet undetermined whether any of them are identical with existing species.

These fossil vegetable remains are accompanied by shells of the genera *Cyclas, Paludina, Planorbis,* and *Cypris.*

Lacustrine marls occupy the west side of the ravine, to the height of perhaps 150 feet above the spot where the fossil plants occur. Their junction with the overlying igneous rocks affords the following interesting series of beds. Pl. I. (Pl. XXXII., Sect. Nos. 1 & 3, Geol. Trans. 2nd ser. v.)

1. Brown porphyritic trachyte capping the hill, and continuous with that which forms the mountain of Cordileon.
2. Alternating beds of tufaceous conglomerate and sand, regularly stratified, and resembling the conglomerate of Mount Perier, near Issoïre, in France, except that the trachytic fragments are less rolled—30 to 40 feet.
3. Tufa, with numerous fragments of decomposed pumice—8 or 10 feet.
4. Beds of rolled pebbles, from the size of an egg downwards, of grey limestone, quartz and schist, but no igneous substances; and similar to the conglomerate which underlies the trachyte on the south side of Smyrna Bay—20 to 30 feet.
5. White and yellow marls, containing, towards the upper part, *Planorbis, Lymnæa, Paludina,* and *Cyclas.* They extend downwards to the bed of the torrent.

All the strata, from No. 2 to 5 inclusive, are conformable, and dip about 20° S.W.

§ 3. Trachytic Rocks.—These extend westward from the torrent of Bournabat to the mouth of the Hermus, and compose, as before stated, the whole mountain of Cordileon. The prevailing rock is a reddish-brown trachyte, identical with that of Smyrna, and, like it, is occasionally divided into laminæ resembling strata.

The junction of these rocks with the lacustrine beds near Bournabat is already described. In following up the ravine to the lakes of Kizghioul and Karaghioul, several varieties of trachyte and tufaceous rocks occur along the junction of the Hippurite limestone and schist. At the lake of Kizghioul, a branch from the igneous rocks extends to the eastward for some distance. The Section, Pl. I. Sect. 5, is here exposed.

About four miles further west, immediately opposite Smyrna (see Map), a long, narrow ridge descends the side of Cordileon, and divides the waters of two torrents. The composition of this ridge differs greatly from that of the brown trachyte of the surrounding hills. It consists of decomposing felspar, principally of a white or yellow colour, yet it presents also various shades of red and brown. It has a soapy feel, and would perhaps be a valuable material for pottery. At the upper end of the ridge, near a village, is a steep, broken escarpment, where some dikes of bluish and reddish trachyte penetrate the light-coloured substance above described.

The soft, decomposing state of this rock has given that smooth, rounded form to the ridge which so strongly contrasts it with the rugged neighbouring hills. Its surface is covered with rounded, erratic boulders of brown trachyte, probably brought down by the torrents, which now flow on each side of the ridge, at a time when they occupied a higher level.

From an inspection of hand specimens, it would not be easy to decide whether this white earthy trachyte has been altered since its ejection, or whether it was originally poured forth in a different state from the ordinary brown trachyte. The latter opinion, however, is the most probable, if we consider, first, the complete dissimilarity between this rock and those which immediately surround it; and secondly, the manner in which this long ridge descends, like a *coulée*, from the higher parts of Cordileon.

These are the principal phænomena which I noticed on the north side of the Bay of Smyrna; and it will be seen that they present a great analogy to those on the south.

The geological events to be inferred from the facts described in this paper may be summed up as follows:—

1. An elevation of several mountains, composed of rocks of the cretaceous age.

2. A long and tranquil deposition of lacustrine matter, with shells and vegetables, in a depression between these mountains.

3. A sudden eruption, at two principal points, of igneous matter, which broke up and overlaid the deposits of the lake, and drained the greater part of its waters. It seems to have been preceded by a commotion in the lake, spreading *non-volcanic* pebbles over the bottom.

4. A more or less sudden cessation of the igneous action, proved positively by the homogeneous character of the erupted rocks, and negatively by the general absence of injected dikes, or of alternate *coulées* of trachyte, tufa, &c., such as are seen in Mont D'Or.

5. A continuance of aqueous deposition in the central and lowest point of the lake after the igneous eruption, proved by beds of marl, alternating with gravel, containing trachytic pebbles.

6. The drainage of this remnant of the lake by the denudation of its outlet, now traversed by the Meles.

List of Fossil Shells noticed in the Lacustrine Basin of Smyrna.

SPECIES.	LOCALITIES.
1. *Helix*, similar to *H. carthusiana*, Drap.	North side of Mount Pagus.
2. *Planorbis*, similar to *P. alba*	Ravine above Bournabat. A point about 4 miles S.W. of Smyrna.
3. *Lymnæa*, resembling the recent *L. peregra*	N.E. side of Mount Pagus. Ravine above Bournabat.
4. *Paludina*, a small turreted species, like the recent *P. acuta*	Ravine above Bournabat. N.E. side of Mount Pagus. South side of the Meles, towards Boudjah.
5. *Cyclas*, a small species, like *C. pusilla*	Ravine above Bournabat, in company with vegetable impressions.
6. *Unio*, an ovate species	North side of Mount Pagus.
7. *Cypris*	Ravine above Bournabat.

The specimens of those fossils which were obtained, are not sufficiently perfect to warrant a greater degree of precision in their determination. Indeed, fossil freshwater shells rarely possess characters sufficiently marked to form the basis of accurate specific distinctions.

---

3. ON THE GEOLOGY OF THE ISLAND OF ZANTE.

[*Read November* 1, 1837.]

THE observations detailed in this paper are the result of a few days' residence in the Island of Zante, during which I was enabled to take a general view of its geological features; but a complete survey of this and the other Ionian Islands is still to be desired. Such an investigation would be a work of labour, for though there is little variety in the rocks, there is much complexity in their arrangement. The structure of Zante, more simple than that of the other islands, presents an epitome of their component rocks in an almost unbroken succession; and it may, therefore, be selected as a type to which the phænomena of the other islands may be referred.

The geological phænomena of Zante may be arranged under the three heads of,—1. Apennine Limestone; 2. Tertiary Deposits; and 3. Mineral Springs. Plate II. (Pl. XXXIII. Geol. Trans. 2nd ser. v.)

1. *Apennine Limestone.*—This is perhaps the most convenient appellation for that deposit of compact white or greyish limestone, which is so largely developed in the South of Europe, and especially on the shores of the Adriatic. It has a uniform character throughout many thousand feet of vertical thickness, and many hundred miles of horizontal extent. The few fossils it contains agree with those of the cretaceous, and in part also of the oolitic series of Northern Europe. It constitutes, in Zante, an anticlinal ridge, extending in a north-north-west and south-south-east direction along the south-western coast, from Point Skinari to Point Cheri, and this ridge is continued through the island of Cephalonia. Along the eastern side the prevailing dip of the strata is from 30° to 45° to the east-north-east; west of Point Skinari an opposite dip commences, and continues, with few exceptions, to near Point Cheri, where the strata again dip to the eastward.

The tertiary beds occur only on the east of this ridge. On the west we find a series of cliffs, upwards of 600 feet high and almost perpendicular, the sea-worn caves and fir-clothed crags of which present highly picturesque scenery. This steepness is continued to a great depth beneath the surface of the sea, as is proved by the deep soundings along this coast.

The Apennine limestone, of which these cliffs are composed, is nearly white, and less compact than usual, often resembling the hard chalk of the north of England. No beds of flint were noticed here, though in Corfu they are not unfrequent. Organic remains are by no means abundant, yet Nummulites, fragments of Hippurites, and indistinct traces of other fossils may frequently be detected by searching.

From its compact, inflexible, and brittle texture, the Apennine limestone generally abounds with faults and fractures, which give rise to numerous caverns (*catavothra*), subterranean rivers, and thermal and mineral springs. In these respects, no less than in its mineral structure, it presents a close analogy to the Carboniferous Limestone of Northern Europe, for which it has often been mistaken by observers, who paid no attention to its fossils.

The frequency and violence of earthquakes in many parts of the South of Europe may perhaps be accounted for by the unyielding texture of this rock, the vibrations being propagated to much greater distances than in countries composed of more loosely aggregated or more elastic materials*.

In Zante the Apennine limestone presents numerous faults, one or two of which will be alluded to in a subsequent page.

2. *Tertiary Beds.*—These occupy the greater part of the island of Zante. Reposing on the eastern flank of the Apennine limestone, they extend to the coast: they rise also in several detached hills through the alluvial plain which forms the centre of the island. They have evidently yielded to the same disturbing force as the limestone range; and they dip from this rock to the eastward. The upper portion is the counterpart of the beds, described by Mr. Hamilton and myself, as occurring near Lixouri in Cephalonia†. In Zante, they are best displayed in the Castle Hill above the town, and in the cliff which extends thence to the eastern coast. Pl. II. (Pl. XXXIII. Sect. 1, Geol. Trans. 2nd ser. v.) The upper strata near the Lighthouse consist of a porous, calcareo-arenaceous stone, of a pale yellow colour, and easily worked. Fossils are rare in it, except on the east coast, where one or two of the strata contain numerous casts of Cerithia and other mollusca.

These strata are succeeded by a thick deposit of blue clay and marl, forming the height on which stands the citadel of Zante. The shells found in it are principally *Pectunculus auritus*, Broch., *Buccinum semistriatum*, Broch., and *Natica glaucina*, Lam. All these species occur also in the middle portion of the Lixouri section, but more abundantly, and associated with many others.

The gypseous beds, which at Lixouri succeed the argillaceous strata, are not visible in this part of Zante, but on the south coast they form

* Dr. Davy, however, draws an exactly opposite conclusion, asserting that the earthquakes in the Ionian Islands are always most severe in the marly tertiary districts—H. E. S. MS.

† See Geological Proceedings, No. 51, and *postea*.

the commencement of a section which carries us much further down in the series than the lowermost beds examined at Lixouri. Pl. II. (Pl. xxxiii. Sect. 2, Geol. Trans. 2nd ser. v.) The gypsum also occurs in white and conspicuous patches on the hill at the south-east extremity of the island. The rest of this hill consists of sand and clay belonging to the upper part of the series; but the beds are much disturbed, and not easily reducible to the regular arrangement seen in Sections 1 and 2.

The uppermost beds in Section 2 consist of gypseous marls and gypsum, sometimes fine-grained and saccharine, but sometimes only a coarse aggregate of selenitic crystals like that at Lixouri. The stratification is occasionally preserved, though in others it appears to have been obliterated by the action of crystallization. Angular fragments of a black marlstone, imbedded in the gypsum, seem to have been derived from strata of stone, broken up by the force of the crystallizing process.

The strata of yellow limestone above the gypsum, exhibited in Section 1, and at Lixouri, clearly belong to the Pliocene epoch, many of their fossils being identical with those of the Subapennine hills. The strata which underlie the gypsum in Section 2, consist of a series of brown sandy clays and marls, but whether they also belong to the Pliocene or to a prior epoch, it is not easy to determine. They extend for about two miles along the coast, and dip about 25° to east-north-east, with a few local interruptions. Fossils are very rare in these beds, and in general they are too much crushed to allow the species to be determined. They were noticed only near the middle of the argillaceous series and near its base. At the former spot are crushed fragments of Echini and obscure bivalves; and at the latter is a bed of indurated bluish marl containing an abundance of the shells of Hyalæa and Creseis, but they are larger than those of the species now living in the Mediterranean (*Hyalæa cornea* and *Creseis spinifera*), and are therefore probably distinct species.

The argillaceous beds are succeeded by yellowish calcareous sandstone and loosely aggregated limestone. A great subsidence appears to have taken place between that point and the range of secondary limestone, about a mile distant. This tract forms the marshy plain of Port Cheri, towards which the tertiary strata dip on both sides. There is consequently no traceable sequence between the argillaceous beds above described and the calcareous strata which we are now considering. The latter dip about 18° south-west, and extend along the east side of the marsh, forming some hillocks at its upper end. They consist in general of calcareous particles, interspersed occasionally with pebbles of secondary limestone; but some of the beds approach the texture of Portland stone. Minute Foraminifera are abundant in it, and the only other fossils noticed were two species of small Pectens.

These calcareous rocks seem referable to a distinct epoch, and may perhaps eventually prove to be of the Meiocene or even of the Eocene age. The fine-grained limestone, extensively quarried near Lixouri in Cephalonia, belongs probably to this part of the tertiary series.

On the west side of Port Cheri is a low cliff of blue marl and clay, the beds of which abut against the secondary or Apennine limestone, and dip about 18° north-east. The only fossils noticed in it were a few scales and vertebræ of fish, and a species of Vermiculum, Mont. (*Quinqueloculina*, D'Orb.)

This small, argillaceous mass has been probably derived from a higher part of the tertiary series, and brought down to its present position by the subsidence, which seems to have formed the valley and bay of Port Cheri. Of this depression, there is further proof in a remarkable fault, which occurs in the Apennine limestone, and is marked by a smooth surface of the rock descending to the sea. It may be traced inland in a direction west-north-west for half a mile or more, rising like a wall above the downcast portion on the north-east side. At the point where it joins the sea the surface is nearly a plane, inclined about 55°. It is scored with numerous striæ, inclined at an angle of 65° to the horizon, the dip of the strata being about 25° north-east.

The enormous friction and pressure of the descending mass have imparted to the surface of rock a remarkable degree of hardness, and a darker colour than usual. This change of character penetrates to the depth of about two or three inches from the surface; the rock below being softer and white, and resembling the compact chalk of Yorkshire. The tertiary beds range from Port Cheri northwards along the foot of the limestone, and reappear on the north shore about two miles beyond the village of Catastari; they are shown in Section 3, which is in some respects a counterpart of Section 2, but presents differences which it is not easy to explain. The porous yellow limestone, which at Port Cheri intervenes between the argillaceous beds and the secondary rocks, is here wholly absent, and the tertiary clay appears to pass gradually into the secondary limestone. The highest beds in the section, consisting of blue marl with shells of Creseis and Hyalæa, are the precise equivalents of that which contains these fossils near Port Cheri; and we are thus furnished with a common point of departure in our comparison of Sections 2 and 3. They are succeeded (Section 3) by numerous beds of blue clay and marl, apparently destitute of fossils, becoming more calcareous in the lower part, and ultimately passing into a white limestone resembling hard chalk. A stratum of conglomerate, used for millstones, occurs here, and consists of rolled pebbles of compact Apennine limestone; beneath this are other strata of compact limestone, undistinguishable from that of the secondary mountain range. The beds above described are conformable throughout, and seem to pass downwards into the secondary limestone; whether this is really the case, cannot, however, be determined from the section before us, for the sequence is again interrupted by an extensive fault (seen at the left-hand of Section 3). It is therefore possible that the limestone beds below the conglomerate may be considerably above the true base of the tertiary series, and that an *hiatus* may exist between them and the real secondary limestone, which they so much resemble. A careful survey of the line of junction between the secondary and tertiary formations throughout the island would perhaps solve the difficulties presented by this section.

§ 3. *Mineral Springs.*—The springs of bitumen, for which Zante has been celebrated from the time of Herodotus, rise in the marsh at Port Cheri: see Pl. II. Sect. 2. The principal one is a well, about 5 feet deep; the bitumen oozing up from the bottom; and above it the well is filled by clear, cool, and tasteless water, which is probably only an accidental accompaniment of the bitumen. Some travellers (Walsh, Chandler, &c.) state that bubbles of gas are given out by the bitumen, but in two visits which I made to the spot, nothing of the kind was observed. The produce has been stated at forty barrels annually*. Bitumen also rises in the Bay of Cheri, some hundred yards from the shore. Pl. II. This circumstance proves that the bitumen is not derived from the peaty soil of the marshy plain, and there is nothing in the composition of the rocks around to induce us to refer its origin to them; we must therefore suppose that this substance is derived from that region of volcanic action which may be almost demonstrated to underlie the Ionian Islands. This supposition derives further probability from the fact, before noticed, that the spot where the bitumen rises has been the site of a vast dislocation.

On the northern coast is another remarkable mineral spring, which seems to have escaped the notice of previous observers. It occurs about half a mile to the north of the junction of the tertiary and secondary rocks, shown at the left-hand of Pl. II. Sect. 3. The Apennine limestone here forms a low cliff descending abruptly to the sea. A spring of turbid water, resembling diluted milk, gushes out at the foot of the cliff beneath the sea-level, and rising to the surface, from its less

* For further details relative to the "tar-springs," see Hawkins in Walpole's 'Travels in the East;' Chandler's 'Travels,' vol. ii. ch. 79, &c.; also Davy's 'Ionian Islands,' vol. i. p. 154.

Dr. Davy states that bubbles of gas (consisting chiefly of carburetted hydrogen) are given out by stirring the bitumen at the bottom, though they do not arise spontaneously. He found the water in the northern spring to be brackish, sp. gr. 1·006, resembling sea-water, but with a little sulphate of lime. The water in the southern spring (described above) was sp. gr. 1·0011, and contained the same ingredients, but in much less degree. The tar consists of naphtha, petroleum and bitumen. Total produce 60 to 100 barrels annually.—H. E. S. MS.

specific gravity, flows away above the sea-water in a stratum a few inches thick; flakes of a slimy white substance (probably glairine) abound in this water, and are seen in the surrounding sea for a considerable distance; a strong smell of sulphuretted hydrogen is diffused around, but no bubbles of gas rose in the water, nor was there any appearance of inflammable gas on the application of a lighted taper*.

On the 30th of May, 1836, the temperature of the air in the shade being 82° Fahr., of the adjoining sea at the surface 73°, and at the bottom in 4 fathoms, 69°, the spring indicated a temperature of 65°. This is so near the mean temperature of the latitude of Zante, that the mineral spring cannot be regarded as thermal. A bottle of the water is now in my possession, but I have not yet ascertained its chemical ingredients; from its close resemblance, however, to the mineral waters of many volcanic regions, as those of the "Aquæ Albulæ" near Rome, we must refer its origin to some analogous cause. Of this we have a further proof in the fact, that this spring rises on the line of a considerable fault which has affected the Apennine limestone at right angles to its strike. The upcast is on the south, and presents a smooth and almost polished surface of rock, rising like a wall at an angle of about 80°, and running in a straight line for about a quarter of a mile inland. See the Plan, Pl. II. It projects some distance into the sea, and the spring above described rises in a recess a few yards to the south. Two smaller springs of similar turbid water issue at the base of the smooth face of rock. There can be little doubt that the outburst of these springs is owing to this fault, which has opened a passage from the abyss in which they originate.

We have, then, in this mineral spring, an additional indication of the existence of a region of volcanic action at some vast depth beneath the Ionian Islands,—of which there is already much presumptive evidence in the springs of bitumen, the frequent earthquakes, and, above all, in the current of sea-water absorbed into a chasm in the neighbouring island of Cephalonia†. It is, however, somewhat remarkable that no rocks of igneous origin exist, as far as is known, throughout these islands.

I cannot conclude this imperfect contribution to the geology of the Ionian Islands, without expressing a wish that some competent geologist

* This spring is described by Dr. J. Davy in 'Notes on Ionian Islands and Malta,' vol. i. p. 148. He made its temperature 62° Fahr. He analysed the Glairine, and found it to consist of sulphur and a substance allied to animal mucus. The specific gravity of the water he made 1·01103. His analysis otherwise nearly agrees with Dr. Daubeny's. He states that this spring is also described by St. Sauveur in his account of the Ionian Islands.—H. E. S. MS.

† See Geological Proceedings, vol. ii. pp. 220, 393.

would undertake an accurate survey of them. Such an undertaking would be well worthy the attention of our Government, and would form an important appendix to the splendid survey which has been already effected by the French Government in the Morea.

Postscript to the Paper on the Geology of Zante.

Since the printing of the paper on the Geology of Zante, I have been favoured by Dr. Daubeny with the following analysis of the mineral spring described at p. 23:—

Specific gravity, 1·020.
Solid contents in one pint, 174 grains.
Ferrocyanite of potash produced no effect, proving the absence of iron.
Barytic salts produced a cloud, proving the presence of sulphuric acid.
Nitrate of silver produced a dense cloud, proving much common salt.
Oxalate of ammonia produced a cloud, proving the presence of lime.
Phosphoric test caused evident indications of magnesia.

From the above analysis it appears that the water of this spring differs but little from ordinary sea-water. Indeed, from the manner in which it rises in the sea, a large quantity of sea-water must unavoidably become mixed with it. At the same time, its inferior specific gravity, its milky colour, the flocculi of glairine, and the strong smell of sulphuretted hydrogen, all serve to characterize it as a mineral spring analogous to those of volcanic regions.

4. ON THE PRESENT STATE OF KNOWLEDGE OF THE GEOLOGY OF ASIA MINOR*.

[The following short paper gives a good view of the state of European information in regard to Asia Minor up to the date of its publication. It was called forth in consequence of an extract of a letter from M. Tchihatcheff which Sir Roderick Murchison read to the Geological Society, and which was afterwards printed in their 'Proceedings†.' M. Tchihatcheff claimed to himself more than he was entitled, and challenged the accuracy of Mr. Hamilton and others; and Mr. Hamilton also found it necessary to remark upon these observations, "as an act of justice to himself and his fellow-traveller‡."]

In the last Number of the Journal of the Geological Society, part 2, p. 74, is a letter from M. von Tchihatcheff, extracted from Leonhard and Bronn's 'Neues Jahrbuch,' 1847. I rejoice to find from it that this gentleman is about to undertake a systematic geological survey of Asia Minor, a country which, from the magnificent scale on which its secondary and tertiary rocks are displayed, and the wonderful diversity of its volcanic phænomena, is probably inferior to none of equal area in geological interest. Our knowledge of the geology of Asia Minor is, in truth, comparatively limited, and we may therefore look for results of the highest value from M. Tchihatcheff's researches. But although much remains to be done by the geologist in Asia Minor, yet we are not wholly without information on this subject; and as it might be inferred from M. Tchihatcheff's silence as to the labours of others that such was the case, I have thought it desirable to give a brief summary of the progress that has already been made in this branch of inquiry.

The existence of a tertiary marine formation on the shores of the Dardanelles was made known nearly half a century ago by Olivier, a scientific zoologist, who recognized in this pliocene deposit many existing species of Mediterranean shells, the names of which he has enumerated.

The slaty rocks of the Thracian Bosphorus, flanked by volcanic rocks on the north, and by tertiary beds on the south-west, have long been known. The occurrence of fossils in the former rocks was noticed by Fontanier ('Voyages en Orient'). In 1836 Mr. W. J. Hamilton and myself proved, by means of these fossils, that the formation was Silurian; and in a paper by myself in the Transactions of the Geological Society, vol. v. p. 385, on the Geology of the Thracian Bosphorus, p. 1, the district between the Sea of Marmora and the Euxine is described in some detail. In the following year the same region was explored by M. de Verneuil, whose researches (published in the 'Bull. Soc. Géol. de France') entirely confirm those which we had previously made.

* London and Edinburgh Philosophical Magazine, vol. xxxii. p. 137.

† Proceedings of the Geological Society, vol. v. p. 360. ‡ Ibid. p. 362.

The vicinity of Smyrna was geologically explored by Mr. Hamilton and myself during the winter of 1835–36, and the results are given in my memoir on that district, p. 9 (Geol. Trans. vol. v. p. 393). In this paper will be found the first attempt at a *classification* of the geological formations of Asia Minor.

The classification is further carried out in a joint memoir which we published, p. 29 (Geol. Trans. vol. vi. p. 1), on the Geology of the Western part of Asia Minor, in which we described the southern shores of the Sea of Marmora, the valleys of the Macestus, the Rhyndacus, the Hermus, the Cayster, and the Mæander, besides giving short notices of Erythræ, Boodroom, Cnidus and Rhodes. The most interesting of the districts here described is unquestionably the Catacecaumene, the volcanic phænomena of which were illustrated by coloured maps, sections and landscapes. After my return to England Mr. Hamilton penetrated to Armenia, and returned through the interior of Asia Minor to Smyrna, during the whole of which lengthened journey he kept careful notes of all the geological phænomena which came in his way. These facts will be found duly recorded in his work, entitled 'Researches in Asia Minor, Pontus, Armenia, with some account of their Antiquities and Geology.' The same journey supplied him with the materials for his memoir in the Geol. Trans. "On the Geology of part of Asia Minor between the Salt Lake of Kodj-hissar and Cæsarea of Cappadocia, with a description of Mount Argæus." The latter mountain was ascended by Mr. Hamilton, and its height ascertained by the barometer.

The geological survey of the neighbourhood of Smyrna was extended westward along both shores of the Gulf, and over the peninsula of Karabournou, by Lieut. Spratt, and the fossils which he collected have been described by Prof. E. Forbes (Journ. Geol. Soc. vol. i. p. 156).

The same gentlemen have given a short notice of the geology of Lycia in the Journ. Geol. Soc. vol. ii. p. 8; and in their joint 'Travels in Lycia,' their observations are to be found in greater detail, and are embodied in the beautiful map of Lycia which accompanies the work.

The geology of Rhodes and of Samos has been described by Lieut. Spratt (Proc. Geol. Soc. vol. iii. p. 774, and Journ. Geol. Soc. vol. iii. p. 65).

Mr. Warrington W. Smyth has described the mining districts of the Eastern Taurus in Journ. Geol. Soc. vol. i. p. 330. And lastly, Dr. Daubeny, in his work on Volcanoes, just published, has devoted an entire chapter to the volcanic phænomena of Asia Minor.

Besides these more elaborate treatises, a variety of scattered hints and notices on Anatolian geology may be collected from the works of Fontanier, Andréossy, Beaufort, Texier, Ainsworth, Fellows, and others. From these multifarious sources I commenced, three or four years ago, to construct for my own use a general geological map of Asia Minor. Of course it is a very fragmentary production, and the numerous blank spaces in it show how much we have yet to learn as to the geology of that country.

I trust, however, that I have now shown that the geology of Asia Minor is not so completely untrodden a field as M. Tchihatcheff's letter would seem to imply; and having thus briefly vindicated the labours of others, I shall look forward with lively interest to the valuable additions to our knowledge which we may expect from that traveller's researches.

---

5. ON THE GEOLOGY OF THE WESTERN PART OF ASIA MINOR.

[This memoir, was compiled from those read by the authors at different times. Mr. Strickland returned from Asia Minor before Mr. Hamilton, and his memoir, entitled "A General Sketch of the Geology of the Western part of Asia Minor," was read before the Geological Society, November 2, 1836, and an abstract was given in the 'Proceedings,' 1836–37, vol. ii. p. 423. Mr. Hamilton's memoir, "On the Geology of part of Asia Minor between the Salt Lake of Kodj-Hissar and Cæsarea of Cappadocia, including a brief description of Mount Argæus," was read February 21, 1838, and published in the Transactions of the Geological Society for 1839–40, vol. v. p. 583.]

CONTENTS.

PLATES.

PREFACE.

IN the spring of 1836 we proceeded from the south coast of the Sea of Marmora to Brusa*, and thence up the previously unexplored course of the river Rhyndacus†, to its sources, whence we crossed to the Plain of Hushak, and descended the valley of the Hermus to Smyrna. In the following year, one of the authors (Mr. Hamilton) followed a track nearly parallel to the above, but from thirty to fifty miles further westward, or from Cyzicus to the river Macestus, and across the Demirji range to Koola (lat. 38° 31′, long. 28° 44′). These two excursions, and a third previously made (by Mr. Strickland) into the valleys of the Mæander and Cayster, together with journeys in the valley of the Mæander, and a voyage from Smyrna to Rhodes (by Mr. Hamilton), have supplied the materials for the following sketch of the geology of a portion of Asia Minor‡.

Our principal object has been to describe correctly the phænomena

* See Map, Pl. III.

† We have in many instances, in the following memoir, used the ancient names of districts, mountain-chains and rivers, in preference to their modern Turkish denominations; the former being more generally understood as well as more exact in their application than the latter. The names of towns and villages, on the contrary, are chiefly Turkish. All the places mentioned in this memoir will be found on the new map of Asia Minor to accompany Mr. Hamilton's forthcoming journal. See also the Map, Pl. III.

‡ The memoirs from which the paper is compiled, although referring to nearly the same line of country, were drawn up by their respective authors in a very different manner; when, therefore, the question of their publication came before the Council, it was considered desirable that the several papers should be embodied in one memoir, to avoid unnecessary repetition, and that the materials should be so arranged as to give at one glance a more enlarged view of the general structure of the country.

which we observed; but as geological facts are dry and uninteresting without arrangement, we have ventured on such a classification as the rocks of Asia Minor appeared naturally to admit. We are, nevertheless, well aware how imperfect such a first attempt at generalization must be in a country which had been only partially explored, and concerning the geology of which we could obtain very little satisfactory information from external sources. We wish, therefore, the general views here advanced to be considered as provisional, and open to correction; our only source of confidence in them being derived from the general simplicity of materials and uniformity of arrangement which mark the rocks of Asia Minor, and which render the task of classification much easier than in many other districts of equal extent. It remains, however, a question of some doubt how far the relative antiquity of the metamorphic rocks can safely be depended upon; and whether they are not in some instances really secondary deposits, which have been acted upon by the volcanic and igneous influences which have operated so abundantly throughout the peninsula of Asia Minor.

Physical Structure of Western Asia Minor.

The western part of Asia Minor is thickly beset with mountains, some of which form chains of considerable extent, but others are more isolated and indefinite in form. These mountains are often alpine in their character, and present many points of interest to the general observer. They rise for the greater part abruptly from horizontal plains, the smooth and verdant surface of which is strikingly contrasted with the rugged outlines of the surrounding mountains*. It will be seen hereafter that this abrupt transition from mountain to plain is the result of peculiar geological conditions.

On the great scale, the prevailing direction of these alpine chains is nearly east and west. This *latitudinal strike* is exhibited in no less than six parallel ranges.

1. The chain of Olympus. This mountain-tract commences with Mount Ida (lat. 39° 40′), whence a range of high land extends eastward to the Mysian Olympus, the snowy summit of which is about 7000 feet above the sea. Beyond this point the same chain reappears under the ancient name of the Bithynian Olympus, and ranges eastward as far as the Halys. The chain sends numerous streams from its northern slope into the Sea of Marmora and the Euxine; and the Macestus, Rhyndacus

* The mountains are often clothed with aboriginal pine forests, and they harbour bears, wolves, jackals and panthers, whose depredations are often productive of loss to the inhabitants of the plains. These forests are the winter retreats of the Turcomans and Eurukes, tribes of nomadic herdsmen, who in summer pitch their tents over the vast unoccupied plains of this thinly peopled empire.—H. E. S. MS.

and Sangarius conduct the drainage of its southern slopes through transverse gorges, the two former into the Sea of Marmora after passing through the Lake Apollonia, and the last directly into the Euxine.

2. The chain of Temnus or Demirji (lat. 39° 5′). This range is furcated at its western end, enclosing the valley of the Caicus. These branches unite to the north-east of Thyatira, and form the chain of Demirji, which constitutes the watershed between the Hermus on the south and the Macestus and Rhyndacus on the north, and terminates in Morad Dagh, near Kutahiyah.

3. The chain of Tmolus commences with Mount Tartali near Smyrna, and forms a narrow and lofty ridge, which, after dividing the valleys of the Hermus and Cayster, passes to the southward of Koola and terminates to the south-east of Takmak.

4. The range of Messogis (lat. 38°), which commences at Ephesus, and divides the fertile plains of the Cayster and the Mæander. At its northern end it unites with the range of Tmolus, and thus completely separates the Caystrian Plain from the surrounding country. It then extends eastward to the south of the Mæander, until it joins the southern end of the chain of Bourgas Dagh, and forms the southern boundary of the great plain of Hushak and Gobek.

5. The range to the south of the Mæander, which, commencing with Mount Latmus (lat. 37° 35′), extends past the lofty height of Cadmus to the eastward, where it merges into the Taurus chain, and continues with a general easterly bearing to the Euphrates. The drainage of the northern flank of this extensive range is effected, in the western portion, by the Mæander. Further eastward is the ancient district of Isauria, the drainage of which is collected into lakes and marshes with no apparent outlet. The waters on the south side of the Taurus range flow by short courses into the Mediterranean.

6. The islands of Crete and Cyprus, though south of the district we are here describing, may be regarded as the crests of a submarine chain of mountains belonging to the same system of elevation as the five parallel ranges above described.

The only important exception to the east and west direction, which prevails in the cases above enumerated, is found in the range which extends from Morad Dagh (lat. 38° 52′, long. 30° 9′) to the south-east, where it assumes the name of Sultan Dagh, and forms the watershed between the sources of the Mæander and the great central plain of Asia Minor. This range connects the eastern end of the Demirji chain in an oblique direction to the middle of the Taurus range.

Classification of the Rocks of Western Asia Minor.

The rocks of this district, as far as we are acquainted with them, admit of being classed into the following geological groups, in ascending order:—

A.—Sedimentary Rocks.

1. Micaceous schist and marble*.
2. Cretaceous System (Hippurite limestone, &c.).
3. Tertiary marine deposits.
4. Tertiary lacustrine deposits.
5. Modern aqueous deposits.

B.—Igneous Rocks.

1. Granitic rocks.
2. Greenstone, and older trap rocks.
3. Trachytic and newer trap rocks.
4. Modern volcanic rocks.

A.—Sedimentary Rocks.

§ 1. *Micaceous Schist and Marble.* Pl. IV. and Pl. V. Sect. 1 to 8, and Sketches 12 and 13.

This series occupies a very important place in Asia Minor, constituting the greater portion of the mountain-chains which intersect the country†. It consists principally of micaceous schist, with which are associated beds of white or bluish crystalline limestone and stratified quartz-rock. The latter deposits seem to occur in no determinate order, but are interstratified with the micaceous schist, into which they often pass imperceptibly. Clay-slate is of very rare occurrence in this group of rocks, and we saw but few instances of *slaty cleavage* distinct from the stratification. Veins of white quartz are as frequent as in similar rocks of most other countries. The formation is very uniform in its characters, and the list of minerals which we found in it is very scanty. On the whole, this group of semi-crystalline deposits bears much analogy to the gneiss and mica-schists of the Scotch Highlands, and may therefore, in the absence of

* The Silurian rocks of the Bosphorus, described in Mr. Strickland's paper, p. 2 (Geol. Trans. 2nd ser. v. p. 385), probably intervene in age between the schistose rocks and the Hippurite limestone. As they have not been noticed to the south of the Sea of Marmora, we have omitted them in the above list. Fossiliferous rocks of this early age are unknown (as far as we are aware) in the whole circuit of the Mediterranean basin; and Constantinople may therefore be supposed to form the southern limit of the great geological basin of Northern Europe, throughout which these protozoic rocks are abundant.—H. E. S.

† Their elevation, however, appears to have taken place at very different periods; for though by far the most considerable portions were elevated into the present mountain-chains at periods of remote activity, and by the agency of granitic action, yet Mr. Hamilton observed many places along the coast, where these rocks had been elevated by outbursts of trachytic rocks, and at periods probably subsequent to the deposition of the cretaceous formation.

definite evidence as to age, be referred provisionally to the primary or transition epoch.

The crystalline marbles of this formation are very generally dispersed over the country *. Sometimes they are pure white, but more commonly they are striped bluish grey in the lines of stratification, or are wholly of the latter colour. Thin seams of mica often pervade the marble, which is then capable of being split into slabs; and they greatly assist the labours of the stone-mason. The vast abundance of marble in Asia Minor enabled its ancient inhabitants to carry architecture and sculpture to their highest perfection; and the numerous antiquities which remain in this interesting country, prove that in these arts the Asiatic Greeks fully equalled their kinsmen of Europe.

The quartz rock of this formation is more local than the crystalline marble. In its purest state it is white with a glassy lustre, and it is identical in appearance with the quartz rock which occurs in veins. Its origin however is very different, for it is distinctly interstratified with beds of micaceous schist. Laminæ of mica are often disseminated through the quartz, and when these predominate, the rock passes into regular micaceous schist.

As might be expected in formations of such antiquity, the stratification of this group is very irregular, both as to the amount and direction of the dip. The strike however is more constant, and commonly coincides with that of the mountain range in which the observation is made.

The following are the principal localities where rocks of this group are exhibited:—

(*a*) *The island of Proconnesus* (*Sea of Marmora*), *north of the granitic ridge of Cyzicus.*—Large quarries of white marble exist here, and have given the modern name of Marmora to the island, as well as to the adjoining sea.

(*b*) *Cyzicus.*—Near Erdek (ancient Artaki), at the south-western extremity of the promontory of Cyzicus, thick beds of argillaceous schist and crystalline marble occur, having a south and south-western dip. The upper portion of the promontory of Melanos, a mile south-south-west of Erdek, consists of crystalline marble dipping south-west and resting upon beds of argillaceous schist. The dip is quâquâversal, and evidently caused by the granite which forms the central mass of the Cyzicene peninsula.

(*c*) *Aidinjik, south of Cyzicus.* Pl. V. Sect. 1.—The strata here present a steep escarpment to the north, rising to the height of 900 or 1000 feet, while the hill slopes much more gradually to the south down to the Lake of Maniyas. The summit of these hills consists of a fine-grained crystalline marble, extensively quarried a little further eastward, whence marble was probably obtained to supply the wants of Cyzicus, one of the most splendid cities of antiquity in point of architectural decoration.

* Although the marbles of this country are in general associated with the mica-schist, yet it is possible that in some cases they may be altered rocks of a more recent date.

The marble is interstratified with beds of schist of various colours, red, black, purple, yellow and white, the whole dipping south and south-east by south at angles of 70° or 80°.

(*d*) *Mount Olympus of Mysia.* Pl. V. Sect. 2.—We did not ascend to the summit of this precipitous mountain. It appears, however, from the evidence of several travellers*, that a part at least of its loftier regions is composed of grey granite. All the lower portions of the mountain which we explored consist of rocks of the schistose group. The ravine above the town of Brusa exhibits the Section No. 2. The upper beds consist of fine-grained crystalline marble, dipping north-east about 25°. The marble passes downwards into micaceous schist, approaching the texture of gneiss; and it is penetrated by quartzose veins, which are sometimes slightly granitic.

A long ridge of schistose rocks runs to the westward from Mount Olympus, between the Rhyndacus and the Lake of Apollonia. The same formation also occupies the south side of the Rhyndacus, till we enter the lacustrine basin of Harmanjik. Throughout this district, the rugged hills of mica-schist and marble scarcely admit of cultivation, and are for the most part covered with forests of pine.

(*e*) *The chain of Demirji.* Pl. V. Sects. 1, 4, 6, and Sketch 11.—We have no evidence of the structure of the western part of this ridge; but at Mumjik, on the Macestus, the pebbles and boulders, brought down from its higher parts, consist of quartz, gneiss, and a large-grained, micaceous granite. On the low part of this ridge, south-east of Simaul, Sect. 6, these older rocks are concealed by younger sandstones. We had no opportunity of visiting the lofty insulated mountain of Ak Dagh; but its appearance so greatly resembles that of the mountains of Argæus and Hassan Dagh, as to warrant the supposition that it may be a vast trachytic mountain, analogous to those of Mont Dor and the Cantal. (See Geol. Proceedings, vol. ii. p. 651.) The ridge connecting Ak Dagh to Morad Dagh is composed of schistose rocks, and appears to be the eastern continuation of the Demirji range. This ridge, which bounds the Plain of Azani on the south, sends off, along the west side of the same plain, a branch in which marble predominates. It probably joins the mountain of micaceous schist which we crossed between Taushanli and Gozuljah. See Pl. V. Section 5.

(*f*) *Morad Dagh and Sultan Dagh.*—The branch ranging westward from Morad Dagh, between the Hermus and the Plain of Hushak, consists of schist with occasional beds of marble. The same formation com-

* Fontanier, 'Voyages en Orient,' Paris, 1829, p. 92. Seetzen, quoted in Walpole's Travels in the East, p. 113. M. Texier's personal communications; also De Verneuil in 'Bull. Soc. Géol. de France,' viii. p. 268 *et seq.*, who gives an interesting description of the geology of Mount Olympus.—H. E. S. MS.

poses Bourgas Dagh, a ridge which bounds the Plain of Hushak on the east, and is connected with Morad Dagh and Sultan Dagh. Along the west foot of this ridge we found an abundance of stratified quartz rock, passing into micaceous schist.

(*g*) *Mount Tmolus.* Pl. V. Sects. 3, 6, Sketches 12 and 13.—The portions of this range which we visited, consist of micaceous schist with occasional beds of marble. At Nimphi, about fifteen miles east of Smyrna, mines of gold and silver are recorded to have been worked in the Middle Ages; and we can hardly suppose that the traditions respecting the golden sands of the Pactolus had not some foundation in fact, although they have been unproductive since the time of Strabo*. At present no metalliferous veins are worked in Mount Tmolus, but this may be owing to the paralysing effects of ignorance and despotism, rather than to the actual scarcity of the precious metals.

The chain of Tmolus expands to the south-west of the Catacecaumene, and a branch from it crosses the Hermus at Adala, Pl. V. Sketch 13, and is probably connected with the range of Demirji. About four miles west of Koola, we found some coarse garnets in the micaceous schist,—the only examples of that mineral which occurred to us in Asia Minor. The schistose rocks near Koola also abound in crystalline marble, Pl. V. Sketch 12, which was extensively quarried in ancient times near the village of Ghieurdiz; and three lateral ridges of the same schistose rocks extend to the northward in the district of the Catacecaumene, constituting, as will be hereafter shown, some of the most remarkable features of that district. At Aktash, between Koola and Takmak, a large crystal of oxide of titanium was found, and veins of the same substance are disseminated through the micaceous schist.

(*h*) *Mount Messogis.* Pl. V. Sect. 3.—The mountains around Ephesus, anciently called Gallesus, Prion and Pactyas, abound in marble; and the ancient cities of Metropolis and Ephesus were built on solid masses of it. The town of Scala Nuova stands upon an insulated rock of blue marble, a portion of that branch of Mount Messogis which stretches toward Ephesus, while the other extends south-west toward Mount Mycale. Marble is less abundant in the ridge of Messogis Proper, and none was noticed in crossing from Aidin to Tireh. The range here consists of pure mica-schist, occasionally penetrated with ferruginous matter. On the north of Tireh, the hillocks, which jut through the alluvial plain of the Cayster, consist of schist and marble.

* We were shown at Smyrna some specimens of rich copper ore, brought from the vicinity of Mount Tartali; but we could not learn whether they came from the schistose formation or from the Hippurite limestone.

For an account of Mount Tmolus and Mount Tartali, see Mr. Strickland's memoir on Smyrna, p. 9 (Geol. Trans. vol. v. p. 393).

(*i*) *Erythræ.*—In the Bay of Ritri, opposite the island of Scio, Mr. Hamilton found a great development of grey and blue crystalline limestone, resembling that of Ephesus already described, and extending in a semicircular line round the ruins of the ancient town. In many places all traces of stratification were obliterated; but near the ruins of Erythræ, he found thin vertical beds of limestone and indurated sandstone, with veins and small masses of quartz. Some of the beds are much contorted, and the strike is nearly north and south. The Acropolis stood upon a lofty rock of red trachyte; and although the exact line of contact could not be traced, the marble was observed to be much shattered near the junction with the trachyte. The anchorage is formed by two large wooded islands, about $1\frac{1}{2}$ mile from the shore. The largest is called Karabagh, and consists of the same blue crystalline limestone; but apparently very thickly bedded, for no trace of stratification could be perceived.

(*k*) *Boodroom* (lat. 37°, long. 27° 25′).—The high hills which rise behind the town of Boodroom (anc. Halicarnassus) consist of blue, compact, semi-crystalline marble belonging to this formation. In some places it is interstratified with thin bands of siliceous limestone, or cherty beds, which being less easily acted upon by the atmosphere, stand out more prominently than the rest. As far as our imperfect knowledge of the physical geography of Caria will enable us to judge, these hills belong to a south-west branch of the fifth great mountain-chain before mentioned (p. 32), diverging off from the lofty range of Mount Cadmus.

(*l*) *Cnidus* (lat. 36° 40′, long. 27° 22′).—The promontory of Cnidus, forming the extremity of the Doric Gulf to the south, Boodroom constituting that to the north, consists of thin-bedded limestone-shale, overlaid by thick-bedded blue marble; the dip being to the south and south-west from 45° to 60°. This formation was observed several miles up the promontory, and is composed of the following beds in a descending order:—

The summit of the peninsula towards the west, consists of thin-bedded calcareous shale and a concretionary rock dipping to the south-west; and of thick-bedded grey marble containing caves and fissures, many of which are covered or filled with stalactitic calcareous deposits, resembling in colour some varieties of oriental alabaster.

To the eastward of the hollow in which stood the ancient city, the marble reappears, dipping in the same direction, and is interstratified with a hard, greenish, sandstone grit.

The hills rise rapidly towards the east-north-east, attaining, at a distance of about two miles, a height of above 2000 feet. This summit is formed by a very narrow ridge of rocks, a quarter of a mile in length from north-west to south-east, consisting of thinly laminated calcareous shales, dipping south-west at an angle of 45°, and presenting an almost precipitous escarpment towards the north-east.

(*m*) *Rhodes.*—The blue crystalline marble occurs frequently on the east coast of the island of Rhodes, near the town of Lindo, dipping 25°

north-west. It forms high hills, against which the tertiary shelly limestone reposes, as shown in Section, p. 43. The Acropolis of Lindo is built upon strata of it, also dipping to the north-west. It occurs likewise further north, between Rhodes and Archangelo, where, besides the high range of hills about two miles from the sea, another low ridge of rocky points of the same formation rises up in the middle of the plain nearer the sea, forming a low ridge parallel to the coast. In one place, between Lindo and Archangelo, this rock was observed to become harder and schistose, and to pass into a hard black crystalline limestone, with a slightly concretionary structure like the Silurian limestone of the Bosphorus; it likewise dips to the north-west, but at a much greater angle.

§ 2. *Cretaceous System.* Pl. V. Sects. 1, 4 and 6.

The fossils found in this group of rocks prove it to be the equivalent of the "Apennine Limestone" of Italy, Dalmatia, the Ionian Islands, and Greece*. The following are the localities at which we have evidence of its occurrence:—

(*a*) *The Northern Flank of the Olympus Chain.*—The south shore of the Lake of Apollonia (lat. 40° 2′, long. 28° 35′) consists of compact, fine-grained, pale yellowish limestone, resembling the Apennine limestone of the Ionian Isles. We did not succeed in finding fossils. It extends for about eighteen miles in a nearly straight line from east-north-east to west-south-west, dipping at a high angle into the lake; and it apparently rests upon the schistose rocks which range from Mount Olympus to the westward. The same limestone occurs on the north side of the hills between the Lake of Maniyas and Susugherli. Sect. 1. It is accompanied by argillaceous shales and micaceous sandstones, which at the latter place form hills on both sides of the Macestus. Higher up that river, between Ildiz and Kebsit, is a range of schistose hills, in some places very micaceous, and in others resembling argillaceous slate. Huge masses of compact white limestone are exposed on the hill-sides in several places, and appear to be allied to the Hippurite limestone rather than to the schistose series.

(*b*) *The Upper Basin of the Hermus.*—An arenaceous deposit which occurs in this district, we are inclined to refer to the newest portion of the secondary series. We first observed it on the north slope of a range of schistose rocks, which extends westward from Morad Dagh, between the Hermus and the Plain of Hushak. Pl. V. Sect. 4. It is well exposed in the ravine east of the village of Makouf, where it consists chiefly of brown micaceous sandstone, finely laminated; but it is not crystalline

* For a description of the Apennine Limestone of Zante, see p. 19 (Geol. Trans. vol. v. p. 403); and for that of Greece, see the 'Expédition scientifique de la Morée,' partie Géologique, par MM. Boblaye et Virlet. Fol. Paris, 1835.

or compact, like the schistose formation above described. Some thin beds of soft white limestone were here noticed; and a little further north, a conglomerate of rolled pebbles of quartz, marble, mica-schist, &c., is interstratified with the sandstone.

On the south of the schistose ridge, between Sorkoum and Hushak, Sect. 4, is a zone of brown sandstone and quartzose conglomerate, dipping from 40° to 60° to south-west, which is probably referable to the same age. A similar sandstone occurs on the north side of the Hermus, between the ridge of Demirji and the Catacecaumene. On the south of Simaul, Sect. 6, it occupies the whole surface of the ridge, which is there very low, and consists of thinly laminated sandstone, containing mica, and breaking into large slabs. It extends thence southward for about five miles, where it is overlaid by a tertiary tufa. About eleven miles further south, the laminated sandstone reappears on the banks of the Aïneh-chai, dipping from 30° to 40° south by west, and is covered unconformably by the horizontal tufa. The accompanying section is well exposed in this deep ravine, the hills on each side of which rise to the height of 600 or 700 feet.

Section, 7 miles north-east of Selendi. Lat. 38° 45′, long. about 29°. (See also the description, p. 47.)

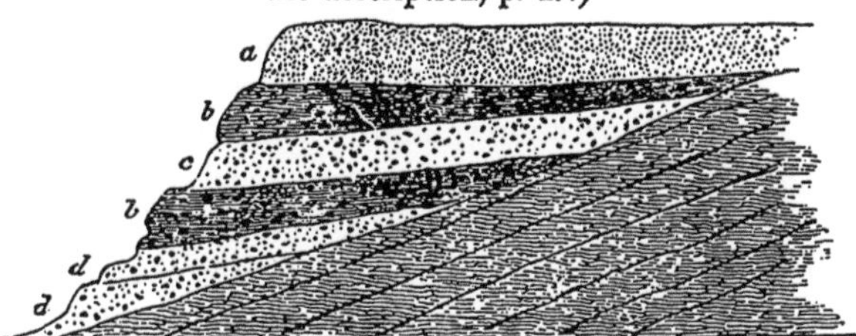

Micaceous sandstone.

a. Peperite. *b.* Trachytic conglomerate. *c.* White marly quartz pebbles. *d.* Peperite, with boulders.

These sandstones appear at intervals on the sides of the river to the distance of six miles below Selendi. They are nearly horizontal, and on the south side of the Aïneh-chai have a slight northerly dip, reposing against a low ridge of saccharine marble, which ranges between that river and the Hermus. On the south of this axis the sandstones again occur, dipping with a southerly inclination towards the Hermus, and are overlaid by horizontal beds of tertiary freshwater limestone. The following section in descending order is exposed about a mile north of the Hermus:—

White marl of the tertiary formation	5 or 6 feet.
Alternate beds of marl and sand	20
Loose beds of sand and gravel, with pebbles of quartz-rock and schist	30

The sandstones above described did not occur to us elsewhere in Western Asia Minor. They appear to be devoid of fossils, but are clearly younger than the schistose group, and older than the tertiary beds; and we have therefore preferred, on account of their comparative unimportance, to class them for the present under the Cretaceous system, than to assign them a separate place.

(*c*) *The vicinity of Smyrna.*—The compact grey limestone with Hippurites and Nummulites, associated with greenish argillaceous and arenaceous shales, which form the mountains around Smyrna, are among the best types of this formation which fell under our notice in Asia Minor. They are described at length in a separate memoir, p. 9. (Geol. Trans. vol. v. p. 393.)

(*d*) The *Island of Scio* and the peninsula of Kara-bornou consist of compact grey limestones, which we refer to this formation. These naked mountains present precisely the same dull grey appearance, which forms so peculiar a feature in the scenery of Greece and Albania, where similar rocks prevail.

(*e*) *Sighajik* (lat. 38° 14′, long. 26° 50′).—The rocky peninsula which extends between Sighajik and Teos, and is connected with the main land by the low marshy plain, on which stand the ruins of Teos, consists of a greyish-white limestone resembling scaglia, but having sometimes a much more earthy character. It dips slightly to the west by south. It is generally thick-bedded, and very nearly resembles the lacustrine limestone near Smyrna. In places it is underlaid by beds of sand and sandstone, containing calcareous concretions; and is sometimes thinly bedded, the strata being separated by slightly micaceous marly waybands.

(*f*) *Syme* (lat. 36° 35′, long. 27° 52′).—The whole of the bare and rugged island of Syme consists of a greyish-white compact alpine limestone or scaglia, with occasional nodules and bands of siliceous limestone. It is in general very thickly bedded, but is sometimes thin-bedded, the strata being separated by thin seams of marl or greensand. One of the authors (Mr. Hamilton), who was several days on the island, sought in vain for organic remains. The general stratification of Syme is horizontal; but toward the north the beds dip 80° north-north-west, whereas to the south the dip is south-south-east. The coast-line is rugged and deeply indented, and the general features and characteristics of the island closely resemble those of the island of Ithaca. The whole of the southern shore of the Gulf of Syme consists of the same formation. The limestone varies a little in colour, being sometimes a pale red. At the eastern end of the Gulf, it alternates with bands of pale red jaspery chert, dipping north-west 50°; and it is in places much contorted. Further eastward the dip is still greater.

(*g*) *The Island of Rhodes**.—The greater part of that portion of the island which has been explored, consists of those rocks which, under the

* See also Spratt on Rhodes in Geol. Proc. vol. iii. p. 773.—H. E. S. MS.

name of Scaglia, or Apennine limestone, have been considered the equivalents of the Cretaceous system of Europe. This formation is composed in the island of Rhodes of—1st, red and brown sandstone and conglomerates; and 2ndly, of whitish-grey or cream-coloured and red scaglia limestone.

1. *Red and brown Sandstone and Conglomerate.*—These appear to constitute the upper beds of the system, and are found near the centre of the island. A red conglomerate sandstone occurs between Apollonia and Embona, dipping south-south-west 50°, and resting conformably upon white scaglia. Indurated red marls and hard sandstone grits also occur in the same locality. At the foot of Mount Atairo, to the north-north-west, is another bed of conglomerate, containing chiefly boulders of greenstone and a greenish granitic rock, together with rounded masses and pebbles of the grey scaglia, of which the neighbouring hills consist. The igneous rocks were not observed *in situ.*

2. *Scaglia Limestone.*—This formation is chiefly developed in the lofty ridges of Mount Atairo, the ancient Mons Atabyrius, which rises to a height of 3500 or 4000 feet above the sea. The summit is a narrow ridge above two miles in length from north-east to south-west, which is nearly the direction of the axis of the island. The beds all dip uniformly to the south-east, at angles of 15° or 20°, and present a steep escarpment towards the north-west. The upper portion consists of thick-bedded grey scaglia without flints, and is underlaid by a thinly laminated schistose limestone with tabular masses or beds of flint, below which again the beds become thicker and the flints are nodular. These formations constitute the uppermost 900 feet of the mountain, below which the scaglia limestone is interstratified with a red calcareous shale or marly limestone, which is again succeeded by the thick-bedded limestone without flints.

The Acropolis of the ancient town of Camiro on the east coast of the island, six miles north of Lindo, is built upon an insulated table mass of white or cream-coloured limestone, very compact, and with rather a conchoidal fracture; and it has every appearance of having once been an island, before the tertiary formations, by which its base is encircled, were raised above the surface of the waves.

(*h*) *Deenair* (lat. 38° 3′, long. 30° 22′).—The same formation occurs at this place near the sources of the Mæander, where it contains numerous Nummulites. It is so extensively developed to the east and south-east, as to constitute the greater part of the Mount Taurus range; and Nummulites have also been found in it near Adalia*.

The secondary deposits here described under one head, may ultimately prove to have been formed at various epochs, but we are at present too imperfectly acquainted with their organic remains to decide this point. The only fossils found by us are Hippurites and Nummulites, and these indicate a very recent secondary date, equivalent to that of the Cretaceous

* Also at Castel Rosso, according to M. Texier, and on the west side of Phineka, according to Mr. Fellows.—H. E. S. MS.

system of Northern Europe. These deposits appear to have been affected by the same movements as the schistose rocks which underlie them; and all our evidence on the subject leads to the inference, that the mountain-chains of this country were elevated above the sea towards the close of the secondary period. It is this circumstance which has caused that peculiar feature in the scenery of Asia Minor before adverted to,—the abruptness with which precipitous mountains rise out of horizontal plains. We here find limestones of the Cretaceous age elevated into mountain-chains, which rival in altitude those of the Carboniferous and Protozoic periods in Northern Europe. Those younger secondary strata, which in Britain form tabular or undulating hills, and conduct us gradually from the plains of Essex to the mountains of Wales, have here undergone disturbances on the most gigantic scale. They have also acquired a compactness of texture much more analogous to the Protozoic than to the Cretaceous rocks of Britain. The tertiary strata which have since been deposited around the bases of these mountains have undergone very little disturbance, and almost invariably retain their original horizontality.

§ 3. *Tertiary Marine Deposits.*

Deposits of this class are of small extent in Asia Minor, and appear to be entirely confined to the vicinity of the sea-coast. We know of their existence at only the two following localities in Western Asia Minor:—

(*a*) *The Troad.*—The west coast of the Troad, and both shores of the Dardanelles, consist of horizontal strata of light-coloured limestones and marls. We had no opportunity of landing there; but as Olivier gives the names of several species of existing shells which he found in this deposit*, we infer that it belongs to the Pliocene group. A sandstone with Pectines is stated by Fontanier† to occur in Tenedos.

(*b*) *The Island of Rhodes.*—The north-east end of Rhodes consists of a yellow, concretionary, calcareous tufa, with beds of sand and gravel almost horizontal, and full of tertiary shells, of which the genera Pecten, Cardium and Venus are the most common. This deposit attains a height of 200 or 300 feet above the sea, forming in many places hanging terraces resting against the steep sides of the blue marble mentioned above‡. (See Woodcut, p. 43.)

* Travels in the Ottoman Empire, vol. ii. ch. 2.

† Voyages en Orient, vol. i. p. 9.

‡ Mr. Hamilton also observed in the ruins of Tripolis ad Mæandrum, nearly one hundred miles from the mouth of the river, several large blocks of a shelly tertiary marine limestone, which could not have been brought from any great distance; but he did not observe the rock *in situ*.

§ 4. *Tertiary Lacustrine Deposits.* Pl. V. Sect. 1, 3 to 6, and 9.

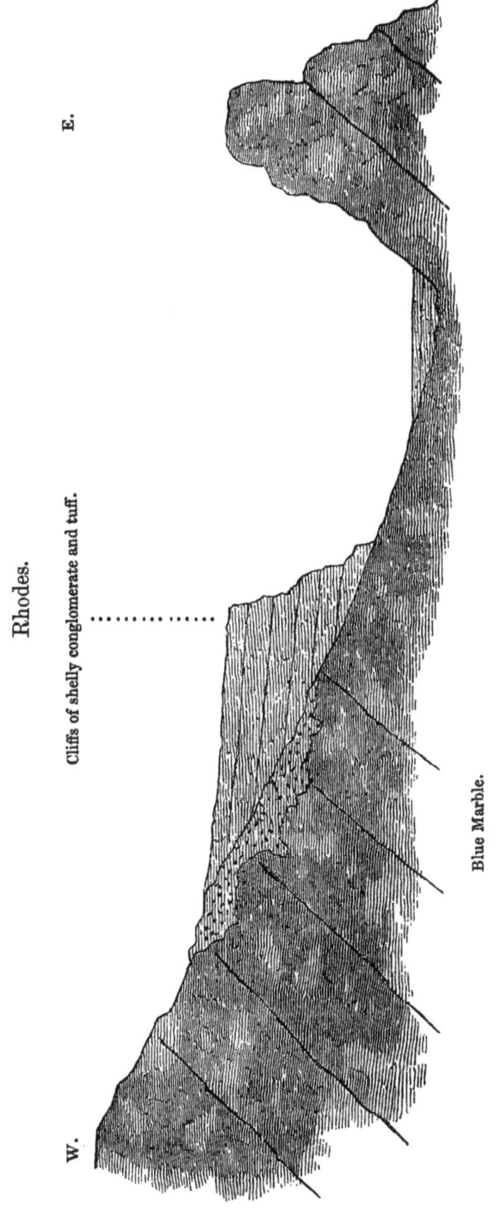

These deposits occupy a large portion of the surface of Asia Minor. We have already stated our belief (p. 42) that the schistose and cretaceous rocks, towards the close of the secondary period, were raised above the surface of the ocean, and formed that tumultuous mass of mountains which still intersects the country. It would seem that the inequalities of surface thus produced gave rise to numerous inland lakes, whose vast calcareous deposits denote a long period of comparative tranquillity. At length, by the slow denudation of their rocky barriers, aided probably by the convulsions incident to a volcanic country, these lakes were drained, and the rivers assumed the geographical positions which they still occupy. In the parts of Asia Minor which we visited, nearly every large valley was found to contain these lacustrine deposits; but in the more narrow ravines of the mountain-chains, nothing of the kind was noticed. This fact may enable us to distinguish the modern valleys of denudation from those older depressions which were caused by the original elevation of the mountains.

The lacustrine formations of Asia Minor consist, in general, of calcareous marl and white limestone. The latter is often identical in composition with the chalk of England, and the resemblance is further increased by the presence of beds and nodules of black flint. Sometimes the limestones are more compact, and approach in character the Italian scaglia or the Bavarian lithographic stone. Towards the margins of the lacustrine basins, the marls and limestones gradually become sandy and then gravelly, and not unfrequently traces may be found of a regular shingle beach *. Pl. V. Sect. 5.

The lacustrine strata are, with few exceptions, nearly horizontal, and seem in general to have remained undisturbed since the period of their deposition. The lapse of time however has subjected these deposits to extensive denudation, and sometimes nothing remains of them but a few detached platforms, which skirt the sides of the valleys at the same level, and show the precise form and extent of these ancient lakes.

The organic remains which occur in these deposits consist of lacustrine shells and portions of vegetables. They approach very closely to existing forms, though we do not venture to assert a completely specific identity in any of them with the present Fauna or Flora of Asia Minor. No remains of Mammalia or other Vertebrata were noticed during our rapid survey of these deposits; and we wish to call the attention of future travellers to this desideratum in the geology of Asia Minor.

We now proceed to enumerate the localities at which we met with deposits of this class, though we do not pretend to assert the absolute contemporaneity of all these detached lacustrine basins.

(*a*) *The basin of Moudania* (lat. 40° 23′). Pl. III.—The country between Mount Olympus and the Sea of Marmora is occupied by a tertiary formation, but no fossils occurred to us in it. Near Moudania it forms hills overlooking the sea, and several hundred feet in height. The similarity of the white limestone of the beds of flint which accompany it, to the lacustrine deposits of Smyrna and other places, induces us to refer it provisionally to the same group. The formation extends to the Lake of Apollonia and reappears at Kirmasli, whence it sends a branch up the valley of the Rhyndacus as far as Kesterlek.

(*b*) *Basin of Dundâr* (lat. 39° 50′).—This basin is of small extent, occurring in a valley on the south-west side of the Rhyndacus. The deposit is a soft, white, sandy limestone, dipping at the village of Bourmâh, about 10° east-south-east. Here also, having found no fossils, we are guided only by analogy in referring it to the lacustrine group.

(*c*) *Basin of Harmanjik* (lat. 39° 47′).—This is bounded on the north by the schistose hills which form the base of Mount Olympus, and on the south-east by the volcanic rocks of Eshén. It consists throughout of white limestone horizontally stratified. Like similar deposits in

* To the eastward of that part of the country now more immediately under consideration, beds of selenite and gypsum occur in some of the upper portions of the formation, as between Sevrihissar and Afiom Karahissar.

Auvergne and the Cantal*, it contains an abundance of black flint in beds and nodules; and considerable quantities of gun-flints are made there and sent to Constantinople. Near Haidár, at the northern limit of the basin, is a picturesque gorge, at least 600 feet deep, traversed by the Rhyndacus, which affords an admirable section of the horizontal strata on each side. The limestone contains a small turreted species of Paludina, resembling *P. acuta* (Michaud), or *P. stagnorum* (Turton). A species, similar in form but double the size, occurs associated with a Limnæa near the south boundary of the basin, about five miles from Harmanjik, and two from Eshén.

(*d*) *Basin of Taushanli* (lat. 39° 35′).—The trappean rocks of Eshén separate this basin from the last. The lacustrine deposit, if such it be, has received considerable modification from the vicinity of the igneous rocks, which appear to have been in action during its deposition. In place of a pure earthy limestone, we have a peculiar kind of tufa or peperite, composed of particles of quartz imbedded in a felspathic cement, and accompanied by hexagonal crystals of hornblende. Near Taushanli this tufa becomes more calcareous, and contains angular fragments of jasperine rocks†. It is capped by a stratum of brown impure jasper, containing a large proportion of iron‡.

The alluvial plain of Taushanli is surrounded by platforms of this tufaceous deposit,—the scattered outliers of a formation once continuous. No organic remains were noticed.

(*e*) *Basin of Gozulját* (lat. 39° 25′).—We crossed this basin in going from Taushanli to Azani. It is filled with white limestone identical in appearance with English chalk, and containing strata of black flint. The scenery also strongly resembles that of our own chalk counties, consisting of rounded hills intersected by dry valleys, and covered with fir and juniper. No organic remains were noticed.

Near the village of Gozulját, Pl. V. Sect. 5, is a section showing the gradual transition of this white limestone into gravel, as it approaches the margin of the ancient lake.

(*f*) *Basin of Azani* (lat. 39° 14′).—The ancient town of Azani, the

* See Scrope on Central France, p. 25. Lyell's 'Principles of Geology,' 3rd edition, vol. iv. p. 106, &c.

† Mr. Hamilton is of opinion that this tufaceous deposit, which he observed in many other parts of Asia Minor, does not belong to the lacustrine group, but is anterior to it, being in fact an accompaniment of the trachytic rocks which everywhere appear in its neighbourhood.

‡ This deposit of tufa and jasperine rocks is probably continuous with a similar district on the north of Kutahiyah, described by Mr. Fellows in his 'Excursion in Asia Minor,' pp. 128, 129.

See also Texier in 'Comptes Rendus,' iv., where he calls the tertiary deposits of the plain of Kutahiyah "CRAIE."—H. E. S. MS.

interesting ruins of which are among the best-preserved in Asia Minor, stood in an alluvial plain near the source of the Rhyndacus, surrounded by mountain-chains of mica-schist and marble. Sect. 4. Around the bases of these mountains are beds of gravel, passing into white limestone. The latter rock approaches within three miles of Azani on the south, and has been extensively used in the construction of its ancient buildings.

The following species of shells were noticed in it:—

1. *Limnæa*, a species closely resembling *L. longiscata*, Sow.
2. ———, a small turreted species like *L. fossaria*, Flem.
3. *Physa*, a small turreted species like *P. hypnorum*, Drap.
4. *Planorbis*, a large species like *P. cornea*, Drap.
5. ———, small, with five rounded volutions, like *P. spirorbis*, Müll.
6. ———, small, subcarinate; allied to *P. nitidus*, Müll.
7. *Pupa*, a small species.
8. *Clausilia*.
9. *Paludina*, a ventricose species like *P. similis*, Drap.

(*g*) *Basin of the Macestus* (lat. 39° 20′).—The valley of the Macestus, between Kebsit and Boghaditza, is occupied by a horizontal formation of calcareous marl and white limestone, thinly bedded and of chalky texture, containing in some parts many nodules of flint. In the southern portion of the basin, the beds are broken up by the intrusion of igneous rocks, which will be described hereafter. Pl. V. Sect. 1. No organic remains were observed.

(*h*) *Basin of Ghiediz* (lat. 39° 2′).—A deposit of white limestone fills the upper valley of the Hermus, and is contained between two ranges of schistose mountains which run westward from Morad Dagh. No fossils were noticed during our rapid survey. Pl. V. Sect. 4 and 9.

(*i*) *Basin of Hushak* (lat. 38° 39′).—This is a lacustrine formation of great extent, bounded on the north and east by branches of Morad Dagh, and on the west by the mountains near Takmak, while to the south it extends to the Mæander, and is bounded by the continuation of the range of Messogis. The whole of this area is a nearly level plain, Pl. V. Sect. 4, drained by tributaries of the Mæander, which flow in narrow ravines, deeply denuded in the lacustrine limestone, and discoverable only on a close approach. A remarkable case of the kind occurs near the town of Gobék. Here, a vast ravine, at least 600 feet in depth, is surrounded by cliffs of limestone in horizontal strata. At the bottom flows a meandering stream, probably the "Kopli-sou" of Tavernier. The time-worn cliffs resemble an assemblage of towers, castles and cathedrals, and present a scene which can hardly be depicted by the pencil or the pen.

In the more southern portion of this plain, where the formations have been cut through by the Mæander, the Banas-chai and the Kopli-sou, the subjacent beds of micaceous and argillaceous schists and crystalline marble of Mount Messogis, together with intermediate beds of sand and gravel, which everywhere underlie the lacustrine limestone, are well exposed along the borders of the ancient lake.

We found the following organic remains near Kalinkesé, in the central part of this basin (lat. 38° 30′) :—

1. *Limnæa*, apparently the same as No. 2 in the Azani list (p. 46).
2. *Planorbis*, ———————— No. 5 ————————.
3. *Paludina*, ———————— No. 9 ————————.
4. Minute seeds of *Chara* ("Gyrogonites").

(*k*) *Basin of the Catacecaumene* (lat. 38° 35′).—The tertiary deposits in this basin consist of horizontal beds of white limestone passing downwards (in the northern part of the basin) into volcanic tufa. The limestones have been deeply denuded by the Hermus and its tributaries, and now form a series of lofty plateaux, frequently capped by tabular masses of basalt, which will be described hereafter. Pl. V. Sect. 6, and Sketch 11.

About nine miles south of Simaul (lat. 39° 5′, long. 29° 5′) the following section in descending order, of the horizontal tufaceous beds, was noticed :—

a. Hard, volcanic tuff, slightly crystalline, with boulders and pebbles of trap, and numerous concretions of green marl	12 feet.
b. Soft, earthy, yellowish tufa, with small fragments of pumice	10 —
c. A hard crystalline bed, which, but for its stratification, might be taken for a slightly decomposed igneous rock................	

These tufaceous beds are capped with white limestone, from beneath which they crop out in many places in the sides of the ravines. They rest unconformably on beds of secondary sandstone, as is shown in the woodcut at p. 39. At the point where that section occurs, seven miles north-east of Selendi (lat. 38° 45′), the lower part of the tufa, resting upon the sandstone, consists of a conglomerate with large masses and boulders of quartz, greenstone, and other primary and igneous rocks, in irregular waved beds. The strata in ascending become finer-grained and more nearly horizontal, dipping 10° south. (See woodcut, p. 39.) These appearances may be owing to the violent disturbance and gradual subsidence of a body of water acted upon by volcanic forces.

On approaching the Hermus from Selendi, the tufaceous beds thin out, and the lacustrine limestone rests on the secondary sandstone. Pl. V. Sect. 6. This limestone covers an extensive tract on both sides of the Hermus, forming table-lands about 700 feet above that river. Between the Aïneh-chai and the ruins of Saittæ, it contains beds of tabular flint about a foot thick; and in one place, near a local outburst of basalt, a mass of hard brown siliceous rock, about fifty yards in extent, was found in the midst of the limestone, taking the place of the beds of flint.

Our attention was too much occupied with the volcanic phænomena of this interesting district to search for organic remains in the tertiary limestone; but from the complete analogy of the formation to those deposits in which freshwater shells were found, we have no doubt of its lacustrine origin*. The drainage of this ancient lake has doubtless been effected by the opening of the gorge in the schistose rocks near Adala, through which the Hermus flows. Pl. III., Pl. V., Sketch 13.

(*l*) *Basin of Sardis* (lat. 38° 30′, long. 27° 55′).—The greater part of this deposit has been removed by denudation, but some low hills of limestone yet remain on the south side of the Gygean Lake, and others to the east and west of Adala. A zone of sand and gravel, perhaps of a different age from the limestone, extends along the north flank of Mount Tmolus, Pl. V. Sect. 3, presenting a succession of broken hills and ravines, which, backed by the snowy heights of Tmolus, form a scene of the richest beauty. This gravelly formation is subject to rapid destruction; and the celebrated Acropolis of Sardis, which could once hold out against a Persian army, is now reduced at the top to an area of a few yards square, which a short lapse of time will probably diminish to a point. Sect. 3.

(*m*) *Basin of Smyrna*.—The lacustrine deposits near Smyrna, with their imbedded shells and vegetables, are fully described in a paper devoted to the geology of that district, p. 9 (Geol. Trans. vol. v. p. 393).

(*n*) *Lower Vale of the Mæander* (lat. 38°).—It is with some doubt that we cite this district in the list of lacustrine basins, as no freshwater limestone was noticed in it; and the masses of gravel and sand which skirt the south side of Mount Messogis might have been produced by the waves of the sea. However, they have so complete an analogy to those on the north side of Mount Tmolus, that we are inclined to refer them to the same origin, whatever it may be. Pl. V. Sect. 3.

In the valley of the Mæander, these accumulations of detritus form a zone of broken hills, many hundred feet in height, which flank the schistose range of Mount Messogis. They appear to have been afterwards modified by water at a lower level than that which originally produced them; for horizontal platforms of gravel rest against the sides of the loftier hills, at a height of about 200 feet above the alluvial plain of

* Mr. Hamilton, who, in the year 1837, succeeded in carrying his barometer across these plains, found that the elevation of the highest parts agreed, within a few feet, with that of the Plain of Gobék and Hushak; he also thinks, that although separated near Koola by the eastern prolongation of Mount Tmolus, these two plains may be connected more to the north-east; their elevation above the sea varying from 2000 to 2400 feet, the latter altitude being the elevation of the highest portion of the lacustrine formation of Koola.

the Mæander. The ancient town of Tralles stood on one of these platforms, defended on three sides by an abrupt escarpment, and on the fourth by its citadel, which was placed on the summit of a lofty peak of indurated gravel and sand. The peculiarity of its situation is most accurately described by Strabo, lib. xiv. c. 1. The gravel consists of rounded pebbles of quartz and mica-schist, derived from the higher parts of Mount Messogis.

It is remarkable that of these tertiary deposits, so common elsewhere, no trace has been noticed in the valley of the Cayster. This extensive vale is completely insulated by the schistose ranges of Tmolus and Messogis, and the only exit for its waters is through a narrow gorge into the sea at Ephesus. The absence of these deposits renders it probable that the vale of the Cayster has been brought into its present condition at a comparatively recent period; or if contemporary with the other valleys of the country, we may suppose that, from the absence of a dam to retain its waters, no lacustrine deposit could accumulate.

§ 5. *Modern Aqueous Deposits.* Pl. V. Sect. 2.

Under this head we have to notice only travertine accumulations and river alluvions. Of the former, a remarkable instance occurs at Brusa. Pl. V. Sect. 2. The hot springs which rise at the foot of Mount Olympus have the high temperature of 184° Fahr.*; and they constantly deposit travertine, a platform of which extends about two miles from the springs, along the foot of Olympus, into the town of Brusa. It is here about half a mile in width and 100 feet high. At this end there are no thermal sources at present, and it is probable that they became stopped up by the calcareous deposit, and found a new exit at the point where they now flow.

At Ilijah (lat. 39° 10′), in the upper valley of the Macestus, about seven miles east of Singerli, are some thermal springs of a very high temperature, which have formed extensive accumulations of fibrous and mammillated travertine. One source is ejected a foot and a half above the top of a mound of travertine which it has accumulated, and which is approached by a natural bridge of the same substance formed across a stream from another spring. In one place the water falls over a cliff 8 or 10 feet high, producing stalactites in its course.

The hot springs of Hierapolis, in the upper valley of the Mæander, flow over a cliff which they have encrusted with white travertine and stalactites. The remarkable appearance here presented has given the

* According to M. de Verneuil, they vary from 42° to 84° Cent. (=107$\frac{3}{5}$° to 183$\frac{3}{5}$° Fahr.). These springs are also described by M. Jouannin in the 'Bulletin' of the Société Géographique de la France, xi. p. 290.—H. E. S. MS.

Turkish name of Bambouk-kalesi, or "Cotton Castle," to the town, and has excited the attention of many ancient as well as modern authors*.

Asia Minor presents some interesting examples of the geographical changes effected by the alluvia of rivers. The detrital deposits of the Mæander were so abundant as to excite the attention of Strabo. Since his time, the island of Lade, near which a naval action was once fought between the Persians and Ionians, has become a hill in the midst of a plain; the Latmic Gulf is changed into an inland lake; and the once flourishing town of Miletus, losing her commerce from this cause, has become a heap of ruins.

The alluvia of the Cayster have reduced Ephesus to the same state as Miletus. Her port, once the seat of commerce and civilization, is now a stagnant pool, separated from the sea by a marshy plain, which infects the surrounding district with malaria.

A similar fate perhaps awaits the now flourishing city of Smyrna. The Delta of the Hermus is fast advancing across the Gulf: a narrow channel only is at present open to shipping; and a few centuries may see that populous city reduced to the condition of Ephesus and Miletus.

The absence of tide in the Mediterranean is doubtless the chief cause of the rapid extension of these alluvial deposits in historical times. The bays and estuaries not being cleansed by the daily reflux of the tide, the alluvial matter brought down by the rivers deposits itself close to their mouths, and their deltas are thus pushed into the sea with a rapidity of which we have no examples in tidal oceans.

Under the head of modern alluvial formations, we may mention a singular lacustrine deposit in the valley of the Rhyndacus, above Kirmasli (lat. 39° 56′). A low range of tertiary limestone crosses the valley at that town, and is cut through by the Rhyndacus, which traverses a narrow gorge. A lake once existed on the upper side of the limestone hills, and has left considerable deposits of mud, sand and gravel. The opening of the gorge having drained the lake, the Rhyndacus has since had time to remove the greater part of the lacustrine deposit; but platforms of it still skirt one or both sides of the valley, at the height of 50 or 60 feet above the more modern plain in which the river now flows. Want of leisure prevented us from searching for organic remains in this deposit, but the appearances which it exhibits seem to refer it to the modern or newest tertiary epoch.

B.—Igneous Rocks.

§ 1. *Granitic Rocks.* Pl. V. Sect. 1 to 3, and 6.

Rocks of this class are rarer in Asia Minor than might have been

* See Strabo, lib. xiii. cap. 4. Vitruvius, lib. viii. cap. 3. Chandler, Trav., vol. i. ch. 68. Laborde, 'Voyages en Orient.' Fellows, 'Excursion in Asia Minor,' p. 283, &c.

expected in a country where schistose formations of great antiquity and igneous rocks of a later period are so largely developed. The only good example of granite which fell under our notice is in the peninsula of Cyzicus (lat. 40° 25′, Pl. V. Sect. 1). This mountainous mass consists of a fine-grained grey granite, decomposing rapidly in consequence of the felspar which it contains, and thereby producing a luxuriant soil. It also contains large masses of hornblende or amphibole, and is sometimes traversed by veins of felspar.

This granite appears to have been elevated subsequently to the formation of the schistose rocks, which dip away from it, at Erdek to the south-west, and at Aidinjik to the south and south-east by south. Sect. 1. The marble of the island of Proconnesus on the north has probably been elevated by the same mass of granite.

From the occurrence of granite pebbles in the detritus at the north foot of the Demirji chain, near Mumjik (lat. 39° 7′, long. 28° 35′), we infer that granitic rocks exist in the higher parts of that lofty range.

We have only to add on this head, that M. Texier, whose interesting researches in Asia Minor are in course of publication, informed us that he had met with granite on the summits of Mount Olympus in Mysia, Mount Tmolus, and Mount Latmus near the mouth of the Mæander*. Pl. V. Sect. 2 and 3.

§ 2. *Greenstone and older Trap Rocks.*

Igneous rocks, which can safely be referred to this class and period, are not of very frequent occurrence in the western parts of Asia Minor. They were observed by us at the following localities:—

(*a*) The promontory of Bozbornou, north of the Gulf of Moudania, consists at its western end of a greenish trap, sometimes forming rounded concretions, which at a distance resembled an artificial wall (lat. 40° 35′).

(*b*) Immediately behind the town of Moudania a compact, green, trappean rock appears above the surface, and its decomposition has in many places produced a fine rich soil. This rock is however associated with tertiary strata, and is probably referable to a late epoch.

(*c*) Another example of greenstone occurred between Kesterlék and Edrenos on the south side of the Rhyndacus. It appears in the midst of the schistose formation, and is accompanied with red and brown jasper. We however saw no sections which exhibited its connexion with the schistose rocks, but we are disposed to think that the jasper is an altered rock intervening between the greenstone and the schist.

On the south side of the basin of Taushanli, above Karjitash, some trappean dikes were noticed in the schistose formation.

* M. Texier, in his 'Sketch of the Geology of Asia Minor' (Comptes Rendus, iv. p. 465), speaks of granite only at Cyzicus and on Mount Olympus.—H. E. S. MS.

§ 3. *Trachytic and newer Trap Rocks.*
Pl. V. Sects. 1, 4, 6, 9, and Sketch 10.

A remarkable feature in the geology of Asia Minor is the vast abundance of trachytic rocks which have been erupted at detached points throughout the country. They appear generally to occur in the sides or bottoms of the valleys, and are frequently in contact with the tertiary formations.

No instance occurred to us of rocks of this class in the higher parts of the schistose mountain-chains. From this circumstance, we consider the trachytic rocks of the country to be all, or nearly all, of the tertiary period; and it is probable that their ejection was consequent on the breaking-up of the superficial crust by the elevation of the mountain-chains at the close of the secondary period. Indeed, in some cases we have direct proof of the date of these eruptions by their having broken up, injected and altered the tertiary formations. It is remarkable, that in such cases the general horizontality of the tertiary strata remains undisturbed, the dislocations produced by the trachyte being confined to a few hundred yards around the focus of eruption. We may hence infer how insignificant have been these volcanic operations in comparison with those forces which could elevate the secondary formation into lofty mountain-chains.

We proceed to enumerate the chief localities at which we noticed this class of rocks:—

(*a*) On the north side of the lake of Apollonia, about three or four miles north-east from the town of Abullionte (ancient Apollonia ad Rhyndacum), the hills consist of red porphyritic trachyte containing imbedded crystals of glassy felspar (lat. 40° 10′).

(*b*) At Hammamlí, near Kirmasli (lat. 39° 56′, long. 28° 25′), is a mass of porphyritic trachyte of various colours. It forms an isolated hill not more than a mile in extent, rising in the midst of tertiary limestones.

(*c*) Around the village of Eshén (lat. 39° 43′, long. 29° 20′), between Harmanjik and Taushanli, is an extensive patch of trappean rocks containing an abundance of iron ore, and occasionally traversed with veins of carbonate of magnesia. On the north side of these igneous rocks, on the road to Harmanjik, a dike of compact basalt, resembling greenstone, traverses lacustrine limestone containing Paludinæ and Limnææ.

Towards Taushanli these trappean rocks assume the character of a brown and grey trachyte, and form some considerable hills. They were probably erupted during the deposition of the tufaceous sandstones which occupy the basin of Taushanli, as before described, the materials of which were probably ejected from the same craters in the state of ashes and cinders.

(*d*) In the valley of the Macestus are several outbursts of trachytic rocks. Pl. V. Sect. 1. About two miles south of Susugherli, a mass of compact porphyritic trachyte stretches down from the hills on the west; and half a mile further south, at Tashkapu, is a considerable extent of white volcanic tufa, dipping slightly to west-south-west, and forming a perpendicular cliff, around which the Macestus flows. Hills of red trachyte with glassy felspar occur a mile to the south of Kebsit, and again at Kalbourja, six miles further south. One of these hills, of a conical form, rises far above the rest; and on its south side is a deposit of stratified volcanic sand and conglomerate, dipping south-west and resting upon trachyte. A mile further south is a very talcose rock resembling serpentine, much decomposed, and breaking with an irregular, lustrous surface. These rocks rise in the midst of tertiary, lacustrine limestone above described, which, on the south side of the trachyte, is much disturbed, and is sometimes almost vertical. About three miles north of Boghaditza the same limestone is again elevated and broken by the protrusion of a small mass of igneous rock resembling a fine-grained granite.

The largest mass of trachytic rocks in the valley of the Macestus commences on the south of Boghaditza, Pl. V. Sect. 1, and extends thence to the foot of the Demirji chain. The northern part of this trachytic mass is of a greenish colour, resembling that of the Bosphorus*. It contains veins of chalcedony, and is accompanied with a trachytic conglomerate. In one place a large mass of contorted, calcareous, thinly laminated marl is caught up in the conglomerate. Further south this trachyte varies in colour from white to green and red, and is mixed with an angular conglomerate, which often presents a stratified appearance. On the east of Singerli is a mass of porphyritic trachyte, with crystals of white glassy felspar and hornblende in a reddish-brown matrix; and it much resembles the trachyte of Smyrna. It flowed in a *coulée* from the hills to the south-east.

The hot springs of Ilijah, described at p. 49, rise in these porphyritic rocks, which extend about eight or ten miles to the east of Singerli.

(*e*) On the south side of the ridge of schistose rocks, between Azani and Ghiediz, Pl. V. Sect. 4, is an outburst of trachyte reposing on pumiceous sand and ashes. Further down, on approaching Ghiediz, several masses of igneous rock jut out amidst the lacustrine limestone. The most remarkable case appears at Ghiediz. Pl. V. Sect. 4 and 9, Sketch 10. This singularly situated town lies in a deep ravine, furrowed in the lacustrine limestone. On the lower side the ravine is blocked up by a conical rock of compact and amygdaloidal basalt, containing green earth and augite, about 150 feet high, and riven to its

* See Geol. Trans. vol. v. p. 388.

base by a deep and gloomy fissure, the width of which at the bottom does not exceed two yards. This fissure affords an exit for the rivulet which descends the valley, and which would otherwise form a lake on the spot where the town stands.

On the north side this basaltic mass has sent forth a *coulée* of columnar, amygdaloidal basalt about 10 feet thick, resting conformably on beds of sand and gravel containing rolled pebbles of trachyte. Sect. 4 and 9. This basaltic *coulée* may belong to a more recent period.

We regret that our short stay at Ghiediz prevented our determining the relations of this volcanic mass to the surrounding lacustrine limestone. It is probably posterior to the lacustrine strata, for the latter contain no traces of volcanic matter. The beds of sand and gravel underlying the *coulée* were perhaps a local alluvium of the same date as the great lacustrine formations*.

(*f*) *Trachytic rocks of Gunay* (lat. 38° 53′). Pl. V. fig. 4.—These rise in the midst of the hills of secondary sandstone between Ghiediz and Hushak, about eight miles south of the Hermus. On the east side of the village of Gunay are some singular varieties of volcanic opal or pitchstone; and on the west is a well-defined *coulée* of grey trachyte with large crystals of felspar, which has flowed from the focus of eruption above, down a small valley in the sandstone hills.

(*g*) *Trachytic hills east of Takmak* (lat. 38° 26′, long. 29° 8′).—This is a considerable cluster of abrupt conical hills standing at the western extremity of the plain of Hushak, and visible from the eastward at a great distance. We passed through them in going from Gobék to Takmak. This eruption of volcanic matter must have taken place anterior to the lacustrine formation of Hushak, for the gravel beds of that deposit, which abut against these hills on the east side, contain many boulders of trachyte mixed with pebbles of schistose rocks.

The hills are chiefly composed of trachyte of various colours, occurring either *en masse* or in angular fragments imbedded in a tufaceous paste. About four miles west of Karadjá Achmet Kieui, is a patch of stratified marl, sandstone and clay, enveloped and surrounded by the trachytic rocks. The beds are altered into the condition of quartz rock and Lydian stone.

(*h*) *The Catacecaumene* (lat. 28° 30′).—The volcanic phænomena from which this district derives its name, belong to a more recent epoch, and will be fully described in a subsequent part of this paper; but there are also igneous rocks of a different character and older date which require to be noticed here. They occur in the northern part of the basin, from eight to sixteen miles north of Selendi, on the road to Simaul, Pl. V.

* In the Sections and Sketches, Pl. V., we have thought it best to colour this basaltic mass like the basalt of the *first period* in the Catacecaumene.

Sect. 6, and consist of masses of trachytic rocks jutting out in the bottoms of the ravines, beneath the lacustrine formation. It was before stated, that the lower portion of the lacustrine beds, in this part of the basin, consists of tufaceous deposits, which are doubtless connected in their origin with these trachytic eruptions. In some cases the tufaceous beds have been broken up by the igneous rock; but in the only instance (thirteen miles north of Selendi) where the superincumbent limestone was seen in contact with trap, its stratification remained undisturbed, and therefore it is probable that these eruptions were anterior to its deposition.

Between the ninth and tenth miles from Selendi, the tufa rests upon protruded masses of decomposing syenitic trap with a nearly vertical cleavage, and half a mile lower down, at the junction of two valleys, is a mass of trachytic conglomerate, with a solid trappean nucleus. Igneous rocks again rise from beneath the tufa at the eighth mile from Selendi, and have assumed, in consequence of decomposition, a great variety of colours.

(*i*) About eight miles from Adala, on the road to Koola, is a small isolated hill of grey trachyte. It rises in the vale of Lydia, at the foot of the schistose mountains which form the north boundary of the valley of the Cogamus. A section, formed by a watercourse, shows that this trachyte is older than the platform of lacustrine gravel which reposes against it.

(*k*) *Trachytic rocks of Smyrna.*—These extensive masses of igneous matter, which break up and overlie the lacustrine deposits of the Smyrna basin, are described at length in a memoir specially devoted to the geology of that district, p. 9. (Geol. Trans. 2nd ser. vol. v. pp. 396, 401.)

(*l*) *Fouges*, anc. *Phocæa* (lat. 38° 40′).—Igneous rocks, which have assumed a great variety of character, porphyritic, trappean and trachytic, form the chief ingredient of the hills which rise to the north of the little harbour of Fouges. They are associated with, and overlaid by various beds of tufaceous and pumiceous sands, which have sometimes assumed the hard, semi-vitrified character of the trachytic rocks themselves, and from which they can be scarcely distinguished. These beds all dip towards the south, and sometimes contain numerous cavities, which are either filled or coated with mammillary concretions of chalcedony*.

(*m*) *Ritri*, anc. *Erythræ* (lat. 38° 18′).—The Acropolis of Erythræ is situated upon an insulated peak of red trachyte, which rises abruptly to the height of two or three hundred feet, and although perfectly crystalline like that in the neighbourhood of Smyrna, it has the appearance of being stratified, and dips to the north. In its range towards the north, the colour passes from red to grey; and the appearance of stratification is

* This locality is mentioned by Texier in Comptes Rendus, iv.—H. E. S. MS.

marked partly by colour and partly by cleavage. The trachyte is in contact with the blue marble, which it has elevated and shattered.

(*n*) *Boodroom*, anc. *Halicarnassus* (lat. 37°).—About five miles to the south-west of Boodroom is a lofty conical hill called Chifoot Kaleh, or Jew's Castle, nearly 1000 feet high, which rises directly from the sea. It consists of a reddish felspathic trachyte. The greater part of the promontory of Karabaghla, to the west of it, is composed of the same rock, as well as the islets of the coast. The hills between Chifoot Kaleh and Boodroom consist chiefly, if not entirely, of trachytic and pumiceous conglomerate, and horizontally stratified beds of volcanic sand, the contained fragments being generally angular, and varying in size and character.

§ 4. *Modern Volcanic Rocks.*

Pl. IV. and Pl. V. Sect. 4, 6 to 9, and also Sketches 10 to 13.

The only part of Western Asia Minor where phænomena strictly analogous to those of active volcanoes are exhibited, is in that district of Lydia known anciently by the name of Catacecaumene, or the Burnt Country*†. Pl. IV. We term these volcanic products *modern*, from the relation they bear to the other formations of the country; but it is certain that, compared with human history, their antiquity must be very great. No record of the activity of these volcanoes existed in the time of Strabo, who however rightly infers their igneous origin from the phænomena which they present. Indeed, these phænomena are so striking as to excite the attention of every beholder, and feed the superstition of the ignorant; and we accordingly find that the ancient inhabitants laid the scene of Typhon's exploits in this region, while the modern Turks refer these mounds of cinders to the agency of Sheitan.

The Catacecaumene is described in a previous part of this paper as a tertiary lacustrine basin, surrounded by hills of schistose rocks. It is drained by the Hermus, which escapes at Adala through a narrow gorge in the schistose formation, the closing of which to a sufficient height would again convert the upper country into a lake. Numerous volcanic eruptions have taken place among the older rocks, which formed the southern margin of the basin; and streams of lava, flowing from these foci, have overspread the lacustrine deposits.

* Strabo has considerably overrated the extent of this volcanic district, in giving it a length of 500 stadia and a breadth of 400. Its real length is about eighteen or nineteen miles, and its breadth seven or eight, if we do not include the lava-stream which has flowed down the valley of the Hermus to Adala, which would increase the length to about ten miles more.

† "North of Hushak, at Insesler-Kaia-Si, which seems to be the ancient Acmonia, is a large crater, full of scoriæ and cinders at the bottom. The sides are of tufa, like that of the Roman Campagna, cut full of caves, formerly inhabited." Texier, 'Asie-Mineure,' p. 81.—H. E. S. MS.

The outbursts of volcanic matter appear to be referable to three great periods. How long may have been their duration, or how long the intervals of repose between each, is buried in the tomb of time. All that we can now assert, is, that long intervals must have passed between each period of eruption; and that the latest eruption occurred antecedently to the commencement of traditional or authentic history.

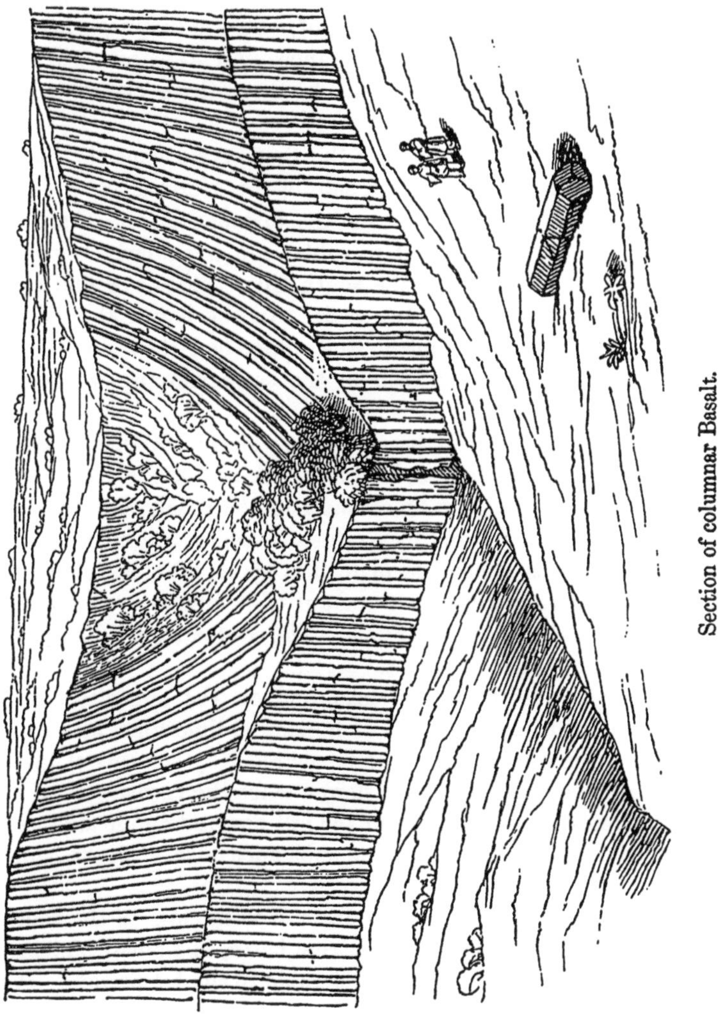

Section of columnar Basalt.

The oldest series of eruptions took place at a time when the bed of the lake presented a nearly level and unbroken surface, and before the first commencement of the excavation of the present valleys; for the basaltic rocks of that period invariably form the capping of the vast horizontal plateaux of tertiary lacustrine limestone. These eruptions could not have occurred long before the drainage of the lake; for if they had, we should find sedimentary deposits overlying the volcanic rocks, which is not the case; nor could they have occurred long after that event, for the valleys of denudation were not then formed. The eruptions of the second period were subsequent to the drainage of the lake, and to the excavation of deep valleys in the lacustrine deposits. Those of the third period are still more recent, and are distinguished by their entire identity of character with volcanoes now in action. We will describe the localities and phænomena of these three classes of volcanic rocks in the order of their antiquity.

1*st period.*—A mass of basalt of the first period caps the platform of lacustrine limestone on the north side of the Hermus for a distance of several miles, and may be seen in the distance in the sketch, Pl. V. fig. 11. The greater part of it splits into irregularly wedge-shaped masses; but in a hollow on the south side, there is a fine section of columnar basalt divided into two beds. In the lower, the columns are uniformly perpendicular; but in the upper bed they diverge to each side with a slight curve, as seen in the accompanying woodcut in the preceding page.

This vast tabular mass of basalt is here above 100 feet thick, but in general it does not much exceed 50 feet. A bed of soft sand intervenes in some places between the limestone and the basalt, but the junction-beds are in general obscured by a talus of basaltic fragments.

The summit of this basaltic plateau is about 800 feet above the bed of the Hermus, proving the vast amount of denudation which has taken place since the period of its eruption. The source from which the basalt has flowed is no longer apparent.

On the south-west of this plateau, and on the opposite side of the Hermus, is a small insulated hill of horizontal limestone, also capped by basalt at the same level. Its base is encircled by streams of lava of the 2nd and 3rd periods, which have flowed since the excavation of the valleys. The limestone at the base of the hill has undergone alteration, and is converted into a yellow jasperine substance with a bright, conchoidal fracture. Pl. IV.

About five miles east by south from Koola, close to the foot of the schistose hills which skirt the lacustrine basin, is a small plateau of imperfectly columnar basalt, which is referable to the 1st period. It is seen in the foreground of the sketch, Pl. V. Sketch 11. The lacustrine beds on which the basalt reposes, being near their junction with the older rocks, consist, in this place, of gravel and sand derived from the schistose formation. Pl. IV. and Pl. V. Sect. 8.

On the south side of the Hermus, about five miles north-north-west from Koola, is another large insulated plateau of basalt, reposing on the lacustrine formation. Pl. IV.

2nd period.—To the second period of volcanic action, we refer the numerous conical hills of scoriæ and ashes which cover the schistose ridges on the south of the lacustrine formation. The range of mica-schist and marble which runs from east to west on the south of Koola*, sends off three nearly parallel ridges towards the north, and may therefore be compared to the letter E. Pl. IV. The volcanic cones of the second period are scattered along this principal ridge and its three lateral branches, and many streams of lava may be traced flowing from them, and descending the valleys of denudation in the lacustrine formation towards the Hermus.

The volcanic products of this period are distinguished by the smoothness of their outlines, and by the vegetation which clothes their surface. The cones of scoriæ are all low and flat, rising at an angle of about 20°; their craters have either disappeared, or are marked by only small central depressions, and all their asperities seem to have been smoothed down by time. The scoriæ which form them are sufficiently decomposed to admit of cultivation, and they are almost invariably covered with vineyards, producing the Catacecaumene wine, celebrated from the time of Strabo to the present day†. The streams of lava which have flowed from them, are level on the surface and covered with turf. The cones of this period are about thirty in number, and their position is indicated on the map, Pl. IV. We proceed to describe some of the more important examples.

On the most eastern of the three schistose ridges above mentioned, are two conical hills of considerable magnitude, which are exhibited in the middle of the Sketch 11. See also Section 7. Remains of a crater are very evident on one of these hills. A broad *coulée* of lava has flowed from each of them towards the north, and uniting has formed a plain about 200 feet above the Hermus, which has cut a passage through the northern extremity of the lava and the subjacent beds. This *coulée* is overlaid by a stream of lava of the third period, which will be afterwards described.

A great number of volcanic cones rise on the top of the central schistose ridge on the west of the plain of Koola. Sketch 12, Sect. 7. Most of them have a slight crateriform depression on their summits. Two or three of these cones, which stand at the eastern foot of the ridge, are partly surrounded by lava of the third period, flowing from Karadewit; and the *coulées* which descend from them towards the Hermus are overlaid in some places by lava of the same age.

* These schistose hills form part of the northern base of the Tmolus range.

† Of the fertility of many volcanic soils, when sufficiently decomposed, we see proofs in the vineyards of Vesuvius, Ischia, the Euganean Hills, &c.

The most western of the three schistose ridges, near the village of Megné, Sect. 7, is also covered with mounds of scoriæ, most of them destitute of lava-streams; but where these exist, they follow the slope of the country towards the north. On the eastern side of the ridge, near the summit, are some stratified beds of volcanic sand and ashes, sloping away from one of the cones.

About six miles north-west from Megné, on the north side of a table-land of lacustrine limestone, is a stream of basaltic lava, which descends nearly to the Hermus. Its lower part is black and vesicular, but it becomes redder and more scoriaceous higher up. The crater from which it has flowed is not apparent. Pl. IV.

3rd period.—The volcanic cones of both the second and third periods have been poured forth since the excavation of the valleys in the lacustrine formation; but their diversity, in point of age, is marked no less by the order of superposition than by the great difference in their state of preservation. The cones of the second period were described as being low and rounded in form, and covered with vegetation. Those of the third period, on the contrary, have all the features of volcanoes now in action. They rise at an angle of 30° or 32°, and the ashes and scoriæ which compose them are so loose as to render the ascent laborious. A few straggling shrubs and plants are the only vegetation which they produce; and the lava which has flowed from them is as rugged and barren as the latest products of Etna or Vesuvius.

The volcanoes of this third period are only three in number, and are nearly equal in size. Pl. IV. and Pl. V. Sect. 7. They stand in a nearly straight line from west by north to east by south, and at the distance of about six miles from each other*. It is remarkable, that each of them rises in the centre of one of the small alluvial plains which alternate with the three schistose ridges before described, therein differing from the cones of the second period, all of which stand upon or near those ridges. We will describe them in the order in which they stand in Section 7.

The easternmost of these three cones is about a mile and a half north-north-east of the town of Koola. Nothing can be more striking than the appearance of this town as viewed from the schistose ridge to the eastward. Pl. IV., Pl. V., Sketches 11, 12. On the right of the Sketch 12, is the cone of scoriæ and ashes, about 520 feet in height, denominated by the Turks *Kara-dewit*, or the "Black Inkstand." An immense sea of black and rugged lava has flowed from its base, and spread over the surrounding plain. This eruption has formed a dam across the valley of Koola, and by stopping the natural drainage has

* These three craters are evidently alluded to by Strabo in his description of the Catacecaumene:—*Δείκνυνται δὲ καὶ βόθροι τρεῖς, οὓς φύσας καλοῦσιν, ὅσον τετταράκοντα ἀλλήλων διεστῶτες σταδίους.*

caused a marsh, which in wet weather is increased into a lake. To avoid the risk of inundation from this source, the town of Koola has been built upon the extremity of the lava current, and is thus raised 15 or 20 feet above the plain.

The cone of Karadewit consists of scoriæ and ashes, principally of a reddish colour. On its north side is a small but very well-preserved crater; and a vast number of small conical hillocks of scoriæ have been thrown up amidst the lava which surrounds its northern base. The lava from this volcano flowed southward as far as Koola; but the larger portion of it descended in a northerly direction to the Hermus, which river has evidently been diverted from its course, and now makes a circuit round the north extremity of an overhanging cliff of lava nearly 50 feet high. The lava is vesicular and scoriaceous in the upper part, for about 6 or 7 feet. The next 20 feet are irregularly columnar, which structure gradually disappears in the lower part of the bed. The marls and limestones on the opposite or north-east side of the river have been here much altered by their contact with the igneous rock, having been converted, in many places, into a yellow jasper-like substance, with a shining lustre and conchoidal fracture.

This lava encircles the base of several cones of the second period, and overlies the older *coulées* which they have sent forth. Pl. IV. In one place, 2½ miles north-north-east from Karadewit, a narrow stream of lava has diverged to the east from the general mass, through a lateral valley in the schistose range, and has spread over the *coulée* from the easternmost cones of the second period before described. The contrast between the lava-streams of the two periods is most striking.

The second of these recent volcanic cones rises in the plain between Sandal and Megné*, and is also called Karadewit. Pl. IV. and Pl. V. Sect. 7. This hill consists of scoriæ and ashes resting on a more solid crust of red scoriaceous rock. The crater is quite perfect, and from 150 to 200 feet deep, the bottom being choked with large stones which have rolled down its sides. Many considerable caverns and hollows exist in the scoriaceous rock. The summit consists of a very narrow ridge surrounding the crater. A few small pine-trees grow on the sides of this cone, which, on the whole, bears more vegetation than the Karadewit of Koola. A stream of lava has issued from the west foot of the cone, and flowed northwards for about five miles to the Hermus; and a small crater, about a mile to the west of the great one, has sent forth a narrow *coulée* which soon joins the general mass. There is also a low flat cone to the south-east, apparently belonging to the 2nd period, from the crater of which a small stream of lava, about half a mile in length, has

* Near the centre of the plain of Megné a well has been sunk through the alluvial soil to the micaceous schist below.

been poured forth, apparently at the same time as the eruption from the large cone. This lava is remarkable from having flowed over the margin of the crater, instead of bursting through the sides, like all the other lava-streams in the district. This is probably owing to its having flowed from the crater of an older and therefore more consolidated cone; whereas the recent cones, being formed of loose scoriæ and ashes, and much higher, were unable to support the mass of heated lava until it should overflow at the summit, and therefore yielded at their bases a passage for the lava.

The third or most western of the three great cones is called by the Turks Kaplan Alan. Pl. IV. and Pl. V. Sect. 7. It is surrounded on all sides for more than a mile by the black and rugged lavas, which have flowed from its base, and render the approach laborious. The cone itself is more clothed with vegetation than either of the other two, though it does not yield to them in the steepness of its sides and the general freshness of its appearance. It possesses a very perfect crater, between 300 and 400 feet deep, surrounded by a ridge only 10 or 12 feet wide at the top, and about half a mile in circumference.

On the north-west side of the great cone, a small one about half its height has been thrown up, with a regularly formed crater, but no lava appears to have flowed from it. On the west, several other small cones of scoriæ have been formed amongst the lava.

The lava from Kaplan Alan has chiefly issued from the east side, whence, spreading round both flanks of the cone, the two streams again united, and flowed down the plain to the west for about three miles; then turning south-west the stream descended a narrow valley for several miles, parallel to the Hermus, from which it is separated by a ridge of lacustrine limestone. At length, finding an opening into the valley of the Hermus, it followed the bed of that river along the narrow gorge in the schistose hills, whence, issuing at Adala, it spread for about a mile over the plain of Sardis, and at length terminated at the distance of at least thirteen miles from its source*. We were able to examine closely only the two extremities of this remarkable *coulée*, and we recommend the future geologist to follow its entire course from Kaplan Alan to Adala, as likely to afford him some interesting examples of the action of the river upon the lava.

The appearance of this rugged mass of lava, issuing from a narrow gorge in the schistose hills, and spreading over the fertile plain at Adala, is very striking. Half a mile above the town is the remarkable scene

* It is remarkable that this is the precise length of the lava-stream of the Puy de Tartaret in Auvergne, which has flowed down the narrow valley of the Couze to Nechers, in a manner strictly analogous to the case here described. See Scrope on Central France, p. 117.

exhibited in the sketch, Pl. V. 13. It appears that the Hermus, when its course was first impeded by the lava, was compelled to flow over the top of the *coulée*, the cavities of which have here been filled by river-worn gravel, forming a stratum on the surface of the lava. In course of time the river has effected a passage between the mica-schist and the lava, and has denuded both rocks to a great depth. It now flows at the base of a cliff of compact lava about 80 feet high, imperfectly columnar in its lower part and scoriaceous above. Higher up in the gorge, the river has cut a channel completely through the *coulée* of lava, and changed its course from the right side to the left. Such are the effects which this river has produced in the lapse of ages. Yet so compact is the substance of the lava, that those parts of the *coulée* which have escaped the action of the running water are as rugged and naked as the latest eruptions of Vesuvius, and exhibit not the slightest tendency to decomposition.

It is evident from the phænomena here exhibited, that however recent may be the appearance of the lava-stream, and of the cone from which it flowed, a very high antiquity must in reality be assigned to them. It is certain, from historical testimony, that no volcanic eruption has taken place in this country for at least thirty centuries; and how many more may have been required to enable the Hermus to denude the solid lava to the extent we now see, it is not easy to calculate. It would not indeed be fair to assume that the whole of the present channel of the river between the lava and the mica-schist has been excavated to the depth of 80 feet since the flowing of the lava; for the rounded margin of a *coulée* commonly leaves a hollow between it and the pre-existing rock, which the river would naturally follow, and by undermining the columnar lava might produce a perpendicular cliff in less time than would at first sight be supposed. But when we find the river cutting *across* the *coulée*, and passing from one side of the lava to the other, we can assign no other satisfactory cause for this phænomenon than the mechanical action of the river, operating during vast periods of time, and gradually wearing down its bed to its present level*.

The freshness of appearance in the three great cones, the vast anti-

* The above observations respecting the agency of running water in cutting through the lava-streams, are Mr. Strickland's. In justice to himself, Mr. Hamilton is obliged to state that he does not agree with the opinion, that the cutting across the *coulée* and the formation of the cliff are at all owing to the effect of running water acting upon the adamantine basalt. The very circumstance of the perpendicularity of the sides is an argument against it. He is rather inclined to attribute it to the fall of the basaltic masses in consequence of their having been undermined by the waters, the operation of which cause may have been hastened by fissures and crevices produced by the numerous earthquakes to which this country has at all times been exposed.—W. J. H.

quity of which we have thus demonstrated, proves how small is the effect of atmospheric agency upon volcanic products when unassisted by running water. Had these volcanoes been ejected within the last ten years, the cones of scoriæ could hardly have been more perfect, the craters better defined, or the streams of lava more black, rugged and barren. Nor is this difficult to explain, when we remember that the lava and scoriæ in this region seem to have scarcely any tendency to chemical decomposition, and that the rain which falls on the cones, being instantly absorbed by their porous soil, exerts no other mechanical force than the mere impact of the drops in falling.

These considerations render it difficult to explain the marked difference of appearance between the volcanic products of the second and of the third periods, on the mere assumption of a difference in their ages. It appears to us that a tenfold greater period of time than has already elapsed since the ejection of the volcanoes of the third period, would barely suffice to reduce them by atmospheric agency alone to the smooth and rounded forms presented by the cones of the second period; but if we suppose the latter to have been submerged beneath a body of water, either at the time of their ejection or subsequently, the explanation will be more easy. The action of water would soon obliterate all the sharp features of these mounds of ashes, and by filling up the cavities in the lava-streams with transported matter, would form that smooth and fertile surface which they now present. It is true that this explanation is not without its difficulties. The lake which existed here before the basalt of the first period was poured forth had long been drained, and deep valleys of denudation had been excavated in its bed before the second period of volcanic eruption commenced. In order that the valley might be reoccupied with water to a sufficient elevation to cover the tops of the volcanic cones, the gorge at Adala would require to be blocked up to a height of perhaps 2000 feet, and the sojourn of this supposed mass of water must have been very transient, as we find no regular sedimentary deposits of any importance which can be assigned to it. Such a supposition involves an extensive flight into the region of conjecture, and many geologists would probably prefer explaining the wasted condition of these volcanic mounds by means of ordinary atmospheric causes acting through an immense period of time*.

We have before alluded to the singular fact, that all the cones of the

* It may assist conjecture as to the relative and perhaps also the absolute antiquity of the three sets of volcanic eruptions above mentioned, if we remind the reader that the valleys have been denuded to the depth of 800 feet since the *first* period of eruption, of 200 feet since the *second*, and 80 feet since the *third*. The different density of the materials acted on in each case must of course be taken into the account.

second period rise out of the ridge of schist and marble on the south of Koola, and its three lateral offsets (p. 58); while the three volcanoes of more recent date occur in the plains which intervene between the several ridges (p. 60). This phænomenon may perhaps be accounted for by supposing that several cracks and fissures were produced in the schistose ridges at the time of their elevation, through which the igneous products at first found vent; that in process of time, lavas having been injected into all the fissures and become cooled, the whole mass was rendered more compact than it was before; and therefore that when, at a later period, volcanic forces were again in action, the points of least resistance would be in the planes which intervened between the consolidated ridges, where the new eruptions would consequently take place.

It may be doubted whether the older series of eruptions supplied the forces which elevated the ridges themselves. The stratification of the latter is very irregular, and seems not to possess an anticlinal arrangement. Moreover, the undisturbed horizontality of the lacustrine beds in the vicinity is scarcely compatible with any great subsequent disturbance of the schistose rocks.

Mineralogy.—To the mineralogist the Catacecaumene presents rather a barren field. The scoriæ and lava are of ordinary character; the latter contains olivine and augite, but no other crystalline minerals were observed.

Thermal springs.—In a valley about a mile north of the Hermus, and six miles north-north-east of Koola, are some remarkably copious hot springs, rising in the lower beds of the lacustrine limestone. The water is perfectly tasteless, and forms no deposit or sediment whatever. The thermometer rose at the lowest spring to 123° Fahr. A short way to the east, in the centre of the ruins of an undetermined ancient city, are two other sources, at which the thermometer rose respectively to 133° and 137°. They rise out of a crevice in the calcareous tufa, and exhibit a slight development of sulphuretted hydrogen gas.

Comparison of the Catacecaumene with Auvergne.

No well-informed geologist can fail to be struck with the numerous and remarkable analogies between the Catacecaumene and the volcanic district of Central France, so elaborately described by Messrs. Scrope, Lyell, Murchison, &c. We will here present, in a tabular form, a few of the chief points of resemblance between the two districts. Some of these analogies may perhaps be only accidental coincidences, but the majority of them afford instructive examples of the universality of Nature's laws, whereby similar causes produce similar effects, even in the most distant regions.

Points of agreement between the Catacecaumene and Auvergne.

	EXAMPLES.	
	Catacecaumene.	Auvergne.
1. In both cases, clusters of extinct volcanoes occur in the interior of continental tracts remote from the sea.		
2. They are referable to several distinct periods, all subsequent to the commencement of the tertiary epoch.	Three periods.	Numerous successive periods.
3. They rise out of rocks apparently of the primary epoch.	Mica-schist.	Granite.
4. The direction of the volcanic zone coincides with the strike of the older rocks through which it is ejected.	E. by S. to W. by N.	N. to S. and N.N.W. to S.S.E.
5. The volcanoes rise near the margin of a tertiary lacustrine basin, which they have partially overspread with basalt and lava.	Basin of the Catacecaumene.	Limagne, d'Auvergne.
6. Eruptions of an early date have poured forth basalt over the lacustrine deposits prior to their denudation.	Basaltic plateaux N.E. of Koola.	Plateau of Chateaugay, Gergovia, &c.
7. Extensive valleys of denudation have been excavated in the lacustrine deposits, which now form tabular hills capped with basalt.	Valleys of the Hermus and Aïneh-chai.	Valleys of the Allier and its tributaries.
8. Eruptions of a later date have poured forth streams of lava, which have flowed down the valleys of denudation.	Cones of the 2nd and 3rd periods.	Puy de Nugère, La Vache, Gravenere, &c.
9. These eruptions have in some cases stopped the drainage of the valleys and produced lakes.	Marsh near Koola.	Lakes of Aidat and Chambon.
10. These newer streams of lava have been deeply excavated by the action of rivers now flowing.	Gorge above Adala.	Lava of Pont Gibaud, Jaujac, &c.
11. The more recent volcanoes, though extinct from the earliest historic times, present all the sharpness of outline and barrenness of surface incident to volcanoes now in action.	The two Karadewits and Kaplan Alan.	Puy de Pariou, Come, La Vache, &c. &c.
12. Thermal springs rise in the vicinity of the volcanic rocks.	Springs six miles N.N.E. of Koola.	Mont Dor, St. Nectaire, &c.

General Conclusion.

It would be rash, without much more extended observations on other parts of Asia Minor not visited by us, to attempt any generalization or theory on the successive formations of the different groups and systems of rocks in the Western part of Asia Minor. Most of those which have occurred to us as safe, in the present imperfect state of our knowledge and the scarcity of organic remains, have been introduced into the foregoing pages. There are, however, a few remarks which result from the observations on the coast of Ionia and Caria, which we think may with propriety be noticed here.

1. That the scaglia, or compact white Alpine limestone, is more abundant in Rhodes and the southern parts of Asia Minor than in the northern or central districts of the peninsula. It is the same formation which constitutes the principal mass of the range of Mount Taurus, of which it may be considered as the western prolongation. Judging from the Nummulites found in it near Adalia, and at Deenair, near the source of the Mæander, it appears to be the same group which extends through the Morea into the Ionian Islands.

2. The micaceous schist and saccharine marble, which are so extensively developed in the mountain-chains extending through the central parts of Asia Minor, as Mount Tmolus, Gallesus, Messogis, and along the coast of Ionia, gradually disappear towards the south, where they are replaced by the scaglia formation* †.

3. The igneous and volcanic rocks, which in the central and northern parts of Asia Minor are found almost universally distributed, become extremely rare towards the south.

4. Trachytic or other igneous rocks are frequently found associated with the crystalline marbles, as at Erythræ and Boodroom, along the coast; and at Cyzicus and in the Catacecaumene in the interior. But in the absence of organic remains, it is impossible to decide whether these marbles may not have had the same origin as the scaglia, and have been altered and crystallized by their proximity to the foci of igneous eruptions; or whether the volcanic outbursts have not raised to the surface formations of a former period, and from a greater depth, in the immediate neighbourhood of their elevation ‡.

We will conclude this imperfect contribution to the geology of West-

* The peninsula of Asia Minor may therefore be regarded as one vast axis of schistose rocks, flanked on the south by the Cretaceous system.—H. E. S.

† M. Texier draws very similar conclusions, except that he considers the rocks of Taurus to be tertiary instead of cretaceous. See Comptes Rendus, vol. iv.—H. E. S. MS.

‡ Mr. Strickland is however of opinion, that the mica-schists and saccharine marbles constitute a distinct and well-marked formation, long anterior in age to the Scaglia. He considers it as an hypogene rock, which owes its present metamorphic character to causes far more extensive and deeper-seated than the volcanic eruptions which appear on the surface. He grounds this opinion on the constancy of character in this rock, whether in the vicinity of trachyte or otherwise; and on the fact, that the alterations produced in the tertiary beds by the injected trachyte are very limited in extent and frequently imperceptible.—H. E. S.

ern Asia Minor by comparing some of its more remarkable features with those portions of North-western Europe which have been most carefully explored. It will be seen that, notwithstanding the close analogy which we have shown to exist between some of the most recent geological phænomena of these two portions of the world, yet on the whole the geology of Asia Minor is more remarkable for its contrast than for its resemblance to that of Northern Europe. These discrepancies do not indeed affect those general principles of the science, which, established on the inductions afforded by our own and the neighbouring countries, have since been found to prevail throughout the world. Thus the crystallized condition of the older sedimentary rocks,—their elevation on either side of granitic axes,—the disruption and alteration of strata by igneous injections,—the parallelism of mountain-chains within limited districts,—the absence of organic remains in the older strata,—their gradual approach to existing forms as we descend the chronological scale,—even the want of continuity between the secondary and tertiary series of deposits,—all these phænomena of Northern Europe find their counterpart in the region here described. But when we enter into greater detail, we shall find that Asia Minor, while it agrees in many respects with the geological types afforded by Greece, Italy, Spain, and the whole circuit of the Mediterranean as far as it is known, differs no less from those parts of Europe which lie to the north of the Alps and Pyrenees. For the sake of brevity and perspicuity we will recur to the tabular form in exhibiting these points of disagreement.

In Western Asia Minor,	*In North-western Europe,*
1. The oldest organic remains belong to the cretaceous system *.	1. An immense series of fossiliferous strata underlies the cretaceous system.
2. The limestones and shales of the cretaceous system are compact, and sometimes semi-crystalline.	2. The limestones and shales younger than the carboniferous system are soft and earthy.
3. The mountain-chains have been elevated at the close of the secondary period.	3. The loftier mountain-chains have been elevated chiefly in the transition or early part of the secondary period.
4. The igneous rocks are chiefly of the tertiary period.	4. The igneous rocks are chiefly of the secondary period †.

It may become a question for the speculative geologist, whether all these discrepancies may not be resolved into a mere difference in the dates at which a similar train of events occurred in both these regions.

* The Silurian rocks of the Bosphorus have been shown to belong to a distinct geological region. See note, p. 33.

† Auvergne and the Eifel form, we believe, the only exceptions to this proposition.

Thus, if we assume that a series of vast subterranean commotions, accompanied by the exhibition of intense heat, has taken place in both countries,—obliterating the organic remains in the lower strata, and indurating the substance of the upper ones, elevating mountain-chains and setting in action a long train of volcanic operations,—we have only to suppose that in Asia Minor this period of subterranean energy occurred at the close of the secondary epoch, and in Northern Europe at its commencement; and the chief geological differences between the two countries may perhaps be explained without much difficulty.

---

6. ON A TERTIARY DEPOSIT NEAR LIXOURI, IN THE ISLAND OF CEPHALONIA*.

[*Read May* 3, 1837.]

THE tertiary deposit here described occupies a considerable portion of the peninsula of Lixouri, on the western side of the Gulf of Argostoli. It forms a series of ridges, extending for two or three miles to the north and south of the village of Lixouri, and running parallel to the strike of the secondary rocks of the island of Cephalonia, as well as of the whole of the Ionian Isles, which conform in their directions to the great mountain systems of the Apennines and of Dalmatia. The beds slope gradually to the eastward, and present a succession of steep escarpments towards the west. The width of this tertiary zone may be about four miles from the sea on the east to the mountainous ridge of secondary rocks against which they rest on the west. The beds are all perfectly conformable, dipping a few degrees to the north of east by compass, at an angle of from 45° to 55°. Their aggregate thickness may be estimated at about 900 feet. They are remarkable for the great number and variety of fossils which they contain, some of the beds being almost wholly composed of shells, in the most perfect state of preservation, many of which belong to species now existing in the Mediterranean. A large proportion of these shells are identical with species figured by Brocchi from the Subapennine beds, indicating that this deposit must be referred to the Pliocene epoch. This locality is also interesting from the great thickness of the beds, and the variety of material of which they consist. The lowest portion is composed of fine-grained white limestones, which, from their resemblance to the rocks of Malta, are probably referable to the Miocene age. Unfortunately, our time was too short to examine this part of the series thoroughly, but it is possible that a transition may here be traced from the Miocene to the Pliocene series.

The accompanying section will show, in descending order, the super-

* This paper was read to the Society so long ago as in the year 1837, before Mr. Hamilton's return to England (Proc. Geol. Soc. vol. ii. p. 545), but its publication has been delayed till the present time in consequence of my having sent a selection of the fossils referred to in it for the examination of M. Deshayes, at a time when that eminent naturalist was absent in Algeria. After his return the specimens were mislaid, and I only received them from Paris a short time ago. M. Deshayes has kindly favoured me with his notes upon these fossils, and they have also been examined by Prof. E. Forbes, whose specific determinations are annexed to those of M. Deshayes.—H. E. STRICKLAND, Jan. 1847.

position of these beds. It was made about two miles to the north of Lixouri, where a road descends from the quarries of Schœno to the sea.

Section about two miles north of Lixouri.

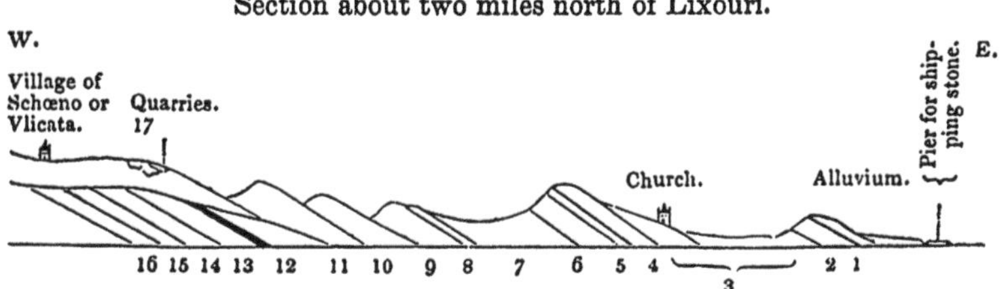

No. 1. Very hard thin-bedded limestone of a reddish-brown colour, containing casts of recent species of shells, chiefly Turritella and Pecten of small size. This bed dips at an angle of 50° towards the east, and at some points rises to a considerable height, with a steep westerly escarpment. The thickness of the bed is from 10 to 20 feet. About three-quarters of a mile north of Lixouri it consists of a hard conglomerate, with small pebbles of flint and quartz in a hard calcareous matrix.

2. Dirty yellow sand, containing in the upper part many irregular concretionary nodules of limestone, containing Pecten, Ostrea and Anomia. The calcareous concretions gradually diminish downwards, and the lower portion consists of fine sand full of shells, principally Pecten, Cerithium, Dentalium, *Isocardia cor*, Turritella, Venus, &c. The thickness of this bed may be about 100 feet. It is best seen in a small hill south of the road from the quarries to the sea.

3. Blue marly clay, which in most places immediately underlies the sands of the last stratum, but in some places a band of hard calcareous marl, consisting almost entirely of shells, from 3 to 5 feet thick, is interposed between the two. These shells appear to belong almost entirely to the same species as occur in the blue clay below. The latter deposit is about 200 feet thick, and contains a great variety of shells, of the genera Dentalium, Fusus, Rostellaria, Buccinum, Murex, Turritella, Cerithium, Cardium, Cardita, Venus, &c. In a lower division of this blue clay Turritellæ are most abundant, scarcely any other shell being found with them, and in an upper portion of the same bed a small oblong species of Sponge is very plentiful.

All the above beds are exposed to the eastward of the road which leads from Lixouri to the quarries of Schœno; the following are to the westward of the same road.

4. Limestone resembling No. 1 in colour and general appearance, but not so hard. This bed is about 8 or 10 feet thick, and contains numerous Pectens of a very large size.

5. Bed of *Lobularia arborea*, about 2 feet thick, the stems and

branches of which are not in the least broken or displaced. Imbedded in them are a few large Pectens and Oysters.

6. Various beds of yellow limestone, sand and marl, with many shells of Pecten, Ostrea, Cardium, Cardita, Terebratula, Pectunculus, and casts of various other bivalves. The Terebratulæ only occur in the very lowest portion of this bed. The whole thickness of these beds may be about 50 feet. They form a hill of considerable height, with an abrupt escarpment on the west, and sloping to the east at an angle of about 45°. A church and the road to Lixouri are at its eastern base, and the road from the quarries to the sea passes to the north. The view from the summit is highly picturesque, and exhibits the structure of the country in a remarkable manner. The alternations of hard and soft strata in this tertiary deposit have produced a series of low parallel hills running north and south, each presenting a steep escarpment on the west, while they slope to the east at an angle equal to the dip of the strata, varying from 45° to 55°.

7. A deposit of blue marl or clay between 200 and 300 feet thick, containing numerous shells, chiefly of the genera Buccinum, Fusus, Turritella, Cerithium, Dentalium and Pectunculus. This bed is well displayed in the broken ground on the south-west and west of the last-mentioned hill.

8. Limestone or calcareous marl, a few feet thick, containing *Pecten varius* and Caryophyllia.

9. Alternating blue and white marls, apparently without fossils, cracking into cubical and rhomboidal fragments at right angles to the stratification. The thickness of these beds is about 100 feet, and they are exposed in a hill on the north of the road to the quarries.

10. Hard yellow marly sandstone without fossils, breaking into irregular fragments. It is exposed in the rivulet on the south of the road.

11. Blue and white marl.

12. Gypsum, varying from 10 to 50 feet thick, composed of an aggregation of large selenitic crystals. Their weathered surfaces have a curious appearance, resembling the crystallization seen on windows during a hard frost, and are partly finely laminated, varying in colour from clear white to grey. In one place the gypsum rises to a high ridge, on the edge of which the village of Vlicata is situated, and is here of considerable thickness.

13. Yellow or white marly sandstone, containing rarely shells of Pecten, Ostrea and Terebratula.

14. Bands of thin-bedded red and grey limestone, very hard, containing no fossils.

15. Blue clay about 20 feet thick, also without organic remains.

16. Gypsum resembling No. 12, and equally devoid of fossils.

Nos. 10 to 16 are seen on the left of the road in ascending towards the quarries. On the right, at the distance of about 100 yards, the rock No. 17 juts out abruptly and unconformably to the tertiary series. It consists of a fine-grained whitish limestone, similar in appearance to that of Malta, and like it probably belonging to the Miocene age. It is here largely quarried, and transported to Argostoli and other places for building. The stone is soft and easily worked, but decays if much exposed to the sea-air. Fossils are rare in it, but we noticed a large species of Pecten, and were shown one or two shark's teeth which had been found here. At the quarries this rock dips about 10° to the south-south-west, but in a valley half a mile to the east it has an easterly dip, and if viewed at this point alone might be supposed to underlie conformably the bed No. 11 of the tertiary series which composes the opposite side of the valley. The relations of these rocks to the secondary limestones, which form the mountain ridge on the west, must be worked out by observers who have more leisure than was at our command.

Appendix by Mr. Strickland.

The Pliocene portion of the beds above described may be classed under three principal divisions,—the calcareo-arenaceous beds, consisting of Nos. 1 and 2 in the section,—the argillaceous, comprising Nos. 3 to 10, —and the gypseous, including Nos. 11 to 16. These divisions are of importance, because a similar arrangement prevails in the tertiary beds of the isle of Zante, of which I have published a description, p. 19 (Geol. Trans. 2nd series, vol. v. p. 403).

The total number of fossil species found in these beds is upwards of ninety; but as some of them are not easily determinable, and as the microscopic species are not yet examined, the number included in the following list is only eighty-four. The remarks of M. Deshayes and of Prof. E. Forbes, which were made independently of each other, are respectively distinguished by the letters D. and F. Where these letters are wanting, the specific names have been determined by myself from the works of Brocchi and others.

A considerable number of the species enumerated are now living in the Mediterranean. A comparison of their distribution as fossils with that of their recent homologues*, as shown in the valuable tables pre-

* The word *homology* having been recently introduced into this country by Prof. Owen as a term of Comparative Anatomy, and being in fact identical with *affinity* as opposed to *analogy* (see Phil. Mag. S. 3. vol. xxviii. p. 357), I venture to recommend the adoption of the word *homologue* in place of the usual but inaccurate term *analogue*, to express those recent and fossil species which are either actually identical or are allied by very close affinity to each other. The term *ana-*

sented by Prof. E. Forbes in his Report on Ægean Invertebrata (Report of Brit. Assoc. 1843), enables us to determine approximately the depths at which the several beds were deposited. The highly fossiliferous beds 1 to 8 are about 105 fathoms thick, and from their organic remains appear to correspond to Prof. E. Forbes's Regions IV., V., VI., VII., and the upper part of Region VIII., representing a depth beneath the surface of from 20 to 125 fathoms. The marly and gypseous beds 9 to 16, in which organic remains are either very rare or wholly absent, are equivalent to the lower portion of Region VIII., which approaches the zero of animal life. We may therefore infer a tranquil condition throughout the deposition of this Pliocene formation, during which the sea gradually became shallower from the filling up of its bed, and the fauna underwent corresponding modifications. The unconformability of the subjacent Miocene rocks indicates a preceding period of disturbance, while the high angle at which the whole Pliocene series is now elevated, proves an enormous dislocation at a very recent geological date.

List of Fossils from Lixouri in Cephalonia.

The figures in the right-hand column refer to the beds in which some of the species are found, as distinguished in the accompanying section. Those species to which no number is attached were chiefly found in the beds 3 and 7, but their precise position was not noted.

The species of which specimens have been presented to the Geological Society are marked G. S.

			Stratum in which found.
G. S.	1.	*Caryophyllia conica?*	No. 8
G. S.	2.	*Fungia*	8
G. S.	3.	*Lobularia arborea*	5
		Balanus balanoides?, living in the Mediterranean.—F.	
G. S.	4.	*Serpula arenaria*, Linn., Brocchi; *Vermetus arenarius*, Desh.—D.	
G. S.	5.	*Serpula.*	
G. S.	6.	*Serpula glomerata*, Lam.	
G. S.	7.	*Amphidesma pubescens*	6
G. S.	8.	*Corbula nucleus.*	
	9.	*Lucina spinifera* (Mont.), living in Med. and Brit. seas.—F. (*Lucina lupinus*, Desh.—D.)	
	10.	*Astarte incrassata*, Brocchi, pl. 14. fig. 7.	
	11.	*Venus paphia*, Linn.	
	12.	*Venus boryi*, nob. fossile de Morée.—D.	
G. S.	13.	*Venus casinoides?*	3

---

logue might still be retained for those distinct groups of animals which discharged *analogous* functions at successive geological epochs; thus the Zoophagous Gasteropoda of the tertiary and recent period are analogues of the Cephalopoda which prevailed in the oolitic series, the existing Chiroptera are analogues of the Pterodactyles, the Decapodous Crustacea of the Trilobites, &c.

Stratum in which found.

14. *Venus radiata*, Brocchi.—D. (*Venus ovata*, Mont.—F.)
G. S. 15. *Cardita aculeata*, Philippi; *Chama aculeata*, Poli.—D. ...... No. 3
16. *Cardita.*
17. *Circe minima* (Mont.), living in Med. and Brit. seas.—F. (*Cytherea an venetiana? jeune.*—D.)
18. *Cardium aculeatum* 6
19. *Isocardia cor* 2
G. S. 20. *Arca antiquata?*
21. *Arca* 2
G. S. 22. *Pectunculus*, appartenant probablement au *P. glycymeris* de Linné.—D. 6
G. S. 23. *Pectunculus auritus*, Brocchi, pl. 11. fig. 9 7
24. *Nucula minuta*, Brocchi, pl. 11. fig. 4.
25. *Nucula.*
26. *Chama.*
27. *Pinna* 6
28. *Pecten* 4
29. *Pecten*, from the quarries at Schœno.
30. *Pecten pleuronectes* de Brocchi, spec. nov. pour moi.—D. (*Pecten cristatus*, Bronn, a fossil in Italy.—F.)
31. *Pecten varius* 8
G. S. 32. *Pecten opercularis* 2
33. *Pecten pusio*, living in Med.—F. (*P. ornatus?*, Lam.—D.)
34. *Pecten dumasii*, Payr., living in Med.—F. (*Mihi incognitus.*—D.)
35. *Pecten.*
36. *Ostrea edulis* 6
37. *Anomia ephippium* 2
38. *Terebratula* (like *T. perovalis*) 6
G. S. 39. *Dentalium elephantinum* 3
40. *Dentalium sexangulum*, Brocchi, pl. 16. fig. 25.
41. *Dentalium fissura*, Lam.
G. S. 42. *Ditrupa coarctata*, Brocchi, pl. 1. fig. 4.
43. *Pileopsis ungarica* 6
G. S. 44. *Natica millepunctata.*
45. *Natica tigrina?*, Defr.
46. *Natica*, an spec. nov.? très voisine du *N. Dilwynii*, Payr.—D.
47. *Pyramidella suturalis*, Sow.
48. *Pyramidella subulata*, Brocchi, pl. 3. fig. 5.
49. *Solarium plicatum*, Lam.
50. *Trochus agglutinans*, Lam.
51. *Turbo rugosus.*
G. S. 52. *Turritella triplicata*, Brocchi, pl. 6. fig. 14. 3
G. S. 53. *Turritella acutangula*, Brocchi, pl. 6. fig. 10.
54. *Turritella ungulina*, Desh.; *Turbo ungulinus*, Linn.—D.
55. *Cerithium vulgatum*, Brug.
G. S. 56. *Cerithium varicosum* 3
57. *Cerithium fuscatum*, var. living in Med.—F. (*C. mediterraneum*, var. nob.—D.)
58. *Pleurotoma reticulata*, var. *spinosa*, living in Med.—F. (*Fusus fusulus*, Brocchi.—D.)

Stratum in which found.

59. *Pleurotoma multiruga*, Bronn, a rare Sicilian fossil.—F. (*P. balteata*, Beck.—D.)
60. *Pleurotoma attenuata* (Mont.), living in Med. and Brit. seas.—F. (*P. vulpeculus*, var. Brocchi, sp. nov. nobis.—D.)
G. S. 61. *Pleurotoma dimidiata*, Brocchi.
62. *Pleurotoma renieri*, Scacchi, a Calabrian fossil.—F. (Nouvelle espèce.—D.)
G. S. 63. *Fusus*, nov. spec.—D. .... No. 7
64. *Fusus** near *F. crispus*, but distinct from any species with which I am acquainted.—F. (Confondu par Brocchi parmi les variétés de *P. dimidiata*, sp. nov.—D.) 7
65. *Fusus vulpeculus*, Brocchi, pl. 6. fig. 11. 7
66. *Fusus muricatus*.
67. *Fusus rostratus*?, Brocchi, pl. 8. fig. 1.
G. S. 68. *Fusus crispus*, Lyell, pl. 1. fig. 8.
69. *Fusus longiroster*, Brocchi, pl. 8. fig. 7. (*F. rostratus*, Oliv., living in Med.—F.)
70. *Triton cutaceus*.
G. S. 71. *Rostellaria pes-pelecani* 3
72. *Buccinum mutabile*.
G. S. 73. *Buccinum prismaticum*, Brocchi, pl. 5. fig. 7 7
G. S. 74. *Buccinum semistriatum*, Lyell, pl. 1. fig. 11 7
75. *Buccinum*, c'est une des espèces confondues avec le *B. reticulatum* de Linné.—D.
76. *Buccinum asperulum*, Brocchi, pl. 5. fig. 8.
77. *Nassa variabilis*, living in Med.—F. (Confondu avec le *B. turbinellus* de Brocchi, spec. distincta, nob.—D.)
78. *Purpura*.
79. *Columbella polita*; *Fusus politus*, Bronn; *Nassa columbelloides*, Brocchi.—F. (*C. subulata*, Brocchi.—D.)
80. *Mitra plicatula*, Lyell, pl. 1. fig. 12 7
81. *Mitra philippiana*, Forbes?, living in Med.—F. (Confondu avec le *plicatula* de Brocchi, pour moi une espèce distincte.—D.)
G. S. 82. *Mitra cupressina*, Brocchi, pl. 4. fig. 6 7
83. *Mitra*†, large and fine species, not known to me. Not *M. zonata*, as I supposed.—F. (Nov. spec. mihi incognita.—D.)
84. *Conus antediluvianus*, Brocchi, pl. 2. fig. 11.

The rest of the island of Cephalonia, as far as we observed it, consists of scaglia or Apennine limestone. Fossils are rare, but about a mile north of Argostoli we observed in it many small spiral univalves, and near the interesting Cyclopian walls of ancient Krani we found specimens of Nerinæa. In crossing the island from Argostoli to Samos, the stratification of these secondary limestones is distinctly developed, dipping for several miles about 25° east, and near the middle of this vast

* Figured in next page, *F. filamentosus*, Strickland.

† Figured in next page, *M. juniperus*, Strickland.

formation we found two beds of a plicated Ostrea, each about a foot thick.

In Cephalonia and in the range of St. Salvador in the north of Corfu, the secondary or Apennine limestone is admirably displayed, having a regular dip to the eastward, and exhibiting an aggregate thickness of many thousand feet. If a careful observer were to make a section at these two points across this great formation, he might establish a series of mineralogical or palæontological subdivisions, and might determine to what extent this vast calcareous deposit of the Mediterranean basin is equivalent to the secondary series of Northern Europe.

Description of two apparently new species mentioned in the foregoing list. By H. E. STRICKLAND, Esq.

Fusus filamentosus, Strickland.—Small, taper, volutions about nine, tolerably rounded, with a fine suture. Ribs twelve on the first volution, very prominent, regular, and rounded; terminating rather abruptly backwards, and leaving a slight space between them and the suture. The intervals are deep, hollowed, and equal to the ribs. Both ribs and intervals are uniformly covered by fine regular thread-like striations, of which the first volution has about twenty-eight, including those which cover the canal; the second volution has ten, and the third six, some of the alternate ones having disappeared. Besides these striations there are about four much finer ones which cover the posterior portion of each volution between the ends of the ribs and the suture. Aperture oblong-ovate, both lips smooth, canal moderately long, and curved to the right. Length ·55 inch, breadth ·15 inch, first volution ·25 inch: angle of spire 30°.

Fusus filamentosus.

Mitra juniperus, Strickland.—Taper, with ten or eleven slightly rounded volutions. On each volution are from twenty to twenty-two slender, rounded, tolerably regular, longitudinal ribs, becoming evanescent towards the suture, which is distinctly marked by a fine line. The intervals between the ribs are in general slightly wider than the ribs themselves, and are shallow, flattish, and furnished with fine irregular longitudinal wrinkles, especially on the first volution. Both ribs and intervals are crossed by nine or ten fine thread-like striæ, producing a reticulated surface; two of these are at the posterior part of the volution near the suture; in front of these is a smooth zone, equal to about one-fifth of the exposed

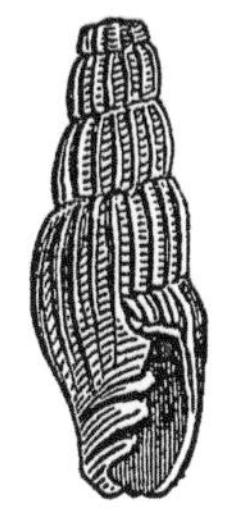
Mitra juniperus.

part of the volution, which is destitute of striæ; the remaining striæ are at the anterior portion of the volution. The anterior portion of the first volution has in addition about twenty more striations, regular and closely compacted, which are concealed by the succeeding volution. Aperture narrow; columella with three strong, rounded spiral folds, the hindermost largest, the anterior one slight, their intervals broad, flat and smooth; between the two anterior ones is a slight trace of another. The medial portion of the outer lip is occupied interiorly by eleven or twelve raised thread-like lines, which penetrate into the mouth as far as can be seen. Length about 1 6 inch, breadth 45 inch, first volution 6 inch: angle of spire 18°.

MEMOIRS ON THE GEOLOGY OF GREAT BRITAIN AND IRELAND.

1. ON THE GEOLOGY OF THE VALE OF EVESHAM.

[From the 'Analyst,' vol. ii. p. 1.]

[So early as 1833, and when not yet elected a member of the Geological Society, Mr. Strickland communicated in letters to his friend Mr. Greenough two papers. The one was "On the Red Marl and Lias of Worcestershire," read December 1833; the other, "On the Occurrence of Freshwater Shells of existing Species beneath the Gravel near Cropthorne, Worcestershire," read 1834.

The first paper was accompanied by a manuscript map, on which was laid down with greater accuracy than had been before attempted, the boundaries of the red marl and lias in the districts adjacent to Pershore, Evesham, Bitford, Alcester, Droitwich and Worcester; and for the first time a line of fault is pointed out ranging from a little north of Bredon Hill in Gloucestershire, to Inkberrow north of the road from Alcester to Worcester. Mention is also made of bones and teeth of Hippopotamus and Deer having been found in the gravel near Cropthorne. (Proc. Geol. Soc. vol. ii. p. 5.)

In the other letter (April 3, 1834), Mr. Strickland states that the surveyor of the roads had opened a new pit, in which the same species of shells had been found under horizontal beds of gravel and sand, which presented no signs of having been disturbed since their deposition; and that the remains of the Hippopotamus, Deer, and, he believes, Ox, had been found in considerable abundance in the same pit. (Proc. Geol. Soc. vol. ii. p. 95.)

Again, in December 1834 a paper was read, "An Account of Land and Freshwater Shells found associated with the bones of Land Quadrupeds beneath diluvial gravel at Cropthorne in Worcestershire," of which the following abstract was given in the 'Proceedings':—"The deposit in which these remains were found is situated on the road from Evesham to Pershore, and on the east side of the small rivulet which flows from Bredon Hill towards the Avon. In May 1834, the deposit presented a section about 70 yards in length, and 8 feet 6 inches high in the middle. The lower part of it consisted of lias clay, on which rested a layer of fine sand, containing twenty-three species of land and freshwater shells, with fragments, more or less rolled, of bones of the Hippopotamus, Bos, Cervus, Ursus and Canis. The sand passes upwards gradually into gravel, which extends to the surface, and differs in no respect from the other gravel of the neighbourhood, being composed principally of pebbles of brown quartz, but occasionally containing chalk-flints, and fragments of lias Ammonites and Gryphites. The bones, though most abundant in the sand, are interspersed also through the gravel; but the shells are confined to the sand. Lists are given of the bones and of the species of the shells, two of which are supposed to be extinct. The author from these phænomena assigns the deposit to the newer Pliocene era; and from the fluviatile habits of some of the shells, he conceives that it occupies the site of an ancient river-bed, and not of a lake. In the course of his paper he points out the inferences which may be drawn from these deposits, respecting the greater change which has taken place in the mammifers of this island than in the molluscs, since the era when the gravel was accumulated; and the little variation which the climate appears to have undergone since the same epoch. In conclusion he notices the published accounts of similar deposits at North Cliff, near Market-Weighton, and at Copford near Colchester, and states that he was informed at Bath that freshwater shells had been discovered under gravel, in sinking for foundations in the lower parts of the city." (Proc. Geol. Soc. vol. ii. p. 111.)

The whole of the information thus obtained was communicated in a lecture read to the Worcestershire Natural History Society in March 1834, and subsequently published as the following "Memoir" in the second volume of the 'Analyst.']

At the request of several members of the Society, I have been induced to draw up an account of the geology of that part of the county of Worcester with which from personal residence I was best acquainted,—an investigation which was the more desirable, as this district does not appear to have been hitherto very minutely examined by the geologist.

Before I proceed to examine the subterranean structure of the Vale of Evesham, I will give a short sketch of its more obvious and superficial characters. This district is bounded on the east by the lofty escarpment of the Cotteswold Range, and on the south by Dumbleton and Bredon Hills, which constitute outliers of the same geological structure as the Cotteswolds, with which they have doubtless at some period been continuous. To the north and west this district has no natural boundaries, and might without impropriety be extended over nearly all Worcestershire, which is in fact one vast vale, of which the south-east angle is denominated the Vale of Evesham.

The Vale of Evesham has an undulating surface, and in some parts of it hills occur of considerable steepness, and from 200 to 300 feet in height. These hills, however, by no means render the term *vale* inapplicable to the district, since from the lofty heights of Broadway and Bredon those minor elevations are barely distinguishable from the plain. From its low and sheltered position, the average temperature of the district is high, and hence in a great measure arises the fertility for which it has been perhaps too highly celebrated. In some parts of the district, indeed, tracts of very good land occur. This is chiefly the case where the surface consists of red marl, diluvial sand, or the alluvial deposits of the Avon. But where the lias clays come to the surface without any foreign admixture, as is the case over a large portion of the district we are considering, a wet tenacious soil is the result, requiring great labour and attention on the part of the farmer, and often rewarding him with very inferior crops. It would seem, indeed, that the lias is better suited for garden ground than for agriculture. Near Evesham and Pershore are some extensive and valuable market gardens, in which the lias clays, by a high degree of manuring and cultivation, are rendered very productive.

The whole drainage of the Vale of Evesham falls by a variety of small brooks and watercourses into the Avon. This river, though immortalized by the poets, has little to recommend it in the eyes of the painter. The clayey soil through which it flows imparts a considerable degree of muddiness to its waters at all seasons, and being kept by means of locks nearly on a level with the surrounding meadows, it loses all the picturesque effect which its neighbour the Severn derives from the steepness of its banks. From being kept constantly full, a moderate quantity of rain suffices to cause a rapid overflow; but as the water seldom remains on the land many hours before it subsides, it deposits but a small portion of the silt which it holds in suspension.

The Avon is commonly flanked on one or both sides by extensive meadows, whose level surface proves them to result from alluvial deposition. These meadows produce heavy crops of hay, which, from its excellent quality, bears generally the highest price in the market.

The bed of the Avon is for the most part deep, with a muddy bottom, with a few exceptions where beds of gravel occur. Its ordinary depth is from 12 to 20 feet, and its breadth from 30 to 50 yards.

Having thus given a sketch of the external features of the Vale of Evesham, we will proceed to examine its internal structure. In so doing it will be most convenient to begin with the highest stratum in the district, and proceed in geological order to the lowest. We shall thus investigate in succession the following formations:—Inferior Oolite, Upper Lias Shale, Marlstone, Lower Lias Shale, and New Red Sandstone.

The Inferior Oolite occupies the brow of Ilmingdon and Broadway Hills, and extends thence with great regularity along the brow of the Cotteswolds, into Gloucestershire and Somersetshire. It also forms a cap on the summit of Meon Hill, and of that magnificent outliėr Bredon Hill, where it rises gradually towards the west and north to the height of about 900 feet above the level of the sea, and may be conveniently examined near the Summer-house, which forms a conspicuous landmark to the surrounding country. On Broadway Hill it reaches the height of 1086 feet, and forms the surface of that elevated table-land. This hill, as well as the whole of the Cotteswold Range, has lately been very carefully surveyed by Mr. Lonsdale, Secretary to the Geological Society, to whose labours those may be referred who wish for further information respecting that district.

The Inferior Oolite is quarried at Bourton-on-the-Hill, and in many other places, as a building-stone, for which it is well suited. In appearance it has so close a resemblance to the Great Oolite, that in many cases it can only be identified by examining its geological position, the mineralogical characters of both formations being the same. It commonly consists of a yellowish limestone, in which the "*ova*," or small globular particles from which it takes its name, are more or less numerous and regular in different specimens. On Bredon Hill, portions of this stone sometimes occur of a brick-red colour, owing probably to the presence of an oxide of iron.

On Bredon Hill the Inferior Oolite appears at some period to have been much disturbed; in the various quarries its strata are seen greatly shattered, and dipping in all directions, often with a high angle of inclination. These dislocations are doubtless of great antiquity, as the present outline of the surface does not seem to be affected by them. The general dip of the oolite of Bredon Hill is to the south, and hence it descends with a gradual inclination much lower down on that side than on the north, where it terminates suddenly in a bold escarpment.

In many parts of England this formation contains an abundance of fossils, but they seem to be comparatively rare in those portions of it which abut on the Vale of Evesham. There has, however, been no scarcity of animal life in the period of its deposit, for a great part of the oolite is composed of fragments of shells and corals, but it is rare to find these remains in a sufficiently perfect state to fit them for cabinet specimens. The most numerous shells are those of the genus Terebratula.

These researches, which have been chiefly confined to Bredon Hill, have as yet produced only the following fossils:—

Ammonita ..	1 species.	*Pecten*	2.	*Terebellaria* ..	1.
Terebratula .	5.	*Cidaris*	1.	*Flustra*	1.
Trigonia ..	1.	*Pentacrinus* ..	1.	*Sarcinula*	1.

Nine genera, 14 species.

The next succeeding formation is the Upper Lias Shale. The traces of this formation in our district are so imperfect, that in many places its existence is rather to be inferred from analogy than proved from ocular evidence. Nevertheless, as there are some places where it certainly exists, it would be improper to omit this stratum in our list, especially as in the North of England it assumes a very important character, both in a geological and commercial point of view. In this and the adjoining counties, the stratum is vastly reduced in thickness, and there are no inducements for the speculator to penetrate its interior, and hence illustrative sections of it are rarely to be met with. Mr. Lonsdale states that he has met with it in many parts of Gloucestershire, and on his authority I have coloured it in the Society's Map, along the side of Broadway and Ilmingdon Hills, where it may be looked for at about three-quarters of the way up. It occupies a similar situation in Bredon Hill, and may be traced round the north side from the height above Aston-under-Hill to Wooller's Hill, its situation being commonly marked by a grassy slope, below the steep brow caused by the inferior oolite, and above the line of the marlstone quarries. Numerous springs are thrown out along the line of its course, as is always the case where clay interstratifies with more porous strata.

This formation being in this district much concealed by grass and vegetable soil, but few fossils have as yet been found in it; but if any good sections of it should occur, it may be expected to reward the geologist amply for the trouble of an examination*.

The next stratum in the descending order is the Marlstone. This formation consists of a series of beds of sandstone, marl and sand, in various degrees of induration. It may be traced along the side of the Broadway Range, at about half the height up; and skirts Bredon Hill in the same manner, where it forms the summit of five or six flat-topped projections, half the height of the main range, and jutting out from it on the north and east sides. In Dumbleton Hill, which is of inferior height, it occupies the summit, proving, by the regularity of its occurrence in these hills, that the intervening valleys have been denuded, and that Dumbleton and Bredon Hills are correctly termed outliers.

The marlstone rarely possesses sufficient solidity of texture to qualify it for a building-stone, but in lieu of better materials it is quarried in many places to repair the roads. The marlstone of our district seems unquestionably to be the equivalent of the bed of the same name on the Yorkshire coast, which separates the upper and lower lias shales. In both localities it contains an abundance of fossils, including several species which are common both to Worcestershire and Yorkshire. In this neighbourhood the most numerous and remarkable fossils are Belemnites, *Gryphæa gigantea*, and *Pecten æquivalvis*. The following is a list of the fossils hitherto noticed in it, and probably many more might be added on a closer examination of the several quarries where it is exposed:—

Ammonita	5 species.	*Lutraria*	1.	*Pinna*	1.
Nautilus	1.	*Unio?*	3.	*Avicula*	1.
Belemnita	1.	*Corbula?*	1.	*Plagiostoma*	4.
Terebra	1.	*Tellina?*	1.	*Pecten*	9.
Turbo	1.	*Astarte*	1.	*Ostrea*	1.
Trochus	1.	*Cardium*	1.	*Gryphæa*	1.
Mya	1.	*Modiola*	1.	*Terebratula*	8.

Twenty-one genera, 45 species.

We now arrive at a formation more important both in thickness and superficial extent than any hitherto described—the Lower Lias Shale, commonly known by the simple name of Lias. This stratum occupies nearly the whole of the Vale of Evesham, and extends from 200 to 300 feet up the sides of Bredon and Broadway Hills. Its total thickness is probably upwards of 500 feet. At Bretforton it has been sunk into more than 300 feet, in quest of coal, without being perforated. This

* An account of this stratum and its organic remains will be found in Mr. Murchison's excellent little work on the Geology of Cheltenham, just published. [A new edition was published in 1845, "augmented and revised by James Buckman, F.G.S. and H. E. Strickland, M.A., F.G.S."]

excavation was commenced three years ago by a sanguine speculator, in spite of the warning advice of geologists, and after a great sum expended, is at last given up as hopeless. All scientific geologists know that true coal is only to be found beneath the New Red Sandstone, and that to seek for it in the lias, which is above that formation, can only end in disappointment. And such is the enormous thickness of the New Red Sandstone, (as shown by Mrs. Brown's excavation, which will be noticed hereafter,) that it is scarcely less chimerical to seek for it there, except, indeed, in the very lowest beds of that formation. If by diffusing a knowledge of the general principles of geology, the Worcestershire Natural History Society shall prevent future speculators from sinking their fortunes underground in places where they will never draw them up again, our infant Society will not be without its use.

The lias formation consists here, as elsewhere, of a series of black or blue shales, producing, by exposure to the atmosphere, a cold, stiff clay soil. At the lower part of the formation, thin beds of limestone occur, from 2 to 8 or 10 inches thick, which produce excellent lime, but when used as a building-stone are apt to shiver with the frost. At Binton, near Bidford, and at Hasler, these beds are thin, smooth, and of fine quality, and are used for flooring and other purposes. Experiments, partly successful, have been made to apply the Hasler stone to lithography. It is well adapted for the lithographic *ink*, but is not suited for *crayons*.

The strata we have hitherto described are very regular and conformable in their arrangement with respect to each other, but the lias presents us with an extensive fault or break in the strata, by which the red marl beneath is unexpectedly exposed on the surface. This fault has been traced from near Netherton on the south to Lower Bentley on the north, a distance of fifteen miles, and is distinguished on the map by a narrow strip of red marl running towards the south, with lias on each side. From Netherton to Radford, on the Worcester and Alcester road, this fault is marked by a shallow valley, from half a mile to a mile in width, crossing the valley of the Avon, and interrupted near the middle of its length by the Cracombe Hills. Throughout this space the eastern side of the valley is the highest and steepest, the rise on the west being very gradual. This valley is one of those which geologists term *valleys of elevation*, being a gap, caused by strata separating and sloping off to either side in consequence, as they suppose, of an elevation of the strata beneath. But as the same effect (as far at least as relates to *level*) would ensue from the *depression* of two neighbouring districts as from the *elevation* of some point between them, it would perhaps be better to give to valleys of doubtful origin a name founded, not on theory, but on facts, and to term them *anticlinal valleys*, that is, valleys in which the strata on either side dip away in opposite directions.

The portion of red marl exposed by the fault in question, is at first a narrow strip, with a regular width of about half a mile, commencing near Netherton, and passing between Cropthorne and Charlton, whence it spreads out to a mile in width, reaching from Cracombe nearly to Chadbury. The Cracombe Hills cause an interruption to the anticlinal valley, and the lias, which is continued uninterruptedly along their summit, forms a kind of bridge, connecting the lias on the two opposite sides of the line of fault. The red marl rises about three-quarters of the height of these hills, and may be traced dipping beneath the lias, both on the east side and the west.

Beyond Cracombe Hill the valley resumes its course as far as Rouse Lench and Radford, with a width of half to three-quarters of a mile. The eastern limit of the red marl follows the brow of a steep bank to Abbot's Lench, and an isolated patch appears in a valley about a mile to the east of that place. The lias caps the hill between Rouse Lench and Abbot's Lench, and then turning to the east appears no more on this side of the line of fault. The western limit of the red marl passes near Bishampton and Abberton, and crosses the Alcester road a short distance west of Radford.

Up to this point the break in our strata has been marked by an anticlinal valley, but from hence to Feckenham it assumes the form of a fault properly so called, the red marl being raised up, and forming a long range of hill, with the lias on the west abutting against it at the base. This is best seen by Section II., taken near Inkberrow. The lias, as before observed, extends no farther than Rouse Lench on the east side, but on the west it continues past Little Inkberrow, Morton-under-Hill and Feckenham, nearly to Lower Bentley, which is its furthest northern point in Worcestershire. At Feckenham, and near Bentley, the red marl is again seen to dip to the westward, and apparently to run under the lias.

Although the extent of the lias in this county may be best seen by inspecting the map in the Society's Museum, yet it may be useful to give a general sketch of the line pursued by the junction of the lias and new red sandstone or red marl.

About three miles from Alcester, on the road to Stratford, is a hill composed of lias, with red marl at the base; the former stratum extending northward, but to what distance has not yet been ascertained. From this point the lower junction of the lias extends towards Bidford, and crosses the Avon about a mile east of that town. It soon afterwards turns westward, and occupies the summit of a steep bank or cliff, with the river at the base, as far as Salford, whence the ridge turns to the southward, and is known by the name of Cleeve Bank. The river still keeps close to the foot of the bank, the sides of which are composed of red marl, and the summit of lias. This continues to near Littleton Ferry, whence the boundary of the two formations is obscurely defined, but appears to pass near Norton, and turning to the north, to follow the eastern foot of the hills on which Atch Lench stands, as far as the Worcester and Alcester road, which it crosses for a short distance; it then sweeps round to the westward and forms the hills above Rouse Lench, before mentioned. Having already described the form assumed by these strata throughout the line of fault, we will resume the line of junction at Lower Bentley, where that description terminated. From near Lower Bentley the lias passes about one mile south-east of Hanbury, and thence at the back of Meer Hall to Goose Hill, and on the north of Trench Wood to Crowle, its course being for the most part marked by a low range of hill. At Crowle is a good section of it, showing very distinctly its junction with the red marl. From Crowle, the line of junction crosses successively the roads to Alcester, Evesham and Pershore, and then turning due south passes close to Pirton, and across Croome Park, where it forms a low bank, with the house at the foot and the gardens upon the top. Beyond Croome the lias runs to the south for about three miles, and turning to the westward past Ripple, it stretches thence to Tewkesbury. The nearest point at which the lias approaches Worcester is on the Pershore road, where it reaches within $3\frac{3}{4}$ miles of Worcester Cross.

The lias of our neighbourhood, though not so productive in organic remains as it is in Dorsetshire and Yorkshire, contains, notwithstanding, in some of its beds, considerable abundance and variety. The vast Saurian reptiles for which Lyme Regis is so famous, though rare, are not wanting in our district. Vertebræ of the Ichthyosaurus have occurred at Coltknap Hill, at Abbey Manor, and at Hasler, where also a fine vertebra of the Plesiosaurus was found, which has been presented by Mrs. Brown to our Museum. These facts suffice to render it highly probable that good specimens of these magnificent reptiles may occur in our neighbourhood, and lead me to hope that the interest excited by our Society in the cause of geology may be the means of saving many valuable fossil specimens from the ruthless hammer of the quarryman.

The most conspicuous fossils of our lias are the *Plagiostoma giganteum*, and an oval bivalve, apparently the *Unio hybridus* of Sowerby (Min. Con. pl. 154), but belonging to a new and undescribed genus, which last is very common in some of the lower beds of the lias. Besides these, the *Gryphæa incurva*, the never-failing attendant of the lias in nearly all countries, is in some parts of our district very abundant. The following is a list of the genera I have hitherto noticed:—

Ichthyosaurus ..	1 species.	*Astarte*	1.	*Gryphæa* ..	2.
Plesiosaurus ..	1.	*Lucina*	1.	*Pecten*	1.
Ammonita	7.	*Unio?*	2.	*Plagiostoma*.	2.
Nautilus	1.	*Pinna*	1.	*Modiola*	2.
Belemnita	1.	*Avicula*	2.	*Terebratula*.	1.
Trochus	1.	*Arca*	1.	*Serpula*	1.
Turritella	3.	*Nucula*	1.	*Cidaris*	1.
Orbicula	1.	*Ostrea*	1.	*Pentacrinus*.	1.
Venus	1.				

Twenty-five genera, 38 species.

Before dismissing the Lias formation, I ought to mention certain substances which in some parts of it are not unfrequent. These are hard masses of stone in the form of a cylinder or truncated cone, from 1 to 4 inches in diameter, and about the same in length. Their surface is rough and uneven, with sometimes faint irregular ridges in a circular direction. When broken they appear composed of a hard marble-like stone, containing fragments of shells. These bodies appear to be the nuclei of nodular concretions, such as are common in many formations, the softer parts of which have been decomposed.

The lowest stratum which occurs in the Vale of Evesham is the New Red Sandstone of geologists. This formation, which composes the greatest part of Worcestershire, is seen to dip under the lias, whose escarpment generally forms a low range of hill along the northern and western borders of the district which it occupies. The New Red Sandstone formation possesses great uniformity of character throughout England, and in our district is not marked by any peculiarities. The uppermost beds consist here, as elsewhere, of a red friable marl, producing a rich strong soil. The highest bed of all, next to the lias, is commonly of a whitish or grey colour, but in texture much resembling the red marl beneath; it is seldom schistose like the lias, but breaks into fragments in all directions. Wherever circumstances admit of a close examination of the union of this formation with the last, the transition appears quite sudden and well defined, but without any marks of violent action. The new red sandstone is quite conformable with the lias, and in the case of the fault before described, the disturbing force has affected both formations alike.

The upper part of the new red sandstone in Britain has in no instance, I believe, supplied the geologist with fossils, and its list of mineral contents is very scanty. The only mineral contained in the red marl of this district is gypsum, which occurs sparingly near Cracombe and at Hasler, but is rare, I believe, in other parts of the county. Of the salt, which at Droitwich and Stoke Prior forms a never-failing source of prosperity, no traces exist in these upper strata.

At Inkberrow, the extensive fault before mentioned brings to the surface the sandstone beds, which in most parts of England underlie the red marl. These beds are there quarried for a variety of uses. Their position with respect to the lias is seen in Section II.

The vast thickness of the New Red Sandstone formation is proved by the shaft sunk in 1804, by Mrs. Brown, of Hasler, in quest of coal. The following is the best account of the strata that can now be procured, and geologists may regret that a more exact account was not kept of so deep and interesting an excavation:—

	feet.	inches.	
Lias beds, about................	75	0	(At 51 feet, a strong spring.)
Red marl, about................	282	0	
Gypsum	1	0	
Grey and red strata	47	0	
Black strata	15	0	
Red and white beds with gypsum	387	9	(At 582 feet, a thin vein of coal.)
Total............	807	9	

Having now traced the strata of the Vale of Evesham in succession, as far as that district is considered to extend, the investigation of the rest of the county is left to others. It only remains to give, by way of an appendix, a sketch of what are commonly called *diluvial* deposits, as far as these occur in our district*.

The deposits of gravel, sand and clay, which in most parts of the world lie in irregular patches upon older rocks, have been by many geologists referred to the Mosaic deluge. But recent observations seem to show that these deposits have not all been simultaneous, and that, granting some of them to originate in the Mosaic deluge, others have been caused by irruptions antecedent to that period. Be this as it may, there is nothing, I think, in our district to prove that the diluvial beds, which are there very abundant, are not the result of the Mosaic deluge alone.

These deposits consist of clay, gravel and sand, in various proportions, and scattered over the country with capricious irregularity. Where the sand predominates, it is often of great service in lightening and fertilizing the otherwise clayey soil. The clay is in many places dug for brick-making, and the gravel is a valuable material for the roads. The latter is composed of a variety of broken rocks, for the most part of older formation than those of this district, but chalk flints are not unfrequent, and the *Echinocorys scutatus* and *Spatangus cor-anguinum*, two well-known chalk fossils, have been met with. In the neighbourhood of the lias, the diluvial beds often contain rolled fragments of the fossils of that formation, such as *Gryphæa incurva*, Ammonites, &c. At the village of Bredon, the *Hippopodium ponderosum* occurs in the gravel in addition to

* [See the paper following this memoir.]

the above fossils. Near the oolite hills the diluvial beds contain, as might be expected, fragments of oolite.

Besides borrowed fossils, the diluvial beds occasionally contain fossil remains peculiar to themselves, consisting of the bones of land animals, which appear to have been living in this country at the time of the catastrophe which caused the deposits in which they are now imbedded. This Society possesses several bones of the Hippopotamus, Ox and Deer, found at Cropthorne*. In a gravel-pit at Chadbury, bones of the Rhinoceros have been found; also, a fine molar tooth of that animal, which has been presented to this Society by W. Perrot, Esq., of Fladbury. Fossil bones of some large animal have also been found in Mr. Day's clay-pit at Bengworth; and the Society is indebted to Mr. Stokes, the Surveyor of the Roads, for a fine tooth of the Elephant from Stratford-on-Avon†.

Thus, then, there is ample evidence of the existence, in our diluvial deposits, of these interesting remains, which carry us back to a period, and, geologically speaking, not a distant one, when the Hippopotamus, Rhinoceros and Elephant roamed, undisturbed, in the valleys of Worcestershire; and hence I beg to recommend to the attention of the Society the numerous pits of gravel, sand and clay which abound in the county, not doubting that many valuable relics may thus be rescued from the workmen, who, unless taught otherwise, will still continue to throw them aside as worthless and unprofitable.

It is now time to close these imperfect remarks, which may suffice to show that this county contains much that is interesting to the geologist, by whom Worcestershire has hitherto been much neglected, but we may hope that, under the auspices of the Natural History Society, it will receive a thorough investigation, and that the treasures it contains will be added to the commonwealth of science.

P.S. With a view to illustrate the order and position of the several formations above described, an imaginary section, No. III., is introduced, commencing at Severn Stoke, and intersecting Bredon and Dumbleton Hills, till it meets the Broadway or Cotteswold range.

* The excavations at Cropthorne have been lately resumed, and have brought to light a considerable number of bones, accompanied by many species of land and freshwater shells, the same as now exist in the neighbourhood. An account of this discovery was read to the Society on November 25, 1834.

† Since the reading of this paper, the Museum has been enriched by two molar teeth of the Rhinoceros, from Sandlin, near Malvern, and the tooth of an Elephant from Powick.

2. ON THE DEPOSITS OF TRANSPORTED MATERIALS USUALLY TERMED *DRIFT*, WHICH EXIST IN THE COUNTIES OF WORCESTER AND WARWICK.

[*Written June* 20, 1835.]

[This is the "paper" referred to by Sir R. Murchison in his 'Silurian System,' p. 555: "I refer to his interesting paper, shortly, I hope, to be published at length, for the details presented at different localities." It is from this also that the list of fossils is given in the note. The memoir was written before starting on the expedition with Mr. Hamilton, and before the exposition of the Glacial Theory by M. Agassiz. It was sent to the Geological Society, as we find by the note of the Secretary attached, in June 1835, but, in consequence probably of Mr. Strickland's absence, it was never read. It is the best record now existing of the beds in the Vale of the Avon then exposed; but most of the localities mentioned are now either filled up, worked out, or become otherwise inaccessible.

Diluvium was used in the title and throughout this paper, and is so referred to in the 'Silurian System,' *l. c.* *Drift* is now substituted in consequence of a memorandum added to the MS. changing the term to the latter.]

Being now on the point of leaving England for a considerable time, I am induced to bring my observations on this subject before the Society in their present state, though they might possibly have been rendered more complete by further investigations which I am not now able to make.

The arrangement which is here attempted of the drifted beds of Worcestershire and Warwickshire, is based solely on the phænomena observed in those two counties. To apply that arrangement to the various drifted deposits which exist in other parts of England, and to reduce those abnormal masses of detritus to a regular system, would require a combination of minute inquiry with judicious generalization which few persons would be competent to devote to the subject.

It may prevent misconception to premise, that by the term *Drifted* deposits, no theoretical notions are implied,—that expression being merely used for the sake of convenience, to denote all those superficial accumulations of transported materials which are so circumstanced that they cannot have been produced by the tranquil causes which are in daily operation in the district.

Drift, *i. e.* superficial erratic detritus of the Midland Counties.

- General.
  - Quartzose*.
  - Flinty.
- Local.
- Fluviatile.

The deposits of this kind in the counties of Worcester and Warwick admit of being divided into three classes, each of which will probably be found to be the produce of a different epoch. For the sake of con-

* This is synonymous with the *Northern* Drift of Sir R. Murchison.

venience, I propose to distinguish these varieties by the terms *General*, *Local*, and *Fluviatile* Drift.

The *General Drift* has been produced by an agency far more powerful and extended than that which has caused the other two varieties. It occurs either with an abundance of chalk-flints, or with very few and frequently no flints, in which case it is chiefly composed of quartz pebbles. The first of these varieties we will call, for the sake of distinction, the *Flinty* drift, and the second the *Quartzose*.

The *Quartzose* general drift is spread over the middle and northern portions of Worcestershire, and extends thence into Warwickshire, Staffordshire, and others of the Midland Counties. It is composed, for the most part, of gravel containing a great variety of pebbles derived from the primitive, transition, igneous and altered rocks of Shropshire, Staffordshire and the North of England. A large proportion of these pebbles appear to have been brought immediately from the conglomerate beds of the New Red Sandstone, so that we must beware of attributing their rounded form to the transitory action which has placed them in their present position. These masses of gravel occur on the tops of considerable hills, at a much greater elevation than is attained by the *local* or *fluviatile* groups of drifted matter.

The drift of which we are now treating is distinguished by an abundance of brown granular quartz pebbles, which have been shown by Sir R. Murchison to be sandstones of the transition series altered by the action of heat. Another peculiarity of this kind of drift may be found in the great rarity, and in most cases total absence of chalk-flints, of fragments of oolite and lias, or of any rock more recent than the New Red Sandstone.

As far as my inquiries have hitherto extended, this kind of drift appears to contain no mammiferous or other remains which can throw light on the era of its deposition. A wider induction must, however, be employed before this negative assertion can be firmly established.

From the northerly position of the rocks which have supplied fragments to the quartzose drift, we may infer that the current which has placed these gravel beds in their present situation came from the northward. This is further proved by the fact noticed by Sir R. Murchison, that as we proceed from Worcestershire towards Cheshire, the boulders are found gradually to increase in size.

The *Flinty* variety of general drift occurs in vast abundance in the eastern parts of Warwickshire, and probably extends northward through Leicestershire towards Derby. In the neighbourhood of Rugby it covers the hills to a great depth, rendering the boundary of the Lias and the New Red Sandstone very difficult to trace. It is composed in part of the same materials as the quartzose gravel, but is distinguished by the great abundance of chalk-flints and pebbles of yellow sandstone, possibly derived from the same tertiary source as the *grey wethers* of the Wiltshire Downs. At Brownsover Hill, and on the north of Princethorpe Nunnery, a patch of stiff brown clay occurs containing rolled masses of soft chalk, which is probably subordinate to this drifted deposit*.

As far as I have been able to ascertain, this group is, like the last, destitute of mammalian remains, except in the low valleys, where it is probably modified by the causes which will be mentioned under the head of *Fluviatile* drift.

There is no appearance of a *transition* from the flinty drift to the quartzose variety in the east of Warwickshire, where the two kinds of gravel approach each other very closely. The boundary between them appears in some degree to correspond with the line which separates the Lias from the New Red Sandstone.

The flinty drift extends from the neighbourhood of Rugby along the base of the oolite hills to the Vale of Shipston, where are extensive beds of gravel with flints which seem to be referable to this deposit. It does not appear to exist in Worcestershire, and the Vale of Shipston is probably its limit on the west. From the great abundance of chalk-flints, we may conclude that the current which brought this gravel into Warwickshire came from some point between north-east and south-east.

It remains for further observation to decide whether these two varieties of drifted matter each mark a different epoch, or whether they are results of the same cause differently modified.

The next class of transported detritus of which we come to speak is that which I have termed *Local* drift. It occurs in the vale at the foot of the oolitic range of Broadway and the outlier of Bredon. It is at once distinguished by the entire absence of any materials which are foreign to the neighbourhood: hence it consists exclusively of fragments derived from the oolite, the marlstone, and the lias. These are more or less rounded by friction, but have evidently not undergone so violent an action as the pebbles in the *General* drift. It lies in patches on the low eminences of lias at the base of the oolite hills, from which it does not extend to a greater distance than two or three miles. These circumstances denote an aqueous action much more circumscribed in extent, and perhaps also in duration, than that which has caused the *General* drift†. The elevated position, however, which it occupies serves clearly to distinguish it from modern drift, and proves that the existing rivulets of the district, in their present state, are quite inadequate to its production.

* A similar deposit of brown clay with chalk pebbles has been noticed by Prof. Sedgwick in Cambridgeshire.

† In the postscript to this paper I have shown that these *local* beds of gravel may be accumulations round the margin of the same sea which brought down the *General* northern drift into Worcestershire.

The *General* drift before mentioned appears, in Worcestershire at least, seldom to approach within two or three miles from the base of the oolite escarpment, its place being there occupied by the *local* deposits. I have not hitherto succeeded in finding the two kinds in contact or passing into each other. A case of this kind might throw light on the relative ages of the two deposits, to which, from the apparent absence of terrestrial organic remains in both, there is at present no clue.

The last variety of drift which occurs in this district is that which I have termed *Fluviatile.* It appears to be a mixture of the other three kinds, modified by the comparatively tranquil action of an ancient river which has followed the same line of drainage which is now traversed by the Avon. The beds of clay, sand and gravel which compose this drift occur in patches on the flanks of the Avon and some of its tributaries, at various heights from 20 to 40 feet above it. The gravel is composed in great measure of the same pebbles as the *Quartzose General* drift, but to them is added a considerable proportion of chalk-flints and frequent fragments of oolite and lias. This drift is commonly stratified more distinctly than the *General* drift, the sand and gravel being disposed in layers a few inches thick, in some cases accompanied by ferruginous clay, and at other times washed very clean of argillaceous matter. These beds may generally be distinguished by their yellow or ferruginous colour from the *Quartzose General* drift, which, from the proximity of the new red marl, is commonly of a dull red colour.

The chalk-flints which occur in this drift in Worcestershire have doubtless been derived from the upper parts of the Avon valley in Warwickshire, where the *Flinty General* drift, which is absent in the former county, exists in great abundance.

The most interesting mark of distinction between the *Fluviatile* drift and the other varieties of transported materials, consists in the presence of organic remains, of which I have found no traces in the other drifts*. These remains consist of the bones of mammalia and the shells of terrestrial and aquatic mollusca, which have been traced along the course of the Avon from Lawford in Warwickshire to Defford in Worcestershire, within a few miles of the junction of that river with the Severn.

In describing this variety of drift, it will be necessary to enter more into detail; beginning therefore with Lawford, the most eastern locality where I have noticed this deposit, we will follow the course of the Avon towards its termination.

Having previously met with fluviatile shells mixed with bones in the lower parts of the Avon valley, I was induced to search for them at

* In the 'Mémoires Géologiques,' vol. i. p. 80, M. Boué arrives at the same conclusion, viz. that the bones of extinct or exotic animals are only found in low situations, and in the vicinity of existing rivers.

Little Lawford, a locality which has supplied an abundance of mammalian remains, and on visiting the spot in March 1835 these expectations were realized. Being assured by the workmen that they occasionally find patches of small shells between the gravel and the subjacent lias, I commenced a search and succeeded in procuring the four following species :—*Lymnæa peregra, Cyclas amnica, C. henslowiana,* and portions of a Unio.

The cavity in the lias which contained the remarkable accumulation of bones described in Buckland's 'Reliquiæ Diluvianæ,' has been long since worked out, but from the extant accounts, it clearly appears to have been a deep pit in the bed of a river or lake, into which the bones had been drifted, and had there become covered with about 15 feet of mud. In this mud one of the workmen assured me that he found many white bivalve shells, very soft and decayed. It is much to be regretted that these shells did not at the time excite the attention of geologists, who seem to have been too much attracted by the remarkable congeries of bones discovered at this place, to notice the less conspicuous but not less interesting molluscous remains which accompanied them.

The drift at Little Lawford in which these remains occur is close to the river Avon, and elevated a few feet above it.

Not having as yet examined the course of the Avon between Lawford and Stratford, I am not aware whether these fluviatile beds exist in this interval. At Shottery near Stratford are beds of gravel in which the teeth of the Elephant have occurred, but whether they are accompanied by freshwater shells is at present undecided.

The fluviatile drift is best displayed in that part of the Avon valley which traverses the Vale of Evesham. The first locality to be noticed is Bengeworth near Evesham, where the drift exists in the form of a ferruginous clay with subordinate beds of sand and gravel. This deposit rests on blue lias clay, which is mixed with the drifted clay for making bricks. At this place were found, a few years ago, a considerable number of bones of Elephants and other mammalia. Very perfect specimens of *Unio ovalis* are occasionally found in the ferruginous clay, but I have not yet procured any other species of shell at this place.

The drift of Bengeworth is distant about a quarter of a mile from the river Avon, and is situated about 40 feet above it.

The valley of the Avon expands about three miles below Evesham, and contains an extensive patch of this variety of drift between Charlton and Fladbury on the left bank of the river, and another reaching from Fladbury to Wyne on the right. These deposits consist of clean washed gravel with a large admixture of sand, in which are occasionally found remains of the Elephant, Rhinoceros and other mammalia. The gravel rests for the most part on lias clay, which holds up the water and renders it very difficult to penetrate to the bottom of the drift, at which part the freshwater shells, when present, are usually to be found. The existence of these shells is ascertained at a gravel-pit, now closed up, in Charlton North Field. Not having seen them myself, I can only give the account of an intelligent workman, who states that at this spot, beneath about 10 feet of gravel, is a level and hard surface of red marl, on which he found, a few years since, numerous river mussels (Anodon), and occasionally bones of various kinds. This spot is about 100 yards to the north of a small rivulet which joins the Avon at Cropthorne Ford, and is raised about 15 feet above it.

The next place at which freshwater shells exist is near the village of Cropthorne. It was here that my attention was first drawn to the existence of these shells in drifted gravel, and I have had more opportunities of collecting and examining the organic remains that occur here than at any other of the localities mentioned. This bed of gravel appears at the side of the turnpike-road from Evesham to Pershore, about a quarter of a mile south of Cropthorne. A small rivulet rising in Bredon Hill, about three miles to the south, follows the course of a shallow valley, and running under the turnpike-road at the place in question, falls into the Avon about half a mile further north. In 1830 the two sides of the valley were lowered several feet, in order to diminish the inequalities of the road, thus exposing on either side a mass of gravel resting on lias clay. On the east side of the valley, between the gravel and the lias, is a stratum of sand containing freshwater shells, of which no traces have appeared on the west side. This stratum of shells was conveniently exposed in 1834 in consequence of a gravel-pit which was opened for some months, but is since closed, and the shells are now only accessible by digging.

A section was thus exhibited about 70 yards in length, 8 feet 6 inches high in the middle, and sloping away towards either end. The lower part of the bank consists of 2 or 3 feet of lias clay, on which rests a stratum of fine sand from 2 to 6 inches in thickness, containing numerous land and freshwater shells amounting to twenty-four species, and bones of the genera Hippopotamus, Bos, Cervus, Ursus, and Canis. This bed of sand merges gradually into the gravel which extends to the surface, and corresponds in character with the other beds of *Fluviatile* gravel in the neighbourhood. The bones, though most abundant in the sand-bed at the bottom, were partially interspersed through the gravel, but the shells seemed entirely confined to the sand. Some of the bones were found imbedded a few inches in the lias clay, but the majority of them rested immediately upon it. They consisted in great measure of detached fragments with the broken surfaces more or less rolled, and even those which were found entire were much damaged in removing them, on account of the compactness of the gravel and the brittle condition of the bones. The rolled fragments occurred principally in the gravel, and the perfect bones were chiefly in the sand and clay at the bottom.

Near the middle of the section the stratum of shells was absent for the space of a few yards, owing to a slight rise of the lias clay at that part, forming a bank with a depression on each side in which the shells had been deposited. It may be further remarked, that both the bones and shells were by far the most abundant at the western or lowest extremity of the section, and became gradually scarcer eastward till they finally disappeared. Towards the western end the stratum of shells, instead of resting on the lias, rose a few inches above it, presenting the following section:—

	ft.	in.
Vegetable soil	1	0
Bluish clay	0	4
Gravel	2	6
Sand and shells	0	8
Clay and shells	0	2
Gravel	0	2
Clay	0	6
Gravel	0	2
Lias clay	1	0
	6	6

The bones which occurred at this place belonged to the following animals:—

Hippopotamus	A few bones and teeth, probably belonging to one individual.
Bos urus	Numerous remains of several individuals.
Cervus	A few bones and fragments of horns.
Canis	A radius.
Ursus	A metatarsal bone.

The following is a list of the shells found here:—v. r. denotes *very rare*; r. *rare*; c. *common.*

TERRESTRIAL.

1.	*Helix virgata*	v. r.
2.	*H. pulchella*	r.
3.	*Pupa marginata*	r.
4.	*P. pygmæa*	v. r.
5.	*Succinea amphibia*	r.

AQUATIC.

6.	*Limnæa palustris*	v. r.
7.	*L. fossaria*	r.
8.	*L. peregra*	r.
9.	*L. auricularia*	v. r.
10.	*Planorbis nautileus*	v. r.
11.	*Planorbis vortex*	r.
12.	*P. complanatus*	r.
13.	*P. lateralis*, Strickl.	c.
14.	*Ancylus lacustris*	v. r.
15.	*A. fluviatilis*	v. r.
16.	*Valvata fontinalis*	c.
17.	*Paludina tentaculata*	c.
18.	*P. minuta*, Strickl.	r.
19.	*Cyclas henslowiana*	c.
20.	*C. amnica*	c.
21.	*C. cornea*	c.
22.	*Anodon anatinus*	c.
23.	*Unio ovalis*, Fleming	c.
24.	*U. antiquior*, Strickl.	r.

In addition to the above shells the valves of a Cypris* also occur.

* [This is now referred to *Candona reptans*, Baird.—Ed.]

The *Anodon anatinus* occurs in considerable numbers at the west extremity of the section, but was not noticed at any other part. This fact seems to show that these shells lie in their original situation, for had they been washed thither from a distance, they would have been scattered indiscriminately over the whole surface.

The shells which are denominated *Paludina minuta* and *Unio antiquior*, appear to be extinct and undescribed; all the rest being existing species indigenous to Britain. Thus, though the majority of these molluscous animals have survived to our times, yet certain species occur in this ancient fluviatile deposit which, like the Hippopotamus that accompanies them, belong to a zoological epoch that has now passed away. This deposit is also distinguished negatively by the *absence* of certain very common species, such as *Helix nemoralis, Planorbis albus*, &c., which are rarely wanting in modern alluvions.

The extinct species may be thus characterized:—

Paludina minuta. P. testâ turritâ, angustâ, anfractibus quinque, rotundatis, apice obtuso, aperturâ ovatâ, staturâ minimâ.

Shell slender, smooth, the volutions five, moderately rounded. Aperture perfectly oval, not interrupted by the body whorl. Outer lip sometimes thickened by an external rib.

Length, $\frac{1}{10}$ inch; diameter, $\frac{1}{20}$ inch.

Differs from *Paludina stagnorum* of Turton, being but half the size, narrower in proportion, with the sides more nearly parallel, and the apex rounded, whereas that shell tapers uniformly from the aperture to the apex* †.

Unio antiquior ‡. U. testâ cordato-ovatâ, crassâ, valvulâ dextrâ unico, sinistrâ duobus dentibus crassis, brevibus, subtrigonis. Lamina postica utriusque valvulæ simplex.

The posterior primary tooth of the left valve is short, strong, and erect, its base being a nearly equilateral triangle. The anterior one is rather narrower and smaller. Between them is a deep triangular pit, receiving the tooth of the opposite valve. The posterior lamina of the left valve (which is double in the other British species) is single in this, the plate next the cartilage being nearly obsolete. Anterior muscular impression very deep and corrugated.

Length, $1\frac{3}{4}$ inch; breadth, $2\frac{1}{4}$ inches.

The outline of this shell bears a near resemblance to that of *Cytherea chione*.

* Closely resembles *Paludina gibba*, Michaud; a species found in brackish estuaries in the South of France, and perhaps belonging to the genus Rissoa rather than to Paludina.—H. E. S. MS.

† Now referred by Morris (Brit. Foss. p. 266) to *P. marginata*, Michaud.—Ed.

‡ Referred by Morris (*l. c.* p. 230) to *U. littoralis*, Lam., but differs much in form from recent specimens.—Ed.

To these I will add the characters of a shell, which, though probably not extinct, does not appear to have been previously described.

Planorbis lateralis *. P. tribus, raró quatuor, anfractibus rotundatis glabris, utrâque facie æquè concavâ, aperturâ subrotundâ.

Volutions uniformly rounded, not in the least flattened or carinated, smooth, with numerous very fine lines of growth; perfectly lateral, hence there is a considerable concavity on either side. Aperture nearly circular, very slightly embracing the body whorl.

Diameter, $\frac{1}{6}$ inch; height, $\frac{1}{25}$ inch.

Resembles *P. alba*, but is less, the volutions do not increase so rapidly as in that shell, they are not decussated, and the aperture is less oblique. It partly agrees with Mr. Jeffreys' description of *P. glaber* in Linn. Trans. vol. xvi. p. 387; but that shell is stated to be flatter on the upper side than below, while this species is never less concave, and is sometimes more so, on the upper side than on the under.

This species is apparently the same with a nondescript Planorbis, *P. lævis*, Ald., found by Mr. Alder in a pond at Whitley, near Newcastle, of which there are some specimens in the British Museum. I have never seen recent specimens of it from any other part of England. The Cropthorne shell-bed is about 50 yards distant from the adjacent rivulet, and from 20 to 24 feet above it. The fall of the rivulet from hence to its junction with the Avon may be estimated at 15 feet.

In the localities hitherto noticed the occurrence of the freshwater shells at the bottom of the drifted gravel, and their absence in the upper parts of it, would seem to denote a period of tranquillity succeeded by one of comparative violence. But at the place which will next be described, the dispersion of the shells through the substance of the gravel, proves that in this instance, at least, the deposition of both has been contemporaneous. This phænomenon appears at Bricklehampton Bank, close to the river Avon, about half a mile west of Cropthorne. A gravel-pit lately opened exposes—Ferruginous clay and sand with freshwater shells, 5 feet. Stratified gravel and sand with interspersed bones and freshwater shells, 8 feet. Blue lias clay, 20 feet.

The shells are of the same species as those in the Cropthorne deposit, which is only about half a mile distant. The bones hitherto found belong to the *Bos urus*, of which animal I lately procured at this place a portion of the skull, with the cores of the horns attached. Their distance from point to point is 3 feet 3 inches, and their circumference at the base 14 inches.

At a brickyard 150 yards further west, the ferruginous clay predo-

* *P. glaber*, Jeffreys = *P. lævis*, Alder. See Forbes and Hanley, Brit. Moll. iv. p. 150.—Ed.

minates, and the gravel and shells, though present, occur only in subordinate layers.

Two other localities yet remain to be noticed, at which freshwater shells are said to exist in the fluviatile drift. As the shells in these cases rest immediately on the lias, below the ordinary layer of the springs, they are only accessible after a continuance of dry weather, and I have not yet had an opportunity of procuring them myself. But from the report of some of the workmen, whose veracity I have no reason to doubt, it appears, that when the lowness of the water permits them to reach the bottom of the gravel, they frequently find "small snail-shells and river-mussels" resting immediately on the blue clay of the lias. The first of these localities is at a pit between the villages of Moor and Wyre, on a platform of gravel elevated about 25 feet above the Avon, from which it is distant about 100 yards. The next place is on a similar platform south of the village of Defford, distant about 300 yards from the Avon, and about 35 feet above it. At this spot a considerable number of bones have been procured. This is the last place in the Avon valley at which the freshwater shells have been found, but there are deposits of fluviatile drift at Bredon, three miles above the junction of the Avon and Severn, which not improbably contain them.

Such are the principal facts connected with that variety of transported detritus provisionally termed *Fluviatile drift,* as exhibited in the valley of the Avon. I will now attempt, with all diffidence, the more difficult task of deducing a few conclusions from the phænomena observed.

These beds have been found to exist at various points along the valley of the Avon from Lawford to Bredon, a distance (following the windings of the river) of about seventy miles. Throughout this distance there is so great a uniformity in their character, that we cannot refuse to assign them a common origin.

There is evidently a close connexion between this drift and the existing drainage of the country. These beds are restricted to certain distances from the Avon and to certain heights above it, yet it is quite certain that this river, ever since it has assumed its present state, can have had no share in their formation. There is no proof of any local disruption having taken place in the surface of the district, by means of which these freshwater beds can have been *elevated* to their present height above the Avon. The lias and new red sandstone on which they rest exhibit no traces of disturbance of so recent a date. And the absence of these beds in other parts of the district, and their uniform occurrence on the flanks of the Avon valley and some of its branches, prove that at the period of their deposition the drainage of the country followed the same lines of depression which still perform that office.

We must then suppose that at this ancient epoch the waters of the

Avon valley flowed at a height of not less than 40 feet above their present level. To point out the specific causes which have lowered the river to its present position would be difficult, and perhaps impossible, but we may form some not improbable conjectures as to their general nature.

Several circumstances tend to show that at the period of these deposits the vale of Avon was occupied by flowing, not by stagnant waters. Had this valley been the bed of an ancient lake, we should find an abundance of silt and argillaceous matter, with little or no admixture of gravel, whereas, in fact, the clay-beds are much less abundant than the gravel, and the latter is in many instances washed quite clean of argillaceous substances. Moreover, among the mollusca whose remains are found in these beds, there are at least three species, which are rarely if ever found to inhabit still water. These are, *Cyclas amnica*, *C. henslowiana*, and *Unio ovalis*.

The former existence of running waters at unusual heights is more easily explained than that of tranquil lakes in the same circumstances. The supposition of a succession of minor obstacles subsequently removed by denudation, will account for a river having occupied a greater elevation formerly than at present. But to explain the occurrence of a once elevated lake, unless the drainage of it has been effected by the elevation of the ground itself, we must assume the former existence of one solid dam, compact enough to be water-tight, and sufficiently strong to withstand the pressure of a vast mass of water. In cases where no vestiges remain of such a dam, or of a gorge through which the lake may have been drained, so extended a flight into the realms of hypothesis should only be used as a last resource. But in the case before us, the fluviatile and non-lacustrine character of these beds relieves us from this difficulty, and enables us to explain the former height and present depression of the river level, by supposing that the sand-banks, rocks, cliffs, or sinuosities which still impede the current of the Severn, and partially of the Avon, in their progress towards the sea, once existed to a much greater extent than they do now, having gradually become modified and removed by the ceaseless action of the flowing waters. We may further suppose that the bed of the Avon itself has become lowered by denudation, for we must otherwise attribute a depth of more than 50 feet to the ancient Avon, viz. 40 feet above the present surface of the water, and 10 or 15 below it*.

It is difficult to form a conjecture as to the relative magnitude of this

* In the Postscript at the end of this paper, I have explained the lowering of the bed of the Avon, by supposing the land to have risen, which would give the river a greater fall, and cause it to denude its bed until it reached a level at which the current became too slow to effect any further denudation.—H. E. S.

ancient river, whose course appears to have nearly coincided with the modern Avon. The beds of freshwater shells which form the chief evidence of fluviatile action are not more abundant or extensive than might have resulted from a river similar to the present one. But the beds of gravel which contain or cover the shells, require for their production a current possessed of much greater physical powers. From the circumstance that the beds of fluviatile shells are subordinate both in amount and extent to the gravel, and usually occur below and not in it, we may suppose that the river itself was not possessed of any remarkable magnitude or force, but that its ordinary action was aided or followed by one or more extraordinary exhibitions of drifted violence.

This fluviatile deposit is evidently of a very recent geological date, but the extinct animals it contains clearly distinguish it from our actual epoch. The relative proportion of the existing and extinct species of shells seem to refer it with some probability to the Newer Pliocene period of Sir C. Lyell, in which a large majority, but not the whole, of the animal kingdom belonged to species which are still in existence.

On a comparison of the mammalia and mollusca of these beds with those which now inhabit our island, it is evident that a much greater change has taken place in the former than in the latter. And as the distribution of mollusca is known to depend really upon climate, we may hence conclude that the temperature of these latitudes has altered less in the interval which has elapsed, than a comparison of the mammiferous remains alone would lead us to suppose.

I have thus endeavoured to describe the phænomena of the fluviatile modification of drift which exists in Worcestershire, and have ventured to deduce a few inferences from the facts. It now only remains to show to what extent these phænomena probably prevail in other parts of our island.

Future investigation must decide whether the apparent absence of mammalian remains in the other varieties of drift, and their supposed exclusive presence in the fluviatile beds, which seems to be the case in Worcestershire, is also true in other districts of Great Britain. One fact seems certain, that the bones of mammalia occur, if not universally, at least much more frequently in the detritus of valleys and low districts, than in that which exists on the tops of hills. In the former situations the deposits of transported matter may have been caused or modified by fluviatile or lacustrine action, but they can only have been placed in the latter position by violent *Débacles* such as must have produced that kind of detritus which I have termed *general drift*.

Several localities have been recorded in various parts of England where the occurrence of freshwater shells and bones of extinct mammalia present phænomena similar to those above described. To facilitate further investigation I will here briefly refer to these published statements.

1. In Geol. Trans. 1st series, vol. i. p. 340, Mr. Parkinson states, on the authority of Mr. W. Trimmer, that bones and freshwater shells were found in calcareous earth beneath gravel, at Kew.

2. In the 'Philosophical Magazine,' New Series, vol. vi. p. 225, the Rev. W. Vernon Harcourt has described a deposit, apparently lacustrine, containing bones of extinct mammalia, and thirteen existing species of land and freshwater shells. This occurs at North Cliff, near Market Weighton in Yorkshire.

3. In Loudon's 'Magazine of Natural History,' vol. vii. p. 274, is a notice by Mr. S. V. Wood, of a freshwater formation discovered by Mr. E. Charlesworth, at Stutton, on the banks of the Stour, in Suffolk. This deposit, which is surmounted by gravel, contains bones of the Elephant and other quadrupeds, and thirty-five species of land and freshwater shells. Of these shells, one species, a Cyrena, is extinct, the rest are all existing species, and this fact furnishes an additional point of resemblance to the fluviatile beds of the Avon valley.

4. In vol. vii. p. 436 of the same work, Mr. J. Brown describes a deposit of this kind at Copford, near Colchester. Bones of ox and deer, mixed with fragmented shells, are covered by 3 feet of drifted gravel.

These are all the published instances that I have hitherto met with; to which I may add, that there is reason to believe that there are many cases of this kind in the low district of Holderness, though some care would be there requisite in distinguishing them from more recent alluvions. Indeed it is probable that if attention was paid to the subject, freshwater or land shells would be found very frequently to accompany mammalian bones in the gravel of our island.

I will not attempt, at present, to draw a comparison between the cases which occur in Britain, and similar ones which may exist on the Continent. I will merely remark, that the *loëss* of the Rhine seems an analogous deposit to the fluviatile beds of the Avon valley, though, from the shells there contained belonging *entirely* to existing species, it may possibly be of a more recent date*.

P.S. Nov. 1836.—Since the date of this paper, Sir R. Murchison has published his views respecting the Northern drift of Worcestershire, &c. He refers the deposits which I have termed *General drift* to the action of a marine current, flowing from north to south across some of the

* Professor Braun, of Carlsruhe, maintains that many of the *loëss* shells belong to extinct species. Having examined the loëss since writing this paper, I have come to the conclusion that it is completely analogous to the fluviatile drift of the Avon.—H. E. S.

Midland counties, and down the valley of the Severn. To this supposition I have no objection to offer. I may remark, that the marine channel through which this current flowed, was probably bounded on the east by the oolite escarpment, as far even as Lincolnshire, for vast accumulations of *Quartzose drift* lie near the foot of the oolite hills, through Lincolnshire, Leicestershire and Warwickshire, to Worcestershire. Branches from this Northern drift have been traced down the valleys of the Evenlode, Windrush, and Charwell, into the London Basin (see the Map in Buckland's 'Reliquiæ Diluvianæ'), so that the chalk and oolite hills of Gloucestershire, Oxfordshire, Buckinghamshire, &c. probably formed at this period several islands.

My *Flinty general drift* was probably formed by that part of this marine current which swept along the eastern side of the chalk hills, and is therefore to be considered contemporaneous with the *Quartzose* gravel produced by the same current further westward.

My *Local drift,* which lies at the foot of the oolite hills in Worcestershire, may be merely a shingle beach, formed at the margin of the sea, where the current was too shallow and powerless to bring the erratic pebbles of the Northern drift. Hence this local gravel consists solely of oolite, derived from the overhanging cliffs of Broadway and Bredon Hill.

The *Local drift* then may be considered contemporary with the two varieties of *General drift,* and it is remarkable that no mammiferous bones, as far as I know, have been found in any of these three varieties.

It is probable that the enormous denudation of the oolite in Worcestershire (proved by the outlier of Bredon Hill) has been caused by the long-continued action of this marine current, which, from the gradual narrowing of its sides, must have set with immense force through the "Straits of Malvern."

Thus far the views which I have advanced in this paper may be made to coincide with those of Sir R. Murchison on the Northern drift. I cannot, however, entirely agree with that part of his paper which supposes the fluviatile gravel of the Avon valley to have resulted from a river flowing into, and therefore contemporary with, the Straits of Malvern. It appears to me that the fluviatile drift, containing bones and shells, must be more modern than the marine drift, because the latter is found in Worcestershire and Warwickshire capping the hills on both sides of the Avon, at a great height above the platforms of fluviatile gravel which skirt the valley below. Take, for instance, Cracombe Hill in Worcestershire, capped with unstratified gravel and quartzose pebbles, without flints or bones, and which may be referred to the Northern drift; while the finely stratified sand and gravel, not exceeding 40 feet above the Avon, contains chalk-flints brought down from Warwickshire, and also bones of extinct mammalia and freshwater shells.

It appears from this, that a spot, once the bed of a marine current, which must have attained the height of at least 200 feet above the modern Avon, afterwards became dry land, and was traversed by a river which flowed at from 20 to 40 feet above that stream.

These phænomena may, perhaps, be explained by supposing—

1. A period when Worcestershire and most of the Midland Counties were at the bottom of the sea, which set in a rapid current through the Straits of Malvern.

2. An elevation of several hundred feet converts this part of England into dry land. A river descends from Warwickshire through the Vale of Evesham, and re-arranges much of the gravel brought from the north by the marine current,—disposing it in thin strata, and imbedding shells and the bones of quadrupeds which lived on its banks. This river flowed into the Bristol Channel, which perhaps extended as far up as Tewkesbury.

3. A further elevation of the land takes place to the amount of 40 or 50 feet, and the Bristol Channel backs out to its present position. The fall between Warwickshire and the sea being thus much increased, the river now called the Avon acquires a greater velocity, and excavates its bed till its general level is lowered 30 or 40 feet. It at last arrives at its present state of equilibrium, when the velocity is no longer sufficient to continue the denuding process. It is by a similar elevatory movement that Sir C. Lyell supposes the loëss of the Rhine to have been denuded.

3. ON THE NATURE AND ORIGIN OF THE VARIOUS KINDS OF TRANSPORTED GRAVEL OCCURRING IN ENGLAND.

[This paper was read before the Meeting of the British Association at Liverpool, in 1837, and an abstract only was printed in its Reports of Sections. It is now printed entire from the original manuscript, and the "Queries respecting the Gravel in the Neighbourhood of Birmingham," which were brought before the British Association at that town, are also added as an Appendix.]

THE object of this paper is to call attention to a much-neglected branch of English geology,—that of the superficial beds of transported materials, commonly termed *Diluvium* and *Drift*. The backward state of our knowledge on this subject is doubtless owing to the obscurity and complication of the phænomena, and the consequent difficulty of reducing them to any satisfactory general rules. Many geologists have referred these multifarious deposits to one universal *cataclysm*, and the Gordian knot being thus cut rather than untied, the further investigation of this subject has been in consequence greatly neglected. But as it has now become evident that nothing but a process of patient and widely-extended induction can lead us to any accurate generalizations on the subject of transported gravel, it is highly desirable that a considerable number of geologists should engage in this field of research. In the hope of advancing this object I have drawn up a slight sketch of the facts which I have observed, and of the conclusions to which they seem to lead, and shall be fully satisfied if others be thereby induced to take up the inquiry, and to confirm or disprove the views here advanced.

A considerable portion of the surface of this island is covered by patches of gravel, sand and clay, transported from distant localities by forces which are no longer in operation. On examining these masses of drifted materials at various points, many remarkable distinctions of character are perceived. The presence or absence of certain varieties of rocks affords a clue to the direction of the forces which have brought them to their present sites. In some cases the pebbles and boulders are derived from the immediate vicinity, in others their origin must be sought for at the distance of many hundreds of miles. Sometimes these beds are found wholly unstratified; in other cases they are finely laminated, and a violent or tranquil state of the transporting current is inferred accordingly. Some varieties of drift occupy the summits of hills, and are independent of the present configuration of the surface; others occur on the sides or bottoms of valleys, with a constant relation to the present lines of drainage. Lastly, some gravel-beds contain remains of mammalia and lacustrine mollusca, indicating the existence of dry land and of fresh water at the time of their formation; others contain only marine remains, while a large portion of them appear, as far as a negative proposition can be asserted, to be wholly destitute of any organic remains. The question now arises whether all these varied phænomena can be referred to one common agency, operating simultaneously in all parts of South Britain,—or whether they are the result of a plurality of causes, distinct in kind, and operating at separate epochs.

From the fact that all these varieties of transported detritus are unconformable to the rocks on which they rest, and from their lying in detached patches which are very rarely brought into contact, it becomes very difficult to determine their respective ages, and it is probable that they have been produced at a greater number of distinct epochs than we can at present venture to assign to them. It will, however, be shown in the sequel, that there is evidence of at least two distinct periods for the formation of these ancient gravel-beds. The erratic matter deposited during the first of these periods may be called *Marine drift*, being the result of submarine currents at a time when the central parts of England were under the ocean. The second variety we will call *Fluviatile drift*, as it seems to have been deposited by ancient rivers (or river-lakes), at a time when the whole, or a great part of England had become dry land.

Geologists are indebted to Sir R. Murchison for having dispelled much of the obscurity which involved the subject of transported gravel. He has shown, by the evidence of marine shells, that the gravel which covers our midland counties from Cheshire to Gloucestershire has resulted, not from a transient cataclysm over the land, but from a marine current flowing from the north, and dividing the oolitic hills of England from the older rocks of Herefordshire and Wales. This view appears more consistent with the phænomena than any other which has yet been proposed, and we may therefore adopt it to explain the origin of that vast mass of gravel which occupies the great central plain of England. We must therefore suppose a large part, and, perhaps, the whole of our island to have been, at the period in question, lower by many hundred feet than it is at present, so as to admit the sea into the interior of the country. Indeed it is possible that it may have been entirely submerged, for the absence of erratic gravel on the oolite hills, and on the mountains of Wales and Herefordshire, does not necessarily imply that those districts were dry land when the gravel was drifted into the Midland Counties, for the erratic pebbles would always seek the lowest levels in the bed of the sea, and districts which lay out of the direction of the current might undergo degradation *in situ*, without any admixture of extraneous matter. And it seems probable that the chalk and oolite hills of England were in great measure if not wholly submerged, from the fact that ramifications from the general mass of gravel in Warwickshire extend through the transverse valleys of the oolitic range, and follow the course of the Thames, even capping considerable hills near Oxford and Henley, as shown by Dr. Buckland in his 'Reliquiæ Diluvianæ.'

The marine shells which have been found in this gravel in Cheshire, Staffordshire, Shropshire and Worcestershire belong chiefly to existing species, and we must therefore assign a very recent epoch to the formation of these deposits. It appears also that the causes which transported the gravel were comparatively transient, for we can hardly suppose the sea to have occupied the central plain of England during a very long period without leaving some traces of regular tertiary strata, especially in small valleys and basins, which might be sheltered from the action of the northern current. The most probable supposition seems to be, that this transport of erratic matter is intimately connected with the elevation of the land, and that, at a very recent geological epoch, the new red sandstone of central England being covered up by younger deposits, a process of elevation and of accompanying denudation commenced, whereby the upper secondary strata were removed, and the new red sandstone exposed to the action of the marine currents. A further rise would elevate England above the sea-level, and the Midland Counties would then present an undulating surface of new red sandstone and other rocks, with scattered patches of erratic gravel, the last relics of the action of the denuding currents.

It is premature to speculate how far this process of elevation and denudation was gradual or otherwise, but I think we may venture to assert that the contact of sea-water with the present surface of the Midland Counties has been too transient, or the destructive agency of the currents has been too incessant, to admit of the formation of tertiary deposits properly so called. It would also be difficult to decide whether the northerly current, which has effected such devastation, was the direct result of the act of elevation (which would imply great violence and consequent short duration in the elevatory force), or whether it was allied to ordinary marine currents, such, for example, as now set through the Pentland Firth.

Such is the theoretical view which in the present state of our knowledge seems best adapted to explain the origin of all those deposits to which it is proposed to give the name of *Marine drift*. The following are the general characters which distinguish them in the Midland Counties. They consist of gravel, sand, or clay, in varying proportions. The gravel contains numerous pebbles of white or brown quartz, with a greater or less admixture of other substances. It is *in general* very imperfectly, or not at all stratified, though to this rule some local exceptions may be found. There is reason to believe that no mammiferous remains occur in this species of drift, and the only genuine fossils it contains are marine shells, which have been found at a few localities in Cheshire, Shropshire, Staffordshire and Worcestershire. The most essential feature of the marine drift is, that it occurs independently of the minor variations of the surface, covering extensive tracts, and capping hills of 400 or 500 feet in height.

The Midland Counties furnish us with three principal varieties of gravel which are referable to the head of *Marine* drift. These are,—1st, *Erratic gravel without chalk flints*; 2ndly, *Erratic gravel with chalk flints*; and 3rdly, *Local* or *non-erratic gravel.*

The gravel without chalk flints covers the country to the north and west of the Warwickshire Avon, while the gravel with flints occurs chiefly between that river and the foot of the oolite hills. The chalk flints indicate an easterly direction in the current which brought them into Warwickshire. The gravel without flints, on the contrary, appears to have come from the north. As no section has yet shown a superposition of one of these kinds of gravel on the other, which alone can prove them to be of different ages, we must, for the present at least, assign them to one epoch, and suppose the difference of direction in the currents to have been caused by the obstacles which they encountered. Thus a current flowing to the south or south-east through the counties of Nottingham, Leicester and Northampton might encounter the chalk hills of Huntingdonshire, and being thus turned to the westward, might carry chalk flints into Warwickshire, and mix them with the quartz and other northern pebbles. Meanwhile the western part of the same current would flow uninterruptedly through Staffordshire and North Warwickshire, depositing pebbles of northern origin in its way, and make its final exit into the Bristol Channel. This hypothesis of the contemporaneous existence in Warwickshire of a northerly and an easterly current is perhaps preferable, in point of simplicity, to the supposition that the gravel with flints, and that without flints, are of different ages.

Another variety of gravel we have termed *Local drift,* being wholly derived from rocks which exist in its immediate vicinity. It occurs in patches along the base of the oolite escarpment in Warwickshire, Worcestershire and Gloucestershire. The evidence of superposition is still wanting to prove any difference of age between these deposits and the erratic gravel before described, and till that defect is supplied, it will be more prudent to refer them to a modification of the same agency, viz. a denuding current of sea-water. It is, I think, very conceivable that while this current was bringing pebbles from great distances into the central and lower parts of its bed, local shingle beaches might be forming round its margin, composed wholly of the rocks which were there undergoing degradation, without any admixture of foreign substances;

just as in the dry bed of a river we may see the margin skirted by the detritus from its banks, and the central part occupied by pebbles which have been washed from a distance.

The *Local drift* which occupies Wales and Herefordshire, has been shown by Sir R. Murchison to be overlaid by the *Northern or Erratic* drift near Shrewsbury, and is therefore referred by him to an antecedent epoch.

The various kinds of drifted gravel above enumerated, appear to have been deposited by marine currents at a time when a large portion of England was under the sea. We now come to consider the deposits of the rivers and lakes which were formed as soon as this region became dry land. These deposits consist of the same materials as the *Marine drift*, and hence are apt to be confounded with them by cursory observers. They differ, however, in the situations which they occupy, and in the organic remains which they contain. We will distinguish this class of deposits by the name of *Fluviatile drift*. They bear a constant relation to the present form of the surface, and are commonly found flanking the sides or covering the bottoms of valleys, often at a definite elevation above the present drainage. They may in general be distinguished from the marine drift by their finer state of lamination, proving a more tranquil deposition. They seem to form the sole depository of mammalian remains, which I cannot learn have ever been found in the marine gravel which occurs on the tops of hills. Freshwater shells of existing species sometimes accompany the bones of land animals, and perhaps if search were made for them they would be found to be generally diffused in these deposits.

I have principally studied these fluviatile deposits in the valley of the Warwickshire Avon. Platforms of gravel, containing bones and freshwater shells, may be traced at intervals down the valley of that river from Rugby to Tewkesbury, at heights of from 10 to 50 feet above the modern stream. The hills which flank the river are often capped with the *Marine drift*, in which the want of stratification, the absence of mammiferous bones, and the mineral character of the gravel are strongly contrasted with the fluviatile gravel in the valley below. An absolute junction of the two kinds of gravel has not yet been noticed, but they occur within a quarter of a mile of each other.

All the recorded instances of the Fluviatile drift which I have been able to collect have occurred in low situations, where ancient rivers and lakes may have existed. The evidence which we now possess on the subject tends to establish the two following conclusions:—

1. The great mass of the erratic gravel which exists in England has been brought by a northerly current, at a time when all or the chief part of England was under the sea, and contains no terrestrial fossil remains.

2. Bones of terrestrial mammalia are only found in the deposits of ancient rivers or lakes, formed after England had been raised above the sea, and had assumed its present form.

To conclude this paper, we will enumerate a few of the points respecting which it is most desirable to obtain information.

1. Attention should be paid to the different *varieties* of the marine gravel, their mineral composition, organic remains, if any, and their relation to the present form of the surface, so as to elucidate the question of their contemporaneous or non-contemporaneous formation.

2. Search should be made for the actual junction of any of these varieties of gravel, in order to decide whether they pass gradually into each other, or whether a distinct order of superposition can be ascertained.

3. Search should be made for marine shells in gravel in all those parts of England where they have not yet been found.

4. It is desirable to decide whether bones of mammalia are ever found in that species of gravel which, from its position or organic remains, is considered to be of marine origin.

5. To ascertain the greatest height above, and the greatest distance from, existing rivers and lakes, at which bones or freshwater shells occur.

6. To search for freshwater shells in all the localities where mammalian bones have been found.

Queries respecting the Gravel in the neighbourhood of Birmingham.

1. Does the gravel near Birmingham ever contain chalk flints, fragments of oolite, &c., which may indicate a southern origin, or is it purely of northern extraction?

2. Does it contain marine shells?

3. Are these shells of existing or extinct species?

4. Does it ever contain bones of land animals or freshwater shells?

5. What are the circumstances of position, material, &c. of the gravel (if any) in which mammalian bones or lacustrine shells are found, and is it distinguishable in any respect from gravel in which marine remains are found?

6. Are mammalian remains ever found in company with marine shells?

---

4. ON THE UPPER FORMATIONS OF THE NEW RED SANDSTONE SYSTEM IN GLOUCESTERSHIRE, WORCESTERSHIRE, AND WARWICKSHIRE; SHOWING THAT THE RED OR SALIFEROUS MARLS, INCLUDING A PECULIAR ZONE OF SANDSTONE, REPRESENT THE "KEUPER" OR "MARNES IRISÉES;" WITH SOME ACCOUNT OF THE UNDERLYING SANDSTONE OF OMBERSLEY, BROMSGROVE, AND WARWICK, PROVING THAT IT IS THE "BUNTER SANDSTEIN" OR "GRÈS BIGARRÉ" OF FOREIGN GEOLOGISTS.

[*Read June* 14, 1837.]

[The following joint Memoir by Sir R. Murchison and H. E. Strickland is now reprinted from the Transactions of the Geological Society, and the notes having the initials A. C. R., are added by Professor Ramsay, with the sanction and approval of Sir Roderick. The notes having H. E. S. MS. are attached to Mr. Strickland's private copy.—ED.]

PREFACE.

IN former communications to the Geological Society*, the System of the New Red Sandstone of Gloucestershire, Worcestershire and Shropshire was subdivided by Sir R. Murchison in the following manner:—

	FOREIGN EQUIVALENTS.
1. Marls with salt, gypsum, and a thin course of sandstone	*Keuper—Marnes Irisées.*
2. Quartzose red sandstone and conglomerate	*Bunter Sandstein—Grès Bigarré.*
3. Calcareous conglomerate = magnesian limestone	*Zechstein*†.
4. Lower new red sandstone	*Rothe-Todte-Liegende.*

The objects of this memoir are to describe with greater precision than had been previously attempted, the two superior members of this group, by showing that they are both fossiliferous, and that they can be separated from each other by their zoological and lithological characters, as well as by geological position. The communication is the result of a joint examination of the district, and of other observations made independently by the authors. In it, they propose to describe the prominent characters of the two superior formations of the New Red System as exhibited in Gloucestershire and Worcestershire, and to extend their remarks concerning those beds into Warwickshire. The strata will be considered in descending order.

§ 1. *Saliferous Marls and Sandstone* (For. Syn. *Keuper—Marnes Irisées*).

Having examined the junction-beds between the lias and the new red

* See Proceedings, vol. i. p. 471, March 27th, 1838; vol. ii. p. 115, January 21st, 1835; and 'The Silurian System.' p. 27 *et seq.*

† This conglomerate is considered by the Geological Survey to belong to No. 4, *Rothe-Todte-Liegende.*—A. C. R.

sandstone from the southern extremity of Gloucestershire through Worcestershire into Warwickshire*, we have only to repeat what was adduced by one of us on previous occasions, that in many positions the lowest calcareous band or "limestone of the lower lias shale" is also underlaid by shale, which passes into the new red system by alternations of whitish sandy marlstone, black shivery schist, and greenish marls, with occasional courses of very thinly laminated, flag-like sandstone†. The uppermost zone of the new red system, beneath these junction-beds, is composed of whitish-green marl, and is succeeded by red marl, which passes downwards, alternating frequently with parti-coloured marls, though red for the most part predominates. As the country, occupied by these marls, presents few steep acclivities or neatly cut escarpments, it is not often easy to estimate with precision the depth beneath the junction lias-beds at which a change of mineral character is first perceived; but we may state, that a zone of sandstone rises, at many places, from beneath the upper portion of these marls. This Keuper sandstone we have detected at intervals, and we have laid it down on the accompanying Map (see Pl. VI.). It lies at a distance varying generally from half a mile to three miles, from the lowest edge of the lias, the distance changing with the angle of inclination of the strata; its relative position being also dependent on dislocations along the boundary-line. There are, however, localities where this sandstone is covered by a considerable thickness of red marl; and by prolonging the course of the strata beneath adjoining platforms of lias, we have been enabled to estimate that this sandstone is separated from the lias by at least 200 feet of red and green marls‡. It was stated, on former occasions, that this rock appears in Gloucestershire, at Tibberton, in the form of sandy marlstone §, and that it reappears at Burge-hill‖ quarries, near Eldersfield, at the southern extremity of Wor-

* This junction-line is correctly laid down in the accompanying Map, though Mr. Greenough has not kept to it in the 2nd edit. of his large Map.—H. E. S. MS.

† The "Bone-bed" occurs in this part of the series at Coomb Hill, Wainlode, Westbury, Aust, Watchet and Axmouth.—H. E. S. MS.

‡ According to Phillips, 250 feet in Gloucestershire. Sections of the Geological Survey, Sheet 13, No. 3. This Keuper sandstone is probably very nearly continuous from Gloucestershire into Warwickshire and Derbyshire, broken however by faults. (See Maps of the Geological Survey.) Though on the same general horizon, it is not to be supposed that the very same bed is constant throughout. In Derbyshire it is nearer the base than the top of the marl, probably in consequence of a thinning out of the lower marl. There are, however, in Derbyshire several thin bands, near the base of the marl, similar to this part of the Keuper strata in Gloucestershire, Worcestershire and Warwickshire. In general, all of these exhibit pseudomorphous crystals of salt.—A. C. R.

§ See 'Silurian System,' p. 29. It occurs also on the south of Hartpury.—H. E. S. MS.

‖ This place, though spelt Burg-hill on the Ordnance Map, is invariably pronounced with the soft *g*.—H. E. S.

cestershire. South-west corner of the Map, Pl. VI. The section of these quarries is,—

1. Red marl, about		4	feet.
2. Greenish grey marl		5	"
3. Thinly laminated marl, with white gritty sandstone	2		"
4. Soft white sandstone	6		"
5. Greenish and grey marly sandstone	8		"
	--	16	
Dip 10° towards the south.		25	

The sandstone beds thin out, and their surface often presents ripple-marks. The marls in the bed fig. 3 were much cracked at the time of their deposition, for the crevices are filled with whitish sandstone, in the same manner as septaria are occupied by calcareous spar*. Sometimes the marl is more broken up, and detached angular fragments are imbedded in the sandstone. The general relations of the Burge Hill sandstone to the surrounding formations is shown in Section, Pl. VI. fig. 1 †.

This sandstone rock, as before stated, is also exposed distinctly at Ripple, three miles north of Tewkesbury. To the north of Ripple it appears gradually to thin out; and in a section at Old House Farm, south of Spetchley, near Worcester, it is represented by only about 12 feet of grey shivery marls with thin courses of sandstone, the whole reposing on red marl ‡. It is also seen half a mile east of Spetchley at the junction of the Alcester and Evesham roads. It is not discernible farther north, being lost or obscured amidst the great mass of red marl in which it is enclosed. We therefore proceed to describe the characters of the rock at Inkberrow, about twelve miles east and by north of Worcester, in which neighbourhood it has long been quarried, and is so clearly exposed, that it appears to us to afford the best lithological type of the subdivision in Worcestershire.

The sandstone of Inkberrow occupies a prominent part of a distinct ridge about three miles in length, and varying in height from 210 to 320 feet (above the sea), and it extends from the quarry pits, north-west of Radford, in a low hummock, one mile west of Inkberrow, by Penhills to Morton Hill, as a culminating point, whence the ground lowers to Feckenham on the north. The same stone reappears, at intervals, in an attenuated form to the north of Feckenham, ranging by Lower Berrow,

* At New Barn, about 2½ miles west of Forthampton, these sandstone veins are as much as 2 inches thick, and the intervening marl about 6 inches, so that the latter must have contracted about one-quarter of its bulk in drying, before the sand entered the cracks.—H. E. S. MS.

† See also 'Silurian System,' pl. 29. fig. 1.—H. E. S. MS.

‡ At this place is a good section of it (disturbed by a fault), on the line of the Birmingham and Gloucester Railway.—H. E. S. MS.

and wrapping round a bay of lias, from beneath which it rises at Wall-house Farm, at an angle of 20° to 25°. (See Section, fig. 2.)* Near Inkberrow the sandstone is quarried at three chief localities, viz. the Quarry Pits, Stone Pits, and Mucklow's Grave. It is also clearly exposed on the sides of any of the lands, descending from the ridge of red ground on the east to the vale of lias on the west†. The following section exhibits the whole of the strata here exposed. Pl. VI., Section, fig. 3.

1. Red and green marls occupying the crown of the ridge, estimated thickness .... 100 to 120 feet.
2. Sandstone, consisting in detail of
   a. A cover of finely laminated flag-like marly sandstone, of delicate greenish and light drab colours, alternating with marls .... 20 to 30 feet.
   b. Thick-bedded, finely laminated, soft, siliceous sandstone, of various colours, the prevailing one being a white, or pinkish white, with occasional tints of green, purple, &c. .... 15 to 30 „
   c. Finely laminated flag-like sandstones, similar to *a.*
3. Red and green marls similar to No. 1.

These sandstones, though usually of a white or lightish colour, exhibit so many lithological varieties, that their characters can with difficulty be rigorously defined.

At the Great Quarries or Inkberrow Stone Pits, there are several excavations exposing surfaces of the rock from 30 to 40 feet in depth. The chief or central beds are brownish red, brownish yellow, rusty pink, delicate pale green and white, the same tint being seldom persistent for more than a few yards in horizontal extension. The quality of the stone also varies very much, in short distances. A sandstone of a fine grain and good quality sometimes thins out and terminates in soft marly beds, whilst wedge-shaped masses of marl are dove-tailed with the sandstone. The very great variety of the rock is well seen in the tombstones piled up for sale. From these the purchaser may select slabs of five or six distinct tints, varying from nearly a pure white to a deep purple; each slab being for the most part of one colour. This uniformity, however, extending only to the length of a large tombstone, and to the depth of 4 or 5 inches, is due to the fine lamination of the beds; and the effect, though evident to the depth of a few inches, disappears

* Owing to the undulations of the red marl in which it is contained, the same rock is again brought out at Love Lane, about two miles to the east.

† This plain of lias, bounded on the east by hills of red marl, indicates a line of fault which is traceable from Bredon Hill on the south towards Bromsgrove Lickey, and has been described by Mr. Strickland, Geol. Proc. vol. ii. p. 5, and 'Analyst,' vol. ii. p. 1.

where the rock is quarried for troughs or building purposes, when various tints and zones occur in the *same* block. It is, however, essential to remark, that although blotches of marl appear at intervals to certain depths, the rock is never spotted red and green like many of the inferior sandstones, whether of the new or of the old red systems. In the quarry at Mucklow's Grave, the upper flag-like beds are underlaid by sandstone, which weathers to brownish yellow colours with dark ferruginous stripes, marking the laminæ of deposition, the edges of the stone being sometimes worn into cavities. The thin way-boards are of green and deep red colours, and they sometimes cover rippled surfaces of the rock, on which are often raised serpentine strings of sandy marl. Occasionally these assume the forms of Septaria, the fine sand penetrating irregularly in minute and sinuous courses through the marly way-boards. In some instances, these layers of clay appear to have been cracked before the sand was accumulated; in which respect they precisely coincide with the beds at Pyle in Glamorganshire, recently described by Dr. Buckland as Keuper sandstone. At this spot the best building material is a white sandstone, composed of rounded grains of quartz, with specks of whitish decomposed felspar, having in some parts a delicate pink tinge. The mass cuts into blocks 2 feet thick.

The same sandstone, subordinate to the marls, is seen near Harvington, south of Inkberrow, where it was formerly worked; also between Ragley and Alcester, and at numerous localities in the western part of Warwickshire. At Oversley Lower Lodge, near Alcester, it caps a platform about one mile and a half from the lias escarpment*. The quarry there presents the following section:—

1. Red marl .... 4 feet.
2. Grey marl .... 4 „
3. White sandstone with laminæ and fragments of greenish marl.. 6 „
4. White sand with veins of gypsum.
5. Red marl to the foot of the hill.

A similar sandstone also caps the great Alne Hills, ranging parallel to a narrow outlier of lias, which extends north-west from the main body of that formation. There are here indications of an upcast of the red marl, ranging from near Stratford-on-Avon towards the north-west†. On the north of this line of fault near Knowle (north boundary of the Map and Section, fig. 4), distant full eleven miles from the chief escarpment of lias on the south, is a kind of basin of red marl, with a small

* Further eastward this platform is crossed by the Alcester and Stratford Road. —H. E. S: MS.

† We may hereafter endeavour to show that this, as well as several other lines of dislocation, terminate in the Lickey Hills.

outlier of lias in its centre. On the east this basin is bounded by a fault which appears to strike northwards from Warwick, while its northern margin is obscured by the great prevalence of quartzose gravel in that direction. The stratum of sandstone above described occurs within this area at the following points:—Mousehill near Tanworth, Lapworth, Knowle, Rowington, Shrewley Common, Barnmoor near Claverdon, and Wolverton. At these localities its mineral character and geological position distinguish it from the sandstone of Warwick, which underlies the marls, and identify it with the sandstone of Inkberrow, Ripple and Burge Hill. One of its most distinguishing lithological characters is the abundance of greenish marl, which separates the sandstone into finely laminated beds, rarely 15 inches thick, and commonly much less. The marl is often intersected by the thin sinuous veins of sandstone resembling septaria, before alluded to, a character which, both in Warwickshire and Worcestershire, seems to distinguish this upper bed of sandstone from the lower one which occurs at Warwick. This sandstone is also more variable in its texture than that of Warwick; for though *hand* specimens may be found much resembling the latter in appearance, yet in some of the beds a coarse grit prevails similar to that of Burge Hill.

The best place for studying this sandstone in Warwickshire is at Shrewley Common, about four miles north-west of Warwick, where it occurs near the summit of an extensive platform of red marl. The great thickness of this underlying marl is evident at Hatton Hill, which is wholly composed of it. At Shrewley Common the sandstone is exposed by a tunnel, on the Birmingham and Warwick Canal, and in some adjacent quarries. The following section is visible:—

Red marl	30 to 40 feet.
Sandstone and green marl	20 "
Red marl	10 "
	70

The stratification is very nearly horizontal.

The sandstone is obtained in large slabs and removed to Warwick, where it is used for tombstones and other purposes. It is whiter than the Warwick sandstone, and considerably harder, thinner bedded, and closer grained. Small nodules of ironstone occasionally occur in it, and their great hardness gives much trouble to the stonemason. Ripple-marks and septaria-like veins of sandstone are very abundant; and this locality is further distinguished by the occurrence of footsteps of animals, bivalve shells, and teeth of fishes, which will be described in the sequel.

The same beds are repeated at Rowington, where a deep cut on the canal exposes the following strata:—

White sandstone and green marl	20 feet.
Red marl	40 „
	60

The stratification horizontal.

At the foot of the hill south-east of Knowle the same stratum of sandstone appears, and though not exposed by quarries, may be seen in one of the locks on the canal. It is surmounted by red marl, which dips beneath lias, about a mile farther north*.

To the south-east of Warwick the red marls are so denuded that there is little hope of finding this band of sandstone clearly developed, though the sandy character of the ground at Radford, about two miles from the lias escarpment, seemed to us to indicate the range of the sandstone: Pl. VI., Section, fig. 4. The undulations and faults near Warwick, by which the younger beds are reproduced as outliers on the north-west, are therefore very important for our present purpose. Having traced this peculiar sandstone through so wide an area, there was no difficulty in identifying it when discovered near Knowle and in the adjacent hills between that place and Warwick, as represented in Pl. VI., Section, fig. 4. Besides a precise mineralogical resemblance with the sandstones of Burge Hill, Ripple and Inkberrow, the rock in Warwickshire contains the *same* peculiar little bivalve shell, *Posidonomya minuta*, Pl. VI.[b] fig. 4, which occurs at the southern extremity of Worcestershire, and thus throughout a course of not less than forty miles we are enabled to mark the position of this thin band of sandstone, and to distinguish it from other sandstones which not only underlie it, but are separated from it by a great thickness of marl.

Again, the exact geological position of this sandstone, which we consider to be the equivalent of the Keuper sandstone of Suabia and Alsace, is 200 or 300 feet below the lowest beds of the lias, a position which coincides well with that of the principal mass of this sandstone in Wurtemberg, where one of the authors has examined it. In Germany, however, the Keuper formation contains several courses of sandstone and grit, but always subordinate to thick masses of marl. In England we have one well-defined band only, which, occurring from 200 to 300 feet below the lias, is completely and distinctly separated from the great red sandstone of the central counties by a vast thickness of red and green marls, which in certain tracts are saliferous. Independently of natural sections, the great thickness of the red marls, or the depth to which they

* The outlier of lias at Knowle is about a mile and a half in length by half a mile broad. It was worked for limestone a few years ago at Waterfield Farm and Copt Heath, and the shafts of the workings still remain. The discovery of this outlier is due to Dr. Lloyd of Leamington, to whom we are indebted for much valuable information on the geology of Warwickshire.

extend beneath the Keuper sandstone, is established by the shafts and borings made at the salt-works of Stoke Prior near Droitwich, where the gypseous marls with masses of rock-salt were penetrated to a depth of nearly 600 feet *without an indication of any bed of sandstone**†.

Combining these facts with the sections exposed in the escarpments of Inkberrow, Knowle, and other places, it appears evident that the Keuper of England (on the whole quite as largely developed as that of Germany) is, like the "Marnes Irisées" of France, a great marly formation, with one principal band of sandstone subordinate to it, which sandstone is separated by at least 600 feet of marls from the great mass of the underlying new red sandstone.

Organic Remains of the Keuper. Plate VI.[a]

When our examination of this tract commenced, it was a prevalent belief, that as no fossils had ever been found in the red marl formation, so was it hopeless to look for them. We have now to announce the existence of fishes, shells, and the impressions of footsteps of probably a Batrachian animal. The locality in which organic remains were first discovered in the Keuper sandstone is at the Burge Hill quarries, before mentioned. Here they are by no means plentiful, but after considerable search a long Ichthyodorulite was found‡, fig. 3, with numerous casts of small bivalve shells, fig. 4. This Ichthyodorulite manifestly belongs to an undescribed species of the genus Hybodus (Agassiz), and it is therefore exceedingly interesting in showing that the same types of organization which prevail in the lower lias existed during the deposition of the upper part of the new red system. The dentated posterior margin of the spine is unfortunately wanting, but sufficient characters remain to warrant us in regarding it as a new species. It is remarkable for the straightness of the anterior margin, which hardly deviates one-eighth of an inch from a straight line. It appears to have been of a remarkably taper form; but from the loss of the hinder margin, the diameter from front to back cannot be determined. The transverse diameter at the larger end is about three-eighths of an inch, and the total length is 5 inches. The longitudinal *costæ* are smooth, rounded, regular and closely packed; at the larger end, six of them cover a quarter of an inch (*i. e.* their diameter is $\frac{1}{24}$th of an inch). We propose to name this species *Hybodus keuperinus*§.

* See Dr. Hastings's (now Sir Charles Hastings) paper in the 'Analyst,' vol. ii. p. 359.

† The salt, however, is not confined to the lower marls, as a strong brine-spring has been found on Defford Common by sinking through the lias into the upper marls.—H. E. S. MS.

‡ By Mr. Strickland.

§ Spines of a Hybodus have also been found, together with vegetable impres-

The bivalve shells, Pl. VI.[a] fig. 4, appear to be the *Posidonomya minuta* of Bronn (Lethæa Geognostica, p. 164. pl. 11. fig. 22), or the *Posidonia minuta* of Goldfuss (Petrefacten, pl. 113. fig. 5) and of Zieten (Versteinerungen Württembergs, pl. 54. fig. 5)*.

In Germany this shell is stated to pervade the new red system from the "Keuper" to the "Bunter sandstein" inclusive, but in this country it appears peculiar to that band of sandstone which we have proved by stratigraphical evidence to represent the upper formation. It is indeed a very characteristic shell; for, as previously stated, we have detected it at Burge Hill and Inkberrow† in Worcestershire, and at Shrewley Common in Warwickshire, where it is very abundant in some of the sandstone beds.

Batrachian.—Our proofs of the existence of probably a Batrachian in this rock are similar to those which have been held sufficient to establish the claims of the sandstones of other countries to a similar distinction, viz. the impressions of the feet of animals. The slabs which we lay before the Society will, we trust, bear out this inference. We found these interesting relics in the sandstone of Shrewley Common. They afford the same proofs as those which were insisted on in the case of the footsteps of tortoises in Dumfriesshire, viz. the same inverted position of the claws—similar raised portions of sandstone behind each impression, caused by the progressive movement of the animal—and similar indentations proving occasional halts in the march of the animal; in short, all the *indicia* by which the Rev. Dr. Duncan was first led to refer the impressions in the red sandstone of Dumfriesshire to the footsteps of animals, and by which Dr. Buckland sustained and established those views. The Warwickshire impressions are further distinguished by a depressed groove, running intermediate between the footsteps, and apparently caused by the tail of the animal dragging in the soft sand. Pl. VI.[a] fig. 1, represents a large slab, now in the Warwick Museum, the counterpart of that which we have presented to the Society. These footsteps, which are given of their natural size in fig. 2, bear some resemblance to one of the Hildburghausen species figured in Buckland's 'Bridgewater Treatise,' pl. 26‴; and supposed by Prof. Owen, Mr. Broderip, and other

sions, in the Keuper sandstone of Ripple. In October 1842, I found a nearly perfect specimen of the dorsal spine of *Hybodus keuperinus* in a quarry at New Barn, about two and a half miles west of Forthampton. The anterior margin is slightly curved *forwards*, the posterior surface rounded, finely striated longitudinally, with a double row of small prominences towards the smaller end only.—H. E. S. MS.

* Bronn has changed the generic name to Posidonomya, the term Posidonia being pre-occupied in botany. Captain Portlock has lately detected this shell in the new red sandstone of Roan Hill near Dungannon, Ireland.

† It also occurs in the thin beds of Keuper sandstone at Severn Stoke.—H. E. S. MS.

zoologists who have examined them, to belong to some genus of crocodilian saurian. Greater accuracy of definition cannot at present be attained; but perhaps, when the science of comparative ichnology shall be more advanced, the nature of the animal, which has left us these faint traces of its existence, may be ascertained with greater precision *†.

Teeth of Fishes.—Two small teeth of Squaloid fishes were found at Shrewley.

§ 2. *New Red Sandstone* (For. Syn. "*Bunter Sandstein*" *and* "*Grès Bigarré*").

The red sandstone of Gloucestershire (as described on former occasions) is comparatively of small dimensions; but as the formation advances to the north of Worcester, it expands, and massive strata of sandstone rise from beneath the thick cover of marls above described. The uppermost beds of this great sandstone formation are distinctly exhibited, at various places, between Ombersley and Bell Broughton. At Ombersley, and the adjoining hamlet of Hadley, Pl. VI. fig. 5, these beds have been much quarried; and they afford a very beautiful lightish-coloured, fine-grained, slightly micaceous, quartzose sandstone, sometimes tinged with slight shades of pink and green; but a delicate olive colour prevails. At Hadley, the quarries expose from 30 to 40 feet of sandstone, covered by red marl. The sandstone is of greenish-white colours, tinged, in parts, with purplish and reddish hues, and it is somewhat micaceous; the upper portion, or that nearest the marls, being a breccia of marl and sandstone. About 8 to 10 feet below the surface, carbonaceous laminæ are very abundant; and on flaking off the beds, their surfaces are found to be covered by numerous impressions of plants, nearly the whole of which have passed into a black powdery charcoal, which readily disintegrates on extraction. Hence it is very difficult to preserve these fossils with all the freshness of form which they exhibit when first disinterred, and when the jet-black colour forms a striking contrast to the light buff-coloured matrix. When the black, incoherent matter has fallen out, the impression of the matrix is usually marked by ferruginous colours. The large slabs, which we procured for

* Professor Owen now supposes these footsteps to belong to the genus Rhynchosaurus, the head of which has been obtained at Grinshill in the Bunter sandstone, 1841.—H. E. S. MS.

† Mr. James Plant has given a minute description of these beds as they occur near Warwick, where they are said to contain remains of plants, viz. "casts of *Echinostachys oblongus* and Equisetæ, Annelid marks, *Estheria minuta* (=*Posidonomya minuta*), teeth of Placoid fishes, Ichthyodorulites, fish-coprolites ?, and fragments of bone."—Journal of the Geol. Soc. vol. xii. p. 373.—A. C. R.

the Society from this locality, have been examined by Professor Lindley, who thus expresses himself concerning their vegetable contents:—

"The only plant among the specimens that has been published, is the Echinostachys, of which slight traces are visible, Pl. VI.[a] fig. 11. I presume they all belong to the same species; but at all events, one of them is *identical* with that published by Adolphe Brongniart, from the *Grès bigarré* (*E. oblongus*). The remainder of the impressions consist of many narrow, monocotyledonous leaves, resembling those of *Grasses*, a portion of a *flabelliform Palm-leaf*, some large moulds of stems of a doubtful character, a longitudinal section of a portion of a *dicotyledonous stem* with the bark on, a considerable portion of a broad leaf of some monocotyledon, and a great multitude of fragments wholly indeterminable. I can detect no trace of a Voltzia, unless a very imperfect stain upon one of the smaller slabs should be of that nature; the genera Æthophyllum and Paleoxyris are equally absent; Convallarites may be present in the form of some of the broken leaves, but it cannot be identified."

Owing to a line of dislocation which runs along the Doverdale valley, the sandstone of Hadley is thrown to the west, and is partly covered by marl; but further west it rises with a gentle easterly dip, and in that position forms the low ridge, at the southern end of which is the village of Ombersley*. By pursuing a transverse section from this little ridge to the Severn, we pass through an unbroken, descending series of sandstones, in the following order. See Pl. VI. Section, fig. 5.

a. Beds of the Keuper or red marl, forming the crest of the Hadley and Ombersley ridge.
b. Thin-bedded, red sandstone.
c. Sandstone of a whitish and yellow colour, with plants similar to those of Hadley, 10 feet below its surface.
d. Thick masses of deep red sandstone. They rise up to the west, and cap the hills on the left bank of the Severn above Holt Bridge. They are underlaid by way-boards of sandy marl.
e. Alternating thin bands of coarse, concretionary, or rather fragmentary marlstone,—gritty small quartzose conglomerate, and soft, thick-bedded, dull red sandstone, the latter much predominating;—the whole dipping east-south-east about 12°. The right bank of the river, at Holt Bridge, offers magnificent sections of the massive red sandstone, some quarries exposing faces of 60 feet and upwards.

This transverse section from Hadley and Ombersley to the river Severn, completely establishes the fact, that the light-coloured *sandstone, with plants, is inseparable from the great mass of red sandstone, and lies beneath the whole of the Keuper marls*, by a large portion of which it is entirely separated from the true Keuper sandstone.

In following the Ombersley ridge upon its strike to Elmley Lovett,

* The quarries from which the pretty church of Ombersley was built, are about half a mile south of the village, but they are now abandoned. This is the point at which the rock first emerges from beneath the marl; and hence it is overlaid by much rubbish.

the same light-coloured sandstone is found, at intervals, for five or six miles, and is largely quarried at the latter place. Here, indeed, the prevalent tint of the rock is *red,* and the plants are again found in some quantity, generally in layers, which separate the principal masses of sandstone. The line of dislocation which passes by these quarries will be noticed hereafter. The tract to the east of this ridge is covered by red, saliferous marls, extending by Droitwich to Stoke Prior. But at Bromsgrove, the peculiar sandstone of which we are speaking rises from beneath the marls, and offers, on the high road from Droitwich, a most instructive section, which confirms, in every respect, the transverse section from Ombersley to the Severn, and proves the light-coloured rock to be an integral part of the great red sandstone formation.

Descending section of Breakback Hill, one mile south-west of Bromsgrove:—

Red marls crowning the hill.

a. Ledge of darkish red sandstone, consisting of Thin-bedded earthy sandstone, alternating with marls, and whitish, thick-bedded, soft sandstone. Top beds partially brecciated like those of Ombersley, and inlaid with blotches of greenish marl. It contains fragments of plants and carbonized wood.

b. Thick-bedded, greenish-white sandstone, full of plants *.

c. Deep red-coloured, solid, massive sandstone, scarcely micaceous, without way-boards; 40 feet are exposed in one quarry. Patches of marl here and there inosculate in the form of wedges or irregular concretions.

d. Light yellowish and brownish sandstone, with wedge-shaped masses of grit and marly calcareous breccia, occasionally cavernous, with traces of plants.

e. Red sandstone, slightly green in some parts.

These last-mentioned beds lie near the turnpike-gate, west of the town of Bromsgrove, the whole mass dipping westerly. They are succeeded, on the opposite side of the valley in which the town stands, by distinct courses of deep red, soft sandstone, which in their turn are underlaid by strata containing irregular concretions of impure limestone or cornstone. The further consideration of these lower beds, in Worcestershire, would carry us beyond the limits of this memoir; and as they have been already described at some length by one of the authors, and proved to be the equivalents of the dolomitic conglomerate and lower new red sandstone†, and that the latter passes down conformably into the coal-measures, they are now merely cited to corroborate the inference, that the overlying red sandstone of Bromsgrove, including the light-coloured rock, with plants, constitutes the true equivalents of the Bunter sandstein‡, Pl. VI. fig. 5.

* I have since obtained a fragment of bone from this stratum.—H. E. S. MS.

† See the 'Silurian System,' p. 46 *et seq.*

‡ Passing over these lower rocks, and also those of much higher antiquity, which constitute the Lickey, we again meet with a fine section of whitish sandstone rock

Sandstone of Warwick.

We now proceed to show that the sandstone of Warwick occupies the same geological position as that of Ombersley and Bromsgrove. In approaching Warwick from the west or south-west, this peculiar rock rises suddenly from beneath the red marl. It extends for several miles to the north-east, and is quarried at Warwick, Guy's Cliff, Leek Wootton, Blakedon Hill, Leamington, Bubenhall, and other places. Its strike appears to be from north-east to south-west; while its abrupt termination at Warwick may perhaps be due to a fault ranging in a northerly direction to the west of Kenilworth*. It is probable that the upcast which has thus raised the sandstone above the marls on the west, has given the more or less dome-shaped structure to the hill on which Warwick stands. On the south, however, the sandstone rises quite conformably from beneath the red marl; while on the other side, the stratification, as far as the irregular bedding allows it to be traced, appears in some places to dip towards the north-west†. See Pl. VI. fig. 4.

at the "Sandhills" near Alvechurch, the dip being reversed to the north-north-west. The section of these quarries consists, in descending order, of

a. Red loam and marl .... 6 to 8 feet.
b. Thin, flag-like sandstone and marl alternating .... 10 to 12 „
c. Solid, whitish sandstone, of a delicate green tinge, void of way-boards, soft under the hammer, and working into any form. Lines of bedding partially indicated by flakes of dark mica, so that when cut and smoothed, the undulations of the dark-coloured materials appear through the light-coloured ground; thus producing the appearance so well known in the *Cipollino* marble. In other parts the mass is freckled with small dark spots (manganese?), which, when the stone is rubbed down, give it a warm brown tinge. We could detect no plants or organic remains in this beautiful sandstone. It is perhaps, therefore, not precisely of the same age as that of Bromsgrove and Ombersley, but a repetition of analogous strata rather lower in the series.

* This inference is correct, having been positively ascertained during the progress of the Geological Survey, and affords a good proof of the accuracy of Mr. Strickland's judgment in matters relating to geological mapping. See Maps of the Geological Survey of Great Britain, 53 N.W.—A. C. R.

† About one mile north of Warwick, and half a mile north of the canal, is a singular knoll of sandstone, partly removed by quarrying, which must be very near the line of this fault; for at the canal bridge, on the south-west, a section of red marl is exposed, and no traces of the subjacent sandstone are visible. This fault, as it passes Kenilworth, is much obscured by the abundance of gravel; but its presence is proved by the fact, that there is no intervention of the buff-coloured or Warwick sandstone, between the red sandstone of Kenilworth and the red marl, which appears to abut against it on the west. To the south-east of Warwick, the hill of red marl called Highdown Hill, with a strip of lias at its eastern base, not improbably indicates the continuation of the fault above described.

At Leamington the sandstone rises very gradually from beneath the red marl on the south. The uppermost beds consist of very soft sandstone and white sand, with alternations of marl. The following section was lately exposed in cutting a large drain on the south side of the Leam:—

a. Gravel with flints	6 to 8 feet.
b. Light-coloured sand with an irregularly denuded surface, containing concretionary masses of sandstone	3 to 5 "
c. Red marl	4 "

The concretionary masses in the sand contain numerous fragments of bone in a better state of preservation than is usual in the Warwick sandstone. They appear, however, to have undergone much attrition; and it is rare to find any traces of their original form. A small tooth of a fish, probably that of a shark, was also found in these sandy concretions, Pl. VI.[b] fig. 7 *a*.

Below these rubbly beds the sandstone assumes a degree of compactness which adapts it for masonry, and many houses in the new town of Leamington are built with the stone which has been extracted from their cellars. Indeed, in beauty of tint, facility of working and durability, the light-coloured sandstone of Warwickshire and Worcestershire, like that of Grinshill near Shrewsbury, with which we shall presently compare it, is probably surpassed by no other rock in the British Isles.

To the north of Leamington the same variety of the red sandstone is quarried at Blakedon Hill, where fragments of bones have also occurred in it. At the northern foot of this hill, the subjacent sandstone, of a deep red tint, commences; but the junction of the two rocks is not exposed.

The quarry which has been most productive in the remains of Vertebrata is at Coton End, on the south-east side of Warwick, where the following section is exposed:—

a. Soft white sandstone and thin beds of marl	8 feet.
b. Whitish sandstone, thick-bedded	12 "
c. Very soft sandstone, coloured brown by manganese, called "Dirt-bed" by the workmen	1 "
d. Hard sandstone, called "Rag," about	2 "
	23

The dip about 3° to the south.

The bones are principally found in the so-called "Dirt-bed." They are in the same rolled and fragmentary condition as at Leamington, but in a state of much greater decomposition. When first taken from the quarry they resemble stiff jelly, with singular hues of blue and red. It is necessary to remove them with great care from the quarry, and, when dry, to saturate them with a solution of gum-arabic as the best means of preserving them. The only specimens which exhibit any distinct zoological characters, are some teeth obtained by Dr. Lloyd, to whose kindness we are indebted for the figures given of them in Pl. VI.[b] figs. 6, 7, 8.

The light-coloured sandstone extends from Warwick, by Guy's Cliff, to Wootton Grange, beyond which it is succeeded by the underlying redder sandstone of Kenilworth. In a quarry in the grounds at Guy's Cliff is the following section:—

Sandstone and beds of marl	8 feet.
Solid sandstone, whitish or grey, occasionally of a reddish tint	12 „
Red, micaceous marl, with wedges of sandstone	8 „
Solid, light-coloured, reddish-tinted sandstone, about	20 „
	48

The bedding very irregular. Ripple-marks occur on some of the beds.

We have now to speak of the redder sandstone which underlies the light-coloured strata. The upper part of this red rock consists of flaggy beds, with marly way-boards. In a quarry south-east of Ashow, on the left bank of the Avon, is the following section:—

a. Thin, flaggy beds of red sandstone	6 feet.
b. Thick-bedded red sandstone	8 „

Calcareous matter pervades the thin beds *a*, in concretionary patches; and where that is the case the sandstone is intensely hard.

One of the best sections of the red sandstone is in a quarry on the north-west of Kenilworth Castle:—

a. Laminated, marly, red sandstone	5 feet.
b. Thick-bedded, reddish-brown sandstone, sometimes discoloured by manganese, with occasional fragments of red marl, and a few rolled pebbles of altered sandstone and porphyry	20 „
	25

The bedding irregular. The general dip about 5° south-east*.

It is needless to give further details respecting this great red sandstone deposit, for its general features are very uniform. We therefore proceed to make a few concluding observations.

The section fig. 4 proves that the light-coloured sandstone of Warwick agrees completely with the rocks of Ombersley and Bromsgrove, fig. 5, in rising from beneath the marls, and in passing downwards into solid

* These beds are only cursorily noticed in this memoir. On closer examination they have since been proved to be Permian. "The error arose from the absence of the pebble-beds and the lower and upper brick-red sandstones of the Bunter series; and thus it happens that between Leamington and the country a little to the south of Tamworth, the white and brown sandstones" (the Keuper sandstone of Buckland at Warwick) "that immediately underlie the new red marl, rest directly on the Permian sandstones and marls, which were thus naturally mistaken for the lower part of the Bunter strata. Having satisfied myself, on purely stratigraphical and lithological grounds, that these were true Permian strata, the truth of this surmise was further confirmed in 1852, when I found fragments of Lepidodendron and Calamites near Exhall, and in the same quarry a few casts of a shell more

red sandstone. Although this rock has been recently described by Dr. Buckland, it is essential to our purpose to state, that, from geological and other evidences, we consider that it cannot be, as he conceives, the equivalent of the German Keuper. The true position of that rock is fortunately indicated in natural sections near Warwick, as well as in many parts of Worcestershire, where it is demonstrated that the *thin-bedded* sandstone, or true Keuper, is separated from the *thick-bedded* sandstone of Ombersley, Bromsgrove and Warwick by a vast thickness of red and green marl. Obedient, however, to geological principles, based on zoological evidences, we should not pretend to set up the classification here suggested, in opposition to the views of so distinguished a geologist as Dr. Buckland, if founded only on the relative gèological position of these rocks. On the contrary, if it could have been shown that the fossils which we have now pointed out as characterizing the upper sandstone, occurred also in the lower,—that the plants in the lower sandstone were similar to the well-known plants of the German *Keuper,*—and that the fragments of Saurians found in the sandstones of Guy's Cliff and Warwick really belonged to the *species* peculiar to the Keuper,—then, indeed, we should willingly allow that the lower sandstone *also* must be grouped with that formation. Seeing, however, that the animal remains of the one sandstone are, as far as we can judge, entirely different from those of the other, and that the plants, so abundant in the lower rock, have none of the characters of the Flora of the Keuper, but, on the contrary, contain one remarkable plant, the Echinostachys, a genus considered by Adolphe Brongniart as peculiarly characteristic of the *Grès bigarré,*—we are compelled to adhere to our opinion, and to contend that the peculiar sandstone of Burge Hill, Ripple, Inkberrow, Alcester, Shrewley Common, &c., which we have been the first to describe, is the true equivalent of the *Keuper Sandstein*; and that the sandstones of Ombersley, Bromsgrove and Warwick, are not the *Keuper*, but a portion of the *Bunter Sandstein.* In respect to the Sau-

allied to Strophalosia than to any other genus. The silicified trees near Allesley and Meriden, and apparently several species of Caulerpites and Breea now in the Warwick Museum, belong to the same rocks (formerly supposed Bunter species); and, in addition to this, it is interesting to know that the beds near Kenilworth in which the *Labyrinthodon bucklandi* was found by Dr. Lloyd belong to the same series. This reptile, previously considered of Bunter date, must therefore be transferred to the Permian period." (Ramsay, Geol. Journ. 1855, vol. xi. p. 197.) These Permian rocks lie on the Warwickshire coal-measures nearly conformably, and the sandstones at the base of the Keuper marls lie unconformably on both. The Permian rocks stretch from the country a little north of Warwick, and overlie the coal-measures between Atherston and Kingsbury. The same rocks overlie the greater part of the coal-measures of South Staffordshire, North Staffordshire, Coalbrook Dale, &c.—A. C. R.

rian of Guy's Cliff, which we have had no opportunity of examining, it is sufficient to state, that Dr. Buckland himself does not contend that it is either of the species of the Phytosaurus (Jäger) of the German Keuper; and he hesitates even to refer it to that genus. Now the mere existence of a Saurian in the Warwick sandstone proves nothing; for geologists are well aware, that various species of the family occur in all the formations, from the lias down to the magnesian limestone inclusive. Indeed, as these animals are not unfrequent in the Bunter Sandstein of Germany, we cannot avoid suspecting that the animal remains of Guy's Cliff may, if ever accurately determined, be assigned to some of those species, mentioned by M. Voltz and others, as occurring in the Bunter Sandstein, or German deposits, which, from the other proofs adduced, we consider to be of the same age*.

In concluding this sketch of the structure and contents of the two upper formations of the New Red System, in the central counties of England, we may observe, that in Shropshire and the adjacent parts of Staffordshire, where these deposits have been described at length by one of the authors, the upper marls, or Keuper, have been so much denuded, particularly near their junction with the lias, that the peculiar band of Keuper sandstein which we have found persistent in Gloucestershire, Worcestershire and Warwickshire, has not yet been met with. But, rising from beneath the whole of the marls, and apparently separating them

* This light-coloured sandstone occurs in great force near Bromsgrove. It is also found rising from under the red marl south of the Malvern Hills, and resting on the upper soft red sandstone of the Bunter series. It ranges from the rising ground, half a mile south of Bromsberrow, to the neighbourhood of Huntley. The same rock, underlying the marl and broken by many faults, ranges from Bromsgrove northward (east of Coalbrook Dale) to Grinshill, New Cliff, the Peckforton Hills, De la Mere Forest, and so into Cheshire on both sides of the Mersey. It also more or less surrounds the South Staffordshire, Warwickshire, and Ashby-de-la-Zouch coal-fields, and again occurs under the marl between North Staffordshire and the country near Derby; the thickest development is from 200 to 300 feet. The strata are frequently interstratified with numerous beds of red marl, and in rare cases (south of Cheadle, for example) the interstratified marl is equal in importance to the beds of sandstone. The Geological Survey originally considered them to be Bunter beds, but latterly they have adopted Dr. Buckland's view, on stratigraphical grounds, that they belong to the Keuper rather than the Bunter series, or in other words, that the white sandstones are of later date than the continental Muschelkalk. With the marl it *overlaps* all the lower parts of the trias (Bunter), and near Newent rests directly on the coal-measures and old red sandstone, in Warwickshire on the Permian rocks and coal-measures, and near Derby on the millstone-grit and upper carboniferous limestone shales. There seems also to be no perfect palæontological reasons why these beds should be classed with the Bunter beds. The Labyrinthodons in these strata are catalogued by Professor Morris as of Keuper species, and the genus Echinostachys has been found by Mr. Plant in undoubtedly higher Keuper sandstone since this memoir was written.—A. C. R.

from the underlying massive sandstones, there does exist a thin course of impure limestone, which, it is presumed, may represent the *Muschelkalk*, and of which an account is given in the 'Silurian System,' p. 36, by Sir R. Murchison*. If this should really prove to be the equivalent of the Muschelkalk, the age of the formation which underlies it will be still more clearly established; for the sandstones of Cheshire and Shropshire, which there rise from beneath the saliferous marls, as in Worcestershire and Warwickshire, correspond in mineral structure with those which we have been describing, particularly in containing, near their upper limits, courses of a stratum (sometimes 70 to 80 feet thick) of whitish or light-coloured sandstone, of which the celebrated quarries of Grinshill, near Shrewsbury, and the picturesque rocks of Hawkstone, are good examples †. These rocks are indeed identical with the sandstone of Ombersley, Bromsgrove and Warwick; and like those we have here described, they pass into, and are inseparable from, the great mass of the new red sandstone of England, which, from geological and zoological proofs, we consider to be the equivalent of the Bunter Sandstein.

In thus identifying the red and green marls, and an included band of sandstone with the Keuper, and separating this marnose formation from the underlying sandstones, we have the direct authority and example of the best French geologists; for M. Elie de Beaumont, in a memoir on the Vosges, has shown that the formation of *Marnes Irisées*, which he places on a parallel with our English red marl, is the true representative of the Keuper; and in the south-western parts of France, M. Dufrénoy has shown us, that these marls and the underlying sandstones are brought together precisely in the same manner as in England, the Muschelkalk, or subdividing limestone, having thinned out ‡ §.

* Sir P. Egerton considers the "bone-bed" of Aust Cliff to represent the Muschelkalk (Geol. Proc. No. 77); but that cannot be the case, as it would throw back the whole of the red marls to the age of the Bunter Sandstein. In my opinion it only proves that certain fish of the Muschelkalk survived to the commencement of the Lias.—H. E. S. MS.

† At Grinshill, eight miles north of Shrewsbury, where the rock is identical with the best building-stone of Leamington and Warwick, the mass of whitish-coloured sandstone (80 feet thick) is exposed between a cap of red marly sandstone and a deep red sandstone of vast thickness, on which it rests; thus perfectly resembling the section at Bromsgrove. The Hawkstone Hills are chiefly composed of the same light-coloured variety of the red sandstone. It is indeed remarkable that a mere lithological distinction of colour should be so very persistent.

‡ See Dufrénoy and Elie de Beaumont, 'Mémoires pour servir à une description géologique de la France,' vol. i. p. 313 *et seq.*

§ Of the tendency of limestones to thin out in short distances, we have many examples in the Wenlock and Aymestry limestones, the carboniferous limestone, the coral rag, &c.—H. E. S. MS.

In a subsequent memoir we may offer some explanation of the lines of dislocation by which these deposits have been effected.

POSTSCRIPT.

Dr. Buckland has described in the Geol. Proc. vol. ii. p. 439, the silicified stems of trees which occur at Allesley, near Coventry, but has not determined the age of the rock in which they are imbedded. Having visited that spot, we have no doubt that these trees occur in that part of the series which we have shown to be the equivalent of the Bunter Sandstein. The red sandstone of Kenilworth may be traced without interruption from that place to Coventry and Allesley, where it is interstratified with beds of quartzose and trappean conglomerate, which are identical with those of North Worcestershire, Staffordshire, &c. (See Sir R. Murchison on the Silurian System, p. 42.) The fossil trees of Allesley are found in a stratum of this conglomerate, containing manganese, and lying between strata of ordinary red sandstone.

The same red sandstone occurs in thick beds at a quarry, lately opened, about three-quarters of a mile north-west of Coventry. It differs in no respect from the sandstone of Kenilworth, and the only remarkable circumstance connected with this quarry, is the discovery of a fossil jaw, represented at Pl. VI.[b] fig. 5. The teeth of this jaw are very irregular both in form and position, and appear to have lost much of their distinctive character by trituration. The teeth are not inserted in alveoli, but are united with the maxillary bone like the teeth of fishes.

This specimen, the only example known of an animal relic in this part of the New Red Sandstone of central England, is deposited in the Warwick Museum.

At the railway station near Coventry, the cuttings have laid bare the upper or light-coloured beds of variegated sandstone, spotted with manganese, dipping to the east-south-east, and covered by the red marl or Keuper, which, on the line to London, is seen to graduate into and pass under the lias. In following the railroad from Coventry towards Birmingham, the underlying or deep red sandstone is traversed; and in the adjoining hill of Meriden, it is again capped by the same light-coloured rock as at Coventry. The red sandstone continues as far as Berkswell, where it is succeeded by the upper marls of the Knowle basin, which extend to within a mile of Birmingham. At Berkswell a singular variety of this sandstone occurs, exhibiting the appearance of a pavement of bricks, 2 or 3 inches square, separated by bands of a whitish cement. The stone splits into thin flags, and the division in the colours extends through each layer. No change of texture is visible in the white bands from the red portions which they enclose; but along the centre of each is a fine raised line, apparently due to a cast in an extremely narrow crack. The following woodcut represents a fragment of this stone.

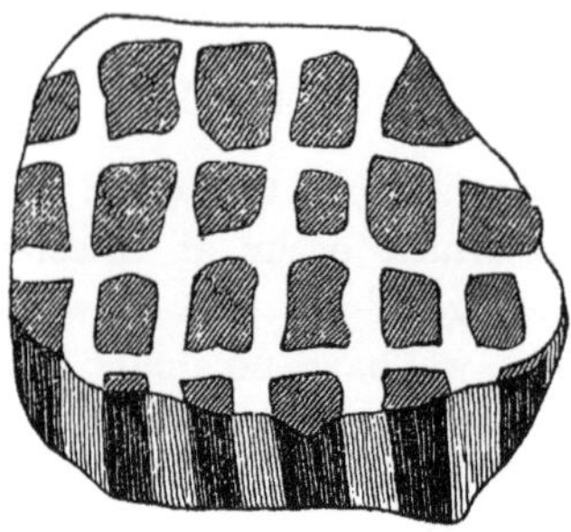

A similar sandstone is stated by Dr. Macculloch to occur in the coal-measures of Arran (Description of the Western Islands, vol. ii. p. 374*).

Plate VI.

Map and Sections: p. 111 *et seq.*

Plate VI.[a]

Fossils from the Keuper Sandstein.

Fig. 1. Ripple-marks and impressions of footsteps on a slab of sandstone from Shrewley Common, Warwickshire, one-third the natural size. The impressions are in relief. The slab is preserved in the Warwick Museum; but a portion of the counterpart is in the Museum of the Geological Society.

The footsteps were formed by an animal which was probably allied to the Batrachians. The toes appear to have been destitute of claws, and the larger or posterior foot exhibits faint traces of having been webbed. In these respects, and in their general form, they have some resemblance to the feet of frogs, but differ from them in having the outer toe of both the hind and fore feet longest, while in frogs the second toe is the longest. It is moreover evident, that this animal possessed a tail, which, dragging in the soft sand, has caused a groove, about ⅜ths of an inch wide, intermediate between the alternate footsteps. This groove meanders slightly from side to side, approaching each pair of footsteps alternately.

The impressions of the feet are repeated twelve times, at equal intervals of about 9¼ inches. To this, however, there are two exceptions,—the distance from *a* to *b* in fig. 1 being only 6 inches, and from *c* to *d* 7 inches. We may conjecture that the animal stopped for a time at this point, and then proceeded, an opinion which is confirmed by the semicircular ridge of sand marked *e*, which may have been formed by the underside of the animal, or by the impulse necessary for the resumption of its motion. Several smaller impressions of similar footsteps may be detected on the surface of the slab: four of them, marked *f*, *g*, *h*, *i*, are in regular alternate succession; the others, marked *k*, are in no regular order, the impressions corresponding to them being too faint to be detected.

* The trees, &c. found at Allesley, and the jaw found near Coventry, are now known to be of Permian date. See note, p. 125.—A. C. R.

A considerable resemblance exists between these impressions and some which have since been discovered (in company with those of the Cheirotherium) in the Bunter Sandstein at Storeton Hill, near Liverpool; also at Grinshill near Shrewsbury, specimens of which latter are in the Museum of that town.

Fig. 2. A pair of footsteps of the natural size marked with a star on fig. 1.

Fig. 3. An Ichthyodorulite of *Hybodus keuperi* (Murch. and Strickl.) from Burge Hill, now in the Museum of the Geological Society.

This Ichthyodorulite was nearly perfect when the description at p. 118 was written, but having been since unfortunately injured, we are able to show only the fibrous structure of the interior in fig. 3, the external costæ on the lower side being concealed by the matrix. The costæ are however seen on the fragment fig. 3 *a*, natural size, and the fibrous internal structure is shown, magnified twice, at fig. 3 *b*, p. 118.

Fig. 3*. A small tooth, probably of a species of Hybodus, from Shrewley Common, natural size and magnified twice. In the Museum of the Geological Society.

Fig. 4. A fragment of sandstone with casts of *Posidonomya minuta*, Bronn, from Shrewley Common. In the Museum of the Geological Society: p. 118.

Plate VI.[b]

Fossils from the Bunter Sandstein.

Fig. 5. Maxillary bone of a fish in the Warwick Museum, from a quarry one mile north-west of Coventry. The left-hand end of the figure seems to be the anterior extremity. The bone, which is here thickened and rounded, gradually becomes thin and flat towards the other end. Much of the thinner portion is wanting, but the impression of the lower surface remains in the sandstone, as shown by the dark shade in the figure. The specimen seems to have been considerably worn, and to have lost several of the teeth, before it became imbedded in the sandstone.

Fig. 6. Tooth of Megalosaurus?: (*a*) natural size, (*b*) magnified twice. From Coton End near Warwick: p. 124.

Fig. 7. A smaller tooth of the same animal, from Coton End. Warwick Museum: p. 124.

Fig. 7*a*. A small tooth, from Leamington. Museum of the Geological Society.

Fig. 8. Tooth of a Saurian?, from Coton End. Warwick Museum: p. 124†.

Fig. 9. A smooth curved tooth, flattened on one side, from Leamington. Warwick Museum.

Fig. 10. A vertebra, from Coton End. Warwick Museum: p. 124.

Fig. 11. *Echinostachys oblongus*, from Hadley near Ombersley. Museum of the Geological Society: p. 121.

Woodcut: p. 130.

Fragment of sandstone from Berkswell.

---

† This specimen having been since sliced and examined by Professor Owen, has formed (in conjunction with a jaw and other fragments since discovered) the groundwork of his genus Labyrinthodon, which he has identified with Phytosaurus, *Jäger*, Salamandroides, *Jäger*, Mastodonsaurus, *Jäger*, and (probably) Cheirotherium, *Kaup*. See Owen, 'Odontology,' and Geol. Trans. n. s. vi. p. 503.—H. E. S. MS.

5. MEMOIR DESCRIPTIVE OF A SERIES OF COLOURED SECTIONS OF THE CUTTINGS ON THE BIRMINGHAM AND GLOUCESTER RAILWAY*.

[*Read June* 10, 1840, and *June* 16, 1841.]

A FEW years ago, when the various lines of railway in this country were first projected, it was usual to hear geologists congratulate themselves on the acquisitions which the science would receive from the numerous opportunities thus afforded for studying the British strata. These anticipations have been in some degree fulfilled, but it is undeniable that during the last ten years many golden opportunities of studying geological phænomena and collecting specimens have been irrecoverably lost, in consequence of the railway sections not having been visited by geologists during the brief period when they were exposed to view. The practice commonly followed by engineers of covering up the "slopes" or sides of the cuttings with vegetable soil as soon as the excavations are completed, detracts greatly from the advantages which the science would otherwise derive from them, for unless some geologist happens to inspect the section at the right moment, the interesting phænomena which it discloses are buried for ever beneath the verdant sod. The only remedy for this evil seems to be for each line of railway to be systematically surveyed by a competent geologist, who should pay repeated visits to every section during the time of its excavation, in order to collect minerals and fossils, and to record the features of its stratification. It would be highly desirable if each railway company had a professed geologist on its staff; but as such an appointment would probably be considered foreign to the objects of these companies, the ends of science might, perhaps, be attained, if the Geological Society were to make applications to the several railway companies, requesting from them a series of their published sections, the colouring of which might be undertaken by some member of the Society. Such sections, even if uncoloured, are still highly valuable, as containing a vast fund of accurate measurements, levelings and other geographical facts. But if to these geometrical sections, the geological phænomena be superadded by the aid of colour, it is evident that such documents will acquire the utmost value. And we must remember that the means of doing this is fast passing away; in a few years all the important lines of railway will have been completed, and future geologists will perhaps sigh for the opportunities which the present generation have neglected.

Anxious to contribute towards so desirable an end, I gladly yielded

* Trans. Geol. Soc. n. s. vi. p. 545.

to a request made to me some time ago by Captain Moorsom, chief engineer of the Birmingham and Gloucester Railway, to undertake a geological survey of that line. To Captain Moorsom the Society are indebted for the accompanying lithographed sections, to which I have added the geological colouring; and I must here express my obligations to that gentleman, as well as to Captain J. Vetch, F.G.S., for much valuable assistance during the survey. The line here described had previously been surveyed by Mr. F. Burr* (see Geol. Proc. vol. ii. p. 593), and I am happy in bearing testimony to the general correctness of his observations; but as none of the excavations were commenced when he made his survey, he had no other data than the "trial-shafts" sunk from time to time, which of course could not exhibit geological phænomena with the same accuracy as open cuttings.

Before entering on the description it is necessary to remark, that the lithographed sections, being made for engineering purposes alone, exhibit a great disparity between their vertical and horizontal scales, the former exceeding the latter in the ratio of about 13 to 1. The inequalities of the surface are in consequence greatly exaggerated, and I have therefore in general abstained from depicting the inclination of the strata, except where the junction of two formations rendered it necessary.

The Birmingham and Gloucester Railway is not, perhaps, one of the most interesting lines, in a geological point of view, because from running nearly parallel to the strike of the strata, it passes through but a small succession of formations. This circumstance, however, causes each stratum to be exposed in greater detail, and affords a wider scope for local variations.

The geological starting-point on this railway does not coincide with either of its locomotive termini. The lowest rock exposed is on the anticlinal axis of the Lickey, about ten miles south-south-west of Birmingham. It will be remembered that in the description given by Sir R. Murchison of the Bromsgrove Lickey (Sil. Syst. p. 569), the effects of that trappean eruption are stated to extend many miles towards the south-south-east, in the shape of a long elevated tract called the Ridgeway, forming the western watershed of the river Arrow. The transverse sections of this ridge afforded by the railway, as well as by the Birmingham and Worcester canal a mile further south†, sufficiently prove an anticlinal arrangement of the strata.

The cutting No. 95‡ on the engineering section§, crosses the anti-

* The sections here referred to, consisting of thirty-five folio sheets, are deposited in the Society's collection of maps, &c.

† See Burr, Geol. Proc. vol. ii. p. 594.

‡ The account of this cutting was read June 16, 1841.

§ See Sheet 24 of the lithographed sections.

clinal ridge of the Lickey, and being excavated to the great depth of 56 feet, it was naturally expected to exhibit some points of interest. About half a mile to the north of it trap-rock appears *in situ* *, elevating and altering the Caradoc sandstone or Lickey-quartz. Neither of these rocks, however, appear in the railway excavation, which exhibits, nevertheless, clear proofs of the disturbance attending the upheaval of the Lickey.

The lowest rock visible in this cutting is a mass of very hard brownish or reddish sandstone, commencing about fifty yards to the east of the post which marks the railway summit, and extending about seventy yards to the eastward, where it attains the height of 20 feet above the railway, and is suddenly cut off by a nearly vertical fault, Pl. VII. figs. 1, 2. Some of the beds of this sandstone are of a grey colour and uniform compact texture, with specks of white decomposed felspar; others are coarser-grained, and of a brown or reddish tint, containing rolled fragments of ferruginous or dark red indurated marl. No organic remains were noticed in it, and it hence becomes difficult to fix the precise age of this rock. I at first considered it to be a portion of the Caradoc sandstone of the Lickey, but from a closer inspection of its mineral characters, I should prefer classing it in the "Lower New Red Sandstone" of Sir R. Murchison.

The strata of this rock dip at the high angle of about 60° to east-south-east, or *from* the great mass of trap composing the Upper Lickey †. The rock above described is overlaid *unconformably* by a vast mass of conglomerate belonging to the "Upper New Red," or Bunter Sandstein. The latter deposit exhibits an imperfectly anticlinal arrangement ‡. On the east of the fault above mentioned, it dips about 5° to east-south-east, while on the western side its dip may be estimated at about 5° to the south-south-east, or south. The bedding is, however, so irregular that great accuracy on this point is not attainable, but it will suffice to state, that the stratification never departs far from horizontality, and is hence most strongly contrasted with that of the highly inclined sandstones above described, on which the conglomerate reposes.

The cutting here described affords a rare opportunity of inspecting the conglomerates of the upper new red sandstone. For the depth of nearly 60 feet the section consists almost exclusively of rounded pebbles imbedded in soft red sand. The resemblance of this deposit to ordinary diluvial gravel is so perfect, that when the excavations were first begun,

* See 'Silurian System,' p. 495. † See Sir R. Murchison's Map.

‡ The arrangement of the new red sandstone on the south side of the Lickey is more properly *mantle-formed* than *anticlinal.* On the line of the railway the dip sweeps round from east-south-east to south-south-east, south, and west-south-west.

I considered it as such, and it was not till the completion of the cuttings that I became undeceived. The gravel is now not only seen to contain numerous wedge-shaped masses of red sandstone and red marl (exhibited in the section, Pl. VII. fig. 2), but it is further shown to underlie the regular thick-bedded sandstone at each end of the section, so that its antiquity is thus most clearly proved.

From this stratum of new red conglomerate a large portion of the superficial or diluvial gravel of the surrounding counties has evidently been derived. At least nine-tenths of the pebbles of the conglomerate consist of quartz, either white and crystalline, or brown and granular, the latter doubtless derived from altered sandstones, such as are still seen *in situ* at the Lickey. The remaining portion of the pebbles includes various trap-rocks, chiefly porphyritic, which are often decomposed into the condition of clay. Boulders of a hard quartzose conglomerate, probably derived from the Old Red system, also occur, together with pebbles of chert, enclosing casts of Spirifers and Crinoidea*. The beds of conglomerate are interspersed and dove-tailed with bands of soft red sandstone, loose sand, and occasionally red marl, in the manner represented in Plate VII. fig. 2. At each end of the section the conglomerates are seen to be overlaid by beds of massive red sandstone, varying in texture from a compact rock to a loose sand, which again appears about a mile to the south-west, in the cutting No. 89, at the top of the inclined plane.

The point which appears of the greatest interest in the section here described, is the *unconformability* of the lowest rock, above mentioned, with the overlying new red conglomerate. Assuming that the lowest rock is correctly identified with the "Lower New Red," it follows that we thus obtain a tolerably exact date for the principal protrusion of the volcanic rocks of the Lickey. The sandstone in question dips at the high angle of 60° directly from the trap-rocks, which exist *in situ* a short distance to the north-west, and it is overlaid by conglomerates which are not far removed from horizontality. It is therefore clear that these trap-rocks must have been erupted *after* the deposition of the lower new red, and *before* the conglomerates of the upper new red, and it is probable that the rolled pebbles of the conglomerate were derived in great measure from the shattered strata which were thus upheaved in the immediate vicinity†. It is further to be inferred, that additional elevations of the region of the

* These fragments of chert appear to indicate a considerable transporting power in the sea which deposited the new red conglomerate, no chert being now known nearer than Derbyshire, a distance of about fifty miles from the Lickey.

† The age here assigned to the trap-rocks of the Lickey coincides with that attributed to them, as well as to those of Abberley and Malvern, by Sir R. Murchison (Sil. Syst. p. 67), though I believe he had nowhere noticed an example of unconformity between the upper and lower new red, such as is here exhibited. The elevation of the Nuneaton district seems also referable to the same epoch.

Lickey took place at a later date, for we find these overlying conglomerates traversed by a fault, and upheaved into an anticlinal position, which may be traced on each side of the ridge as far up in the series as the saliferous marls. Indeed it is probable that some of the dislocations connected with the Lickey were later than the age of the lias, for in the south of Worcestershire and Warwickshire the lias and red marl are affected by several extensive faults, the directions of which have an appearance of *radiating* from the Lickey (see p. 111).

In Groveley Hill (Pl. VII. fig. 1), on the north-east of the Lickey ridge, the red sandstone passes occasionally into a hard conglomerate of quartz pebbles, in a calcareous paste, recurring in nodular masses and thin strata *. Similar conglomerates are described by Sir R. Murchison as occurring in the upper new red sandstone in several parts of Worcestershire, Staffordshire and Warwickshire †. A considerable dislocation appears to traverse this hill, for at the north end of Groveley Tunnel the strata dip about 30° north-east, while at the south end the inclination is only about 3° south-east (Plate VII. fig. 1).

At Finstal, on the south-west flank of the Lickey ridge, the upper portion of the sandstone becomes light-coloured, and contains obscure vegetable impressions. It is, in fact, a prolongation of the stratum of light-coloured sandstone with vegetables exposed at Breakback Hill on the west of Bromsgrove ‡. These decomposed fragments of plants form the only examples of organic remains exposed by the railway cuttings in the new red sandstone.

The new red sandstone dips beneath the red or Keuper marl on either side of the Lickey ridge. On the north-east, it is traversed by the line of railway from Groveley Hill to Birmingham, which town stands on the new red sandstone, but is closely skirted by red marl on the south and east. The same formation extends from Birmingham along the London Railway as far as Berkswell, forming in North Warwickshire a basin of red marl, with the small lias outlier of Knowle in the centre §. This great extension of red marl was not known at the time of the construction of Sir R. Murchison's Map, in which the new red sandstone of Warwickshire is carried too far to the southward. The true boundary of the marl and sandstone ranges from Hewell Grange nearly north by Cofton Hacket to Northfield, and thence north-east to the south suburbs of Birmingham. The course of the railway is nearly parallel to this boundary-line.

* In the portion of these sections introduced in the Plate, this disparity is diminished one-half.

† See 'Silurian System,' p. 52. Coll. Papers, p. 127.

‡ Found also at Hadley. See Geol. Trans. vol. v. p. 341. Coll. Papers, p. 121.

§ See Geol. Trans. 2nd series, vol. v. p. 336. Coll. Papers, p. 116.

On the south-west side of the Lickey ridge the new red sandstone becomes marly and thin-bedded in the upper part, and eventually passes into the incumbent red marl. The latter still retains an inclination of 6° to 8° south-west, resulting from the elevation of the Lickey ridge, till we reach the salt-works at Stoke Prior. The deep excavation made at this place is fully described by Sir R. Murchison*, and I will therefore only remark, that, from the dip of the strata exposed by the railway, it is evident that the position of the salt rock must be near the bottom of the marl immediately above the sandstone.

The lowest rock exposed is a small mass of very hard sandstone, extending about 100 yards on the east of the summit. It dips to east-south-east at the high angle of 60°, and seems to belong to the lower new red, very near the coal. It has been elevated by the trap of the Lickey, before the conglomerates were deposited, for they overlie it unconformably.

The red marl presents no feature of importance till we reach the neighbourhood of Hadsor, where the railway crosses a promontory of lias projecting from the main body of that formation (Pl. VII. fig. 3). On the north side the marl is cut off by a fault, but on the south, at Dunhamstead, is an interesting section of the junction of the two formations.

The following section is here exposed:—

a.	Lias clay with contorted beds of lias limestone, containing Saurian bones.	
b.	White micaceous sandstone	2 ft.
c.	Lias clay	6 ft.
d.	Grey marl	35 ft.
e.	Red marl.	

The whole dipping 5° north-north-east.

The sandstone, *b*, has much resemblance to the "Keuper Sandstone" described in p. 111, but it occurs considerably higher in the series. It here contains numerous specimens of a smooth oval bivalve, larger and more oblong than the *Posidonomya minuta* of the Keuper, Pl. VI.[a] fig. 4, but too imperfect to exhibit generic characters†.

In the hill south of Dunhamstead is a fault which causes the grey marl to abut against the red, as seen in the railway section, sheet 17, Pl. VII. fig. 3. For the next five miles the railroad runs through red marl, in a valley between the escarpment of the lias and a ridge caused by the "Keuper Sandstone." On the south-east of Spetchley ‡ is a dislocation which causes the Keuper sandstone to change its strike from

* Silurian System, p. 31.

† In a paper read December 1841, I have shown that this band of sandstone is the true equivalent of the "bone-bed" of Somerset and East Devonshire.

‡ See sheet No. 14, and Pl. VII. fig. 4.

south by east to south-west, forming a projecting angle, which is intersected by the railroad*. The stratum is here but a feeble representative of the Keuper sandstones of Burge Hill, Inkberrow and Shrewley, p. 112, consisting chiefly of greenish marl with thin laminæ of white sandstone, the whole forming a deposit about 20 feet thick, with red marl both above and below. The bands of sandstone vary rapidly in thickness, and one wedge-shaped mass of solid sandstone, about 30 feet long, is 2 feet thick in the middle, and thins out entirely at each end.

About a mile further south, at Norton, the railway traverses the lias escarpment, which presents a section exactly analogous to that at Dunhamstead, showing the same succession of lias limestone, clay, white thin-bedded sandstone, grey marl and red marl. The sandstone also contains the oval bivalve met with at Dunhamstead. About a mile south of this point the lias clay contains many calcareous concretions abounding with shells, including *Plagiostoma giganteum*†, *P. duplicatum*, *P. hermanni*, Goldf., *Terebratula ornithocephala*, *Modiola minima*, and a *Caryophyllæa*, which is remarkable, from the general scarcity of corals in the lias formation. Further south, near Abbots Wood, the beds of fissile sandstone, at the base of the lias, are again exposed on the railway, being brought up by a fault. Thence the lias clay presents little interest till we reach Defford, Pl. VIII. fig. 5, where numerous specimens of *Cardinia listeri* (Stutchbury) (*Unio listeri*, Sow.), *Gryphæa incurva*, *Astarte lurida*, *Plagiostoma punctatum*, *P. duplicatum*, and several apparently new species of Ammonites, Modiola, &c., occurred in the cuttings. The same shells are found also at Eckington, fig. 5, on the south of the Avon; but at Bredon, fig. 5, we reach a higher stratum of the lias clay, and meet with an almost entirely distinct set of organic remains, among which *Pleurotomaria anglica*, *Hippopodium ponderosum*, *Gryphæa maccullochii*, *Nautilus striatus*, *Pholadomya ambigua*, *Modiola scalprum*, *Plagiostoma punctatum*, *P. duplicatum*, *Plicatula spinosa*, *Spirifer walcotti*, *Corbula cardioides* (Phill.), *Amphidesma donaciforme* (Phill.), *Pecten sublævis* (Phill.), *Ammonites planicosta*, *A. obtusus*, *A. turneri*, *A. conybeari*, and *A. birchi* have been identified. Between Bredon and Cheltenham the ground is very level, and but few sections of importance occur; but at Cheltenham, and between that town and Gloucester, the excavations in the lias clay are very extensive, and have supplied large collections of organic remains to the cabinets of the Cheltenham geologists. Several of these fossils are enumerated in Sir R. Murchison's 'Memoir on the Geology of Cheltenham‡,' but a considerable number of

* These Keuper sandstones are interesting from their fossils. See Pl. VI. & VI.[b]

† All the shells included in this and the following lists are described in Sowerby's 'Mineral Conchology,' unless a different authority is given.

‡ See also 'Silurian System,' p. 18.

species, especially of Ammonites, appear to be new, as I have been able as yet to assign names to only the three following:—*Ammonites armatus* (Sow.), *A. ovatus* (Young and Bird), and *A. lenticularis* (Young and Bird). With the exception of the *Hippopodium ponderosum* and *Gryphæa maccullochii*, and one or two others, the Cheltenham fossils are wholly distinct from those of Bredon, proving how small a difference of vertical elevation will effect an almost total change of organic remains. At Hewlitts, east of Cheltenham, the lias near the base of the marlstone presents another series of distinct fossils, consisting of *Hippopodium ponderosum* (rugose variety, perhaps a distinct species), *Modiola scalprum, Spirifer granulosus* (Goldf.), *Terebratula rimosa* (Bronn), *Perna ventricosa, Cardinia attenuata* (Stutchb.), *Littorina imbricata* (*Trochus imbricatus*, Sow.), *Ammonites henleyi* (*A. striatus*, Rein.), *A. heptangularis* (Young), *A. cheltiensis* (Murchison), &c.*, so that in the lower lias alone we have evidence of at least five well-marked successions of molluscous faunæ, ranging through a vertical height of about 400 or 500 feet, and unaccompanied by any change in the mineral character of the deposit.

Having now described the secondary formations along the line of the railway, we will recommence at the Birmingham end, and examine the deposits of superficial detritus. The phænomena exhibited by the railway cuttings are entirely confirmatory of the views which I have announced elsewhere †, respecting the distinction between those ancient terrestrial alluvia in which mammalia occur, and the general mass of submarine "drift" which covers most parts of the island. In pursuance of these views, the ancient superficial detritus of this district may be divided according to its efficient cause, into *fluviatile* and *marine*, the latter according to its origin into *local* and *erratic*, and this again, according to its composition, into gravel *with flints*, and *without flints*.

§ 1. *The marine erratic gravel without flints* occurs at intervals along the line of railway from Birmingham till it approaches the Valley of the Avon. Vast accumulations of this detritus (the "Northern drift" of Sir R. Murchison) occur on all sides of Birmingham. My own personal inquiries addressed to resident geologists, railway engineers and excavators, aided by printed queries circulated in the Geological Section of the British Association at Birmingham ‡, all tend to prove the *utter absence* of mammalian remains in the deposits of this class in that neighbourhood.

Chalk flints, though not absolutely wanting in the Birmingham gravel, are yet so extremely rare as to prove that the current which transported

* Outline of the Geology of the Neighbourhood of Cheltenham, 1834.

† See Report of British Association, vol. vi. Sections, p. 61. Coll. Papers, p. 105.

‡ See Report of British Association, 1839, Sections, p. 71. Coll. Papers, p. 110.

it came from the north and not from the east, and furnish a well-marked distinction from the flinty gravel described below.

At Mosely the railway cuts through a vast deposit of this gravel, upwards of 80 feet thick, reposing upon red marl. It is composed of rolled pebbles, rarely exceeding 4 inches in diameter, of various granitic and quartzose rocks and altered sandstones, imbedded in clean ferruginous sand, devoid of argillaceous matter. A stratum of sand about 30 feet thick, free from pebbles, occurs in the middle of the gravel.

Between Mosely and the Lickey the railway line is in general free from gravel. The only mass of stone of sufficient size to deserve the name of an erratic block, occurred on the line of the railway between Cotteridge and Wytchall*. It is a shapeless mass, about 5 feet by 4, with the angles partially rounded, reposing on the red marl, and consists of a greyish porphyritic trap.

Patches of gravel of this class occur on each flank of the Lickey ridge, though none, as before shown, were found on its summit at the part traversed by the railway. The singular manner in which the gravel reposes on an irregular surface of new red sandstone is shown in the cutting at the summit of the inclined plane, Plate VII. fig. 1. The superficial drift closely resembles the genuine new red conglomerate seen in the cutting on the Lickey ridge, as it consists in great measure of the same materials, but it may be distinguished by containing, in addition, many fragments of slaty rocks, and by the sand in which the pebbles are imbedded being freer from red argillaceous matter, and consequently of a whiter colour. This gravel attains on the line of the railway a height of 544 feet above the sea (not 387, as misprinted, or 587, as corrected in the Proc. Geol. Soc. vol. iii. p. 316). I may here remark, that the gravelly soil of the Lickey Beacon, 900 feet high, which has been quoted as an example of superficial gravel at a great elevation, may very probably not be derived from these recent or "diluvial" deposits, but from the genuine new red conglomerate, though its existence *in situ* cannot be determined in consequence of the want of sections. For the present, therefore, the above-mentioned elevation, 544 feet, is the greatest which can be assigned with certainty to the superficial gravel or "Northern drift" of this part of England.

The next locality where gravel occurs is at Sugars Brook †. This deposit consists of quartzose pebbles, commonly less than 3 inches diameter, and rarely attaining 6. No stratification is observable. The surface of the bed is about 12 feet above the brook, and from its low position it appeared likely to belong to the fluviatile class, and to contain mammalian bones; but though many thousand tons of the gravel

* Sheet No. 28.

† See Pl. VII. fig. 1, upper line.

have been raised for the use of the railway, I have not been able to learn that any organic remains have been found in it.

From this point no more gravel occurs on the line of the railway for the next sixteen miles; but a few hundred yards east of it, at Abbots Wood, is an extensive deposit of quartzose gravel mixed with ferruginous sand, devoid of flints, and resting upon lias clay. Here also large quantities of gravel have been quarried, but the evidence of the engineers and workmen is unanimous as to the absence of organic remains.

§ 2. *Marine erratic gravel with flints.*—It was stated in a paper read at the Liverpool meeting*, that the gravel to the south-east of the Avon in Warwickshire and Worcestershire abounds in flints, indicating a current from the chalk district lying to the eastward; and accordingly no sooner does the railroad pass that river, than we find a very large per-centage of flints in the gravel. The village of Bredon stands on a platform about 70 feet above the ordinary surface of the Avon, capped with an extensive deposit of this kind of gravel, about 10 to 15 feet thick†. It reposes on an uneven surface of lias, and is mixed with much ferruginous clay. Its elevation above the Avon valley, though slight, has been sufficient to protect it from the modifying effects of fluviatile agency, and hence the extensive excavations near Bredon have furnished no example of mammalian bones.

§ 3. The only example of *fluviatile gravel* on the railway line is on the two flanks of the Avon at Defford and Eckington‡. We have here on either side of the river a tabular platform whose surfaces do not exceed 45 feet above the Avon. They are capped by about 10 feet of gravel, precisely similar in composition to the flinty gravel of Bredon; but with this important difference, that they contain an abundance of mammalian remains: they were found chiefly in the cutting on the north of the village of Eckington. The bones occur chiefly in the lower part of the gravel, and often on the surface of the subjacent lias clay, about 35 feet above the river. They are accompanied with numerous freshwater shells, agreeing in species with many of those enumerated in Murchison's 'Silurian System,' p. 555, as found in a similar position higher up the Avon at Cropthorne. The most abundant species are the *Cyclas amnica* and *C. cornea*. Great numbers of the bones found here have been broken and lost by the carelessness of the workmen; but those which have been preserved, and which are in the cabinets of the Worcestershire Natural History Society, of Mr. Fowler of Cheltenham, Mr. Dudfield of Tewkesbury, and my own, are referable to the following species:—*Elephas primigenius*, *Hippopotamus major*, *Bos urus*, and *Cervus giganteus*?.

* See Report of British Association, vol. vi., Sections, p. 62.

† See sheet No. 7, and Pl. VII. fig. 5.

‡ See sheet No. 9, and Pl. VII. fig. 5.

On the north or opposite side of the Avon, bones of *Elephas primigenius, Rhinoceros tichorhinus* and *Hyæna spelæa,* accompanied with freshwater shells, occur under similar circumstances to those at Eckington, beneath a few feet of gravel, and about 20 feet above the river.

In endeavouring to account for the presence of bones and freshwater shells at this part of the railway line, and their absence in all other parts of the district described, I can offer no other explanation than that formerly proposed*, viz. that *after* the beds of marine gravel had been deposited where we now find them, and had been laid dry by the elevation of the land, a large river or chain of lakes extended down the Valley of the Avon at a height of from 20 to 50 feet above its present course; and that the gravel previously brought into the district by marine currents was remodified by the river-stream and mixed up with remains of the Mammalia and Mollusca which tenanted its banks or its waters.

§ 4. *Local gravel.*—This occurs abundantly at Cheltenham†, and consists exclusively of detritus from the oolite and lias of the vicinity‡. The composition varies from the state of gravel to fine calcareous sand. No bones or other terrestrial remains have occurred in it, and it is therefore referred, in the absence of other evidence, to a marine origin.

Modern alluvia.—No formations of any importance belonging to this class occur on the line of the railway, except the peaty deposits on the banks of the Avon and its tributary streams. In sinking through this peaty soil for the foundation of the bridge over the Avon at Defford, a human skeleton was found at the depth of 18 feet from the surface.

On a general review of the sections afforded by the Birmingham and Gloucester Railway, although they do not present us with any very striking or novel phænomena, yet we may justly attach some value to the evidence they afford of geological facts, whether as confirmatory or as corrective of those previous researches which were undertaken without the assistance now furnished by the railway cuttings.

* Report of British Association, vol. vi. Sections, p. 64. Coll. Papers, p. 105.

† See Railway Section, sheet No. 1.

‡ See Murchison's 'Geology of the Neighbourhood of Cheltenham,' p. 28.

6. DESCRIPTION OF CUTTINGS ACROSS THE RIDGE OF BROMSGROVE LICKEY, ON THE LINE OF THE BIRMINGHAM AND GLOUCESTER RAILWAY*.

[Abstract.]

WHEN Mr. Strickland laid before the Society in June 1840†, a description of a series of coloured sections on the Birmingham and Gloucester Railway, the cuttings on the Lickey not having been completed, he was prevented from detailing the phænomena exhibited on this part of the line. The present communication is therefore supplementary to the former memoir.

Where the cutting crosses the ridge, it has been excavated to the depth of 56 feet, and exhibits clear proofs of the disturbance which attended the elevation of the Lickey. The lowest rock which has been exposed is a mass of hard, grey, brownish or reddish, compact or coarse-grained sandstone, occupying a horizontal distance of about 70 yards, and rising gradually to the north-east to the height of 20 feet. At the point where it attains this visible thickness, it is suddenly cut off by a nearly vertical fault. The strata dip about 60° to the east-south-east, or from the trap composing the Upper Lickey. No organic remains having been noticed, it is difficult to fix the precise geological position of the deposit; but he is inclined to assign it, on mineral characters, to the lower new red sandstone of Sir R. Murchison‡.

These highly-inclined strata are overlaid unconformably by a vast mantle-shaped mass of conglomerate, belonging to the "upper new red" or Bunter Sandstein. The bedding of this deposit is so irregular that great accuracy of dip is not attainable; but to the east of the fault the inclination is about 5° to the east-south-east, and to the west about 5° to the south-south-west or south; and the slight departure from horizontality is strongly contrasted with the high inclination of the lower sandstone. The resemblance of this deposit, consisting of rounded pebbles in soft red sandstone, to ordinary gravel is so perfect, that Mr. Strickland was at first induced to consider it as superficial detritus; but its true nature is proved by its containing numerous wedge-shaped masses of red sandstone and red marl, and by its being overlaid at each extremity of the cutting by the regular thick-bedded sandstone, which again is surmounted by red or Keuper marls. At least nine-tenths of the pebbles consist of white and crystalline, or brown

* Proc. Geol. Soc. vol. iii. p. 446.

† See *antè*, p. 132.

‡ Silurian System, p. 54.

and granular quartz, the latter doubtlessly derived from such altered sandstones as exist *in situ* in the Lickey. The remainder of the pebbles are composed of various traps, chiefly porphyritic, and often decomposed into clay. Boulders also occur of a hard quartzose conglomerate, derived, the author believes, from the Old Red system; likewise pebbles of chert, containing casts of Spirifers and Crinoidea.

Patches of gravel overlie the red sandstone on the flanks of the Lickey, sometimes filling up considerable irregularities in its surface, but none were exposed on the summit of the ridge. The gravel resembles the conglomerate of the new red sandstone, as it consists chiefly of the same materials, but it may be distinguished by containing many fragments of slaty rocks, and by the whiter colour of the pebbles. It attains on the line of the railway a height of 544 feet; and as the gravelly soil which has been stated to occur on the Lickey Beacon at an elevation of 900 feet, may, Mr. Strickland says, belong to the new red conglomerate, the gravel on the line of the railway occupies the highest position which can with certainty be assigned to the northern drift of that part of England.

The point of greatest interest exposed in this cutting is the unconformability of the lowest rock to the overlying conglomerate. Assuming that the former is correctly identified with the "lower new red," it follows, Mr. Strickland observes, that a tolerably exact geological date is obtained for the principal protrusion of the volcanic rocks of the Lickey* †; and that they must have been erupted after the deposition of the lower new red, and before that of the upper new red. It is also probable, he states, that the pebbles of the conglomerate were in great part derived from the shattered upheaved strata in the immediate vicinity. The author further infers, from the fault in the upper conglomerate beds, that additional elevations of the Lickey region took place at a later date, and threw the superior strata into an anticlinal position. He also suggests that some of the dislocations connected with the Lickey may have occurred subsequently to the deposition of the lias, as the faults which have affected that formation and the new red sandstone in Worcestershire and Warwickshire appear, he says, to have radiated from the Lickey‡.

* The age here assigned to the trap rocks of the Lickey coincides, Mr. Strickland says, with that attributed to them, as well as to the trap rocks of Abberley and Malvern, by Sir R. Murchison, though the want of unconformability between the upper and lower new red strata was apparently unknown to that gentleman. (Silur. Syst. p. 67.)

† The "volcanic rocks" of the Lickey, Abberley and Malvern country mentioned in this memoir have since been shown to be brecciated trappoid rocks of aqueous origin.—Geol. Journal, vol. xi. p. 186.—A. C. R.

‡ See Geol. Trans. 2nd series, vol. v. pp. 333, 335.

7. ON THE ERRORS WHICH MAY ARISE IN COMPUTING THE RELATIVE ANTIQUITY OF DEPOSITS FROM THE CHARACTERS OF THEIR IMBEDDED FOSSILS *

GEOLOGISTS are indebted to Mr. Charlesworth and Dr. Richardson for having pointed out two important sources of error in the application of Sir C. Lyell's rules for fixing the age of tertiary deposits. The following paper is intended to show that these errors do not affect the *principle* of Sir C. Lyell's test, but its *practical application,* and that, although they may diminish, they do not destroy its utility.

Sir C. Lyell's proposition may be stated thus:—The ratio of extinct to existing species of animals in the tertiary deposits is proportionate to their antiquity. To this Mr. Charlesworth objects, first, that naturalists differ greatly in their estimate of specific characters, and therefore that we cannot make accurate enumerations of species; and secondly, that fossils of different ages are liable to be mixed in the same stratum, and that we may thus be misled in inferring their contemporaneous existence. We will consider each of these objections separately.

The *essential* point in Sir C. Lyell's test is this, that fossil forms recede from existing ones in a degree proportionate to their antiquity. This general proposition is admitted by all who are conversant with geological facts. Observations made in all parts of the world upon the whole fossiliferous series have uniformly tended to establish this truth, and though a few isolated exceptions may exist, they do not affect its general application. The difficulty is to find a correct *measure* of the resemblance or disagreement between extinct and living forms. In the secondary strata, where the fossil species are admitted to be wholly extinct, we have at present no other measure than the general estimate of resemblance and disagreement which persons conversant with organic forms may be capable of making. We cannot, I think, venture upon *numerical* comparisons of extinct and existing *genera,* for these groups are as yet little more than vague human generalizations to which arithmetical calculation cannot be correctly applied.

But in the tertiary series, where existing species are supposed to make their first appearance, Sir C. Lyell conceived that a measure of zoological resemblance might be found, in the numerical proportion of these existing species to the extinct ones which accompany them. Now those persons who, unlike the Lamarckian school, believe in the reality and permanency of specific distinctions, must admit that this is a very logical path to the desired truth. For on comparing the "fauna" or entire

* Ann. Nat. Hist. n. s. vol. i. p. 234.

assemblage of animals of two geological periods or geographical regions, we shall commonly find that the numerical ratio between the species which are common to both, and those which are peculiar to each, will be proportionate to their total resemblance between the two faunæ, or in other words, the amount of *identity* will be proportionate to the amount of *similarity*. Hence the degree in which the fauna of any tertiary deposit resembles the existing creation may be arithmetically expressed by the number of living species which it contains, and its relative antiquity may be inferred accordingly.

Mr. Charlesworth, however, objects to this numerical test of Sir C. Lyell, because he finds that hardly any two naturalists agree in their estimate of specific differences. This is certainly an important, but I trust not a fatal objection. It seems to be rather a proof of human infirmity than of imperfection in the *per-centage* principle, which is shown above to be not essentially unphilosophical. I would rather suppose that this variation of opinion proceeds from our imperfect acquaintance with specific distinctions and our inaccurate estimate of their importance, than from those characters being themselves really mutable and indefinite. In proportion, therefore, as our knowledge of zoology advances, the utility of the per-centage test will increase; and though mathematical *accuracy* is not to be expected from it, it will probably be found to indicate the zoological similarity of different deposits, and consequently their relative ages, with more certainty than any other rule which can be laid down.

I now proceed to Mr. Charlesworth's second objection to Sir C. Lyell's principles, grounded on the liability of fossils of different ages to become mixed up in the same deposit. This is a subject which has not yet excited the attention which it deserves, and important errors may arise from inattention to this point. Mr. Charlesworth considers that many fossils of the coralline crag have in this way become enveloped in the red crag, whence they are again removed by the waves, and mingled with the present inhabitants of the German Ocean. Dr. Richardson also has described* the manner in which bones and shells are daily removed from the gravel and London clay of Kent, and deposited in the estuary of the Thames. Most geologists must have had opportunities of witnessing the same phænomenon.

Among instances which have come under my own observation, I may mention the highly interesting, though little-known, *sponge-bed* of Faringdon. This deposit, which has many analogies to the coralline crag, is supposed to be of the age of the greensand. It is a thick bed of sand and fine gravel resting on hills of coral rag. It contains multitudes of beautiful Sponges and other Zoophyta, also various shells, especially Terebratulæ, which are commonly empty, and exhibit the curious bony

* Mag. Nat. Hist. n. s. vol. i. p. 122.

apparatus of the inside in a rare state of perfection. Along with these delicate and beautiful fossils are rolled fragments of Belemnites and other remains, which, to all appearance, are derived from the subjacent coral rag.

I lately saw a remarkable instance of the same kind in the cabinet of M. Nicolet at Chaux-de-fond, near Neufchâtel. Most of the valleys in that part of Switzerland are partly filled with *molasse,* a Miocene deposit containing Ostreæ, Pectines and various tertiary shells. Intermixed with these occur numerous secondary fossils, such as Ammonites, Belemnites, Trigoniæ, Terebratulæ, &c., derived from the surrounding hills, which correspond to the English series, from the Oxford clay to the chalk inclusive.

Superficial gravel or ancient alluvium exhibits the same intermixture of fossils of widely different periods. In the gravel of Worcestershire may be found Corals from the Silurian rocks, Plants from the coal, Gryphites from the lias, Terebratulæ from the oolite, and Spatangi from the chalk, mixed with bones of Rhinoceros and Hippopotamus, and freshwater shells of existing species.

A person not conversant with geology, on a review of these facts, might exclaim, " How vain then must be the attempt to class formations by their organic remains, and how gratuitous the assumption that the fossils of any stratum are contemporaneous with its deposition !" This, however, would be jumping too fast to a conclusion, and the following considerations will show, that though geologists should always be on their guard against errors arising from the intermixture of fossils of different ages in the same stratum, yet the possibility of these errors is so rare as not to affect the general deductions of geology.

In the first place, an aqueous action of considerable violence is necessary to remove fossils from one formation and to deposit them in another. Hence all fine-grained deposits are beyond the limits of the above error, and they form the largest portion of the whole stratified series. No one can doubt that the fossils of the carboniferous limestone, the lias, the chalk, and the London clay are contemporary with the deposition of these strata respectively, for a current sufficiently strong to bring these fossils from a distance would not have allowed the fine-grained matter which now envelopes them to be deposited.

Moreover, the fossils of an older formation can hardly become enveloped in a newer one, unless the two deposits are unconformable. The older deposit must be raised up so as to form sea-cliffs and projecting rocks before its destruction can take place, for we can hardly suppose that the sea ever denudes its own bed to such a depth as to lay bare and extract the fossils of a preceding and extinct creation. Hence, when we have a long series of conformable deposits, as in the Silurian and oolitic rocks of England,

L 2

there can be no doubt that the various groups of fossils are contemporary with the beds which they characterize, even though the sandy or gravelly structure of some of these beds may indicate the flow of submarine currents. I am not aware whether the red crag can be shown to be unconformable to the coralline crag; but if it cannot, I should strongly suspect that they belong to the same geological period, or at least, that all the tertiary fossils found in the red crag are contemporaneous with its deposition.

Again, in the cases where an admixture of fossils of different ages has unquestionably taken place, there can rarely be any difficulty in distinguishing the genuine ones from the erratic.

To take the examples above cited in the Faringdon *sponge-bed*, the Belemnites, which I suppose to be derived from the coral rag, are all worn and broken, while the Sponges and Terebratulæ indigenous to the spot are in the most delicate state of preservation. Again, the secondary fossils found by M. Nicolet in the molasse, though very slightly worn, are easily distinguished by their solidity and petrified condition, independently of their zoological characters, from the tertiary shells which accompany them. In the beds of ancient alluvial gravel there can be no difficulty; the secondary fossils there found are not only petrified and worn, but are commonly attached to portions of their original matrix, while the mammiferous bones and freshwater shells are both uninjured and unaltered, except by the subtraction of part of their animal matter. There can, therefore, be no doubt that these bones and freshwater shells are contemporary with the deposition of the beds of gravel.

To the above arguments it might be added, that the removal of fossils from one stratum to another must always be a local and partial occurrence, and therefore that when any species is found to characterize a given formation over a large area, we cannot err in regarding it as an inhabitant of the sea which deposited the stratum. There is, however, no occasion to enlarge on this topic; enough has been said to prove that the errors likely to arise from the removal of fossils from one stratum to another, though not to be lost sight of by geologists, are yet confined within very narrow limits.

Dr. Richardson concludes his interesting account of the manner in which shells from the London clay and bones from the gravel become mixed with existing shells in the estuary of the Thames, with a speculation as to the geological appearances which would be presented if that estuary became dry land, and the errors into which an examination of it might lead a future geologist. He says, "The sedimentary and tranquil character of the formation, consisting of alternating bands of sand and clay, and the total absence of extraneous materials, give assurance to his (the supposed geologist's) conclusions; and the synchronous existence of the organic contents is instanced with unhesitating confidence. He meets with no fact that can excite suspicion or create distrust; the intrusive and extinct fossils are linked with the recent by the closest of all ties: they lie in peaceful juxtaposition, and upon undisturbed beds of oysters." This hypothetical statement is probably not quite accurate. If the formation was of a "sedimentary and tranquil character," the bones of large mammalia could not be transported into it, or if they could, gravel and extraneous materials would accompany them, and would "excite suspicion and create distrust." Indeed, Dr. Richardson himself states that the interior of the bones fished up on the Whitstable oyster-beds is "still filled with the yellow loam and the small flints of the gravel." The *probable* appearances presented would be these:—In the central and tranquil parts of the estuary, the geologist would find shells which are *now* existing in a perfect state, and mixed with these would be the lighter and fragile portions of gravel bones and London clay shells. The broken condition of the latter would excite his suspicions, which would be confirmed as he approached the margin of the estuary. He would there find a shingle beach of various erratic pebbles, and containing entire those bones and shells of which he had before found only small fragments. They would, however, be in general more or less rolled, and easily distinguishable from the Mammalia or Mollusca which were living in A.D. 1837. Rolled masses of London clay would probably occur, and the shells derived from that formation would often contain portions of it in their interior different from the more recent matrix in which they were now imbedded. All these circumstances would soon lead a geologist to a true conclusion, even if no cliffs of London clay and gravel remained to prove that those formations were anterior and unconformable to the one under investigation.

---

8. ON SOME REMARKABLE DIKES OF CALCAREOUS GRIT, AT ETHIE IN ROSS-SHIRE. [*Read March* 7, 1838.]

[Trans. Geol. Soc. n. s. vol. v. p. 599.]

I AM desirous of calling the attention of the Society to some remarkable dikes which penetrate lias schist at Ethie, near Cromarty. The beds of lias at this place have been described by Sir R. Murchison in the Transactions*, and their relations are given in the accompanying woodcut. These beds are only exposed at low water, and dip towards the sea at an angle which gradually increases in approaching high-water mark, where they become perpendicular. Cliffs of gneiss and red conglomerate rise immediately from the shore, but the broken fragments which are scattered near high-water mark, conceal the actual junction of the lias with these older rocks.

The dikes, which are here seen to penetrate the lias, are remarkable for their mineral character, which so precisely resembles that of certain altered quartzose sandstones, that it is impossible to refer them to a purely igneous origin.

The substance of which they are composed is intensely hard. On examining it with a lens, its component particles are distinctly seen, and bear the closest resemblance to ordinary water-worn grains of sand. The stone is penetrated by carbonate of lime, which produces a slight effervescence on the application of an acid; and it has assumed a pseudo-crystalline structure, which exhibits, when held in certain lights, the peculiar lustre of the calcareous sandstones of Fontainebleau and other places.

A careful examination of the locality sufficed to show, that however much this rock might resemble an aqueous product, yet that it forms genuine intrusive dikes, penetrating the lias shale in all directions. A and B are two dikes (see the woodcut) which are parallel to the stratification of the lias shale, and their injected origin is not consequently

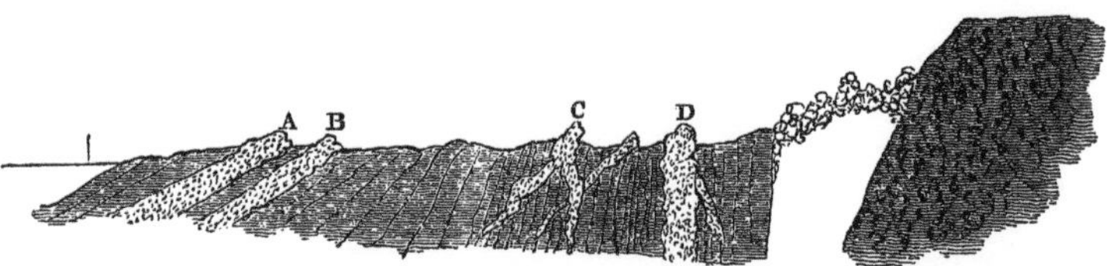

immediately apparent; but C is a dike which sends off branches in

* Trans. Geol. Soc. n. s. vol. ii. p. 308.

various directions, and in no part of its course is parallel to the strata which it penetrates. The thickness of this dike is about 1 foot, and that of its lateral offsets from 3 to 6 inches. D is the largest dike of all, being 3 feet thick, and ranging parallel to the shore for at least 200 yards. In one place it also gives off a lateral branch, which extends a few yards and then rejoins the main dike.

These several dikes exhibit no variation whatever in their texture or composition, except being occasionally penetrated by small veins of carbonate of lime. They show no signs of lamination, but are frequently fractured transversely to their own direction. The transition is instantaneous from the dike to the lias shale, which seems to have suffered neither alteration nor contortion by the intrusion of this extraneous matter. The shale, from its greater softness, has been washed out between the dikes, leaving them to project like walls from 1 to 3 feet in height*.

I have been the more exact in describing this locality, because the identity of the specimens here exhibited with ordinary stratified sandstone is so perfect, that the clearest evidence was necessary to prove that they had been inserted into fissures of the lias subsequently to its deposition. The sedimentary structure of this rock forbids us to refer it to igneous injection from below; and notwithstanding the complete resemblance of these intrusive masses to ordinary plutonic dikes, we have no resource left but to refer them to aqueous deposition, filling up fissures which had been previously formed in the lias. We have no clue to the period at which this insertion of sand into fissures of the lias took place, no fossils having been noticed in the substance of the dikes themselves.

In speculating on the causes of this phænomenon, we should bear in mind the total absence of trappean rocks on the eastern coast of Ross-shire and Sutherland, and the presence of vast masses of granitic and syenitic rocks, which have been shown by Sir R. Murchison to have been erupted subsequently to the deposition of the oolitic series.

* Sir R. Murchison, in his paper on Brora (Geol. Trans. 2nd series, vol. ii. p. 304), mentions a dike of quartz rock as occurring at Kintradwell, on the coast of Sutherland. That gentleman has further informed me, that in company with Professor Sedgwick he noticed similar dikes, both at Ethie and at other places on that coast.

Something of the kind of these dikes is indicated in Arran by Mr. Ramsay. See his Memoir, p. 23.—H. E. S. MS.

9. ON A FOSSIL DRAGON-FLY FROM THE LIAS IN WARWICKSHIRE.

[Mag. Nat. Hist. new ser. vol. iv. p. 301. 1840.]

A LIBELLULINE insect has lately been found in the lias of Warwickshire. It is the property of Mr. J. Gibbs, of Evesham, who has kindly lent it to me for examination and description. It exhibits a very distinct impression on the surface of a slab of blue lias limestone, the wing being of a pale brown colour, and the nervures a darker tint of the same. The opaque spot which exists at the anterior margin of the wing in most of the Libellulidæ is here distinctly marked, being of a much darker brown than any other part of the wing. This specimen appears to be the left anterior wing of the insect. On comparing it with recent species of Libellulidæ, it exhibits a close resemblance to them in the general arrangement of the nervures. It is well known that the insects of this family present certain generic peculiarities in the nervures at the base of the wings. In this respect the specimen before us exhibits characters most nearly allied to the genus Æschna, Fab., but approximating also to the structure of Libellula. The dimensions of the fossil are about one third greater than those of *Æschna grandis*, one of the largest of our British species, its length being 2 inches 10½ lines, and its greatest breadth 8½ lines.

It is proposed for the sake of distinction to denominate this specimen *Æschna liassina**.

The specimen before us furnishes, I believe, the first example of an insect of this family in so old a formation as the lias. It is well known that Libellulæ occur in the lithographic stone of Solenhofen †, which belongs to the upper part of the oolitic series, and is the lowest rock in which these insects have hitherto been found. The present specimen, therefore, is unquestionably of great geological interest, especially when we contrast its close resemblance to existing forms with the extraordinary Saurian, Piscine and Molluscous structures which were its contemporaries.

This specimen was found in the neighbourhood of Binton, near Bidford,

* "In the description of *Æschna liassina*, Strickland, a fossil Dragon-Fly found in Warwickshire, figures of the wings of the recent *Æschna grandis* and *Libellula depressa* are also given to show the difference in size and structure. The stigma on the wing shows it to be nearer to Cordulegaster, and still more to Petalura (Zool. Misc. vol. ii. pl. 94), where it is stated to be a native of New Holland; and I have a ♀ of the same from Mr. Children's cabinet, which on comparison bears a close resemblance to the fossil fly."—J. C. Dale, Ann. Nat. Hist. vol. ix. p. 257.

† See De la Beche, Geol. Man. p. 345, &c.

in Warwickshire, where the beds of limestone near the base of the lias are extensively quarried for flooring, &c. These beds of limestone, besides the usual Ammonites of the lias, occasionally contain specimens of Ichthyosauri, Plesiosauri, three or four species of fish, crustaceans, and two or three kinds of ferns. The latter circumstance indicates the proximity of land at the time of the deposition of the strata, a supposition which is further borne out by the fossil insect above described.

One of the species of fish found here appears to be a Cycloid, and furnishes an exception to the generalization of M. Agassiz, that no Cycloidian fish occur below the chalk.

In the first series of this Magazine, vol. v. p. 549, is a figure of a fossil fish from Wilmcote (misspelt Wilments), near Binton. This specimen is now in the Warwick Museum, and has been figured by M. Agassiz under the name of *Tetragonolepis angulifer*.

The rock in which these fossils are found is a fine-grained blue limestone divided into thin slabs, resembling in texture the Solenhofen stone, and like it adapted to lithographic purposes.

---

10. ON THE OCCURRENCE OF THE "BRISTOL BONE-BED" IN THE LOWER LIAS* NEAR TEWKESBURY.

[An abstract only of this paper was published in the 'Proceedings' of the Geol. Soc. iii. pp. 585, 732. The entire paper, as read to the Society, is now printed.]

In the neighbourhood of Bristol a peculiar stratum occurs at the base of the lias, which, from consisting almost wholly of the fossil remains of fish and saurians, has long been known to collectors by the name of the "Bone-bed." In that district it has been found at various places between Westbury on the north and Watchet on the south, also at Gold Cliff, Glamorganshire, and St. Hilary, near Cowbridge (Geol. Trans. i. n. s. p. 305), which points have been hitherto supposed to form the geographical limits of the stratum. Its superficial area has, however, been lately shown to extend much further in contrary directions. On the south it has recently been detected on the coast near Axmouth, and I have now to record the occurrence of this bed at Coomb Hill between Tewkesbury and Gloucester, a locality many miles further north than any before described. We have thus evidence of the existence of this remarkable stratum over an area at least ninety miles by twenty-five, and if we contrast this great superficial extent with the extreme thinness of the bed itself, and consider also the regularity of its position in the series at distant localities, and the vast abundance and variety of its organic remains, we may regard the "Bone-bed" as a highly interesting geological phænomenon.

I proceed to describe the stratigraphical position of the bone-bed at this locality, and of its equivalents in other places, and shall afterwards consider its organic remains.

1. *Coomb Hill.*—This place is about four miles south of Tewkesbury. A deep cut for a road has here been made through the lias escarpment, and the section exposed has been briefly described by Sir R. Murchison in his 'Geology of Cheltenham,' Ed. 1. p. 24, Ed. 2. p. 47, and Sections, fig. 1 (see also Sil. Syst. pp. 20, 29, pl. 29. fig. 1). At the time of Sir R. Murchison's visit, the excavation had been made many years, and the banks were so far obscured that the thin stratum forming the bone-bed did not attract his attention. Last summer, however, the road being

* British geologists are still of different opinions as to the Keuper or Liassic age of these beds, some believing, with Sir Philip Egerton, that their fish indicate their relation to the triassic, while Mr. Strickland and others believed that their shells and Saurian remains united them with the lias. For a brief account of their probable continental equivalents, see the 'Supplement to the Fifth Edition of a Manual of Elementary Geology,' Lyell, pp. 28-31.—A. C. R.

lowered several feet, a considerable surface of the bone-bed was exposed, which was rescued from destruction by Mr. Dudfield of Tewkesbury, and many specimens from it were added to his interesting collection of fossils from that neighbourhood. The following section is now visible at this place, Pl. VIII. fig. 2.

	ft.	in.
1. Yellow clay	2	0
2. Lias limestone	0	3
3. Yellow clay	5	0
4. Nodules of lias limestone	0	6
5. Brown clay	14	0
6. Impure pyritic limestone with Pectens and small bivalves	0	6
7. Black laminated clay	8	0
8. Hard grey pyritic sandstone	0	2
9. Black laminated clay	1	0
10. Greyish sandstone	0	2
11. Black laminated clay	1	6
12. *Bone-bed*	0	1
13. Black laminated clay	3	6
14. Compact angular greenish marl	25	0
15. Red marl	3	0
	64	8

Dip about 12° to the east.

The stratum No. 12, which is identified by its position and organic remains with the "Bristol Bone-bed," rarely exceeds an inch in thickness, and frequently thins out in short distances to one-fourth of an inch or less. It consists chiefly of a dense mass of scales, teeth, bones, and small coprolites, cemented by pyrites, the golden colour of which contrasts beautifully with the jet-black of the animal remains. These osseous fragments have the appearance of having been washed into the hollows of a previously rippled surface of clay, in the same manner as we often see patches of coal-dust and small shells on the sea-beach. They have evidently been subjected to a gentle mechanical action, as the fragments often present broken and worn surfaces. The former existence of gentle currents is further proved by small rounded pebbles of white quartz, a substance of very rare occurrence in the liassic series*. In some places the bones and coprolites compose nearly the whole substance of the bed, in other parts they thin out rapidly, and are replaced by whitish micaceous sandstone. The only mollusc occurring in this bed is a smooth bivalve, too imperfect to be further identified.

2. *Wainlode Cliff.*—About three miles west-south-west from Coomb Hill, the Severn has excavated the side of a hill, and formed the interest-

* In this respect, as well as in its other mineral characters, there is a perfect identity between the bone-bed of Axmouth and that of Coomb Hill.

ing section called "Wainlode Cliff." The following beds are here exposed, Pl. VIII. fig. 1.

	ft.	in.
1. Black laminated clay, with a band of lias limestone with Ostreæ near the top	22	0
2. Bed of slaty calcareous sandstone with a peculiar species of Pecten	0	4
3. Black laminated clay	9	0
4. *Bone-bed*, passing into white sandstone	0	3
5. Black laminated clay	2	0
6. Light green angular marl	23	0
7. Red marl, with zones of greenish	42	0
	98	7

Dip very slight to the south.

The "Bone-bed" at this place is far less rich in organic remains than at Coomb Hill. Its prevailing character is that of a fissile white micaceous sandstone, sometimes acquiring a flinty hardness, and presenting at rare intervals accumulations of osseous and coprolitic fragments similar to those of Coomb Hill. The upper surface of this bed is covered with ripple-marks, and in some cases with mechanical impressions apparently produced by the feet of Crustacea. The specimens of these impressions hitherto found, are, however, too irregular to determine their nature with certainty. A small bivalve is the only species of shell in this stratum.

The bed No. 2 contains an abundance of a small species of Pecten, with numerous fine ribs like the recent *P. varius*. This bed is evidently the same with No. 6 of the Coomb Hill section.

3. *Bushley.*—Near Bushley, about two miles and a half west of Tewkesbury, the Ledbury road cuts through the escarpment of the lias, and presents us with the following section, Pl. VIII. fig. 3*.

	ft.	in.
1. Black laminated clay, about	10	0
2. Lias limestone	0	4
3. Black laminated clay	6	0
4. Compact slaty beds with numerous small bivalves and a Pecten (the same species as at Wainlode and Coomb Hill)	0	3
5. Black laminated clay	9	0
6. *White micaceous sandstone* with impressions of two species of bivalve shells	1	0
7. Black laminated clay	2	6
8. Greenish marl, about	20	0
9. Red marl.		
Dip about 8° east.	49	1

* This section is given (with some slight discrepancies) by Sir R. Murchison in his 'Geology of the Neighbourhood of Cheltenham,' p. 23, Ed. 1.

It will be observed that the sandstone bed No. 6 occupies precisely the position of the bone-bed at Coomb Hill, but as no fragments whatever of osseous remains have been found in it, a geologist might hesitate in pronouncing it the true equivalent of that stratum. The section at Wainlode Cliff, however, removes all the difficulty, for there we see the same stratum assuming in some parts the character of the bone-bed, and in others that of a white sandstone identical with No. 6 of the Bushley section. It is therefore evident that this band of white sandstone, which is interpolated in so singular a manner between the beds of black clay at the base of the lias, is the true representative of the bone-bed at Coomb Hill.

On further comparing these sections, we recognize at Bushley, about 9 feet above the representative of the bone-bed, the same band of slaty stone with Pectens which at Wainlode and Coomb Hill occurs within a very few inches of the same position.

If we now refer to the railway section at Dunhamstead, near Droitwich, given in the 'Proceedings Geol. Soc.' vol. iii. p. 314, we again recognize a band of white micaceous sandstone, amounting in thickness to 2 feet, and at the height of 6 feet above the top of the green marl. This sandstone contains the same small bivalve which occurs at Bushley and Wainlode, and as its mineral character and position also correspond, we may infer that we are again on the precise horizon of the "Bone-bed." I have examined other sections of the lias escarpment at Norton near Kempsey, and at Cracombe Hill near Evesham, and always with the same result: a thin band of white sandstone invariably occurs a few feet above the base of the lias clay, charged with a small bivalve shell, but destitute (in Worcestershire) of any Saurian or Piscine remains.

The "Bone-bed," as it occurs at Axmouth, Watchet, Aust, Westbury and other southern localities, occupies precisely the same position as at the places above described, being always near the base of the lias clay, and a very few feet above the top of the greenish marls which terminate the New Red system. It appears, however, that this stratum, which in East Devon, Somerset and Gloucestershire is so highly charged with organic remains, loses its ossiferous character when we enter Worcestershire. Its identity, however, is not lost; and when it is considered that from Axmouth on the south to Dunhamstead on the north is a distance of about 112 miles, we have a remarkable instance of the continuity of a very thin stratum over a great distance.

This great continuity of extent, combined with the prodigious abundance of organic remains in some parts of this stratum, render it probable that a much longer period may have elapsed during its deposit than in the case of an equal thickness of the less fossiliferous clay-beds above and below. Like those laconic sentences of Jewish history which merely state that "the land had rest forty years," so here, the geological records of years, or even of centuries, may be condensed into a few inches. Generations of fishes and of saurians may have added their remains to the common mass, while from the clearness of the water, or from the existence of a gentle current which prevented the deposit of muddy particles, scarcely any mineral matter was added to the bottom of the sea.

Organic Remains.

I now proceed to enumerate the organic remains met with in the "Bone-bed" at Coomb Hill.

1. Scales of *Gyrolepis tenuistriatus* ?.
2. Scales of Amblyurus ?.
3. Teeth of *Saurichthys apicalis*, a species which occurs also in the bone-bed at Axmouth, and in the continental Muschelkalk.
4. Teeth which much resemble in form the incisors of Mammalia. They are, however, proved to belong to fish, both by their organic structure, and by their analogy to those of Sargus and other existing genera of fish.
5. Portion of a tooth with two finely serrated trenchant edges. This tooth appears to belong to a Saurian allied to the genus Palæosaurus, discovered in the magnesian conglomerate near Bristol.
6. Tooth of *Hybodus delabechei*, Charlesworth, Mag. Nat. Hist. ser. 2. vol. iii. pl. 4. It seems to be the same as *H. medius*, Ag. There has also occurred a portion of skin similar to that figured by Charlesworth, *l. c.* pl. 4. fig. 5*a*, and a dorsal spine similar to his fig. 9.
7. Teeth of *Acrodus minimus*.
8. Teeth of *Hybodus minor*.
   (The last two species are abundant in the Bone-bed at Axmouth.)
9. A small vertebra of a fish.
10. Ichthyodorulite of *Nemacanthus monilifer*, Agass. This species occurs also at Aust Passage.
11. Teeth of Pycnodus ?.
12. Small coprolites.
13. Bones of an Ichthyosaurus.
14. Imperfect casts of a small, smooth, oval bivalve, occurring also at Wainlode, Bushley and Dunhamstead.

Sir P. Egerton has lately communicated an able memoir on the Bone-bed of Somerset and East Devon*, in which he shows that though most of the species of fish found in it are exclusively characteristic, yet a certain proportion of them also occur in the continental Muschelkalk. He hence infers that this stratum should be referred, not to the liassic, but to the triassic or pecilitic series. On this point I would remark in the first place, that the bone-bed is clearly not of the same age with the Muschelkalk itself, because it overlies the red and green marls which have been shown, I think satisfactorily, to be equivalent to the German Keuper Sandstein. All that we can therefore assert is, that certain species of fish survived

* Geol. Proceedings, vol. iii. p. 409.

from the epoch of the Muschelkalk to that of the English "Bone-bed." It is further to be observed, that these Muschelkalk fish are accompanied by various species of Hybodus, Acrodus and Ichthyosaurus which strictly belong to the lias, and there are, therefore, as good zoological data for referring the bone-bed to the lias as to the triassic system. But there will appear yet stronger grounds for placing the bone-bed in the liassic series, if we advert to the remarkable change of mineral character which takes place at the horizontal line marked AB, Pl. VIII. Below that line the beds consist of compact angular marl, greenish in the upper part and red below; and there is no trace of anything like black laminated clay till we descend through the New Red system into the coal-measures. But at the line AB there is a complete change from angular green marl to fissile black clay, and the transition is so sudden that it may be defined within the eighth of an inch. No more marl occurs above this point, but the clay-beds (interrupted only by thin bands of limestone, Pl. VIII. L.) extend upwards for many hundred feet till we reach the arenaceous rock termed "marlstone." It is therefore evident that at the epoch marked by the line AB, a great and sudden change took place in the deposits of the ocean, accompanied by great alterations in animal life, and we are therefore justified in taking the line AB as the base of a new system of geological formations. And although in the case of the bone-bed an arenaceous deposit, similar to the Keuper sandstone below, recurred for a limited period, accompanied by some of the surviving animal forms of the triassic series, yet this does not invalidate the evidence of the commencement of a new order of things,—of that series of argillaceous deposits which we recognize by the name of Lias. The evidence, therefore, preponderates in favour of placing the bone-bed within the limits of the liassic series, although it certainly affords an interesting example of transition from the triassic or pecilitic system below.

Before quitting the subject, it may be well to remark that two other examples of "bone-beds," precisely analogous to the one here described, occur in other parts of the geological series. One of these is the well-known bone-bed of the Upper Ludlow rock, so fully described by Sir R. Murchison in the 'Silurian System,' p. 198. The other has been lately found in Caldy Island, near the junction of the carboniferous limestone with the subjacent old red sandstone. It is singular that in all these three instances the bone-beds occur near the passage from one great geological system of rocks to another. It may deserve inquiry whether this is a mere accidental coincidence, or whether it may arise from some extraordinary mortality among the inhabitants of the ocean in consequence of those great geological changes which produce a new series of deposits.

Postscript to the Memoir on the Occurrence of the "Bristol Bone-bed" in the neighbourhood of Tewkesbury.

Since the reading of my paper on this subject in December last, I have met with evidences of the further extent in a northerly direction of the ossiferous portion of this stratum. The locality in question is at some old salt-works belonging to the Earl of Coventry on Defford Common in Worcestershire, about half a mile *within* or to the eastward of the lias escarpment. At this place is a shaft sunk about seventy years ago to the depth of 175 feet. It is filled with moderately strong brine, a small stream of which constantly overflows at the surface. This shaft was emptied a few months ago, when it was found that at the bottom is a tunnel which follows the dip of the strata to the eastward for about 160 yards. From the specimens of rock brought up, it appears that this shaft descends through the lias into the *grey marl* which forms the top of the triassic series, but without reaching the *red marl*, which commences a few feet lower down. The shaft consequently cuts through the horizon of the "Bristol bone-bed," and accordingly we find among the rubbish on the surface considerable quantities of the peculiar white sandstone with bivalves which I have before shown to represent in Worcestershire the "bone-bed" of Aust Passage and Axmouth. I now exhibit a large slab covered with the bivalve in question, which appears to be a species of Posidonomya, distinguished by its larger size and more oval form from the *P. minuta* of the Keuper sandstone. It is also an interesting circumstance, to find that at this locality the sandstone occasionally assumes the true character of a bone-bed, some portions of it being charged with the same kinds of teeth, scales and coprolites which are so abundant at Coomb Hill, as before described. Now, as Defford Common is about ten miles further north than Coomb Hill, we have evidence of the extent of the ossiferous portion of the bone-bed through a distance of about 104 miles from Defford Common to Axmouth.

The occurrence of an abundance of pure salt water within the area of the lias is also an interesting phænomenon*. The saliferous beds having been ascertained to lie at the bottom of the red marl, probably 500 feet below the lias, we must suppose that the saline waters percolate upwards from that depth to the lower beds of the lias, in the manner suggested by Sir R. Murchison in his work on the 'Geology of Cheltenham.'

* An imperial gallon of the brine from Defford Common is stated to contain 5817·6 grs. of chloride of sodium, 195·8 grs. of sulphate of lime, and a trace of magnesia. See Dr. Hasting's paper in the 'Analyst,' vol. ii. p. 384.

11. ON THE DISTRIBUTION AND ORGANIC CONTENTS OF THE "LUDLOW BONE-BED" IN THE DISTRICTS OF WOOLHOPE AND MAY HILL. [*Read December* 1, 1852.]

[Quart. Journ. Geol. Soc. vol. ix. p. 8.]

In a paper read to this Society in June last*, I noticed the occurrence at Hagley, four miles north-east of Hereford, of that remarkable deposit the "Ludlow Bone-bed," a stratum interesting not only for its wide extension, as contrasted with its very slight vertical thickness, but also as presenting nearly, if not quite, the earliest known indication of vertebrate life on the surface of our planet. I showed its close conformity, at Hagley, to the type of the same deposit as first described by Sir R. Murchison near Ludlow ('Silurian System,' p. 198), and enumerated certain fish and Mollusca which it there contained. I also briefly referred to certain seed-like bodies which seemed to indicate the commencement, or at least the first appearance, of terrestrial vegetation.

The interest of the subject has since induced me to trace out the same deposit at various points towards the south-east. Professor Phillips had already indicated the existence of Ichthyic fragments near the boundary-line of the Silurian and Devonian systems at two or three points in this direction (Mem. Geol. Surv. vol. ii. pp. 178, 191), but had not gone into much detail as to their structural character. Having succeeded in tracing this stratum at additional localities, and having obtained in it some fossil remains of considerable interest, it seemed desirable to communicate these results to the Society.

It will be remembered that at Hagley the Ludlow bone-bed occurs as a thin stratum of fish bones, scales and coprolites, mixed with carbonaceous fragments, not more than from 1 to 2 inches thick. The same bed occurs around the north-west margin of the Woolhope district, between Stoke Edith and Prior's Frome, where vegetable fragments were noticed by Professor Phillips, though the bone-bed has not yet been there detected. At Prior's Frome Mr. Scobie has lately found specimens of the same seed-like body as that found at Hagley, which will be hereafter described.

At the locality now to be mentioned the bone-bed is much increased in thickness. This is at a point described by Professor Phillips (*l. c.* p. 178) between Lyne Down and Gamage Ford, on the south-west side of the Silurian area of Woolhope. The bed crops out in the side of a lane, and presents a stratum nearly a foot thick, containing the remains of fish in immense profusion. Professor Phillips describes it as "a layer

* See Quart. Journ. Geol. Soc. vol. viii. p. 381.

of fish-bones and pebbles in a loose blackened state," but though it is evidently a drifted deposit, I did not succeed in finding pebbles in it. It is literally a bone-bed, the great bulk of the deposit consisting of osseous and coprolitic matter, mingled with a very small proportion of sand and mud, barely enough to cement it together, and apparently containing no erratic fragments except a few marly nodules. Some portions of it have the appearance of a coarse brown sandstone, but on close examination it is found that the apparent grains of sand are really the so-called teeth of *Thelodus parvidens*, Agassiz ('Silurian System,' pl. 4, figs. 34, 35, 36), which compose its entire substance. These curious little bodies, composed of two flattened pieces connected by a narrow neck, and which in form much resemble a common shirt-stud, are considered by Agassiz to be the teeth of a fish, though it appears to me not unlikely that they may be the placoid scales, and not the teeth, of the animal. It would be difficult otherwise to account for the extraordinary profusion of this minute organism, which forms by far the most abundant fossil of the Ludlow bone-bed, wherever it occurs, and in this locality constitutes the principal bulk of the stratum.

The only other ichthyic fossils which I have noticed at this locality are the striated spines of *Onchus tenuistriatus*, Agassiz (Sil. Syst. pl. 4. fig. 58).

Coprolites are abundant in the deposit. They are usually much rolled and water-worn, presenting but little regularity of form, though from their organic contents and the phosphate of lime which they contain *, there can be no doubt as to their nature.

The following Molluscous remains occur here, either imbedded in masses of coprolite, or in the form of casts filled with coprolitic matter:—

Orbicula rugata, Sow. Sil. Syst. pl. 4. figs. 47, 48.
Lingula cornea, Sow. Sil. Syst. pl. 4. fig. 49.
Turbo ?
Bellerophon expansus ?, Sow. Sil. Syst. pl. 4. fig. 50.
Orthoceras semipartitum, Sow. Sil. Syst. pl. 4. figs. 52, 53 ? Detached casts of chambers with the siphuncle *lateral.*

At Gamage Ford, as at Hagley †, rolled masses of carbonaceous matter accompany the animal remains. These are in the state of coal, and usually present no trace of their original structure. There are, however, some remarkable seed-like bodies, evidently identical with those before noticed at Hagley and at Prior's Frome. They are usually of an almost perfectly spherical form, and present no trace of any point of attachment to the parent plant. Their diameter varies from 0·1 to 0·2 of an inch. The surface when in a perfect condition is very smooth.

On breaking open these globular bodies, a central cavity is seen,

* See 'Silurian System,' p. 607, for the analysis of those in the bone-bed at Ludlow.

† *Loc. cit.* p. 382.

which is in some cases empty or filled with mineral matter; in others it contains a powdery carbonaceous substance in which no organization is perceptible. The diameter of this cavity varies from one-half to one-third of that of the entire body, so that the wall which encloses it is from one-fourth to one-third of that diameter in thickness. It is wholly composed of straight regular fibres, of even thickness, radiating from the internal cavity to the external surface (see fig. 1). The fibres have flat sides, and being closely packed together, they are necessarily, on the average, hexagonal. On rubbing down a portion of the external surface, the ends of these hexagonal fibres are exposed to view. They do not appear to be tubular.

In one specimen the extremities of these fibres are exhibited on the external surface by numerous small dark-coloured dots on a pale ground. Their outlines are here no longer hexagonal, but round or oval, often confluent, with a tendency to form parallel lines, yet subject to some irregularity (see figs. 2, 3)*. A further notice, by Dr. Hooker, of these curious fossils, and of a fragment of carbonized wood found with them, is appended to this paper.

Figs. 1, 2, & 3. *Showing the Structure of the globular vegetable bodies found in the Ludlow Bone-bed.*

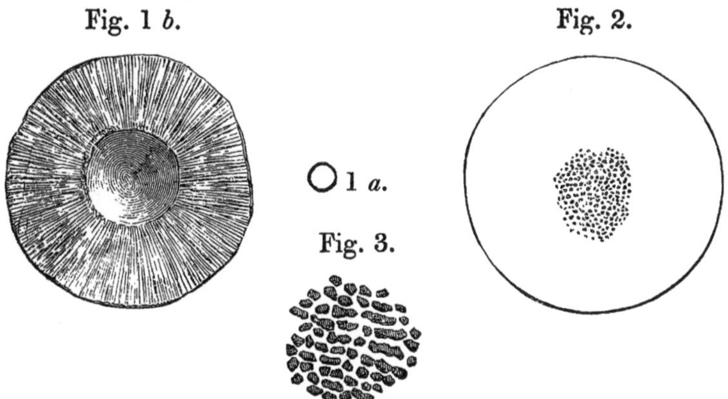

Fig. 1 *a*. Natural size.
Fig. 1 *b*. Transverse section; magnified.

Fig. 2. Dotted markings on the external surface; magnified.
Fig. 3. The same, more highly magnified.

In crossing the axis of the Woolhope elevation from Gamage Ford to Much Marcle, I again detected the bone-bed, in the side of the road,

* The object figured in Sil. Syst. pl. 4. figs. 65, 66, and described as the "palatal bone of Sclerodus," has every appearance of being one of these seed-like bodies split in half. Somewhat similar, though imperfect, specimens were found by Mr. Scobie and myself in the Cornstone of Llanfihangel, near Abergavenny.

M 2

near the base of the old red sandstone. It here presents numerous rolled coprolitic fragments mixed with the supposed scales of *Thelodus parvidens,* imbedded in brown micaceous sand. I also noticed casts of an Atrypa in it.

The beds of yellow sandstone quarried at Gorstley Common, between Woolhope and May Hill, belong to the same part of the series as those which contain the Ludlow bone-bed, though no remains of fish have yet been found in them. They contain, however, numerous vegetable remains, chiefly amorphous fragments of carbon in the state of coal. Among these frequently occur globular seed-like bodies identical with those above described, and exhibiting the same internal structure.

Proceeding six miles further south, nearly to the southern extremity of the May Hill elevation, there is an interesting section just laid open by the works of the Hereford, Ross and Gloucester Railway, at a place called the Velt-house. A cutting, of about 100 yards long, here exposes the lower beds of the old red sandstone, the yellow sandstones termed "Downton Sandstones" by Sir R. Murchison (which are here about 20 feet thick), and a considerable extent of the Upper Ludlow shales. The whole series is conformable, and dips about 75° to the west. Near the base of the Downton sandstones, and 4 or 5 feet above the Ludlow shales, I detected the bone-bed, not more than 2 inches thick, containing numerous rolled fragments of black coprolitic matter imbedded in a grey micaceous sandy shale. The scales of *Thelodus parvidens* are frequent, and I also found a spine of *Onchus murchisoni**. The only other fossils noticed were casts of a depressed spiral shell, which appears to be the *Trochus helicites,* Sowerby, Sil. Syst. pl. 3. fig. 1 *e,* but whose true generic position is doubtful. This shell has been also found in the Tilestones near Kington, and in the Llandeilo district. Its resemblance to some of the commoner forms of recent Helix, such as *H. hispida* or *H. cantiana,* is very striking, though from its collocation it was probably a marine mollusc. The diameter is about $\frac{1}{2}$ an inch, the height $\frac{3}{8}$†.

I have thus traced this remarkable deposit of ichthyic, molluscous and vegetable remains from Hagley to the Velt-house, a distance of about seventeen miles. Professor Phillips speaks of the equivalent of the Ludlow bone-bed as occurring at Pyrton Passage, about seven miles

* By *Onchus murchisoni* I mean a genuine Ichthyodorulite, distinct from *O. tenuistriatus,* and agreeing with figs. 9 and 11 in plate 4 of the 'Silurian System,' but not with fig. 10, which Professor M'Coy has shown to be a Crustacean. [Jan. 6, 1853.]

† The genus Helix does not, I believe, make its appearance until the tertiary period; otherwise I should be disposed to speculate on the possibility of these molluscs having been drifted down from the same land whence the seed-like fossils above described were probably derived. [Jan. 6, 1853.]

further south (Mem. Geol. Survey, vol. iii. part 1. p. 191). From this point to the banks of the Teme, near Ludlow, where the bone-bed was originally detected, is a direct distance of forty-five miles. It is not a little remarkable to find a stratum of such trifling thickness thus persistent in character through so wide an area. In this respect, no less than in the nature of its osseous and coprolitic drift, and in the fact that it occurs just at the boundary-line between two great geological systems, its analogies with the well-known bone-bed at the base of the lias are deserving of notice. In both cases we have indications of an extensive mortality among the fish then in being, a phænomenon not improbably connected with the great physical changes which are proved to have taken place at the periods respectively indicated by these deposits.

On the Spheroidal Bodies, resembling Seeds, from the Ludlow Bone-bed.

By JOSEPH HOOKER, M.D., F.R.S., F.G.S.

The fossils in question consist of spherical bodies varying in size from 1 line to $\frac{1}{4}$ of an inch; they have suffered no compression nor mutilation during their conversion into mineral matter. The surface is nearly smooth and uniform to the naked eye, but seen under the microscope it appears to be covered with circular or hexagonal areolæ, placed in contiguity in the latter case.

Fractured specimens show them to be hollow spheres, whose walls are fully twice as thick as the cavity they enclose. Their contents consist of a little loose white powder, of mineral matter, displaying no organic structure.

The walls (or integument) consist of a single series of narrow hexagonal cells, placed side by side, radially. The cell-walls are very thin, and exhibit no markings, nor are there any intercellular spaces. No vascular tissue is observed in any part, nor remains of any outer integument.

This simple structure of spore-case is very characteristic of the natural order Lycopodiaceæ and of the allied fossil genus Lepidostrobus, and I am not aware of any other order to which these fossils may more safely be referred. In their spherical form they differ from any known spore-case of this alliance; but mere form is a character of very minor importance in such organs, suggestive of specific only and not of generic difference.

In the great thickness of the walls, and consequent length of the radiating cells forming the latter, they differ conspicuously from any recent or modern spore-case with which I am acquainted.

The great difference in size of the specimens, unaccompanied by any other character, is remarkable, as is the apparent absence of any point of attachment. The latter is probably due to contraction of the tissues, and cannot be regarded as any evidence of these fossils having been seeds or spores rather than the cases in which such are contained; for in some of the best specimens of Lepidostrobus which I have examined, there is no evident attachment between the spore-case and the modified leaf or scale which supports it, and through which it was nourished in its progress to maturity.

The accompanying fossil-wood presents very obscure traces of structure, and none on a cursory examination that throw any light upon the origin of the spherical bodies; but I have not had time to adopt the usual means for making a satisfactory examination.

---

12. ON CERTAIN IMPRESSIONS ON THE SURFACE OF THE LIAS BONE-BED IN GLOUCESTERSHIRE. [*Read November* 30, 1842.]

[Ann. Nat. Hist. xi. p. 511. Proc. Geol. Soc. iv. p. 16.]

[In 1842 a paper with the above title was read before the Geological Society; and as bearing on the subject of the two previous memoirs, the following abstract, printed in the 'Proceedings' of the Society, is now copied.—Ed.]

[Abstract.]

THE singular markings described, which the author in a former communication suggested might be caused by the crawling of Crustacea, but which further opportunities and observations have induced him to refer to a different cause, have been noticed only at Wainlode Cliff on the Severn. There they occur on the uppermost surface of the band of micaceous sandstone which represents the "bone-bed," and which appears to have consisted of a fine-grained muddy sand, capable of receiving the most minute impressions, while the pure black clay which forms the superincumbent stratum has preserved this ancient surface in the most unaltered condition. The ripple-marks produced by currents on the surface of this bed of sand are very interesting, from their perfect preservation, and from often exhibiting two sets of undulations oblique to each other, indicating two successive directions in the currents, such as would result from a change of tide.

The impressed markings were evidently produced by living beings, probably by fish or invertebrate animals. To determine their nature Mr. Strickland observed the progression of two species of Littorina among Gasteropodous Mollusca, and of *Carcinus mænas* among Crustacea, but the impressions produced were very different from those under consideration.

The fossil impressions are of four kinds:—

1st. Lengthened and nearly straight grooves, about one-tenth of an inch in width, and several inches long, very shallow, with a rounded bottom. These, Mr. Strickland considers as caused by some object striking the surface of the sand with considerable impetus. They may often be seen to cut through the ridge of one ripple-mark, and after disappearing in the depressed interval, they are again seen pursuing their former direction across the next ridge. They may have been caused by fish swimming with velocity in a straight direction, and occasionally touching the bottom with the under part of their bodies.

2nd. Small irregular pits averaging one-fourth of an inch wide and one-eighth of an inch deep. These might have been caused by some small animal probing the mud and turning up the surface in quest of food. Mr. Strickland conjectures that some of the numerous species of fish found in the bone-bed may have produced them, the heterocene form of tail common to most of which, Dr. Buckland has suggested, enabled them to assume an inclined position with the mouth close to the ground.

3rd. Narrow deep grooves, about one-twelfth of an inch in width, the sides forming an angle at the bottom, irregularly curved and often making abrupt turns, apparently formed by a body pushed along by a slow and uncertain movement, such as might arise from the crawling of molluscs. Mr. Strickland refers them to the locomotion of Acephalous Mollusca, and supposes that the only shell found in this bed, a small bivalve named by him *Pullastra arenicola,* might have produced them*.

4th. A tortuous or meandering track consisting of a slightly raised ridge about one-tenth of an inch wide, with a fine linear groove on each side. These tracks are analogous to those formed by the crawling of small annelidous worms, as may often be seen on the mud of the sea or fresh water.

About 11 feet above the stratum which presents the impressions above described, a second ossiferous bed occurs at Wainlode Cliff, which escaped Mr. Strickland's notice in the section formerly given (Geol. Proc. vol. iii. p. 586). It is a band of hard, grey, slightly calcareous stone, about an inch thick, containing a plicated shell resembling a Cardium, and scales and teeth of *Gyrolepis tenuistriatus, Saurichthys apicalis, Hybodus delabechei, Acrodus minimus* and *Nemacanthus monilifer,* all of which occur in the true "bone-bed" below. On the upper surface of that bed are numerous impressions, termed by Mr. Strickland fucoid, consisting of lengthened wrinkled grooves, variously curved, about three-quarters of an inch wide, one-eighth of an inch deep, and of variable length. The bone-bed seems to be a local deposit, not being met with in the other localities examined by the author, and being confined to a portion only of Wainlode Cliff, where it constitutes No. 9 in the following corrected section:—

	ft.	in.
1. Blackish lias clay	3	6
2. Limestone, with *Ostrea* and *Modiola minima* (the bottom bed)	0	4

---

* Mr. Strickland describes this species as follows:—"Its form is nearly a perfect oval, depressed, nearly smooth, but with faint concentric striations towards the margin. The apex is about halfway between the middle of the shell and the anterior end. The general outline closely resembles that of the recent *Pullastra aurea* of Britain. Maximum length 7 lines, breadth 4½ lines, but the ordinary size is less."

		ft.	in.
3.	Yellowish shale	1	0
4.	Limestone, with remains of insects	0	4
5.	Marly shale and clay	5	3
6.	Yellowish limestone nodules, with occasional remains of Cypris	0	6
7.	Yellowish marly clay	6	0
8.	Black laminated clay	3	6
9.	Stone, with scales and bones of fish, and on the upper surface fucoid impressions	0	1
10.	Black laminated clay	1	6
11.	Slaty calcareous stone, with Pectines	0	4
12.	Black laminated clay	9	0
13.	BONE-BED and white sandstone, with casts of *Pullastra arenicola*	0	3
14.	Black laminated clay	2	0
15.	Greenish angular marl	23	0
16.	Red marls with greenish zones	42	0
		98	7

---

13. ON SOME REMARKABLE CONCRETIONS IN THE TERTIARY BEDS OF THE ISLE OF MAN. [*Read November* 16, 1842.]

[Ann. Nat. Hist. xi. p. 507. Proc. Geol. Soc. iv. p. 8.]

[Abstract.]

THE north extremity of the Isle of Man consists of an arenaceous pleistocene deposit, occupying an area of about eight miles by six, bounded on the west, north and east by the sea, and on the south by the mountains of Cambrian slate which occupy the greater portion of the island. The arenaceous formation attains in some parts a height of about 200 feet above the sea, though the undulations of its surface prove that considerable portions of the deposit have been removed by denudation. This district, comprising about fifty square miles, furnishes perhaps the most extensive example in the British Isles of a marine newer pliocene or pleistocene deposit. In the Isle of Man the sea-cliffs on each side of this tertiary district afford a good insight into its structure and composition. On the north of Ramsey the cliffs average about 100 feet in height, and consist principally of irregularly stratified yellowish sand, sometimes clayey, with interspersed bands of gravel and scattered pebbles. The gravel is chiefly composed of slate-rock, quartz, old red sandstone, granites, porphyries and chalk flints, all of which occur *in situ* in the island except the last two, which may have been drifted, the former from Scotland, and the latter from the north of Ireland. About four miles north of Ramsey the cliffs attain 150 feet. Here the lowest portion, only visible at intervals, is a brownish clay loam, and the remainder of the cliff is sand and coarse gravel, less distinctly stratified than is the case near Ramsey, and containing rudely rounded boulders, some of which are upwards of a ton in weight. They consist of granite, and occasionally of carboniferous limestone.

Organic remains are sparingly diffused in this deposit: Mr. Strickland enumerates twenty species. Of these, five, viz. *Crassina multicostata, Natica clausa, Nassa monensis, Nassa pliocena,* and *Fusus forbesi,* are not known in the British seas. *Crassina multicostata* and *Natica clausa* are found living in the Arctic ocean, but the two species of Nassa and the Fusus are unknown in a recent state*.

* Mr. Strickland gives the following characters of three species of shells found in the newer pliocene beds of the Isle of Man; specimens of which have been examined by several eminent conchologists in London, who all concur in believing them to belong to extinct species.

"1. *Nassa monensis*, Forbes, in Mem. Wern. Soc. vol. viii. p. 62. Small; volu-

Between three and four miles north of Ramsey, the beds of this deposit occasionally exhibit a very remarkable concretionary structure. The sand has here been cemented into masses, which are extremely hard, and even sonorous when struck, though the sand in which they are imbedded is perfectly loose. The cementing ingredient, which the application of acid proves to be carbonate of lime, seems to have been influenced in its operations partly by the planes of stratification, and partly by the direction in which the sand has been originally drifted by currents. In the former case the concretions are in the form of flat tabular masses parallel to the stratification, often mammillated on their surfaces, or perforated obliquely by tubular cavities. In the latter case they assume a subcylindrical or pear-shaped form, and occur parallel both to the stratification and to each other. A pebble is frequently attached to the larger

tions about six, rounded; suture deep; ribs, nine on the first volution, straight, rather distant, strong, subacute and slightly oblique. The first volution has thirteen, and the second six, distinct, regular, thread-like, spiral striæ, crossing alike the ribs and their interstices. Aperture orbicular-ovate; canal very short and oblique; pillar-lip simple; outer lip with about five slight marginal denticles on the inside, and an external rib slightly more developed than the ordinary ribs. Total length, 7 lines; first volution, 3¼ lines; breadth, 4½ lines; angle of spire, 40°.

"*Obs.* Resembles the recent *N. macula*, but is larger, more ventricose, has fewer ribs, and the terminal rib is less suddenly developed.

"2. *Nassa pliocena*, Strickland, 1843. Large; volutions about seven, rather flat, with a distinct thread-like suture; ribs, twelve on the first volution, straight, distant, rounded, very slightly oblique; the interstices flat, exceeding the width of the ribs by one-half. The first volution with thirteen, and the second with about nine fine spiral striæ, only visible in the interstices, the ribs being smooth; but this may be due to attrition. Aperture ovate; canal very short and oblique; pillar-lip with about five obscure denticles, and a spiral groove immediately behind the canal, continued into the interior of the shell; outer lip with about eight internal marginal denticles; no rib at the back. Total length, 1 inch 8 lines; first volution, 8 lines; breadth, 9 lines; angle of spire, 40°.

"3. *Fusus forbesi*, Strickland, 1843. *Fusus*, nov. sp., Forbes, Malacologia Monensis, pl. 3. fig. 1. Middle-sized; volutions about six, slightly rounded, suture distinct; ribs, eleven on first volution, straight, rounded, smooth (perhaps from attrition); interstices concave, and hardly wider than the ribs. First volution with about fifteen, and second with about seven distinct, rather irregular spiral striæ, of which those on the first volution are alternately large and small; they are only visible in the interstices of the ribs. Aperture ovate, double the length of the canal, which is straight, and rather oblique to the left; pillar-lip smooth, with one obscure denticle at the posterior end; outer lip with about ten small linear denticles within, continued a short way into the mouth, and a well-marked external rib remote from the margin. Total length, 1 inch 3 lines; first volution, 7 lines; breadth, 8 lines; angle of spire, 43°.

"*Obs.* This species belongs to a group of Fusus which seems closely allied to Nassa. First described by Mr. E. Forbes, from a worn specimen found on the coast of the Isle of Man, and supposed by him to be an existing species; but the discovery of additional specimens *in situ* proves it to be a genuine fossil."

end of the concretion, which springs from it as from a root, to the length of a foot or more, and gradually terminates in an obtuse flattened point. All these varieties are sometimes combined together into vast clusters of several tons weight, resembling masses of stalactite, the component portions being nearly parallel to each other. Mr. Strickland supposes that currents of water (or possibly of wind, operating during ebb tide), flowing in a certain direction, may have disposed the sand in ridges parallel to that direction, and the carbonate of lime may have afterwards been attracted into these ridges in preference to the intermediate portions. This view is confirmed by the fact, that these concretions have frequently a pebble attached to the larger end, as though it had protected a portion of sand from the current, and caused it to accumulate in a ridge on the lee side, a circumstance which may frequently be observed where sand is drifted by the wind or water.

14. ON CARDINIA, AGASSIZ, A FOSSIL GENUS OF MOLLUSCA CHARACTERISTIC OF THE LIAS. [*Read before the Meeting of the British Association at Plymouth*, 1841.]

[Report Brit. Assoc. vol. x. Sections, p. 65. Ann. Nat. Hist. xiv. p. 100.]

THERE are few groups of fossils which, both in their generic and specific relations, have been involved in greater confusion than the very natural and characteristic genus of which I am about to speak. Having resided for some years in a locality where several species of this genus abound, and having, by the examination of many hundreds, I might say thousands of specimens, aided by the kindness of Mr. J. Morris, author of the valuable 'Catalogue of British Fossils,' been enabled to trace them through their several varieties, and thus to circumscribe the boundaries of the species, I hope to correct some of the errors into which other authors have fallen.

The genus of Mollusks in question is evidently most nearly allied to Astarte, Sow. (Crassina, Lam.), a genus which most authors agree in placing among the Veneridæ. From the great strength of the shell, single valves are often preserved in a perfect state, and we are thus enabled to ascertain all its characters with an accuracy that is rarely attainable in fossil bivalves, especially of the older formations. The genus may be described in general terms as an Astarte with the addition of very strong lateral teeth. The shell is longitudinally oval, very thick, equivalve, inequilateral, perfectly closed; the hinge very strong; the right valve with two oblique converging cardinal teeth as in Astarte, but these teeth are flat, and only divided by a slight groove, which is sometimes obsolete. Below these teeth and immediately behind the lunule is a depression extending in front of the anterior lateral tooth, with a corresponding elevation in the left valve, in which the true cardinal teeth are almost wholly obsolete. Above the cardinal teeth in both valves is a deep narrow groove, evidently for the reception of an external ligament, as in Astarte. In front of the hinge is a deep and distinct lunule. The lateral teeth are remote and very strong; the anterior one of the right valve obtusely conical, the posterior one of the left valve elongated, and both mutually entering deep pits in the opposite valves. Umbones approximate. Muscular impressions very deep, placed immediately below the lateral teeth, their surfaces smooth; the posterior impression round, the anterior one ovate. Above the latter in both valves is a small oval detached muscular impression placed on the hinder surface of the lateral tooth, for the insertion of the retractor muscle of the foot. Pallial impression entire, parallel to the margin, which is not crenated. External surface of the shell more or less irregularly imbricated by the lines of growth. The geographical distribution of this genus is as yet confined to Northern Europe; its geological range is from the base of the lias up to the inferior oolite.

Several species of this genus were described by Sowerby in his 'Mineral Conchology,' under the genus Unio. They differ, however, from the whole of the Unionidæ in many respects, especially in the want of the small accessory muscular impression *behind* the anterior one (which occurs in the Unionidæ, and to which a branch of the retractor muscle of the foot is attached), in the presence of the lunule, in the shell not being nacreous, and in the habitat having been marine, as is sufficiently proved by the other fossil animals whose remains invariably accompany these shells.

M. Goldfuss has been no more successful than Mr. Sowerby in detecting the true generic relations of these shells, having in his 'Petrefacten' referred different species of them to the genera Unio, Cytherea and Lucina, without detecting the essential characters which distinguish them from all these genera.

M. Agassiz was the first to combine the different species of this group into one genus, though he failed to perceive that they are much more closely allied to the Veneridæ than to the Unionidæ. To this genus he gave the name of Cardinia, in a paper read to the Helvetic Society at their meeting at Basle in 1838*, and in 1840 he published the characters of the genus in his translation of Sowerby's 'Mineral Conchology.' In 1840 Mr. J. E. Gray gave the name Ginorga to this genus in the 'Synopsis of the British Museum,' p. 154; but this mere name, destitute alike of etymology and of definition, can have no claim for adoption. In January 1841, M. de Christol defined a genus Sinemuria in the 'Bulletin de la Société Géologique de France,' which from the characters assigned is evidently identical with the genus before us, though he errs in supposing the ligament to have been internal instead of external. Lastly, in March 1842 Mr. S. Stutchbury described this group in great detail in the 'Annals of Natural History,' and bestowed on it the name Pachyodon, a name which had been used four years before by M. von Meyer for a genus of Mammals.

It appears from this historical statement, that as M. Agassiz was the first to publish the characters of the genus, so his generic name Cardinia must supersede all later ones.

Some authors have been disposed to extend the geological range of this genus, by including in it those numerous species from the coal-measures which Sowerby and most other palæontologists have regarded

* *Thalassides,* Berg. 1833, without generic characters. *Thalassites,* Quenst. 1843.

as true Unionidæ. Whether Agassiz originally proposed this extension of the genus I am not aware, having never yet been able to meet with his translation of the 'Mineral Conchology,' in which the group is first defined; but in his last work on the subject, the 'Etudes critiques sur les Mollusques Fossiles,' he seems to regard Cardinia as exclusively confined to the lias and lower oolite. De Koninck, however, in his 'Description des Animaux Fossiles du Terrain houillier de la Belgique,' classes these coal-measure shells as Cardiniæ, and prefixes a definition of the genus which seems to be chiefly copied from De Christol's definition of Sinemuria, and we may therefore conclude that De Koninck had not been able to examine the *interior* of the fossils which he describes. He seems to have made a compromise between the real characters of Cardinia and the erroneous statement of De Christol as to the *internal* ligament; for he says that the shell had *two* ligaments, one *internal* and the other *external*, a statement which I believe to be wholly incorrect.

Captain Thomas Brown also seems to regard the coal-measure fossils as generically identical with the lias ones, since he has described, under Mr. Stutchbury's name Pachyodon, no less than twenty-six species of shells from the coal-measures, which he has illustrated with very accurate figures in the 'Annals of Natural History' for Dec. 1843, and in his own 'Fossil Conchology of Great Britain,' plate 73.

There are, however, many reasons for regarding as doubtful the supposed affinity between the Unioniform shells of the coal-measures and the true Cardiniæ of the lias, although it must be admitted that there is much general resemblance in their external forms. In the *first* place, I believe no author has yet seen or described the *interior* of any of the coal-measure shells, and there is consequently no positive evidence whatever as to the structure of their hinges. *Secondly*, although the general characters of the muscular and pallial impressions, as exhibited by the casts in both these sets of species, are very similar, yet in the coal-measure shells the muscular impressions are much smaller and shallower than in those of the lias, and the lateral teeth, if present at all, are evidently much less developed. *Thirdly*, in conformity with this greater feebleness of the connecting muscles, we find that the shells of the coal-measure fossils are much thinner and weaker than in those from the lias. *Fourthly*, the shells from the coal-measures rarely exhibit any trace of a lunule, and when present it is more diffused and indistinct than in the liassic species. *Lastly*, the Cardiniæ from the lias were wholly marine in their habits, while there are strong grounds for believing that the species from the coal-beds inhabited fresh, or at most brackish water. This is shown by the fact that these Unio-like shells are almost invariably found in the beds of shale accompanying the coal, and not in the really marine formations of the same age. Now whether we suppose the coal to have grown *in situ* like peat, or to have been washed by currents into certain localities (both which theories are no doubt true in certain cases), we cannot deny the coal to be a terrestrial production; and therefore when we find a particular family of mollusks constantly, and almost always exclusively, accompanying the beds of coal, we have a very strong presumption that these animals had a lacustrine or estuarine habitat.

It is true that in some cases, as in Coalbrook Dale, at Halifax, at Glasgow and in Belgium, the coal-measures contain an admixture of these bivalves with various marine genera; but this does not necessarily prove them to be marine species, for they may either (as suggested by Mr. Prestwich in his memoir on Coalbrook Dale, 'Geol. Proceedings,' vol. ii. p. 405) have been washed down into an estuary and there become mixed with marine shells, or by a depression of the land the sea may have washed the marine shells into the marshes tenanted by these supposed freshwater species. And it is important to remark, that in the carboniferous limestone, a strictly marine formation immediately preceding, and in some cases alternating with the coal-measures, these peculiar bivalves rarely if ever occur.

For these reasons I think we ought to abstain from classing the shells of the coal-measures with the well-marked and clearly-defined genus Cardinia of the lias. I do not indeed mean to assert that the carboniferous group of shells really belong to the Unionidæ, where they were formerly classed, for they want the supplementary anterior muscular impression which distinguishes that family*; but I think they may be for the present regarded as a distinct family, probably lacustrine, and possibly allied to Unionidæ, but the precise characters of which, and especially the structure of the hinge, are as yet unascertained. Perhaps Dr. Carpenter, whose researches on the microscopic structure of shells have opened to us a new element for the determination of fossil Mollusca, may be able to throw further light on the affinities of these ambiguous yet characteristic fossils†.

Confining our attention therefore to the shells of the lias and lower oolite, we will proceed to examine the species of Cardinia which really exist in nature, as well as those which have been described in books.

* Mr. G. B. Sowerby, in his 'Genera of Recent and Fossil Shells,' states that he could find *no difference* between the casts from the coal-measures and those which he made from the inside of recent Unios, but he had perhaps overlooked the supplementary muscle of the latter.

† These palæozoic shells have been described by King (1844), under the name of *Anthrocosia*. Woodw. Man. p. 303.

I. *Ascertained species of Cardinia.*

1. CARDINIA LISTERI, *Sow.* (sp.)

Donax? Park. Org. Rem. pl. 13. f. 7.
Unio listeri, Sow. Min. Con. pl. 154. f. 1, 3, 4.
Pachyodon listeri, Stutchb. Ann. Nat. Hist. vol. viii. pl. 9. f. 1, 2.

Var. 1. Subelongate.

Cytherea latiplexa, Goldf. Petref. pl. 149. f. 6.
Unio hybrida, Sow. Min. Con. pl. 154. f. 2.
Pachyodon hybridus, Stutchb. Ann. Nat. Hist. vol. viii. pl. 9. f. 3, 4.
Cardinia hybrida, Agass. Et. Crit. Moll. pl. 12.

Var. 2. Subcompressed.

Cytherea lamellosa, Goldf. Petref. pl. 149. f. 8.

Var. 3. Lines of growth very numerous.

Pachyodon imbricatus, Stutchb. Ann. Nat. Hist. vol. viii. pl. 9. f. 5, 6.

Var. 4. Small-sized (probably young).

Pachyodon cuneatus, Stutchb. Ann. Nat. Hist. vol. viii. pl. 10. f. 11, 12.

Var. 5.

Cardinia amygdala, Agass. Et. Crit. Moll. pl. 12. f. 10–12.
Formation.—Lower lias.
Localities.—Whitby Yorkshire; Grantham; Langar Nottinghamshire; Cropthorn, Defford and Eckington Worcestershire; Frethern Gloucestershire; Wurtemburg.

In Worcestershire and Gloucestershire this species is very abundant in a zone of the lower lias, about 150 feet above the base of that formation. Single valves are frequent. It is subject to much variation in the thickness of the shell, the frequency and regularity of the imbrications, and the length or shortness of the posterior extremity. Having examined a very extensive series of specimens, I have little doubt of the correctness of the above synonyms.

2. CARDINIA CRASSISSIMA, *Sow.*

Unio crassissima, Sow. Min. Con. pl. 153.
Pachyodon crassissimus, Stutchb. Ann. Nat. Hist. vol. viii. pl. 9. f. 7.
Lower oolite.—Dundry; Wick near Bath.
Marlstone.—Dumbleton Worcestershire.

3. CARDINIA CRASSIUSCULA, *Sow.*

Unio crassiusculus, Sow. Min. Con. pl. 185; Zieten, Verst. Wurt. pl. 60. f. 1.
Pachyodon crassiusculus, Stutchb. Ann. Nat. Hist. vol. viii. pl. 9. f. 8.
Pullastra antiqua, Phill. Geol. Yorksh. pl. 13. f. 16.

Var. 1. Small-sized, perhaps young.

Cardinia elliptica, Agass. Et. Crit. Moll. pl. 12. f. 16, 17.

Var. 2.

Cardinia similis, Agass. Et. Crit. Moll. pl. 12. f. 23.
Formation.—Lias.

Localities.—Pocklington and Robin Hood's Bay, Yorkshire; Nottinghamshire; Gloucestershire; Somersetshire.

Wurtemburg; Stuttgard. Var. 1. Argovie; var. 2. Soleure.

After a careful comparison of specimens, I have little doubt of the specific identity of the above references.

4. Cardinia lanceolata, *Stutchb.*

Pachyodon lanceolatus, Stutchb. Ann. Nat. Hist. vol. viii. p. 484.

Formation.—Lower lias.

Locality.—Robin Hood's Bay, Yorkshire.

The figure intended for this species by M. Agassiz was taken from a specimen of *C. attenuata* which I sent him.

5. Cardinia attenuata, *Stutchb.*

Pachyodon attenuatus, Stutchb. Ann. Nat. Hist. vol. viii. pl. 10. f. 13, 14.
Cardinia lanceolata, Agass. Et. Crit. Moll. pl. 12″. f. 1–3.

Formation.—Top of lower lias, just below the marlstone.

Localities.—Hewlets near Cheltenham; Bourton-on-the-Water, Gloucestershire.

M. Agassiz's figure above-quoted is taken from a specimen which I sent him, and I am therefore satisfied that it belongs to the present species.

6. Cardinia concinna, *Sow.* (sp.)

Unio concinnus, Sow. Min. Con. pl. 223. f. 1, 2; Zieten, Verst. Wurt. pl. 60. f. 2–5; Goldf. Petref. pl. 132. f. 2; Bronn, Lethæa Geogn. p. 361.
Pachyodon concinnus, Stutchb. Ann. Nat. Hist. vol. viii. pl. 10. f. 15, 16.
Cardinia concinna, Agass. Et. Crit. Moll. pl. 12. f. 21, 22.

Formations.—Marlstone and lias.

Localities.—Yorkshire; Langar Nottinghamshire; Daventry Northamptonshire; Saltford and Weston near Bath; Wurtemburg, Fachsenfeld; Mogglingen; Staffelegg in Argau.

This is the largest species of the genus. I have a specimen from the marlstone of Byfield in Northamptonshire which is 5½ inches long by 3 inches broad.

7. Cardinia ovalis, *Stutchb.*

Lucina lævis, Goldf. Petref. pl. 146. f. 11.
Pachyodon ovalis, Stutchb. Ann. Nat. Hist. vol. viii. pl. 10. f. 17, 18, 19.
Cardinia unionides, Agass. Et. Crit. Moll. pl. 12″. f. 7–9.

Var. 1.

Cardinia cyprina, Agass. Et. Crit. Moll. pl. 12″. f. 4–6.

Formation.—Lower lias.

Localities.—Dunhamstead and Coltknap Hill Worcestershire; Ashleworth and Frethern Gloucestershire; Watchet Somersetshire; Blumenroth, Coburg.

M. Goldfuss's specific name *lævis* is prior to the other two; but as it is founded on an erroneous identification with the *Corbis lævis* of

Sowerby, which is a very different shell, I retain Mr. Stutchbury's name *ovalis*. The two supposed species figured by M. Agassiz are both founded on specimens which I sent to that learned naturalist myself, and I am therefore able to identify them positively with the present species. In Worcestershire this fossil abounds at about 100 feet above the base of the lower lias. Single valves are very rare.

8. CARDINIA SULCATA, *Agass.*

Cardinia sulcata, Agass. Et. Crit. Moll. pl. 12. f. 1–9.
Formation.—"Calcaire à Gryphites."
Locality.—Soleure.

Judging from the figure and description, the above seems to be a distinct species.

9. CARDINIA APTYCHUS, *Goldf.*

Cytherea aptychus, Goldf. Petref. pl. 149. f. 7.
Formation.—Lias.
Locality.—Amberg.

I have seen and examined specimens of all the above species except nos. 8 and 9.

II. *Species referable to this genus, but whose specific characters require further investigation.*

1. *Pachyodon abductus*, Stutchb. Ann. Nat. Hist. vol. viii. pl. 9. f. 9, 10.

I think this is probably one of the numerous varieties of *C. listeri*. I agree with M. Agassiz that it is not the *Unio abductus* of Phillips.

2. *Cardinia oblonga*, Agass. Et. Crit. Moll. pl. 12. f. 13–15.

From the lower oolite of Normandy. Described *from a cast*, an authority on which it must be very unsafe to found *specific* distinctions.

3. *Cardinia lævis*, Agass. Et. Crit. Moll. pl. 12″. f. 13–15.

From Mulhausen. It is not the *Lucina lævis* of Goldfuss. Perhaps a variety of *C. listeri* or *crassiuscula*.

4. *Cardinia securiformis*, Agass. Et. Crit. Moll. pl. 12″. f. 16–18.

From Soleure; described from a cast, and perhaps only a variety of *C. concinna*.

5. *Sinemuria dufrenii*, De Christol, Bullet. Soc. Géol. de France, Jan. 11, 1841.

From "fer oligiste" of Semur. It is impossible to say from the brief description given, whether this shell be a distinct species or not.

6. *Unio depressus*, Zieten, Verst. Wurt. pl. 61. f. 1.

From Dejerloch near Stuttgard. Probably referable to variety 1 of *C. listeri*.

III. *Species apparently referable to other genera.*

1. *Venulites trigonellaris*, Schloth. Petref. p. 198; *Cytherea trigonellaris*, Goldf. Petref. pl. 149. f. 5.

From the lias of Alsace; perhaps not a Cardinia.

2. *Unio abductus*, Phillips, Geol. of Yorksh. pl. 11. f. 42.

From inferior oolite of Glaizedale. Possibly a Cardinia, but M. Agassiz regards it as a Gresslya.

3. *Cardinia quadrata*, Agass. Et. Crit. Moll. pl. 12''. f. 10–12.

From lias of Lower Rhine. The above figure appears to represent an Astarte, and much resembles *A. lurida*, Sow.

4. *Unio listeri*, Goldf. Petref. pl. 132. f. 1.

This seems to be the *Amphidesma donaciforme* or *rotundatum* of Phillips, and belongs to the genus Gresslya, Agassiz.

5. *Unio uniformis*, Sow. Min. Con. pl. 33. f. 4.

6. *Unio acutus*, Sow. Min. Con. pl. 33. f. 5, 6, 7.

The last two species, said by Sowerby to be from the middle oolite, are referred to Cardinia by Agassiz, in his translation of the 'Mineral Conchology.'

7. *Pachyodon hamatus*, Brown in Ann. Nat. Hist. vol. xi. pl. 16. f. 6.

From Oxford clay of Gristhorpe Bay, and certainly not a Cardinia.

8. *Pachyodon vetustus*, Brown in Ann. Nat. Hist. vol. xi. pl. 16. f. 7.

From shale at Gristhorpe Bay, and probably not a Cardinia.

9. *Unio striatus*, Goldf. Petref. pl. 132. f. 3.

From coral rag, Nattheim.

10. *Unio liasinus*, Zieten, Verst. Wurt. pl. 61. f. 2; Bronn, Lethæa Geogn. pl. 19. f. 17.

From Fildres near Stuttgard. This is evidently a Gresslya, allied to *Amphidesma rotundatum*, Phillips.

15. ON CERTAIN CALCAREO-CORNEOUS BODIES FOUND IN THE OUTER CHAMBERS OF AMMONITES.

[From Journ. Geol. Soc. vol. i. p. 232, 1845.]

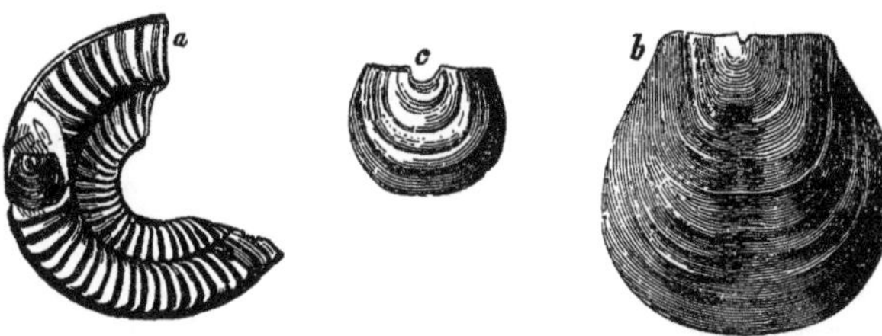

a. A portion of the broken shell of an Ammonite, with the calcareo-corneous body *in situ*, reduced to one-fourth natural size.
b. The body in (*a*), of the natural size.
c. Another similar body of a different species, also of the natural size.

In 1841, Miss Anning, of Lyme Regis, drew my attention to some black-coloured substances which she had occasionally met with in the interior of the *Ammonites bucklandi*, and which she considered to indicate the presence of an ink-bag in the animal of the Ammonite, corresponding to that of the Sepiadæ. From these and other specimens, it appeared to me evident that these substances had constituted, not an ink-bag, but a laminar appendage to the animal, adapted to discharge some unascertained function. The specimens presented the appearance of a very thin concave shell, glossy on its outer surface, with irregular concentric undulations, crossed by longitudinal striæ and fine irregular oblique wrinkles. In the middle of the external margin is a large undulation or sinus. The inner surface, as exhibited by its cast, is of a dull black, the outer surface of the shell being of the colour of horn. Miss Anning informed me that these bodies generally occur about the middle of the outer chamber of the Ammonite, whence they are obtained by breaking the fossil; but as this process more frequently destroyed than exposed the object of search, I was unable, during my stay at Lyme Regis, to procure any tolerably perfect specimens, or to arrive at any satisfactory conclusion as to their nature.

In 1843 my attention was again called to the subject, by finding in a bed of lias limestone, at Temple Grafton and Bickmarsh, near Bidford in Warwickshire (a bed remarkable for the variety of fish, plants, insects and crustacea which it contains), several anomalous bodies whose characters were difficult to define. These substances are of a nearly semicircular form, very thin, slightly concave, presenting a small notch at the middle of the straight side, and having their surface covered with irregularly wrinkled lines of growth, concentric to the notch above mentioned. From the same point of departure also proceed fine radiating lines, visible only with the help of a lens. The colour is usually black, but they sometimes present a browner tint, as if from a mixture of calcareous and carbonàceous matter. The usual diameter is from half to five-eighths of an inch. (See fig. *c.*)

In speculating on the nature of these bodies, although the black colour seemed to indicate a vegetable origin, yet the concentric lines of growth appeared so evidently allied to the structure of molluscous shells, that I could not hesitate to seek for their affinities among the latter class of animals. Indeed the general aspect is so much like that of an Orthis, that had they been found in a Palæozoic rock, I should probably have referred them to the Brachiopoda. But on closer examination it was evident that these bodies were very little, if at all, calcareous, and that though their mode of growth was similar to that of shells, yet their composition was in great measure corneous, and probably elastic, like the plate in the genus Laplysia. It seemed therefore likely that they were part of the internal organization of some mollusk, and on comparing them with the bodies before mentioned, as occurring in the Ammonites of Lyme Regis, it seemed not improbable that they were of a similar nature. Now the bed of lias in which these substances occur, contains two species of Ammonites, the *A. planorbis,* Sow., and another allied to *A. conybeari;* and the dimensions of these Ammonites are such as would very well permit the bodies in question to be contained in their outer chamber. The form, too, of the bodies is nearly that of a transverse section of the chamber of the Ammonite, so that they might easily close it in the manner of an operculum. From these considerations, the most probable supposition seemed to be, that the detached substances in the Bidford lias were portions of the animals of the Ammonites which occur in the same stratum.

This conjecture has been recently verified by finding a very interesting specimen. It is a species of Ammonite, allied to *A. turneri,* Sow., but as yet, I believe, unnamed, which occurs in a bed of clay at Defford Worcestershire, near the middle region of the lower lias. By a fortunate fracture, this specimen exhibits, imbedded in the stone which fills the outer chamber, a substance evidently identical in its nature with those just described. It lies with the convex surface outwards, and the straight side turned towards the mouth of the shell. The portion exposed to view is the cast of the interior surface, which is somewhat irregularly waved, but exhibits distinct concentric lines of growth. The whole of this inner surface is black like the Bidford specimens, but portions of the substance itself, which still adhere to the cast, are white and calcareous, showing that in this species, at least, the body was of a shelly nature. The slight portions which remain of the outer surface of this thin calcareous plate exhibit fine lines, radiating rather irregularly from the centre of the straight side, in which there is a very small but deep emargination or notch. (See the figure *b*.)

Judging from the specimens thus repeatedly obtained within the outer chamber of several species of Ammonite, there can be no reasonable doubt that these bodies were appendages of the cephalopodous mollusk which inhabited those shells. I leave it to more expert comparative anatomists to pronounce as to the precise nature of these corneo-calcareous appendages, which were possibly the representatives of the horny girdle described by Professor Owen as occurring in the recent Nautilus, and which aids in the attachment of the animal to the shell. They may also possibly be the equivalents of that "ligamento-muscular disc" which protects the head of the recent Nautilus.

These singular bodies may perhaps throw light on the nature of that much-disputed fossil the Trigonellites or Aptychus. I am aware that Professor Forbes has recently seen some reason for referring the latter fossil to the existing Holothuriadæ. But as this supposed affinity is as yet far from being demonstrated, I may be allowed to remark that the two valves of Trigonellites, when expanded, closely resemble in appearance the univalve disc which I have been describing; and when we recollect that Trigonellites have hitherto only been found in formations which also contain Ammonites, and that they have in several instances been found in the interior of Ammonites precisely as in the case of the bodies before us, there is, I think, a fair presumption that these singular bodies are allied in origin and in function to the remarkable fossils here described.

On referring to a paper communicated by M. Voltz to the Natural History Society of Strasburg, which will be found in the 'Institut' for 1837, p. 48, it appears that he was acquainted with fossils similar to those before us, and that he also considered them to be allied to Trigonellites or Aptychus. He divides the Aptychi into three groups, *A. cornei*, *imbricati*, and *cellulosi*, the former of which differs from the two latter (which are *calcareous* and *bivalve*) in being *corneous* and *univalve*, both which characters are applicable to the fossils which I have above described. He supposes that in the corneous species a certain degree of motion was effected in the two halves of the body by means of its own elasticity, while in the calcareous groups the same end was obtained by means of a bivalve structure. He enumerates five species of the corneous group, all of which are from the lias and inferior oolite, and which, like the imbricate and cellulous species, are occasionally found in the interior of Ammonites, occupying a symmetrical position, and corresponding in their dimensions to the shell in which they are found. From these and other reasons, M. Voltz regards the whole of this group of fossils as appendages to the animals of Ammonites, a view which is confirmed by the facts adduced in the present communication.

16. ON THE GEOLOGY OF THE OXFORD AND RUGBY RAILWAY.

[From Proc. Ashmol. Soc. p. 192, 1848.]

This railway follows the valley of the Cherwell, which is deeply denuded through the oolitic strata into the lias, which last formation is thus brought within twelve miles of Oxford. The railway cuttings exhibit in succession (beginning at the north) the lower lias, marlstone, inferior oolite, great oolite, cornbrash, and Oxford clay. The *lower lias* is exposed in a cutting about three-fourths of a mile north-west of Heyford Warren. It here consists of blue clay, with occasional nodules of limestone, containing beautiful Ammonites, Plicatula, Coprolites, and other fossils. The *marlstone* is a formation of sandy or marly rock of a rusty-brown colour, which is extensively spread over the country about Deddington and Banbury, and forms the summit of Edge Hill. It is well exposed in a cutting in Steeple Aston parish. Two strata of stone are here exhibited, the upper about 2, the lower 3 or 4 feet thick, separated by a bed of bluish sandy clay. The lower bed consists of enormous roundish flattened concretions 10 or 12 feet in diameter, and exceedingly hard. Many hundreds of tons of this rock have been removed by blasting, and are now lying on the sides of the railway. They contain an abundance of fossil shells, especially the *Gryphæa gigantea,* characteristic of the marlstone, also Pecten, Terebratulæ, Helicinæ, Dentalium, Fucoids, &c. About thirty-two species of shells were here procured, and many more might doubtless be obtained by searching. The clayey beds occasionally contain septaria, the veins in which are filled with sulphuret of iron, which is remarkable for displaying the most beautiful iridescent colours. The next stratum above the marlstone is the *upper lias,* which probably exists about Rousham, but is not exposed on the line of the railway. About a mile south of Rousham the railway again cuts into a hill, and exposes some beds of siliceous sand of various colours. Above this is a bluish clay, with numerous small oblong Ostreæ. These beds are probably referable to the *inferior oolite,* for in the next cutting, a little north of Nethercot, we find some light-coloured beds which seem to belong to the *great oolite.* They abound with fossils, especially Pholadomya, Terebratula and Modiola. Eighteen species of fossils were here found. The next locality described is the railway-cutting at Enslow Bridge, more commonly called Gibraltar, from a public-house adjacent. At this place the *cornbrash,* or uppermost bed of the great oolite, rises from beneath the *Oxford clay,* and presents the following section:—1. Bluish clay, about 6 feet. 2. Shaly oolitic stone, about 3 feet. 3. Hard oolitic stone, blue in the middle, 1 to 3 feet. 4. Dark blue shaly clay, 6 feet. 5. Hard oolitic rock, 5 feet, passing downwards into soft white limestone with numerous shells, especially a large smooth Terebratula. The stony beds here contain Corals, Serpulæ, fragments of wood, and numerous fossil shells. The clay-bed No. 4 contains an abundance of fossil wood, and is further remarkable for the gigantic bones of Saurians occasionally found in it. They belong to the Cetiosaurus, a genus of extinct reptiles, probably allied to the crocodile, but equal in dimensions to a whale. This genus was established by Professor Owen, on the authority of some bony fragments procured partly at this very locality, and partly near Chipping Norton, and now in the Geological Museum of the University. But a larger and more perfect bone has lately been obtained. It is an entire femur, 4 feet 3 inches in length, and is probably one of the largest fossil bones in existence*. It is accompanied by a smaller bone, found at the same time, and probably belonging to the same individual. The large bone has undergone great fracture and compression, from the enormous pressure to which it has been subjected. To comprehend this amount of pressure, we must conceive a time when the oolite of Oxfordshire was covered up by the incumbent beds of the oolitic, wealden, cretaceous and tertiary series, which have since been removed by denudation. The vertical thickness of this mass of strata cannot have been less than one-third of a mile, and as this fossil bone presents a surface of about 4 square feet, it must have sustained the pressure of at least 250 tons. Had it been exposed on a hard surface to such a weight, the bone would have been crushed to powder, but being packed in a dense stratum of clay, it merely underwent the same amount of compression with the stratum in which it was imbedded.

* This bone was presented by Mr. Strickland to the Geological Museum in Oxford.—Ed.

17. ON TWO SPECIES OF MICROSCOPIC SHELLS FOUND IN THE LIAS.

[From Journ. Geol. Soc. vol. ii. p. 30, 1845.]

a. Orbis infimus, Strickland *. *b. Polymorphina liassica*, Strickland.

N.B. The figures are greatly magnified, the small dot being the natural size.

THE shells of the microscopic order Foraminifera, which occur so abundantly in the cretaceous and tertiary series, are found much more rarely as we descend through the secondary formations. Examples of them have indeed occurred both in the oolitic and the palæozoic series, but I believe that these fossils have not hitherto been met with in the lias.

The Rev. P. B. Brodie, while pursuing his interesting and successful researches into the fossil insects of the lower lias, was the first to discover these minute objects. In a bed of yellowish shaly stone, a few feet above the "insect limestone" of Wainlode Cliff, Gloucestershire, he detected small white dots, about $\frac{1}{50}$th of an inch in diameter, which when examined by a powerful microscope prove to be discoid spiral shells, apparently unattached, with five or six smooth, rounded, narrow volutions, devoid of striations or any other distinctive characters. As there are no traces of concamerations, we perhaps ought to refer them to the Serpulidæ rather than to the Foraminifera, although their extreme minuteness would point to the latter family as a more probable clue to their affinities. It has been suggested to me that their characters resemble those of the genus Orbis of Lea, and I will therefore denominate the fossil provisionally *Orbis infimus* (see figure *a*). These fossils also occur in a similar bed of shaly limestone near the base of the lias at Cleeve Bank, between Evesham and Bidford. Along with them I here obtained one specimen of an equally minute fossil, which exhibits more decidedly the characters of the group Foraminifera, and is referable to the family Stichostega. It is an oval body, pointed at both ends, smooth and glossy, convex, divided into three concamerations, the largest of which extends all the length of the shell. As the aperture is not visible, it cannot be referred with certainty to any of the known genera of Foraminifera, but in general form it approaches sufficiently to the genus Polymorphina of M. d'Orbigny, and we will therefore denominate it *Polymorphina liassica* (see figure *b*).

* Now *Spirillina infima* (Strickl.).

18. ON THE RESULTS OF RECENT RESEARCHES INTO THE FOSSIL INSECTS OF THE SECONDARY FORMATIONS OF BRITAIN*.

[Read at the Meeting of the British Association in Cambridge, 1845.]

Fossil insects were, till recently, very little known in the secondary rocks of Britain, and the only examples were those from the Stonesfield slate, one from the lias, and a few from the coal-measures. The very large additions to our knowledge of fossil entomology, made by the Rev. P. B. Brodie, have been derived from two principal groups, the Wealden and the Lias. In the Wealden no less than seventy-four insect forms have been described and figured by Mr. Westwood from Mr. Brodie's specimens. These are generally remarkable for their small size, from which, and from their zoological characters, Mr. Westwood infers that they belong to a temperate climate. The gigantic beetles, locusts and Cicadæ of our modern tropics are here wanting, and the specimens consist, with very few exceptions, of small Curculionidæ, Tipulæ, Libellulæ and Aphides, such as swarm at this moment in European climates.

This then is a very remarkable fact, when taken in connexion with the gigantic reptiles and remarkable forms of vegetable life which occur in the Wealden formation, and which by analogy we must refer to a tropical climate. We must either suppose, what is scarcely conceivable, that insects of European forms could coexist with tree-ferns and other tropical productions, or what is perhaps more probable, that the insects of a cooler climate floated down some vast river into the great Wealden estuary, just as the insects of Upper Canada or the Rocky Mountains might be carried by the Mississippi, in the present day, into juxtaposition with the alligators and palm-trees of the Gulf of Mexico. A similar anomaly is presented by the insects first discovered by Mr. Brodie, and afterwards collected by Mr. Hope, the author, and others, in the lower lias of Gloucestershire and the adjacent counties. Of many hundred specimens examined by Mr. Westwood, the whole present indications of a temperate climate, a conclusion wholly opposed to that which we are accustomed to draw from the vertebrate and molluscous fauna of the same epoch. We must here, as in the case of the Wealden insects, reconcile this apparent discrepancy by supposing that the insects were drifted from cooler climates to the spots where we now find them. There are probably no organic bodies of such delicate structures which are capable of floating to so great distances as insects; their extreme lightness and the strong materials of which their corneous parts consist, would enable them to float down rivers, and to be diffused far and wide

* Reports Brit. Assoc. vol. xiv. Sects. p. 58.

over the sea, there to be imbedded with truly marine products. In conformity with this view, we find that the insects of the Wealden, and still more so of the Lias, consist chiefly of Coleoptera and other strongly compacted forms, that they most commonly present only detached portions of the entire insect, and such portions (chiefly wings and wing-cases) as are the most compact and durable. There is therefore no doubt that these insect-remains have been drifted from the land into the sea, in other words from higher ground to lower, and we have only to suppose that the original habitat of these insects was sufficiently elevated to supply them with a cool or temperate climate, and the whole difficulty is removed.

Another very unexpected result of the examination, by a skilful entomologist, of these fossil insects, is the remarkable affinity which they present to existing forms: even in so ancient a deposit as the lias, we find no insects of decidedly new types of organization; they are in almost every instance referable to families, and frequently to genera, which belong to the existing fauna. In one instance only has Mr. Westwood ventured to propose a new generic name, and it is remarkable that the peculiar form so indicated is common both to the Wealden and the Lias. It would appear, therefore, that from the time of the Lias to the present day, the class Insecta has undergone a far less amount of alteration, either by the extinction of old forms or the introduction of new, than any other large group of the animal or vegetable kingdom with which we are acquainted. It was indeed well known that the different classes of the animal kingdom vary greatly in what we may call their amount of durability; that the higher groups of Vertebrata, for instance, present a rapid succession of forms as we descend the chronological scale, while certain molluscous and infusorial structures are continued with little or no change during vast geological periods; but perhaps there is no other instance of so remarkable a persistency of character in a whole class of animals, as that which is presented to us in comparing the insects of the Lias and Wealden with those of the existing fauna.

---

19. NOTES ON A SECTION OF LECKHAMPTON HILL*.

[From Journ. Geol. Soc. vol. vi. p. 249, 1850.]

THE well-known promontory of the Cotswolds, called Leckhampton Hill, affords probably the best locality in Gloucestershire for studying the relative position of the various beds of the inferior oolite and subjacent lias. From its great height and steepness, the entire series of the oolite is admirably exposed to view, and the extensive quarries from which Cheltenham has been mainly built, afford every facility for examining the formation. The geologists of Cheltenham and the neighbourhood are well acquainted practically with the subdivisions of the strata and their organic contents, but no exact definitions or precise measurements of these strata have, I believe, ever yet been made. Mr. Buckman's 'Chart' contains a section of Leckhampton Hill, and is valuable for its descriptions of the mineral character and organic remains of some of the beds. But it does not attempt to exhibit all the subdivisions of the oolite, or to show their absolute and relative thicknesses. It appeared therefore desirable that a more elaborate survey should be made, and with this object I gladly availed myself of the aid of the Rev. A. D. Stacpoole of Oxford, whose skill in the use of surveying instruments was of great service. Where the strata were exposed in a vertical escarpment and were accessible, they were accurately measured, but in other cases their thickness could only be ascertained by means of the sextant. By putting together the varied observations thus obtained, we constructed the accompanying section†. It is drawn on a uniform scale of height and distance, so that it exhibits the precise profile of Leckhampton Hill, without those exaggerations seen in others where two scales are adopted. The height of this hill above the sea has been ascertained by the officers of the Ordnance Survey to be 978 feet (see 'Annals and Magazine of Natural History,' second series, vol. v. p. 257). I will now proceed to enumerate the strata here observed, adopting, as far as they go, the names employed in Sir R. Murchison's 'Geology of Cheltenham.'

* The sands classed in this paper in the inferior oolite are now by some geologists considered as forming the upper strata of the upper lias (Wright, Journal of the Geol. Soc. vol. xii. p. 292). Mr. Strickland's was the first accurately measured section of these strata.—A. C. R.

† This section was given with Mr. Brodie's paper "*On certain beds in the Inferior Oolite near Cheltenham, by the Rev. P. B. Brodie, M.A., F.G.S. With Notes on a section of Leckhampton Hill, by H. E. Strickland, M.A., F.G.S.*" Printed in the Quarterly Journal of the Geol. Soc. vol. vi. p. 239[1].

[1] See Introduction to collected papers.—ED.

(1.) Trigonia grit: so called from the abundance of Trigoniæ which it contains. It is better exhibited in other localities, especially at Cold Comfort Farm, but it may be seen in the quarries on the highest part of Leckhampton Hill, its thickness being about 7 feet.

(2.) Gryphite grit: characterized by the very peculiar and conspicuous shell "*Gryphæa cymbium.*" Extensively quarried on the summit of the hill, and used for roads, building walls, &c. Thickness 7 feet.

(3.) Beds of brown rubbly oolite: not employed for any purpose, but exposed between the inclined planes and the Cirencester road. It contains a considerable variety of fossils. Thickness 24 feet.

(4.) Oolitic freestone: too fragmentary to be of much value for building, and apparently not distinguished by any peculiar fossils. 26 feet.

(5.) Oolite marl: consisting of soft, whitish oolitic stone containing much aluminous matter and some beautifully preserved fossils, and especially characterized by the very peculiar *Terebratula fimbria.* 17 feet.

(6.) Beneath the marl, the strata for more than 100 feet consist of compact light-coloured oolite. The upper portion forms the best building-stone, and has been extensively quarried; hence it is more especially distinguished by the name of "freestone." Its thickness at the point where we measured it was 31 feet 6 inches. The lower part of this oolitic mass is coarser and more variable in texture, and is hence more rarely quarried. Numerous and beautiful fossils may be procured in it, but only by great patience and perseverance. The resemblance of these and their matrix to those of the great oolite on Minchinhampton Common is very remarkable, and caused considerable discussion among geologists until the distinctness of the two strata was absolutely demonstrated. This part of the series is 75 feet thick, and is locally known by the name of "Roestone."

(7.) Ferruginous beds, consisting of coarse oolite in the upper part, and of the very peculiar, large-grained oolite or pisolite ("pea-grit") in the lower. A few miles to the south the pisolite disappears, and is replaced near Painswick and at Haresfield Hill by strata containing ferruginous oolitic grains in a brown paste. This is the precise equivalent of the well-known oolite of Dundry near Bristol, which may be recognized as far off as Bridport on the Dorset coast. At Leckhampton the pisolite rests on a few feet of ferruginous oolite and sand. The total thickness of this portion of the series is 42 feet.

(8.) Immediately below this sand we find in the gravel-pits on the side of the hill, an abrupt transition to beds of bluish clay. This is the uppermost portion of the upper lias, the thickness of which cannot be easily measured for want of sections, but which may be estimated at 180 feet.

(9.) Marlstone: this stratum, so rich in fossils at Bredon and

Alderton Hills, is not exposed to view on Leckhampton Hill, but is probably indicated by the low ridge which rises to the south of Leckhampton Church. Its thickness may be estimated at 50 feet.

(10.) Lower lias: it would be a matter of great difficulty to determine accurately the thickness of this formation, from the base of the marlstone to the top of the new red marls. Its depth is undoubtedly very great, and it is divided into several distinct zones, well marked by peculiar fossils. From an examination of numerous sections made during many years, I am disposed to estimate its total thickness at probably between 500 and 600 feet.

By adopting Leckhampton Hill as our standard of comparison with other places where the inferior oolite is exposed, we may trace (as Mr. Brodie has already done) the enlargement of some strata, the thinning out of others, and the introduction of new ones. We shall find also considerable variations in the organic contents of the same stratum at different localities, whereby we may be enabled to study the causes which operated in the distribution of animal life in the ancient seas.

20. ON THE ELEVATORY FORCES WHICH RAISED THE MALVERN HILLS.

[From the Philosophical Magazine, 4th ser. vol. ii. p. 11. pl. 1, 1851.]

PROFESSOR PHILLIPS has already pointed out (Mem. Geol. Survey, vol. ii. p. 5) that the syenitic ridge of the Malvern Hills forms a part of a great line of dislocation, extending for at least 120 miles from Flintshire on the north to Somersetshire on the south. He shows that this line of disturbance forms the eastern boundary of that vast region of elevation which includes the whole of Wales and part of Southern Ireland, and that the principal movement which caused this elevation took place between the Carboniferous and Triassic epochs*. He further shows that this line of fracture, bounding the elevated region on the east, partakes throughout the greater part of its course of the nature of a fault; that this fault is on an enormous scale in its vertical and horizontal dimensions, and that it is much concealed by the thick deposits of new red sandstone which have covered it up on the downcast side, and followed the sinuosities of its course.

The demonstration of so vast a line of disturbance, evidently due to one set of operations acting at a very remote epoch, enormous in dynamic amount, yet comparatively limited in their duration, is one of the grandest generalizations at which British geologists have arrived. The nature of the movement which has produced these results seems consequently to deserve a fuller investigation than it has yet received.

These disturbing forces appear to have been partially continued during, and even after, the deposition of the new red sandstone. Both that and the incumbent lias show proofs of elevation and of dislocation, which may be regarded as the expiring efforts of those vast forces which raised the mountains of Wales above the plains of England. Indeed the general south-easterly inclination of the whole secondary series of Southern England is a further proof of the continuation of these elevating move-

* We cannot speak more precisely as to the date of a convulsive movement which perhaps extended over a considerable period. According to the researches of Sir R. Murchison in other regions, an entire geological epoch—that of the "Permian System"—intervened between the Carboniferous and Triassic systems. But deposits of this age are scarcely, if at all, traceable in the region here described; and we cannot therefore assert whether the Malvern ridge was elevated at the beginning, the middle, or the end of the Permian epoch. From the conformability, however, of the "Lower New Red Sandstone" to the coal-measures in Staffordshire and Shropshire, and its unconformability to the Triassic or Upper New Red Sandstone, we may consider the conclusion of the Permian epoch as the probable date of this event. (See Murchison's Silur. Syst. p. 131.)

ments down to a late geological date. But all these more recent changes of level were so feeble in amount compared to the vast convulsions of the pre-triassic period, that we may eliminate them altogether from our present inquiry. We shall gain clearer notions by supposing the new red sandstone and all the superior formations entirely removed, and by endeavouring to decipher the state of things which immediately preceded the deposition of those strata.

Of the whole line of dislocation above mentioned, the ten or fifteen miles which include the Malvern and Adderley Hills probably afford the best information on this subject. The syenitic axis of Malvern, eight miles long, about half a mile wide, and almost perfectly straight, naturally suggests the idea of a vast dike of injected trap rock. But Professor Phillips has successfully shown, from the absence of lateral ramifications of syenite, from the rare and slight indications of metamorphic action, and from other phænomena, that this plutonic ridge must have been elevated in a solid state. Indeed the fact that it occurs, not on a line of simple fissures, but on a line of fault, is conclusive of its having been elevated as a solid; for the downcast side being lower by several thousand feet than the upcast, the syenite, if fluid, could not have been raised to its present position, but would have overflowed the downcast side to a great distance.

Admitting this wall-like mass of syenite to have been forced up from below in a solid state, we at once obtain a clue to the vertical or highly inclined (sometimes reversed) position of the sedimentary strata on the west, or upside cast of the Malvern ridge.

It appears, then, that the Malvern district, though forming part of a great line of fault, yet exhibits the phænomena of a fault under a very complicated aspect. To explain this I must refer for a moment to a few elementary principles.

In the simplest form of a fault, when one portion of a horizontal stratum is elevated by an equally diffused pressure from below, while the other portion remains at rest, the stratum preserves its horizontality up to the very plane of separation; or, more frequently, the friction of the two masses causes the strata to bend slightly *towards* each other on the opposite surfaces of the fault. Again, if the upward pressure be confined to a *line* instead of being spread over a *surface*, the strata are thrown in opposite directions, and an anticlinal is the result.

But the Malvern region presents us with a combination of both these kinds of forces, and of both their resulting phænomena. There has been an elevatory force diffused more or less equally under a vast area, which has heaved up in a mass the entire region for hundreds of miles to the westward of the Malvern axis. And there has also been a local force applied immediately beneath this axis, which has given an extra amount of elevation to the marginal portion of the upcast area.

It is this excessive development of motive force *at the very margin* of an elevated region, and in immediate contact with a non-elevated tract, that renders the phænomena of the Malvern Hills peculiarly anomalous. Under ordinary circumstances, when an upward force is applied locally along a line, it acts equally on both sides of that line, elevating the strata, as already shown, into an anticlinal position. If, however, the resistance be greater on one side of the axis than the other, a certain amount of displacement ensues, and the anticlinal arrangement is combined with that of a fault. The Malvern elevation is probably an extreme and unusually exaggerated instance of the last class of phænomena. If we could strip off the thick mantle of new red sandstone which conceals the eastern side of this axis, we should probably find the strata from the Caradoc sandstone up to the coal-measures more or less upturned at their edges. Plate IX. So vast a force as was required to elevate the syenitic axis could hardly have failed to shatter and twist up the margin of the deposits on its eastern or downcast side, although their amount of statical resistance was such as to forbid any general elevation of them *en masse*.

Assuming that such was the condition of things in this region before the deposition of the new red sandstone, let us endeavour to trace the mode of action of the forces which produced it.

There is evidence that elevatory movements have taken place along the axis of the Malvern chain before, as well as since, that great and transient outburst which dates between the Carboniferous and Triassic epochs. A mass of syenitic rock had been elaborated by igneous agency beneath this tract in very remote geological times. It had become solidified, and had been elevated above the oceanic surface before the Upper Silurian formations were deposited. The sections on the west side of the Malvern Hills show that the mollusca and corals of the Caradoc sandstone lived and flourished in immediate contact with the plutonic rock, and that pebbles of the latter were rolled into the sea of that period, and were there imbedded in company with the animal remains. (See Mem. Geol. Surv. vol. ii. p. 33.) We may therefore suppose that at this period a state of things prevailed such as is here represented.

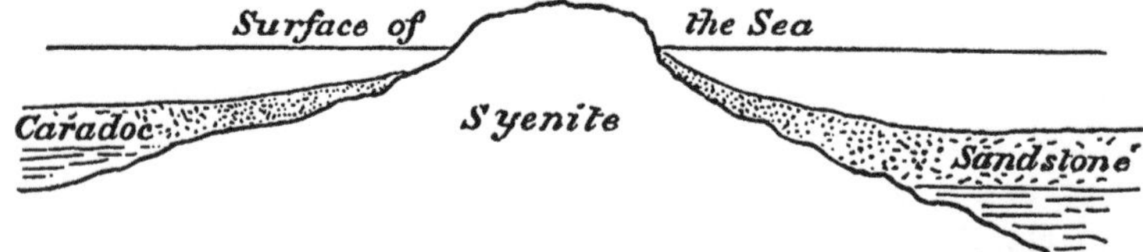

In other portions of the Welsh region we find similar proofs of elevations having taken place in remote palæozoic times. Thus at Bishop's Castle, and in the country to the north-west of it, the Caradoc sandstone is found to lie unconformably to the subjacent rocks; and the Wenlock shale in the same way overlaps the Caradoc sandstone near Bishop's Castle and Builth. (See Ramsay in Journ. Geol. Soc. vol. iv. p. 296.) Some of these ancient disturbances were probably connected with those referred to in the Malvern district. But this most ancient elevation of the syenite seems to have been comparatively small in amount, and was wholly covered up by the formations which succeeded the Caradoc sandstone, and which contain no fragments of syenitic rocks. In order to explain the changes which now took place, it may be legitimately assumed that the floor of solidified syenite on which the sedimentary deposits rested was itself underlaid by igneous rock in a fluid and active state. Let it be further granted, that the present *breadth* of the Malvern syenite (averaging half a mile) approximately represents the *thickness* of the upper or solid portion of the plutonic rock. Such I assume to have been the condition of things when that great elevatory movement commenced which upheaved the westernmost side of our island. It is irrelevant here to inquire whether this general upheaval was effected by the mere expansion caused by increased temperature, or by the introduction from other quarters of vast masses of fluid matter beneath the elevated area. It will be sufficient to admit that a special volcanic focus existed beneath the syenitic axis of Malvern, and that its energies were called into action simultaneously with the more general movement which elevated the area of Herefordshire and Wales.

We may now suppose that the elevatory forces beneath the Cambrian region had accumulated so as to overcome the superincumbent weight; while the region to the eastward, either from its greater rigidity, or from the less amount of subjacent force, remained in a quiescent state. A separation would now take place between these two areas; a long and sinuous line of fracture would divide them; and the region where force had overcome resistance would begin to rise higher and higher above the area which remained unmoved.

The previous elevatory movement which has been shown to have existed along the Malvern axis probably rendered this a weak point in the earth's crust, and caused the line of fracture to coincide with that axis. As soon as one side of this line began to rise and a fault to be produced, the volcanic forces which had been pent up beneath the syenitic axis would now find, or endeavour to find, a vent. Struggling to escape along the line of fault, they would thrust up the solid syenite above them, raising it into a lofty cliff above the downcast area, and elevating, overturning, or crumpling up the edges of the Silurian, Devonian and carboniferous strata which rested upon it. Plate IX.

In the above diagram I have taken as a basis the section across the Worcestershire Beacon published by the Geological Survey, and have endeavoured to supply conjecturally those portions of the strata which have been removed by denudation, or which lie too deep to be visible. I have supposed that a vast mass of Devonian and carboniferous rocks has been upheaved bodily, while the lower strata nearer the syenite are more or less fractured, crushed and contorted. The thickness of the strata which have been since denuded may appear enormous; but it is founded on the careful measurements of the Geological Survey, which give about 5500 feet for the old red sandstone of Herefordshire, and 3500 for the incumbent carboniferous series at the nearest point (Dean Forest), where the undulations of the beds have saved them from denudation. As, however, the coal-fields of Wyre Forest and the Clee Hills on the north present a less development of the series than is seen in Dean Forest, I have reduced the thickness of the coal-measures and carboniferous limestone which once existed on the west of the Malvern Hills to about 2300 feet. Adding these amounts to the thickness of the Upper Silurian, Caradoc sandstone and syenite, we obtain a total of at least 13,000 or 14,000 feet for the amount of dislocation between the two sides of the great fault; an amount greater, perhaps, than can be paralleled in any other instance of a single fault which the world can produce. Nearly one-half of this amount may, however, be assigned to the more local forces which elevated the Malvern syenite; so that about 7000 or 8000 feet would represent the difference of level between the strata in the less disturbed parts of Herefordshire west of Malvern, and their equivalents now buried beneath the new red sandstone of Worcestershire, allowing about 1000 feet for the thickness of the latter down to the subjacent coal-measures.

The fluid matter which I suppose to have thus forced up the solid syenite may itself have never reached the surface. The plutonic axis of Malvern seems only to exhibit its upheaving effects, and shows no signs of fluid ejections contemporaneous with the elevation. It is possible, however, that volcanic matter may have poured out over the downcast area, where it is now concealed by the new red sandstone. And the laterally injected dike of Brockhill, as well as the trappean masses in the black shales on the west of Ragged Stone and Midsummer Hills, are not improbably connected with the volcanic forces which thrust up the syenite. This supposition appears to me at least equally probable with that of Professor Phillips (Mem. Geol. Surv. vol. ii. p. 56), that these greenstone eruptions were contemporaneous with, and overlaid the black Caradoc shales with which they are in contact. In the arrangement of the strata around Eastnor Park, we seem to have indications of a crater of elevation, caused by an incipient volcanic eruption whose focus never reached the surface. The great expansion and crumpled condition of the Silurian rocks at Ledbury, and their general semicircular arrangement round a central point, indicate a local and special development of volcanic energy beneath. But the syenitic axis itself affords no more signs of eruptive force at this point than at any other. The efficient force seems to have acted not in, but at the west side of this axis. A mass of basaltic matter ejected beneath the Caradoc sandstone will explain these phænomena. Its ramifications would be likely to select the black shales as being less resisting than the sandstones above and below them, and would produce that series of trappean dikes which Professor Phillips was the first to describe. By penetrating the shales (as trap-dikes often do) in the planes of stratification, they would produce an appearance of contemporaneity, though their real dates might be long subsequent.

The district here referred to seems to be exactly analogous to the well-known *elevation crater* of Woolhope, distant only seven or eight miles to the westward, in which we also see the ineffectual struggles of a focus of volcanic energy to burst through the incumbent strata. Here also the concealed volcano has left a collateral proof of its existence in the single basaltic dike of Bartestree Chapel.

These detached indications seem to show that the volcanic matter which underlies, and which has elevated this region, is different in mineral character from the more ancient syenite of Malvern, and is probably more allied to greenstone or basalt. The trap-rocks of Wyre Forest, north of Abberley, further corroborate this view.

In tracing to the north or south that long line of dislocation of which the Malvern Hills form a part, we find a continuation of analogous phænomena more or less modified by local circumstances. The Abberley range of hills is, as is ably shown by Professor Phillips (Mem. Geol. Surv. vol. ii. p. 145), completely analogous to the Malvern district; the chief difference being, that the syenitic axis which upheaved the Silurian rocks is here almost wholly concealed from view, and (with one small exception) is only known by its effects. The Silurian and Old Red formations are here, as on the west of Malvern, *overturned* for a distance of several miles. This remarkable phænomenon, may, I think, be explained in a simpler mode than either of those proposed by Professor Phillips. All that is requisite is to resolve a certain portion of the vertical uplifting force into a lateral direction. Now it is certain that an enormous fault-line runs along the eastern side of all the disturbed and elevated district, and that the downcast region on the east has remained relatively rigid and unmoved. In accordance with the well-known law that the plane of a fault (almost invariably) dips *towards the downcast side*, it is evident that this oblique surface would act mechanically as an inclined plane or wedge, in reference to a vertical uplifting of the strata on the west, and would force them over to a certain distance in a lateral direction. Plate IX.*

* A very analogous case occurs at Hohnstein in Saxony, where a mass of granite, upheaved in a solid state, has not only elevated but *overturned* the contiguous

The same lateral force would explain the sharp anticlinal curves into which some parts of the Ridge Hill near Abberley are compressed. (Mem. Geol. Surv. vol. ii. p. 151.)

At numerous other points, as we proceed northwards along the eastern limit of the elevated district, or southwards by May Hill to Tortworth, we find indications of the same great line of fault. Sometimes, as at Oswestry and Higley, these faults have affected the lower new red sandstone as well as the carboniferous rocks, proving that here at least the elevatory movement was subsequent to the commencement of the Permian epoch. Generally the great marginal fault seems to have formed a nearly vertical cliff, against which the upper or triassic portion of the new red sandstone was deposited, as in the Shropshire coal-field, at Bewdley, Abberley, Malvern, May Hill and Pyrton Passage.

The Cambrian and Herefordshire area having now become elevated many thousand feet above the eastern region, and the volcanic forces having spent their energy in thrusting up and overturning the syenite and incumbent strata of Malvern, a period of comparative tranquillity ensued. The elevated region had become dry land, while the downcast area remained beneath the sea. The sands and marls of the triassic series filled up the bed of this sea, while its littoral waves, beating against the syenitic cliffs of Malvern, formed accumulations of conglomerate such as those of Rosemary Rock and the Berrow and Woodbury Hills. The oolitic, cretaceous and tertiary formations were successively piled upon the triassic rocks, and may possibly have raised this downcast area to the same level as the upcast portion, though there is no evidence that they ever overlaid the latter in the region west of the Severn.

The elevated area meanwhile was undergoing a vast amount of denudation. During the long ages of the triassic and oolitic systems, it was doubtless exposed to atmospheric degradation, and supplied the adjacent ocean with much of its sedimentary matter, as has been ably shown by Prof. Ramsay (Mem. Geol. Surv. vol. i. p. 297). The denuding forces which were so active in the pliocene period terminated these vast operations, and gave to this rugged and dislocated area those smooth undulating outlines which it now generally presents.

I trust that I have now in some degree confirmed and extended the proofs adduced by the geological surveyors of the elevation and subsequent denudation of the Cambrian region, and that I have shown how the peculiar phænomena of the Malvern district may be explained by the supposition of a local development of plutonic energy superadded to a more general upheaving force *.

strata, causing beds of the Jurassic series to repose upon cretaceous ones. (See Cotta, 'Geognostische Wanderungen.' Dresden, 1838.)

* The opening of the Malvern Tunnel has confirmed the truth of Mr. Strickland's views regarding the "Malvern fault."

21. ON A PROTRUDED MASS OF UPPER LUDLOW ROCK AT HAGLEY PARK IN HEREFORDSHIRE*.

[*Read June* 16, 1852.]

So laboriously minute have been the researches of the officers of the Geological Survey of Great Britain, that it is only where some fresh sections have been subsequently exposed by the operations of nature or of man that any material additions or corrections of that Survey can be looked for. A case of the kind has lately occurred in Herefordshire, revealing a small protrusion of Silurian rocks in the midst of the old red sandstone, and accompanied by circumstances of some geological interest.

A quarry having been opened near the base of the old red sandstone a few hundred yards west of Hagley House, near Lugwardine, and a deep drain having been cut from the quarry towards the south-east, the junction-beds of the old red sandstone and of the Upper Ludlow rock were unexpectedly exposed. This circumstance attracted the notice of M. J. Scobie, Esq., of Hereford, to whom I am indebted for having my attention drawn to the spot, and for many interesting organic remains and geological details which his residence in the vicinity enabled him to collect. I must also express my obligations, and those of the other geological friends who accompanied me, to Robert Biddulph Phillips, Esq., the owner of the land, who kindly caused part of the quarry to be re-opened for our inspection.

The area of Silurian rocks here exposed on the surface does not exceed three or four acres; it consists of yellowish sandstones referable to the "Downton Sandstones" of Sir Roderick Murchison, resting on grey micaceous Upper Ludlow schists, and dipping on all sides beneath the sandstones and marl of the Old Red series. They seem to form a portion of a very flattened dome, and the quarry, which extends about seventy yards from north-west to south-east, cuts through this dome on its south-western slope. Such at least is the conclusion drawn from the dip of the beds, which at the north end of the quarry is about 10° north-west by west; at the middle of the quarry, 5° west-north-west; about twenty yards further south, 8° west-south-west; and at the southern extremity, 7° south-south-west. The following section is here exposed in descending order, as far as the irregularities of the stratification permit them to be measured.

* Journ. Geol. Soc. vol. viii. p. 381.

		ft.	in.
Old Red Sandstone ..	1. Red marls and clays, containing bands of whitish sandstone, not calcareous, about	12	0
Downton Sandstones..	2. Hard brownish sandstone	2	0
	3. Flaggy, slightly micaceous, brown sandstone..	2	0
	4. Highly micaceous, thin-bedded, brown sandstone..	2	0
	5. Band of clay and rubble, about	0	6
	6. Micaceous yellowish sandstone, with traces of carbonized plants.........................	2	0
	7. Clay and rubble	0	6
	8. Micaceous yellowish sandstone, with numerous fragments of carbonized plants	4	0
Upper Ludlow Rocks.	9. Bones, teeth, and scales of fish, about........	0	1
	10. Grey micaceous shale, effervescing with acid, full of fossils, about......................	4	0
		29	1

The vegetable remains in the beds Nos. 6 and 8 are interesting from their extreme antiquity, but in general present no traces of their organic structure. They are merely rounded and water-worn fragments converted into a coaly mass, which cracks in drying. When ignited, these fragments burn like anthracite, without smoke or flame, and remain ignited until they are reduced to a light white ash. The occurrence of vegetable remains in the corresponding beds at Downton Castle is noticed by Sir R. Murchison* and near Stoke Edith, and in the May Hill district by Professor Phillips†.

The bed No. 9 is interesting as being unquestionably the representative of the "Ludlow Bone Bed," described by Sir R. Murchison‡. His description of this deposit near Ludlow, as "a mass of scales, ichthyodorulites, jaws, teeth and coprolites of fishes, united by a gingerbread-coloured cement," is precisely applicable to the stratum at Hagley. The cement which unites the bones is calcareous and imperfectly crystalline, exhibiting a *chatoyant* lustre when the eye catches the light reflected from the cleavage-planes. This singular deposit of ichthyic remains occurs as a thin band, in some places no thicker than a wafer, and gradually increasing at other points to about an inch and a half in thickness, as if deposited by eddies in shallow depressions of the sea-bottom §. These minute osseous fragments are mostly much water-worn and highly polished by mutual friction. Some of them are black, but the majority are of a yellowish or ferruginous tint. As very few of the bones or teeth are

* Silurian System, p. 197.

† Mem. Geol. Survey, vol. ii. pp. 176, 188, 312.

‡ Sil. Syst. p. 198.

§ Precisely the same conditions exist in the case of the well-known bone-bed at the base of the lias, and are doubtless due to the difference between the specific gravity of the fish-bones and that of the arenaceous grains of the sea-bottom, causing the former to be separated from the latter by the action of the currents.

sufficiently perfect to indicate generic or specific characters, we are only able to enumerate the following:—

Spines of *Onchus murchisoni*, Agass., Sil. Syst. pl. 4. fig. 10.
Teeth of *Thelodus parvidens*, Agass., Sil. Syst. pl. 4. figs. 34–36.
Teeth resembling that figured in Sil. Syst. pl. 4. fig. 37, but serrated at the margin.
Ganoid scales.

The only molluscous remains in the fish-bed are the *Orbicula rugata*, Sil. Syst. pl. 5. fig. 11, and an Orthis.

In some places fragments of coaly matter, similar to that in the bed No. 8, are mixed up with these osseous remains. One of these carbonaceous pellets seems to be the seed of some terrestrial plant. It is globular, about a quarter of an inch in diameter, and being broken across exhibits a central cavity, the parietes of which are about one-tenth of an inch thick, and composed of fibres radiating to the external surface*.

The bed No. 10 corresponds in character with the uppermost strata of the Ludlow rocks wherever they are visible in the neighbourhood. It is a fine-grained sandy shale, of a greenish or greyish colour, abounding with small particles of mica, and effervescing with acids, although not sufficiently calcareous to deserve the name of a limestone.

The following organic remains occur in it at Hagley Quarry:—

Cyathophyllum ?.
Favosites polymorpha, Goldf., Sil. Syst. pl. 15. fig. 2.
Cophinus dubius, König, Sil. Syst. pl. 26. fig. 12.
Crinoidal stems (pentagonal).
Cyathocrinites macrostylus, Phillips, Mem. Geol. Surv. ii. p. 384.
Serpulites longissimus, Murch., Sil. Syst. pl. 5. fig. 1.
Homalonotus knightii, König, Sil. Syst. pl. 7. figs. 1, 2.
Calymene blumenbachii, Brongn., Sil. Syst., pl. 7. fig. 5.
Rhynchonella semisulcata, Dalm. (*Terebratula lacunosa*, Sil. Syst. pl. 5. fig. 19).
—— *wilsoni*, Sow., Sil. Syst. pl. 6. fig. 7 *a*.
—— *nucula*, Sow., Sil. Syst. pl. 5. fig. 20.
Orthis orbicularis, Sow., Sil. Syst. pl. 5. fig. 16.
—— *lunata*, Sow., Sil. Syst. pl. 5. fig. 15.
Strophomena filosa, Sow., Sil. Syst. pl. 13. fig. 12.
Leptæna sarcinulata, Schloth. (*L. lata*, Sil. Syst. pl. 5. fig. 13).
Orbicula striata, Sow., Sil. Syst. pl. 5. fig. 12.
—— *rugata*, Sow., Sil. Syst. pl. 5. fig. 11.
Lingula minima, Sow., Sil. Syst. pl. 5. fig. 23.
Orthonota amygdalina, Sow., Sil. Syst. pl. 5. fig. 2.
—— *retusa*, Sow., Sil. Syst. pl. 5. fig. 5.
Avicula ampliata, Phillips, Mem. Geol. Surv. ii. pl. 23. fig. 1.
Orthoceras bullatum, Sow., Sil. Syst. pl. 5. fig. 29.

* I propose on a future occasion to give a fuller description of these singular bodies, which have since been detected in the same stratum at several other localities.

Orthoceras perelegans, Salter, Mem. Geol. Surv. ii. pl. 13. fig. 2.
—— *ibex*, Sow., Sil. Syst. pl. 5. fig. 30.
—— *gregarium*?, Sow., Sil. Syst. pl. 8. fig. 16.

Many of these fossils, especially the Orthocerata, are penetrated with sulphuret of iron, which gives them a bright metallic gloss.

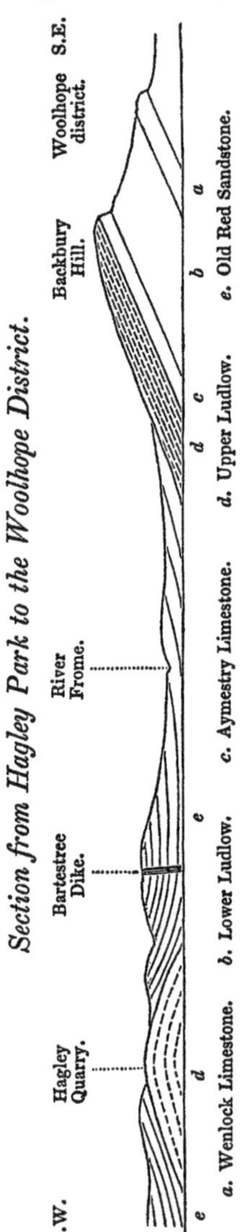

Section from Hagley Park to the Woolhope District.

In addition to the above-mentioned invertebrate forms, an interesting portion of a crustacean has been forwarded to me by Mr. M. J. Scobie from the Upper Ludlow shale underlying the bone-bed, of which Mr. Salter has kindly undertaken the description*.

Traces of ichthyic remains, especially the minute teeth of *Thelodus parvidens*, are occasionally found interspersed in the bed No. 10, but never in the same abundance as in No. 9.

Having now enumerated the strata of Hagley Quarry and their organic contents, I must make a few remarks on the geological phænomena attending them.

It was stated above that the beds here exposed assume the form of a flattened dome. By reference to the Map it will be seen that this protruded dome is about half a mile to the north-west of the well-known dike of greenstone at Bartestree, which cuts through horizontal strata of old red sandstone, and runs in an east-north-east direction towards the south edge of the protruded Silurian mass of Shucknall Hill†. About a mile and a half further to the south-east, we come to the elevated region of Woolhope, the axis of which runs for more than ten miles still in a south-east direction. It appears, therefore, that the ejection of the trap-dike at Bartestree, together with this protrusion of Silurian rocks at Hagley Park, occur exactly on the axial line of the great elevation of Woolhope.

But though this coincidence of position deserves notice, yet the forces which have caused these minor protrusions have in fact acted nearly at right angles to the direction indicated. The Woolhope region, though possessing an axis from north-west to south-

* Description of *Pterygotus problematicus*, Agass., by J. W. Salter, Esq. Quart. Journ. Geol. Soc. vol. viii. p. 386. pl. 21.

† See Murchison, Sil. Syst. p. 185; and Phillips, Mem. Geol. Surv. vol. ii. p. 180.

east, is essentially an *area,* and not a *line,* of elevation. Its pressures have been distributed, not in two opposite directions from an axis, but in every direction from a centre. In conformity with this view, we find that in its north-western portion it is cut through by the "great Mordiford fault," running north-east by east (or nearly at right angles to the major axis), and causing the strata about Dormington and Stoke Edith to assume the same strike. Beyond the Woolhope area we find the valley of the River Frome, the Bartestree Dike, and the protruded Silurian mass of Shucknall Hill assuming almost exactly the same east-north-easterly direction. And in the small dome-like protrusion of Hagley Park, which lies parallel to Bartestree Dike, and precisely in the axis of Shucknall Hill, we find a further proof of the same movement.

It appears probable, then, that the pressure caused by the elevation of the central dome of the Woolhope area, acting in every direction, has on the north-west side caused great undulations in the Silurian and Devonian strata which lie beyond the region of actual elevation. In two instances, that of Shucknall Hill and of Hagley Park, the denudation of the old red sandstone has exposed to view the subjacent Silurian rocks on the summits of these undulations. Great shattering and dislocation would of course accompany these movements, and in the Bartestree Dike it is interesting to find one of the great crevices thus formed, and filled with eruptive matter derived from the plutonic region where all these great movements originated.

The accompanying section will serve to show the relations in which the Hagley protrusion stands to the Bartestree Dike and the Woolhope elevation.

22. ON PSEUDOMORPHOUS CRYSTALS OF CHLORIDE OF SODIUM IN KEUPER SANDSTONE*.

[*Read December* 1, 1852.]

About one-third of a mile south-south-west of the village of Blaisdon in Gloucestershire, I lately noticed in the side of a road some thin flaggy beds of the peculiar white sandstones, alternating with greenish marl, which have long been recognized in this country as the equivalents of the Keuper sandstone of Germany. At this locality they are elevated at the unusually high angle of about 45°, dipping to the east by south, a position which they doubtless owe to their proximity to the elevated Silurian mass of Blaisdon and Huntley Hills, distant only a few hundred yards on the west, and connected northwards with the May Hill range. For though the chief elevation of those Silurian rocks doubtless took place prior to the deposition of the Triassic series, yet there are abundant proofs along the east flanks of the May Hill and Malvern ranges that additional upheavals were communicated to those masses subsequently to the Triassic period.

These beds of Keuper sandstone thus elevated, are ripple-marked on their surface, and present the usual characters of the stratum as it occurs in Gloucestershire. But what especially attracted my attention was that the surfaces, both upper and lower, of some of the sandstone beds, which alternate with laminæ of marl, are studded over with small bodies, resembling crystals, of cubical or nearly cubical forms. On examination, these apparent crystals prove not to be crystalline in their interior, but to be wholly composed of white sandstone, passing into, and inseparable from, the stratum from whose surfaces they project. It is also evident that the grains of sand which compose them are not held together by crystalline matter, like those in the well-known sandstone crystals of Fontainebleau, as they present no trace of cleavage-planes, and their cubical forms prove them not to be due to carbonate of lime.

It appears then that these crystal-like bodies belong to the class termed pseudomorphous, and to the lowest subdivision of that class. In some cases, where one substance fills by chemical infiltration the cavity which has been formed by another, the product is still a crystalline body, though it has assumed an external form foreign to its real nature. But in the case before us the operation is purely mechanical;—the cavities formed by pre-existing crystals being merely filled with sand, poured into them from above, and taking the form of whatever cavities it might meet with.

* Journ. Geol. Soc. vol. ix. p. 5.

The question next arises, what was the nature of the original crystals which are now replaced by sand? On examination we find that the majority of them are cubes, or modifications of cubes. In some of them indeed a slight obliquity in their angles is perceptible, but this seems evidently owing to the effect of pressure, which has crushed and distorted the original form of the crystals, after they were replaced by the sand. We may therefore regard them as having originally conformed to the cubical type.

Of substances which crystallize in cubes, the only ones which usually occur in the Triassic formations are sulphuret of iron or iron pyrites, and chloride of sodium or common salt. It is hardly possible that sulphuret of iron can have supplied the moulds into which the sand was afterwards poured, as it would require a considerable time both for the formation and for the removal of crystals of that mineral, whereas it is evident that the crystals in question must have been formed, and must have afterwards been removed, leaving an empty cavity, in the short interval between the deposition of one bed of sand, and of the one immediately superimposed. All these conditions, however, are supplied in the most satisfactory manner by supposing chloride of sodium to have been the material which formed the moulds for these pseudomorphous crystals. The ripple-marks,—the cracks formed by desiccation in the argillaceous beds, and afterwards filled with sand poured in from above,—and the not unfrequent impressions of the feet of air-breathing reptiles, all of which phænomena especially characterize the Keuper sandstones of our English counties, seem to point to a very shallow state of the sea, abounding with sand-banks, and extensive salt-water marshes, often laid bare in the intervals of the tides. If now we suppose that at the locality in question, a sandy marsh existed, which at high spring tides was covered by the sea, we can easily conceive that in the interval between two spring tides, or in the still longer one between two equinoctial tides, the sea-water, ponded up in such a marsh, had time to evaporate and to deposit its crystals of chloride of sodium, which being slowly and tranquilly formed would assume their normal shape of cubes. As the desiccation proceeded these crystals would be enveloped by the fine muddy sediment which usually forms the last deposit of water as it evaporates to dryness. When, after a given interval, the tide again overflowed the spot, the returning sea-water (not being saturated) would dissolve these saline crystals, leaving cubical cavities in the mud which contained them. The tide would bring with it a fresh deposit of fine sand, a portion of which would pour into the cavities formed by the crystals, and the remainder would form a homogeneous stratum immediately above. Such seems to me the probable explanation of the phænomenon in question, as seen on the *under* side of the slabs of sandstone.

There are, however, examples on one of the slabs of similar crystals on the *upper* surface, similarly connected with the mass of sandstone, and less easy to be explained. In the former case we suppose the sand to have poured by simple gravitation into the subjacent cavity. But here the sand rises above its ordinary level, filling a cavity in the superincumbent stratum of marl. I can only suppose that here the crystals having formed between the sand and the mud which covered it, and being afterwards dissolved away by the returning tide, their places were filled by a portion of the subjacent sand, which being in a soft state, and intermixed with water, might without much difficulty be pressed upwards through the short space, not exceeding $\frac{1}{10}$th of an inch, in which these crystals project above the surface.

That chloride of sodium was the original substance that gave form to these bodies is further shown by the fact that in one example we find that peculiar concave figure so often seen in crystals of common salt, and which is due to the original cube floating at the surface of the brine, when a succession of smaller cubes form round its margin, and ultimately give it a kind of basket-shaped form, which is distinctly seen in the specimen in question.

These pseudomorphous crystals of salt therefore supply us with an additional evidence of those subaërial agencies affecting littoral deposits, of which the Keuper sandstone has already furnished examples in the ripplings, the rain-drops, the mud-cracks and the reptilian footprints so often presented on its surface.

I may add, that on showing these specimens to Professor J. Phillips, he informed me that he had seen similar examples of pseudomorphous crystals in the Keuper sandstone at Spetchley, in Worcestershire. There are also some specimens of the same phænomenon in the Museum of the Birmingham Philosophical Institution, presented by Miss Jukes, the locality of which I have not been able to ascertain. And Dr. Percy has kindly communicated a specimen from Clifton Grove, Nottinghamshire, which seems to prove the extension of Keuper sandstone, affected by similar conditions, over a considerable area in the Midland Counties. It is a greenish micaceous sandstone, closely resembling the Gloucestershire specimens,—ripple-marked on its upper surface, and covered below with cubic pseudomorphous crystals, the largest of which is about $\frac{1}{4}$ inch in diameter, and exhibits a trace of the basket- or hopper-shaped form, analogous to those above described.—[Jan. 21, 1853.]

23. ON GEOLOGY, IN RELATION TO THE STUDIES OF THE UNIVERSITY OF OXFORD.

[The substance of the following Essay is taken from the Introductory Lecture of a course on Geology, delivered in Michaelmas Term, 1850, with such alterations as appeared requisite.—H. E. S. 1852.]

It has always appeared to me that there is a great deal of truth in an epigrammatic sentence which I once heard uttered,—that "a well-educated man ought to know something of everything and everything of something." In other words, he ought to have a certain general acquaintance with the principles and outlines of all, or at least of a great many, branches of knowledge, and he ought also to select *some one*, or at most *some few*, subjects of study, of which he should endeavour to obtain the entire mastery. Without the former, most of the ideas which circulate in general literature and general conversation become to him a dead letter, as unsuggestive as the inscriptions of Assyria or Etruria; without the latter, he possesses no detailed or systematic knowledge to exercise his judicial or discursive powers. Both general and particular knowledge are necessary to complete the mental structure;—the man who *only* knows "something of everything" is superficial, while he who *only* knows "everything of something" is narrow-minded.

So intense in some minds is the appetite for special knowledge, that they waste their energies in striving to master the entire details of every subject that comes before them, forgetting the shortness of life, and the limited powers of the human mind. Of such men it has been said that science is their *forte*, but omniscience is their *foible*. It often demands no little judgment to make a wise selection of special subjects of study, and great self-denial in adhering steadily to them. The particular duties, talents and tastes of each individual must be consulted. The subjects which have a practical bearing on his social and professional duties must of course have the first claim for his selection. But there are few men who could not, if they would, spare with advantage a portion of their leisure from their daily employments, in order to refresh their minds with some more abstract subject of study, and nothing can more conduce to preserve the *mens sana in corpore sano* than such a change of intellectual occupation.

The principles above stated have been instinctively recognized, with greater or less precision, by the founders of Universities in modern times. They have endeavoured, theoretically at least, to give the chief prominence to those studies which concern all men as members of a Christian and a civilized community, and they have also not forgotten to provide for the acquisition of all or most other accessible branches of knowledge. Their idea of a University, as implied by its very name, was that of a microcosm, or epitome of universal knowledge, as far as it is attainable by the faculties of man. The entire Cosmos, the *omne scibile* of all external things, was supposed to be concentrated and reflected within our collegiate walls, as the features of a boundless landscape are condensed into the narrow limits of a *camera obscura.* All men were thus enabled to enlarge their minds by acquiring the general principles of every science, while each individual had the means of mastering the profundities of such especial subjects as best suited his tastes and talents.

It is needless to say that no academical body has ever yet thoroughly carried out the details of this theoretical scheme. Yet every institution deserving of the name of University has embodied the idea above explained with greater or less success. Our own Oxford exhibits some deficiencies and some redundancies in her machinery for instruction, but all the most essential parts of the apparatus are established and in working order. The chief thing wanted is a more general appreciation, among our members, of the many pleasurable and invigorating sources of knowledge, both subjective and objective, which the universe supplies, and of the ample facilities which our University possesses for conducting these streams of knowledge from their fountains in external nature to the mental soil which they are ever ready to irrigate. This place furnishes libraries, museums, professorships and lecture-rooms in abundance;—the trees are loaded with fruit, but too many of us neglect to gather it.

Although my especial purpose in this Essay is to point out the claims of GEOLOGY to a conspicuous place among the studies of this University, yet I have no wish to assign to it any greater importance than it really possesses. I fully admit that Moral Science has a higher claim on our attention than Physical Research, but I only maintain the expediency of superadding to our graver studies a general acquaintance with the principles of physical science. Every man who goes forth from our Colleges into the outer world ought to carry with him a certain amount of these general elements of knowledge,—or he cannot be called a thoroughly educated man. And if his personal talents and tastes incline him to plunge deeper into any particular branch of science, to aim at knowing the "everything" of this especial "something," let him not lose the golden opportunity which the facilities of this place and the intervals of his academical leisure afford. Many a man, when immersed in after-life in the ceaseless bustle of the world, has regretted the hours of unprofitable idleness which he might have devoted to fruitful study in our lecture-rooms and libraries.

The claims of physical science as an important part of the studies of Oxford are the more urgent when we consider the great dignity and influence to which it has attained in other places. It is highly inexpedient and even dangerous for us practically to ignore many matters upon which vast masses of able and energetic minds, both British and foreign, are actively employed. Now, if we look around us we shall find, that though in past times metaphysical or mathematical science was almost exclusively pursued, and physical knowledge either misunderstood or repudiated, the present age is doing its best to restore the balance of these sciences. When compared with past ages, the living generation may, to some, appear too exclusively absorbed in physical inquiries; but if this be true of the practical men of our day, it is surely not so in the case of our philosophers. These fully appreciate the excellence of mental, moral and abstract science, and only differ from the great minds of other days in also allowing a duly adjusted influence to the sciences which treat of external and tangible objects. And as physical science is taking its rightful place among its compeers, so are its several subdivisions daily becoming more justly estimated, as they become better understood. By the employment of strict observation and sound logic, chemistry has become developed out of alchemy, and astronomy out of astrology. Sciences such as zoology and geology, whose names were unknown a century ago, because their practical utility was less obvious than that of some others, are now cultivated and honoured. Flourishing societies are established for their promotion, and authorized teachers for their diffusion, while vast libraries and museums are the fruit and the proofs of the amount of scientific energy which has been thus put in motion. Sciences once despised by the vulgar because they were supposed to be useless and unprofitable, and only pursued by a few self-devoted and earnest truth-seekers, are now heaping a thousand benefits on the many in return for the disinterested labours of the few. The steam-engine, the railway, the electric telegraph, and countless other temporal blessings which we now enjoy, would never have existed if the *cui bono* cavillers of past times (a race not yet extinct) had succeeded in annihilating those physical investigations, the results of which no one could foresee.

The great extent to which the different physical sciences mutually illustrate each other, has been a further cause of their being more equitably appreciated in modern times. Sciences of minor interest or dignity themselves are found to throw an important reflected light upon others, and the more the laws of nature are studied the more evident becomes the complex network of relations which connects each department of knowledge with all the rest. The meanest objects cannot be despised or ignored without detriment to the highest.

Such, then, being the direction in which multitudes of active minds are ceaselessly working elsewhere, it surely behoves the University of Oxford to give a little more attention to these matters than she has hitherto done. With the means at her command, there is no reason why her fame should not be increased by original researches and physical discoveries such as those which throw lustre on London, Edinburgh, Berlin, Paris, and countless other foci of civilization. And possibly, while sharing in the intellectual reputation which has been earned by the physical labourers of other Universities, Oxford might acquire an additional renown by giving a right direction to these studies;—guiding them by the mechanism of her logic and by the light of her ethics, and thus avoiding alike the mysticism of Germany and the materialism of France.

Having said thus much on the motives for a general cultivation of the physical sciences in this place, I will now proceed to speak more especially of geology. This is a science which occupies a peculiarly central position in relation to the rest. Chemistry, zoology, botany, mineralogy, crystallography, electricity, magnetism, hydrostatics, dynamics, astronomy, geography, may all derive light and life and strength from geological research, or may communicate them in return. And if from the sciences we proceed to the arts, we find that agriculture, metallurgy, mining, engineering, architecture, to say nothing of the more æsthetic arts of painting and sculpture, may be aided by a knowledge of geological principles, or may suffer from their neglect.

Geology is the science which treats of the present structure of the earth, and of the past causes which have produced that structure. It is, therefore, a science at once of observation and of inference, of induction, and of deduction. It begins by observing, collecting, and generalizing facts, and it proceeds to reason from effect to cause, and to draw conclusions as to the former conditions through which our planet has arrived at its present state. Geology may, therefore, be divided into two distinct departments, positive or descriptive; and inferential or dynamical. Positive geology, or the observation of the actual phænomena presented to us by nature, is of course the most fundamental branch of the subject. To learn what to observe, and how to observe it, to classify under general and well-selected heads the facts thus collected, is the main business of the geologist. But it is impossible for the human mind to stop here. *Cognoscere res* is not enough for man,—he is irresistibly impelled a step further, and endeavours, if he does not always succeed, *rerum cognoscere causas*. In the case of geology this is often a difficult undertaking, and one which requires in the reasoner an extensive knowledge of facts, and a patient and unprejudiced examination of them in all their bearings, before he ventures to draw conclusions. The phænomena are palpable to our senses, but the causes which produced them operated chiefly in the bowels of the earth or in the depths of the ocean. Even had man existed contemporaneously with those causes, they would have been in great measure concealed from his sight and consequently inexplicable by his reason. But in point of fact the vast majority of the phænomena which the geologist investigates are due to operations which took place long anteriorly to man's existence, at periods when the physical conditions of the earth were very dissimilar to those of our own day. Our conclusions as to these past events in the earth's history are therefore frequently incapable of demonstration; they are at best but theories, which we maintain and believe on the understanding that they are liable to occasional modification from the accession of fresh observations.

In some respects, however, we are far better able to infer the nature of causes now extinct, than a contemporaneous observer could have been. Preadamite geologists, had such existed, could neither have penetrated to the foci of volcanoes nor to the abysses of the sea. The earthquakes which elevated the Alps would have brought death to the bystander, and the most ardent zoologist would have quailed in the presence of a living ichthyosaurus or iguanodon. But in the present more tranquillized state of our planet these obstacles do not exist. On the cool summits of Dartmoor or the Grampians we may examine at our ease the mode in which the molten granite was elaborated far in the interior of the earth. In Headington quarries and in Shotover Hill we may walk dry-shod on the ancient bed of the sea, and investigate the growth of coral-reefs and the accumulation of detritus. The voracious monsters of former days are now the tenants not of our forests, but of our museums, where we may reason in security upon the structure of their skeletons, and infer the forms of their living bodies, together with their food, their habits, and their affinities to other creatures. So circumstantially minute, so irresistibly convincing, are many of the evidences afforded by geology, that we may often derive from them valuable conclusions as to what is now taking place in regions of our globe inaccessible to us. By examining the dikes and intrusive veins of trap rocks, the columnar structure of basalt, and the alterations made by igneous matter on contiguous sedimentary strata, we may pronounce with approximate certainty as to the actual condition of things at vast depths below Vesuvius or Etna. By examining the rock surfaces and the gravel ridges in the valleys of Wales or Cumberland, we learn of mechanical processes which take place on the lower surfaces of Alpine glaciers. And in our stone-quarries or our railway-cuttings we gain a clearer idea of the deposition of drifted materials, the aggregations of living animals, and many other submarine phænomena, than we can ever hope to do by actual inspection of what is now taking place. In this way the past and the present mutually illustrate each other.

The theoretical department of geology has fallen into undue discredit from the absurd inventions, the crude guesses, and illogical inferences of Whiston, Burnet, and a host of cosmogonists of former centuries. But all who have deserved the name of geologists in modern times have pursued a totally different method from those visionaries. By a strictly Baconian induction of facts, and by a comparison of the effects produced by causes now in operation with those resulting from agencies once active but now extinct, they have succeeded in giving to their conclusions an amount of probability which practically amounts to positive certainty. When arrived at by these sound principles, the theories of geology become valuable auxiliaries to the practice. Ours were indeed a dry and mentally unprofitable pursuit, if it were restricted to the mere collecting of specimens, the measuring of strata, and the delineation of sections. But when from these data we proceed to generalize, a far higher and nobler series of faculties are called into action. From the examination of the effect we are led to study the cause, and the modes and conditions under which it has operated. A comparison of causes, and a consideration of their various aspects and conditions, lead to the discovery of laws of nature which were previously unknown; laws which have been in force since the earliest ages of our planet, but which have been hidden from man until these latter days—and lastly, from these laws of nature we are led to the perception and appreciation of design. We discover as an absolute fact, demonstrable to our senses, that creative wisdom and providential benevolence is unchanged by the lapse of time. In the long succession of races of organized beings which have peopled the earth through past epochs, we see variety of action, but unity of purpose; the aspect of nature changing, but not her laws; the exits of old actors, and the entrances of new on the stage of organic existence, but the adaptation of their structures to the surrounding circumstances remaining the same. An immense and an unlooked-for accession has thus been made to the unanswerable phalanx of arguments by which natural theology is defended. Dr. Buckland's Bridgewater Treatise has completed the task which Paley had begun, and a quibbling mysticism is now the only resource of the sceptic and the atheist.

Nor is it going too far to assert, that not merely natural theology, but even revealed religion, may derive support from the evidence which geology unfolds. Paley's argument that the proofs of design which nature exhibits, demonstrate the existence of a Designer, will serve to confute the atheist, but are not equally efficacious with the deist. There are men who admit that a creation implies an original Creator, yet who refuse to allow that any fresh manifestations of the Creator's power have operated in the lapse of time or in the changes of circumstance. They can only conceive a creation called into being by the divine fiat, and then left for ever to the guidance of eternal laws. They cannot believe in a providence watching over the working of His vast machine, and correcting its irregularities by special acts of interposition. They refuse all assent to recorded testimonies of such interpositions, however well authenticated, because, say they, miracles are contrary to experience.

Now, geology alone, it would seem, of all the physical sciences, is commissioned to declare to man, what revelation declares in another language, that miracles are *not* contrary to experience, or at least *not* contrary to fact. The other sciences deal with the properties of matter as they are, irrespectively of past time. But geology treats not only of the present, but of the past—not merely of phænomena which exist, but of events which have successively taken place. It is in fact a department of history, and only differs from what is commonly so termed, in treating chiefly, though by no means entirely, of events anterior to man's existence. Now, although in the inorganic department of geology we perhaps find that the past is strictly conformable to the present, and that the laws of nature, though they may once have operated on a greater scale than now, present no signs of change or even of suspension, yet when we turn to organic nature, a very different result is exhibited. We there see unmistakeable proofs of frequent suspension of the laws of nature, and of special interpositions of creative power. It is a law of nature, and, as far as mere human experience goes, a constant one, that organic beings, animal and vegetable, are produced only from a pre-existing parent. Equivocal generation, or the spontaneous conversion of inorganic matter into an organic tissue, is now exploded as a fallacy, and "omnia ex ovo" is established as an every-day truth. Yet unquestionably true as this principle is, within the limited span of man's observation, it is no less certain that in the vastly larger cycles of geological time, this law has undergone frequent suspensions. There have been occasions when whole races of animals and of plants have ceased to propagate their kind, and have become extinct. And to meet these losses in the treasury of nature, new species, new genera, new orders and classes of animals and plants, which had no previous existence whatever, were suddenly called into being, to fill existing voids, and to provide for new complications of external circumstances. Such direct interpositions of Almighty Power are in every sense supernatural and miraculous. How grateful Butler would have been, could the state of science of his day have supplied him with this important addition to the analogies of natural and revealed religion! This is no *à priori* or *ex parte* conclusion, drawn with a view to the support of natural or revealed religion. It is simply the logical result of a great physical problem, a result to which nearly all the greatest physical philosophers who have grappled with the subject have been compelled, willingly or unwillingly, to arrive. I need only mention the names of Cuvier, of Agassiz, of Owen, of Lyell, of Sedgwick, as a few examples of the host of witnesses who might be cited in support of this proposition.

An unsound, if not a morally perverted, school of reasoners may try to reduce these manifestations of infinite power to the domain of natural law, but in vain; their plausible sophistry vanishes before the appliances of sound logic*. The records of a mighty Creator acting independently of law, at successive epochs, are engraven on the strata of the everlasting hills, and it is the high privilege of the geologist to interpret their testimony. Beyond this point geology does not go. She has reached the limit of her own domain, and she leaves it to the theologian to pursue the subject further. On this point I may quote the words with which Dr. Whewell concludes his able discussion of the relations in which physical science stands to theology:—

> "From what has been said it follows, that geology and astronomy are of themselves incapable of giving us any distinct and satisfactory account of the origin of the universe or of its parts. We need not wonder, then, at any particular instance of this incapacity, as for example the impossibility of accounting by any natural means for the production of all the successive tribes of plants and animals which have peopled the world in the various stages of its progress, as geology teaches us. That they were, like our own animal and vegetable contemporaries, profoundly adapted to the condition in which they were placed, we have ample reason to believe; but when we inquire whence they came into this our world, geology is silent. The mystery of creation is not within her legitimate territory; she says nothing, but she points upwards."

It is also no slight enhancement of the interest which geology possesses in our eyes, that it is pre-eminently a progressive science. Far be it from me to say that any branch of knowledge is not progressive. The human faculties have not yet, and perhaps never will, attain the extreme limits of any field of inquiry, nor if they reach the margin, will they ever wholly exhaust the soil of the interior. Still, some sciences have come into existence, and have attained to excellence at very different periods from others. Geology is one of the youngest born, and is but just attaining to the vigour of maturity. Men are now living in whose youth geology, viewed as a science and not as a speculation, had no real existence. Its grandest discoveries, and its sublimest generalizations, are the work of our own generation, and are yet far from completion. This, which may be called an accident and not a property of the science,

* In our own country these doctrines have recently gained some popularity from that specious, but most shallow publication, the 'Vestiges of Creation.' To those who may have been in the slightest degree influenced by its perusal, I would strongly recommend the study of the *new edition* of Prof. Sedgwick's discourse on the Studies of the University of Cambridge, and the equally able 'Footprints of the Creator' of Mr. Hugh Miller.

is yet unquestionably a source of its great vitality at the present moment. Our intellectual faculties derive both pleasure and instruction from contemplating a science growing before our eyes, receiving and reflecting back new lights from surrounding sciences, and pointing the way to undiscovered truths.

It is in this respect especially that geology claims a position among the appropriate studies of this place. Erudition, and not discovery, necessarily forms the staple employment of our youth, and the physical sciences are therefore valuable, precisely because of the contrast they present to classical learning. A change of diet is as necessary for mental as for bodily health. If our chief business in this University is to acquire a knowledge of the results at which the master minds of antiquity have long since arrived, it must also be a beneficial, though subordinate, exercise to learn the right methods to discover new truths, and to assist in discovering them for ourselves.

If mathematical studies discipline the mind to sound reasoning, physical science is no less useful in supplying a habit of exact observation. A correct syllogism is worse than useless if its premises be false, and the greater the powers of reasoning in a man, the more dangerous they become, unless the reasoner possesses a sufficient stock, a mental capital of general and particular truths. Now, though physical science does not supply man with all the truths that he requires, it disciplines him in the art of acquiring the rest. It teaches him to be ever striving for the greatest attainable amount of certainty, to apply his powers of observation to the utmost, and to appreciate the minutest details of quantity, quality, or relation, in their influence on the general result.

These advantages attach more or less to the study of any physical science, and to none more so than to geology. A man cannot become a good geologist without acquiring at least a general knowledge of most other branches of physics, and there is, therefore, no fear of his mind becoming narrowed by an exclusive devotion to one pursuit. To be a geologist, he must study mineralogy, and this implies an acquaintance with the principles of chemistry, and of crystallography, with its attendant train of geometrical investigations. Hypogene geology, or the department which treats of volcanic operations, of earthquakes, and the elevation or depression of the earth-crust, leads to a series of dynamical researches highly interesting to the mathematician, such as those which have recently been pursued with great success by Mr. William Hopkins of Cambridge. The connexion between geology and astronomy is also evident, for while astronomy studies the forms and motions of *all* the heavenly bodies, geology investigates that *one* of them which happens to be the most accessible to ourselves. Nor, indeed, is geology wholly confined to our own planet, for the telescope supplies us with some highly important information regarding the geological structure of the moon, and to a certain extent of other bodies of the solar system. The other branches of physics, such as electricity, terrestrial magnetism, and that highly interesting subject, physical geography, which the labours of Humboldt, of Ritter, of Mrs. Somerville, and of Guyot, have recently elevated to the rank of a science, have all an intimate connexion with geology. If we turn to the sciences which treat of organic life, zoology and botany, we find (what we had no reason to expect *à priori*) that they actually form a part of geology, and geology of them. The fossil remains which have been disinterred from the cemeteries of past epochs, have almost doubled our actual knowledge of the forms of animal and vegetable life, and have filled up innumerable gaps which were otherwise apparent in the chain of organized existences. In short, there is hardly a department of physical science which does not in some way or other illustrate geology, or receive illustration from it.

This wide extent of scientific relations which geology exhibits, may perhaps deter some from the pursuit. They may suppose that a long course of physical science must be gone through before a man is fit to commence this study. The fact is, however, that geology rather leads to, than presupposes, a knowledge of the other sciences. A thorough acquaintance with chemistry, zoology, and botany, is doubtless the best groundwork for the study of geology, yet it is equally certain that a sufficient knowledge of these may be acquired, collaterally with an advancement in geological science. No one can have studied the latter subject diligently, without gaining incidentally, a considerable knowledge of the principles of the other physical sciences; and should he afterwards wish to pursue any one of them more in detail, he will be all the better qualified for so doing.

Much advantage may also be gained by means of the division of labour, a principle to which we owe so much of our modern refinement and civilization. If the geologist wishes to analyse a specimen, he may take it to a chemist, whose practised hand will at once supply the required information. So if he is in doubt about the name or structure of a fossil shell he consults one naturalist; of a fossil coral, another; of a fossil plant, a third. In this way he readily acquires an amount of information, to arrive at which independently a lifetime would be too short. With zoologists like Owen, Mantell, Forbes, Lonsdale, Egerton, Waterhouse, with botanists like Lindley, Bunbury, and Hooker, with such chemists as Faraday, Playfair, or Graham, and with mathematicians like Herschel, Hopkins, or Babbage, we possess in this country a phalanx of scientific referees, competent to the solution of any practicable geological problem.

Geology is, indeed, a science of a peculiarly comprehensive range of vision. It is not, like many other sciences, confined to mere *matter*, but extends its flight through *time*, and through *space*. It directs a telescopic, as well as a microscopic glance upon its subject-matter, just as a painter, after labouring at the minute details of a portrait, finds it necessary from time to time to retire to a distance, and take a general survey of his whole design. Thus a geologist finds his advantage in sometimes laying aside his hammer, or his microscope, and retiring to the mountain top, where the distant prospect not only gives him that refined pleasure which is common to all men, when contemplating the majesty of Nature, but also supplies positive geological information of the most valuable kind. Or even if travelling rapidly by railway, he may still cultivate his favourite science, and acquire general views, which the stone-quarry does not supply. The geographical and picturesque features of a country, the forms of the hills and valleys, the direction of the streams, the botanical and agricultural produce, the social and moral condition of the people, are all closely connected with the geological structure of the district which he traverses, and will therefore legiti mately engage his attention.

I need not enlarge on the important relations in which geological science stands to the practical employments of life, which in the present state of our civilization have become so widely multiplied. There is no profession which may not derive advantage from combining a knowledge of this subject with its more especial studies. The connexion of geology with mining and engineering operations is self-evident, and is now generally recognized. The architect will render his structures all the more durable, if he possesses a scientific knowledge of the strata on which he builds, and of the material which he employs. The medical man will acquire more enlarged, and more correct views of the influence of localities upon diseases, if he comprehends those geological idiosyncrasies on which the configuration of the surface depends. Military men will be the better able to fortify their positions, and to comprehend the natural defences of a country, by acquainting themselves with its geological structure. The artist will view with additional delight the picturesque features of the landscape, if he understands the geological agencies of which that landscape is the result. The agriculturist will be more likely to derive profit from his draining and manuring, by an acquaintance with the composition, the inclination, and the superposition of the strata which underlie his fields. The politician and statistician must be informed as to the geological structure of a country, before they can explain or remedy many anomalies in the wealth, the industry, the morality, and other social phænomena of particular districts. And, lastly, the clergy, who form the larger portion of the students annually reared and sent forth from this place, will derive much benefit, and no detriment, by an acquaintance with the principles of geological science. Natural Theology is an appropriate adjunct to Revealed Religion, and a study of the works of the Deity in antecedent ages of the earth, cannot be inconsistent with a devout contemplation of the dealings of God with man. It is also no small advantage to a studious or laborious ecclesiastic to have an inducement for taking exercise in the open air, and refreshing the mental faculties by the contact of external nature. Many of our clergy too are destined to pass their lives in remote parishes, far from educated neighbours, and where their parochial duties are too light to occupy their whole time. The monotony of such a position will often, in spite of the best principles, react upon the nerves, and render a person listless, if not discontented. To a mind thus diseased, I can prescribe no better remedy than this :—" Make a geological map of your parish. Form a collection of all its animal, vegetable, and mineral productions. Read the books which will teach you their names and classification. When this is done, extend your researches to the neighbouring parishes, or to the whole county. If this alterative treatment be combined with an earnest discharge of your sacred duties, the result cannot fail to be beneficial."

But there are higher reasons which make it incumbent on our clergy to acquire some knowledge of physical science, and especially of geology. This science has demonstrated in a manner which cannot be gainsayed, that the earth is of a vast and untold antiquity, and that man is but as a thing of yesterday, compared with the successive races of animated beings which have peopled the ancient world. This unexpected discovery has naturally surprised, if not alarmed, many sincere champions of revelation. As long as no proof appeared to the contrary, men acquiesced from ancient times in that which appeared the most obvious interpretation of the Mosaic History, namely, that the creation of the Heavens and of the Earth was nearly coincident with that of Man, and that these wondrous manifestations of Almighty Power bear a date of some 6000 years ago. Now, however, when it can be proved that vast epochs of animated existence preceded the advent of the human race, it is evident that some modified interpretation of the few brief words which commence the book of Genesis must be admitted.

All truths must be consistent with each other, and in order to retain the Mosaic narrative as a portion of inspired truth, a re-interpretation of the passages in question appears indispensable. Those, therefore, to whose keeping the inspired records are especially entrusted, are bound to make themselves acquainted with both sides of the argument.

The different forms in which Universal Truth is exhibited to our view, such as Physical, Moral, and Religious, may be compared to a cluster of hill fortresses, all in view from one another, each one serving either as a protection to the rest, or as a point of attack against them, according to the motives of the men who garrison it. Now it is unquestionable that there have been, and that there are, men who would gladly seize on the eminences of physical science, as a vantage ground from which to direct their batteries against revealed religion. But if the champions of revelation will pre-occupy those eminences, they may so fortify them as to protect the Church of Christ from her foes.

An able writer of the present day speaks thus of the expediency of combining physical science with theological studies:—

"The mighty change which has taken place during the present century in the direction in which the minds of the first order are operating, though indicated on the face of the country in characters which cannot be mistaken, seems to have too much escaped the notice of our theologians. . . . Judging from the preparations made in their colleges and halls, they do not now seem sufficiently aware,—though the low thunder of every railway, and the snort of every steam-engine, and the whistle of the wind amid the wires of every electric telegraph, serve to publish the fact,—that it is in the department of physics, not of metaphysics, that the greater minds of the age are engaged,—that the Lockes, Humes, Kants, Berkeleys, Dugald Stewarts, and Thomas Browns belong to the past, and that the philosophers of the present time—tall enough to be seen all the world over—are the Humboldts, the Aragos, the Agassizes, the Liebigs, the Owens, the Herschels, the Bucklands, and the Brewsters. In that educational course through which, in this country, candidates for the ministry pass, in preparation for their office, I find every group of great minds which has in turn influenced and directed the mind of Europe for the last three centuries, represented, more or less adequately, save the last. It is an epitome of all kinds of learning with the exception of the kind most imperatively required, because most in accordance with the genius of the time. The restorers of classical literature—the Buchanans and Erasmuses—we see represented in our Universities by the Greek and what are called the humanity courses, and the Lockes, Kants, Humes and Berkeleys by the metaphysical course. But the Cuviers, the Huttons, the Cavendishes and the Watts, with their successors the practical philosophers of the present age,—men whose achievements in physical science we find marked on the surface of the country in characters which might be read from the moon,—are *not* adequately represented;—it would perhaps be more correct to say that they are not represented at all; and the clergy as a class suffer themselves to linger far in the rear of an intelligent and accomplished laity,—a full age behind the requirements of the time. Let them not shut their eyes to the danger which is obviously coming. The battle of the evidences will have as certainly to be fought on the field of physical science, as it was contested in the last age on that of the metaphysics. And on this new arena the combatants will have to employ new weapons, which it will be the privilege of the challenger to choose. The old, opposed to these, would prove but of little avail. In an age of muskets and artillery, the bows and arrows of an obsolete school of warfare would be found greatly less than sufficient in the field of battle, for purposes either of assault or defence."—Hugh Miller, 'Footsteps of the Creator,' p. 21.

The physical and practical tendencies of the present age, the importance of which, as a matter of fact, is so impressively urged in the above passage, have received a very marked impulse even in the short period since those words were written. The Great Exhibition of 1851 has tended in a very great degree to enhance and diffuse the social influence of practical industry, and of the physical sciences which lie at its foundation. The millions who visited that marvellous display of human energy have carried home a fund of objective ideas which will influence their future lives. And many of the raw materials and manufactured results of industry there assembled have since been placed in permanent repositories for the information of future generations.

The Museum of Practical Geology in Jermyn Street has not only received many accessions of this kind from the Great Exhibition, but has within these few months assumed the form of a geological college, in which the several branches of Chemistry, Mechanical Science, Natural History and its Applications, Geology and its Applications, Mining, Mineralogy and Metallurgy, are now regularly taught by Professors of the highest scientific eminence. This noble institution, which no visitor to London should omit to inspect, together with the collections of the British Museum, of the College of Surgeons in Lincoln's Inn, and of the Geological Society in Somerset House, furnish a strong evidence of the high estimation in which the physical sciences, and especially those which cluster round geology as a centre, are now held by the Government and by the nation.

I fully grant that for a people so busied with practical matters as ourselves, it is a high privilege to possess, by way of counterpoise to these utilitarian tendencies, a few ancient Universities like Oxford, where classical learning, moral science, and pure theology may be cultivated for their own sakes. But I see no reason why our Universities should exclusively devote themselves to these abstract studies. It is surely beneficial both to our intellectual and moral faculties, as well as conducive to our social interests, to inform ourselves of many matters which exert a great practical influence on the external world, and among these geological science occupies a conspicuous place.

In regard to the influences of geology as a moral and intellectual discipline, I may quote the words of Dr. Whewell, who, in his 'History of the Inductive Sciences,' thus speaks of geological studies:—

The labours of the members of the Geological Society "have shown that there are no talents and no endowments which may not find their fitting employment in this science. Besides that they have united laborious research and comprehensive views, acuteness and learning, zeal and knowledge, the philosophical eloquence with which they have conducted their discussions has had a most beneficial influence on the tone of their speculations; and their researches in the field, which have carried them into every country and every class of society, have given them that prompt and liberal spirit, and that open and cordial bearing, which results from intercourse with the world on a large and unfettered scale. It is not too much to say that in our time, practical geology has been one of the best schools of philosophical and general culture of mind."—Whewell, Hist. Ind. Sc. vol. iii. p. 524.

Nor should we forget how greatly the pleasures of travelling are enhanced by a knowledge of physical science, and especially of geology. Oxford sends forth annually scores of active-minded and strong-bodied men who seek to enliven the Long Vacation, or the interval between their B.A. degree and their assumption of a profession, by exploring distant lands. With the present facilities of locomotion, neither the North Cape, nor Abyssinia, nor the Mississippi are too remote for their enterprise. Possessed of such opportunities, what gratification would they not derive, and what additions would they not make to human knowledge, were they but grounded, even in a moderate degree, with the elements of physical science! Instead of merely gratifying a thirst for adventure and excitement, they would acquire the purer and more wholesome love of knowledge, and would return enriched with a stock of mental acquirements which would supply them with material for pleasing thought and for active usefulness during the rest of their lives.

To such men any branch of natural history would furnish an invaluable resource, and none more so than geology, a science which so admirably combines the general with the special—the enlarged views of the physical geographer with the minute researches of the zoologist, botanist, and mineralogist.

Such are some of the many grounds which recommend geology to the attention of the Members of this University. I may add that there are few places possessed of greater local advantages for studying the science. The Geological Museum in the Clarendon, for which the University is mainly indebted to the untiring energy and great liberality of Dr. Buckland, presents an immense and very precious collection of specimens from formations of every age and from all parts of the world. It is true that the space now assigned to this collection is very inadequate for its proper display, but this defect we may hope ere long to see remedied by the erection of a suitable museum for these and other materials connected with physical science. Meanwhile the geological collection, though confined in space, yet presents a quite sufficient amount of systematic arrangement for purposes of general instruction, with which view it is opened gratuitously to Members of the University three days in every week. Two courses of Lectures on Geology will henceforth be delivered in this Museum, while to those who wish to go deep into the subject, the recently established School of Natural Science will afford the means of academic distinction. In the Radcliffe Library we possess nearly every standard work on geological science, accessible without fee or reward to every student who may wish to consult it. And, lastly, in the district around Oxford we may read in the Book of

Nature one of the most instructive chapters of geological history. In the short distance between Steeple Aston on the north, and Shotover or Cumnor Hills on the south, we may examine in succession the Lias, the Marlstone, the Inferior Oolite, the Stonesfield Slate, the Great Oolite, the Cornbrash, the Oxford Clay, the Coral Rag, the Kimmeridge Clay, the Portland Stone, and the Wealden Sands;—each formation marked by its peculiar mineral and physical characters, and still more so by its numerous and beautifully preserved organic remains, differing in every stratum, and telling us of the lapse of countless ages, and of wondrous changes in the animal population of the earth. Or if we take the rail and travel an hour's journey to the east, we ascend to the Greensand, the Chalk, and the London Clay, while by passing west to Bristol we may descend from the Lias and New Red Sandstone to the lower depths of the Coal-measures, the Carboniferous Limestone, and the Devonian series. All this vast succession of formations is now brought within the easy access of a few hours' excursion from the doors of our Colleges.

Let me hope, in conclusion, that the junior Members of the University will cease to overlook the treasures of knowledge which lie beneath their feet, and will be induced to give a larger portion of their leisure and of their energies to the cultivation of some of those branches of knowledge, of which no man ever yet repented the acquisition.

ORNITHOLOGICAL MEMOIRS.

1. LIST OF THE BIRDS NOTICED OR OBTAINED IN ASIA MINOR IN THE WINTER OF 1835 AND SPRING OF 1836†.

[From Proceed. Zool. Society, 1836, p. 97.]

THE winter of last year was one of unusual severity in all parts of Europe. At Smyrna, where I resided from November to February, the weather, which had been mild in the early part of December, underwent a sudden change about Christmas-day. A north wind and violent storms of snow brought vast flocks of northern *Birds* to take shelter in Smyrna Bay. A frost of more than three weeks followed, a circumstance almost without parallel at Smyrna, which is situate close to the sea, and in the low latitude of 38½°. This statement will explain the occurrence in the following list, of many *Birds* whose usual abode is in high northern latitudes.

In the month of February I visited Constantinople, and returned overland to Smyrna, which I reached at the end of April. A great change had now taken place in the ornithology of that neighbourhood. The spring was now at its height, and numerous summer birds had arrived, of a more exotic race than those which had been observed during the winter. I was now, however, compelled to return to Europe; but the few days which passed before I left Smyrna, served to give me a taste of the rich ornithological harvest which might be reaped by a summer's residence in Asia Minor.

Vultur, Illig. }
Aquila, Briss. }

Two or three species of each of these families frequent the neighbourhood of Smyrna, but all my endeavours to procure specimens of these wary birds were unavailing.

*1. *Falco æsalon*, Smyrna; rare.

*2. *Falco tinnunculus*, Linn. Smyrna; rare.

---

† Of those species in the following list which have an asterisk attached, specimens had been obtained by Mr. Strickland and were exhibited.

*3. *Falco tinnunculoides*, Temm. Very abundant in Asia Minor during the spring. It frequents the Turkish villages, and builds in the roofs of the houses. Its mode of hovering is similar to that of the common Kestrel, but it is more gregarious in its habits than that bird.

*4. *Accipiter fringillaria*, Ray. Smyrna.

*5. *Buteo vulgaris*, Bechst. Smyrna.

*6. *Circus cyaneus*, Flem. Smyrna.

*7. *Circus rufus*, Briss. Smyrna.

8. *Otus brachyotus*, Cuv. Smyrna.

*9. *Ulula stridula*, Selby. Smyrna.

*10. *Bubo maximus*, Sibb. Smyrna.

*11. *Noctua nudipes*, Nilss.† Very common in the Levant.

*12. *Lanius minor*, Linn. Smyrna, in April.

*13. *Lanius rufus*, Briss. Smyrna, in April.

*14. *Lanius collurio*, Linn. Smyrna, in April.

15. *Turdus merula*, Linn. Smyrna.

16. *Turdus solitarius*, Linn. Frequents the rocks and hills near Smyrna.

17. *Turdus viscivorus*, Linn. Smyrna, during the winter.

18. *Turdus pilaris*, Linn. Smyrna, during the winter.

19. *Turdus musicus*, Linn. Smyrna, during the winter.

20. *Turdus iliacus*, Linn. Smyrna, during the winter.

21. *Cinclus aquaticus*, Bechst. Rivulets near Smyrna. I cite this bird with some doubt, not having been able to obtain a specimen. It is possible that the Smyrna Cinclus may be the *C. pallasii*, Temm., though I am inclined to refer it to the former species.

*22. *Oriolus galbula*, Linn. Smyrna, in April.

*23. *Saxicola rubicola*, Bechst. Winters at Smyrna.

*24. *Saxicola aurita*, Temm. Arrives at Smyrna in April. Its habits are similar to those of our Wheatear, and from its shy and restless motions it is very difficult to procure.

*25. *Saxicola œnanthe*, Bechst. Smyrna, in April.

26. *Saxicola rubetra*, Bechst. Common at Smyrna during the winter.

27. *Phœnicura suecica*, Selby. I believe that I saw this bird near Smyrna in April.

*28. *Phœnicura tithys*, Jard. and Selb. This bird is common on the bare rocky hills near Smyrna, where it remains during the winter.

29. *Philomela luscinia*, Swains. First heard on the 5th of April at Hushak in the interior.

30. *Salicaria phragmitis*, Selby. Seen at Smyrna in December.

31. *Curruca cinerea*, Bechst. Smyrna, in April.

*32. *Curruca melanocephala*, Bechst. This delicate little bird, which is only found in the most southern parts of Europe, remains through the winter in the neighbourhood of Smyrna. It is a retired solitary bird, frequenting sheltered ravines thickly beset with various evergreen shrubs.

*33. *Sylvia rufa*, Temm. Shot near Smyrna in November.

*34. *Sylvia brevirostris*, Strickl. Also killed in November near Smyrna. This species, which I believe to be new, may be thus characterized.

---

† This specimen from Smyrna was considered by Prince Bonaparte as the true *Athene noctua* (Retz.).

SYLVIA BREVIROSTRIS, Strickland.

Sylv. corpore suprà olivaceo brunneo, subtùs albido; pedibus nigris.

Plumage closely resembling that of *S. trochilus.* Above brown with a tinge of olive. A pale yellow streak over the eye. Throat and breast pale fulvous with a slight tinge of yellow; belly whitish. Inner wing-coverts of a pale yellow. *Remiges*: the 4th and 5th longest and equal; the 2nd equal to the 8th. Beak dusky; legs black.

Long. tot. poll. $4\frac{3}{4}$; *rostri,* $\frac{1}{4}$; *caudæ,* $2\frac{1}{8}$; *alæ,* $2\frac{2}{5}$; *tarsi,* $\frac{3}{4}$.

Differs from *S. rufa* in its greater size, and from *S. trochilus* in the shortness of the beak, and the dark colour of the legs.

Habitat prope Smyrnam. Hyeme occisa.

*35. *Accentor modularis,* Cuv. Killed near Smyrna in the winter, but is rare.

*36. *Regulus ignicapillus,* Cuv. Frequents the olive groves near Smyrna.

*37. *Troglodytes europæus,* Linn. Common near Smyrna. Undistinguishable from English specimens.

38. *Motacilla alba,* Linn. Smyrna.

39. *Motacilla boarula,* Linn. Smyrna.

*40. *Anthus pratensis,* Bechst. Common at Smyrna.

*41. *Anthus aquaticus,* Bechst. Killed on the coast near Smyrna.

42. *Hirundo rustica,* Linn. I believe that all the British species of *Hirundinidæ* frequent the Levant, but have only ascertained the above species.

*43. *Alauda arvensis,* Linn. Immense flocks of this bird arrived from the northward at the commencement of the severe weather at Christmas.

*44. *Alauda cristata,* Linn. Very common.

*45. *Alauda arborea,* Linn. Smyrna; common.

*46. *Alauda calandra,* Linn. Arrived during the cold weather.

*47. *Parus major,* Linn. Smyrna.

*48. *Parus cæruleus,* Linn. Smyrna.

*49. *Parus lugubris,* Natt. Smyrna.

*50. *Emberiza miliaria,* Linn. Common.

*51. *Emberiza cia,* Linn. Frequents the rocky hills near Smyrna.

*52. *Emberiza cirlus,* Linn. Haunts the vicinity of streams. It seems to replace the *E. citrinella,* which I never noticed in Asia Minor.

*53. *Emberiza palustris,* Savig. The habits of this species of Reed Bunting exactly resemble those of *E. schœniclus.* The beak is rather less gibbous than in the Dalmatian specimens.

*54. *Emberiza cæsia,* Cretzsch. Killed at Smyrna in April. It is frequent in Greece and in the Ionian Islands.

*55. *Emberiza hortulana,* Linn. Smyrna, in April.

*56. *Emberiza cinerea,* Strickl. This new species is thus characterized:—

EMBERIZA CINEREA, Strickland*.

Emb. capite viridi-flavescente; corpore suprà cinerascenti, subtùs albo.

Male. Crown of the head greenish-yellow, becoming cinereous at the nape. Back cinereo-fuscous, with an obscure streak of brown in the middle of each feather. Rump cinereous; tail dark brown; the two lateral pairs of feathers white on the inner webs for near half their length towards the extremities.

Wings dark brown, the covers and quills margined with whitish, the scapulars with fulvous. Chin and throat yellow, becoming greenish on the cheeks.

---

* See Plate in Memoir, Chapter VI.

Breast cinereous; abdomen white, sides cinereous.
Bill dusky; legs flesh-coloured.
Long. tot. poll. 6; *rostri*, $\frac{2}{5}$; *alæ*, $3\frac{1}{2}$; *caudæ*, $2\frac{3}{4}$; *tarsi*, $\frac{3}{4}$.
The beak of this species most nearly resembles that of *Emberiza cia.*
Habitat in collibus juxta Smyrnam. Mense Aprili occisa.

57. *Pyrgita domestica*, Cuv. This is the common House Sparrow of the Levant.

*58. *Pyrgita hispaniolensis*, Cuv. A single specimen was obtained in April at Smyrna.

*59. *Linaria cannabina*, Swains. Common.

60. *Carduelis elegans*, Steph. Common.

*61. *Fringilla cœlebs*, Linn. Very common in the Levant.

62. *Fringilla montifringilla*, Linn. Occurred during the winter.

*63. *Fringilla serinus*, Linn. Gregarious during the winter. Assembles in large flocks, which chirp incessantly in a small low note.

64. *Coccothraustes chloris*, Flem. Common.

65. *Sturnus vulgaris*, Linn. Smyrna.

66. *Corvus corax*, Linn. Smyrna.

67. *Corvus cornix*, Linn. Common near Smyrna.

68. *Corvus monedula*, Linn. Common near Smyrna.

Obs. The *common Rook* was not noticed, and I do not believe that it exists in the country.

69. *Pica caudata*, Ray. Common in the Levant.

*70. *Garrulus melanocephalus*, Bonelli. This bird was first described by M. Gené in the Memoirs of the Academy of Turin, vol. xxxvii. p. 298, pl. 1, from specimens in the Turin Museum, received from Lebanon. It is common in the vicinity of Smyrna, and its note and habits are identical with those of the European Jay, whose place it supplies.

*71. *Sitta syriaca*, Ehrenb. Frequents the open hills near Smyrna, where it is seen climbing up the masses of rock, or perched on their summits. It never is seen on trees. The note is a loud clear warble.

*72. *Sitta europæa*, Linn. Inhabits the groves of aged olive-trees which abound in the bottoms of the valleys. The specimens are smaller than British ones, but not otherwise distinguishable.

73. *Upupa epops*, Linn. Seen at Hushak in April.

*74. *Alcedo ispida*, Linn. Common.

*75. *Alcedo rudis*, Linn. This bird may often be seen in the salt-water marshes west of Smyrna. It never seems to follow the rivers, but always remains near the coast. It sometimes hovers for several minutes, about 10 feet above the water, and then drops perpendicularly on to its prey.

76. *Picus martius*, Linn. I saw a specimen of this bird in the possession of Mr. Zohrab at Broussa. It was shot in the pine forests of Mount Olympus.

*77. *Picus major*, Linn. Common near Smyrna.

*78. *Cuculus canorus*, Linn. Smyrna, in April.

79. *Phasianus colchicus*, Linn. Common near Constantinople on both sides of the Bosphorus. It has probably migrated thither spontaneously from Colchis, its native country.

80. *Francolinus vulgaris.* Occurs in the marshes of the Hermus and the Cayster, whence it is sometimes brought to market at Smyrna.

*81. *Perdix saxatilis*, Meyer. Abundant on the hills round Smyrna.

82. *Coturnix dactylisonans.* Remains near Smyrna during winter.

83. *Columba palumbus*, Linn. Smyrna.

84. *Columba œnas*, Linn. Smyrna.

*85. *Columba turtur*, Linn. Smyrna, in April.

*86. *Columba cambayensis*, Lath. This bird inhabits the Turkish burial-grounds at Smyrna and Constantinople, which are dense forests of cypress-trees. It is strictly protected by the Turks, and it was with some difficulty that I obtained a specimen. It was perhaps originally introduced by man, but now seems completely naturalized.

87. *Otis tarda*, Linn. Frequents the plains south of Smyrna. It is called *Wild Turkey* by the European residents.

*88. *Otis tetrax*, Linn. Abundant during the winter in the poultry-shops at Smyrna.

89. *Œdicnemus crepitans*, Temm. Said to occur in this part of Asia Minor.

90. *Vanellus cristatus*, Meyer. Appeared in vast flocks at the commencement of the cold weather.

91. *Grus cinerea*, Bechst. A flock seen in the plain of Sardis at the end of April.

*92. *Ardea egretta*, Linn. Frequents the sea-marshes west of Smyrna.

*93. *Botaurus stellaris*, Steph. Smyrna.

*94. *Ciconia alba*, Bellon. Very abundant in Turkey during summer. It swarms in every village, and is protected with the same strictness by the Turks as by the Dutch. It is said to have quite deserted Greece, since the expulsion of its Mahometan protectors.

95. *Numenius arquatus*, Cuv. Smyrna.

96. *Scolopax rusticola*, Linn. So abundant were *woodcocks* at Smyrna during the severe weather, that many were killed in small gardens in the midst of the town.

97. *Scolopax gallinago*, Linn.
98. *Scolopax gallinula*, Linn. } Abundant in the marshes near Smyrna.

*99. *Tringa variabilis*, Meyer. Common on the coast.

*100. *Tringa temminckii*, Leisl. Smyrna, in winter.

*101. *Totanus glottis*, Bechst. Smyrna, in winter; rare.

102. *Totanus calidris*, Bechst. Common in the marshes.

103. *Totanus ochropus*, Temm. Seen on the coast.

*104. *Recurvirostra avocetta*, Linn. Smyrna; rare.

*105. *Rallus aquaticus*, Linn. Smyrna.

106. *Crex pratensis*, Bechst. Smyrna, in winter.

*107. *Crex porzana*, Bechst. Smyrna, in winter.

108. *Gallinula chloropus*, Lath. Smyrna, in winter.

109. *Fulica atra*, Linn. Smyrna, in winter.

*110. *Glareola torquata*, Meyer. A pair of these birds were brought to me at Smyrna in April.

*111. *Podiceps cristatus*, Lath. The young of this bird is abundant in the harbour at Constantinople, where, in common with all other waterfowl, it is strictly protected.

*112. *Puffinus anglorum*, Ray. Flocks of this bird are constantly seen flying up and down the Bosphorus. They are rarely seen to alight, and from their unceasing restlessness, the Franks of Pera have given them the name of *âmes damnées*. I am not aware that this bird has before been noticed in the southern parts of Europe.

*113. *Larus ridibundus*, Linn.

*114. *Larus argentatus*, Brunn. These two species of *Gull* frequent the Golden Horn at Constantinople, where they are so tame that they may easily be struck with an oar.

115. *Pelecanus onocrotalus*, Linn. Frequents the marshes near Smyrna, where it remains during the winter.

*116. *Phalacrocorax carbo*, Briss. Abounds in the harbour of Constantinople, and roosts on the roofs of the houses.

*117. *Phalacrocorax pygmæus*, Briss. Shot near Smyrna in winter.

118. *Cygnus olor*, Linn. Visited Smyrna Bay in the winter.

119. *Clangula vulgaris*, Leach. Smyrna, during the winter.

120. *Fuligula ferina*, Steph. Smyrna, during the winter.

121. *Fuligula cristata*, Steph. Smyrna, during the winter.

*122. *Rhynchaspis clypeata*, Shaw. Smyrna, during the winter.

123. *Tadorna vulpanser*, Flem. Smyrna, during the winter.

124. *Querquedula acuta*, Selby. Smyrna, during the winter.

125. *Anas boschas*, Linn. Smyrna, during the winter.

126. *Mareca penelope*, Selby. Smyrna, during the winter.

127. *Tadorna rutila*, Steph. Frequent in the poultry-shops at Smyrna, but owing to the Turkish practice of cutting the throats of birds as soon as shot, I was unable to obtain a perfect specimen.

128. *Querquedula crecca*, Steph. Smyrna, in the winter.

*129. *Mergus albellus*, Linn. Smyrna, in the winter.

---

2. ON SOME NEW GENERA OF BIRDS.

[From Proc. Zool. Soc. 1841, p. 27.]

It is not without some unwillingness that I venture to point out some new generic groups of birds, because I am of opinion that the process of naming and defining new genera has been in many cases carried too far already. The class of Birds probably does not contain more than 6000 species, and these have already been distributed into upwards of 1000 genera*, and I think therefore there can be no doubt that systematic ornithologists are now fast approaching the point beyond which it will not be expedient to carry the subdivision of the older groups into new genera. For we must not lose sight of the fact, that expediency or practical convenience *does* form an element in the construction even of a natural system. In such a system the *materials* which constitute any group must be naturally allied, and they must be placed in such *order* as will best show their natural affinities, and yet the number of nominal genera into which such a group is to be divided may be a question of mere expediency. For the *species*, which are the only *real* ingredients in a family or subfamily, often pass from one form of structure to another without any hiatus, so that it becomes a mere matter of opinion whether the so-called genera into which they are to be classed shall be many or few. Nature draws no line by which the rank or extent of genera can be determined. As a general rule, varieties of *form* are considered to constitute genera, and varieties of colour, species; but this criterion is far from infallible, for we very rarely find two species, however closely allied, possessing precisely the same form and proportions, so that if every difference of structure be held to be generic, we shall end in having as many genera as there are species. Take, for instance, the two very natural groups, Corvus and Parus, as now restricted; it will be found on examination that there are marked differences of both structure and habits which characterize almost every one of the species. It is plain then that we have in general no other guide in the definition of new genera than a mere *opinion* as to the amount of structural variation which is considered to authorize their adoption, and I think there can be no doubt that in many of the genera recently established in ornithology, the standard has been reduced too low; in other words, these genera are based on diversities of structure of so little value as to be practically inconvenient. All genera profess to be of equal rank, and we should therefore aim at making them as nearly so as possible, and at the same time not inconveniently numerous.

Granting, however, that many of the existing genera of birds are based on insufficient characters, and may require to be reincorporated with the

* The number of genera now, 1857, is about 3000!—Ed.

groups from which they have been divided, it is equally certain that other groups exist in our cabinets, whose characters, as yet undefined, are so marked, as to demand in fairness, and with the view of producing equality of rank, to be defined and named as genera. A few of these groups I now bring forward, illustrated by specimens from my own cabinet, and the meeting will be able to judge how far the structures here exhibited appear to authorize generic distinction.

I also take this opportunity of exhibiting a specimen of that rare bird the *Glyphorhynchus cuneatus* (Licht.), afterwards named *Xenophasia platyrhyncha* by Mr. Swainson. This bird has the tail of Dendrocolaptes and the general form of Xenops, while the beak is altogether anomalous, being *com*pressed at the sides and *de*pressed at the apex.

Fam. SYLVIADÆ.

Subfam. MALURINÆ.

Genus SPHENŒACUS *, n. g.

< *Motacilla*, Gmel., < *Malurus*, Swains., < *Sphenura*, Licht.

Diff. Char.—*Beak much compressed; tail cuneate.*

Rostrum mediocre, compressum, ad basin elevatum, culmine subrecto, juxta apicem deorsum, gonyde pariter sursum curvato, tomio maxillari emarginato, commissurâ leviter deflexâ. Nares oblongæ, membranâ suprà tectæ. Vibrissæ nullæ.

Alæ breves rotundatæ, remigibus 4â, 5â, 6â, 7â, æqualibus, primam duplo excedentibus.

Cauda longiuscula, maximè cuneata, rectricibus strictis, acutis, subdecompositis, intermediis exteriores triplò superantibus.

Pedes fortes, tarsis longiusculis, acrotarsiis scutellatis, paratarsiis integris, digitis lateralibus æqualibus.

Ungues modicè curvati, acuti.

Ptilosis rigida, pennis subdecompositis.

Habitat in Africâ.

Species unica, *S. africanus* (Gmel.), Levaill. Ois. d'Af. pl. 112. f. 2. (*Sphenura tibicen*, Licht.)

Fam. MUSCICAPIDÆ.

Subfam. FLUVICOLINÆ.

Genus COPURUS†, n. g.

< *Platyrhynchus*, Spix, < *Muscipeta*, Cuv.

Diff. Char.—*The two medial rectrices greatly prolonged.*

Rostrum triangulare (desuper spectanti), paulo longius quàm latum, modicè depressum. Culmen rotundatum, ad basin lentè, versus apicem citiùs decurvans. Nares ovatæ, patulæ. Tomia maxillaria modicè emarginata. Commissura leniter decurvata, gonys leniter ascendens. Vibrissæ rictales mediocres.

Alæ longiusculæ, remige primâ breviore, 2â, 3â, 4â, ferè æqualibus.

Cauda mediocris, quadrata, nisi quòd maris rectrices duæ intermediæ graciles spathuliformes reliquas magis duplo excedunt.

Tarsi mediocres, acrotarsiis paratarsiisque scutellatis. Digitus externus interno

* Σφὴν, a wedge; οἴαξ, a helm.

† Κώπη, an oar; οὐρὰ, the tail.

longior, ad basin pauló cum intermedio coadunatus. Ungues longiusculi, satis curvati, acuti, graciles.

Habitat in Americâ meridionali.

Species unica, *C. filicaudus* (Spix), Av. Bras. ii. pl. 14. (*Muscicapa leucocilla*, Hahn.)

Obs. This bird bears much resemblance to Pipra in the colours of its plumage and in the elongate rectrices, as Mr. Swainson has remarked, 'Classif. Birds,' vol. ii. p. 90. The depressed beak, however, rounded culmen, shorter tarsus, and slender claws, sufficiently prove its true place to be among the Flycatchers, near Alectrurus.

Fam. PIPRIDÆ?

Subfam. PARDALOTINÆ?

Genus PRIONOCHILUS *, n. g.

< *Pardalotus*, Temm.

Diff. Char.—*Margins of the beak minutely serrated.*

Rostrum longiusculum, subcompressum, mandibulis subæqualibus; culmine subcarinato, juxta basin recto, deinde usque ad apicem gradatim decurvato; commissurâ modicè decurvatâ; gonyde sursum curvatâ. Maxilla haud emarginata, sed tomia mandibulæ utriusque per medietatem externam minutissimè serrata. Nares oblongæ, membranâ suprà tectæ.

Alæ mediocres, remige 1â spuriâ, 3â, 4â et 5â subæqualibus.

Cauda brevis, rectricibus æqualibus.

Pedes subbreves, gressorii, acrotarsiis subscutellatis, paratarsiis integris. Digitus externus interiore longior, ad medium per longitudinis dimidium coadunatus.

Habitant in Malasiâ.

Species: 1. *P. percussus* (Temm.), Pl. Col. 394. fig. 2. 2. *P. thoracicus* (Temm.), Pl. Col. 600. figs. 1, 2. 3. *P. maculatus* (Temm.), Pl. Col. 600. fig. 3.

Obs. The nearest affinity of this group is Calyptomena, Raffl., to which it approaches in the structure of the beak and feet much more nearly than to Pardalotus. The serrations of the tomia appear not to have been hitherto noticed.

Fam. LANIADÆ?

Subfam. ———?

Genus ÆTHIOPS†, n. g.

Diff. Char.—*Beak subconical, slightly emarginate, dilated at the base.*

Rostrum subconicum, subelongatum. Maxilla ad basin paulo dilatata, juxta apicem compressa, leviter emarginata, apice paulo deorsum curvato, tomiis inflexis. Culmen subcarinatum, gradatim à basi ad apicem decurvatum. Nares ovatæ. Commissura leviter decurvata, gonys ascendens.

Alæ mediocres, rotundatæ, remige 1â spuriâ, 3â, 4â et 5â subæqualibus.

Cauda breviuscula, rotundata.

Tarsi breves, acrotarsiis scutellatis, paratarsiis integris. Digiti mediocres, medius tarsum æquans, postico longior; externus internum paulo superans. Ungues breviusculi, modicè curvati, ad latera sulcati.

Habitat in Africâ occidentali. Species unica.

* Πρίων, a saw; χεῖλος, a lip.

† Αἰθίοψ, a negro, in reference to the colour and habitat.

ÆTHIOPS CANICAPILLUS, Strickland.

Æth. vertice, cervice dorsoque canescente-cinereis, verticis lateribus uropygioque canescente-albidis. Fronte, genis, gulâ partibusque inferioribus omnibus nigerrimis. Alæ nigræ, tectricibus minoribus omnibus, majoribusque dorso proximis maculâ subapicali rotundatâ albâ. Cauda unicolor nigra, tectricibus superioribus nigrescente-plumbeis. Rostrum pedesque nigri.

Long. tot. 5 poll. *Rostrum* ad rictum 7½ lin., ad frontem 6 lin., latum 3½ lin., altum 3 lin. *Ala* 2 poll. 8 lin. Rectrices medii 1 poll. 11 lin., externi 1 poll. 7 lin. *Tarsus* 8 lin. Digitus intermedius cum ungue 8 lin., externus 6 lin., internus 5½ lin., posticus 6 lin.

Habitat in insulâ Fernando Po. Mus. Strickl.

Obs. This singular generic form is very difficult to classify. The beak is somewhat similar in form to that of a Tanager, but its other characters and the African habitat forbid such a collocation. The beak also exhibits some resemblance to that of Artamus, but the shortness of the wings makes a marked contrast to that genus.

Fam. FRINGILLIDÆ.

Subfam. TANAGRINÆ.

Genus STEPHANOPHORUS*, n. g.

< *Tanagra*, Temm., < *Pyrrhula*, Vieill.

Diff. Char.—*Beak very short, tumid, of equal height and length.*

Rostrum breve, subconicum, mandibulis subæqualibus, intumidis; culmine gradatim deorsum, gonyde sursum incurvatis; commissurâ subrectâ, leviter deorsum curvatâ; maxillâ juxta apicem obsoletissimè emarginatâ. Nares subrotundæ.

Alæ mediocres, rotundatæ, remigibus 3â, 4â (hâc longissima) et 5tâ ferè æqualibus.

Cauda mediocris, rectricibus subæqualibus.

Pedes mediocres, acrotarsiis scutellatis, paratarsiis integris. Digitus externus interiorem paulo excedens. Ungues mediocres, leviter curvati.

Ptilosis cærulescens, nitore sericeo. Vertex colore igneo insignis.

Habitat in Americâ meridionali.

Species unica, *S. cæruleus* (Vieill.), (*T. diadema*, Temm.), Pl. Col. 243.

Obs. The beak is more tumid and the under mandible more developed in this well-marked type than in any other of the Tanagrinæ, and it has hence been referred to the genus Pyrrhula. The marginal notch, however, together with the blue and silky plumage, and the geographical distribution, sufficiently prove the true place of this bird to be among the Tanagers, and in the vicinity of Tanagra, Linn. (restr.), and Calospiza, Gray (Aglaia, Swains.).

Fam. PICIDÆ.

Subfam. CELEINÆ.

The genus Brachylophus, as defined by Mr. Swainson, includes three very distinct groups: first, the Green Woodpeckers, which had previously been named Gecinus by Boié; secondly, the crimson-winged species, *miniatus, puniceus,* and *mentalis,* to which I propose to restrict Swainson's name Brachylophus; and thirdly, the short-thumbed Woodpeckers, which are here characterized.

* Στέφανος, a crown; φέρω, to bear.

Genus BRACHYPTERNUS *, n. g.

< *Picus*, Linn., < *Brachylophus*, Swains.

Diff. Char.—*Hind toe and claw very short, almost obsolete.*

Rostrum longitudine caput æquans, apice obtusè securiformi, culmine paulatim decurvato, acie laterali nullâ, commissurâ rectissimâ, gonyde paulo ascendente.

Alæ mediocres, remige 1â subspuriâ, 4â, 5â (hâc omnium longissimâ) et 6â subæqualibus.

Cauda mediocris.

Tarsus mediocris. Digitus intermedius ac versatilis tarsum æquiparantes, posticus cum ungue brevissimus, propemodum obsoletus.

Ptilosis: dorsum alæque aurantia aut rubra, facies et partes inferiores albido nigroque variegatæ.

Habitant in Indiâ, Malasiâ.

Species: 1. *B. aurantius* (Linn.), (*P. bengalensis*, Gmel.; *P. nuchalis*, Wagl.; *B. hemipodius*, Swains.). 2. *B. goensis* (Gmel.), (*P. peralaimus*, Wagl.). 3. *B. philippinarum* (Lath.), (*P. palalacca*, Wagl.). 4. *B. hæmatribon* (Wagl.). 5. *B. erythronotus* (Vieill.), (*P. neglectus*, Wagl.).

Fam. CHARADRIADÆ.

Subfam. CHARADRIANÆ.

The group of Plovers affords an instance in addition to those furnished by the genera Ceyx, Alcyone, Jacamaralcyon, Tiga, Tridactylia, Halodroma, and others, that the presence or absence of the hind-toe in birds becomes, under certain circumstances, a character of very small value in the natural arrangement. The fact seems to be, that when in any group the hind-toe becomes so slightly developed as to be unable to perform those functions of prehension or of progression which are its usual duties, the transition from the abortive state of this organ to its total disappearance becomes very unimportant. In the group of the Plovers too much weight has hitherto been attached to the presence or absence of the hind-toe; it has been made the groundwork of divisions into families and subfamilies, whereas the utmost value that can justly be assigned to it amounts only to that of a generic character. This is proved by the fact that the absence of the hind-toe is not coincident with the other and more extensive changes of structure in the group, so that it becomes indicative of analogy rather than of affinity, as the following table of the genera of Charadrianæ will show:—

A. Acrotarsia reticulate; wings pointed; plumage spotted.

Three-toed .... *Charadrius.*
Four-toed .... *Squatarola.*

B. Acrotarsia reticulate; wings pointed; plumage black, white, and grey, in large masses.

Three-toed........ *Eudromias. Hiaticula.*

---

* Βραχὺς, short; πτέρνα, a heel.

C. Acrotarsia scutate; wings rounded; plumage black, white, and grey, in large masses.

I. Face unwattled.

Three-toed. *Philomachus*, Mœhr. (*Hoplopterus*, Bonap.)
- *a.* wing-spine short, *P. coronatus* (Gmel.).
- *b.* wing-spine long, *P. spinosus* (Linn.).

Four-toed. *Vanellus*, Temm. ....
- *a.* wing-spine short, *V. cristatus* (Linn.).
- *b.* wing-spine long, *V. cayennensis* (Gmel.).

II. Face wattled.

Three-toed.................. *Sarciophorus*, Strickl.
Four-toed .... *Lobivanellus*, Strickl.

The last two groups which have hitherto been united, the one with Charadrius, the other with Vanellus, are now for the first time defined.

Genus SARCIOPHORUS *, n. g.

< *Charadrius*, Gmel.

Diff. Char.—*Three-toed, lores wattled.*

Rostrum ut in Charadriis.

Membrana loris affixa, nuda, erecta, in anticum protensa.

Alæ elongatæ, caudam vix superantes, remigibus tribus primariis subæqualibus. Spina pollicaris brevis, obtusa.

Cauda modicæ longitudinis, rectricibus æqualibus.

Pedes elongati, graciles, tridactyli, acrotarsiis scutellatis.

Habitant in Africâ, Asiâ, Australiâ.

Species: 1. *S. pileatus* (Gmel.), Pl. Enl. 834. 2. *S. tricolor* (Vieill.), (*Charadrius pectoralis*, Wagl.). 3. *S. bilobus* (Gmel.), Pl. Enl. 880.

Genus LOBIVANELLUS †, n. g.

< *Parra*, Gmel., < *Tringa*, Lath., < *Charadrius*, Wagl., < *Vanellus*, Cuv.

Diff. Char.—*Four-toed, lores wattled.*

Rostrum ut in Charadriis.

Membrana loris affixa, nuda, erecta, in anticum protensa.

Alæ elongatæ, caudam vix superantes, remigibus tribus primariis subæqualibus. Spina pollicaris valida, acuta.

Cauda modicæ longitudinis, rectricibus æqualibus.

Pedes elongati, graciles, tetradactyli, acrotarsiis scutellatis.

Habitant in Africâ, Asiâ, Australiâ, (Americâ?).

Species: 1. *L. goensis* (Gmel.), Pl. Enl. 807. 2. *L. gallinaceus* (Wagl.), Jard. Ill. Orn. ser. 1. pl. 84. 3. *L. ludovicianus* (Gmel.), Pl. Enl. 835. 4. *L. senegalus* (Linn.), Pl. Enl. 362. 5. *L. albicapillus* (Vieill.), Swains. B. W. Af. vol. ii. pl. 27. 6. *L. tricolor* (Horsf.), (*Ch. macropterus*, Wagl.). 7. *L. dominica* (Gmel.), (*Ch. brissonii*, Wagl.). 8. *L. albiceps* (Gould), Proc. Zool. Soc. pt. ii. p. 45. 9. *L. cucullatus* (Temm.), Pl. Col. 505.

---

* Σαρκίον, a caruncle; φέρω, to bear.

† *Lobus*, a caruncle; *vanellus*, a Lapwing.

3. DESCRIPTIONS OF SEVERAL NEW OR IMPERFECTLY-DEFINED GENERA AND SPECIES OF BIRDS.

[From Annals and Magazine of Natural History, vol. xiii. June 1844.]

THE details of zoology are now diffused over so wide a field of literature, that it is next to impossible to pronounce with certainty that any given specimen belongs to an undescribed species; and although confusion is often caused by the too hasty and careless definition under new names of species previously described, yet, on the other hand, science may be retarded by too great backwardness in making known new species and groups. With this feeling I now venture to describe a few out of many species of birds which have long remained unnamed in my cabinet; and though it is very possible that some of them may be already described in works to which I have not had access, yet having searched carefully through a large number of ornithological publications without meeting with any notice of these species, I am disposed to believe that the majority of them are really nondescript.

FALCONIDÆ, ACCIPITRINÆ.

Genus ISCHNOSCELES, Strickland (ἰσχνοσκελής, exilia habens crura).

Rostrum asturinum, subexiguum, elevatum, compressum, cera longiuscula, culmine satis curvato, dertro hamato, commissura subrecta, vix sinuata, dertrum versus subito deflexa, mandibula debili, denticulo obtuso versus apicem instructa, gonyde vix ascendente. Nares ovatæ, obliquæ. Alæ mediocres, caudæ trientem attingentes, rotundatæ, remigibus graduatis, 5ª et 6ª longissimis. Cauda elongata, rotundata. Tarsi gracillimi, acrotarsiis paratarsiisque scutatis, scutis lævigatis, subobsoletis. Digiti graciles, digitus medius elongatus, *externus interno multum brevior*. Ungues curvati, acuti, subtus complanati, externus longe minimus, alii subæquales.

Typus *Ischnosceles gracilis* (Falco gracilis, *Temm.* Pl. Col. 91).

The slenderness of the tarsi in this bird, and the remarkable proportions of the toes, seem to justify its generic separation from Astur and from Accipiter, where it has been hitherto classed. The external toe (exclusive of the claw) falls short of the extremity of the second phalanx of the middle toe, and the end of the inner toe is parallel with the middle of the third phalanx, while in most other raptorial birds the outer toe is longer than the inner.

SYLVIIDÆ, SAXICOLINÆ.

PRATINCOLA PASTOR, Strickland.

Le Pâtre, Levaill. Ois. Af. pl. 180.

P. ptilosi omnino *Pratincolæ rubicolæ* (Linn.), nisi uropygio, abdomine, caudæque tectricibus omnibus (etiam in fœmina) pure albis, pectore intense rufo.

Habitat in Africâ meridionali.

Several authors have mentioned the common Stonechat of Europe

(*Pratincola rubicola*) as occurring in South Africa, but I believe that all the specimens which have been so considered will be found to belong to the present nearly allied species. It was first indicated as a distinct species by Levaillant, but as later writers have persisted in uniting it with *P. rubicola,* it has never yet received a systematic name. The plumage is identical with that of *P. rubicola,* except that the rufous of the breast is more intense, and the belly and upper and lower tail-covers in both sexes are uniform pure white. The dimensions are moreover rather larger than in *P. rubicola.*

Total length 5¼ inches; beak to gape 7½ lines, to front 5 lines; wing 2¾ inches; medial rectrices 2 inches 4 lines, external 2 inches 2 lines; tarsus 11 lines.

TURDIDÆ, PITTINÆ.

Pitta cucullata, Hartlaub.

P. cucullata, Strickl. Ann. Nat. Hist. vol. xiii. p. 410. pl. 11.
P. nigricollis, Blyth.—H. E. S. MS.

P. summo capite ferrugineo, loris, mento, gutture, genis, auricularibus nuchaque nigerrimis dorso, scapularibus, remigibus tertialibus tectricibusque dorso proximis obscurè viridibus; tectricibus majoribus externis obscurè viridi-cæruleis, minoribus et uropygio vividè lazulinis; remigibus primariis nigris, 1ª et 2ª macula alba mediana in latere interiore, quatuor sequentibus utrinque similiter notatis; remigibus secundariis nigris, apicem versus in latere exteriore viridi-cæruleis; caudæ tectricibus superioribus nigris, plumis uropygii lazulinis obtectis, rectricibus nigris, apicibus obscurè cæruleis. Pectus, venter et hypochondria pallidè viridia, nitore cærulescente; abdomen, crissum, tectricesque caudæ inferiores coccinea, rostrum fuscum, pedes pallidi.

I had described and figured the above bird under the impression that it was a new species, when I found that it was already described by M. Hartlaub of Bremen in the 'Revue Zoologique,' 1843, p. 65. As however the species is rare in collections, the present delineation will make it better known. It inhabits Malacca.

The crown is deep ferruginous; the chin, throat and sides of the head deep black, forming a collar on the nape. Back, scapulars, tertials and covers next the body dark green; the outward greater covers dark greenish blue, the lesser covers and rump bright glossy azure; primaries black, the first and second with a medial white spot on the inner web, the four next with a white bar crossing both webs; secondaries black, broadly margined externally with greenish blue towards their extremities. Upper tail-covers black, concealed by the blue feathers of the rump; tail black, tipped with greenish blue. Breast, upper belly and sides pale sea-green with azure reflexions; abdomen, vent and lower tail-covers crimson. Beak fuscous, legs and claws pale.

Total length 7 inches; beak to gape 1 inch, to front 9 lines; wing 4 inches 2 lines; medial rectrices 1 inch 6 lines, external 1 inch 4 lines; tarsus 1 inch 4 lines.

TURDIDÆ, PYCNONOTINÆ.

CRINIGER ? ICTERICUS, Strickland.

Trichophorus indicus, Jerdon.—H. E. S. MS.

C. corpore supra olivaceo-viridi, remigibus fuscis, extus ferrugineo-flavido, intus stramineo-marginatis, rectricibus olivaceo-viridibus, intus stramineo-marginatis, loris, superciliis, genis corporeque toto inferno lætè flavis, rostro pedibusque cinerascentibus.

This bird differs from the type of Pycnonotus only in the beak and rictal bristles being somewhat longer. As it possesses nuchal bristles, I refer it for the present to Criniger, Temm., though that character is common to most of the true Pycnonoti. In fact, there seems no very good ground for separating these two genera at all. The present bird resembles Brisson's description of his *Merula olivacea indica* (Turdus indicus, *Gmel.*), but its dimensions are considerably less. I believe it to have been brought from the East Indies, but do not know the precise habitat.

Upper parts olive-green; quills fuscous, margined externally with ferruginous yellow and internally with straw-colour, as are the rectrices. Lores, circuit of eye, and whole lower parts bright yellow, with a slight olive tinge on the breast. Beak and legs cinereous.

Total length 7 inches; beak to gape 9½ lines, to front 7½; rictal bristles ½ an inch; wings 3½ inches; medial and lateral rectrices 3¼ inches; tarsus 8 lines.

PYCNONOTUS FINLAYSONI, Strickland.

Brachypus finlaysoni, Horsf. MSS.

P. fronte, genis guttureque flavis (plumarum scapis flavissimis); pileo olivascente-cinereo; loris nigris, dorso, alis, caudaque obscurè olivaceis, remigibus rectricibusque extus olivaceo-flavescentibus, rectricibus lateralibus strictissimè flavido terminatis; pectore et abdomine cinereo-olivaceis, ventre imo, crisso et alarum tectricibus infernis lætè flavis. Rostrum pedesque corneo-brunnei.

I am not aware that this bird has been yet described, but as it has received from Dr. Horsfield the MS. name of *finlaysoni*, I think it right to adopt that appellation. It is a typical Pycnonotus; the form of the beak agrees with that of *P. capensis*, and the feathers of the rump are very long and downy. It is probably from some of the Malasian islands, but I am unacquainted with the precise habitat.

Front, cheeks and chin yellow, brightest down the middle of each feather; lores velvety black; upper parts obscure olive, greyish on the crown, and yellowish on the wings and tail; three or four pairs of lateral rectrices narrowly tipped with pale yellowish. Below dirty olive; lower belly, tail-covers and lower wing-covers bright yellow. Beak and legs corneous.

Total length 6¾ inches; beak to gape 7½ lines, to front 6½ lines;

wing 3 inches 1 line; medial rectrices 3¼ inches, external 2 inches 10 lines; tarsus 9 lines.

PYCNONOTUS CROCORRHOUS, Strickland.

Yellow-vented Flycatcher, Brown, Ill. Zool. pl. 31. fig. 1.
Muscicapa hæmorrhousa, β, Gmel. Syst. Nat. p. 941.
Turdus hæmorrhous, Horsf. in Linn. Trans. vol. xiii. p. 147.
Ixos hæmorrhous, Vigors in Raffles's Life, p. 661.
Trichophorus virescens, Jerdon.—H. E. S. MS.

P. capite subcristato, facie, mento, nigerrimis; dorso alisque fuscis, marginibus pennarum pallidioribus, caudæ tectricibus superioribus albis; cauda fusco-nigricante, tenuiter albido terminata; regione parotica alba; partibus infernis cinerascente-albidis, crisso lætè aurantio-croceo. Rostrum pedesque nigri.

This bird, which has been long known, requires a new specific name, being quite distinct from the true *Pycnonotus hæmorrhous* (Gmel.) of Ceylon, in which the vent is crimson. If we regard the genus Pycnonotus of Kuhl to be typified by *Turdus capensis*, Linn., we must refer to it all the species of Hæmatornis, Swains., and the present bird among the number. It closely agrees in form with *Pycnonotus capensis*, Linn., but is at once distinguished by the vent being orange instead of yellow, and by other characters.

This bird inhabits Java. The head and chin are black; back and wings dusky, with paler margins; rump white; tail black-brown, narrowly tipped with whitish; ears white; lower parts dirty white; under tail-covers bright saffron-colour; beak and legs black.

Total length 7¾ inches; beak to gape 10 lines, to front 7½ lines; wing 3¾ inches; medial rectrices 3 inches 7 lines, external 3 inches 5 lines; tarsus 10 lines.

PYCNONOTUS FLAVIRICTUS, Strickland.

Galgulus philippinensis, Kittlitz.—H. E. S. MS.

P. striga superciliari a naribus excurrente, alteraque suboculari albis, loris nigris, macula in mandibulæ basi mentoque flavis; capite corporeque toto supra obscurè olivaceis, remigibus secundariis rectricibusque basin versus flavido-olivascente limbatis, hisce strictissimè albido terminatis; corpore inferno cinerascenti-albido, flavido pallescente strigato, crisso pallidè flavo. Rostrum pedesque corneo-brunnei.

I purchased this bird from a dealer, who informed me it was from Madras. It is a typical Pycnonotus, with the rump-feathers very downy, nearly allied to *P. goiavier* (Scop.) (Muscicapa psidii, *Gmel.*, Turdus analis, *Horsf.*), but is distinguished by the gonys being slightly curved upwards, by the yellow rictal spot, &c.

A white superciliary streak from the nostrils is separated from one below the eye by the black lores. The tip of the chin and a spot at the base of the lower mandible are yellow. Upper parts obscure olive; secondaries and rectrices margined with yellowish-olive, the latter narrowly tipped with whitish. Lower parts dirty white; feathers margined laterally with very pale yellow, producing a streaked appearance. Vent and lower tail-covers pale yellow.

Total length 7½ inches; beak to gape ¾ of an inch, to front 7 lines; wing 3 inches 5 lines; medial rectrices 3¼ inches, external 3 inches 1 line; tarsus 11 lines.

HYPSIPETES PHILIPPENSIS, Strickland.

H. pileo cinereo-fusco, dorso alisque fusco-olivaceis, remigibus fuscis extus fusco-olivaceo limbatis, rectricibus fuscis; genis gulaque fusco-ferrugineis, scapis pennarum albidis, pectore et abdomine olivascente-albidis, crisso albido. Rostrum pedesque corneo-fusci.

This species agrees with the type of Hypsipetes, except in the tail being slightly rounded. It was brought by Mr. Cuming from Manilla. The feathers of the crown and chin are pointed. Three or four nuchal bristles project half an inch beyond the plumage. Rump-feathers downy.

Crown cinereous brown. Upper parts dark olive; remiges and rectrices fuscous, the former margined with olive; cheeks and chin obscure ferruginous, the shafts of each feather whitish. Lower parts dirty white with an olive tinge; lower tail-covers whitish. Beak and legs corneous.

Total length 8¼ inches; beak to gape 11½ lines, to front 10 lines, height 3 lines, breadth 3½ lines; wing 3 inches 8 lines; medial rectrices 3 inches 6 lines, external 3 inches 4 lines; tarsus 8½ lines.

MUSCICAPIDÆ, TYRANNINÆ.

SUIRIRI? ICTEROPHRYS (Vieill.).

Suiriri obscuro y amarillo, Azara, Pax. Par. ii. p. 118. No. 183.
Muscicapa icterophrys, Vieill. Nouv. Dict. d'Hist. Nat. xxi. p. 458.
M. chrysochlores, Max.
M. capistrata, Licht.—H. E. S. MS.

S. fronte, pileo, nucha et dorso toto olivaceo-viridibus, alis fuscis, tectricibus omnibus largè, remigibus secundariis tertiariisque strictè, cinerascente albido terminatis, caudæ tectricibus supernis fusco-olivaceis, cauda subfurcata, fusco-atra, rectricibus externis extus albido marginatis; linea superciliari lætè flava a naribus oriente; loris plumisque paroticis olivaceo-fuscis; corpore toto inferno lætè flavo. Rostrum pedesque atri.

Inhabits Buenos Ayres.

Of the numerous species of American birds which have been classed in the genera Tyrannula and Elania, and which exhibit much variety in the modifications of the beak, the present one has that organ the most elongate; indeed it approaches in form the beak of the Sylvicoline genus Myiodioctes, though the straight culmen, the comparatively short black tarsi, covered with seven or eight short scuta, the slender toes and sharp claws, show the true place of the bird to be among the Tyranninæ. I had intended making this the type of a new genus, under the name of Satrapa (*quasi* a petty tyrant), but perceiving that M. D'Orbigny has included it in his genus Suiriri, I retain that generic name for the present. M. D'Orbigny admits that it differs in its smaller head, more slender beak and longer tarsi from the Suiriri of Azara, No. 179, which is the type of his genus; but not having examined the latter bird, I am fearful of creating a new genus without sufficient reason. M. D'Orbigny's *second* species of Suiriri belongs to Gould's genus Pyrocephalus, a very distinct form from the present bird.

Above olive-green, beneath bright yellow; wings dusky, the covers broadly, the secondaries and tertiaries narrowly, edged with greyish white. Tail very dark brown, outer rectrices margined externally with whitish. A bright yellow streak from the nostrils over the eye. Lores and ear-covers dusky olive. Beak and legs black.

Total length 6¼ inches; beak to gape 8 lines, to front 6 lines, width 2½ lines, height 2 lines; wing 3¼ lines; medial rectrices 2½ inches, external 2¾ inches; tarsus 9 lines; middle toe and claw 8½ lines, hind ditto 6 lines; outer toe slightly longer than the inner.

EUSCARTHMUS CINEREUS, Strickland.

E. capite supra nigro, plumis medianis basin versus albis, dorso cinereo, in uropygio dilutiore, tectricibus minoribus cinereis, mediis et majoribus fusco-nigris, cinereo terminatis; remigibus fusco-nigris, tertiariis cinereo limbatis; rectricibus fusco-nigris; gula pectoreque dilutè cinereis, abdomineque crissoque albidis. Rostrum pedesque atro-fusci.

Inhabits Chili.

Beak slightly broader than in *E. parulus* (Kittlitz), and proving the affinity of this genus to Tyrannula. Feathers of the crown rather lengthened, forming a crest.

Crown black, with a concealed white vertical spot. Upper parts grey, palest on the rump; middle and greater wing-covers dusky black, tipped with grey; remiges dusky, tertials margined with grey; tail dusky black; chin, throat and breast pale grey; belly and lower tail-covers nearly white; beak and legs blackish.

Total length 4 inches; beak to gape 6 lines, to front 4 lines; wing 2 inches 2 lines; medial rectrices 2 inches, external 1 inch 10 lines; tarsus 8 lines.

LANIIDÆ, FORMICARIINÆ *.

Genus HOLOCNEMIS, Strickland. (ὅλος, integer, κνημὶς, ocrea.)

Rostrum elongatum, ad basin subdepressum, apicem versus subcompressum, mandibulis juxta apicem leviter emarginatis. Culmen rectum, dertro deflexo, commissura recta, ad apicem deflexa, gonys elongata, subrecta, leviter ascendens. Vibrissæ nullæ. Nares ovatæ, nudæ, a plumis lori subremotæ. Alæ mediocres, rotundatæ, remige 4ª vel 5ª longissima, remige 1ª dimidio breviore. Cauda sub-

* In this subfamily I include the genus Thamnophilus, as it cannot possibly be separated from the American Ant-thrushes in any natural arrangement.

brevis, rotundata. Tarsi elongati, acrotarsiis et paratarsiis integris. Digiti sublongi, graciles; externus phalange prima ad medium annexus.

The two birds which I propose to distinguish under the above generic name are distinguished from the genera Formicivora, Swains., and Myrmeciza, Gray, by the elongate beak, short tail, and entire acrotarsus; the latter character, as well as their greater length of tail, distinguishes them from Urotomus, Swains., and the greater freedom of the external toe separates them from Pithys, Vieill. Possibly the present genus may be referable to Leptorhynchus, Menetries; but as the latter name is preoccupied, Holocnemis may in that case take its place.

HOLOCNEMIS FLAMMATA, Strickland.

H. corpore supra olivaceo-fusco, alis fuscis, tectricibus omnibus olivaceo limbatis, scapis et gutta subtriquetra apicali albis; remigibus olivaceo limbatis; cauda obscurè fusca, rectricibus obtusè acuminatis, lateralibus albido strictè terminatis; gula alba; genis et partibus infernis pallidè olivaceis, litura in singulis plumis longitudinali acuminata (in pectoris plumis latissima), alba. Rostrum albidum, maxillæ basi fusca, pedes unguesque albidi.

Habitat unknown, though doubtless American.

The middle toe and claw are about equal in length to the tarsus; hind toe shorter; outer toe slightly longer than the inner. Claws considerably developed, compressed, moderately curved.

Above olive-brown; wings fuscous, the covers edged with olive, and with the shafts and a subtriangular terminal spot white. Remiges fuscous, margined with olive; tail dark fuscous, the feathers obtusely pointed, the external ones slightly tipped with whitish. Throat white; cheeks and lower parts pale olive-brown, each feather with a pointed white streak, very broad on the breast and narrowest on the sides. Beak whitish, basal half of upper mandible brown; legs and claws very pale yellowish white.

Total length $5\frac{1}{4}$ inches; beak to gape 1 inch 1 line, to front 10 lines, breadth $3\frac{1}{2}$ lines, height 3 lines; wing 2 inches 10 lines; medial rectrices 2 inches, external $1\frac{3}{4}$ inch; tarsus 11 lines; middle toe and claw 11 lines, hind ditto 8 lines.

HOLOCNEMIS CINNAMOMEA, (Gmel.).

Wall-creeper of Surinam, Edw. ♀.—H. E. S. MS.
Turdus cinnamomeus, Gmel. Pl. Enl. 560. f. 2.

H. fronte, pileo, dorso toto caudaque tectricibus supernis rufo-ferrugineis, tectricibus alarum nigris, minoribus albo-, mediis majoribusque ochraceo-, terminatis, remigibus rectricibusque fuscis, extus fusco-ferrugineo marginatis; linea superciliari albo a naribus ad pectoris latera descendente et abdomine albo confluente; loris, oculorum ambitu, genis, gula et pectore toto nigerrimis, hypochondriis crissoque ferrugineis. Rostrum pedesque cornei.

Taking *H. flammata* as the type of Holocnemis, the present bird is somewhat aberrant, the tail being rather longer and more rounded, the tarsi longer, the lateral toes equal, and the claws shorter and less curved. The style of plumage bears much resemblance to *Myrmeciza loricata* (Licht.), and to *Urotomus ? formicivorus* (Gmel.).

The upper parts are deep ferruginous; wing-covers black, the lesser tipped with white, and the middle and greater with ochraceous, forming two bars; remiges and rectrices fuscous, margined with ferruginous. A narrow white line commences at the nostrils, and descending the side of the neck and breast blends into the white of the abdomen. The whole space enclosed by this white line is deep black. Sides and lower tail-covers ferruginous. Beak and legs corneous.

Total length 5½ inches; beak to gape 1 inch, to front 10 lines; breadth 3 lines, height 2½ lines; wing 2 inches 7 lines; medial rectrices 2¼ inches, external 1¾ inch; tarsus 1 inch; middle toe and claw ¾ of an inch, hind ditto 7½ lines.

MYRMECIZA MELANURA, Strickland.

M. capite, nucha, dorso alisque obscurè ferrugineo-fuscis, uropygio obscuriore, cauda fusco-atra, loris fusco-cinerascentibus, gula albida, pectore et abdomine dilutè fuscis, rufescente tinctis, crisso fuliginoso-atro. Maxilla pedesque fusci, mandibula albida.

In the form of the beak and general proportions this species agrees with *M. leuconota*, Spix, Av. Bras. vol. ii. pl. 39. f. 2 (*Drymophila atra*, Swains.). The acrotarsia are divided into five scuta, the paratarsia entire. Habitat unknown.

Upper parts dull ferruginous brown, darker on the rump; tail dusky black; lores dusky cinereous; throat whitish; breast and belly pale rufous brown, sides darker; lower tail-covers sooty black; upper mandible and legs brown, lower mandible whitish.

Total length 7 inches; beak to gape 10 lines, to front 8 lines; wing 3 inches; medial rectrices 3 inches, external 2¼ inches; tarsus 1 inch 1 line; middle toe and claw 1 inch, hind ditto 8 lines; outer ditto 8½ lines, inner ditto 7½ lines.

LANIIDÆ ? TIMALIINÆ.

In this group I would include the genera Timalia, Brachypteryx, Malacopteron, Eyton (Trichostoma, Blyth), Macronus, Jard., and several of the East Indian "Myiotheræ" of Temminck. We do not know enough of their habits to decide whether they are most allied to the Formicariinæ or to the Malurinæ; all that can be said of them is, that they form a natural group inhabiting the Malasian region, and that they appear to approach the Laniidæ in structure sufficiently to warrant their collocation for the present in that family. Mr. Eyton's genus Malacopteron exhibits this Laniine structure to the greatest degree, the beak being precisely that of a Thamnophilus, with the addition of strong rictal bristles.

I now proceed to describe an apparently new species,

MALACOPTERON MACRODACTYLUM, Strickland.

M. capite, nucha et dorso superiore rufo-brunneis, plumis fusco marginatis, uropygio, alis caudaque fusco-ferrugineis, hac obscuriore, loris albidis, genis fuscis, mento gulaque albis, plumis fuliginoso terminatis, abdomine obscurè albido, hypochondriis et crisso dilutè rufo-brunneis, rostro pedibusque fuscis, digito medio laterales multum superante.

Agrees with Mr. Eyton's type-species, *M. magnum*, in the form of the beak, wings and tail, in the scale-like structure of the coronal feathers, and the loose downy plumage of the rump, but differs in the greater strength of the hind toe and the remarkable length of the middle one. The lateral toes are equal, and the bases of their claws are parallel with the distal end of the second phalanx of the middle toe, their extremities reaching about two-thirds the length of its third phalanx. The claws are less curved than in *M. magnum*, that of the middle toe being nearly straight. My specimen was brought from Malacca.

The feathers of the crown and upper back are rufous brown, margined with dusky; rump, wings and tail ferruginous brown, the last darkest. Lores white; cheeks fuscous; chin and throat white, the feathers of the latter largely terminated with sooty black. Lower parts dirty white; sides and lower tail-covers pale rufous brown. Beak and legs horn-coloured.

Total length $6\frac{1}{4}$ inches; beak to gape 1 inch, to front $\frac{3}{4}$ of an inch, breadth 4 lines, height 3 lines; wing $3\frac{1}{4}$ inches; medial rectrices $2\frac{1}{2}$ inches, external 2 inches 4 lines; tarsus 1 inch 2 lines; middle toe and claw 1 inch 2 lines, hind ditto 10 lines, lateral ditto $8\frac{1}{2}$ lines.

FRINGILLIDÆ, PLOCEINÆ.

SPERMOPHAGA MARGARITATA, Strickland.

S. fronte, capite summo, nucha, dorso alisque obscurè ferrugineis, unicoloribus; primariis intus fuscis; caudæ tectricibus supernis, rectricumque marginibus externis obscurè vinaceo-rubris; rectricibus in reliqua parte nigris; loris, superciliis, genis, gutture pectoreque vinaceo-rubris; partibus reliquis infernis nigerrimis, pectus versus et ad latera maculis magnis rotundis caryophyllaceis (binis in singulis pennis) punctatis. Rostrum nitidè cyaneum, pedes (exsiccati) albidi.

This beautiful little bird was purchased at Cape Town, and was said to have been brought from Madagascar. The beak is less developed than in *Spermophaga hæmatina*, Vieill., and the first quill is barely one-third the length of the fourth, fifth and sixth (which are equal), but in other respects it accurately accords with the type of *Spermophaga*. The arrangements of its colours show its affinity to *S. guttata*, Vieill., and the peculiar blue colour of the beak is common to both, as well as to *S. hæmatina*, Vieill., the specific distinctness of which from *S. guttata* is at present undecided.

Mr. G. R. Gray has changed Mr. Swainson's name *Spermophaga* to *Spermospiza*, because the name *Spermophagus* is already used in entomology; but as I am by no means prepared to concede that mere *similarity* affords a sufficient ground for cancelling generic names, I have retained Mr. Swainson's appellation.

The whole upper parts of this bird are rich ferruginous brown, except the quills, which are dusky within; the upper tail-covers and outer margins of the rectrices dull vinous red, and their inner webs and apical portions black. The circuit of the eyes, cheeks, throat and breast pale claret-red; rest of lower parts deep black, spotted next the breast and on the sides with large pearl-like spots the colour of peach-blossom, of which two are placed transversely and subterminally on each feather.

Total length $4\frac{3}{4}$ inches; beak to gape $5\frac{1}{2}$ lines, to front 5 lines; width $2\frac{1}{2}$ lines, height 3 lines; wing 2 inches 1 line; medial rectrices 2 inches, external $1\frac{3}{4}$ inch; tarsus $7\frac{1}{2}$ lines; middle toe and claw 7 lines, external 5 lines, internal 4 lines, hind 5 lines.

FRINGILLIDÆ, TANAGRINÆ.

TACHYPHONUS SAUCIUS, Strickland.

Tachyphonus phœniceus, Swains.—H. E. S. MS.

T. corpore toto atro-chalybeo, tectricibus alarum supernis minoribus juxta humerum albis, juxta carpum sanguineis, infernis omnibus albis. Rostrum nigrum, mandibula medio alba, pedes nigri.

Allied to *Tachyphonus nigerrimus*, Gmel., but differs in the smaller size, shorter beak, and sanguineous spot near the carpus. I presume it to inhabit Columbia or Central America*.

Entirely black, with a purplish gloss, except the lesser wing-covers next the humerus, which are white and pass into bright orange-red as they approach the carpus. Lower wing-covers white. Beak black, middle of lower mandible whitish; legs black.

Total length $5\frac{1}{2}$ inches; beak to gape 7 lines, to front 6 lines; wing $2\frac{3}{4}$ inches; medial rectrices $2\frac{3}{4}$ inches, external $2\frac{1}{2}$ inches; tarsus 9 lines.

TACHYPHONUS RUFICEPS, Strickland.

T. fronte, mento summo, loris oculorumque ambitu nigris, capite toto reliquo gulaque intensè rufo-castaneis, corpore toto plumbeo-cinerascenti, alis caudaque obscurioribus. Rostrum nigro-cinereum, tomiis albidis, pedes brunnei.

The upper mandible is smaller than the lower, somewhat like that of an *Emberiza*, but the culmen is more arched. It agrees however sufficiently with the structure of *Tachyphonus quadricolor*, Vieill. (T. auri-

* Interior of Brazil, Borba (Natterer). Sclater, Synop. Av. Tanag. p. 40.

capillus, *Spix*) to warrant its collocation in the same genus. I am unacquainted with the habitat*.

Front, lores, upper chin and circuit of eyes black; head, cheeks and throat deep chestnut-red; rest of plumage leaden grey, passing to fuscous on the wings and tail. Beak blackish, margins of mandibles whitish; legs brown.

Total length 5½ inches; beak to gape 7 lines, to front 6 lines; wings 2½ inches; medial rectrices 2½ inches, external 2¼ inches; tarsus 10 lines.

CALLISTE THALASSINA, Strickland.

C. capite cyaneo, loris et oculorum ambitu nigris, mento summo genisque thalassino-viridibus, collo toto et gula nigris; dorso, scapularibus caudæque tectricibus supernis cyaneis, nitore thalassino, tectricibus alarum minoribus intensè cæruleis, mediis, majoribus remigibusque secundariis nigris, viridi-thalassino marginatis; primariis, rectricibusque nigris cæruleo limbatis, pectore, abdomine crissoque albis, thalassino imbutis. Rostrum nigrum, pedes cinerei.

A typical *Calliste*, believed to be brought from Mexico†. It is nearly allied to *Tanagra nigroviridis*, Lafr., Mag. Zool. pl. 43, but that has more of yellow and green and less of blue in its plumage than the present bird.

Head pale azure; chin and cheeks vivid sea-green; lores and eyelids black; circuit of the neck black; back, scapulars and upper tail-covers pale blue, with a gloss of sea-green. Lesser wing-covers deep vivid blue; middle and greater covers and secondaries black, margined with sea-green; primaries and tail black, margined with blue; lower parts white, with a delicate gloss of sea-green and azure. Beak black, legs cinereous.

Total length 4¾ inches; beak to gape 6 lines, to front 5 lines; wing 2 inches 7 lines; medial and external rectrices 2 inches; tarsus 7½ lines.

NEMOSIA FULVESCENS, Strickland‡.

N. capite toto supra aurantio, loris gulaque flavis, corpore supra cinereo-olivascenti, remigibus rectricibusque obscurioribus, corpore inferno dilutè fulvescenti, abdomine albido. Rostrum cinereum, tomiis albidis; pedes fusco-cinerei.

Allied to, but sufficiently distinct from, *Nemosia ruficapilla*, Vieill., Gal. Ois. pl. 164. Inhabits Brazil?

Head orange, passing into yellow on the lores and throat; upper parts greyish olive, darker on the wings and tail; lower parts pale fulvous or cream-colour, almost white on the belly. Beak cinereous, margins whitish. Legs dark cinereous.

* Brazil.—S. Paolo.—Paraguay. Sclater, Synop. Av. Tanag.—ED.

† This will stand as *C. nigro-cincta*, Bonap. 1837. We have specimens from Bogota, and Mr. Sclater gives New Grenada, Province of Quixos, Ecuador, &c. as habitats.—ED.

‡ Mr. Sclater makes this=*N. sordida*, Lafr. and d'Orbig. He has compared the type specimens in the Parisian, and the *N. fulvescens* in the Strickland collections. *N. sordida* is from Bolivia.—ED.

Total length 5¾ inches; beak to gape 6 lines, to front 5 lines; wing 2 inches 7 lines; medial rectrices 2 inches 7 lines, external 2 inches 5 lines; tarsus 9½ lines.

EMBERNAGRA LONGICAUDA, Strickland.

E. linea a naribus ad oculos, horumque ambitu albis, corpore toto supra, alis caudaque cuneata viridi-olivaceis, scapis in vertice nigris; loris, genis lateribusque colli olivaceo-cinereis, pectore dilutè cinerascenti, gula abdomine crissoque dilutè fulvis, maxilla fusca, mandibula flavida, pedibus brunneis.

Inhabits South America.

Closely allied to *Embernagra platensis*, Gmel. (Emberiza platensis, *Gmel.*, *Azara*, no. 90), but differs in the lower mandible being yellowish instead of orange, the white line over the eye, the longer and more cuneate tail, and the shorter tarsus and claws.

A white line from the nostrils surrounds the eye. Upper parts greenish olive, brightest on the wings and tail; shafts of the feathers on the crown black. Lores, cheeks and sides of neck olive-grey. Lower parts pale fulvous, tinged with grey on the breast. Upper mandible fuscous, lower yellowish; legs pale brown.

Total length 8¼ inches; beak to gape 9 lines, to front 8 lines; wing 3 inches; medial rectrices 4 inches, external 2½ inches; tarsus 1 inch; middle toe and claw 11 lines, hind ditto 8½ lines.

ORTHOGONYS, Strickland. (*ὀρθὸς*, rectus; *γωνὺς*, gonys.)

Rostrum elongatum, compressiusculum, culmine obtusè carinato, a basi ad apicem curvato, commissura satis decurvata, tomiis paulo inflexis, maxilla vix emarginata, gonyde rectissima, nec ascendente. Nares ovatæ subbasales, pilis raris frontalibus tectæ. Alæ mediocres, remigibus 2, 3, 4, subæqualibus, 1ª paulo breviore. Cauda mediocris, rotundata. Tarsi subbreves, acrotarsiis scutellatis, paratarsiis integris; digiti mediocres, externus internum paulo superans, ungues satis curvati, acuti.

The general habit of this bird suggests the idea of a Tanagrine form, but it is distinguished from all the genera which I know by its elongate beak, much curved culmen, and perfectly straight gonys. The beak is somewhat like that of *Lamprotes*, but is not so high and compressed.

ORTHOGONYS VIRIDIS (Spix).

Tanagra viridis, Spix, Av. Bras. pl. 48. f. 2.

O. genis et corpore supra olivaceo-viridi unicolore, subtus lætè flavo, pectore et hypochondriis olivaceo tinctis, rostro nigro, pedibus brunneis.

Inhabits Brazil.

Body wholly olive-green above, beneath yellow, tinged with olive on the breast and sides. Beak black; legs light brown.

Total length 8 inches; beak to gape 10 lines, to front 9 lines, breadth and height 3½ lines; wing 3¾ inches; medial rectrices 3½ inches, external 3 inches 2 lines; tarsus 10 lines; middle toe and claw 10 lines, hind ditto 8 lines.

4. REPORT ON THE RECENT PROGRESS AND PRESENT STATE OF ORNITHOLOGY.

[Report Brit. Assoc. vol. xiii. p. 170. 1844.]

Introduction.

The object of this Report is to give a sketch of the recent progress, present state, and future prospects of that branch of zoology which treats of the class of Birds. As the chief, indeed the only method by which his study can be developed into a science, consists either in describing and depicting the character and habits of this class of animals in *books*, or in preserving and arranging the objects themselves in *museums*, I shall review in succession the progress which has been made in these two departments of the subject, and shall conclude with a few remarks on the *desiderata* of ornithology.

In treating of the bibliography of ornithology, however, it is not necessary to go into much detail respecting the works of older date than about fifteen years ago. The ornithological works of the last and the earlier part of the present century are well known to most naturalists, and the reader will find ample and for the most part just criticisms respecting them in Cuvier's 'Règne Animal,' vol. iv., Temminck's 'Manuel d'Ornithologie,' Swainson's 'Classification of Birds,' and his 'Taxidermy and Bibliography,' Wood's 'Ornithologist's Text Book,' Wilson's article *Ornithology* in the 'Encyclopædia Britannica,' Rev. L. Jenyns's 'Report on Zoology,' 1834, Burmeister's article *Ornithologie* in Ersch and Gruber's 'Encyclopädie der Wissenschaften,' and other sources. I shall therefore only give such a cursory notice of some of the earlier writers on ornithology as will serve to introduce the more legitimate subject of this Report.

It may perhaps surprise those who are not very conversant with the subject to be told that ornithology is in a less advanced state than many other departments of zoology. Persons who are accustomed to regard "stuffed birds" as constituting the most usual and most attractive objects of a public museum, will not readily admit that the various species of Mammalia, Fish, Insects, Mollusca, and even Infusoria, are more accurately determined and more perfectly methodized than the class of Birds. Such is, however, the case, and although in the last few years ornithology has certainly made a very marked progress, yet it is still considerably in the rear of its sister sciences.

This backward condition of ornithology must be attributed in great measure to the pertinacity with which its followers during many years adhered to the letter instead of to the spirit of Linnæus's writings. In this country the venerable Latham, who for half a century was regarded as the great oracle of ornithology, persisted so late as 1824 in classifying his 5000 species of birds in the same number of genera (with very few additions) as were employed by Linnæus for a fifth part of those species. The consequence was that many of the genera in Latham's last work contain each several hundred species, frequently presenting the most heterogeneous characters, and massed together without any, or with only very rude attempts at further subdivision. Shaw's 'General Zoology' was, in a great measure, a servile copy of Latham's 'Ornithology,' and these two works formed for many years almost the only text-books on the subject. On the Continent meanwhile, those who were not disciples of Linnæus, transferred their allegiance to Buffon, and often exceeded that author in their contempt for systematic arrangement and uniform nomenclature.

Cuvier, indeed, as early as 1798, had sketched out an improved classification of birds in his 'Tableau Elémentaire de l'Histoire Naturelle,' repeated with amendments in his 'Anatomie Comparée' in 1800. The main features of his arrangement correspond with that which he afterwards adopted in his 'Règne Animal.' About the same period also, Lacépède published a system, arranged on a new plan and containing the definitions of several new genera. Another outline of an improved ornithological system was published in 1806 by M. Duméril in his 'Zoologie Analytique.' But these attempts at progress seem to have been made before the scientific world was able to appreciate them, and several years elapsed before their influence was generally felt.

The logical and accurate Illiger was the next who endeavoured to introduce sounder principles into ornithology; his admirable 'Prodromus Systematis Mammalium et Avium,' published in 1811, after long years of neglect, has now become an almost indispensable handbook to the studier of mammals and birds. But this young reformer died at an early age, and ornithology again relapsed under the drowsy sway of the Linnæan and Buffonian schools.

The next effort in advance was made in 1817, when Cuvier, having previously arranged the Paris Museum according to his own views of the natural system, embodied the results in the 'Règne Animal.' In the ornithological portion Cuvier was anticipated by Vieillot, who having access to the galleries of the museum, is charged with having appropriated the labours of Cuvier by attaching names of his own to the groups there pointed out. Be this as it may, the 'Analyse d'une nouvelle Ornithologie Elémentaire' of Vieillot, and the ornithological portion of the 'Règne Animal' of Cuvier, contain many new generalizations based upon highly important but previously neglected structural characters, and their publication indicated a vigorous effort at transferring the subject from the domain of authority to that of observation.

Temminck, who in his 'Histoire des Pigeons et des Gallinacés,' 1813–15, had introduced several new generic groups into the Rasorial order, published in the second edition of his 'Manuel d'Ornithologie,' 1820, the outline of a general system of ornithology, containing many important additions to the arrangements of Cuvier and Vieillot.

The method of De Blainville, completed in 1822, deserves notice, from his having introduced as a new element of classification the structure of the sternum and of the bones connected with it. The distinctive characters thus deduced are now generally admitted as forming valuable auxiliaries in the search after a natural arrangement.

The improved methods of classification, thus originated on the Continent, made a gradual but slow progress into this country. Dr. Leach seems to have been the first British naturalist who duly appreciated the labours of Cuvier, and in the concluding volumes of Shaw's 'Zoology,' published under his superintendence, the new generic groups of the continental authors were successively introduced, and engrafted upon the stock of Linnæus and Latham. Dr. Horsfield also entered thoroughly into the spirit of the reformers of zoology, and in his valuable memoir on the Birds of Java in the 'Linnean Transactions,' vol. xiii., he adopted the arrangements of Cuvier and of Leach, with many excellent additions of his own. Dr. Fleming's 'Philosophy of Zoology,' 1822, also contributed to render the naturalists of Britain familiar with the improved systems of the Cuvierian school.

The late Mr. N. A. Vigors gave, in 1823, a great impulse to the study of ornithology by his elaborate memoir in the 'Linnæan Transactions,' vol. xiv., on "The Natural Affinities that connect the Orders and Families of Birds." This treatise abounds with original observations and philosophical inferences, but unfortunately they are applied in support of a theory which the most careful inductions and the most unprejudiced reasonings of subsequent naturalists have shown to have no claim to our adoption as a general law. Without entering further upon the *vexata quæstio* of the "Quinary System" than as regards its application to ornithology, I may remark that if we can show that this supposed universal principle fails in its application to any one department of the animal kingdom, it loses its character of universality, and a presumption is raised against its truth even as a *special* or *local* law. The quinary system in fact includes several distinct propositions, the truth of any one of which does not imply that of the remainder. First, it is laid down that all natural groups, if placed in the order of their affinities, assume a circular figure; secondly, that these circles are each subdivided into *five* smaller circles; thirdly, that two of these are *normal*, and the remaining three *aberrant*; and fourthly, that the members of any one circle represent analogically the corresponding members of all other circles. I shall have occasion to recur to these points in speaking of Mr. Swainson's writings, and at present will merely remark, that the application by Mr. Vigors of these novel and singular doctrines to the class of Birds contributed in no small degree to the advancement of ornithological science; for however erroneous a theory may be, yet the researches which are entered upon with a view to its support or refutation invariably advance the cause of truth. Alchemy was the parent of chemistry, astrology of astronomy, and quinarianism has at least been one of the foster-parents of philosophical zoology. Another debt of gratitude which we owe to the quinarians is the broad and marked distinction which they were the first to draw between AFFINITY and ANALOGY—between agreements in *essence*, and agreements in *function* only and not in essence, the one constituting a *natural*, and the other an *artificial* system. And although their foregone conclusions sometimes led them to mistake the one for the other, yet by their clear definitions on the subject they enabled others to detect the errors which in such cases they could not see themselves*.

In 1824 Vieillot presented a new edition of his system, with but slight alterations, in his 'Galerie des Oiseaux,' and in the following year

* The distinction between affinity and analogy is as yet but imperfectly established on the Continent, or at least the terminology employed is very vague. French writers continually use the term *analogie* to express what we call *affinity*, a defect in their scientific language which they might easily remedy by making use of the word "*affinité*," and by restricting *analogie* to its true meaning. The same inaccuracy also exists in the language of geologists, British as well as foreign, when they speak of the *recent analogue* of a fossil, meaning thereby that recent species which has the strongest affinity to the extinct one. They might term it with more propriety the *recent affine*. A similar alteration would also introduce greater precision into the terminology of comparative anatomy. The parts which in different groups of animals are *essentially equivalent*, though often differing in function, are commonly termed *analogous members*, but it would be more correct to call them *affine members*, and to restrict the term *analogous* to those organs which resemble in function *without being essentially equivalent*. Thus the *tooth* of Monodon, the *nose-horn* of Rhinoceros, the *intermaxillaries* of Xiphias, and even the *rostrum* of a Roman galley, all perform a similar function, and are therefore *analogous* organs, but the relation between the weapon of offence in Monodon and the masticatory teeth of other Mammalia is an agreement in essence but not in function, and is therefore not an *analogy* but a real *affinity*. There is yet a third kind of relation between organic beings which does not deserve the name of *analogy*, but which may be simply called *resemblance*, consisting of a mere correspondence in form, but not in function or essence, such as the *resemblance* between *Murex haustellum* and a Woodcock's head, between *Ophrys apifera* and a Bee, &c., a relation which is in every sense *accidental*, though the advocates of the quinary theory have often regarded it as a true analogy.

Latreille proposed another arrangement, which however differs very little from that of Cuvier as finally left by him in the second edition of his 'Règne Animal,' 1829. The celebrity of its author caused the latter work to be speedily translated into other languages, and it soon became the text-book for classification in most of the museums of Europe. The 'Règne Animal' will ever remain a monument of the industry of Cuvier and of his extraordinary powers of generalization, but it would be vain to expect that all parts of so vast an undertaking should be equally perfect, and it is therefore no matter for surprise that the class of Birds, which do not seem to have been a favourite branch of Cuvier's studies, should present many defects in their arrangement. Certain it is that, not to mention many proofs of haste in the citation of species and of authors, the series of affinities is in this work often rudely broken or arbitrarily united. In his arrangement of birds Cuvier seems to have too closely followed the old authors, in adopting an isolated character as the basis of his classification, a practice which inevitably leads to arbitrary and artificial arrangements. He places, for example, the Tanagers, Philedons and Graculæ in the midst of the Dentirostres, Dacnis, Coracias and Paradisea among the Conirostres, Sitta and Tichodroma among the Tenuirostres, Furnarius in Nectarinia, &c. Many of these defects were pointed out by Prince C. L. Bonaparte in an admirable critique published at Bologna in 1830, entitled 'Osservazioni sulla seconda edizione del Regno Animale,' and which is an indispensable appendage to Cuvier's work. Another valuable accompaniment to the 'Règne Animal' is the series of plates published by Guérin under the title of 'Iconographie du Règne Animal de Cuvier.'

This slight preliminary sketch of the progress of ornithological classification has now conducted us to a period when it becomes necessary to enter into greater detail.

I propose, as far as I am able, to notice all the more important ornithological works, which have been published since 1830, and which have contributed to bring the subject to its present state, not indeed of *perfection*, but what is more interesting to those engaged in it, of *progress*. I must however regret, that from the difficulties of obtaining access to many rare continental publications, especially to the almost innumerable annals of scientific societies, this attempt at a general survey of the subject will unavoidably be somewhat incomplete. I shall of course pass over such works as are devoid of *scientific* merit, as well as those mere compilations, which from their want of any new or original matter tend only to diffuse and not to advance the science.

In entering on so large a field it becomes necessary to subdivide the subject, which may be treated of under seven heads, viz.—1. General systematic works. 2. Works descriptive of the Ornithology of particular regions. 3. Monographs of particular groups. 4. Miscellaneous descriptions of species. 5. Pictorial Art as applied to Ornithology. 6. The Anatomy and Physiology of Birds; and 7. Fossil Ornithology.

1. *General Systematic Works.*

Lesson, who in 1828 had published a useful little 'Manuel d'Ornithologie,' based chiefly upon Cuvier's classification, brought out in 1831 a more extended work, entitled 'Traité d'Ornithologie.' This book, which professes to enumerate all the species of birds in the Paris Museum, is upon the whole a very unsatisfactory performance, presenting all the marks of great haste and consequent inattention. Many professed new species are named without being described, others are described without being scientifically named; no measurements are given, and the descriptions are often so brief and obscure, that it is impossible to determine a species by their means. The work, nevertheless, contains the definitions of many new generic groups which are now adopted into our systems, and M. Lesson is therefore entitled to the credit of these original generalizations. The classification followed in this work is very complex, and in some of its portions very artificial, the genera being arrived at through a numerous and irregular series of successive subdivisions, founded in many cases upon arbitrary and isolated characters. Perhaps the most valuable portions of the work are the generic definitions, which are worked out with greater care than the specific descriptions.

Professor Eichwald gave a synopsis of the class of Birds, with brief descriptions of the Russian species, in his 'Zoologia Specialis,' Wilna, 1831. Prefixed to it is a good general *résumé* of the characters, external and internal, of the ornithic class.

The arrangement of birds proposed by Wagler, 'Systema Amphibiorum,' and by Nitzsch, 'Pterylographia,' have not yet fallen under my inspection.

In 1831 the Prince C. L. Bonaparte published his 'Saggio di una Distribuzione Metodica degli Animali Vertebrati,' exhibiting a system of ornithology, of which he had previously given a sketch in the 'Annals of the Lyceum of New York,' vol. ii. 1828. As this arrangement seems in its main features to approach more nearly to the system of nature than any contemporary method, it will be worth while to enter into some detail respecting it. The author divides the class of Birds in the first instance into *two* great groups or subclasses, Insessores or Perchers, and Grallatores or Walkers, the first including the orders Accipitres and Passeres, and the second the Gallinæ, Grallæ and Anseres. Most other zoologists, from the time of Linnæus to the present day, unconsciously prejudiced by the size, rapacious habits and celebrity of the birds of prey, have attached too much importance to their characters, and have made them into one of the primary divisions of the class Aves. But on an unbiassed estimate of their characters it will appear that the Accipitres form merely a division of the great group of Perchers, agreeing with them in all essential points of organization, and not differing more than some of the subdivisions of the Perchers do from each other. It was therefore a justifiable act to lower the Accipitres from the lofty place which they had long occupied, and to subordinate them to the Insessores. I even think that the learned author might have gone a step further, by making his subclass Insessores to consist of *one* order, Passeres only, while the Accipitres would stand on a level with his Scansores and Ambulatores, as a tribe or subdivision of Passeres.

The primary division of all birds into Perchers and Walkers, though professedly based on the position and development of so unimportant an organ as the hind-toe, and therefore liable at first sight to be termed arbitrary and artificial, is yet confirmed by so many other important and coextensive characters to which the structure of the hind-toe serves as an external indication, that we cannot doubt of this arrangement being conformable to nature. No person acquainted with the difficulty of defining the larger groups of zoology, will, of course, expect logical exactness in the application of these or of any other set of characters to the orders of ornithology. But allowing for such exceptions as occur in all zoological generalizations, it is certain that by this arrangement two great groups of birds are pointed out: the one arboreal, with perching feet, monogamous, constructing elaborate nests, and rearing a blind, naked, and helpless offspring; while the others are terrestrial, with ambulatory feet, frequently polygamous, displaying no skill in the form of their nests, and producing young which are clothed and able to see and to run as soon as hatched.

The classification of Vertebrata, which Prince C. L. Bonaparte sketched out in the above work, is further developed in a paper which he communicated to the Linnean Society (Transactions, vol. xviii.). The diagnostic characters of all the families and subfamilies are here worked out with elaborate exactness, as they are also in his 'Systema Ornithologiæ,' published in the 'Annali delle Scienze Naturali di Bologna,' vols. iii. and viii. In these latter essays the author introduces several modifications, the most important of which is, that he removes the Psittacidæ from the other Scansores, and places them as a separate order at the commencement of the system, before the Accipitres. This arrangement, which was first proposed by Blainville, is grounded on the curvature of the beak, the presence of a cere, and the reticulation of the tarsi, which are supposed to connect the Psittacidæ and Accipitres. I must be allowed however to differ from this opinion, as the Parrots appear to me to be much more closely allied to the other Scansores, with which they are usually classed. In the nature of their food, the prevailing red and green colours of the plumage, the structure of the tongue in some genera (Trichoglossus), and of the beak in others (Nestor, &c.), they seem really allied (though somewhat remotely) to the Rhamphastidæ, and through them to the Bucconidæ and Picidæ.

An arrangement of the chief families and genera of birds, with definitions of their distinctive characters, will be found in the 'Elémens de Zoologie,' by M. Milne-Edwards, 1834 (2nd ed. 1837), and in similar introductory works by Oken and Goldfuss.

Professor Sundevall published a new classification of birds in the 'Kongl. Vetensk. Acad. Handlingar,' Stockholm, 1836. He divides them into two large groups, nearly corresponding with the Insessores and Grallatores of the Prince of Canino. He agrees with Mr. Swainson in attaching a real importance to the analogical representation of groups, but appears not to insist on their numerical uniformity.

Mr. Swainson had, in 1831, given a sketch of his ornithological system in Dr. Richardson's 'Fauna Boreali-Americana,' but as his plan is more fully developed in the 'Classification of Birds,' forming part of Lardner's 'Cyclopædia,' published in 1836–37, we will confine our attention to the latter work. Of all the authors who have followed the quinary arrangement, Mr. Swainson has carried it to the greatest extent, having in various volumes of Lardner's 'Cyclopædia' endeavoured to apply it not only to the whole of the Vertebrata, but also to the Mollusca and Insecta. In speaking of Swainson as a *quinarian* author, it should be explained that he divides his groups in the first instance into *three*; but as one of these is again divided into *three*, these last, with the two undivided groups, make up the number *five* (see 'Geog. and Classif. of Animals,' p. 227). His method is therefore only a modification of the quinary theory, originally propounded by MacLeay and further developed by Vigors. In following Mr. Swainson into the details of his method, we miss the philosophical spirit and logical though not always well-founded reasoning of the two last-named authors. Firmly wedded to a theory, he is driven, in applying it to facts, to the most forced and fanciful conclusions. Compelled to show that the component parts of every group assume a *circular* figure, that they amount in the aggregate to a *definite number*, into which each of them is again subdivisible, and that there is a system of *analogical representation* between the corresponding members of every circle, which forms the sole test of its conformity to the natural arrangement, we need not wonder at the difficulties with which our author is beset; and we may certainly admire the ingenuity with which he has grappled with the Protean forms of nature, and forced them into an apparent coincidence with a predetermined system. I need not follow out the details of this Procrustean process, having already treated of it elsewhere ('Ann. Nat. Hist,' vol. vi.

p. 192). With all its faults the 'Classification of Birds' is a very useful elementary work, containing numerous details of structural characters, and many just observations on the affinities of particular groups. A large number of new genera are here defined, although many which Swainson considered to be new had been anticipated by continental authors, with whose writings he was unacquainted.

Although the quinary theory, properly so called, has made but little progress beyond the British Islands, yet there is a school of zoologists in Germany whose doctrines are of a very similar character. The most eminent of these authors is Oken, who has explained his ideas on classification in several of his detached works, as well as in that valuable periodical the 'Isis,' and who communicated an outline of his theory to the Scientific Meeting at Pisa in 1839. We find in his system the same arbitrary assumption of premises, the same far-fetched and visionary notions of analogy, and the same Procrustean mode of applying them to facts, which distinguish the writings of Swainson. He professes to deduce as a conclusion, what is in fact the *à-priori* assumption on which his whole theory is based,—that the animal kingdom is analogous to the anatomy of man, that is to say, that each of the organs which, when combined in due proportion, constitute the human body, are developed in a predominant degree in the several classes of animals, which represent those organs respectively. This doctrine is far too fanciful to stand the test of common sense, but it is certainly very ingenious, and we may admit that *se non é vero é ben trovato*. The subkingdom Radiata he considers to represent the egg, Mollusca the sexual organs, Articulata the *viscera*, and Vertebrata the *essentially animal*, or *motive organs*. The subdivisions of these groups represent not only individual anatomical organs, but also each other, in a mode somewhat like that asserted by Mr. Swainson, but even more complex and ingenious, and which I have not space to develope*.

The work which most nearly represents in Germany, the quinarian school, is the 'Classification der Säugthiere und Vögel' of Kaup, 1844. This author, like Oken, compares the Animal Kingdom to the human anatomy, but he extends the analogy of the "five senses" over every part of the system (except his subkingdoms, which are three), so as to

* The author having assumed not only that the class Mammalia represents the organs of sense, but that the genera of each family represent the individual senses, and these latter being commonly (though not correctly) enumerated as five, it results that, as far as the Mammalia are concerned, Oken's system is, like Swainson's, a *quinary* one. This coincidence of number is, however, proved to be arbitrary, and not real, by the fact that these two authors, who seem to have been wholly unacquainted with each other's writings, have in no one instance adopted the same subdivisions for their corresponding groups.

form a uniformly quinary arrangement. Thus, though Kaup agrees with Swainson in adopting the number *five*, these authors are guided by different principles of analogy, the former looking to the development of the organs of sense, and the latter to points of external structure connected with habits. Hence these two quinary arrangements are very far from being coincident: Swainson, for instance, makes the Raptores one of his primary orders, while Kaup makes them a subdivision of his Water-birds! Again, Swainson makes Corvus the essential type of all birds, while Kaup gives the same dignified position to Hirundo. I need only add, that Kaup's arrangement, like all *à-priori* systems, is replete with conjectures and fallacies.

The fundamental error which appears to pervade these and many similar modes of classification, is the assumption of a *regularity* and, as it were, *organization* in that which is a mere abstraction, the system of nature. The point at issue is this,—whether or not it formed a part of the plan of Creative Wisdom, when engaged in peopling this earth with living beings, so to organize those beings that, when arranged into abstract groups conformably with their characters, they should follow any regular geometrical or numerical law. Now such a proposition appears, when tested by reason, to be improbable, and when by observation, to be untrue. The researches of the comparative anatomist universally lead to this result,—that all organized beings are examples of certain general types of structure, modified solely with reference to external circumstances, and consequently that the final purpose of each modification is to be sought for in the conditions under which each being is destined to exist. But these conditions result from the infinitely varied arrangements of unorganized matter, they are consequently devoid of any symmetry themselves, and the wild irregularity of the inorganic is thus transmitted to the organic creation. Geology has revealed to us that in all ages of the world new organic beings have been from time to time called into existence whenever the changes of the earth's surface presented a new field for the development of life, and, judging from analogy, we cannot doubt that if a new continent were hereafter raised by volcanic agency in the Southern Ocean, a new fauna and flora would be created to inhabit it, adapted to the new set of influences thus brought into action. Such a supposition appears, as far as man can presume to reason on a subject so far above him, to be more consistent with the benevolence of an all-wise Creator, than the theory which would consider the final purpose for which certain groups of organic beings were created, to be the fulfilment of a fixed geometrical or numerical law. The supporters of the latter view appear to consider that in many cases whole tribes of animals have been made, not because they were wanted to perform certain functions in the external world, but merely in order to complete the circularity of a group, to fill a gap in a numerical arrangement, or to *represent* (in other words, *imitate*) some other group in a distant part of the system. But, from what is above advanced, irregularity, and not symmetry, may be expected to characterize the natural system, and to form, like the features of a luxuriant landscape, not a defect, but an element of beauty.

If this be true, it follows that the natural system cannot be arrived at in any part of its details by prediction, but only by the process of induction. The quinarian authors have themselves suggested a method by which the affinities of organic beings may be worked out inductively, and exhibited to the mind through the medium of the eye. Having observed that the true series of affinities cannot be expressed by a straight line, and having assumed from a few instances of groups returning into themselves, that the circular arrangement was universal, they proceeded to draw these circles on paper, and thus gave the first idea of *zoological maps*. For this idea we may be grateful to them, as it indicates a process, which if pursued inductively, and not syllogistically, seems likely to be of great use in arriving at the natural classification. This process consists in taking a series of allied groups of equal rank, and placing them at various distances and positions according to a fair estimate of the amount of their respective affinities. If this be done with care and impartiality, the traces of a symmetrical arrangement, if any such existed, would soon begin to show themselves; but I am not aware that any indications of such a law are apparent in the cases in which this method has yet been used.

In 1840 I endeavoured to apply this process to the natural arrangement of birds, and exhibited to the Glasgow Meeting of the Association a map of the family Alcedinidæ arranged upon this principle (Annals of Nat. Hist. vol. vi. p. 184). Last year I extended it to the Insessores, and have brought to the present Meeting a sketch of the whole class of Birds exhibited by the same method*. I do not of course guarantee the accuracy of any part of the arrangement in its present state, as the subject is too vast to be perfected by a single individual; but the specimen now shown may nevertheless serve to illustrate a *method* which I believe to be sound in principle, and which I would gladly see tested in other departments of organic creation†.

* See Chart in Memoir, Chapter vi.

† Mr. Waterhouse communicated to the Cork Meeting of the Association an arrangement of Mammalia which is on very nearly the same principle as that above referred to. His groups are all *drawn* as circular, of equal size, and placed in contact, whereas in my map of birds the groups of the same rank are of irregular form and dimensions, and are placed at greater or less distances according to the amount of their affinities. I believe, however, that Mr. Waterhouse does not lay any stress on these points of difference, and that his method is in fact reducible

M. de Selys Longchamps, in the Appendix to his 'Faune Belge,' 1842, has given a sketch of an ornithological system, in which the order of succession differs little from that generally adopted. He divides the class into eleven orders, some of which, as the Inertes, Chelidones, Alectorides, and Struthiones, can hardly be said to be of equal rank with the rest. He adopts the plan proposed by Nitzsch, and followed by Keyserling and Blasius, of including with the zygodactyle Scansores several other groups allied to them in many points of structure, and differing from the remaining Insessores in having the paratarsus scutate instead of entire. It is doubtful how far this last character affords a good ground for the diagnosis of orders, and it may be objected that by adhering to this distinction we separate the Trochilidæ from the Nectariniidæ, Phytotoma from the Tanagridæ, and Menura from Turdidæ. On the other hand, this arrangement has the advantage of bringing into juxtaposition the unquestionably allied groups of Alcedinidæ and Galbulidæ, as well as the Bucerotidæ and Rhamphastidæ. The scutation of the paratarsus, therefore, may form a useful auxiliary to natural classification, although, if too rigidly adhered to, it would produce in some cases an artificial arrangement.

Few more valuable contributions have been made of late years to general ornithology than Mr. G. R. Gray's 'Genera of Birds,' which passed through two editions in 1840 and 1841. It is a list of all the generic groups which had been proposed by various authors, exemplified by reference to a *type-species* in each case, and classed according to Mr. Gray's ideas of the natural system. This work is deserving of praise on several distinct grounds. The author has exercised a rare degree of industry in collecting his materials from numerous sources difficult of access; he has applied the "law of priority" in nomenclature with great fairness and impartiality, and he has sought after a natural arrangement without any theoretical bias, and with very considerable success. Although professedly including in his list *every* genus proposed by others, yet he does not pledge himself to adopt them all, indeed he distinctly asserts that many so-called genera are too trivial for practical utility. With this limitation, the 'Genera of Birds' is by far the best manual extant for the purpose of arranging collections scientifically, and of guiding the student to more hidden and scattered sources of information.

In a compilation of such a nature as Mr. Gray's, many errors of detail are unavoidable, and being sensible of the general value of the work, I ventured to point out some of them in a series of commenta-

almost to an identity with mine. A somewhat similar mode of exhibiting affinities by diagrams has also been recently adopted by Milne-Edwards (Ann. Sc. Nat. 1844), De Selys Longchamps, and others.

ries upon the two editions of the 'Genera of Birds,' which will be found in the 'Annals of Natural History,' vols. vi. vii. viii. Some critiques on the second edition were also made in the 'Revue Zoologique,' 1842, by Dr. Hartlaub, a skilful ornithologist of Bremen, who is understood to be preparing a general work on ornithology, including the distinctive characters of the species.

Mr. G. R. Gray is now engaged in issuing the 'Genera of Birds' in a much more complete and extended form, including the essential characters of the various groups, and full lists of the species and their synonyms. In this work he endeavours to reduce the various genera to an equality of rank, and is consequently compelled to reunite such genera as appear to have been separated by other authors on insufficient grounds. This task requires much judgment as well as industry, but with the resources which the galleries of the British Museum supply to Mr. Gray, he has been enabled to execute it with great success. The lithographic plates which accompany the work exhibit the essential characters of every genus, and of a large number of new or rare species, and the admirable mode in which they are executed by Mr. D. W. Mitchell confers a high degree of excellence upon this publication.

I may here be allowed to mention an undertaking of my own which has occupied the leisure of several years, but which is not yet sufficiently matured for publication,—a complete synonymy of all known species of birds, with full references to all the works where they are figured or described*. This undertaking requires considerable labour and much careful comparison of specific character, as exhibited both in nature and in books; but there is probably no department of natural history in which, from the multiplication of nominal species, and the wide dispersion of the materials, such an analysis of the whole subject is more wanted than in ornithology.

Works of reference connected with ornithology, though not strictly systematic, may be briefly mentioned here. The 'Dictionnaire des Sciences Naturelles,' the 'Dictionnaire Nouveau d'Histoire Naturelle,' the 'Encyclopédie Méthodique,' and the 'Dictionnaire Classique d'Histoire Naturelle,' were all useful works, though now more or less superseded by the progress of science. The best and most recent work of the kind is the 'Dictionnaire Universel d'Histoire Naturelle,' now publishing at Paris, and edited by M. C. d'Orbigny. The ornithological articles have been, till recently, written by M. de Lafresnaye, whose

* The first volume of this work, containing the whole of the Accipitres, has been published: "*Ornithological Synonyms*, by the late H. E. Strickland, M.A., F.R.S., &c. Edited by Mrs. H. E. Strickland and Sir W. Jardine, Bart." The remaining portions of MS. are in preparation.—Ed.

name is a sufficient guarantee for their accuracy. The illustrative plates are engraved with care, but in a stiff and mechanical style, and the colouring is frequently too vivid. Our own country has been less prolific in dictionaries of natural history than France, but zoological subjects are adequately treated of in more comprehensive works of reference, such as the 'Encyclopædia Britannica' and 'Metropolitana,' and the excellent 'Penny Cyclopædia,' in which the ornithological articles are very carefully compiled. The same remark applies to the 'Allgemeine Encyclopädie,' published at Leipzig by Ersch and Gruber.

An indispensable index to ornithology, as indeed to every other branch of natural history, is the 'Nomenclator Zoologicus' of Professor Agassiz, which is a list of all the names of groups, with references to the works where they were first proposed. The portion relating to birds has undergone careful revision, and is believed to present a near approach to accuracy.

While speaking of general methods of classification, I may refer to a new and unlooked-for source, from which a reflected light may in some cases be thrown upon doubtful points of ornithic affinity. The parasitic insects of the order Anoplura, which abound on almost every species of bird, have been till recently most unduly neglected, but that able entomologist Mr. Denny has lately taken up this branch of zoology, and after publishing, with the aid of the British Association, a beautiful work on British Anoplura, is now occupied with the exotic species. He finds that these parasites constitute numerous species, and exhibit many well-marked generic forms. The remarkable fact is further deduced, that several genera of Anoplura frequent certain groups of birds exclusively, so that there is a sort of parallelism between the affinities of birds and those of their insect parasites. Hence we are able to infer the probable position in the natural series of an anomalous bird by investigating the structure of the almost microscopical parasites which infest its plumage, and this apparently paradoxical method has been successfully applied by Mr. Denny, who has shown that the Anoplura inhabiting the genus Tallegalla are allied to those of the Rasores, and the parasites of Menura to those of the Insessores, an arrangement entirely confirming the views recently obtained as to the affinities of these singular birds (Ann. Nat. Hist. vol. xiii. p. 313).

2. *Ornithology of particular regions.*

Europe.—The most important work ever published on the ornithology of our own quarter of the globe is unquestionably the 'Birds of Europe' of Mr. Gould. This gigantic undertaking, consisting of more than 400 beautifully coloured plates, would have sufficed, independently of his other elaborate works, to stamp the author as a man of genius and of enterprise. Nor should it be forgotten that the talents of Mr. Gould were most ably seconded by his amiable partner, who, up to the time of her decease, executed the lithographic department of his various works. The extensive patronage which the 'Birds of Europe' received on the Continent as well as in Britain, is a proof both of the excellence of the work itself and of the scientific taste of the present age.

The long-expected supplements to Temminck's 'Manuel d'Ornithologie' made their appearance in 1835–40, and bring down our knowledge of European birds to the latter date. Although the author hesitates too much in adopting the generic groups of modern science, and does not sufficiently value the law of priority in nomenclature, yet the exactness of his descriptions and the general soundness of his criticisms will long render his work a valuable handbook of European ornithology. The series of illustrative plates, published at Paris by Werner, are a useful accompaniment to Temminck's work. The 'Histoire Naturelle des Oiseaux d'Europe,' now publishing by Schlegel, aided by several zoologists, and superintended by Temminck, may be regarded as an improved and enlarged edition of the 'Manuel d'Ornithologie.' The plates, by Susemihl, are of a superior order. Delarue's 'Galerie Ornithologique' forms another set of illustrations to the birds of Europe.

The 'Wirbelthiere Europa's' of Count Keyserling and Professor Blasius is a well-digested synopsis of European vertebrate zoology. The first part, with which alone I am acquainted, and which is devoted to Mammals and Birds, contains an exact catalogue of their species, with their synonyms and localities, and a statement of the diagnostic characters of the several groups from the class down to the species. These characters are stated in an antithetical mode very similar to the dichotomous method used in Fleming's 'British Animals,' a method which, when viewed in its true light, as an artificial index to specific characters, and as a means of calling attention to the presence or absence of certain structures, is probably superior to any other. Indeed, when the characters employed for the subdivisions are *really essential*, and are placed in successive subordination according to a just estimate of their functional importance, as seems to be generally the case in the work before us, this method is quite compatible with a natural classification. The authors have avoided the error of adopting indiscriminately every genus which other authors have proposed, and by carefully estimating the value of their groups, reducing the less important ones to the rank of sub-genera, they have endeavoured to bring the standard of their generic groups to an approximate state of equality.

As a mere Catalogue of the Birds of Europe, the most full and the most accurate is that by the Prince of Canino, published in the 'Annali delle Scienze Naturali di Bologna,' 1842. It is an improved edition of that contained in the 'Geographical and Comparative List of the Birds of Europe and North America,' London, 1838, containing all the additional results at which the labours of its author have arrived. The names, synonyms, and localities of the species are given with the greatest accuracy, and by rigidly adhering to sound principles of nomenclature, the author has introduced a series of scientific names which there is reason to hope will be permanently adopted.

There remain some recent works on the ornithology of Europe, which I have not had an opportunity of consulting, such as Gloger's 'Naturgeschichte der Vögel Europas,' and others.

Britain.—Prior to 1828, the only complete handbooks of British ornithology were the valuable but somewhat obsolete 'Ornithological Dictionary' of Montagu, and the fascinating, though not always accurate, 'British Birds' of Bewick. In the above year appeared the 'British Animals' of Dr. Fleming, a work which had no small share in introducing into this country the improved systems of modern zoology. The genera adopted are for the most part those of Cuvier's 'Règne Animal,' and the specific descriptions and remarks, though brief, are in general accurate.

A somewhat similar work, the 'Manual of British Vertebrata' of the Rev. L. Jenyns, is one of the best examples of a *handbook* that I am acquainted with, containing every fact of importance connected with each species, and being totally free from superfluous verbiage.

Of the magnificent plates to Mr. Selby's 'Illustrations of British Ornithology,' I shall speak elsewhere. The letter-press, in two volumes, 8vo, 1833, is very complete in its details, which are founded in great measure on the personal observations of the author, and the synonymy has been worked out with very great attention.

In 1836 Mr. T. C. Eyton published a 'History of the rarer British Birds.' It is intended as a supplement to the work of Bewick, containing the species which had been added to the British fauna since his time, and it is illustrated with woodcuts, into which the artist has infused much of the spirit of that celebrated engraver.

Meyer's 'Illustrations of British Birds' are a series of coloured plates very neatly executed.

It remains to notice three other works on British ornithology, the nearly simultaneous appearance of which is an evidence of the popularity of the subject.

Professor M'Gillivray, in 1836, published an account of the 'Rapacious Birds of Great Britain,' which was followed in 1837 by his 'History of British Birds,' in 3 vols. The author, who is an active field naturalist, as well as an expert anatomist, gives very full descriptions of the external and internal structure, as well as of the habits, of the several species and groups. These are interspersed with matter of a more miscellaneous nature, in the style of Audubon's 'Ornithological Biographies,' which render the work an entertaining though voluminous production. The classification is novel, but cannot be regarded as successful, the terrestrial birds being classed in two large sections, one of which consists of the Fissirostral and Raptorial birds, and the other includes the remaining Insessores, together with the Rasores. The remarks on Classification and Nomenclature in the Introduction are, for the most part, sound and judicious, though the author has not always adhered to his own rules.

Professor M'Gillivray has given a condensed abstract of his larger work in two small volumes, entitled 'A Manual of British Ornithology,' 1840–42.

Sir W. Jardine's 'History of British Birds,' forming three * volumes of the 'Naturalist's Library,' is a well-illustrated work, and embodies a great mass of original observations, forming a cheap and excellent work for the student of British ornithology.

The most elegant work on British Birds recently published, is that of Mr. Yarrell. From the beauty of the engravings and of the typography, it may rank as an "*ouvrage de luxe,*" while the correctness of the descriptions, and the many details of habits, geographical distribution and anatomy, render it strictly a work of science. A second edition of this work is in preparation †.

The birds of Ireland are treated of by Mr. W. Thompson in an elaborate series of papers, commenced in the 'Magazine of Zoology and Botany,' and continued in the 'Annals of Natural History.' The author has collected from his own observations and from external sources, much valuable information on habits, migrations, and other subjects connected with Irish ornithology. Being the most western portion of temperate Europe, Ireland presents some interesting peculiarities in its fauna, among which may be mentioned the occasional occurrence of American terrestrial birds in that country, though the nearest point of America is 1500 miles distant ‡. The results of Mr. Thompson's labours are incorporated in his excellent 'Report on the Fauna of Ireland,' read to the British Association in 1840, in which careful comparisons are made between the species of Ireland and of Great Britain.

The subject of British ornithology is now so nearly complete, that the works above enumerated will probably long remain unsuperseded, and we may hope that students and collectors will now extend their attention to the far more neglected department of exotic ornithology.

* Completed in four volumes.

† Mr. Yarrell's work has reached a third edition.

‡ These papers have since been collected, and are published in *three volumes,* forming a complete History of the Ornithology of Ireland.—Ed.

North and Central Continental Europe.—Many useful works on the ornithology of Northern and Central Europe were published between 1820 and 1830, by Brehm, Nilsson, Faber, Boié, Naumann, Walter and others, but as these are prior in date to the period to which I have more particularly limited this Report, and as their various merits are reviewed with candour by M. Temminck, in the introduction to his 'Manuel d'Ornithologie,' part 3, I need not enlarge upon them here.

Of the voluminous works of M. Brehm, his last, the 'Handbuch der Naturgeschichte aller Vögel Deutschlands,' 1831, is perhaps the least valuable, on account of the immense number of so-called new species which he has introduced, based upon the most trivial and inappreciable variations of size, form, or colour. This view of the subject, if carried out, would upset the whole fabric of systematic zoology, the very foundation of which is a belief in the reality, the permanence, and the distinguishableness of species. This author still continues his predilection for imaginary diagnoses in the memoirs which he publishes in the 'Isis.'

Nilsson's 'Skandinavisk Fauna,' Lund, 1835, contains a very complete, and apparently very accurate summary of the ornithology of Scandinavia, but unfortunately the Swedish language renders it a sealed book to the majority of British naturalists. The ornithology of Scandinavia has received some recent additions and corrections from a memoir by Professor Sundevall in the 'Kongl. Vetenskaps Academiens Handlingar,' 1842.

M. de Selys Longchamps, well known by several valuable monographs of European Mammals and Insects, has published the first part of his 'Faune Belge,' Liège, 1842, containing a systematic arrangement of the Vertebrata of Belgium. The specific descriptions are postponed to the sequel of the work, which is nevertheless valuable for its critical remarks on structure, habits and distribution. In the preface are some very judicious observations on the subject of systematic nomenclature, the law of priority, the limitations of species, and the still more difficult, because more arbitrary question, of the due limitation of genera. It is very satisfactory to find that the majority of European zoologists are now making considerable approaches to unanimity upon these general principles, which form the groundwork of philosophical zoology.

Dr. Gloger's 'Schlesiens Wirbelthier-Fauna,' Breslau, 1833, contains a list of the birds of Silesia, with remarks on their habits and migrations.

M. Brandt of Petersburg has published a work, entitled 'Descriptiones et Icones Animalium Rossicorum novorum,' in which several of the natatorial birds of Russia are illustrated by full descriptions and accurate figures.

France.—The ornithological portion of the 'Faune Française,' by M. Vieillot, is a useful manual, though the author has made many unnecessary changes of nomenclature. The descriptions are accompanied with figures on copper, stiffly designed, but delicately engraved.

The 'Ornithologie Provençale' of M. Roux is a respectable work on the birds of Southern France, the text being carefully drawn up, though we may regret that the author has adopted the objectional nomenclature of Vieillot.

Italy.—The ornithological researches of Savi, Bonelli, Ranzani, Costa, and many others, prepared the way for the magnificent 'Iconografia della Fauna Italica' of the Prince of Canino, a work which, after ten years' labour, has recently been completed. It consists of elaborate descriptions and beautiful coloured plates of all the new or imperfectly elucidated Vertebrata of Italy. The birds of that country having been previously more fully investigated than the other classes, occupy in this work the least prominent place, yet several new species are there figured, and our knowledge of others is enriched with much interesting information. The introduction to the work contains an excellent summary of the whole subject of Italian Vertebrata. The noble and philosophical author, who pursues with steady devotion the paths of science, unallured by the manifold attractions of rank and fortune, has devoted the best part of his life to the advancement of zoological knowledge. His elaborate researches on North American ornithology, his classification of vertebrate animals, his critique on the 'Règne Animal' of Cuvier, his comparisons of the European and American faunæ, are all works of the highest value, and we may now congratulate him on the completion of this admirable digest of the vertebrate zoology of Italy. Nor let it be forgotten that he was the first to establish beyond the Alps that great *mental,* no less than *physical* barrier, a peripatetic congress of scientific men, similar to that at which we are now assembled. This *Italian Association for the Advancement of Science* has met in the plains of Piedmont and Lombardy; it has crossed the Apennines into the happy region of Tuscany, and it will next year pass over the Papal dominions, to diffuse the light of knowledge in the distant kingdom of Naples.

An unpretending little volume by Sig^r L. Benoit, entitled 'Ornitologia Siciliana,' published at Messina in 1840, contains many interesting details on the habits and migrations of the birds of Sicily. A work of greater value is the 'Faune Ornithologique de la Sicile' of M. Malherbe, Metz, 1843, in which about fifty species are added to the list of Benoit, making a total of 318. The work abounds with important observations on the geographical distribution of species, not only in Sicily, but in other parts of South Europe and North Africa. As the island of Sicily serves as a sort of *stepping-stone* between these two continents, it affords an interesting station for observing the habits of migratory species.

A similar *catalogue raisonnée* of the birds of Liguria was published at

Genoa in 1840, by the Marquis Durazzo, and is entitled 'Notizie degli Uccelli Liguri.' Catalogues of the birds of the Venetian provinces have been published by Catullo, Basseggio, and Contarini, the latter of whom enumerates no less than 339 species.

A brief notice of the birds of Sardinia will be found in the 'Voyage en Sardaigne,' 2nd ed. 1839, by Count de la Marmora, in which it is announced that Professor Géné is about to publish a complete fauna of that island.

The island of Malta possesses an able ornithologist in Sig^r Schembri, who has published a 'Catalogo Ornitologico del Gruppo di Malta,' 1843. His other work, the 'Quadro Geografico Ornitologico,' is a highly useful volume, showing in parallel columns the ornithology of Malta, Sicily, Rome, Tuscany, Liguria, Nice, and the department of Gard. These form almost the first works on zoology ever printed in the island of Malta, and they show that, even in the most insulated localities, an active naturalist will always find abundant occupation. The author enumerates about 230 species of birds in Malta, nearly the whole of which are migratory.

Several new species of birds have been added to the fauna of the South of Europe by Dr. Rüppell, in the 'Museum Senckenbergianum,' 1837.

Greece.—But little has been done in Greece to illustrate ornithological science. The 'Expédition Scientifique de la Morée' contains a summary of sixty-six species there observed, but without adding much to our knowledge. A few new species (which however require further examination) are described by M. Lindermayer in the 'Isis' and 'Revue Zoologique,' 1843. The most complete work on the subject is the 'Beiträge zur Ornithologie Griechenlands,' by H. von der Mühle, Leipzig, 1844, in which no less than 321 species are noticed, and are accompanied with many original observations of great value. The researches of this author have added several species to the European fauna.

The birds of the Ionian Islands and of Crete are enumerated, and accompanied with some valuable remarks on their migrations and habits, by Captain H. M. Drummond, 42nd R.H., in the 'Annals of Natural History,' vol. xii. p. 412.

Spain.—The ornithology of the Spanish peninsula is as yet but imperfectly known. A list of some of the birds is given in Captain Cooke's (now Widrington) 'Tour in Spain.' (See also his 'Spain in 1843.') That gentleman was, I believe, the first discoverer of the *Pica cyanea* in Spain, a species which, if it be really identical with the *Garrulus cyaneus* of Pallas, found in Siberia and Japan, presents a most unusual instance of the existence of the same species in two remote regions, without occurring in the intervening space. M. Temminck has described several new species brought from the South of Spain by Parisian collectors, and from the proximity of that region to Africa, it is probable that further additions to the European fauna may be there made.

Of the birds of Madeira there is a brief notice by Dr. Heineken in the 'Zoological Journal,' vol. v.; and several species are described by Sir W. Jardine in Ainsworth's 'Edinburgh Journal of Natural and Geographical Science.'

The Canary Islands present a fauna more allied to that of Europe than the southern position of these islands and their proximity to the African continent would have led us to expect. The 'Histoire Naturelle des Isles Canariennes,' a splendid work lately published at Paris by MM. Webb and Berthelot, contains a list of birds, the whole of which, with the exception of a very few terrestrial species peculiar to the islands, are included in the ornithology of Europe.

Asia Minor.—The 'Proceedings of the Zoological Society' contain lists of the birds of Trebizond and Erzeroum, by Messrs. Abbot, Dickson, and Ross, and of those of Smyrna by myself. There is also a short list of those obtained by Mr. C. Fellows in the 'Annals of Nat. Hist.' vol. iv. The greater part of the birds hitherto found in this country are also common to Europe, which may in part be attributed to their having been chiefly collected in the northern districts, or in my own case at Smyrna, during the winter season. An ornithologist who would visit the regions south of the Taurus during the spring, would doubtless meet with many interesting species, a foretaste of which we have in the beautiful *Halcyon smyrnensis*, discovered more than a century ago by the learned Sherard, and restored to science in 1842 by Mr. E. Forbes*.

I may here allude to the 'Catalogue of the Birds of the Caucasus' by M. Ménétries, in the 'Mémoires de l'Acad. Imp. des Sciences de St. Pétersbourg.' Although several of the supposed new species have been reduced to the rank of synonyms, yet this list supplies some valuable information on the geographical distribution of species. For the ornithology of Southern Russia, the student may also consult M. Eichwald's summary of the Caucasian and Caspian birds in the 'Nouveaux Mémoires de la Soc. Imp. des Naturalistes de Moscou,' 1842, and Demidoff's 'Voyage dans la Russie Méridionale,' the zoology of which is edited by Professor Nordmann.

Siberia.—The zoology of Northern Asia was long retarded by the delays which attended the publication of the 'Zoographia Rosso-Asiatica' of that Humboldt of the 18th century, the celebrated Pallas. This posthumous work, though printed in 1811, was not published till 1831, when it at once added to our knowledge a large number of new species. Many commentaries upon Pallas's work, and additions to his species, have been made by various authors, especially by M. Brandt, the learned and indefatigable curator of the Imperial Museum at St. Petersburg, in the 'Bulletin' of the Academy of that city, and by Nordmann in Erman's

* See Annals of Nat. Hist. vol. ix. p. 441.

'Reise um die Erde.' There are also some valuable "Addenda" to the work of Pallas from the pen of Dr. Eversmann, in the 'Annals' of the distant University of Casan, and further additions have been recently contributed by that author to the Petersburg Academy. We may hope that the labours of these and other equally active Russian zoologists will soon make us fully acquainted with the natural history of Asiatic Russia.

A few of the birds of Behring's Straits are elaborately described, though indifferently figured, in Eschscholtz's 'Zoologischer Atlas,' to Kotzebue's second Voyage, Berlin, 1829.

Japan.—Drs. Von Siebold and Burger, who were attached for several years to the Dutch mission in Japan, devoted their leisure to the zoology of that little-known country, and the results have now been published by the Dutch government in a handsome work, entitled 'Fauna Japonica.' A remarkable fact established by their researches, is the great amount of coincidence between the ornithological faunæ of Japan and of Europe. In Temminck's 'Manuel d'Ornithologie' (Introd. to part 3) is a list of the species common to these two regions, amounting to no less than 114.

British India.—It is only within a very recent period that any really original and trustworthy researches have been made into Indian ornithology. Twenty years ago the utmost that was done by the numerous British officers in that country to illustrate this science, was to collect drawings of the species which attracted their notice. These drawings were in most cases made by native artists, who, being utterly ignorant of any scientific principles, executed them in a stiff mechanical style, and neglected the more minute but often highly important characters. Such designs are useful as aids to scientific research, but ought not to usurp its place; yet from these materials the too undiscriminating Latham described and named a great number of so-called species, many of which have not yet been identified in nature. The largest collection of these drawings was made by the late General Hardwicke, a selection of which were engraved and published in 1830; but though carefully edited by Mr. J. E. Gray, the number of nominal species there introduced shows the danger of founding specific characters on the sole authority of drawings.

A better day dawned about 1830, when several British officers in India became interested in the study of scientific ornithology; and we may hope that natural history in this and all its other branches will now become a general pursuit with our countrymen in that region. The first *original* contribution to the ornithology of India in recent times was made by Major Franklin, and was speedily followed by a valuable paper from Colonel Sykes, both of which are inserted in the 'Proceedings of the Zoological Society,' 1831–32. About the same period appeared the first effort of Mrs. Gould's pencil, the 'Century of Birds from the Himalaya Mountains,' a work the plates of which at once established the fame of this admirable artist, while the scientific characters were carefully prepared by Mr. Vigors. In 1832 was also commenced that most valuable repertory of oriental knowledge, the 'Journal of the Asiatic Society of Bengal,' which is still published with regularity at Calcutta. In this journal, and in others of a similar nature, as the 'Asiatic Researches,' the 'Gleanings in Science,' Corbyn's 'Indian Review,' the 'Quarterly Journal of the Calcutta Medical and Physical Society,' the 'Calcutta Journal of Natural History,' are contained the valuable but unfortunately too scattered and inaccessible zoological researches of Hodgson, Hutton, Pearson, Tickell, M'Clelland, W. Jameson and others. Mr. Hodgson, who by his residence in Nepal has been so favourably circumstanced for zoological pursuits, has long since promised to include in an entire work his scientific researches in that country, but various delays have hitherto impeded the undertaking. He has recently, with the utmost liberality, presented the whole of his precious materials to the British Museum and other public collections, and we may hope that the facilities of comparison thus afforded will enable him shortly to commence this very desirable publication.

The Indian species of Coturnix and Turnix have been described with minute exactness by Colonel Sykes in the 'Transactions of the Zoological Society,' vol. ii. This paper is of great service in clearing up the characters of these obscure and ambiguous birds, which however are still far from being thoroughly investigated.

Professor Sundevall, in his valuable Report on recent Zoological Researches, Stockholm, 1841, refers to a paper on the Birds of Calcutta in the 'Physiographisk Tidskrift,' Lund, 1837, a work which has not yet fallen into my hands.

A great impulse has recently been given to Indian zoology by the appointment of Mr. Blyth to the care of the Asiatic Society's Museum at Calcutta. Most of the previous workers in that field were civil or military officers, who took up zoology as an afterthought, and as a relief from more important duties. But Mr. Blyth went to India a ready-made zoologist, who had long devoted himself to the study as a science, and was well acquainted with its literature and its principles. Of the zeal and success with which he is now bringing into order the heterogeneous materials of Indian zoology, the pages of the 'Journal of the Asiatic Society of Bengal' bear ample testimony. Besides many detached memoirs, the monthly reports which Mr. Blyth presents to the Asiatic Society contain a mass of interesting observations, and present an example which the curators of European museums would do well to imitate. By comparing complete lists of the species comprised in each successive accession to the museum, accompanied by critical remarks on the more novel or interesting specimens, previous to their being incorporated into the general collection, a number of important observations on structure, habits and geographical distribution are preserved from oblivion. In the midst of these active and useful labours Mr. Blyth retains his interest in European science, and occasionally sends communications of great value to the 'Annals of Natural History.'

While treating of Northern India I may mention the Catalogue of the Birds of Assam, by Mr. M'Clelland, in the 'Zoological Proceedings,' 1839. The author avoided the too common error of describing as new every species which was *unknown to him*, by the judicious plan of attaching provisional names and descriptions to such species, and then sending them to a highly competent naturalist in England, Dr. Horsfield, to be revised prior to publication.

The presidency of Madras can boast of a 'Journal of Literature and Science,' and of zoologists, Messrs. Jerdon and Elliott, equal in activity and scientific attainments to those of Bengal. The various memoirs of these gentlemen on the characters and habits of the birds of Southern India are of high value. Mr. Jerdon has commenced the publication of a series of 'Lithographed Drawings of Indian Birds,' illustrating many rare species in a style which does credit to the artists of India.

A few species of Indian birds have been described by Professor Jameson in the 'Memoirs of the Wernerian Society,' vol. vii., and several others are figured in Royle's 'Botany of the Himalaya Mountains,' and in the zoological part of Jacquemont's 'Voyage dans l'Inde,' Paris, 1843, the plates of which are beautifully executed. Mr. Blyth has drawn up a notice of the species received from the British officers in Tenasserim, and of the desiderata which remain to be sought for in that province. The zoological portion of M. Belanger's 'Voyage aux Indes Orientales,' 1834, contains descriptions and figures of many of the birds of Pegu and Java, among which are several novelties. Some of the species of continental India are also described in the same work. Ornithological information will also be found in Delessert's 'Souvenirs d'un Voyage dans l'Inde.'

Malasia.—Under this name may be included the peninsula of Malacca and the islands of the Indian archipelago, which taken collectively form a well-marked zoological region, whose fauna, though for the most part agreeing *generically* with that of continental India, presents an almost wholly distinct series of *species*. The first contributor to the ornithology of this region was Brisson, who described, with an exactness that may serve as a model even at the present day, many new species of birds from the Philippine Islands. Sonnerat described some more species in 1776, but scarcely anything has since been added to our knowledge of the vertebrate zoology of that particular group of islands; and it is to be regretted that a considerable collection of birds recently brought thence by Mr. Cuming, were dispersed before any scientific examination of them had been made. The zoology of western Malasia was first investigated by Dr. Horsfield and Sir Stamford Raffles, the first of whom described the birds of Java, and the second those of Sumatra, in the 'Linnean Transactions,' vol. xiii. These are very valuable memoirs, though it is to be regretted that from the brevity of the specific characters some of the species are rendered difficult to recognize. A selection of Dr. Horsfield's species is however more fully described and illustrated by figures in his 'Zoological Researches in Java,' and the original specimens collected by him are preserved in the Museum of the East India Company. The species of Horsfield and of Raffles were arranged into one series by Mr. Vigors in the Appendix to the 'Life of Sir Stamford Raffles.'

Between 1820 and 1830 several Dutch and German naturalists visited the Malasian Islands, and enriched the continental museums with their collections. A considerable number of the species thus obtained are figured in the 'Planches Coloriées' of M. Temminck, who however too frequently described as new the species which had been long before characterized by Horsfield and Raffles.

For two centuries past the Dutch have been famed for their love of collecting rarities, and the numerous settlements of that people in all parts of the world have tended to the gratification of this taste. It is therefore not to be wondered at that the national museum of Holland at Leyden should have become one of the richest collections of natural objects in the world; and it is gratifying to find that the information which its treasures convey is in the course of being diffused abroad. The Dutch government are now publishing a complete zoology of their foreign colonies, under the title of 'Verhandelingen over de Natuurlijke Geschiedenis der Nederlandsche overzeesche Bezittingen.' This superb work contains figures and descriptions of many new species from the remote islands of the Malay archipelago; and it is only to be regretted that so valuable a publication should be compiled in a language with which few men of science out of Holland are acquainted.

A considerable number of ornithological specimens have recently been sent to Europe from the peninsula of Malacca, and indicate a fauna closely allied to, though often specifically distinct from, that of the adjacent islands of Java and Sumatra. Mr. Eyton has described several of these Malacca birds in the 'Proceedings of the Zoological Society,' 1839, and Mr. Blyth has characterized others which had been sent to the Calcutta Museum.

The great island of New Guinea presents features in its zoology which entitle it to be considered a distinct region from the Malasian archipelago, and connected rather with the Australian fauna. We here first meet with that extraordinary group of birds the Paradiseidæ, whose affinities it is impossible to assign with certainty until their anatomy and habits are better known. In this group will probably be ultimately included (as they were originally by the earlier writers) the genera Seleucides, Ptilorhis, Epimachus, Phonygama and Astrapia, which are at present arranged, from conjecture rather than induction, in many widely-separated families. These genera all agree with the Paradiseidæ in the very peculiar structure of their plumage, and what is of no less importance as an indication of zoological affinity, they all (with the exception of Ptilorhis, which is found in the adjacent Australian continent) inhabit the same island of New Guinea; and I think it not improbable that the anomalous Australian genera Ptilonorhynchus, Calodera and Sericulus, may be also referable to the Paradiseidæ. These questions, however, must be resolved by the anatomist and not by the studier of dried skins; and we may therefore regret that New Guinea has hitherto been so inaccessible to naturalists. The specimens from thence are mostly obtained in a mutilated state from the savage inhabitants, and I believe the only zoologists who have seen the Birds of Paradise in a state of nature are M. Lesson, who made some interesting observations upon them during the few days which he spent in the forests of New Guinea ('Voyage autour du Monde de Duperrey,' and Lesson's 'Manuel d'Ornithologie'), and MM. Quoy and Gaimard, whose observations, recorded in the 'Voyage de l'Astrolabe,' 1830–33, were still more limited.

Polynesia.—The ornithology of the innumerable islands of the Pacific Ocean is as yet very imperfectly investigated. From the small size of most of these islands they cannot individually be expected to abound in terrestrial species, though in the aggregate they would doubtless furnish a considerable number, while of aquatic species an interesting harvest might be collected. At present much of our information is derived from no better source than the incomplete descriptions made by Latham of species collected during Captain Cook's voyage. Some of the birds collected by the Rev. A. Bloxam in the Sandwich Islands are described in Lord Byron's 'Voyage'; others were made known by Lichtenstein in the 'Berlin Transactions,' 1838; and the 'Zoology of the Voyage of the Sulphur,' now in course of publication, contains some further materials which have been examined and described by Mr. Gould. A few Polynesian birds are described by MM. Hombron and Jacquinot among the scientific results of the Voyage of the Astrolabe and Zelée (Ann. Sc. Nat. 1841), and several new species from the Philippine, Carolina and Marian Islands are characterized by M. Kittlitz in the 'Mémoires de l'Acad. Imp. de St. Pétersbourg,' 1838. The recent American voyage of discovery will extend our knowledge of Polynesian zoology, and its researches will be made known by Mr. Titian Peale, who is said to have discovered among other rarities a new bird allied to the Dodo, which he proposes to name Didunculus.

Australia.—Shaw's 'Zoology of New Holland,' 1794, was the first work devoted to the natural history of the Australian continent, but its publication was soon discontinued. It was followed by the 'Voyages' of Phillips and White, in which many of the birds of that country were figured and described. The next additions were made by Latham, who in the second 'Supplement to his Synopsis,' 1802, described and named many species on the authority of a collection of drawings belonging to the late Mr. A. B. Lambert. These drawings, however, were very rude performances, and being unaccompanied by descriptions, it is no wonder that Latham was led by them into many errors of classification and synonymy. Fortunately, however, they passed at Mr. Lambert's death into the possession of the Earl of Derby, who liberally entrusted them for examination to Mr. Gould, Mr. G. R. Gray, and myself. By carefully studying these designs and comparing them with Australian specimens, we have been able to identify almost the whole of the species which Latham founded upon them, and by this process many corrections have been introduced into the synonymy of the Australian birds. (See Ann. Nat. Hist. vol. xi.)

It is to be regretted that Messrs. Vigors and Horsfield had not access to this collection of drawings when they prepared their valuable paper on Australian birds in the 'Linnæan Transactions,' vol. xv. They would there have recognized several of the species which, from having failed to identify them in the brief description of Latham, they described as new. Their memoir is notwithstanding a very important contribution to Australian ornithology, especially on account of the many generic forms peculiar to that region which they defined with logical precision.

The above, together with the brief but original work of Lewin (Birds of New South Wales), and a few species described by Quoy and Gaimard in the 'Voyage de l'Uranie,' 1824, and in the 'Voyage de l'Astrolabe,' 1830, and by Lesson in the 'Voyage de la Coquille' and the 'Journal de la Navigation de la frégate Thetis,' 1837, formed the chief materials for Australian ornithology until the expedition of Mr. Gould to that country made a vast accession to our knowledge, which is embodied in his great work, the 'Birds of Australia.' Among those splendid publications of science and art which the liberality of governments have given to the world, there are few which in point of beauty or completeness are superior to this unassisted enterprise of a single individual. Regardless of expense and risk, Mr. Gould proceeded to Australia for the sole purpose of studying Nature in her native wilds, and after spending two years in traversing the forests and plains of that continent, he returned home with a valuable collection of specimens, and a still more precious one of *facts*. These he is now engaged in bringing before the public, and the many new and interesting details of natural history which his work contains, indicate powers of observation and of description which will place the name of Gould in the same rank with those of Levaillant, Azara, Bewick, Wilson, and Audubon.

Of the artistic merits of this publication I shall hereafter speak, and shall refer to it at present merely as a work of science.

Among the new generic groups proposed by Mr. Gould, some, as Pedionomus, Sphenostoma, &c., possess sufficiently well-marked characters; but others, as Donacola, Erythrodryas, Erythrogonys, Synæcus, Geophaps, appear hardly to deserve generic separation. These so-called genera seem to be founded upon slight peculiarities of form, habit, or colouring, to which, however interesting in themselves, we ought not, I think, to attach a generic value, unless we are prepared to reduce all our other genera to the same low standard, a step which would increase the number of genera and diminish their importance to an extent that would be highly inconvenient. I may also remark that some of the birds which Mr. Gould regards as distinct *species* appear to possess insufficient diagnostic characters. Peculiarities of climate and food will always exert a certain influence on the stature and on the intensity of colour in the same species, and so long as the proportions and the distribution of the colours remain unaltered, we should hesitate in raising the local varieties thus produced to the rank of species, unless we are ready to go the same length as M. Brehm, who by this means has trebled the number of European species. As instances of Australian birds the real specific distinctness of which appears to me doubtful, I may mention Mr. Gould's *Malurus cyaneus* and *longicaudus*, *Amytis textilis* and *striatus*, *Astur approximans* and *cruentus*, *Hylacola pyrrhopygia* and *cauta*.

Passing over these slight defects, it is certain that the facts brought for the first time to our knowledge by Mr. Gould have cleared up many doubtful questions respecting the true affinities of the anomalous forms so prevalent in Australia. Being now informed as to their habits, and, in many cases, their anatomy, we are enabled to classify with certainty the once ambiguous groups, Talegalla, Psophodes, Menura, Falcunculus, Artamus and others. In other cases, as in the genera Ptilonorhynchus and Calodera, the observed habits of the birds are even more anomalous than their structure, and rather increase than diminish the difficulty of classifying them.

Mr. Gould's work is also valuable for its critical examinations of the labours of other authors, the synonyms being for the most part carefully elaborated, and a due regard paid to the principle of priority in nomenclature. It is to be hoped that this delightful and truly original work will be hereafter republished in a more portable form, as its present costly style of illustration necessarily restricts it to a small number of readers.

This publication has tended to create a taste for natural history in the Australian colonies, which will advance the cause of morality and civilization. Among recent proofs of an improved tone of mental cultivation, I may mention the 'Tasmanian Journal of Natural Science,' commenced at Hobart Town in 1842, and which is a publication highly creditable to the southern hemisphere. One of its chief contributors is the Rev. T. J. Ewing, who is ardently devoted to science, and who has already increased our knowledge of Australian ornithology.

The tropical parts of the Australian continent exhibit, as might be expected, many new and beautiful forms. A few of these were made known in Captain King's 'Survey of Intertropical Australia,' 1827; and the labours of Mr. Gould's collector, Mr. Gilbert, will now render the zoology of Northern and Western Australia as familiar to us as that of New South Wales.

New Zealand.—The earliest information on the ornithology of New Zealand was obtained by Forster, during the voyage of Captain Cook, of which we shall learn more particulars in Professor Lichtenstein's forthcoming edition of Forster's MSS. A few additional species are described in the Voyage of the Coquille, 1826, and of the Astrolabe, 1830; but little was subsequently added till 1842, when Dr. Dieffenbach submitted his collection to the examination of Mr. G. R. Gray, and the result will be found in the interesting 'Travels in New Zealand' of the former gentleman. As in most oceanic islands remote from a continent, the terrestrial ornithology of New Zealand is somewhat limited; but some interesting representatives of the Australian fauna are there found, and the extraordinary structures of those anomalous birds, the Apteryx and Dinornis, atone in point of interest for the general paucity of species.

The aquatic ornithology of the Southern Ocean and its isles has been hitherto in a state of the greatest neglect and confusion; but some valuable materials for its elucidation will be supplied by the 'Voyages of the Erebus and Terror,' now in course of publication, as well as by many details introduced in Gould's 'Birds of Australia.'

Africa.—The zoology of Lower Egypt has received but few accessions since the French expedition to Egypt; but that of Nubia and Abyssinia, the foundations of which were laid by Bruce and by the present Earl of Derby, who added a valuable appendix to Salt's 'Voyage,' has been since greatly extended by the labours of Rüppell and Ehrenberg. The 'Atlas zu der Reise in Nordlichen Afrika,' and the 'Neue Wirbelthiere' of the former author, are especially valuable for the fulness and accuracy of the descriptions, and for the critical remarks with which they are accompanied. The lithographic plates, though rather coarsely executed, are sufficiently characteristic. The author has made further additions to this subject in his 'Museum Senckenbergianum.' The 'Symbolæ Physicæ' of Messrs. Hemprich and Ehrenberg contain some accurate information on the ornithology of Abyssinia, Egypt and Syria, and we may regret that this excellent work was never completed. Besides much original matter, the authors have added many careful criticisms on the works of other authors who have written on the zoology of those countries. Some additions to Abyssinian ornithology have also been made by M. Guérin-Méneville, 'Revue Zoologique,' 1843.

No special work has been produced on the ornithology of Western Africa, except the useful little book by Swainson, which forms two volumes of Sir W. Jardine's 'Naturalist's Library.' Many new species are there defined and figured with care.

The birds procured during the late unfortunate expedition to the Niger are described in the 'Proceedings of the Zoological Society' by Mr. Fraser, who accompanied the party as naturalist.

The ornithology of South Africa is now far advanced towards completeness. The 'Oiseaux d'Afrique' of Levaillant formed an admirable groundwork for the study, and through the labours of subsequent naturalists, there is probably little more to be added to our knowledge of the subject.

The enterprising Burchell characterized several new species in his 'Travels in South Africa,' and others collected by Sir J. Alexander were described by Mr. Waterhouse in the Appendix to that traveller's 'Expedition of Discovery into the Interior of Africa,' 1838. But we owe the largest additions to South African ornithology to the energy of Dr. Andrew Smith, who, in 1832, planned and executed an expedition of discovery into the remote interior, northwards of the Cape colony. The zoological results of this expedition were first published by Dr. Smith in the 'South African Quarterly Journal,' a scientific periodical printed at Cape Town, and less known in Europe than it deserves to be. They will also be found in a pamphlet entitled 'Report of an Expedition for Exploring Central Africa,' Cape Town, 1836. By the liberality of Her Majesty's Government Dr. Smith has since been enabled to publish these new and precious materials, under the title of the 'Zoology of South Africa,' in a style and form corresponding to the 'Zoology of the Voyage of the Beagle' and of the 'Sulphur,' and forming a standard work for the library of the naturalist.

Of the birds of Madagascar but few have been described since the days of Brisson. M. I. Geoffroy St. Hilaire has made known some remarkable forms from that island in Guérin's 'Magasin de Zoologie,' 'Comptes Rendus,' 1834, and 'Ann. des Sciences Naturelles,' series 2, vol. ix.

North America.—The ornithology of North America (exclusive of Mexico) is now more thoroughly investigated than that of any other quarter of the globe, except Europe. The fascinating volumes of Wilson, and the invaluable continuation of his work by Prince C. L. Bonaparte, contributed to produce in the United States a great taste for natural history, and for ornithology in particular. The works of Wilson and Bonaparte have been made more accessible in this country by means of smaller editions, one of which was edited by Sir W. Jardine, and another by Professor Jameson. A small edition has also been published in America by Dr. T. M. Brewer, Boston, 1840. Foremost among the successors of Wilson is the indefatigable Audubon, whose life has been spent in studying Nature in the forest, and in depicting with pen and pencil her manifold beauties. The plates of his 'Birds of America,' more than 400 in number, are the work of an enthusiastic naturalist and a skilful artist, though the designs are sometimes rather *outré*, and their size is inconveniently gigantic. The latter evil is however remedied in a smaller edition with lithographic plates, which the author has recently published in America. The text to these plates, entitled 'Ornithological Biography,' is an amusing as well as instructive work, though written in a too inflated style. Mr. Audubon has since published a 'Synopsis of the Birds of North America,' Edinburgh, 1839, containing condensed descriptions of the genera and species, and forming a very useful manual of reference. Several of the species of Sylvicolinæ had been unduly multiplied by Audubon, and their synonymy has been rectified by Dr. T. M. Brewer in Silliman's 'Journal of Science,' vol. xlii.

Mr. Nuttall's 'Manual of the Ornithology of the United States,' published at Cambridge, U.S., 1832–34, is a very convenient hand-book, containing a compendium of the labours of Wilson, Bonaparte and Audubon, accompanied with many original observations on the habits of the species. The work is illustrated with woodcuts, which, though not equal to the works of Bewick, are executed in a similar style and with considerable success.

Several of the States of the American Union have adopted the truly enlightened policy of making regular scientific surveys of their respective territories. Of these the State of New York has already published several handsome volumes on other branches of natural history; but the ornithological portion is not yet issued. A list of the birds of Massachusetts will be found in Professor Hitchcock's Report on the Geology of that State. This list has been further extended by Dr. Brewer and by the Rev. W. Peabody in the 'Boston Journal of Natural History,' 1837 and 1841. The latter gentleman has given much valuable information on the manners and migrations of the species. Some popular notices of the birds of Vermont are given by Mr. Z. Thompson in his 'History of Vermont,' Burlington, 1842.

A mass of interesting observations on the zoology of the Arctic portion of North America is contained in the appendices to the narratives of Ross, Parry, Franklin and Back, and in the 'Memoir on the Birds of Greenland' by our respected Secretary Colonel Sabine (Linn. Trans. vol. xii.). These enterprising explorers found the means, during their arduous and protracted expeditions, to add greatly to our knowledge of Arctic zoology, and the results of their labours were brought together and reduced to system in the volumes of the 'Fauna Boreali-Americana,' of which the volume on Birds is the production of Dr. Richardson, assisted by Mr. Swainson. The specific descriptions by the former gentleman are a model of accuracy and precision, and the lithographic plates are executed with Mr. Swainson's usual skill.

In his able 'Report on North American Zoology,' read to the British Association in 1836, Dr. Richardson has presented us with a full catalogue of the birds of North America, including Mexico. He enters at some length into the subject of migration, and has incorporated with his own observations those of the Rev. J. Bachman in Silliman's 'American Journal of Science,' 1836.

His Highness the Prince of Canino continues to take a lively interest in the zoology of North America, where so many years of his life were spent. In 1838 he published a very elaborate 'Comparative List of the Birds of Europe and North America,' exhibiting in parallel columns the species which, whether by identity or by close affinity, represent each other in the two countries. This work exhibits some interesting results connected with the geographical distribution of species and of forms. The region between Mexico and the Polar sea approaches in its fauna much more to the European, and less to the tropical American type, than might have been expected. Of 471 North American species of birds, no less than 100 are identical with European kinds. This is due not merely to similarity of climate, but to the comparatively short interval between western Europe and eastern America, which enables nearly all the marine and some of the terrestrial species to pass from the one continent to the other. Another cause is the proximity of north-western America to Siberia, which has extended the migrations of certain essentially arctic species, and caused them to spread completely round the world to the north of about lat. 50°.

The Prince is at present engaged on an improved edition of the 'List of North American Birds,' in which he now proposes to include the birds of Mexico. This addition will materially modify the numerical results of the former work, as it will introduce a large number of species of a more tropical character than most of those of the United States. It will form a valuable addition to our knowledge, the birds of Mexico being as yet but imperfectly determined, and their descriptions scattered through many remote sources. Some of them have been described by Mr. Swainson (Philosophical Magazine, ser. 2, vol. i., and Animals in Menageries), others by Wagler and Kaup (Oken's 'Isis,' 1832), and Lesson (Ann. Sc. Nat. ser. 2, vol. ix.). Not a few of the nominal species in Latham's 'Index Ornithologicus' are said to be from Mexico, some of which, taken from the original work of Hernandez, might doubtless be regained to science; others, described from the worthless 'Thesaurus' of Seba, are probably altogether apocryphal.

The voyage of Captain Cook supplied the earliest materials for the zoology of North-western America. A few Rasorial birds were brought from that country by the botanist Douglas, and others are described by Mr. Vigors in the 'Zoology of Captain Beechey's Voyage,' 1839. We may regret that no note was taken of the localities of many species brought home by that expedition, and which are described and figured with exactness in the above work. M. Lichtenstein's memoir in the 'Berlin Transactions,' 1838, and the recently published 'Zoology of the Voyage of the Sulphur,' have also furnished some additions to the ornithology of that remote part of the American continent, and twelve species from the Columbia river are described by Mr. Townsend in the 'Journ. Acad. Sc. Philadelphia,' 1837.

Mr. J. P. Giraud has described several new species of birds from Texas in the 'Annals of the Lyceum of New York,' of which he has given coloured figures in a folio form, under the title of 'Description of Sixteen New Species of North American Birds,' New York, 1841.

Central America.—Of this region of tropical forests (in which Honduras and Yucatan may be geographically included) the zoology is almost unknown. Two or three beautiful birds from that country have found their way into Temminck's 'Planches Coloriées,' a few more are described by M. Lesson in the 'Revue Zoologique,' 1842, and Dr. Cabot, an American naturalist, who accompanied Mr. Stephens in his interesting expedition in Yucatan, has enumerated some of the birds which he collected, in the work of the latter gentleman (Incidents of Travel in Yucatan). He considers many of them to be identical with species of the United States, but it is not stated how far this identification rests on a rigorous comparison of specimens from the two countries. Dr. Cabot has given an interesting account of the habits of that beautiful bird the *Meleagris ocellata* in the 'Boston Journal of Natural History,' and the habits of *Trogon pavoninus*, another splendid bird of that country, are recorded by M. Delattre in the 'Revue Zoologique,' 1843.

Galapagos Islands.—This small group of islands illustrates that remarkable law which establishes a general coincidence between geographical distribution and zoological affinity. These islands of the Pacific, though several hundred miles distant from the American coast, are yet much nearer to it than to the numerous islands of the Polynesian archipelago, and in conformity with this position we find that the birds of the Galapagos, though belonging to species exclusively confined to these isles, are altogether referable to an American and not to a Polynesian type of organization. This result is derived from the researches of Mr. Darwin, who, in the 'Zoology of the Voyage of the Beagle,' has described several new species from these remote islands.

West Indies.—The ornithology, and I may say the natural history of the West Indies, is far less known than, from the long connexion of those islands with Europe, might have been expected. Of the birds of Cuba a few were described by Mr. Vigors in the 'Zoological Journal,' vol. iii. This island has since been scientifically surveyed by Ramon de la Sagra, in his 'Histoire Physique, Politique et Naturelle de l'Isle de Cuba,' in which a considerable number of new species of birds are accurately characterized. Many of the birds of St. Domingo were long since described by Brisson, Buffon and Vieillot, and few if any additions to our knowledge of its productions have been made of late years. The natural history of our own island of Jamaica has experienced a degree of neglect which reflects but little credit upon the energy of individuals or of the government. Almost the whole of our knowledge of its ornithology is derived from the obscure descriptions and wretched figures in Sir Hans Sloane's 'Natural History of Jamaica,' published in the beginning of the last century. A few stray species have since been described by various authors, but nothing like a regular scientific survey of that beautiful and interesting island has yet been, or, judging from appearances, is likely to be, undertaken. The smaller West Indian islands have been equally neglected by naturalists; but few of their natural productions ever reach our museums, and these are too often consigned to the cabinet without being scientifically described or published.

South America.—The birds of Columbia were till a recent period wholly unknown (with the exception of a few brief notices by Humboldt in his 'Recueil d'Observations de Zoologie,' 1811), but a considerable supply of specimens has been lately sent to Europe from the province of Bogota, which have added greatly to our knowledge. Many new species thus obtained have been described by MM. De Lafresnaye, Boissonneau, Bourcier and De Longuemare in the 'Magasin de Zoologie' and 'Revue Zoologique,' and by Mr. Fraser in the 'Proceedings of the Zoological Society.' Many of the birds of that country are beautiful and interesting representatives of the better-known species of Brazil, and the family of Tanagers in particular has lately received large additions from that quarter.

The ornithology of British Guiana is not yet so fully worked out as it deserves to be. Mr. Schomburgk has collected many species during his various journeys in the interior, some of which have been characterized in miscellaneous works; but there is no collective publication of the natural history of that colony.

The ornithology of Brazil, on the other hand, is now very fully known, many species having been described by the older authors, and many more in recent times by Prince Maximilian of Neuwied, Spix, Swainson, and others.

The costly work of Spix, 'Avium species novæ in itinere per Braziliam collectæ,' is valuable rather for the amount of new materials which the travels of that author supplied, than for the skill or diligence with which those materials were digested. A sounder criticism was applied by Prince Maximilian of Wied, who has done much to illustrate the ornithology of Brazil, not only in his travels in that country, and his 'Recueil de Planches Coloriées d'Animaux du Brésil,' but in his 'Beiträge zur Naturgeschichte von Brazilien,' Weimar, 1832. A great number of species are there described in detail, and the work is especially valuable as a supplement and commentary to the writings of Azara and Spix. About 1833 Mr. Swainson commenced an illustrated work on the Birds of Brazil, entitled, 'Ornithological Drawings,' but it only attained to about seventy plates. The figures are well drawn and carefully coloured; but they labour under the defect of being unaccompanied by descriptions, without which even the best designs are often insufficient for specific identification. M. Schreiber of Vienna commenced, in 1833, the 'Collectanea ad Faunam Braziliæ,' but only one number of the work was ever published. Several Brazilian birds are also described by Nordmann in the Atlas to Erman's 'Reise um die Erde,' 1835.

Since the publication of the invaluable work of Azara, nothing has been added to the ornithology of Paraguay; but as that country is intermediate to Brazil, Chili and Patagonia, most of Azara's species have been procured by naturalists who have visited the three last-named countries. Many of the birds of Patagonia, Terra del Fuego and the Falkland Isles, are described by Mr. Darwin in the 'Zoology of the Voyage of the Beagle,' and by Captain King (Zool. Journal, vol. iii. and Zool. Proceedings, 1831).

After the publication of Molina's not very accurate 'Saggio sulla storia naturale del Chili,' fifty years elapsed without any addition being made to the zoology of western South America. About 1831 M. Kittlitz published a short paper on the Birds of Chili in the 'Mémoires de l'Académie Impériale de St. Pétersbourg,' in which several new and curious generic forms are for the first time indicated. Descriptions of a few Chilian birds will also be found in the 'Journal de la Navigation de la Frégate Thetis,' 1839, and in papers by M. Meyen in the 'Nova Acta Ac. Leop. Car.,' vol. xvi., and by M. Lesson in the 'Revue Zoologique,' 1842. Subsequently the 'Voyage dans l'Amérique Méridionale,' by M. D'Orbigny, and the 'Zoology of the Beagle,' by Mr. Darwin, have greatly extended our knowledge of this region. Nor ought I to omit the brief but very interesting notes on the Birds of Chili by Mr. Bridges, in the 'Proceedings of the Zool. Soc.' 1843, or the full list of Peruvian birds lately published at Berlin by M. Tschudi, in which many new species are described. Most of the species originally described by Molina are now identified with accuracy, and the long and narrow tract extending the whole length of South America, between the Andes and the Pacific, is shown to possess a peculiar and a highly interesting fauna.

M. A. D'Orbigny, who prosecuted his scientific researches for several years in South America, traversing the interior from Buenos Ayres to Columbia, has reaped a rich harvest of zoology, which is embodied in his 'Voyage dans l'Amérique Méridionale.' Besides discovering many new species of birds, he has identified most of those described by Azara. The plates of his work are however not so perfect as the text, the colouring being too vivid, and the figures unnecessarily reduced in size, when the natural dimensions might have been more frequently retained. He has drawn some interesting conclusions respecting the distribution of species through various zones, of southern latitude, and through zones in some degree corresponding to these, of elevation. Such generalizations, when carefully made, never fail to throw light on philosophical zoology.

3. *Ornithological Monographs.*

No method is so effective in advancing zoological science as that by which an author gives his whole attention to some special group or genus, examines critically all the works of previous writers that relate to it, adds his own original observations, and publishes the result in the shape of a Monograph. I will briefly notice the works of this kind which have appeared of late years.

The different species of Vultur known up to 1830 were critically analysed by M. Rüppell in the 'Annales des Sciences Naturelles' for that year, and his remarks must be studied by all who attempt to define the species of that intricate group.

The characters of the family Strigidæ and of its subdivisions are treated of by M. I. Geoffroy St. Hilaire in 'Ann. Sc. Nat.,' 1830.

Mr. Swainson published a monograph of the genera Tachyphonus and Tyrannus in the 'Quarterly Journal of Science,' London, 1826. Although several species have been discovered since, and new genera proposed, yet these papers still possess considerable value. An essay on the Cuculidæ by the same author is inserted in the 'Mag. of Zool. and Botany,' vol. i.

M. Ménétries has published in the 'Mém. de l'Acad. Imp. de St. Pétersbourg,' 1835, a monograph of the Myiotherinæ, preceded by an historical account of the authors who have treated of this complicated group. This memoir is a valuable contribution to our knowledge, though the series of natural affinities would perhaps have been better exhibited if the Thamnophili had been included among the Myiotherinæ (passing, as they do, almost imperceptibly into Formicarius), and if the so-called Myiotherinæ of the East Indies had been formed into a separate section.

We owe to M. L'Herminier some interesting particulars respecting that anomalous and little-known bird, the Steatornis of Humboldt (Ann. Sc. Nat. vol. vi. p. 60, and Nouv. Ann. Mus. Hist. Nat. vol. iii.). It appears that this nocturnal bird, which inhabits the caverns of Venezuela and Bogota, can only be classed among the Caprimulgidæ, though it differs from all its congeners in its frugivorous habits, while it approaches the Strigidæ in many points of structure (as has been well insisted on by M. Des Murs, 'Rev. Zool.,' 1843).

The same indefatigable naturalist has thrown much light on the structure of the genera Sasa, Palamedea, Turnix and Rupicola, in the 'Ann. Sc. Nat.' vol. viii. p. 96, and 'Comptes Rendus,' 1837. The first of these he shows to be a connecting link between the Insessores and Rasores; the second he places between the Rallidæ and Ardeidæ; the third he considers to have more affinity to the Grallatores than to the Rasores; and the last he retains among the Ampelidæ.

M. Lesson's monographs of the Trochilidæ, entitled 'Histoire Naturelle des Oiseaux Mouches,' and 'Histoire Naturelle des Colibris,' are valuable works for the illustration of species, but the generic subdivisions are not carried into sufficient detail. M. Lesson has elsewhere proposed several generic groups of Trochilidæ, and M. Boié has added others; but many of these appear difficult to define satisfactorily. In fact there is no family of birds whose classification is more imperfect and more in want of careful elucidation than the beautiful but bewildering group of Humming Birds. The two volumes of 'Humming Birds' in Sir W. Jardine's 'Naturalist's Library' contain a synopsis of most of the species, but without professing to form a complete monograph.

Other volumes of the 'Naturalist's Library' are devoted to particular groups, but as they only contain selections, and not entire lists of the species, they do not strictly constitute monographs. Such are the useful volumes by Mr. Selby on the 'Pigeons and Gallinaceous Birds,' and by Mr. Swainson on Muscicapidæ. A more complete work is the volume by Sir W. Jardine on the Nectariniidæ, or rather on the genus Nectarinia, containing a very full synopsis of the species of that extensive and beautiful group.

The 'Histoire Naturelle des Oiseaux de Paradis' by M. Lesson, is a useful monograph of an obscure and difficult group of birds, and is worked out with more care and just criticism than is to be found in many others of M. Lesson's publications.

M. Malherbe of Metz is at present engaged on a general history of the Picidæ, a work much wanted on account of the many genera and species introduced into this family since Wagler's monograph of Picus was published.

Several attempts have been made to compile monographs of the numerous family of Psittacidæ, but the subject is yet far from being exhausted. Levaillant in 1801 had figured and described all the species then known, and Kuhl in 1820 published a valuable monograph in the 'Nova Acta Acad. Leop. Car.' Another and a more complete monograph of the Psittacidæ, by the industrious Wagler, will be found in the 'Abhandlungen der Baierischen Akademie der Wissenschaften,' 1832. Although some of the author's generic divisions have been criticised as being artificial, yet this paper has a great value for its discrimination of species. Lear's 'Illustrations of the Psittacidæ,' 1832, is intended as supplementary to Levaillant's great work 'Les Peroquets.' The lithographic plates are beautifully executed, but as they are unaccompanied by letter-press they hardly belong to the class of monographs.

Another continuation to the work of Levaillant is the 'Histoire Naturelle des Peroquets,' by M. Bourjot St. Hilaire, Paris, 1835–38, folio. Many of the plates are original, others are copied from Spix, Temminck, or Lear; they are executed on stone, and though inferior to the works of Gould and Lear, they are perhaps the best ornithological lithographs which have issued from the French press. The text of this work is prepared with considerable care, but the nomenclature wants precision, the Latin names being often wrong-spelled, and the principle of binomial appellations departed from. Thus the genus Palæornis is in one instance designated Psittacus, in another *Psittacus sagittifer*, and in a third *Conurus sagittifer*, with the addition in each case of a specific name. What can we say of an author who designates a species as "*Psittacus platycercus viridis unicolor*," but that he is deserting that admirably concise and effective method of nomenclature introduced eighty years ago by the great Linnæus, and is resuming the vague and unscientific generalizations of the ancient naturalists?

I only know by name the 'Monograph der Papageien,' published in Germany by C. L. Brehm.

Some interesting details on the genera Crotophaga and Prionites were published by Sir W. Jardine in the 'Annals of Natural History,' vols. iv. and vi., and I last year communicated to the same work a paper on the structure and affinities of the genera Upupa and Irrisor (Promerops of some authors), showing that these genera are really allied, though M. Lafresnaye had maintained that they are widely separated (Proc. Zool. Soc. 1840).

Mr. Vigors communicated to the earlier volumes of the 'Zoological

Journal' several papers of a monographic character, entitled 'Sketches in Ornithology,' which are distinguished by close research and careful induction.

Among the ornithological works of this class which have appeared of late years, Mr. Gould's 'Monographs of the Trogonidæ and of the Rhamphastidæ occupy a conspicuous place. Of these I need only say that they are executed in the same form and with the same excellence as his other superb publications. Mr. Gould has also published a short monograph of Dendrocitta in the 'Zoological Transactions.' He is now collecting materials for monographs of other families, including the Odontophorinæ, the Caprimulgidæ, and the Alcedininæ. Of the Odontophorinæ, or American Partridges, the first number has already appeared; and though they are a less gaudy tribe of birds than many others, yet the admirable taste with which Mr. Gould has depicted them renders the work peculiarly attractive. A translation with reduced plates of Gould's 'Monograph of Rhamphastidæ' has been published in Germany by Sturm.

Prof. C. J. Sundevall has described some species of Euphonia in the 'Kongl. Vetenskaps Akademiens Handlingar,' Stockholm, 1834. This paper is supplementary to the monograph 'De genere Euphones,' by Dr. Lund, published at Copenhagen in 1829.

Dr. Rüppell's work, entitled 'Museum Senckenbergianum,' Frankfort, 1836, contains some admirable monographs of the genera Otis, Campephaga, Colius, Cygnus, &c. They combine laborious bibliographical research with close observation of structure, and are accompanied by excellent illustrative figures.

Mr. Swainson published in the 'Journal of the Royal Institution,' 1831, an essay on the Anatidæ, which, though founded on peculiar theoretical views, deserves to be consulted even by those who do not agree in the author's conclusions. This memoir prepared the way for Mr. Eyton's 'Monograph of the Anatidæ,' 1838, which is in many respects a valuable and accurate work, and is especially useful for its details of anatomical structure. The Latin specific characters might however have been drawn up with more care; and an appendix should have been added, containing the numerous species described by Latham and the old authors, which had not come under Mr. Eyton's observation. No monograph can be considered complete which does not, in addition to the *ascertained* species, enumerate also the *unascertained*, that is to say, those nominal species which for the present exist only in books and not in museums, many of which however will no doubt be again restored to science as real species, while others will be recognized as peculiar conditions of the species we now possess. In this respect the collection of monographs published by Wagler under the title of 'Systema Avium,' and continued afterwards in Oken's 'Isis,' affords a useful model. It was his custom, after describing those species of a genus with which he was himself acquainted, to append two lists, one of "*species a me non visæ*," and the other of "*species ad genera diversa pertinentes.*"

MM. Hombron and Jacquinot have communicated to the Académie des Sciences a memoir on the habits and classification of the Procellariidæ, of which an abstract is given in the 'Comptes Rendus,' March 1844, and in which several new subgenera are proposed. Mr. Gould has also extended our knowledge of this obscure group in the 'Annals of Nat. Hist.,' May 1844.

M. Brandt, of St. Petersburg, who has made the Natatores his peculiar study, has monographed the family Alcidæ, and the genera Phaëton and Phalacrocorax, in memoirs contributed to the Imperial Academy of Sciences at St. Petersburg.

Professor Sundevall states that there is a monograph of the genus Dysporus (Sula) in the 'Physiographisk Tidskrift,' Lund, 1837.

Many monographic summaries of different genera will be found in Temminck's 'Planches Coloriées,' Rüppell's works on Abyssinia, and Smith's 'Zoology of South Africa.'

Besides monographs of the larger groups, there are many valuable memoirs on individual species, such as that by M. Botta on *Saurothera californiana* (originally described by Hernandez as a Pheasant, and now properly termed *Geococcyx mexicanus*, Gmel. (sp.)), in the 'Nouv. Ann. Mus. Hist. Nat.,' vol. iv.; that by Dubus on *Leptorhynchus pectoralis* and other new generic types, in 'Bullet. Acad. Roy. de Bruxelles;' by De Blainville on Chionis (Ann. Sc. Nat. 1836); by Lesson on Euryceros (Ann. Sc. Nat. 1831); by Mr. Yarrell on Apteryx (Trans. Zool. Soc. vol. i.), &c.

4. *Miscellaneous Descriptions of Species.*

Among recent works of this class, Guérin's 'Magasin de Zoologie,' commenced in 1831, demands notice. This publication, which for the excellence of its scientific matter and its moderate price deserves every encouragement, is rendered the more convenient to the working naturalist by being sold in separate sections. The ornithological portion of this periodical contains valuable papers by Isidore Geoffroy St. Hilaire, Lafresnaye, D'Orbigny, Eydoux, Gervais, L'Herminier, Delessert and others. Many new and important forms are there described and figured with great exactness, and although the authors are not in all cases sufficiently conversant with the writings of British ornithologists, yet they duly estimate the claims of the latter when brought before them.

Upon the whole, the 'Magasin de Zoologie' must be regarded as a work highly creditable to French science, and it is much to be regretted that since the discontinuance of our own 'Zoological Journal' no similar periodical has been set on foot in this country. Such a work might however be easily reproduced if our Zoological Society would attach illustrative plates to their very valuable 'Proceedings,' and give them the form of a Journal, as has lately been done by the Geological Society.

A work closely connected with the 'Magasin de Zoologie' is the 'Revue Zoologique de la Société Cuvieriennne,' the object of which is to assert without loss of time the claims of any zoological discovery, by publishing brief but adequate descriptions of new species. The multitude of labourers now at work in the same field, and the importance of adhering to the rule of priority as the basis of systematic zoological nomenclature, render it necessary to publish rapidly and diffuse widely the first announcements of new discoveries. The delays incident to the engraving of plates and the printing of memoirs in scientific transactions have often robbed original discoverers of their due credit, and introduced confusion and controversy into science; and it is to remedy this evil that the valuable though unpretending 'Revue Zoologique' was established.

Original descriptions of new species are scattered so widely that it is impossible to notice all the recent works in which they occur, and I must therefore confine myself to simply enumerating the more important. Of regular periodical works devoted to natural history in general, and including original contributions to ornithology, I may mention (in addition to those above noticed) the 'Zoological Journal;' Ainsworth's 'Edinburgh Journal of Natural and Geographical Science,' 1829; Loudon's and Charlesworth's 'Magazine of Natural History;' Sir W. Jardine's 'Magazine of Zoology and Botany;' Taylor's 'Annals of Natural History;' and the popular rather than scientific 'Field Naturalist's Magazine' of Professor Rennie; the 'Naturalist' of Mr. Neville Wood; and the 'Zoologist' of Mr. E. Newman. Among foreign periodicals are Oken's 'Isis;' Wiegmann's 'Archiv;' Kroyer's 'Naturhistorisk Tidskrift;' Van der Hoeven's 'Tijdschrift fur Natuurlijke Geschiedenis;' Wiedemann's 'Zoologisches Magazin;' 'Physiographisk Tidskrift,' Lund;' Rohatzsch's 'Munich Journal;' the 'Annales des Sciences Naturelles;' Müller's 'Archiv für Anatomie;' Silliman's 'American Journal of Science;' 'Boston Journal of Natural History,' and the scientific journals of India, Tasmania, and South Africa, which I mentioned when speaking of the ornithology of those regions. Among the authorized publications of scientific societies, ornithological details of greater or less amount will be found in the 'Philosophical Transactions;' the 'Proceedings and Transactions of the Zoological Society;' the Transactions of the Linnean, the Cambridge Philosophical, the Newcastle and the Wernerian Societies; the 'Bulletin de la Société Philomathique des Pyrénées orientales;' 'Actes de la Soc. Linnéenne de Bordeaux;' 'Mémoires de la Soc. Linnéenne de Calvados;' 'Bulletin de l'Académie Royale des Sciences de Bruxelles;' 'Mémoires' and 'Comptes Rendus de l'Académie Royale de France;' 'Annales du Musée d'Histoire Naturelle;' 'Annales de la Soc. Linnéenne de Paris;' 'Mémoires de la Soc. d'Emulation d'Abbeville;' 'Mémoires de la Soc. Académique de Falaise;' 'Mémoires de la Soc. Royale de Lille;' 'Mémoires de l'Académie de Metz;' 'Mémoires de la Soc. des Sciences Naturelles de Neufchatel;' 'Mémoires de la Soc. de Physique de Genève;' 'Jahrbuch der Naturforschenden Gesellschaft zu Halle;' 'Nova Acta Academiæ Cæsareæ Naturæ Curiosorum;' 'Abhandlungen der Baierischen Akademie der Wissenschaften;' 'Abhandlungen der Akademie der Wissenschaften zu Berlin;' 'Kongl. Vetenskaps Akademiens Handlingar,' Stockholm; 'Mémoires' and 'Bulletins de l'Académie Impériale des Sciences de St. Pétersbourg;' 'Annales Universitatis Casanensis;' 'Mémoires' and 'Bulletins de la Soc. des Naturalistes de Moscou;' 'Annale delle Scienze Naturali di Bologna;' 'Nuovo Giornale de' Litterati di Pisa;' 'Memorie della Accademia delle Scienze di Torino;' 'Atti dell' Accademia Gioenia de Catania;' 'Journal of the Academy of Natural Sciences of Philadelphia;' 'Annals of the Lyceum of Natural History of New York;' 'Transactions of the American Philosophieal Society,' and many others.

Of recent works specially devoted to the description and illustration of new objects of zoology in general or of ornithology in particular, the following British ones may be mentioned:—Swainson's 'Zoological Illustrations,' 1st and 2nd series, 1820–33; Donovan's 'Naturalist's Repository;' Jardine and Selby's 'Illustrations of Ornithology,' an excellent work, which I regret to say is now discontinued; Wilson's 'Illustrations of Zoology,' fol. Edinburgh, 1827, an accurate and well-illustrated volume; J. E. Gray's 'Zoological Miscellany,' 1831, containing concise descriptions of new species; Swainson's 'Animals in Menageries,' 1838 (in Lardner's Cyclopædia), comprising descriptions of 225 species, many of which however had before been published; Bennett's 'Gardens and Menagerie of the Zoological Society,' 1831, valuable for its observations on the habits of living individuals; and Gould's 'Icones Avium,' equal in merit and beauty to his other works.

Among foreign works of the same kind are Temminck's 'Planches Coloriées,' whose merits are too well known to be here dwelt on, and the text of which, if carefully translated and edited, would form an acceptable volume to the British naturalist; Lesson's 'Centurie Zoologique,' containing eighty miscellaneous plates; those relating to ornithology respectably executed, and exhibiting several new forms, especially of Chilian Birds; the 'Illustrations de Zoologie' form a second volume of the same character as the 'Centurie;' Kuester's 'Ornithologische Atlas der auseuropaischen Vögel,' Nuremberg; Dubois' 'Ornithologische

Galerie,' Aix-la-Chapelle (the last two works I know only by name); Lemaire, 'Hist. Nat. des Oiseaux exotiques,' Paris, 1836, a collection of brief descriptions and very gaudy figures; and Rüppell's 'Museum Senckenbergianum,' a work of first-rate excellence.

5. *Progress of the Pictorial Art as applied to Ornithology.*

The preceding criticisms have chiefly referred to the claims of the descriptive or classificatory portion of the several works noticed, but it may be useful to make a few special observations on the success which has attended the various methods of representing the forms and colours of birds to the eye. In this branch of zoology, as in all others, the pencil is an indispensable adjunct to the pen. The minute modifications of form which constitute the distinctive characters of genera, and the delicate shades of colour by which alone the specific differences are in many cases indicated, are of such a nature as to be frequently beyond the power of language to define without the aid of art, and it is consequently indispensable that the zoological artist should combine a scientific knowledge of the subject with a perfect command of his pencil. In no branch of zoology are these peculiar talents more requisite than in ornithology, where the varieties of habit and of attitude, the unequalled grace and elegance of form, the remarkable modifications of structure in the plumage, and the endless diversities of colouring demand the highest resources of the painter's skill.

The three principal modes of engraving, namely, wood-engraving, metallic plate-engraving, and lithography, have all been applied in turn to the illustration of ornithology.

1. *Wood-engraving.*—For such illustrations of birds as are not intended for colouring, this method is not only the cheapest, but for works of smaller size it is the best. The works of the immortal Bewick have shown us with what complete success the structure and arrangement of the feathers, the relative intensities of the colours, and the characteristic expression of the living bird may be transferred to a block of wood by the hand of original genius. Many recent wood-engravers have approached Bewick, but none have yet equalled him. Among the most successful of these the Messrs. Thompson of London must be especially mentioned. Their woodcuts in Yarrell's 'British Birds' are beautiful works of art; in delicacy of execution they often exceed the engravings of Bewick; but the occasional stiffness of attitude in the birds, and a conventional *sketchiness* in the accompaniments, indicate the professional artist and not the self-taught child of Nature.

The beauty of Yarrell's 'British Birds' is much enhanced by improvements in the preparation of paper and ink, and in the mode of taking off the impressions which have been introduced since Bewick's time. It is probable that if the wood-blocks of Bewick, now in the possession of the great engraver's family, were entrusted to one of our first-rate London printers, an edition of Bewick's 'Birds' could be now produced, far superior in execution to any which was issued in the lifetime of the author.

2. *Metallic plate-engraving.*—Line-engravings or etchings on copper or steel have been at all times extensively applied to the illustration of ornithological works. Such engravings, if uncoloured, are certainly inferior in effectiveness to good woodcuts, as an example of which I may mention the numerous plates of birds in Shaw's 'Zoology' and Griffith's 'Cuvier,' which though often respectably executed, are almost useless for the purpose of specific diagnosis; and even when carefully coloured, engraved plates rarely approach in excellence, and in my opinion never equal the best examples of lithography. The greater stubbornness of the material involves almost of necessity a certain constraint in the attitudes represented: just as the statues of ancient Egypt which were carved out of hard basalt, never attained the grace and animation which has been conferred upon the tractable marbles of Greece, and the still softer alabaster of Italy. In proof of this I may refer to Temminck's 'Planches Coloriées,' and to the recent works of Lesson, Quoy, D'Orbigny and other French ornithologists. The figures of birds in these plates, though delicately and even beautifully engraved, are often exceedingly stiff and unnatural, a defect owing partly no doubt to too great a familiarity with *stuffed specimens*, but in part also to the unyielding material on which they are engraved. If the Parisian ornithological artists have not the means of studying living nature, they might at least take for their models the designs of Nature's best copyist—Gould.

The defects shown to be incident to *line-engraving* attach indeed in a less degree to *etching*. The resistance to the tool being diminished in the latter process, the lines are drawn with greater ease and freedom. Here the main difficulty is to avoid *hardness* and *coarseness* in the delineation of the plumage. Many etchings which are otherwise meritorious, have failed in this point, and the lines which were intended to represent the smooth soft plumage of birds, resemble rather the scales of a fish or the wiry hair of the Sloth or Platypus.

The plates of Mr. Selby's 'Illustrations of British Ornithology' are certainly the finest examples extant of ornithological etchings, though they are nearly equalled by some of the plates etched by Sir W. Jardine, Mr. Selby and Captain Mitford in the 'Illustrations of Ornithology.'

In the plates of Audubon's 'Birds of America' line-engraving is combined with aquatint, a method which, when well executed, may be used with advantage to increase the depth and softness of line-engravings or etchings.

3. *Lithography*.—We have next to consider that style of illustration which is beyond all question the best adapted to ornithology. Lithography possesses all the freedom and facility of *drawing* as contrasted with the laborious mechanical process of *engraving*, and is hence peculiarly fitted to express the graceful and animated actions of birds. Another merit is the expression of *softness* which it communicates to the plumage, and the power of showing the roundness of the forms by a homogeneous shading, instead of the parallel lines and cross hatchings employed in engraving. The lines introduced to represent the individual feathers possess just that amount of indistinctness which we see in the living object, and which adds so much to its beauty.

It is a matter of some pride to us, that while in certain other departments of natural history (especially in fossil conchology) the British lithographers must yield the palm to foreigners, yet in ornithology our own artists have never been equalled. Lithography was, I believe, first applied to the delineation of birds by Mr. Swainson, who soon attained great excellence in the art. His 'Zoological Illustrations,' his plates to the 'Fauna Boreali-Americana,' and his 'Ornithological Drawings of the Birds of Brazil,' possess great merits both of design and execution, as does also Mr. Lear's great work on the Psittacidæ. But all these productions are eclipsed by the pencil of Gould, whose magnificent and voluminous works exhibit a gradual progress from excellence to perfection. Temminck, who in 1835 said of Gould's 'Birds of Europe,' "Ils sont d'un fini si parfait, tant pour le dessin, la pose, et l'exacte vérité de l'enluminure, qu'on pourrait, avec de si beaux portraits, se passer des originaux montés," would, I am sure, pass even higher encomiums on the 'Birds of Australia,' which Mr. Gould is now publishing*. One little fault, and one only can I find in these beautiful drawings, and that is, that the hallux, which in all the Insessores is essential to the steady support of the bird, is too often represented as projecting backwards, instead of firmly clasping, as it ought, the perch. Mr. Richter and Mr. Waterhouse Hawkins, both of whom have been employed in executing on stone the designs of Mr. Gould, have attained great excellence in the art, as has also Mr. D. W. Mitchell, the able coadjutor of Mr. G. R. Gray in the 'Genera of Birds.' The latter has successfully applied the new art of "lithotinting" to the representation of smooth and hard surfaces, such as those of the beak and legs of birds. He has also in some cases executed the whole plumage in lithotint, producing a beautiful and delicate finish, the effect of which is intermediate between lithography and engraving.

Lithography has never been applied extensively to ornithology upon

* Now complete in seven volumes; a supplementary part being from time to time published in continuation, as new materials are discovered.—Ed.

the continent. The plates in Vieillot's 'Galerie des Oiseaux,' and in the Atlas to Erman's 'Reise um die Erde' are very indifferent, those in Werner's 'Atlas des Oiseaux d'Europe' a shade better, and in the 'Petersburg Transactions' they are tolerably good. The Prince of Canino's 'Fauna Italica,' Nilsson's 'Illuminade Figurer till Skandinaviens Fauna,' and Rüppell's 'Museum Senckenbergianum,' are the only continental works which I have seen, in which the lithographs at all approach to the excellence of the British artists.

The lithographic plates in Spix's 'Avium species novæ in itinere per Braziliam collectæ,' are tolerably executed, but in rather a peculiar style; the legs and beaks of the birds, and in some instances the whole body, being first covered with black, and the lighter parts afterwards scraped off with a sharp point. Examples of this style also occur in some of Mr. Mitchell's plates. In particular cases, especially in representing the scuta of the legs and feet, and the details of black plumage, this method may be adopted with great advantage.

There is a real though somewhat paradoxical cause of the superior excellence of the drawings of Gould and of Swainson, which should not be overlooked. It is, that these artists have in almost every case (when the living bird was not accessible) made their designs from *dried skins*, and not from *mounted specimens*. In the skin of a bird, dried in the usual mode for convenience of carriage, the natural outlines and attitudes are nearly obliterated, and the artist is consequently *compelled* to study living examples, to retain the images thus acquired in his memory, and to transfer them to his design. By the constant habit of thus re-animating as it were these lifeless and shapeless corpses, he acquires a freedom of outline and a variety of attitude unattainable by any other means. But when an artist attempts to draw from a stuffed specimen, he beholds only a fabric of wire and tow, too often a mere caricature of nature, exhibiting only the caprices and mannerisms of an ignorant bird-stuffer. Knowing that the object before him is *intended* to represent nature, he is unconsciously and irresistibly led to copy it with all its deformities. Such is no doubt one cause of the stiff and lifeless designs which we see in the French works, drawn as they mostly are from mounted specimens in the Paris Museums.

6. *Anatomy and Physiology of Birds.*

The most complete general treatise on the anatomy of birds that I am acquainted with is the article *Aves* by Prof. Owen, in Todd's 'Cyclopædia of Anatomy and Physiology.' The author's original investigations on this subject are here combined with those of others, and the whole forms an excellent monograph of the structural peculiarities of the class, as well as of many differential modifications which mark particular groups. Much indeed remains to be added to our knowledge of individual organizations, but those anatomical arrangements which distinguish Birds from the other classes of Vertebrata can hardly be described with greater precision or reasoned upon more philosophically than in the work in question. We may indeed regret that this treatise of Prof. Owen is not published in a separate and more accessible form, especially if we consider how essential a knowledge of comparative anatomy is to the scientific zoologist, and what peculiar interest attaches to the anatomy of Birds, as indicating their affinities to Reptiles and to Mammals, and as exhibiting the wonderful arrangements by which their muscular bodies are sustained in a medium at least one thousand times lighter than themselves. We shall however be soon put in possession of Prof. Owen's most recent researches on the anatomy of birds, by the publication of that portion of his 'Hunterian Lectures' which relates to the Vertebrata, and which will doubtless be of equal value with the excellent volume already issued on the Invertebrata*.

Another carefully-prepared summary of ornithic anatomy is that by Prof. M'Gillivray, in the Introduction to his 'History of British Birds.' The author has evidently bestowed much labour, both mental and manual, upon this subject, and has successfully vindicated the claims of comparative anatomy to be considered not an adjunct to, but a part of, scientific zoology. The above work is particularly valuable for its details respecting the organs of digestion, a part of the system to which the author justly attributes great importance, and which he has treated of in a special article in the 'Magazine of Zoology and Botany,' vol. i. *Résumés* of the anatomical peculiarities of birds will also be found in the 'Elémens de Zoologie,' by Milne-Edwards, 1837, and in the 'Encyclopædia Britannica' and 'Penny Cyclopædia.' The article *Zoology* in the 'Encyclopædia Metropolitana' also contains a useful treatise on the subject, though it is damaged by the affectation of using new English terms in place of the received Latin terminology of anatomy.

In Dr. Grant's 'Outlines of Comparative Anatomy,' the structure of birds is described with the same accuracy as that of the other classes of animals; but as the work is arranged anatomically and not zoologically, the details of ornithic anatomy are necessarily intermixed with those of the other classes of animals.

Prof. Rymer Jones has given, in his 'General Outline of the Animal Kingdom,' a careful abstract of the anatomy of birds, including more especially the structure of the eye and the important subject of the development of the ovum. The excellent mode in which the generalities of the subject are treated of, makes us regret that the limits of Prof.

* The last volume of this important work was published in 1846, and completed the comparative anatomy and physiology of Fishes only.—Ed.

Jones's work prevent him from giving a fuller statement of the anatomical characters of the several orders and families.

An excellent synopsis of this subject is contained in Wagner's 'Comparative Anatomy,' of which Mr. Tulk has just published an English translation.

Of special treatises, either on the anatomy of particular organs throughout the whole class, or on the general anatomy of particular groups, many are to be found scattered over the field of scientific literature, and I shall notice some of the more important.

The general subject of the *pneumaticity* or circulation of air through the bodies of birds is ably treated of by M. E. Jacquemin in the 'Nova Acta Acad. Cæs. Leop. Car.' 1842. See also 'L'Institut' and 'Comptes Rendus,' 1836. After minutely describing the modifications of the aërating system in different forms of birds, the author deduces a series of conclusions, and shows that this structure, peculiar to the class of birds, performs the fourfold office of oxidizing the blood,—of enlarging the surface of the body, and consequently the points of muscular attachment,—of diminishing the specific gravity, and of producing a general elasticity which favours the act of flight.

The structure of the ear in birds is treated of in great detail in a memoir by M. Breschet, in the 'Annales des Sciences Naturelles' for 1836, and in a detached treatise on the same subject. After giving an historical sketch of the researches of previous authors, he enters upon an elaborate description of the characters of this organ in various groups of birds. He shows that of the three bones of the tympanum, the *stapes* alone is osseous in birds, while the *malleus* and the *incus*, which in Mammalia are composed of bone, are here represented by cartilaginous processes, and he points out many other minute but important characters which appear to distinguish the ears of birds from those of other Vertebrata.

Dr. Krohn has treated on the organization of the iris, and Dr. Bergman on the movements of the *radius* and *ulna* in Müller's 'Archiv für Anatomie,' 1837–39.

The structure of the *os hyoides* in birds, and the affinities of its several parts to the corresponding organs of the other Vertebrata, are explained in a memoir by M. Geoffroy St. Hilaire, in the 'Nouvelles Annales du Mus. d'Hist. Nat.' 1832.

M. Müller has described the modifications of the male organs of birds in the 'Abhandlungen der Akad. Wissenschaften zu Berlin,' 1836.

M. Cornay, in 'Comptes Rendus,' 1844, p. 94, has announced that he finds an important character to exist in the anterior palatine bone, the modifications of which in the various orders he considers to form a more correct basis of classification than any one hitherto employed.

Until more attention be paid to this organ than it has yet received, it would be premature to pronounce as to the value of it.

The gradual development of ossification in the sternum of young birds, and the relations of its several parts to the skeletons of other Vertebrata, were treated of by M. Cuvier (Ann. Sc. Nat. 1832) and by M. L'Herminier (Ann. Sc. Nat. and Comptes Rendus, 1836–37). These essays involved theoretical views which gave rise to controversies in which MM. Serres and Geoffroy St. Hilaire also took part. The structure of the pelvis and hinder extremities was described by M. Bourjot St. Hilaire in a memoir read to the Académie des Sciences, 1834.

The osteology of the feet of birds is treated of by M. Kessler in the 'Bulletin de la Soc. de Naturalistes de Moscou,' 1841.

The internal temperature of various species and groups of birds is treated of in a general memoir on the subject of Animal Heat, by M. Berger, in the 'Mémoires de la Société de Physique de Genève,' 1836. Dr. Richard King has also published some observations on this subject.

Mr. Eyton has contributed some interesting information on the anatomy of Menura, Biziura, Merops, Psophodes and Cracticus, which throw much light on the affinities and classification of those genera (Annals of Natural History, vol. vii. *et seq.*).

Amidst the numerous profound researches of Prof. Owen on the comparative anatomy of various portions of the animal kingdom are many original investigations into the structure of such rare birds as have fallen under his scalpel. In the 'Transactions of the Zoological Society' he has described the anatomy of *Buceros cavatus*, showing the points of affinity which the Bucerotidæ bear towards the Rhamphastidæ on the one hand and the Corvidæ on the other. He has also suggested that the probable design of the gigantic beak in the Hornbills and Toucans is to protect the eyes and head while penetrating dense thickets in quest of the nestling birds on which they feed. Another memoir, of still greater importance, is the elaborate description of the anatomy of the Apteryx (Trans. Zool. Soc. vol. ii.), for which our successors even more than ourselves will be grateful to Prof. Owen, seeing that but few years will probably elapse before that rare and extraordinary species will be erased from the list of animated beings. He has also contributed to the 'Proceedings of the Zoological Society' excellent anatomical monographs of the genera Sula, Phœnicopterus, Corythaix, Pelecanus, Cathartes, and Talegalla. The invaluable descriptive catalogues of the Museum of the Royal College of Surgeons, which are in great measure the work of Prof. Owen, contain a mine of information on the anatomy of every class, and not least on that of birds. The volume which relates to the Fossil Mammalia and Birds is now in the press.

We are indebted to Mr. Yarrell for several accurate notices on the more remarkable structures of certain birds, among which are papers on the anatomy of the Raptores, on the xiphoid bone and its muscles in Phalacrocorax, and on the muscles of the beak in Loxia, published in the 'Zoological Journal;' memoirs on the convolutions and structure of the trachea in Numida, the Gruidæ, and the Anatidæ, which will be found in the 'Linnean Transactions;' and notices on the anatomy of Cereopsis, Crax, Ourax, Penelope, Anthropoides and Plectropterus, in the 'Proceedings of the Zoological Society.'

A very elaborate account of the anatomy of *Aptenodytes patachonica*, by Mr. Reid, is published in the 'Proceedings of the Zoological Society,' 1835, and we may regret that this gentleman has not made more such contributions to anatomical science.

There are some very interesting remarks by Mr. Blyth on the osteology of *Alca impennis*, in the 'Proceedings of the Zoological Society,' 1837, showing that in this bird (which is wholly unable to fly) the bones of the extremities are nearly solid and filled with marrow, while in the volatile species of Alcidæ the air-cavities of the bones are highly developed, in order to compensate for the shortness of the wings. He adds the important remark, that "when once the object of aërial flight is abandoned, the wings are reduced to exactly that size which is most efficient of all for subaquatic progression; *species of an intermediate character of course never occurring*." This principle of the *necessity of hiatuses* in the natural system (of which numerous other examples might be adduced), is one which I have long regarded as conclusive against that continuity of affinities and symmetry of arrangement which some writers have endeavoured to demonstrate.

Mr. T. Allis of York (whose beautifully prepared ornithic skeletons, now in the York Museum, are so highly creditable to his skill as an anatomist) has made some observations on the connexion between the furculum and sternum, showing that in certain birds possessing powers of long-continued flight these bones are connected by an intimate symphysis, which in Pelecanus and Grus amounts to an actual anchylosis (Zool. Proc. 1835).

The anatomies of Pelecanus, Dicholophus and Corythaix are described in detail by Mr. W. Martin in the work last quoted.

A paper on the anatomy of *Corvus corone* by M. Jacquemin, will be found in the 'Isis,' 1837, and the osteology of the Trochilidæ is described by M. J. Geoffroy St. Hilaire in 'Comptes Rendus,' 1838*.

Several points of ornithic anatomy are treated of by Prof. Wagner in

* The 'Disquisitiones Anatomicæ Psittacorum,' by M. Thuet, Turin, 1838, and Kuhlman's dissertation, 'De Absentiâ Furculæ in Psittaco pullario,' Kiel, 1842, are works which I have not seen.

the 'Abhandl. der Baierischen Akad.' 1837, and the osteology of the genera Crypturus, Dicholophus, Psophia and Mycteria is fully described. The structure of the Struthionidæ is beautifully portrayed by D'Alton in his 'Skelete der Straussartigen Vögel,' 1827.

There is a paper by M. Schlegel on the supposed absence of nostrils in the genus Sula, in the 'Tijdschrift voor natuurlijke Geschiedenis,' 1839, of which, from being unacquainted with the Dutch language, I regret my inability to give a summary.

The osteology of several groups of Natatores is treated of by M. Brandt in an elaborate and highly important paper in the 'Mémoires de l'Acad. Imp. de St. Pétersbourg,' 1839. The researches of this author throw great light upon the classification of many obscure groups, and nothing can be more exact than his figures and descriptions of ornithic osteology.

Mr. Yarrell has paid considerable attention to the subject of *hybridity* (Zool. Proc. 1832, 1836, &c.). The result of his observation seems to be, that hybrid birds will occasionally propagate with the pure race on either side, but rarely, if ever, with each other, thus indicating a special provision of nature to preserve the distinctness and permanency of species. Mr. Eyton and Mr. Fuller have also made notes on the same subject (Zool. Proc. 1835). See also a paper by Mr. W. Thompson in the 'Mag. of Zool. and Bot.' vol. i.

Mr. G. Gulliver, who has made a series of microscopic researches into the blood-corpuscles of the Vertebrata, taking exact measurements of these minute bodies in different genera and species, has in the course of this inquiry given a fair share of attention to the corpuscles of birds, and his labours are recorded in the 'Proceedings of the Zool. Soc.' 1842, &c.

The difficult question of the influence of climate in producing permanent varieties of species is discussed by Dr. C. L. Gloger in a treatise published at Breslau, 1833, and which deserves translation for the use of British naturalists, although the author carries his theory to too great an extent.

The arrangement of the feathers on birds, to which attention was first called by Nitzsch in his 'Pterylologie,' is briefly treated of in a memoir read to the Académie des Sciences by M. Jacquemin (Ann. Sc. Nat. 1836, p. 227), who points out several facts which have not been sufficiently attended to by previous ornithologists.

The various modes by which the changes of plumage in birds at different seasons are effected, whether by actual moulting, by the shedding of a deciduous margin to the feather, or by a change of colour in the feather itself, have been investigated by Cuvier, Temminck, Yarrell (Trans. Zool. Soc. vol. i.), and others. Dr. Bachman of Charleston has made some very interesting observations on this subject in the case of many of the North American birds, which will be found in the 'Transactions of the American Philosophical Society,' 1839.

The subject of moulting, and especially of that remarkable tendency in old female birds to assume the male plumage, is treated of by M. I. Geoffroy St. Hilaire (Ann. Sc. Nat., and Essais de Zoologie Générale, 1841). See also papers by Dr. Butler in the 'Memoirs of the Wernerian Society,' and by Mr. Yarrell in the 'Philosophical Transactions.'

M. De la Fresnaye published in the 'Mémoires de la Soc. Acad. de Falaise,' 1835 (L'Institut, 1837), a paper on *melanism*, or a supposed abnormal tendency in the Raptores to acquire a dark plumage, analogous to *albinoism* in other birds. The examples cited are few in number, and not very conclusive, but the subject is deserving of investigation.

Many writers have written descriptive works on the eggs of birds, especially of the European species. Of the older authors on this subject, as Klein, Wirsing, Sepp, Naumann, Schintz, Donovan, Roux, and Thienemann, I need not here speak. In the 'British Oology' of Hewitson the eggs of our native birds are accurately described and figured, and the second edition now publishing attests the popularity of the subject. An 'Atlas of Eggs of the Birds of Europe' is just commenced by A. Lefevre at Paris, the figures of which are well executed. Of the eggs and nidification of exotic birds our information is very incomplete, and almost the only contributor to this branch of ornithology is M. D'Orbigny, who in his 'Voyage dans l'Amérique Méridionale' gives many figures of eggs and details of nidification, which may aid in clearing up the affinities of certain doubtful forms of the South American continent.

Mr. Gould brought home from Australia a large and interesting collection of eggs and nests, of which we may regret that he has not introduced the figures into the plates of his 'Birds of Australia.' We may hope, however, that when he has completed that great work he will publish an 'Australian Oology,' and perpetuate the knowledge which his unique collection of eggs supplies.

Dr. Carlo Passerini has given an account of the nidification and incubation of *Paroaria cucullata* in a domestic state, in a memoir published at Florence in 1841.

The subject of ornithic oology has been treated of in a philosophical manner by M. Des Murs (Revue Zoologique, and Mag. de Zool. 1842, 1843). By carefully studying the peculiarities of form, nature of shell, and colour, in the eggs of various birds, he finds a correspondence between these peculiarities and the structural characters of the several groups, and thus obtains an additional element in the process of classification.

The *number* of eggs laid by birds of different groups and species is the subject of a paper by M. Marcel de Serres (Ann. Sc. Nat. ser. 2. vol. xiii. p. 164), and the author deduces some interesting generalizations upon this subject.

There is a learned treatise on the structure of the egg prior to incubation by Prof. Purkinje, under the title of 'Symbolæ ad Ovi Avium Historiam,' Leipzig, 1830. The structure of the *vitellus* has been investigated by M. Pouché (Comptes Rendus, 1839), and that of the umbilical cord by M. Flourens (Institut, 1835, p. 324), while M. Serres has described the branchial respiration of the embryo of mammifers and birds in the 'Ann. Sc. Nat.' ser. 2. vol. xiii. p. 141.

Closely connected with oology is the subject of nidification, one of the most interesting branches of ornithological observation, and one which often throws important light on questions of natural affinities. I am not aware of any special work on this subject except the 'Darstellung der Fortpflanzung der Vögel Europa's,' by Thienemann, and the popular 'Architecture of Birds' by the late Prof. Rennie, but the details of the nidification of European birds are contained in most of the works which treat upon them. The nests of the majority of exotic species are still unknown, though Wilson, Audubon, Gould and others have in some measure supplied this deficiency in our knowledge.

The songs and call-notes of birds are very important in their relation to habits and affinities, though, from the imperfect mode of indicating these sounds by alphabetical or musical characters, there is much difficulty attending their study. In some cases, such as the relation of *Phyllopneuste rufa* to *P. trochilus*, or of *Corvus corone* to *C. americanus*, the notes of the living birds present clearer specific distinctions than are shown by their physical structure, and the melody of the woods thus becomes no less interesting to the scientific zoologist than it is fascinating to the unlearned lover of nature.

External Terminology.—The series of terms employed by Brisson, Linnæus and Latham, in describing the external parts of birds, were greatly improved in precision and accuracy by the 'Prodromus Systematis Mammalium et Avium' of Illiger. His series of descriptive terms are still generally current, and have undergone comparatively little change. Definitions and figures illustrative of the terms employed in ornithology will be found in most general treatises on the subject, among which Lichtenstein's 'Verzeichniss der Doubletten,' Berlin, 1823, Stephens's 'General Zoology,' Swainson's 'Classification of Birds,' Wilson's article *Ornithology* in 'Encyclopædia Britannica,' the article *Birds* in the 'Penny Cyclopædia,' and M'Gillivray's 'History of British Birds,' may be mentioned as being useful guides to the language of descriptive ornithology.

There is an excellent summary of the different characters used for ornithological classification, and of the due value to be attached to them, by M. I. Geoffroy St. Hilaire, in the 'Nouv. Ann. Mus. Hist. Nat.' 1832, and in the 'Essais de Zoologie Générale' of the same author, 1841. He shows that the value of the emarginated upper mandible, of the feathers and of the caruncles has been much overrated, and points out that the structure of the tongue, the wing and the toes, furnishes characters which have not been duly appreciated. The importance of the feet, as indicating natural affinities by their structural details, is further insisted on by M. De la Fresnaye in the 'Magasin de Zoologie.'

7. *Fossil Ornithology.*

Our knowledge of Birds has received a less amount of extension from the discoveries of palæontology than perhaps that of any other class of the animal kingdom. Not only are the fossil remains of birds of considerable rarity, and confined principally to the most recent deposits, but when found, they seldom present characters of such a nature as would enable us to predicate generic, much less specific, differences. The generic characters of birds being mostly drawn from the structure of the corneous appendages of the skin, such as the beak, tarsal scuta, claws, remiges and rectrices, are of course effaced in a fossil state, and the study of the bony skeleton has not yet been carried into sufficient detail (except in the case of some very isolated groups) to serve as the basis of generic definitions. The fossil skeletons of birds will nevertheless often guide us to the *family* or even the *subfamily* to which the specimens belong, and as the science progresses a greater amount of precision will no doubt be attained.

Birds, like Mammalia, appear not to have generally "multiplied and replenished the earth" until the commencement of the Tertiary epoch. Examples of their existence at an earlier period do indeed occur, but though the evidence of this fact is indisputable, yet the information it conveys is vague and obscure, and we look in vain for such grand palæontological discoveries as those which in the classes Reptilia, Pisces, Mollusca and Crustacea, have added whole families and even orders to the zoological system.

Many geologists have supposed that the rarity of fossil mammals and birds in the secondary rocks is owing to the improbability of their becoming imbedded in marine deposits, and not to their non-existence altogether. So far, however, as it is possible to draw a conclusion from negative evidence, there seem very strong reasons for believing that, in the European hemisphere at least, neither birds nor mammals were called into existence prior to the middle of the oolitic period. Let us take the case of the coal-measures, a formation of vast extent, and which is proved to have been in some cases a terrestrial deposit, and in others to have been formed in the immediate vicinity of dry land. Yet this vast series of beds, which has been quarried by man to a greater extent than any other, and which contains the remains of plants and even of insects in the most perfect state of preservation, has never yet afforded the slightest indication of a mammal or a bird. When we contrast this fact with the frequent occurrence of bones of these animals in recent peat-bogs, and in deposits, both marine and lacustrine, of the Tertiary epoch, we can hardly attribute the absence of such remains in the coal-measures to any other cause than to the non-existence at that period of the two highest classes of Vertebrata. The triassic or new red sandstone series leads in the European quarter of the globe to the same conclusion. We there find, in Germany and in Britain, evidences of ancient shores and sand-banks, exposed (probably during the recess of the tide) to the sun and the rain, and presenting the footprints of numerous reptiles which walked upon their surfaces. Now these are the localities to which aquatic birds, as well as certain mammals, love to resort, yet no traces of such animals have yet been met with in any ascertained triassic rock of the eastern hemisphere. The lias and lower oolite again, though strictly marine deposits, contain in many places the remains of plants or of insects which have floated from adjacent shores, but invariably unaccompanied by any fragments of birds or of mammals. In the Stonesfield slate we find the *first* and the *only* indication of mammalian remains in the whole secondary series; but the bones from that formation, which were once referred to birds, have been proved to belong to Pterodactyles, and no unequivocal examples of birds occur till we reach the horizon of the Wealden beds, where they are exceedingly rare, and apparently unaccompanied by Mammalia.

In the American continent, however, a remarkable case occurs, which seems to prove the existence of birds at a period long anterior to their first appearance in our hemisphere. I allude to the now well-known instance of Ornithichnites, or birds' footmarks, in the sandstone of the Connecticut valley, first discovered by Dr. J. Deane, and described by Prof. Hitchcock in the 'American Journal of Science,' 1836–37. (See also Buckland's 'Bridgewater Treatise,' pl. 26 *a* and *b*, and 'Ann. Sc. Nat.' ser. 2. vol. v. p. 154.) Two questions arise in connexion with these impressions: first, whether they are really produced by birds; and secondly, what is the age of the rock in which they are found. The first question seems to be now finally settled in the affirmative, some of the impressions being so nearly identical with those of certain existing Grallatores and Rasores as to convince the most incredulous. The footmarks are evidently due to birds of several distinct genera, some of which present structures as anomalous as those found in the reptiles and fish of the same remote epoch. The greater part, however, appear clearly referable to Wading Birds allied in structure to the Charadriidæ or Scolopacidæ. Some are of such a gigantic size that we can only seek their affinities among the Struthionidæ, and others appear to have had the tarsi clothed with feathers or bristles, a character which would exclude them from the Grallatores as at present defined, though, judging from the impressions made by living birds in snow, I think this appearance may possibly be due to the *trailing* action of the foot before it takes its hold of the ground. One very remarkable form (if really belonging to a bird) has the outer and middle toe united as in the so-called Syndactyles of Cuvier, and is further distinguished by all the four toes pointing forwards (neither of which characters are in the existing fauna ever found in ambulatory birds). Such anomalous structures, however (reasoning from the analogy of the fish and reptiles of the older rocks), appear rather to confirm than to disprove the genuineness and antiquity of these Ornithichnites; and as there is no other known class of animals to which they can by possibility be referred, it would be very unphilosophical to deny them to be the footmarks of birds, to which they bear so strong a resemblance.

In his 'Report on the Geology of Massachusetts,' Dr. Hitchcock has described no less than twenty-seven species of these footmarks, and in the 'Reports of the American Association of Geologists and Naturalists, 1843,' he has added five more. (See also Silliman's Journal of Science, Jan. 1844.) One of these much resembles the footprint of a Fringilla, others are similar to those of Fulica. In all these impressions, the phalanges of the toes obey the same numerical law which prevails, with hardly an exception, in the feet of existing birds*. They are accompanied in some cases by reptilian footmarks resembling those of Chirotherium, which are at once distinguished from the ornithic impressions by being *quadruped*, and by the forward position of the thumb.

Granting then that we have here the genuine indications of an ancient ornithological fauna, of which no other traces than these footmarks have been found, we have next to consider the geological age at which they were formed. Now it appears that the phænomena of superposition merely show that this deposit is intermediate between the Carboniferous and Cretaceous series. Could we have availed ourselves of such a latitude for speculation, the analogy of the oldest fossil birds found in the eastern hemisphere, would lead us to adopt the *latest* period within the

* The remarkably simple law referred to is this: that if we consider the metatarsal spine of certain Rasores (and which is wanting in all other birds) as the first toe, the hind toe as the second, and the inner, middle and outer toes as the third, fourth and fifth, the number of phalanges is found to progress regularly from one to five. The only exceptions are in the Caprimulgidæ, Cypselus, and one or two others.

above limits for fixing the age of these impressions. It has been announced, however, both by Dr. Hitchcock and by Sir C. Lyell (Proc. Geol. Soc. vol. iii. p. 796), that the only recognizable organic remains discovered in this deposit are Fish belonging to the genera Palæoniscus and Catopterus, and as these genera have never been found above the Triassic series, we are compelled to follow Dr. Hitchcock in referring the sandstone of Connecticut to the New Red system. These Ornithichnites therefore, abounding in this ancient formation, and separated by so vast an interval of time from the oldest traces of fossil birds in our own hemisphere, remain as one of those anomalies which serve to curb the eager spirit of generalization, and to teach us that nature fulfils her own designs without regard to human theories. Let us hope that the American geologists will never rest till they have discovered some osseous remains of the *raræ aves* whose footprints have given rise to such perplexing questions*.

The rest of the subject of Fossil Birds may be briefly noticed. The oldest example which I can meet with of their actual occurrence is mentioned in Thurmann's 'Soulèvemens Jurassiques' (as quoted by Von Meyer, 'Palæologica'), who remarks, however, that the statement seems to require confirmation. It is there stated that the fossil remains of Birds occur, in company with those of Saurians and Tortoises, in the limestone of Soleure, which is considered equivalent to the Portland beds.

A better authenticated instance is recorded by Dr. Mantell (Fossils of Tilgate Forest, p. 81; Geol. Trans. vol. v.; Proc. Geol. Soc. vol. ii. p. 203), who describes certain bones from the Wealden beds of Sussex, which he shows (and his opinion is backed by that of Cuvier and of Owen) to belong to Waders, and probably to Ardeidæ. Other bones from the same locality apparently belong to birds, yet present a nearer approach to the reptilian type than any known existing genus.

Another example of a fossil bird from the secondary series is mentioned by Dr. Morton (Synopsis of Cretaceous Rocks of United States), who procured a specimen which he refers to the genus Scolopax in the ferruginous sand of New Jersey. This formation he considers to represent the Greensand of Europe, and though its precise equivalents may be somewhat doubtful, there is no doubt of its belonging to the Cretaceous series.

In the "Glaris slate" of Switzerland, a member of the lower portion of the Cretaceous system, a nearly entire skeleton of a bird resembling a Swallow has been found by Professor Agassiz.

The chalk of Maidstone has supplied Lord Enniskillen with some

* M. Agassiz has of late expressed doubts whether any of the footmarks imprinted on the American beds of the New Red system belong to birds, and considers them all Reptilian.—Ed.

fragments of the skeleton of a large natatorial bird, considered by Professor Owen to be most nearly allied to the Albatros (Proc. Geol. Soc. vol. iii. p. 298; Geol. Trans. vol. vi.).

Proceeding to the Tertiary series, we find that ornitholites begin to appear in greater abundance. Here, as in every other department of the animal kingdom, we perceive a rapid approximation to the fauna which is characteristic of the period in which we now live.

The Eocene clays of the Isle of Sheppey have produced the bones of a bird affording almost the only example of a decidedly new ornithological form which has been rescued from the ruins of past geological ages. The sternum of this bird is fortunately preserved, and Professor Owen having worked out its affinities to all known genera with his usual sagacity and success, has arrived at the conclusion that it forms a new genus among the Vulturidæ, which he has denominated Lithornis (Proc. Geol. Soc. vol. iii. p. 163). This interesting specimen will soon be described in Professor Owen's work on 'British Fossil Mammalia and Birds' now in course of publication.

In Kœnig's 'Icones Fossilium sectiles,' fig. 91, some fragments of bones from the Isle of Sheppey are delineated, which the author considers to belong to a natatorial bird, and which he designates *Bucklandium diluvii*. If the original specimens are in existence they would well deserve further examination.

The remaining instances of fossil birds from the Tertiary formations call for but little remark. The fragments which have been found are either undistinguishable, or at any rate have not yet been distinguished, from the genera and species of the existing creation, though it is highly probable that new forms might in some cases be detected if they were subjected to rigid examination. In the Tertiary, and for the most part Eocene strata of the continent, birds' bones have been found in Auvergne, at Pont du Chateau and Gergovia, overlaid by beds of basalt, and in one instance accompanied by fossil eggs; in the Cantal, at Perpignan, Montpellier, Wiluwe, St. Gilles, Sansan (where eggs have also been found), Montmartre, Monte Bolca, Œningen, Kaltennordheim, Ottmuth in Upper Silesia, Westeregeln near Magdeburg, and Neustadt in the Hardt, and are recorded in the writings of Dufrénoy, Bravard, Croizet, Jobert, Marcel de Serres, Karg, Cuvier, Mösler, Germar, Von Meyer, &c. Birds' feathers have been found fossil at Monte Bolca, Aix and Kanstatt.

Proceeding to the newer Tertiary beds, we meet with remains of birds in the Crag of Suffolk and in the Pleistocene fluvio-lacustrine beds at Lawford (Buckland). M. Lund, whose researches into the bone-caverns of Brazil have already very greatly extended our knowledge of fossil Mammalia, has announced that he has also obtained a considerable variety of fossil birds, including a Struthious species larger than the existing Rhea of America; but these remains have not as yet I believe been fully investigated. The same remark also applies to the ornithic remains found by Dr. Falconer in that mine of palæontology the Siwalik Hills of India. Amidst the extraordinary remains of Mammals and of Reptiles obtained by that gentleman, the bones of several species of Birds were found, mostly referable to the Grallatorial order, and exhibiting in some cases very gigantic proportions. As Dr. Falconer's collections are now in course of arrangement at the British Museum, we may hope soon to learn more particulars of these interesting ornithic fossils.

The *Gryphus antiquitatis* of Schubert, a supposed colossal ornitholite from Siberia, appears to be either altogether apocryphal, or to be founded on the cranium of a Rhinoceros, mistaken for that of a bird.

In bone-caverns fossil birds have been found in company with extinct Mammalia at Kirkdale (Buckland), Bize in the south of France (Marcel de Serres), Avison, Sallèles, Poudres near Sommières, and Chokier near Liège (Von Meyer).

The bones of birds are of frequent occurrence in the osseous brecciæ which fill the fissures of limestone on the coasts of the Mediterranean, but these are probably referable in many cases to the recent epoch. They are recorded as occurring at Gibraltar (Buckland), Cette, St. Antoin and Perpignan (Cuvier), Nice (Risso), and Sardinia (Wagner, Nitzsch and Marmora).

I may here mention the remarkable instances of birds which belong to the existing epoch of the world, but have become extinct in recent times. The first is the well-known case of the Dodo, a bird insulated alike in structure and in locality, and which being unable to fly, and confined to one or two small islands, was speedily exterminated by the thoughtless pioneers of civilization. Most fortunately a head and foot of this bird still exist in the Ashmolean, and another foot in the British Museum; and with these data, aided by the descriptions of the old navigators, we are in some degree informed as to the structure and natural history of this anomalous creature. The memoirs on the Dodo by Mr. Duncan in the 'Zoological Journal,' vol. iii., and by M. De Blainville in the 'Nouvelles Annales du Mus. d'Hist. Nat.' vol. iv., are highly interesting, and there is an admirable synopsis of the whole subject from the pen of Mr. Broderip in the 'Penny Cyclopædia,' article *Dodo* *.

The bird described by Leguat (Voyage to the East Indies, 1708) as inhabiting the island of Rodriguez so recently as 1691, and termed by him Le Solitaire, appears evidently to have been another lost species of

* The whole of the information regarding the Dodo and extinct birds of the Mauritius and adjacent isles will be found in the 'Dodo and its Kindred,' by the author and Dr. Melville. This Report perhaps gave the first impulse to examine into and collect their history.

terrestrial bird distinct from the Dodo, and more allied in its characters to existing species of Struthionidæ. It is therefore probable that the supposed bones of the Dodo, described by Cuvier as found beneath a bed of lava in the Mauritius, but which M. Quoy states to have been in fact brought from Rodriguez, as well as the bones from the latter island presented by Mr. Telfair to the Zoological Society (Proc. Zool. Soc. part i. p. 31), but which have been unfortunately mislaid, belonged, not to the Dodo, as Cuvier supposed, but to the Solitaire. On this supposition we can the better account for a fact which threw doubt at the time upon Cuvier's identification of the bones at Paris, namely, that the sternum in this collection presented a mesial ridge, indicating strong pectoral muscles. Now Leguat tells us that the Solitaire, though unable to fly, had its wings enlarged at the end into a knob, with which it attacked its enemies, a structure which would require large pectoral muscles and a sternal crest. These bones and others, said to be from the Mauritius, in the Andersonian Museum at Glasgow and at Copenhagen, require further investigation, and every additional fragment that can be recovered from the caverns or alluvial beds of Mauritius, Rodriguez, or Bourbon, ought to be most carefully preserved.

The island of Bourbon appears to have been inhabited at a recent date by two species of birds allied to, but distinct from, the Dodo of Mauritius and the Solitaire of Rodriguez. I lately found in a MS. journal given by the late Mr. Telfair to the Zoological Society, an exact and circumstantial account of two species of Struthious birds which inhabited Bourbon in 1670 (Zool. Proceedings, April 23, 1844, Ann. Nat. Hist., and Phil. Mag., Nov. 1844). It appears then that this small oceanic group of islands possessed several distinct species of this anomalous family, the whole of which were exterminated soon after the islands became tenanted by man*.

Evidence of the recent existence and probable extinction of another Struthious bird has very lately come to light in New Zealand, where its bones are occasionally met with in the alluvium of rivers. The first portion that was brought to this country was a very imperfect fragment of a femur, which Prof. Owen did not hesitate to assign to an extinct gigantic bird allied to the Emeu (Trans. of Zool. Soc. vol. iii. p. 29). This bold conclusion, which from the imperfection of the data seemed prophetic rather than inductive, was speedily confirmed by the arrival of fresh consignments of bones, and we are now in possession of a considerable portion of the skeleton of this ornithic monster, which has been appropriately named by Prof. Owen Dinornis. That skilful anatomist has even been enabled, from the materials already received, to

* The *Epyornis* of Madagascar is another fossil bird, discovered since the publication of this Report.

point out no less than *five* species of this genus, differing in stature and the proportions of their parts (Proc. Zool. Soc., Oct. 1843). These birds, *if extinct*, must have become so in very recent times, and probably through human agency; but it is as yet by no means certain that they do not still inhabit the unexplored interior of the middle island of the New Zealand group. See notices by Rev. W. Cotton in 'Zool. Proc.,' 1843, and by the Rev. W. Colenso in the 'Tasmanian Journal,' reprinted in the 'Annals of Nat. Hist.' vol. xiv.

Another very interesting bird of the same region, the Apteryx, is now threatened with the fate which has befallen the Dodo and (as presumed) the Dinornis. Civilized man has already upset the balance of animal life in New Zealand. It is stated by Dieffenbach that Cats, originally introduced by the colonists, have multiplied greatly in the woods, and are rapidly reducing the numbers of the Apteryx, as well as of other birds, so that unless some Antipodean Waterton will disinterestedly enclose a park for their preservation, these extraordinary productions of the Creator's hand will soon perish from the face of the earth *.

8. *Ornithological Museums.*

The conservation of specimens for the purpose of reference is no less essential to the progress of zoology than the description of species in books, and in the case of ornithology there certainly is no scarcity of collections, both public and private, of illustrative specimens. Unfortunately, indeed, *classification*, which is no less important, though far less easy, than *accumulation*, is too often wanting or imperfect in such repositories, and their scientific utility is thus very greatly diminished. I may congratulate the zoological world, however, that this is no longer the condition of our great national collection, the British Museum. Without adverting to the immense improvements introduced in the last few years into all its other departments, I need only remark that the ornithological gallery, from the beauty of its arrangements and the extent of its collections, rivals, if not exceeds, the first museums of the continent. The scientific classification of the specimens is making great progress, under the able superintendence of the two Messrs. Gray, and ornithologists will soon possess in this collection a standard model which may be applied with advantage to other museums. This latter object will be greatly aided by the recent publication of catalogues, scientifically arranged by Mr. Gray, of all the species contained in the museum.

These catalogues, which are brought out in an accessible form, are calculated to be of great service to science. The classification and the scientific nomenclature are based on sound principles, and are corrected

* The same fate, if it has not already overtaken it, seems to await the *Gnathodon* of the Samooa Islands.

by the latest observations of zoologists, and every specimen is separately enumerated, with its locality and the name of its donor, which is especially important in a collection containing the *type-specimens*, from which original descriptions have been made. The zoological catalogues of the British Museum will now become standard works of reference, exhibiting both the riches and the desiderata of our national collection, and setting an example which we may hope to see followed by the great public museums abroad. The catalogue of the Mammalia was published last year: of the Birds, the Accipitres, Gallinæ, Grallæ and Anseres are already issued, and the other portions will speedily follow. Dr. Hartlaub has been the first to profit by this spirited example, and has published an excellent catalogue of birds in the Bremen Museum.

Another collection, of almost equal value, is that of the Zoological Society, now in progress of arrangement in a new building at the Society's Gardens*. Among private cabinets I may mention Mr. Gould's Australian collection as one which possesses a peculiar scientific value. It consists of selected specimens of the entire ornithology of Australia, the sexes, dates and localities of each being indicated, and as these specimens form the standard authorities for the accuracy of Mr. Gould's figures and descriptions, we may hope that this unique collection may be preserved for reference in some permanent repository †. But I must abstain from further details, as it would be impossible to give anything like a fair report on the individual merits of the numerous ornithological museums now extant without a far more extended personal inspection of them than I have had opportunity to make. It may however assist the student to be furnished with a list of all the more important collections of birds which have come to my knowledge (though many others doubtless exist); and I shall venture on no other criticism of them than merely to distinguish those general collections which are of first-rate importance by CAPITALS, and those which are confined to British ornithology by *Italics*.

ENGLAND:—Public Museums.—London (1. BRITISH MUSEUM; 2. ZOOLOGICAL SOCIETY; 3. EAST INDIA COMPANY; 4. Linnean Society; 5. United Service Institution; 6. College of Surgeons; 7. London Missionary Society); Newcastle-on-Tyne; Carlisle; Kendal; Durham; Scarborough; Leeds; York; Lancaster; Manchester; Liverpool (Royal Institution); Nottingham; Derby; Chester; Shrewsbury; Ludlow; Hereford; Burton-on-Trent; Birmingham (School of Medicine); Warwick; Cambridge; Norwich; Bury St. Edmunds; Saffron Walden; Oxford; Worcester; Cheltenham; Bristol; Plymouth; Bridport; Gosport (Haslar Hospital); Chichester; Rochester; Chatham (Fort Pitt); Canterbury; Margate.

* This has now been dispersed. Many of the typical specimens were purchased by the British Museum.

† This has been sold to Mr. Wilson, of Philadelphia, and forms a part of that gentleman's splendid ornithological collection.

Private Museums.—EARL OF DERBY, Knowsley *; Lord Say and Sele, Erith; Earl of Malmesbury, Christchurch, Hants; Messrs. Hancock and Dr. Charlton, Newcastle; P. J. Selby, Twizell; *Dr. Heysham*, Carlisle; — Crossthwaite, Keswick; J. R. Wallace, Distington, Cumberland; — Newell, Littleborough, Lancashire; A. Strickland, Bridlington Quay; *J. Hall*, Scarborough; C. Waterton, Walton Hall; *W. H. R. Read*, York; *G. S. Foljambe*, Osberton; *Rev. A. Padley*, Nottingham; H. Sandbach, Liverpool; *Rev. T. Gisborne*, Yoxall, Staffordshire; T. C. Eyton, Eyton Hall, Shropshire; *J. Walcot*, Worcester; H. E. Strickland, Oxford; *Rev. Dr. Thackeray*, Cambridge; J. H. Gurney, Earlham Hill, Norfolk; R. Hammond, Swaffham; Rev. G. Steward, Caistor; *E. Lombe*, Melton Hall, Norfolk; Rev. C. Penrice, Plumstead; *J. R. Wheeler*, Wokingham; — Dunning, Maidstone; *C. Tomkins*, M.D., Abingdon; W. V. Guise, Elmore Court; T. B. L. Baker, Hardwicke, Gloucester; *Rev. A. Mathew*, Kilve, Somerset; Dr. Roberts, Bridport; Dr. E. Moore, Plymouth; J. H. Rodd, Trebartha, Cornwall; H. Doubleday, Epping; *W. Yarrell*, J. Gould, J. Leadbeater, and G. Loddiges, London.

WALES:—Private.—L. L. Dillwyn, Swansea.

SCOTLAND:—Public.—Edinburgh; Glasgow (1. Hunterian Museum; 2. Andersonian Museum; 3. King's College); Aberdeen; St. Andrew's; Kelso; Dumfries.

Private.—Sir W. Jardine, Jardine Hall; Capt. H. M. Drummond, Meggineh Castle, Errol; *E. Sinclair*, Wick; Duke of Roxburgh, Fleurs; Dr. Parnell, Edinburgh.

IRELAND:—Public.—Dublin (1. Royal Dublin Society; 2. Natural History Society; 3. *Ordnance Collection*; 4. Trinity College); Belfast Museum.

Private.—*Dr. Farran* and *T. W. Warren*, Dublin; *Dr. Burkitt*, Waterford; *Dr. Harvey*, Cork; *J. V. Stewart*, Rockhill, Donegal; *R. Davis*, Clonmel; *Rev. T. Knox*, Toomavara; *W. Thompson*, Belfast.

FRANCE:—Public.—PARIS; STRASBURG; Bordeaux; Clermont; Lyons; Boulogne; Caen; Rouen; Metz; Epinal; Marseilles; Avignon; Arles; Nismes; Montpellier.

Private.—Prince Massena, Paris †; MM. Baillon and De Lamotte, Abbeville; Lesson, Rochefort; Allard, Montbrisson; Baron de Lafresnaye, Falaise; Fleuret, Bifferi, Bourcier, and Jourdan, Lyons; Crespon, Nismes; Degland, Lille; Bequillet, Toulouse.

BELGIUM:—Public.—BRUSSELS; Ghent; Louvain; Liège; Cologne (Jesuits' College); Tournay.

Private.—M. Kets, Antwerp; L. F. Paret, Ostend; M. Dubus, Brussels.

HOLLAND:—Public.—LEYDEN; Haarlem.

DENMARK:—Public.—Copenhagen.

NORWAY:—Public.—Christiania; Bergen; Drontheim.

Private.—Prof. Esmark, Christiania.

SWEDEN:—Public.—Stockholm; Lund; Upsal; Gottenburg.

Private.—Mr. R. Dann, Sioloholm, Gottenburg.

RUSSIA:—Public.—ST. PETERSBURG; Moscow; Casan; Odessa.

PRUSSIA:—Public.—Berlin.

AUSTRIA:—Public.—VIENNA; Trieste; Laibach.

WESTERN GERMANY:—Public.—Bonn; Mannheim; Mayence; FRANKFORT-ON-MAIN; Darmstadt; Heidelberg; Karlsruhe; Freiburg; MUNICH; Stutgart; Dresden; Göttingen; Griefswald; Bremen.

---

* Presented at Lord Derby's decease to the town of Liverpool.

† Since dispersed.

Private.—Prince Maximilian Neuwied; C. L. Brehm; J. A. Naumann, Dessau; Dr. Hartlaub, Bremen.

SWITZERLAND:—Public.—Basle; Neufchatel; Berne; Soleure; Geneva; Fribourg (Jesuits' College); Sion (Jesuits' College).

ITALY:—Public.—TURIN; Pavia; Parma; Bologna; FLORENCE; Rome (Accademia della Sapienza); Genoa; Nice; Pisa; Naples.

Private.—Prince of Canino, Rome*; Prince Aldobrandini, Frascati; Marchese Costa, Chambery; Marchese Breme, Turin; Signor Passerini, Florence; C. Durazzo, Genoa; Count Contarini, Venice; Contessa Borgia, Velletri; Signor Antenori, Perugia; Signor Costa, Naples.

SPAIN:—Public.—Madrid; Gibraltar.

IONIAN ISLANDS:—Public.—Corfu.

GREECE:—Public.—Athens.

MALTA:—Private.—Signor Schembri.

NORTH AMERICA:—Public.—Montreal; Cambridge; Salem; Philadelphia (1. Academy of Sciences; 2. Peale's Museum); Charleston; New York; Mexico.

Private.—Signor Constancia, Guatemala.

AFRICA:—Public.—Cape Town.

INDIA:—Public.—Calcutta.

Private.—T. C. Jerdon, Nellore.

AUSTRALIA:—Public.—Sydney; Hobart Town.

In connexion with Museums, the subject of Taxidermy may be briefly noticed. Although in acquiring the somewhat difficult art of preparing the skins of birds for collections, practice is far more important than precept, yet useful hints may often be obtained from the treatises which have been published on the subject. Among the best of these may be mentioned Mrs. Lee's 'Taxidermy,' Swainson's 'Taxidermy' in 'Lardner's Cyclopædia,' Waterton's 'Wanderings,' and his 'Essays in Natural History,' Boitard's 'Manuel du Naturaliste Préparateur,' Brehm's 'Kunst Vögel als Balge zubereiten,' &c., Weimar, and Kaup's 'Classification der Säugthiere und Vögel,' Darmstadt, 1844.

Ornithological Libraries.—It is needless to enumerate all the scientific libraries in which the subject of ornithology is adequately represented, especially as the museums above-mentioned are in most cases accompanied with appropriate collections of books. Of libraries unconnected with museums I may notice, as especially useful to the ornithological student, the Radcliffe at Oxford, the Royal Societies of London and of Edinburgh, and the fine collection of zoological works formed by Mr. Grut of Edinburgh †, to whom I am indebted for access to several rare works.

9. *Desiderata of Ornithology.*

Having now given an account of the recent progress and present state of Ornithology, I will conclude with pointing out the *desiderata* of the science, showing the deficiencies which require to be supplied in order to

* Dispersed. † Since sold and dispersed.

refine the crude mass of knowledge already extracted from the mine, and to make further researches into the storehouses of Nature.

1. There is a great want of increased precision and uniformity in the value of the genera, and of the superior groups which various authors have introduced into ornithology. All groups of the same rank are supposed in theory to possess characters of the same value or amount of importance, and the object of the naturalist should be to bring them as nearly as possible to this state of equality. It must indeed be admitted, that no certain test seems to have been yet discovered for weighing the value of zoological characters. The importance of the same character manifestly varies in different departments of nature, and must therefore be estimated by moral rather than by demonstrative evidence. The real test of the value of a structural character ought to be its influence on the economy of the living animal, but here we too often have to lament our ignorance or our false inductions, and in many cases we are wholly unable to detect the relations between structure and function. More definite principles of classification may hereafter be discovered, and meantime all that we can do is to arrange our systems according to sound reason and without theoretical prepossession. By care and judgement much may be done to give greater regularity and exactness to our methods of classification, either by introducing new groups where the importance of certain characters requires it, or by rejecting such as have been proposed by others on insufficient grounds. At the present day many authors are in the habit of founding what they term "*new genera*" upon the most trifling characters, and thus drowning knowledge beneath a deluge of names. As this is a point of great importance to the welfare of zoology in general, I may be excused for dwelling on it for a few moments.

In the subdividing of larger groups into genera, even in the strictest conformity with the natural method, there is evidently no other rule but *convenience* to determine how far this process shall be carried. However closely the species of a group may be allied, yet as long as any one or more of them possess a character which is wanting to the remainder, it will always be in the power of any person to partition off such species and to give them a generic name. Take the very natural group Parus for instance, as restricted by most modern authors (*i. e.* Parus of Linnæus, deducting Ægithalus and Panurus). First we may separate the *long-tailed* species, and follow Leach in calling it generically Mecistura. Of the remaining Pari, we may make a genus of the *crested* species (*P. cristatus*), then another of the *blue* species with short beaks (*P. cæruleus*, &c.), a third of the *black and yellow* group (*P. major*, &c.), and a fourth of the *grey* species (*P. palustris*, &c.). [N.B. Generic names have actually been given to these groups by Kaup in his 'Skizzirte Entwickelungsgeschichte der Europaïschen Thierwelt.'] But another author may go still further, and may again subdivide the groups above enumerated, a process which would lead to the absurd result of making as many genera as there are species, or in other words, of giving to each species *two specific* names and *no generic* one. Therefore genera should not be subdivided further than is *practically convenient* for the purpose of fixing really important characters in the memory; and seeing that there are already more than 1000 genera provided for the 5000 species of birds (which are probably all that can be said to be *accurately* known) it seems evidently inexpedient to increase the number of genera, except in the comparatively rare cases where new forms are discovered, or really important and peculiar structures have been overlooked.

The precise rank in the scale of successive generalizations which shall be occupied by those groups which we term *genera* is then a matter of *convenience*, and consequently of *opinion*. Nature affords us no other test of the just limits of a genus (or indeed of any other group), than the estimate of its value which a competent and judicious naturalist may form. The boundaries of genera will therefore always be liable in some degree to fluctuate, but this is unavoidable, and it is a less evil than to give an unlimited license to the subdivision of groups and the manufacture of names. The only remedy for this excessive multiplication of genera, is for subsequent authors who think such genera too trivial, not to adopt them, but to retain the old genus in which they were formerly included*.

We may obtain a great amount of fixity, in the position at least, if not in the extent of our groups, by invariably selecting a *type*, to be permanently referred to as a standard of comparison. Every family,

* It is usual where this is done to retain the groups, which are thus deprived of a *generic* rank, under the title of *subgenera*. There appear to me, however, to be great objections to the adoption of *subgeneric names* in zoology. First, it would introduce into a science already overloaded by the weight of its terminology, an additional set of names whose rank is not (like that of families, subfamilies and genera) indicated by the *form* of the word, but which are undistinguishable to the eye from real generic names, and would therefore be perpetually confounded with them. Secondly, subgenera would greatly interfere with the harmonious working of the "binomial method," that mainspring of modern systematic nomenclature; for one author would habitually indicate species by their *generic* and another by their *subgeneric* names, and the same word would be sometimes used in a *generic*, sometimes in a *subgeneric* sense, so that instead of a uniformity of language being adopted by zoologists, nothing but a vague and capricious uncertainty would result. If it were possible to establish a uniform system of *trinomial* nomenclature, so as always to indicate every species by its generic and subgeneric as well as by its specific name, the use of subgenera might indeed be tolerated, but such a method would be far too cumbrous and oppressive for practice, and I must therefore enter my humble protest against subgeneric names altogether. Not that I object to the subdividing large genera for convenience of reference into *defined* though *anonymous* groups; but let not these groups be designated by proper names, unless their characters be sufficiently prominent to warrant *generic* distinction.

for instance, should have its *type-subfamily*, every subfamily its *type-genus*, and every genus its *type-species*. But it must not be supposed, with some theorists, that these types really exist as such in nature; they are merely examples or illustrations selected for convenience to serve as permanent fixed points in our groups, whatever be the extent which we may give to their boundaries. By adhering to this notion of types we may often indicate these groups with greater precision than it is possible to do by means of definition alone.

2. Another desideratum in ornithology is to discover some sure mode of distinguishing *real species* from *local varieties*. The naturalists of one school are disposed to attribute nearly all specific distinction to the accidental influence of external agents, while others regard the most trivial characters which the eye can detect as indicating real and permanent species. Between these two extremes, the judicious and practised naturalist has seldom much difficulty in keeping a middle course, and perhaps in ornithology the cases of ambiguity are less frequent than in many other departments of nature; still the student will be sometimes at a loss to distinguish between those characters which were impressed on a species at its creation, and those which may be reasonably attributed to external agents, and we must look for further research to solve these difficulties.

3. We are greatly in want of more information as to the habits, anatomy, oology, and geographical distribution of the majority of exotic species. With no other data than are furnished by dried skins, we are too often compelled to guess at, rather than to demonstrate, the true affinities of species. However essential may be the arrangement of specimens in museums, they supply only a portion of the requisite evidence, and a vast and fascinating field of research awaits the naturalist who shall devote himself to *observing*, as well as *collecting*, the ornithology of foreign regions*. The anatomy of many genera and even families of birds is wholly unknown, and it would be well if some student would devote himself especially to this department, and endeavour to make a *classification of birds by their anatomical characters alone*. If such a system were found to coincide with the arrangements which have been based on external characters, the strongest proof would be furnished of its reality and truth.

4. There yet remain many extensive regions of the world, of whose ornithology we know little or nothing. Great as have been the zoological collections made of late years by individuals and governments, there is still much virgin soil for the naturalist to cultivate. The birds of the vast Chinese empire are only known by the rude paintings of the natives,

* Collectors would double the value of their specimens if they would invariably attach to them a small label, stating at least the sex, date, and locality, and adding any other observations which they may be able to make.

though nothing would be easier than to instruct those ingenious people in the art of collecting specimens. We obtain, too often indeed in a mutilated state, the gaudy Paradisiidæ of New Guinea, but the less attractive birds of that country, as well as of the whole Polynesian archipelago, are almost unknown. From Madagascar a few remarkable species have been occasionally sent to Europe, but the peculiarly insulated fauna of that island, partaking neither of an African nor an Asiatic character, is still very imperfectly explored. Even our own colonies of the West Indies and Honduras have been regarded only with a commercial, and not with a scientific eye, and their ornithology affords to this day—with shame be it spoken—an almost untrodden field of inquiry. Morocco, Eastern Africa, Arabia, Persia, Ceylon, the Azores, and the rocks and billows of the Southern Ocean, present ample materials for the future researches of the ornithologist, and will doubtless furnish many new generic and specific forms.

5. Besides the collecting of new species, the correct determination of those already described is no less important. The names and characters of species are scattered through such an infinity of works, and are often so vaguely defined, that the apparent number of known species far exceeds the real one, and much critical labour is required to reduce the nominal species to their actual limits. Having myself devoted much time to this department of ornithology, I have found that the number of synonyms is nearly threefold that of the species to which they refer, and it is important that the further growth of this evil should be checked by the publication of exact lists of species and their synonyms.

6. This vast multiplication of nominal species mainly results from the great number of scientific periodical works now issuing in all parts of the civilized world, and which it is almost impossible for any one person to consult. This is an unavoidable consequence of the great diffusion of knowledge at the present day, but the inconvenience which results from it might be much diminished if some method were adopted of centralizing the mass of scientific information which is daily poured forth. It is much to be wished that some publication like the excellent but extinct 'Bulletin des Sciences' were again established, containing abstracts of all the important matter in other scientific works; or if this were found too great an undertaking, a periodical which should merely announce the titles of the articles contained in all other scientific journals and transactions as they are published, would be a most useful indicator to the working naturalist. Perhaps the nearest approach towards supplying this desideratum at present, is made by the French scientific newspaper 'L'Institut,' and in Germany by Oken's 'Isis,' and Wiegmann's 'Archiv.' We shall shortly too possess an alphabetical index to all works and memoirs on zoology, through the praiseworthy efforts of Prof. Agassiz,

whose gigantic undertaking, the 'Bibliographia Zoologica,' is now ready for the press.

7. The science of ornithology would be much advanced if a greater number of persons would devote themselves to the *general subject*. The majority of those who now study it, or form collections, confine themselves to the birds of their own country, under an impression that general ornithology is too wide a field for them to enter upon. They often are not aware at how small an expenditure of money or space a very large general collection may be formed. By adopting the plan first recommended by Mr. Swainson, of keeping the skins of birds in drawers, instead of mounting them in glazed cabinets, the collector may arrange many thousand specimens in a room of ordinary size, and have them at all times ready for reference and study. Or if the ornithologist considers a general collection too cumbrous, he may devote himself to the study and arrangement of particular groups, and supply the science with valuable monographs. Such a course would be of far greater service to zoology, as well as more interesting to the student, than if he were to confine himself to the almost exhausted subject of European or British ornithology.

8. The last point which I shall notice is the prevailing want of scientific arrangement in our ornithological museums, both public and private. I have seen few collections in this country in which anything more is attempted than a general *sorting* of the specimens into their orders and families, and fewer still in which the generic and specific distinctions are indicated by systematic arrangement and uniformity of labelling. It is needless to remark how essential classification is to the scientific utility of a museum, but some excuse for the general want of it may be found in the scarcity of suitable works to serve as guides in arrangement. Now, however, by following the code of zoological nomenclature adopted by this Association (Report for 1842), and by taking as models the excellent 'Catalogues of the British Museum,' and Mr. G. R. Gray's 'Genera of Birds,' the scientific curators of museums can be no longer at a loss, and we may hope soon to see a great reform effected in the arrangement of our ornithological collections.

In concluding this sketch of the progress and prospects of Ornithology, I must apologize for many imperfections and omissions which are unavoidable in treating of so extensive a subject. A person with more time at command and more favourably circumstanced for consulting authorities, would doubtless have rendered this Report more complete, but I trust that it may be of some use in guiding the student to the sources of his information, and in pointing out the best methods of advancing this fascinating department of scientific zoology.

5. ON THE REDISCOVERY OF *HALCYON SMYRNENSIS* (LINN.) IN ASIA MINOR *.

[Read at the British Association at Manchester, 1842.]

ALBIN, in his 'Natural History of Birds,' published about a century ago, describes a bird under the name of the "Smyrna Kingfisher," and gives a figure of it (vol. iii. pl. 27) from a specimen preserved in spirit, which he states was shot by Consul Sherard "in a river of Smyrna." This species, to which Linnæus gave the name of *Alcedo smyrnensis*, has been retained ever since in our catalogues, though from the time of Sherard to the present day, no further evidence has been adduced, as far as I am aware, of the occurrence of this bird on the coasts of the Mediterranean.

The succeeding plate (vol. iii. pl. 28) of Albin's work gives, under the name of the Great Bengal Kingfisher, an indifferent representation of a well-known Indian bird, which was afterwards more correctly depicted by Buffon in his 'Planches Enluminées,' No. 894. This bird, being evidently closely allied to Albin's Smyrna Kingfisher, was classed as a variety of it by Gmelin and Latham, and stands as *Alcedo smyrnensis*, γ, in their catalogues. Later writers on Indian ornithology agree in terming this bird (which appears to be common throughout India from Ceylon to Assam) *Halcyon smyrnensis*, thus implying a belief of its specific identity with the Smyrna Kingfisher of Albin. But as Albin's figure presented certain differences from the Indian birds, and as no specimens from the Mediterranean shores were at hand for comparison, it was impossible to decide this point with certainty, and it seemed probable that the species from India might prove to be distinct, and thus require a new appellation.

The differences in question were these:—The adult Indian bird has the *lesser* wing-covers rufous, the *middle* ones black, and the *greater*, together with the quills, back and tail, bright greenish blue, changing in certain lights to green. Younger specimens from India retain the rufous *lesser* covers, but have the *middle* covers blue-green, like the rest of the wing and upper parts. In Albin's Smyrna Kingfisher, however, the *whole wing* is represented green, changing only into a bluish tinge on the middle covers, but without the rufous on the lesser covers. This discrepancy appeared to indicate a specific distinction, but was neutralized by the circumstance that Albin, or his colourist, has also given *green lesser covers* to the Bengal Kingfisher on plate 28, a mistake,

* Report Brit. Assoc. xi. Sect. p. 70; Ann. Nat. Hist. ix. p. 441.

indeed, which might easily arise from the rufous ridge of the wing being concealed beneath the azure-green feathers of the upper back.

It was clear then that these doubts could only be solved by searching in Asia Minor for the original species described by Albin. During my residence at Smyrna in the winter of 1835–36, I failed in meeting with any traces of this bird, although two other species of Alcedinidæ, viz. *Alcedo ispida*, Linn., and *Ceryle rudis*, Hasselq., were not unfrequent. The *Halcyon smyrnensis*, however, belonging to an insectivorous genus which is rarely met with far beyond the tropics, could hardly be expected to occur so far north as Smyrna in the depth of winter. Failing in this attempt, I took occasion at a later period, when supplying that ardent and philosophic zoologist, Mr. Edward Forbes, with a list of ornithological desiderata to be sought for in the Levant, to call his particular attention to the long-lost "Smyrna Kingfisher," and I am happy to say that his researches have at last been crowned with success. In a letter from him dated Macri, on the coast of Lycia, at the end of February last, he says,—"One of the sailors has just shot a large Kingfisher, which I take to be the one wanted; three or four have been seen, but not got at. The common Kingfisher is also very abundant, or something like it. The large bird was brought alive; its plumage is very beautiful. I have drawn it, and Graves is at this moment busy skinning it. We shall send the skin to you by an early opportunity." Through the kindness of Captain Graves this specimen has since been forwarded to me, and on comparing it with a series of specimens from India, it turns out to be in every respect specifically identical with them. It is in the full adult plumage, possessing the rufous *lesser* and black *medial* covers, which distinguish the perfect bird in India. We may therefore henceforth, without hesitation, retain the original specific name of "*smyrnensis*" for the specimens from India no less than for those of Asia Minor; and from the proximity of the latter country to Crete and the Morea, we may anticipate the future admission of this beautiful and interesting species into the fauna of Europe.

The specimen of the Smyrna Kingfisher depicted by Albin must be regarded as an immature individual, and we must suppose that the uppermost series of wing-covers in his plate were either coloured green instead of rufous, through an oversight, or that they indicate a still earlier stage of development than the Indian specimen which I have had opportunities of examining.

Description of Macri specimen.—Rich chestnut-brown on head, cheeks, back and sides of neck, lesser wing-covers, under wing-covers, sides of breast, abdomen and lower tail-covers; deep black on middle wing-covers; greenish blue, changing in certain lights to verdigris-green, on upper back, scapulars, spurious wing, greater and primary wing-covers, secondaries, tertials and rectrices. On the rump and upper tail-covers this blue assumes a purer tint. Terminal half of primaries black, basal half greenish blue externally, and white within, gradually increasing till the ninth primary is almost wholly white. Inner margin of secondaries and rectrices blackish brown. Chin, throat and middle of breast white. Beak and legs vermilion-red.

Long. tot. $11\frac{1}{4}$; beak to front $2\frac{1}{4}$, to gape $2\frac{3}{4}$; breadth 8 lines; height 8 lines; wing to primary $4\frac{3}{4}$, to tertials, ditto; middle rectrices 3 inches 7 lines; outer 2 inches 11 lines; tarsus 7 lines; middle toe and claw $1\frac{1}{4}$; hind ditto 7 lines.

---

6. ON THE OCCURRENCE OF *CHARADRIUS VIRGINIACUS* (BORKH.) AT MALTA.

[Ann. Nat. Hist. 2 ser. v. p. 40. 1850.]

I HARDLY know whether the occurrence of a new or unrecorded species of bird at Malta is to be regarded as forming an addition to the European fauna, because geographers are I believe not yet agreed as to whether Malta belongs to Europe or to Africa. But in either case the discovery of *Charadrius virginiacus* at Malta is not the less interesting, for this species has not as yet, I believe, been noticed in either of those two quarters of the globe to which that island is intermediate.

I have lately found an accidentally mislaid letter, addressed to me in 1846 by Captain H. M. Drummond, 42nd R.H., whose valuable papers on the Birds of Corfu, Crete, Macedonia and Tunis are well known to the readers of the 'Annals.' In this letter he mentions having procured at Malta "a little Golden Plover, which, on comparing with *C. pluvialis*, I find quite distinct, being only the size of *C. morinellus*, and much longer in the tarsus. It was shot in company with another of the same species in March 1845. They are occasionally observed in Malta every second or third year, generally early in spring, and have never been noticed in company with *C. pluvialis*, but generally solitary or in pairs. They have not been observed with black on the breast. The man who shot it informs me that he has frequently killed them, and that he can immediately recognize them by the note, which is peculiar, differing from that of *C. pluvialis*, and more resembling that of *C. hiaticula*."

Captain Drummond has subsequently been in England, and showed a specimen of this bird to Mr. Yarrell, who ascertained it to be the *C. virginiacus*.

This species possesses a far more extensive geographical distribution than the better-known *C. pluvialis*. The latter occurs throughout Europe, and is recorded as far east as Trebizond and Siberia. But *C. virginiacus* not only frequents the whole of North and South America, but extends over the Polynesian Islands to the Malay Archipelago and India, as well as to Australia and New Zealand*. We have now evidence of its visiting Malta for a short time early in spring, a

* The Australian *C. xanthocheilus* of Jardine's 'Illustrations of Ornithology,' plate 85, and of Gould's 'Birds of Australia,' vol. vi. plate 13, is certainly identical with *C. virginiacus*. The true *C. xanthocheilus* of Wagler inhabits New Zealand (in company with *C. virginiacus*); and, according to Mr. Gray's Catalogue, there are three specimens of it in the British Museum from Van Diemen's Land, though it seems to be omitted by Mr. Gould.

fact which clearly proves that it must winter in Africa, and, occasionally at least, pass the summer in some part of Europe, though it has never yet been obtained in either of these continents. This has probably been owing to the resemblance of its plumage to that of *C. pluvialis*, which bird is recorded by Malherbe in his 'Faune Ornithologique de la Sicile,' by Schembri in his 'Catologo Ornitologico del Gruppo di Malta,' and by Von der Mühle in his 'Beiträge zur Ornithologie Griechenlands,' but without any indication of their having noticed the *C. virginiacus*.

The distinctions between *C. pluvialis* and *C. virginiacus* are numerous, and are carefully pointed out by Sir W. Jardine in his edition of 'Wilson's American Ornithology,' vol. ii. p. 362. It will therefore suffice to mention here that *C. virginiacus* is rather smaller than *C. pluvialis*, has rather longer tarsi, and has the under wing-covers and axillary feathers of a *grey-brown*, while in *C. pluvialis* they are *pure white*.

American specimens of *C. virginiacus* are somewhat larger than the Indian and Maltese ones. Both varieties however have been recently found by Captain Drummond in Bermuda. In a list of the Birds of Bermuda by Mr. H. B. Tristram, which is on the point of being published by Sir W. Jardine in his 'Contributions to Ornithology,' these two varieties are regarded as distinct species, as appears from the following passage: "No. 46, *Charadrius marmoratus* [i. e. *virginiacus*], American Golden Plover. No. 47, *Charadrius* . . . . . ?, an unnamed species smaller than the American and perfectly distinct. Not unfrequent here. It has been also found in Malta by Capt. Drummond, 42nd R.H."

7. PAPERS, &c. FROM THE CONTRIBUTIONS TO ORNITHOLOGY*.

[Strickland was an active assistant to the 'Contributions,' both by advice and communications. From the intention of that work being to record Ornithological information as early as possible, many short notices and descriptions of new species of birds were published, which could not bear the character of papers or memoirs. These were also frequently illustrated by plates, which it would be impossible to repeat here; but in addition to the papers now printed entire, a complete list of all his contributions is also given, with such extracts as relate to scientific arrangement, or where the alliance and affinity of the species is pointed out, or new generic characters have been drawn.]

1848. Descriptions of three birds from the Indian collection of Captain Boys. *Pericrocotus erythropygius*, Jerdon.—*Muscicapa hemileucura*, Hodgson.—*Heterura sylvana*, Hodgson.

Of the last it is remarked:—"Mr. Hodgson has been misled by the lengthened tertials and the striated plumage, into placing this bird, at present a unique species, among the Larks, in which arrangement Mr. Blyth follows him. Mr. Gray seems to me equally wide of the truth in placing this bird among the Emberizinæ. It appears to be a nearly typical genus of the Malurinæ, and consequently belongs to the Dentirostres, not the Conirostres. The flattened forehead, the compressed emarginate beak, the wiry plumage of the head, the narrow and worn rectrices, the strong feet, lengthened toes and curved claws, as well as the striated coloration, are precisely what we find in the Malurine genera, Megalurus, Cinclorhamphus, Sphenura, Sphenœacus, Malacocercus, Prinia, &c.; the arched and indurated membrane which overhangs the nostril is repeated in Sphenœacus, Sphenura, Malacocercus and Prinia, and the pointed and lengthened tertials occur in Megalurus and Cinclorhamphus. The only generic peculiarity of this bird seems to be, that the first four primaries are of equal length, while in the Malurinæ generally, the wing is much rounded by the graduation of the first two quills."

Descriptions of *Scops cristata*, Daudin, var., from the collection of L. L. Dillwyn, Esq., of Swansea:—

"I follow Mr. G. R. Gray in uniting Lophostrix to Scops, from which it seems to differ only in size. Mr. Gray, however, adopts for this united genus the later name Ephialtes in place of Scops, because the name Scops was originally given by Mœhring in 1752 to the Numidian Crane. After a mature consideration of this question, which I discussed in 1842 (Ann. Nat. Hist. vol. viii. p. 368), I am still of opinion that Mœhring's generic names should not be adopted. The binomial system of nomenclature was first introduced by Linnæus in the tenth edition of the 'Systema Naturæ,' published in 1758, and that year ought consequently to be taken as the *datum line*, beyond which no claim for priority of nomenclature can be entertained. It follows, that

* Contributions to Ornithology, by Sir W. Jardine, Bart. 1848—1852.

if we reject the name of Scops in the Mœhringian sense, on the ground of its having been proposed previously to 1752, we must retain the term Scops, as proposed by Savigny, in 1809, for the genus of Owls before us."

Tityra surinama, Linnæus. A good species, and since named by De Filippi *Pachyrhamphus dimidiatus.*—*Timalia leucotis*, Strickland. From Malacca. These six birds were all figured on plates drawn by the anastatic process, and were the first attempts to apply this art to ornithology.

1849. Descriptions of *Hirundo albigularis*, Strickland.—*Momotus gularis*, Lafresnaye.

"This elegant little Motmot conducts us at once from the larger species to the diminutive *Hylomanes momotula* of Lichtenstein, and seems to justify us in reuniting under one genus the closely allied groups of Momotus and Hylomanes. It also serves to illustrate the close and indisputable affinity between the Motmots and Bee-eaters, an affinity which is by no means so generally recognized as it ought to be. I should be inclined to regard Momotus as merely the American form of the subfamily Meropinæ, just as the Cuckoos, the Barbets, the Trogons and the Parrots, have each their peculiar generic forms in the old world and the new. In Momotus, and in Merops, we have the same lustrous sea-green plumage, the prevalence of a black streak through the eye, and of a black spot on the breast, an almost identical form of foot, and a similar prolongation of the medial rectrices. Momotus is mainly distinguished by the rounded form of the wing, in accordance with its more indolent habits, and by the serration of the mandibles, a character of little weight in questions of affinity, as it breaks out, *pro re natâ*, in many remote groups of birds, unaccompanied by any other peculiarity of structure."

Holocnemis nævius, Gmelin, 1789.—Wall-creeper of Surinam, Edw. B. pl. 346.—*Sitta nævia*, Gmel. Syst. Nat. p. 442. Lath. Ind. Orn. p. 263; Gen. Synop. ii. p. 654. Shaw, Zool. viii. p. 114.—*Holocnemis flammatus*, Strickl. Ann. and Mag. Nat. Hist. xiii. p. 415, pl. 13 (female).*

Saxicola opistholeuca, Strickland. From Northern India. Allied to *S. picata*, and *S. leucuroides*, Guérin.

Pachycephala macrorhyncha, Strickland. From Amboina.

"The discovery of a bird of the Australian genus Pachycephala, so far to the north as Amboina, is a very interesting circumstance; the more so, as its peculiar form appears to furnish a clue to the true affinities of what has hitherto been an anomalous and puzzling genus. The

* The figure, Pl. Enl. 823. fig. 1, the type of Gmelin's *Turdus lineatus*, will also rank as a synonym. *Holocnemis*, Strickland, 1844, was used in 1829 by Schilling as a genus of Coleoptera, and on that account Mr. Sclater has proposed to change it to *Heterocnemis*, 1855 (Proc. Zool. Soc. 1855, p. 146).—Ed.

small group of birds comprising the genera Pachycephala and Eopsaltria, has been classed quite at random by most previous writers, who seem to have had no idea of its real affinities, and have been content to place it, from some fancied resemblance, in the utterly remote American families, Ampelidæ and Vireoinæ. The bird before us, though unquestionably a true Pachycephala, is distinguished by a beak considerably longer and more compressed than in the other species. In this respect it offers so much resemblance to certain genera of Laniidæ, as to leave scarcely any doubt that the Pachycephalinæ ought to stand as an Australian subfamily of that extensive group. This view is confirmed by the observations of Mr. Gould, who has shown that their habits are similar to those of the Shrikes, and who was the first to class them in that family. It is more especially the African subfamily of Laniidæ, comprising Laniarius, Telophonus, &c., to which the Pachycephalinæ show an affinity; and its relationship is indicated, not merely by the peculiar form of the beak, but by great similarity of colouring, as will be evident on comparing the African *Telophonus zeylonus* (Linn.), or *Laniarius olivaceus* (Shaw), with their Australian representatives."

Brachypteryx poliogenis, Boie. From Borneo.

Dr. Hartlaub wrote in regard to this description :—" A fine specimen in the Bremen Museum was received direct from Malacca; and I described it in my 'Catalogue of the Bremen Collection,' p. 20, under the name of *B. malaccensis*. This name has the priority."

Pericrocotus minutus (Verreaux, MS.). From Borneo. Considered by Blyth as equal to his *P. igneus*.

Cyanocorax nanus, Dubus?

"The Mexican bird described by Vicomte Dubus agrees in every respect with the specimen here figured, except as regards his expression, 'gutture albo-cærulescente;' while in my bird the throat is deep blue-black. Should it prove distinct, I would propose for it the name of *Cyanocorax pumilo*."*

Phylloscopus trivirgatus (Verreaux, MS.).

Pycnosphys grammiceps (Verreaux, MS.).

"This curious little bird appears to belong to a new genus, which I propose to designate *Pycnosphys*, from πυκνὸς, thick, and ὀσφὺς, the rump or loin, in reference to the length and density of the uropygial feathers. Its affinities are not easily determined. With the general form and coloration of a Sylviine bird, it unites the beak of the Muscicapidæ and the dorsal plumage of the Timaliinæ. It must probably be regarded as belonging to the Muscicapidæ, in which family we find at least one genus (Philentoma) possessing the same development of the

* In 1855, Prince Bonaparte examined this specimen, and judged it to be distinct; "*pumilo*," Strickl., will therefore stand. It was received from Guatemala.—Ed.

dorsal feathers, though in other points of structure they are very distinct, the beak of Philentoma being more depressed, the rictal bristles longer, and the tarsi proportionally shorter.

"*Generic character.*—Beak moderately depressed, the length almost double the breadth; sides nearly straight, very slightly concave; the culmen rather sharp, straight for its basal third, then gradually curved to the tip, which is emarginate and slightly overhangs the lower mandible; margin of upper mandible very slightly decurved; lower mandible nearly straight; gonys gradually curved upwards; nostrils ovate; rictal bristles about half the length of the beak; wing of moderate length, rounded; first quill short, fourth and fifth longest; uropygial feathers very long, thick and decomposed; tail rather long, narrow, the rectrices pointed; tarsi (for a Muscicapine bird) long; acrotarsia and paratarsia entire; toes moderately long, external toe very slightly longer than the inner; claws moderately curved, slender, acute." Inhabits Java.

Trichostoma umbratile, Temminck.

T. celebense, Strickland.

Drymocotaphus capistratoides, Blyth.—*Goldana capistratoides,* Gray.—*Myïothera capistratoides,* Temm.—Verr. MS.

1850. *Monasa flavirostris,* Strickland.—*Todirostrum chrysocrotaphum,* Strickland.—*Euphonia bicolor,* Strickland. All from the upper branches of the Amazon.

Tachyphonus rufiventer, Spix. From the eastern provinces of Peru.

Xanthornus prosthemelas, Strickland. From Guatemala.

Elænia linteata, Strickland. From the upper branches of the Amazon.

Cæreba nitida, Hartlaub. Upper branches of the Amazon.

Paroides flammiceps, Burton. North-western Himalaya.

1851. Ornithological Notes, by H. E. Strickland.—*Dacnis melanotis,* Strickland. The type is Buff. Pl. Enl. 669. figs. 1, 2*.

Two new species of Euphonia, allied to *E. chlorotica*:—*E. trinitatis,* Strickland, from Trinidad, &c.; and *E. strictifrons,* Strickland†, from Surinam and Cayenne.

1852. *Coccyzus pumilus,* Strickland. From Trinidad.

Todirostrum granadense, Hartlaub. *T. multicolor,* Strickland‡.

Nectarinia albiventris, Strickland. Procured by James Daubeny, Esq. at Ras Hassoun§, on the eastern coast of Africa.

See also a list of the species collected by Mr. Daubeny, of which this Nectarinia formed a part, drawn up by Philip L. Sclater, assisted by H. E. Strickland.

* Mr. Sclater now considers this = *D. angelica,* De Filippi.—Ed.

† *E. pumila,* Bonap., and the ♀ of the same bird is *E. minuta,* Caban.—P. L. S.

‡ *T. ruficeps,* Kaup, and is from New Grenada.—P. L. S.

§ This is probably Ras Hafoun of Arrowsmith's map,—the eastern point of the African continent.—H. E. S.

On the distinctness of *Monasa fusca* (Gmel.) from *M. torquata* (Hahn).

On *Parus ignotus*, Gmelin.

"Contributions to Ornithology may be quite as usefully effected by the sweeping away of error and confusion, as by the collecting of positive knowledge. Now there is a nominal species of Parus said to inhabit Norway, which has haunted our systems of Ornithology for nearly a century, but which no recent observer has recognized in nature. Brünnich prophetically called it *Parus ignotus*; Latham named it after its supposed discoverer, *Parus stromei*; and Vieillot went so far as to found a genus for it, under the name of Megistina. Still, all this book-learning was of no avail in adding to our knowledge; the *Parus ignotus* still refused to fall before the gun of the naturalist, or to make its appearance on the shelves of our museums.

"This ornithological phantom is, however, at last dissipated by Prof. Sundevall of Stockholm, who, in a recent letter to myself, has thus replied to my queries about *Parus ignotus*. As the original is in Swedish, I have thought it best to translate his observations:—

"'As regards *Parus ignotus*, it is well known that Gmelin had it from Latham's 'Synopsis,' vol. iv. p. 537; Latham took his description from Brünnich's 'Ornithologia Borealis,' p. 73; and Brünnich merely transcribed it from Hans Ström's 'Physisk og œkonomisk Beskrifvelse over Fögderiet Söndmör, begliggende i Bergens Stift i Norge.' 4°, Soröe, 1762. If, now, we read the original description in Ström, we shall at once perceive that Ström, who was not a scientific ornithologist, has given a pretty good description of *Anthus pratensis*, in the autumnal plumage. This description Brünnich translated into Latin; and in so doing, introduced some little variation into the meaning, so that it has been misunderstood. (See especially the phrase 'abdomen cæruleum propè anum flavescens.') Ström himself places the bird next after the Pari, and adds, that it was *unknown to him*, and thus the name *Parus ignotus* originated. I have not Ström's work now by me; but several years ago I made this investigation into the origin of *Parus ignotus*, and am quite satisfied of its correctness. I presume that the expression 'abdomen cæruleum' arose from a slip of the pen in Ström, who wrote 'bugen mörk blaæ' instead of 'bugen mörk graæ,' which last word means *cinereous*. All the rest of his description applies admirably to *Anthus pratensis* in autumnal livery. In the index to Gloger's 'Handbuch der Vögel Europas,' it is said that *Parus ignotus* was originally a specimen of *Motacilla flava*, but artificially made up,—a supposition which I cannot regard as correct, for the description applies far better to *Anthus* than to *Motacilla*.'"

Description of *Iridisornis dubusia*, Bonaparte.

ON A PECULIAR STRUCTURE IN THE RECTRICES OF *VIDUA PARADISEA* (LINN.).

THE group of Ploceine birds which are distinguished by the assumption, during the breeding season, of greatly lengthened rectrices, and to which the generic name Vidua was given by Cuvier, have been in more recent times divided into three genera. M. Rüppell retained the name Vidua for the form of which *V. regia* (Linn.) is the type, and proposed the name Coliuspasser for another group, typified by *Fringilla macrocerca*, Licht. (*C. flaviscapulatus*, Rüpp.). Mr. G. R. Gray proposed a third division, named Chera, for the "*Emberiza longicauda*" of Gmelin, distinguished by its longer and more rounded wings, in which the fifth primary is the longest, while in the other two divisions it is the third that is longest. I am disposed to retain all these generic groups as distinct, although they are very closely allied to each other. I must remark, however, that M. Rüppell has founded his genus Coliuspasser on a misconception of its true characters. He states, in his 'Neue Wirbelthiere zu der Fauna von Abyssinien gehörig,' that in Vidua it is only the upper tail-covers that are lengthened, the rectrices remaining of ordinary length, while in Coliuspasser the true rectrices are extended far beyond the tail-covers. The fact however is, that if we carefully examine the tails of all these three groups, we shall find that they agree with other Passerine birds in possessing the normal number of twelve rectrices, and that it is by the prolongation of certain of these rectrices, and not of the tail-covers, that the Whidah birds acquire their peculiar character. But though Coliuspasser and restricted Vidua agree in this respect, they are nevertheless well characterized by the different modes in which their rectrices are prolonged. In Vidua the four external rectrices on each side are nearly equal and of ordinary length, while the four middle ones, which Rüppell mistook for tail-covers, are abruptly and very greatly lengthened. Of these four middle feathers, the two on one side have their *under* surfaces turned *towards* the under surfaces of the other two. This group will include the species *V. regia* (Linn.), *V. principalis* (Linn.), and *V. paradisea* (Linn.). *V. superciliosa* (Vieill.), said to have only two rectrices elongate*, may probably be also referred to this genus.

In Coliuspasser, on the contrary, all the rectrices are more or less elongate. Some of them, as *C. macrura* (Gmel.), *C. macrocerca* (Licht.), have the tail simply graduated, the middle rectrices being of moderate length, and the lateral ones gradually shorter. Other species, in which the elongation is carried to a greater extent, have the relative length of the rectrices variously modified. In *C. ardens* (Bodd.), (*V. panayensis*, Gmel.), the tail may be termed furcate, the medial rectrices being shortest, and the outer ones gradually lengthened, though the most

* The Abyssinian bird is considered distinct from the S. African *V. paradisea*, and has been called *V. verrauxii*, Cassin, Pr. Ac. Phil. 1850. *V. sphænura*, Bonap. Conspect. Av.—ED.

external pair is (in my specimens at least) somewhat shorter than the penultimate one. In this species the rectrices are hollowed on their upper side; and it is probable that in the living bird the rectrices on each side of the centre have their *upper* surfaces turned towards those of the opposite side, as in the American "boat-tailed Grackles" (Scaphidurus). This structure of the tail, the exact reverse of that which prevails in the majority of birds, is still more decidedly shown in the *Chera progne* (Bodd.) of Gray, a bird, which, were it not for the different form of the wing, would rank with Coliuspasser, and not with Vidua.

A series of specimens of *Vidua paradisea*, in various stages of moult, collected in Kordofan by Mr. J. Petherick, has revealed a curious fact in the development of the elongated rectrices, which appears not to have been before noticed*. The tail of this bird is a very anomalous one, the submedial pair of rectrices being very greatly elongate, while the medial pair are much shorter, very broad, with a smooth hair-like shaft projecting nearly two inches beyond the webs. If we now examine the longest (or submedial) rectrices, we shall see that they differ from all the other tail-feathers in presenting a serrated appearance at their margins. In the other feathers the barbs end in fine points, so that the webs which they compose terminate in an acute margin. But in this particular pair of rectrices, the barbs terminate abruptly, with an obliquely flattened disk at the extremity of each; and the webs composed of those barbs are consequently both blunt and serrated at the margin.

On examining these feathers when in a half-grown state, a singular hair-like filament is seen to spring from their base, which explains the cause of these marginal serrations. This filament is narrow, flat and thin, much resembling in appearance the barbs of the feathers, but reaching to three or four inches in length. Its distal extremity is free; but towards the base of the half-formed feather it is seen to adhere to the extremities of all the barbs on one (generally the outer) side of the feather, forming a continuous margin or "selvage" to the web. Towards the base of the feather, where the imperfectly formed barbs are collected, as in all young feathers, into a cylindrical bundle, and inserted into a membranous sheath, the barbs belonging to both webs of the feather are seen to be connected at their extremities to the opposite sides of this intermediate filament. As the feather grows and the barbs become mature, their tips are gradually released from this connecting filament, those of the lower or exterior web first, and those of the interior one subsequently. Hence the distal portion of the filament becomes free, and waves loosely in the air. It is probable that when the whole feather reaches maturity, the filament is shed altogether.

The cicatrices, or points of junction between the tips of the barbs and

* About three months after my notice of this remarkable structure was printed, I was not a little surprised to find that this peculiarity had been long since described and figured by that exceedingly accurate observer Brisson, Orn. iii. p. 123.—H. E. S.

the flat surface of the filament, produce a succession of slight indentations on both sides of the latter, and give it a serrated appearance, which is further increased by the alternate tufts of barbules which fringe its margin. Hence also arise the corresponding serrations on the margins of the feather, which have been before referred to.

It is remarkable that these filaments, though apparently formed for some temporary purpose in the development of the feather, should exhibit a structure as highly complex as that of the feather-barbs themselves. I allude to the double row of *barbules* ("*Strahlen*" of German authors) which fringe the outer margin of the filament; they are not continuous as on the barbs ("*Aeste*"), but in little tufts, alternating with the surfaces of attachment of the barb-tips (see fig. 2). These barbules further exhibit those ultimate fringes to which the name *barbicels* has been given.

In ordinary feathers, the barbules on the distal side of the barb are, as is well known, furnished with hooked barbicels ("*Häkchen*"), while those on the proximal side are simple. But it is remarkable, that in these deciduous filaments which I am describing, *both* series of barbules are furnished with hooked barbicels. The object of these is obviously to embrace the barbules of the feather-barbs, during the attachment of the latter to the filament, and as these barbules are attached to both sides of the filament, it is requisite that the filament should be provided with a double series of hooks.

The object of this singular structure is probably the protection of the feather-barbs during the course of their development. But why so complex and elaborate an arrangement should be confined to two feathers only in the bird, and to one species of bird only (as far as is yet known), is one of the many questions of natural science which must probably remain unanswered.

Explanation of the Plate.

Fig. 1. Under view of basal portion of one of the submedial rectrices of *Vidua paradisea*, in a half-developed state.

a, interior web; *b*, exterior web; *c c*, serrated margin; *d*, free portion of filament; *e f*, portion of filament attached to margin of interior web; *f g*, portion attached to both webs; *h*, barbs of exterior web recently detached from the filament and not yet incorporated into the web; *i*, membranous sheath surrounding the immature barbs.

Fig. 2. Portion of filament magnified, showing the alternating tufts of barbules on each side.

Fig. 3. *A A*, portion of filament highly magnified, with portions of barbs attached, as in fig. 1 *f g*.

a a, filament-barbules, forming two series of tufts, both of which are furnished with hooked barbicels; *b b b*, feather-barbs, belonging to the two opposite webs of the feather, connected at their distal extremities to opposite sides of the filament; *c c c*, barbules of the distal side of the barbs, furnished with hooked barbicels; *d d d*, barbules of the proximal side of the barb, devoid of barbicels, or furnished with only a very few simple ones.

NOTES ON SOME BIRDS FROM THE RIVER GABOON IN WEST AFRICA.

A SMALL collection of birds from this new locality has lately been purchased from M. E. Verreaux of Paris, by E. Wilson, Esq., who has kindly submitted them to my examination before sending them to their destination in the Philadelphia Museum. The river Gaboon, or as the French write it, Gabon, is situate exactly under the equator, and is considerably farther south than most of the localities where West African birds have hitherto been collected. It has consequently afforded several new species; and in order to show the geographical distribution of others, I have thought it best to give a list of the whole collection. The specimens were labelled by M. Verreaux, whose MS. names I have retained in all cases where it was practicable.

The species which are *additional* to Dr. Hartlaub's list of West African Birds in the 'Verzeichniss der Vorlesungen' of the Hamburg Gymnasium, 1850 (see Contributions to Ornithology, 1850, p. 129), are marked with an asterisk.

*1. *Hirundo melbina,* Verreaux, MS.* A curious little Swallow, combining the typical structure of Hirundo with the sombre colouring of Chelidon or Cotyle. Crown and sides of head, rump and upper tail-covers, fuscous brown; upper back and wings black, with a steel-blue gloss; tail blackish, deeply furcate, the external rectrices narrow, and 7 lines longer than the next; chin and lower parts white with a pale brownish tinge. Beak black, legs pale brown.

Total length, 5 inches 6 lines; beak to front, 2 lines; wings, 3 inches 9 lines; medial rectrices, 1 inch 6 lines; external, 3 inches; tarsus, 4½ lines.

2. *Platystira leucopygialis,* Fraser, Zoologia Typica, pl. 34. The *P. castanea* of Fraser is the ♀ of this bird.

*3. *Hyliota violacea.* (*Muscicapa violacea,* Verreaux, MS.) This bird is interesting as affording a second species of a genus of which one species only, the *H. flavigaster,* Swains., of Senegal, was hitherto known. It much resembles *H. flavigaster,* but differs in its broader beak and the less extent of white on the wing. Whole upper parts black with a steel-blue gloss, of a rather more purple hue than in *flavigaster.* Three or four of the greater wing-covers next the body are white (in *flavigaster* the whole of the middle and the basal half of the greater covers are white). Lower parts pale cream colour; tibiæ black; beak and legs fuscous.

* Verreaux places this Swallow in Atticora. This genus seems to be little more than an artificial receptacle for several stray Hirundinidæ from all quarters of the globe, presenting the form but not the coloration of Hirundo.—H. E. S.

Total length, 5 inches; beak to front, 5 lines; to gape, 7 lines; breadth, 2½ lines; wing, 3 inches; medial rectrices, 1 inch 9 lines; external, 2 inches; tarsus, 7 lines.

*4. *Dicrurus coracinus*, Verreaux, MS. Whole plumage deep velvety blue-black. Quills internally black. Total length, 8½ inches; beak to front, 6 lines; to gape, 1 inch; height, 3½ lines; width, 3 lines; wing, 4 inches 9 lines; medial rectrices, 3 inches 5 lines; external, 4 inches 3 lines; tarsus, 7 lines.

5. *Chaunonotus sabinei*, Gray. (*Hapalophus melanoleucus*, Verr. MS.)

*6. *Pycnonotus ashanteus*, Bonaparte? (*Ixos ashanteus*, Bonap., Verr. MS.) This bird differs from *Pycnonotus barbatus* (Desfont.), (*Turdus arsinoë*, Licht., *Ixos obscurus*, Temm., *Ixos inornatus*, Fras., *Hæmatornis lugubris*, Less., all of which names belong, in my opinion, to *one* species, though Prince Bonaparte makes *two* of them), *only* in having a very slight wash of yellow on the lower tail-covers, which in *P. barbatus* are white, and also in the lighter tint of the head. Though labelled *Ixos ashanteus*, I doubt its being the true *ashanteus* of Bonaparte, 'Conspectus Generum Avium,' p. 266, as he says it only differs from his *I. obscurus* in being much less, whereas the present specimen agrees in size with *I. obscurus* (*P. barbatus*, Desfont.).

*7. *Ixonotus guttatus*, Verreaux, MS. A curious form belonging to the subfamily Pycnonotinæ. At first sight it resembles in general appearance the genus Chrysococcyx. Beak longer than in Pycnonotus. Rump-feathers very long and thick.

*8. *Pratincola salax*, Verreaux, MS. Resembles *P. rubicola* of Europe, and still more *P. pastor*, Strickland, of S. Africa, but the beak is wider, and the *breast* only is rufous, the *sides* being white.

*9. *Lamprocolius purpureiceps*, Verreaux. Whole head and throat glossy violet-purple; whole back, breast and belly glossy bluish-green; wings glossed with steel-blue, the primaries externally with violet; tail black, slightly glossed with purple; beak and legs black.

Total length, 7 inches 2 lines; beak to front, 6 lines; to gape, 8 lines; wing, 4 inches 3 lines; medial rectrices, 2 inches 8 lines; external, 2 inches 9 lines; tarsus, 7 lines.

10. *Nigrita canicapilla*, Strickland. This bird, hitherto found only in Fernando Po, is now shown to extend to the adjacent African continent.

*11. *Nigrita latifrons*, Verreaux, MS. A typical species of this very limited genus, resembling *N. canicapilla* in the general arrangement of colour, but smaller, and wanting the black front and the white wing-spots.

12. *Malimbus rubricollis* (Swainson). (*Euplectes rufovelatus*, Fras.)

*13. *Anthreptes aurantius*, Verreaux, MS. Closely allied by the stout, straight form of beak to *A. longuemarei* of Senegal, though the arrangement of the colour, and especially the orange tufts on the sides of the breast, connect it with Nectarinia.

14. *Nectarinia chloropygia,* Jardine.

15. *Nectarinia stangeri,* Jardine.

16. *Nectarinia fuliginosa* (Shaw)? Differs from *N. fuliginosa* as usually described, in having the upper tail-covers *purple* instead of *brown* like the rest of the upper parts. In this state it is described by Lesson in his 'Description de Mammifères et d'Oiseaux récemment découverts,' p. 271; but whether it be a peculiar state of plumage, or a distinct species, we have as yet no evidence.

17. *Nectarinia superba* (Vieillot)? This splendid bird is labelled by Verreaux "*Cinnyris sanguineus,* Less. *C. superbus,* Vieill." It agrees with the descriptions of *N. superba* (Vieill.), except in wanting the golden-green band which separates the violet-purple of the throat from the dull red of the breast. The vent and lower tail-covers are deep black. Crown metallic greenish-blue, a minute purple spot behind each supercilium; ear-covers, nape, back, lesser and middle covers, scapulars, rump and upper tail-covers, metallic green, with a golden gloss on the upper back and ear-covers. Rest of wings and tail black, with a faint purplish gloss.

Total length, 5 inches 6 lines; beak to front, 1 inch 2 lines; wing, 3 inches; medial rectrices, 2 inches; external, 1 inch 9 lines.

I can find no reference to the "*Cinnyris sanguineus,* Less." quoted by Verreaux, so I conclude that it is a MS. name.

*18. *Nectarinia johannæ* (Verreaux). The ♀ is dark olive-brown above, very pale yellow on throat and deeper yellow on breast, belly and lower tail-covers; the whole lower parts with a brown streak down the middle of each feather.

19. *Alcedo leucogaster,* Fraser.

*20. *Alcedo quadribrachys,* Bonaparte. This beautiful Kingfisher presents a remarkable similarity to the well-known *Alcyone azurea* of Australia, from which however it is at once distinguished by having four toes, a character which places it in Alcedo. The whole upper parts, cheeks and sides of neck are deep Prussian-blue, barred with black on the crown; the lores, patch behind the ears and throat, are pale fulvous; lower parts deep ferruginous. Remiges blackish; rectrices black with a tinge of blue; beak black; legs red.

A young specimen differs in having a blue tinge spread over the breast.

Length, 6 inches 5 lines; beak to front, 1 inch 7 lines; wing, 3 inches 1 line; tail, 1 inch 5 lines; tarsus, 3 lines.

I have had a specimen of this bird in my collection since 1838, but, from not knowing its locality, I have been unable to identify either its name or habitat till now.

21. *Halcyon cancrophaga* (Latham)? This bird was so named in a list sent me by Mr. Wilson, but Verreaux's label has been lost *. It seems, however, to be different from *H. cancrophaga,* as described by Latham after Buffon.

Head, cheeks, upper back, scapulars and lesser and middle wing-covers, deep chestnut; greater covers blackish. Lower back and upper tail-covers azure; primaries black; secondaries azure, externally tipped with black; tail deep blue, tipped with black; chin and lower parts white; beak red; legs reddish.

Length, 7 inches 5 lines; beak to front, 1 inch 1 line; wing, 3 inches 7 lines; tail, 1 inch 5 lines.

22. *Melittophagus variegatus* (Vieillot). (*M. lefebvrei,* Desmurs. *M. cyanipectus,* Verr. MS.)

23. *Chrysococcyx smaragdineus,* Swainson? The specimen from Gaboon is barred with green, and rufous on all the upper parts, and with green and yellowish-white below.

*24. *Trachyphonus purpuratus,* Ranzani. An interesting bird, almost intermediate between Trachyphonus and Pogonius. There is a slight notch about the middle of the margin of the upper mandible, but no projecting tooth as in Pogonius. The beak, however, is much stouter than in the typical species of Trachyphonus.

25. *Barbatula atroflava* (Sparrmann). (*B. erythronotus,* Cuv.)

26. *Barbatula subsulphurea,* Fraser. (*B. flavimentum,* Verr. MS.)

*27. *Barbatula leucolæma,* Verreaux, MS. Closely allied to *B. subsulphurea,* and only differs in having the chin and throat white, with a pale grey tinge, while in *subsulphureus* the whole lower parts, including the throat, are sulphur-yellow.

* Named by Verreaux *Halcyon badius,* as distinct from Latham's bird, the type of which is Buffon's Pl. Enl. 334. *Alcedo cancrophagus,* Bodd.

LIST OF A COLLECTION OF BIRDS PROCURED BY MR. C. T. ANDERSSON IN THE DAMARA COUNTRY IN SOUTH-WESTERN AFRICA*.

MR. FRANCIS GALTON, an active member of the Royal Geographical Society, set out in the autumn of 1850, with the intention of penetrating the interior of Southern Africa. He started from Walfisch Bay, on the south-west coast, in latitude 23° south; and passing through the Damara country, reached the longitude of 21° east—a distance of about 500 miles in the interior. An account of his journey will be found in the Journal of the Geographical Society, vol. xxii. p. 140†. He was accompanied by Mr. C. T. Andersson, a Swede, who formed a considerable collection of birds, which were consigned to Mr. A. D. Bartlett of London for sale. Unfortunately, as too often happens in such cases, many of these birds were dispersed before any catalogue was made of them. Some were purchased for the British Museum; others were bought by Mr. Frank, a dealer in Amsterdam; and of the residue, about 100 specimens have passed into my collection, and a like number into that of Sir W. Jardine, Bart.

As the Damara country is intermediate between the regions of Southern and Western Africa, the ornithology of each of which has been pretty fully investigated, it is the more to be regretted that no complete list of these birds is now attainable, as it would have thrown much light on the geographical distribution of species. Mr. Sclater and I have done our best to supply this loss, by compiling a list of the species purchased by myself, including also some of those which are now in the British Museum.

The Namaqua Land, where Levaillant collected many of his birds, being adjacent to the Damara country on the south, it is interesting to recognize in this collection several of his species, which are unknown in the Cape colony, and some of which have been hitherto recorded on his authority alone. Dr. Andrew Smith also penetrated in the same direction; and we accordingly find many of the Damara birds delineated in his 'Illustrations of the Zoology of Southern Africa.'

It is remarkable, that among the sixty-two species of birds from Caffraria, described by Professor Sundevall in the 'Öfversigt af Kongl. Vetenskaps-Akademiens Förhandlingar,' 1850, p. 97, there is not one which I have been able to identify in the Damara collection. This fact shows, that, as in the case of South America, the birds of Southern Africa are to a great extent limited to special localities; a circumstance which of course greatly increases the numerical richness of the fauna. Professor Sundevall estimates the total amount of the ornithology of South Africa at 700 species.—H. E. S.

* The notes are conjointly by H. E. Strickland and P. L. Sclater.

† And in 'The Narrative of an Explorer in Tropical South Africa.' Murray, 1853.

1. *Milvus parasiticus*, Daudin; Levaill. Ois. Af. pl. 22.

2. *Accipiter gabar*, Daudin, Ois. Af. pl. 33.

3. *Accipiter niger*, Vieillot, Gal. Ois. pl. 22. (*A. carbonarius*, Licht.) Vieillot describes it as only 9 inches (French) long, but my specimen (probably a ♀) measures 12 inches (English).

4. *Tinnunculus rupicolus* (Daudin).

5. *Scops leucotis*, Temminck, Pl. Col. 16.

6. *Scops senegalensis*, Swainson, Birds West Af. vol. i. p. 127. This species is quite distinct from *Scops zorca* of Europe. The wing measures only 5 inches 1 line, while that of *S. zorca* is 6 inches, and the differences in the length of the primaries, indicated by Swainson, appear to be constant.

7. *Athene licua* (Lichtenstein), Verz. Säug. u. Vög. aus dem Kaffernlande, p. 12.

8. *Caprimulgus pectoralis*, Vieillot; Levaill. Ois. Af. pl. 49. (*C. rufigena*, Smith, Zool. S. Af. pl. 100.)

9. *Caprimulgus lentiginosus*, Smith, Zool. S. Af. pl. 101.

10. *Caprimulgus damarensis*, Strickland. Size, small. Ground colour of crown and upper parts pale grey, minutely speckled with fuscous; each feather on the crown has a conspicuous medial black streak, broad on the front and occiput, narrower on the hind head; cervical collar fulvous, each feather margined with black dots; feathers of back, rump and upper tail-covers with a black longitudinal streak about $\frac{1}{20}$th of an inch broad on each; external row of scapulars with a broad pointed medial black streak, their inner margin fulvous, speckled with black, the outer plain fulvous, contrasting strongly with the black medial stripe; middle and greater covers fuscous, with a large squarish or circular fulvous spot at the tip of the outer web, and some broken bars of fulvous near the base of the same web; primaries with three or four square fulvous spots on the basal two-thirds of their outer webs, separated by equal intervals of fuscous, on their inner webs are three or four larger fulvous spots united at the margins of the web; terminal third of the primaries fuscous, speckled with grey at the tips; secondaries regularly barred with four or five fulvous bars, divided by the black shaft, and separated by equal intervals of fuscous; rectrices pale grey, speckled and barred with fuscous; the medial pair have about nine narrow fuscous bars, which become broader and more numerous externally, so that on the two outer pair they amount to fourteen or fifteen, and the ground colour on these outer remiges is obscure fulvous. Over the eye is a pale superciliary streak; the cheeks dark fulvous, speckled with fuscous; the ear-covers pale fulvous; chin and throat pale fulvous, obscurely barred with fuscous; a nearly white spot on each side of the throat; upper breast grey speckled with fuscous, with a medial fuscous streak on each feather; the feathers of the lower breast have a large subtriangular

fulvous patch, surrounded by fuscous; belly pale fulvous, with narrow transverse fuscous bars, three on each feather; vent and lower tail-covers plain pale fulvous.

Total length 9 inches 5 lines; beak to front 3 lines, to gape 9 lines; wing 6 inches 4 lines; medial rectrices 4 inches 5 lines, external 4 inches 4 lines; tarsus 7 lines, feathered for half its length; middle toe and claw 9 lines; outer and inner ditto 5 lines.

11. *Hirundo rustica*, Linnæus. Identical with British specimens.

12. *Platystira pririt* (Vieillot); Levaill. Ois. Af. pl. 161.

13. *Platystira albicauda*, Strickland*. Size, large. Front pure white, extending laterally as far as the eyes; crown, lores and cheeks deep glossy black; a white spot on the nape surrounded by black; back slaty-grey; scapulars black externally, slaty-grey within, and obscurely tipped with white; wings black, the margins varied with white; the middle and greater covers next the body pure white; basal third of primaries and the extreme tips of the three first, white; secondaries and tertials black, tipped with white, the four inner secondaries next the tertials white for one-fourth from the base; rump and upper tail-covers thick and downy, cinereous like the back, each feather with an elongate sub-terminal tear-like white spot, the extreme tips blackish; tail wholly pure white except an elongate tear-like spot of black on the medial pair, nearly bisected longitudinally by the white shaft. Chin, throat and sides of neck pure white, below which is a black pectoral collar; sides of breast cinereous; lower wing-covers black; middle of breast, abdomen and lower tail-covers pure white; feathers of the tibiæ white at the tips, black at the base; beak and legs black.

Total length 5 inches 3 lines; beak to front $6\frac{1}{2}$ lines, to gape 9 lines, width $2\frac{1}{2}$ lines, height 2 lines; wing 3 inches 5 lines; medial rectrices 1 inch 8 lines, external 1 inch 7 lines; tarsus 1 inch 2 lines.

This is the largest species of *Platystira* that I have seen. The beak is stronger and more compressed at the sides than in the other species, the tail shorter in proportion, and the first primary longer, being nearly two-thirds the length of the fourth.

14. *Dicrurus divaricatus*, Lichtenstein. Agrees with specimens from Kordofan.

15. *Lanius subcoronatus*, Smith, Zool. S. Af. pl. 68.

16. *Lanius collaris*, Linnæus; Levaill. Ois. Af. pl. 62.

17. *Lanius minor*, Gmelin. Differs from specimens from South Europe only in the base of the lower mandible being whitish, and in the rather smaller extent of black on the third pair of lateral rectrices.

* This bird was described by Waterhouse in Appendix to Alexander's Travels in S. Africa as *Lanioturdus torquatus*; and it has been lately named by Prince Bonaparte *Moquinus tandonus*, after M. Moquin Tandon!!!—ED.

18. *Enneoctonus anderssoni*, Strickland. Front, crown and ear-covers ferruginous; a cream-colour streak from the nostrils over the eye; upper parts deep ferruginous, tinged with greyish on the hind neck and rump; remiges pale fuscous, margined with fulvous; tail obscure ferruginous; chin, throat and lower parts very pale cream-colour, with a darker shade on the breast; beak horn-colour; base of lower mandible pale; legs horn-colour.

The specimen before me is not quite adult, and the feathers of the wing-covers, rump and tail are bordered by a submarginal fuscous band, and those of the breast and flanks have two bars of the same colour.

Total length 6 inches 3 lines; beak to front 5 lines, to gape $7\frac{1}{2}$ lines, height $2\frac{1}{2}$ lines; wing 3 inches 6 lines; medial rectrices 2 inches 8 lines, external 2 inches 7 lines; tarsus $9\frac{1}{2}$ lines.

Allied to *E. melanotis* (Valenciennes) of India, but differs in the ferruginous ear-covers, longer wing, shorter tail, &c.

19. *Dryoscopus cubla* (Latham); Levaill. Ois. Af. pl. 72.

20. *Laniarius atrococcineus* (Burchell), Zool. Journ. vol. i. pl. 18.

21. *Telophonus senegalus* (Linnæus); Levaill. Ois. Af. pl. 70.

22. *Nilaus brubru* (Latham).

23. *Eurocephalus anguitimens*, Smith.

24. *Turdus strepitans*, Smith.

25. *Pycnonotus capensis* (Levaillant), Ois. Af. pl. 105.

26. *Crateropus bicolor*, Jardine. The young differs from the adult in having the front, eyebrows and cheeks hoary brownish-white; crown and upper parts umber-brown, palest on the crown and rump, which last may almost be termed dirty-white; wings and tail deep umber-brown; remiges margined internally with fulvous; throat white; feathers of breast pale greyish-brown, broadly margined with white; sides and lower wing-covers pale fulvous; belly, vent and lower tail-covers whitish; beak black; gonys straight; commissure and culmen considerably decurved; legs horn-colour.

Total length 9 inches 5 lines; beak to front 9 lines, to gape 1 inch 1 line; wing 4 inches 8 lines; all the rectrices 4 inches; tarsus 1 inch 3 lines.

Allied to *Crateropus leucocephalus* (Rüpp.), but differs in not having the whole head white, the fulvous colour of the flanks, &c.

27. *Saxicola leucomelæna*, Burchell, Travels in S. Africa, vol. i. p. 335. This bird precisely agrees with Mr. Burchell's description, except that he makes no mention of the white which occupies three-fourths of the length from the base of the four external pairs of rectrices. Mr. Gray refers *S. leucomelæna* of Burchell to *S. cursoria*, Vieill., Levaill. Ois. Af. pl. 90, but that differs in its larger size, black abdomen, &c. It is however possible that the bird before us may be an immature state of

S. monticola, Vieill., such as is represented in Levaill. Ois. Af. pl. 185. fig. 2. The entire crown and hind neck are white, with a faint brownish tinge; the lesser and medial wing-covers, upper tail-covers, abdomen and four outer pair of rectrices, except the tips, also the basal portion of the outer web of the fifth pair, pure white; lower tail-covers black, tipped with white; rest of the bird deep black.

28. *Saxicola hottentotta* (Gmelin). (*Sylvia pileata,* Lath.; *Œnanthe imitatrix,* Vieill.; Levaill. Ois. Af. pl. 881.) There seems no doubt that this is the 'Grand Motteux du Cap de Bonne-espérance' of Buffon, on which Gmelin founded his *Sylvia hottentotta*; though Buffon erroneously states it to be 8 inches long. A specimen which has long been in my collection measures 7 inches in total length, and the wing 4 inches. Mr. Andersson's specimen from Damara is still smaller, being only about 6 inches 3 lines long, and the wing 3 inches 4 lines. Its coloration is identical with that of the larger specimen.

29. *Saxicola.* Light fulvous brown above, pale cream-colour beneath; wing and tail feathers fuscous, margined with fulvous; beak horn-colour; legs black.

Total length 7 inches 3 lines; beak to front 5½ lines, to gape 8 lines, height 2 lines; wing 4 inches 4 lines; all the rectrices 3 inches 1 line; tarsus 1 inch.

This is probably ♀ or immature, and I do not at present venture to name it. It approaches *S. pallida* of Rüppell, Atl. Nord. Af. pl. 34, fig. *a,* but has a much stouter beak, darker tail, &c.

30. *Monticola brevipes,* Waterhouse, Alexander's Exped. Int. Af. vol. ii. p. 263. Front and crown hoary white; lores blackish; chin, throat, ear-covers, upper back and wing-covers, deep slaty-grey; remiges blackish, narrowly margined with whitish; breast, abdomen, lower wing-covers, rump, upper and under tail-covers and tail, bright ferruginous; beak and legs blackish.

N.B.—The two medial rectrices are wanting in my specimen, but probably are more or less marked with fuscous, as in other species.

Total length 6 inches 8 lines; beak to front 7½ lines, to gape 9 lines; wing 4 inches 2 lines; tail 2 inches 9 lines; tarsus 1 inch.

♀ Above greyish fuscous, wing-feathers margined with pale brown; rump and upper tail-covers ferruginous; tail deep ferruginous chestnut, the middle pair of rectrices fuscous, the rest fuscous on the outer webs near the tips; feathers of cheeks, chin and throat whitish, margined with fuscous; lower parts ferruginous, feathers of breast with a subterminal fuscous bar on each; wing only 3 inches 8 lines.

I adopt Boie's original name Monticola, given in 1822, for this genus, because I know no reason for changing it to Petrocossyphus or Petrocincla.

31. *Erythropygia galtoni,* Strickland. This bird agrees with a speci-

men in Sir W. Jardine's collection, procured by Dr. A. Smith, except in being of a rather paler tinge above and below. This latter specimen is labelled "*Saxicola familiaris,*" a name founded on the "Tracquet familier" of Levaill. Ois. Af. pl. 183. But the latter is described as having the outer webs of the lateral rectrices fringed with rufous, whereas in the present bird the whole of the rectrices except the middle pair are rufous, tipped with brown for about 4 lines. The upper parts are light fulvous brown; wing-covers and remiges fuscous, narrowly margined with whitish-brown; whole lower parts pale cream-colour; rump, upper tail-covers and tail deep rufous; medial rectrices fuscous, rufous at the base, the rest tipped with fuscous.

Total length 6 inches; beak to front 5 lines, to base 7 lines; wing 3 inches 4 lines; medial rectrices 2 inches 8 lines, external 2 inches 7 lines; tarsus 1 inch.

32. *Drymœca* (commonly misspelled *Drymoica*) *levaillanti,* Smith, Zool. S. Af. pl. 73. In reference to this species I may mention, that the *Cisticola* (misspelled *Cysticola*) *campestris* of Gould, afterwards named by him *C. magna,* and figured in his 'Birds of Australia,' vol. iii. pl. 41, is evidently identical with the South African *Drymœca levaillanti.* Mr. Gould's figure was taken from a specimen which has been for many years in my own possession, the locality of which was unknown, but which being stuffed with wool, as is frequent in specimens from Australia, I conjectured to be from that country. Mr. Gould accordingly figured it as a new species in his 'Birds of Australia.' The occurrence of a specimen in the Damara collection has enabled me to detect this error.

Mr. Gould's specific name *campestris* was given in 1845, and must therefore yield to *levaillanti,* Smith, 1842.

33. *Drymœca flavicans,* Vieillot; Levaill. Ois. Af. pl. 127.

34. *Drymœca flavida,* Strickland. Front, crown and cheeks pale cinereous; back and wings yellowish-olive; remiges fuscous, margined with clear olive-yellow; rectrices olive, margined with olive-yellow and tipped with pale yellow, the external pair wholly pale yellow; chin whitish; throat and breast pure pale yellow; belly silvery-white; sides, under wing-covers and under tail-covers pale yellow; beak horn-colour; base of lower mandible pale; legs pale brown.

Total length 4 inches 2 lines; beak to front 4 lines, to gape 5 lines, height 1 line, breadth 1½ line; wing 2 inches; medial rectrices 1 inch 6 lines, external 1 inch 2 lines; tarsus 8 lines.

The cuneate tail and form of beak refer this little bird to the vicinity of Drymœca or Prinia, though the coloration is more like that of Phylloscopus. It seems allied to the '*Malurus pulchellus,*' Rüpp. Atl. Nord. Af. pl. 35. fig. *a*, but that has a longer tail, tipped with white, and otherwise differs.

35. *Sylvietta brachyura*, Lafresnaye, Rev. Zool. 1839, p. 258.

36. *Sphenœacus pycnopygius*, Sclater. Capite et dorso superiore nigro brunneoque striatis: lineâ utrinque à rostro super oculum, gulâ et pectore albis; maculis crebris in pectore nigris; ventre toto, crisso, tibiis et tectricibus caudæ superioribus et inferioribus rufo-brunneis; alis caudâque nigricantibus, illis brunneo limbatis; hujus rectricibus lateralibus pallidè brunneo terminatis; mandibulâ inferiore albâ, superiore nigrâ; pedibus brunneis.

Total length 6 inches 8 lines; beak to front 6½ lines, to gape 8 lines; wing 2 inches 9 lines; medial rectrices 3 inches, external 2 inches 6 lines; tarsus 1 inch.—(P. L. Sclater.)

Allied to *Sphenœacus africanus* (Gmel.), but differs in the longer beak, shorter tail, broader and more developed webs of the rectrices, &c.

37. *Parisoma subcæruleum* (Vieillot); Levaill. Ois. Af. pl. 126. (*Parisoma rufiventer*, Swains.) This bird evidently belongs to the Sylviinæ, not to the Parinæ, and is closely allied to Curruca.

38. *Ægithalus smithi*, Jardine.

39. *Parus cinerascens*, Vieillot; Levaill. Ois. Af. pl. 138.

40. *Juida australis* (Smith), Zool. S. Af. pl. 47. (*Lamprotornis burchelli*, Smith.)

41. *Spreo nabouroup* (Daudin); Levaill. Ois. Af. pl. 91. (*Lamprotornis fulvipennis*, Swains.)

42. *Spreo bispecularis*, Strickland; Levaill. Ois. Af. pl. 90. Whole plumage glossy bluish-green; head, rump and tail with a purplish tinge when held *towards* the light; lesser wing-covers violet-purple, with a coppery gloss towards their tips; middle and greater covers greenish, with a small indistinct velvety-black spot at their tips; primary covers violet-purple, but without any coppery gloss.

Total length 8 inches; beak to front 8 lines, to gape 1 inch 1 line, height 2½ lines; the gonys nearly straight; wing 5 inches; medial rectrices 3 inches 3 lines, external 3 inches 1 line; tarsus 1 inch 5 lines.

Closely allied to the true *Spreo nitens* (Linn.) of Angola (*Merula viridis angolensis* of Brisson), of which I possess a specimen, but it differs in the more slender beak, and in the primary covers being purple instead of greenish like the rest of the wing. Sir William Jardine has some specimens from Southern Africa, in which the beak is intermediate between the Damara bird and the true *nitens*.

43. *Dilophus carunculatus* (Gmelin).

44. *Buphaga africana*, Gmelin.

45. *Textor niger* (Smith), Zool. S. Af. pl. 64. (*Textor erythrorhynchus*, Smith.)

46. *Vidua regia* (Linnæus).

46*. *Steganura paradisea* (Linnæus).

47. *Plocepasser mahali*, Smith, Zool. S. Af. pl. 65. (*Leucophrys pileatus*, Swains.)

48. *Ploceus abyssinicus*, Gmelin; Smith, Zool. S. Af. pl. 7. (*Euplectes taha*, Smith.)

49. *Ploceus sanguinirostris* (Linnæus).

50. *Philetærus squamifrons* (Smith), Zool. S. Af. pl. 95.

51. *Estrilda granatina* (Linnæus).

52. *Estrilda lipiniana*, Smith, Report Exped. S. Af. p. 49.

53. *Estrilda astrild* (Linnæus). The specimen from Damara is rather paler on the upper parts and tail than specimens in Sir William Jardine's collection from South Africa, and in my own from Mauritius.

54. *Pytelia melba* (Linnæus); Edw. Birds, pl. 128; Buff. Pl. Enl. 203. fig. 1; Vieill. Ois. Chant. pl. 25. *Fringilla elegans*, Gmel.; *Fringilla speciosa*, Bodd.

There has been much confusion between this species, which seems to inhabit Southern and South-western Africa, and the species figured by Edwards, pl. 272, lower figure, from Abyssinia, Kordofan and Senegal. The present bird is evidently the same as plate 128 of Edwards, upon which Linnæus, in his 10th edition of the 'Systema Naturæ,' founded his *Fringilla melba*, though afterwards Brisson, and Linnæus himself in his 12th edition, united with it Edwards's plate 272, lower figure, an error followed by most succeeding authors. The first author who in modern times has distinguished these two species is I believe the Prince of Canino, in his 'Conspectus Generum Avium,' p. 461, where he gives their respective diagnoses, but erroneously assigns the specific name *melba* to the species of Edwards's plate 272. The latter species in fact requires a new name, and I propose to term it *Pytelia citerior*, in allusion to its less distant habitat.

The *P. melba*, as above indicated, is distinguished by having the front (not including the eyes), chin, and whole of the throat deep scarlet; breast yellowish-olive; crown, cheeks and nape deep cinereous; belly and sides fuscous, barred and spotted with white, each feather having two medial white bars and two subterminal transversely ovate white spots, separated by fuscous bars nearly equal in breadth to the white ones; lower tail-covers cream-colour; back and wings yellowish-olive; upper tail-covers brick-red; rectrices fuscous, margined externally with brick-red; beak reddish; legs pale brown.

The *P. citerior*, Strickl. (Edw. 'Birds,' pl. 272. fig. inf., *Estrilda elegans*, Rüpp., and *Pytelia elegans* of my list of Kordofan birds in Proc. Zool. Soc. 1850, p. 218), differs in having the front (including the eye), fore-part of cheeks and chin, scarlet; throat and breast light yellow; crown and hind neck pale cinereous; belly and sides white, with narrow bars of pale brown, three or four on each feather; lower tail-covers white; back, wings, rump, tail, beak and legs, as in *P. melba*.

55. *Colius erythropus*, Gmelin. (*C. leuconotus*, Lath.)

56. *Colius macrurus*, Linnæus. (*C. senegalensis*, Gmel.)

57. *Crithagra*. Brown, streaked with deep fuscous above, dirty white below.

Length 4 inches 4 lines; wing 2 inches 7 lines; beak and legs pale. Probably a ♀, but it is identical with a specimen in my collection from Kordofan.

58. *Fringillaria capensis* (Linnæus), Buff. Pl. Enl. 158. fig. 2. (*Fringilla nævia*, Gmel.; *Fringillaria vittata*, Swains.)

59. *Alauda erythrochlamys*, Strickland. Whole upper parts and wings nearly uniform ferruginous, becoming paler on the upper tail-covers and medial pair of rectrices. All the wing-feathers narrowly margined with pale cream-colour externally; the remiges light fuscous on their shafts and inner webs. A streak above and a spot below the eye pale cream-colour; ear-covers pale rufous; chin white; breast and lower parts pale cream-colour; beak long, gonys nearly straight, flesh-coloured; legs apparently flesh-coloured; hind-claw short, straight.

Total length 6 inches 5 lines; beak to front 7 lines, to gape 8 lines, height 2 lines; wing 3 inches 7 lines; rectrices 2 inches 8 lines; tarsus 1 inch 1 line; hind-claw $2\frac{1}{4}$ lines.

A well-marked species, leading by its lengthened beak to Certhilauda. The *A. ferruginea* of Lafresnaye approaches it in coloration, but is said to have the beak shorter than *A. calandra*, the breast streaked with black, &c.

60. *Alauda spleniata*, Strickland. Lores and superciliary streak whitish; front and crown chestnut; feathers of back and wing-covers fuscous in the middle, broadly margined with pale greyish-brown; margins of the greater covers and tertials nearly white; remiges light fuscous, narrowly margined externally with a pale impure rufous tinge, their tips and inner webs narrowly margined with whitish; upper tail-covers pale chestnut, tipped with whitish; rectrices deep fuscous, narrowly margined externally and slightly tipped with white, outer webs of the external pair wholly white; ear-covers light brown; cheeks, chin, throat and lower parts whitish; a faint brownish tinge on the middle, and a large chestnut patch on each side of the breast; beak subconical, the margin nearly straight, the gonys slightly curved upwards, the culmen rather more curved downwards, the tip brownish, the base flesh-colour; tarsi flesh-colour; toes and claws brownish, hind-toe short, slightly curved.

Total length 6 inches; beak to front $4\frac{1}{2}$ lines, to gape 6 lines, height 2 lines, width 2 lines; wing 3 inches 5 lines; medial rectrices 2 inches 4 lines, external 2 inches 1 line; tarsus 8 lines; hind-claw 3 lines.

Allied to *A. ruficeps*, Rüpp. Neue Wirbelth. pl. 38. fig. 1, but that has a black, not a rufous patch on each side of the breast. Specimens of this bird were collected in South Africa by Dr. Smith, and labelled by him '*Alauda ruficapilla*,' which is Stephens's name for the 'Calotte Rousse' of Levaill. Ois. Af. pl. 198 (*A. rufipileus*, Vieill.), a very distinct species.

61. *Alauda nævia*, Strickland. Crown greyish-brown, back of neck paler, back and wing-covers fulvous—the feathers of all these parts with a broad, distinct, longitudinal fuscous streak on each; remiges fuscous, margined externally and internally towards the base with pale fulvous; upper tail-covers rufo-fulvous; rectrices deep fuscous, the middle and external pairs broadly, the rest narrowly margined with fulvous; a whitish streak above and below the eye; cheeks and ear-covers light brown; chin whitish; breast pale cream-colour, with a small longitudinal fuscous streak on each feather; belly and vent pale cream-colour; beak horn-colour, paler towards the base; margin nearly straight, gonys curved upwards, and culmen equally so downwards; feet and claws flesh-coloured, tinged with light brown; hind-claw short, slightly curved.

Total length 6 inches; beak to front 6 lines, to gape 7 lines, height 2½ lines, breadth 2½ lines; wing 3 inches 4 lines; all the rectrices 2 inches 3 lines; tarsus 9 lines; hind-claw 3 lines.

62. *Nectarinia anderssoni*, Strickland. Head, back and lesser wing-covers metallic-green, the crown with a coppery gloss; upper tail-covers bluish-green; rectrices black, margined with bluish-green; greater wing-covers and remiges deep fuscous, margined externally with greyish-brown; chin bluish-green; cheeks and throat bright coppery-green; a broad zone on the breast of violet-purple, followed by a narrow one of dull greyish-brown; axillary tufts gamboge-yellow; abdomen, sides and lower tail-covers dirty-white; beak and legs black.

Total length 4 inches 3 lines; wing 2 inches 3 lines; medial rectrices 1 inch 6 lines, external 1 inch 5 lines; tarsus 7 lines.

Not unlike the East African species which I lately described as *N. albiventris* *, but has a longer wing, no orange on the axillary tuft, &c. My specimen has lost the extremity of the beak, so that I cannot assign its dimensions.

63. *Nectarinia fusca* (Vieillot); Levaill. Ois. Af. pl. 296.

64. *Nectarinia senegalensis* (Linnæus); Levaill. Ois. Af. pl. 295. fig. 2.

65. *Nectarinia bifasciata* (Shaw).

66. *Halcyon damarensis*, Strickland. Crown and wing-covers greyish-brown, streaked with black; upper back and scapulars sooty-brown; lower back and upper tail-covers vivid blue; remiges fuscous, margined externally with blue; primaries white at the base on both, secondaries on the inner webs; tail bluish; chin white; cheeks, hind neck, breast

* Collected Papers, p. 324.

and belly yellowish-white, with a black line down the centre of each feather; beak fuscous, base of lower mandible red; feet brown.

Total length 7 inches 3 lines; beak to front 1 inch 3 lines, to gape 1 inch 6 lines; wing 3 inches 5 lines; rectrices 2 inches 2 lines; tarsus 5 lines.

Almost identical in colouring with *H. chelicuti*, Stanl. (*Alcedo striolata*, Licht.; *Dacelo pygmæus*, Rüpp.), from Abyssinia and Senegal, but much larger in size. A specimen of the present bird in the British Museum is erroneously labelled '*A. striolata*,' which name, as defined by Lichtenstein, refers to the smaller species.

67. *Coracias caudata*, Linnæus.

68. *Melittophagus hirundineus* (Lichtenstein). (*Merops furcatus*, Stanl.; *M. taiva*, Cuv.; *M. hirundinaceus*, Swains.; *M. chrysolæmus*, Jard. and Selby, Ill. Orn. ser. i. pl. 99.)

69. *Irrisor erythrorhyncus* (Latham)?

The Damara specimen has the beak 2 inches 2 lines long to the gape, considerably curved and red, as in specimens from the Cape. But the white band on the primaries is broad, as in *I. senegalensis*, the shaft is nearly white, and extends over both webs of the fourth to the tenth primary; while in the Cape bird the band is narrow, commences on the outer web with the fifth primary, and is distinctly divided by the black shaft. Another distinction may also be pointed out, in the greater extent of white at the tips of the *primary covers* (not the *bastard wing*, as inadvertently stated) in the Senegal than in the Cape species. In this respect also the present specimen agrees with the West African, and not with the South African species. The Damara bird thus approaches in plumage the *red-beaked* specimens of *I. senegalensis* from Kordofan described by me in Proc. Zool. Soc. part 18. p. 216, but in these the beak is short and nearly straight, as in the *black-beaked* specimens from Senegal. Whether these variations indicate a plurality of ill-defined species, or a single but very variable one, must be decided by future inquiry. Sir W. Jardine considers the colour of the beak to depend on age or season, but I am not aware that any *red-beaked* specimens have ever been brought from Senegal *.

70. *Rhinopomastus cyanomelas* (Vieillot), Zool. Journ. vol. iv. pl. 1. (*Upupa purpureá*, Burch.; *Rhinopomastus smithi*, Jard.)

71. *Upupa minor*, Shaw.

72. *Buceros nasutus*, Linnæus; Levaill. Ois. Af. pl. 236. This is clearly the same species as Levaillant's bird, though it differs in possessing an elevated casque, with a sharp though obtuse-angled keel along its ridge. This casque is about 2 lines high, and extends to within 1 inch 2 lines

* We have red-billed specimens collected by Mr. L. Fraser at Abomey.—Ed.

of the tip of the beak, its anterior extremity being abruptly truncated. Levaillant's bird had no casque, being probably immature.

73. *Pogonius leucomelas* (Boddaert); Buff. Pl. Enl. 688. fig. 1. (*Bucco niger*, Gmel.; *Trogon luzoniensis*, Scop.; *Pogonius stephensi*, Leach.)

74. *Dendrobates namaquus* (Lichtenstein); Levaill. Ois. Af. pl. 251. (*Picus mystaceus*, Vieill.; *P. diophrys*, Steph.; *P. biarmicus*, Wagl.)

75. *Dendrobates fuscescens*, Vieillot; Levaill. Ois. Af. pl. 253. (*Picus fulviscapus*, Licht.; *Colaptes capensis*, Steph.; *P. chrysopterus*, Less.)

76. *Campethera capricorni*, Strickland. ♂ Front, crown and a broad streak from the base of the lower mandible along each side of the chin crimson; a broad streak of white from the nostrils, under the eye and across the ear-covers; hind neck, back, scapulars and tertials olive-brown, with two or three half-line-wide bars of yellowish-white on each feather (five or six bars on the tertials); wing-covers similar, but barred on the outer webs only, and with a small roundish terminal spot of whitish; remiges blackish internally, yellowish-olive externally; the shafts golden-yellow, the outer webs with five or six marginal spots of yellowish-white, the inner with as many but larger; rump and upper tail-covers yellowish-white, each feather with a subterminal black heart-shaped spot, and two or three medial transverse interrupted black bars; tail fulvous-brown, with six or seven pale fulvous bars, the shafts golden-yellow, the tips black; chin and throat white; breast, belly and lower tail-covers yellowish-white, middle of belly plain; the other parts with a longitudinally ovate tear-like black spot on each feather, those on the breast largest, and about 1 line in diameter; beak and feet horn-colour.

Total length 8 inches 5 lines; beak to front 1 inch 1 line, to gape 1 inch 3 lines; nasal ridge ½ line distant from culmen; wing 4 inches 9 lines; medial rectrices 3 inches, external 2 inches 3 lines; tarsus 1 inch.

Near *C. benneti* (Smith)—(*Picus guttatus*, Licht.; *C. variolosa*, Gray) —but differs in having a stouter beak, smaller spots on the breast, and the rump *spotted* instead of *barred*.

77. *Campethera abingoni*, Smith (*C. smithi*, Malh. Rev. Zool. 1845, p. 403). ♀ Feathers of front and crown blackish, with a round white spot on each; hind head crimson; back and wings olive, barred and spotted with yellowish-white; rump barred with fuscous and greenish-white; remiges fuscous, spotted on both shafts with whitish, the shafts golden-yellow; rectrices similar, but the marginal spots are obscurely fulvous; ear-covers and stripe down sides of neck white; chin, throat and upper breast brownish-black, with a subterminal roundish white spot, and sometimes a second medial one on each feather; lower parts yellowish-white, with a conspicuous longitudinal streak of black on each feather, broadest on the breast, becoming transverse and heart-shaped on the thighs and vent; beak and legs horn-colour.

Total length 7 inches 5 lines; beak to front 1 inch 1 line, to gape 1 inch 3 lines; nasal ridge nearly 1 line distant from culmen; wing 4 inches 7 lines; tail 2 inches 5 lines; tarsus 8 lines.

78. *Pæocephalus meyeri* (Rüppell), Atl. Nord. Af. pl. 11.

79. *Pæocephalus rüppelli* (Gray), Proc. Zool. Soc. ♂ Olive-brown; lesser wing-covers, lower wing-covers and tibiæ yellow; rump, lower belly, upper and lower tail-covers blue. ♂ like the ♀, but wanting the blue colour.

Total length 8 inches; wing 5 inches 7 lines.

80. *Agapornis roseicollis*, Vieillot; Bourg. St. Hil. Perroq. pl. 91.

81. *Columba guinea*, Linnæus; Levaill. Ois. Af. pl. 265.

82. *Turtur vinaceus* (Gmelin); Levaill. Ois. Af. pl. 268.

83. *Peristera afra* (Linnæus); Levaill. Ois. Af. pl. 271.

84. *Francolinus swainsoni*, Smith, Zool. S. Af. pl. 12.

85. *Francolinus gariepensis*, Smith, Zool. S. Af. pl. 83, 84.

86. *Pterocles variegatus*, Burchell; Smith, Zool. S. Af. pl. 10.

87. *Pterocles bicinctus*, Temminck, Pig. et Gall. iii. pp. 247, 713. There has been much confusion in ornithological works between various African and Indian species of Pterocles, to which the specific names *bicinctus*, *tricinctus* and *quadricinctus* have been given. In the Proceedings of the Zoological Society, part 18. p. 220, I have pointed out the distinctions between the true *P. quadricinctus*, Temm., of Africa (*P. tricinctus*, Swains.), and the *P. fasciatus* (Scop.) of India, erroneously named *P. quadricinctus* by Indian ornithologists. Mr. Andersson's collection has now enabled me further to clear up the subject, by affording specimens of the *P. bicinctus*, Temm. (to which Mr. Gray erroneously referred the *P. tricinctus* of Swainson). This rare species, first discovered by Levaillant in the Namaqua country, has never been figured. Though closely resembling the *P. quadricinctus* of Senegal and Kordofan in the colours of its head and under parts, it is at once distinguished by the absence of the black stripes on the back and wings, the feathers on those parts being of a greyish-brown, with a terminal subtriangular white spot. It also wants the chestnut band on the breast, but possesses a very distinct band of white, succeeded by one of black between the fulvous colour of the breast and the finely rayed black and white of the abdomen.

I may add that this, the *P. bicinctus*, Temm., is not the *P. bicinctus*, Licht., the latter being the *P. lichtensteini*, Wagl. and other authors.

88. *Coturnix dactylisonans*, Meyer ?. Plumage darker than European and Indian specimens, and of a deeper rufous on the cheeks and breast.

89. *Turnix lepurana*, Smith, Zool. S. Af. pl. 16. My specimen has the breast bright rufous, in which respect it agrees with the "pectore nitidè rufo" of Smith's Latin diagnosis, but not with the "pale Dutch orange" of his description, nor with the faint yellowish tinge of his figure.

90. *Otis ruficrista,* Smith, Zool. S. Af. pl. 4.

91. *Squatarola helvetica* (Linnæus).

92. *Charadrius hiaticula,* Linnæus.

93. *Charadrius damarensis,* Strickland. Crown, nape, ear-covers, back and wings uniform "hair-brown" (of Syme's nomenclature), paler on the rump and upper tail-covers; a dark fuscous mark at the anterior and lower margins of orbits; front, broad superciliary streak, cheeks, chin and throat dirty-white; breast pale brown, the shafts darker; belly, axillary feathers, vent and lower tail-covers white; primaries fuscous, darkest at the tips, the shafts white in the medial portions, the sixth to the ninth white at the base of the outer web; secondaries pale fuscous; rectrices pale fuscous, darker towards the tips, which are margined with white; outer pair narrow, margined externally with white; beak black; legs dark brown; acrotarsia scutate.

Total length 8 inches; beak to front 8 lines, to anterior termination of nasal groove 4 lines, to gape 1 inch; wing 5 inches 3 lines; medial rectrices 2 inches 3 lines, external 2 inches 1 line; naked part of tibia 7 lines; tarsus 1 inch 5 lines.

94. *Charadrius pallidus,* Strickland. Crown and upper parts pale greyish-brown, darker on rump; front, supercilium, cheeks and cervical collar white; ear-covers pale brown; greater wing-covers tipped with whitish; primaries fuscous, darkest at tips, shafts medially white, the fifth to the ninth white at base of outer web; secondaries tipped with white; all the remiges margined internally with white; medial pair of rectrices deep fuscous, the two next pairs paler, margined with white; the three outer pairs pure white; whole lower parts pure white; beak and legs black.

Total length 6 inches; beak to front 5 lines, to gape 6½ lines; wing 4 inches; medial rectrices 1 inch 6 lines, external 1 inch 5 lines; tarsus 1 inch 2 lines.

95. *Charadrius nivifrons,* Lesson. Agrees with Pucheran's description, Rev. Zool. 1851, p. 280, except in the isabelline or cream-coloured tinge on the lower parts, probably the breeding dress. Differs from *C. varius,* Vieill., *C. pecuarius,* Licht., in the larger patch of white on the front, the absence of the black patch on the side of the neck, the fulvous tinge of the upper parts, the pale brown (not black) rump, &c. Resembles *C. alexandrinus,* Linn. (*C. cantianus,* Lath.), but differs in having an isabelline tinge on the breast and belly, and in having the three proximal secondaries (next the tertials) pure white, instead of pale grey with white margins.

Total length 7 inches; beak to front 6½ lines, to gape 8 lines; wing 4 inches 1 line; medial rectrices 2 inches 3 lines, external 2 inches 1 line; tarsus 1 inch ½ line.

96. *Hoplopterus coronatus* (Gmelin), Pl. Enl. 800.

97. *Machetes pugnax* (Linnæus).

98. *Pelidna subarquata* (Gmelin). The Damara specimen is remarkable for the shortness of its beak, which only measures 1 inch 3 lines to the front, while specimens from Kordofan and various parts of Europe measure 1 inch 6 lines.

99. *Pelidna minuta*, Leisler; Gould, Birds of Europe, pl. 332.

100. *Glottis canescens* (Gmelin).

101. *Totanus glareola* (Linnæus).

102. *Rhynchæa capensis* (Linnæus).

103. *Himantopus melanopterus*, Meyer.

104. *Strepsilas interpres* (Linnæus).

105. *Ciconia* ——? Atra, viridescens; dorso imo cum ventre toto albis.—(P. L. Sclater.)

106. *Phœnicopterus minor*, Geoffroy.

107. *Nyroca brunnea*, Eyton, Monog. Anat. pl. 23.

108. *Podiceps minor* (Linnæus).

109. *Xema phæocephala*, Swainson. Swainson calls this species *pæocephala* (misspelled by him *poiocephala*), a name which would imply green-head, not grey-headed. He no doubt meant to have written *phæocephala*, and I have altered the name accordingly.

Differs from *X. brunneicapillum* of India in having the white at the base of the remiges confined to the *outer* web.

110. *Sternula balænarum*, Strickland. Front, round the eyes and crown black; back, wings and tail very pale grey; primaries hoary fuscous, shafts and inner margins white; below wholly white; beak black; legs brownish flesh-colour.

Total length 8 inches 5 lines; beak to front 1 inch 2 lines, to gape 1 inch 4 lines; wing 6 inches 6 lines; medial rectrices 1 inch 8 lines, external 2 inches 4 lines; tarsus 6 lines.

Remarkable for the shortness of its tarsus.

8. ON THE STRUCTURAL RELATIONS OF ORGANIZED BEINGS.

[Read before the Ashmolean Society of Oxford, March 10, 1845.]

I PROPOSE to make a few observations on the relations which subsist between different organized beings in respect of the *similarities* of their physical structures. This limitation will exclude—first, the relations between individuals, such as that of parent to offspring, for in individuals of the same species the essential points of structure are not *similar*, but *identical*; and secondly, the relations between an organized being and the external circumstances of soil, climate, or food, to which it is adapted, in other words, between structure and function; for these adaptations of the one to the other, however interesting and admirable in themselves, are not relations of *similarity*.

On comparing together the innumerable species of organized beings, we find their structures to present every possible degree of variation, from an almost perfect identity to the utmost amount of difference which the mind can conceive any two organized bodies to possess. These agreements and differences are not, however, devoid of laws and principles; they admit of being classed under certain general heads, and we thus discover the traces of Divine workmanship not merely in the structure of an individual organism, but in the mutual relations of those organisms, the due combinations of which constitute the natural systems of botany and zoology.

When the human mind first began to observe and to compare the structures of organic life, to generalize the points of agreement, and thus to lay the foundation of the science of natural history, no inherent principles of classification were even suspected to exist, characters were compared and generalized at random, and the arrangements which resulted were of the rudest and most unphilosophical kind. The most superficial and arbitrary characters were selected as the basis of classification, and no man was able to give a reason why one mode of arrangement should not be as correct and as true to nature as another. Thus we find the older naturalists classing Lizards, Tortoises and Frogs with terrestrial Mammalia, under the name of "four-footed beasts," while Serpents were made into a distinct class: and Whales, whose physiological organization is as highly developed as in any other mammal, were dismissed among the cold-blooded class of Fish, into which the humble Lobster and the Oyster entered from the other side to keep them company. By some authors we find the Echinus and the Hedgehog approximated, because both are covered with spines; the Ammonite and the Rock-crystal were described in the same chapter "de lapidibus;" Shrew-mice and Spiders were classed together, because both were supposed to be venomous: Bats were referred to Birds, Corals to Plants, and so on.

In the course of the seventeenth century, the few who cultivated natural science began to be conscious that these crude arrangements were not satisfactory, or consistent with the realities of nature; and in the works of Ray and of Lister, we perceive many instances of an instinctive preference for essential instead of arbitrary characters. But it was Linnæus who first pointed out in express terms the great principle of the *subordination of characters.* This principle teaches us to give to each point of structure its due weight, and to attach more value to those peculiarities whose immediate influence on the mysteries of life often renders them the most difficult for our senses to appreciate, than to those external characters which, though most conspicuous to the eye, are but remotely connected with the real essence of the creature. This principle has been further developed by later naturalists, especially by Cuvier, and accordingly we now find that in the modern systems of zoology the *primary* divisions of the Animal Kingdom are based on characters derived chiefly from the *nervous* system, as being the most important feature in organization, the *secondary* subdivisions are grounded on the *organs of respiration,* groups of a lower rank on the *digestive* system, and so on, the most superficial peculiarities, such as external form and colour, being reserved to characterize the ultimate groups of genera and species. These improved principles of classification are gradually bringing the systems of zoology and botany into a state of permanence, consistent with nature, and satisfactory to that truth-seeking instinct which is inherent in the human mind.

A further advance of philosophical classification has shown that the characters of organized beings require not only to be subordinated according to their importance, but subdivided according to their kinds. There are many instances of correspondence of structural characters in organic beings which can never by any process of subordination become elements in a natural classification, and it is important to distinguish those which *can* from those which *cannot* be so employed. Zoologists had long been aware that certain sets of characters produced an arbitrary or artificial method if employed for classification, while others seemed to lead to a natural system, but the question was involved in obscurity till the time of MacLeay, who was the first to give us clear definitions on the distinction between Affinity and Analogy. He applied his views indeed in support of a theory, the *quinary system,* which few naturalists are now disposed to support, and with which we are not now concerned; but his elucidation of affinities and analogies is not the less valuable on that account. Although I am not disposed to take the same view of these principles as that of Mr. MacLeay, yet as the principles themselves are at the foundation of all sound classification, whether in zoology or botany, I may be allowed to make a few further remarks upon this subject.

It appears to me that the instances of resemblance or agreement of structure between any two species of organized beings should be reduced, not into *two*, but into *three* distinct classes, *Affinity*, *Analogy*, and a third, for which I propose to adopt the name of *Iconism**.

I. The highest class of these structural agreements is that of *Affinities*, which appear to be the direct result of those laws of organic life which the Creator has enacted for his own guidance in the act of creation. Affinity consists in an *essential* and physiological agreement in the corresponding parts of organic beings, resulting from a uniformity of plan which pervades the system of nature†. These essential agreements of parts consist rather in a similarity of organic *composition* and of *relative situation*, than of *form*. A microscopic examination of the primary tissue, or a chemical analysis of its substance, will often demonstrate the true affinities of a structure when its external form would only mislead us. And when we have proved an affinity to subsist between the structures of two organic beings, we then apply the term to the beings themselves, and say that an affinity subsists between them, greater or less, according to the number and importance of the organs in which such affinity is shown. Take for example the long, straight weapon of offence in the Narwhal: its general appearance is that of a *horn*, and such the vulgar accordingly call it; but if we examine its organization and its chemical composition, we find that both are utterly unlike those of real horns, but correspond to the structure of teeth. Further, if we examine the mode of its connexion with the skull, we find that it is inserted into a socket like other teeth, instead of being attached in the manner of horns, and we accordingly pronounce it to be not a horn, but a tooth, developed for purposes of offence to an extraordinary extent. And having thus shown that the weapon of the Narwhal has no affinity to real horns, we no longer appeal

* This term, suggested by the Rev. Dr. Ingram, President of Trinity College, appears preferable to *Mimesis*, which I had originally proposed to use.

† We may suppose, for instance, that it was a law of organic creation that all Birds should have the anterior extremities modified into the form of wings; and in obedience with this law we find that there is no bird which is absolutely without wings, though there are several kinds in which the wings are perfectly incapable of flight. Again, it is a law that Mammalia have neither more nor less than seven cervical vertebræ, and we find this law to hold good, without an exception, through the whole class of mammals, from the slender-necked Giraffe to the Whale, which can hardly be said to have any neck at all. The above, out of countless other examples, will show what is meant by *laws of organization*.

to this structure in proof of any affinity between the Narwhal and the truly horned animals. Again, the Narwhal in its external form much resembles a fish; but when we look to its nervous, circulatory, and reproductive organizations, which rank much higher in the scale of characters than external form, we find that it is no fish, but a true mammal, agreeing in every essential point with the warm-blooded quadrupeds of the land, to which its affinities are real and direct. Similar instances of the discordance between outward form and real affinity might be multiplied to a great extent, and it forms a constant employment for the scientific zoologist to distinguish real affinities from apparent ones, and thus to refer every organized being to its true position in the natural system.

It will thus be seen that every instance of asserted affinity between two organic beings is merely a corollary deduced from an observed affinity between the corresponding organs in each; and though it is not usual to apply the term *affinity* to the similarities between parts, yet as the similarity between the wholes results from the similarities of their parts, the word *affinity* may be as correctly applied to the one as to the other. In works of comparative anatomy it is customary to speak of those members which are essentially equivalent in two organic beings as *analogous* organs, but we shall soon see that the word *analogy* has a very different sense; and as the relation between equivalent organs is one of real *affinity*, and forms the sole ground on which we assert the affinity of the whole beings, we may introduce the adjective *affine* or *homologous* in place of *analogous*, when referring to structures which essentially correspond in different organic beings.

When we say that affinity consists in an essential agreement of structure resulting from a fixity of purpose in the mind of Creative Wisdom, it must not be supposed that all affinities are equally strong, direct, and palpable. Any agreement, however slight, or however concealed by more palpable differences, which forms part of the plan of organic existence, is a true affinity, and the principle of subordination of characters before referred to is merely the arranging of these affinities in the true order of their proximities. The proximity of affinities is in the inverse ratio of their essential importance, the most important agreements of characters being those which have the widest extent, and which therefore form affinities between the remotest points in the system of organized beings. We will illustrate this by an example showing the successive series of affinities which the same species bears to others, commencing with the most remote, and proceeding to the closest affinity which can subsist between two distinct species. We will take as an example the species Raven (*Corvus corax*) :—

	A Raven has an Affinity to an	it is the same Affinity which exists between	and is derived from the Affinity between their respective	supplying the diagnostic characters of the
1.	Oak-tree;	all Animals and all Plants,	organic life, &c.	*Organic* EMPIRE.
2.	Locust;	Vertebrata and Insects,	nervous systems, &c.	*Animal* KINGDOM.
3.	Salmon;	Birds and Fish,	vertebral columns, &c.	PROVINCE, *Vertebrata.*
4.	Swan;	Insessores and Natatores,	circulatory systems, &c.	CLASS, *Birds.*
5.	Humming Bird;	Conirostres and Tenuirostres,	structure of feet, &c.	ORDER, *Insessores.*
6.	Sparrow;	Corvidæ and Fringillidæ,	conical beaks, &c.	TRIBE, *Conirostres.*
7.	Jay;	Corvinæ and Garrulinæ,	structure of nostrils, &c.	FAMILY, *Corvidæ.*
8.	Magpie;	Corvus and Pica,	short elevated beaks, &c.	SUBFAMILY, *Corvinæ.*
9.	Carrion Crow;	one species of Corvus and another,	even tails, black plumage, &c.	GENUS, *Corvus.*

The affinities in this series are seen to accumulate successively as we proceed from the remotest organism to the approximate species. The Raven and Carrion Crow not only possess that superficial resemblance of form which constitutes their generic character, but they have in addition all the other points of affinity which extend from them to a greater or less distance into the realms of organic existence. Thus we find that

The Raven has

organization	in common with all	Organized Beings.
a nervous system	...	Animals.
a vertebral column	...	Vertebrata.
a peculiar circulatory system	...	Birds.
perching feet	...	Insessores.
a conical beak	...	Conirostres.
the nostrils covered by feathers	...	Corvidæ.
ridge of the beak arched	...	Corvinæ.
an even tail, and	...	Corvus.
a wholly black plumage	...	Carrion Crow.

It will be seen from the above example, that the whole process of classification consists in observing the affinities of structure in different beings, in estimating their importance, and in arranging them according to that estimate. It follows that a clear comprehension of *affinities*, as distinguished from the other kinds of resemblance, is essential to the objects of the scientific zoologist.

Although affinity consists in an essential and intimate agreement in the structure of certain organs, yet it by no means implies an identity of function in those organs. The modifications of external form are so various that we frequently find the same organ applied by different animals to purposes the most remote from its normal function; and on the other hand we see very different organs applied to discharge the same function. Thus, as a general proposition, it is certain that the proper function of wings is flying, of legs walking, of fins swimming; and yet we find examples where each of these organs is applied to any other function but its own, as in the case of the Bat, Seal, Ostrich, Penguin, Gurnard and Flying Fish. Hence, although it is generally true that certain organs are destined to perform certain definite functions, yet the exceptions are so frequent as to make us attach a minor degree of importance to *function*, while we give the fullest weight to those essential properties which form the only test of real affinity.

II. We have next to consider that class of structural agreements known by the name of *Analogies*. These consist in a similarity of external form and of function connected with it, but without that agreement of essence which constitutes *Affinity*. These analogous agreements are equally the result of natural laws, but of laws of a different class from the former. Agreements of affinity are produced in conformity with the laws of the organic creation, while analogies have a reference to the laws and properties of external and often inorganic matter. In obedience to these laws, it follows that whenever an instrument is required to produce a given effect upon external objects, or to resist their influences in a given manner, there is in general one method, and one only, of effecting the object in the best and most effectual way. Accordingly, whatever be the organ or instrument employed, that organ must have a certain and definite mechanical structure bestowed upon it to obtain the desired end. As a general rule, the same end is attained in different organic beings by means of the same set of organs; but when those organs are required for any other purpose, or are so modified as to be unfit for that special end, then some other set of organs are endowed with the requisite external structure, and are called upon to act as substitutes for the legitimate instruments. Examples of this adaptation of organs to purposes remote from their normal destination are numerous and well known, and I cannot do better than refer to the late Mr. John Duncan's work 'On the Analogies of Organized Beings,' where there are numerous examples of such analogies arranged in a tabular and highly perspicuous form. We need only take the Elephant as an instance. We may suppose that this animal required horns for the purpose of defence, but it belongs to an order, the Pachydermata, in which horns are uniformly absent, and the laws of affinity forbade their introduction. To supply this defect, the incisor teeth are removed from their usual duties of mastication, and are so developed as to assume the form and discharge the function of horns. Further, the great size and weight of these lengthened tusks required a great strength and shortness of neck, and the animal was consequently unable to reach the ground with his mouth. A hand was therefore required to convey the food to the mouth, but the vast weight of the animal required a massive structure in the feet, which forbade them to be adapted to the purpose of hands. To supply this want then the *nose* is lengthened out, furnished with muscles, divided at the end into a finger and thumb, and in this proboscis behold a hand! almost equal in delicacy of manipulation to the hand of Man. And thus we see the Elephant, endowed in one respect with an analogy to the Ox, and in another respect to Man, yet having no immediate affinity with either.

As then Analogy consists in an agreement of function, and only of form so far as it tends to discharge that function, it follows that real and genuine *Analogies* may take place between the works of Nature and the works of Man, while no such relation of *Affinity* can possibly exist. When, for instance, the inventive powers of Man are called upon to imitate any of the operations of Nature, the external matter to be acted on being in both cases the same, a similar arrangement of form is adopted by both. If the problem be to make a floating body adapted for rapid motion through water, Man either by practical experiment or mathematical calculation produces the form of a boat, and thus unconsciously imitates the structure of the Whale and Seal among Mammals, the Penguin among Birds, the Ichthyosaurus and Turtle among Reptiles, the Fish among Vertebrata, the Dytiscus among Coleoptera, the Notonecta among Hemiptera, Sepia among Mollusca, Physalia among Acalephæ, &c. Nor is the analogy between a ship and a fish confined to the external form only: the keel of the one represents the spine of the other, the "ribs" of both agree in name as in nature, the rudder coincides with the tail, the oars with the fins, the masts with the spinous processes, the running rigging with the tendons, the seamen with the muscles, the *look-out man* on the forecastle with the eye, and the captain in the cabin with the mental faculties in the fish's brain. Again, what can be more striking than the analogy between a locomotive steam-engine and a living animal? We see in both an analogous respiratory and digestive system, the same necessity for food and drink and oxygen to sustain that internal combustion which is the source of the vital action, the same obedience of the organs of motion to the impulse of the governing mind, and the same wear and tear of the system, terminating in old age and sudden or gradual death. Yet in all these cases there is no set purpose on the part of Man to imitate the works of Nature, he merely applies the faculties which God has given him to elicit the properties which the same God has given to matter; and by this process alone he often arrives at the same or similar results to those at which Creative Wisdom had arrived before him. It appears to me, therefore, that relations of Analogy, that is to say, agreements in structure in consequence solely of an agreement in the function to be performed, may be as truly and as correctly asserted to exist between artificial and natural productions, as between one object of the latter class and another. It is clear from this how much lower *Analogies* ought to stand in our estimation than *Affinities*. The latter form an essential part of that magnificent plan of Creation, which, notwithstanding the amount of attention which Man has given to it, is of so transcendental a nature, that it may almost be said to be yet "to us invisible or dimly seen." Analogies, on the contrary, appear not to form any element whatever in the great System of Nature, but are merely examples of the recurrence of certain mechanical forms whenever the production of a certain mechanical action called for them; and so far from their being at or beyond the verge of human comprehension, we have seen that Man enjoys the high privilege of copying by these analogies, at a humble distance, the far transcendent works of his Maker.

It would be an improvement in the language of comparative anatomy, if the term *analogous organs* were limited to the sense above defined. The serrations in the beak of a Duck, for instance, are *analogous* in form and in function to *teeth*, but in their essential nature they are only a corneous modification of the *lips*. Most anatomists, however, would habitually say that the beak of a bird is *analogous* to the lips of a mammal, though it must be evident how much more precise their language would become if they spoke of this *essential* relation as an *affinity*, and applied the word *analogous* to *formal* or *functional* relations only. A similar inaccuracy is committed by geologists in speaking of the *recent analogue* of a fossil species, meaning thereby that living species which has the nearest affinity to the extinct one. It would be more correct if they would term it the *recent affine* or the *recent homologue*.

III. There is yet a third species of relation of structural similarity between organized beings which has usually been confounded with Analogy, but which appears to me to be distinct from it in kind, as well as far inferior to it in importance,—I refer to those cases where a resemblance in form or configuration exists, but without any perceptible identity either of essence or of function. Such, for example, are the resemblances between the flower of the Bee Orchis and a Bee, between the shell of *Murex haustellum* and a Woodcock's head, between a Fungia and a Fungus, Ovulum and an egg, Haliotis and an ear, &c. To this class also belong the numerous instances of *similarity of colour* between birds whose affinities are remote, such as the resemblance of Oriolus to Xanthornus, of Dicrurus to Corvus, of Cissopis to Pica, of *Agelaius phœniceus* to *Campephaga phœnicea*. Many errors of classification have been caused by mistaking these similarities for true affinities.

Not only are such cases of external resemblance unconnected with any agreement in the essential structures of the bodies compared, but there is no conceivable similarity in the functions which they are created to discharge. I think therefore that it is not going too far, nor departing from that veneration which the true naturalist will always feel for Nature's God, to call such superficial coincidences of form *accidental.* They seem to arise from the exuberant variety of the works of Nature which causes an occasional recurrence of similar forms, without any express design for such coincidences. Nothing can be inferred from such resemblances, either as to essential affinity or functional design; and they would almost have been beneath our notice, were it not that some authors have regarded them as examples of real analogies. The advocates of the Quinary theory of classification, who regard *Analogies* to be as important an element in the natural system as *Affinities,* often speak of these mere resemblances in the light of true Analogies, and appeal to them in confirmation of their views. Regarding, however, as I do, those views to be erroneous, I think it important that the distinction between *functional Analogy* and *mere resemblance* should be clearly pointed out; and to render the distinction more marked, I would distinguish the latter by the new term *Iconism.*

We must beware indeed of too hastily pronouncing an instance of resemblance to be an *Iconism,* merely because we cannot immediately detect any functional analogy. There may be real reasons for these resemblances, real agreements in the functions to be discharged, which we have not yet detected, and perhaps may never discover. A person might say, for instance, that the species of Mantis called the "walking-leaf" presents a mere Iconism or accidental resemblance to true leaves; whereas it is highly probable that this very resemblance is given to the animal to enable it to remain concealed from its foes amid the verdant foliage. Such at least is undoubtedly the intention of numerous instances in which animals present an analogous colour to the surrounding surface, and even undergo corresponding changes with it, such as that of the Ptarmigan, which during summer is of a speckled grey plumage, like the lichen-covered rocks which it frequents, while in winter it becomes a pure white when those rocks are covered with snow.

I have now endeavoured to show that the relations of resemblance in organized beings are of three kinds, diminishing successively in importance; that *Affinities* are expressions of the real and elementary and esoteric Plan of Creation which the Author of Nature has been pleased to follow; that *Analogies* are coincidences of structure consequent solely upon an identity of external physical conditions; and that *Iconisms* are merely accidental recurrences of similar forms resulting from the exuberance of Nature's riches. It is evident that these distinctions must be clearly understood before we can make any progress in Natural History as a Science, and the remarks above offered may perhaps aid in drawing attention to the subject, or removing the difficulties which surround it.

9. LIST OF BIRDS PROCURED IN KORDOFAN BY MR. J. PETHERICK. WITH NOTES BY H. E. STRICKLAND, M.A., F.G.S.

(Read November 26th, 1850, and published in the Proceedings of the Zoological Society for the same year, p. 214, with plates of *Buteo rufipennis*, *Mirafra cordofanica*, and *Alauda erythropygia*.)

[SPECIES not enumerated in Rüppell's 'Systematische Uebersicht der Vögel Nord-Ost Afrika's,' 8vo, Frankfurt a. M. 1845, are marked N.

Species common to the West Coast of Africa are marked W. These are chiefly determined by reference to Dr. Hartlaub's valuable list of West African birds in the 'Verzeichniss der öffentlichen u. Privat-Vorlesungen am Hamburgischen Gymnasium,' 4to, Hamburg, 1850.]

1. *Neophron percnopterus* (Linnæus).
2. *Vultur occipitalis*, Burchell.
3. *Otogyps auricularis* (Daudin).
4. BUTEO RUFIPENNIS, Strickland, n. s. Upper parts cinereo-fuscous, nearly black on the crown; feathers of back and wing-covers with black shafts; cheeks cinereous, a black line below them from angle of mouth; chin whitish with a medial dark streak; breast and sides ferruginous brown, with a conspicuous medial black streak one-sixteenth of an inch wide on each feather; belly, thighs and vent plain fulvous; primaries and secondaries bright ferruginous, tipped for about an inch and a half with black, and from three to five distant transverse black bands on the inner web; tail cinereo-fuscous, with five dark fuscous bands, each about a quarter of an inch wide, the distal one about half an inch, beyond which the extremity is cinereo-fuscous and the extreme tip white; cere and legs yellowish; beak and claws black.—Mus. H. E. S.

Length 17 inches; wing $12\frac{1}{4}$; medial rectrices $7\frac{1}{2}$; external ditto $7\frac{1}{8}$; tarsus $2\frac{1}{4}$.

Habitat Kordofan.

5. *Aquila nævia* (Gmelin).
6. *Aquila pennata* (Gmelin).
7. *Circaëtus gallicus* (Gmelin). W.
8. *Helotarsus ecaudatus* (Daudin).
9. *Falco biarmicus*, Temminck. (*F. peregrinoides*, Temm.; *F. chiqueroides*, Smith; *F. feldeggi*, Schleg.; *F. lanarius*, Schleg.; *F. rubeus*, Thien.; *F. cervicalis*, Kaup.)

After a careful examination of many specimens, I feel justified in uniting the above synonyms under one species. This is essentially an African bird, extending from the Cape of Good Hope to Egypt, whence it has probably spread into Greece and Dalmatia, to which portions of

Europe it is chiefly confined, though a single straggler has occurred in Germany. It is at once distinguished from *F. peregrinus* by the shorter toes, and the fulvous patch on the crown. The *Falco jugger*, Gray (*F. luggur*, Jerdon), of India is closely allied, but seems to differ constantly in the plumes of the tibia being uniformly dark brown, while in *F. biarmicus* they are cream-coloured or white, like the rest of the under parts, with a small brown spot on the centre of each feather. This is one of the many species to which the name *Falco lanarius* has been given, under the supposition that it may be the *Lanner* of the old works on falconry; but as the original *F. lanarius* of Linnæus is now admitted to be the young of *F. gyrfalco*, and as systematists are generally agreed not to trace binomial titles further back than Linnæus's Systema, of course the specific name *lanarius* must be dropped altogether, and the oldest binominal name, *Falco biarmicus*, Temm., adopted for the present species.

10. *Tinnunculus alaudarius* (Gmelin). W. This widely diffused species extends, without variation of form or colour, from Britain southwards to Central Africa and eastwards to India.

11. *Nauclerus riocouri*, Vieillot. N. W.

12. *Accipiter sphenurus*, juv.? Resembles *A. sphenurus*, Rüpp., in the cuneate form of the tail. Head and neck rufescent, with a fuscous medial stripe on each feather; belly white, barred with brown; back cinereous brown with rufous margins; *upper tail-covers white*; tail cinereous, with three broad fuscous bars, outer feather white, with five bars.

13. *Accipiter carbonarius* (Lichtenstein). N. Two specimens agree with Lichtenstein's description (in his 'Verzeichniss einer Sammlung von Saügethieren u. Vögeln aus dem Kafferlande,' 8vo, Berlin, 1842, p. 11), except in having only *three* or *four* white bands on the tail instead of *five*. With the exception of these bands, and the numerous light and dark brown bands on the remiges, the plumage is wholly black; cere and legs yellow.

Total length 12 inches; wing 7; tarsus $1\frac{6}{10}$.

14. *Melierax gabar* (Daudin). W. (*Accipiter erythrorhynchus*, Swains.)

15. *Melierax polyzonus*, Rüppell. United by Mr. Gray to *M. canorus*, Rislach (*M. musicus*, Daud.), but differs in its smaller size, and in having the upper tail-covers banded grey and white, while in *M. canorus* they are pure white. The wing in *M. polyzonus* measures 12 inches, in *M. canorus* 15 inches.

16. *Polyboroides radiatus* (Scopoli). W. (*Falco gymnogenys*, Temm.)

17. *Circus pallidus*, Sykes. N.

18. *Scops leucotis* (Temminck). W.

19. *Scotornis climacurus* (Vieillot). W.

20. *Caprimulgus infuscatus*, Cretzschmar, female. Agrees with Rüppell's plate, but wants the white wing- and tail-spots of the male bird.

21. *Eurystomus afer* (Latham). W. (*E. orientalis*, Rüpp.; *E. rubescens*, Vieill.; *Collaris purpurascens*, Wagl.)

22. *Coracias abyssinica*, Gmelin. W. (*Coracias caudata*, Wagl.)

23. *Coracias nævia*, Daudin. W. (*C. levaillanti*, Rüpp.; *C. nuchalis*, Swains.)

24. *Ceryle rudis* (Linnæus). W. (*Ispida bicincta*, Swains.; *I. bitorquata*, Swains.) Identical with specimens from Smyrna and S. Europe. The individuals with two pectoral bands (*I. bicincta*, Swains.) are the males.

25. *Merops albicollis*, Vieillot. N.W. (*M. cuvieri*, Licht.; *M. savignyi*, Swains.)

26. *Merops nubicus*, Gmelin. W. (*M. superbus*, Shaw; *M. cæruleocephalus*, Lath.)

27. *Merops lamarcki*, Cuvier. W. (*M. viridissimus*, Swains.; *M. ægyptius*, Kittl.; *M. viridis*, Rüpp.) Closely allied to *M. viridis*, Linn., of India, but smaller, with a larger mixture of golden yellow in the plumage, the throat not blue as in *M. viridis*, and the remiges are rufous on *both webs*, with scarcely any tinge of green externally.

28. *Merops erythropterus*, Gmelin. W. (*M. minulus*, Cuv.; *M. collaris*, Vieill.; *M. lafresnayei*, Guérin.)

29. *Irrisor senegalensis* (Vieillot)? The Kordofan specimens agree, in the shortness and nearly straight form of their beak, with the black-beaked species of W. Africa, *I. senegalensis*, Vieill. (*Nectarinia melanorhynchus*, Licht.), but in the red colour of this organ they agree with the Cape species (*I. erythrorhynchus*). It is well known that the females of the latter have the beak much shorter and straighter than the males, yet in these Kordofan specimens the beak, though of the same length, is considerably straighter than in the female birds from the Cape. Like *I. senegalensis* they have a broad white bar crossing the inner webs of the first three, and both webs, shaft included, of the remaining primaries; while in *I. erythrorhynchus* the white bar of the primaries is much narrower, and divided by the black shaft.

30. *Nectarinia metallica*, Ehrenberg.

31. *Nectarinia pulchella* (Linnæus). W.

32. *Phylloscopus trochilus* (Linnæus). Identical with British specimens.

33. *Saxicola deserti*, Temminck.

34. *Saxicola œnanthe* (Linnæus).

35. *Saxicola isabellina*, Cretzschmar. This is probably the *Sylvia leucorrhoa*, Gmel., in which case it extends to Senegal. It resembles *S. œnanthe*, but is paler on the upper part, and has less white on the lateral rectrices, the terminal black portion being $1\frac{1}{10}$th of an inch in length, while in *S. œnanthe* it is only about $\frac{3}{4}$ inch.

36. *Motacilla capensis*, Linnæus.

37. *Budytes melanocephala* (Lichtenstein).

38. *Anthus* (undetermined species).

39. *Melænornis? erythropterus* (Gmelin). W. (*Turdus erythropterus*, Gmel.) This bird approaches nearly to the type of *Melænornis*, Gray (*Melasoma*, Swains.), though the beak is rather more elongated, and the rictal bristles less developed, than in *M. edoliolides*, Swains. Rüppell refers it to Boie's genus Cercotrichas, which is synonymous with Copsychus, Wagl. Dr. Hartlaub places it in Argya, Less., which is synonymous with Chætops, Swains.

40. *Pycnonotus barbatus* (Desfontaines). W. (*Turdus barbatus*, Desfont. in Mém. Ac. Sc. 1787; *Turdus arsinoë*, Licht.; *Ixos obscurus*, Temm.; *I. inornatus*, Fras.; *Hæmatornis lugubris*, Less.)

41. *Oriolus galbula*, Linnæus.

42. *Dicrurus divaricatus*, Lichtenstein. W. (*D. lugubris*, Ehrenb.; *D. canipennis*, Swains.) Nearly allied to the *D. musicus*, Vieill., of S. Africa, but has the tail less deeply forked, the culmen of the beak more acute, and the primaries pale internally.

43. *Lanius algeriensis*, Lesson in Rev. Zool. 1839. This is probably the species termed *L. excubitor* by Rüppell. It differs from the true *excubitor* of N. Europe in the greater extent of white on the primaries, and in the two external pairs of rectrices being wholly white (except the shafts). It closely approaches *L. lahtora* of India, and only differs in wanting the narrow band of black across the front.

44. *Lanius nubicus*, Lichtenstein. (*L. personatus*, Temm.)

45. *Lanius collurio*, Linnæus. A young male specimen appears referable to this species.

46. *Lanius isabellinus*, Ehrenberg, Symb. Phys. fol. *e*. N. This species is pale fulvo-cinereous above, cream-coloured below; rump and tail rufous; a broad blackish band from the nostril to the ear-covers, margined above by a whitish streak. It much resembles *L. arenarius*, Blyth, Journ. As. Soc. Beng. vol. xv. p. 304, but is of a more cinereous tinge above, and is distinguished from that and all the allied Asiatic species by possessing a conspicuous white band at the base of the fourth to the ninth primaries. The specimen from Kordofan has an obscure dark transverse band near the tips of the rectrices.

47. *Telophonus senegalus* (Linnæus). W. (*Lanius erythropterus*, Shaw.)

48. *Corvus scapulatus*, Daudin. W. (*C. leuconotus*, Swains.)

49. *Corvus umbrinus*, Sundevall. Distinguished by the length and curvature of the beak, and by the grey-brown tint of the head and neck.

50. *Juida rufiventris*, Rüppell. W.

51. *Juida chalybea*, Ehrenberg. W. (*Lamprotornis cyanotis*, Swains.)

52. *Ploceus luteolus*, Lichtenstein. W. (*P. personatus*, Vieill., Jard. Contrib. to Ornith. 1849, p. 35. pl. 7.)

53. *Ploceus sanguinirostris* (Linnæus). W.

54. *Pyromelana ignicolor* (Vieillot). W.

55. *Vidua paradisea* (Linnæus). W. The series of immature specimens in the collection have enabled me to detect a curious structure connected with the development of the tail-fathers, which will be treated of in a separate paper. See Sir W. Jardine's 'Contributions to Ornithology,' 1850, p. 88, pl. 59.—Coll. Papers, p. 326.

56. *Vidua principalis* (Linnæus). W. The specimen from Kordofan, like those from Senegal, has a black spot on the chin, but it is not yet proved whether the presence of this spot amounts to a specific distinction.

57. *Pytelia elegans* (Gmelin). W.

58. *Amadina fasciata* (Gmelin). W. (*Fringilla detruncata,* Licht.)

59. *Amadina cantans* (Gmelin). W. A perfectly typical Amadina, though M. Rüppell makes it an Estrilda.

60. *Philetærus nitens* (Gmelin). W. (*Amadina nitens,* Swains.) From the peculiar form of the beak I am disposed to refer this species, as well as *Estrilda squamifrons,* Smith, *E. musica,* Gray, and *Loxia frontalis,* Daud., to the genus Philetærus.

61. *Crithagra lutea* (Lichtenstein), Temm. Pl. Col. 365.

62. *Passer simplex,* Lichtenstein. W. (*Pyrgita swainsoni,* Rüpp.)

63. *Emberiza striolata,* Rüppell.

64. *Galerida cristata* (Linnæus) ? This is probably the bird so designated by Rüppell, who states it to be abundant in the whole of North Africa. It precisely agrees with European specimens in form, but is of a much paler colour, which however may be easily explained by the bleaching effect of the sun's rays in the scorching deserts which this bird frequents.

65. MIRAFRA CORDOFANICA, Strickland, n. s. N. Above ferruginous, the feathers of the crown and back with an indistinct medial dusky streak, and margined on their inner side with rusty white; tertials broadly margined with whitish, that colour being separated from the ferruginous of the medial portion by a narrow dusky line; secondaries ferruginous, margined externally with whitish; primaries ferruginous at the base, their distal half being pale rufo-fuscous; medial pair of rectrices ferruginous, the next pair pale rufo-fuscous, the two following pairs deep fuscous, with a very narrow rufescent margin, the penultimate pair deep fuscous internally; the external web, and part of the inner at the tip, white; external pair white, the inner web fuscous towards the base; cheeks pale rufo-fuscous, chin and throat white, breast and lower parts pale cream-colour, the former with a few pale rufo-fuscous subtriangular spots; lower wing-covers and sides rufescent; beak, feet and claws pale yellowish.

Total length $5\frac{1}{4}$ inches; beak to front $\frac{1}{2}$, to gape $\frac{6}{10}$ths; wing $\frac{3}{10}$ths;

medial and external rectrices $2\frac{7}{10}$ths; tarsus $\frac{9}{10}$ths; middle toe and claw $\frac{7}{10}$ths; hind toe $\frac{3}{10}$ths; hind claw $\frac{2}{10}$ths.

This, which seems to be a typical *Mirafra*, is remarkable for the predominance of a pure ferruginous tint on its upper parts. The hind claw is remarkably short, though not more so than in some of the Indian species of Mirafra. The single specimen that occurred of this bird is now in the British Museum.

66. ALAUDA ERYTHROPYGIA, Strickland, n. s. Upper parts deep fuscous brown, the feathers narrowly margined with rufo-fulvous; upper tail-covers ferruginous; remiges deep fuscous, almost black on both webs; secondaries narrowly tipped with pale fulvous; tail fuscous black, the middle rectrices narrowly margined with ferruginous, the bases of all ferruginous, extending obliquely nearly to the tips of the outer pair. Lower parts pale fulvous, the chin, throat and breast with a broad medial fuscous streak on each feather; lower wing-covers black, margins of wing fulvous; beak fuscous; legs flesh-colour; hind claw short and slightly curved.

Length $7\frac{1}{2}$ inches; beak to front $\frac{6}{10}$ths, to gape $\frac{11}{10}$ths; wing $4\frac{1}{4}$; medial and external rectrices 3; tarsus 1; hind claw $\frac{3}{10}$ths.

Habitat Kordofan. Mus. H. E. S.

67. *Colius macrurus*, Linnæus. W. (*C. senegalensis*, Gmel.)

68. *Tokus erythrorhynchus* (Kuhl). W.

69. *Palæornis torquatus*, Vigors. W. (*P. cubicularis*, Wagl.) This species, which extends across Africa from Abyssinia to Senegal, is identical with specimens from India.

70. *Pogonius vieilloti*, Leach. W. (*P. senegalensis*, Licht.; *P. rubescens*, Temm.) This generic name was originally written Pogonia by Leach (Zool. Misc. vol. ii. p. 45), in which form it had been preoccupied by a genus of plants. Illiger's name, Pogonias, had also been preoccupied by a fish-genus; but Leach afterwards corrected it to Pogonius, which form had never been used before, and I therefore retain it instead of Mr. G. R. Gray's name Læmodon (erroneously written Laimodon).

71. *Trachyphonus margaritatus*, Rüppell. (*Tamatia erythropygia*, Ehrenb.)

72. *Yunx torquilla*, Linnæus. Identical with specimens from Britain and from India.

73. *Oxylophus serratus* (Sparmann). N. This Cape bird has never before, I believe, been obtained to the north of the equator. The nearly allied *O. jacobinus* (Bodd.) of India (*Cuculus melanoleucus*, Gmel.; *C. passerinus*, Vahl) has the lower parts constantly white. Ehrenberg, in his 'Symbolæ Physicæ,' fol. *r*, describes a Nubian species under the name of *Cuculus pica*, which from the description seems to be identical with the white-bellied *O. jacobinus* of India. Rüppell erroneously refers this

C. pica of Ehrenberg to the *Oxylophus afer*, Leach (Levaill. Ois. Afr. pl. 209), of S. Africa, which differs in having dark streaks on the throat, and which appears from Rüppell's observations to be also an Abyssinian bird.

74. *Oxylophus glandarius* (Linnæus). W.

75. *Columba guinea*, Linnæus. W. (*C. trigonigera*, Wagl.)

76. *Numida ptilorhyncha*, Lichtenstein.

77. *Francolinus clappertoni*, Vigors. Mr. G. R. Gray has separated the *F. clappertoni* of Rüppell as a distinct species, under the name of *F. rüppelli*; but the specimens from Kordofan seem to agree equally well with Rüppell's plate of *F. rüppelli*, and with Gray's plate of what he regards as the true *clappertoni*, between which I can see no difference.

78. *Coturnix dactylisonans*.

79. *Pterocles quadricinctus*, Temminck. N. W. (*P. tricinctus*, Swains.) This African species has long been confounded with the closely allied *P. fasciatus* (Scop.), (*Perdix indica*, Lath.), of India, figured by Mr. Jerdon, in his 'Illustrations of Indian Ornithology,' pls. 10 and 36. Specimens sent by Mr. Jerdon have now enabled me to prove their distinction. The general arrangement of colour is almost identical in these two species, the chief distinction being in the feathers of the back, scapulars, tertials and greater wing-covers, which in *P. fasciatus* are marked transversely with bars of a dull iron-grey (or "inky hue," as Mr. Jerdon well describes it), while in *P. tricinctus* these bands are of a deep glossy black. In *P. fasciatus* the wing-covers next the body have two or three of these dark bands alternating with white ones of equal breadth, the subterminal one being dark, and the tip of the feather ochreous yellow. In *P. quadricinctus* the wing-covers have only one black band (or a very faint trace of a second), narrowly margined on *both* sides with a fine white line, the terminal and basal parts of the feather being ochreous. Temminck's original description of *P. quadricinctus* is evidently taken from the African bird, but he erroneously gives India as its habitat, in consequence of having confounded it with *P. fasciata*. Vieillot has increased the confusion by *figuring* the *quadricinctus* in his 'Galerie des Oiseaux,' pl. 220, under the specific name of *bicinctus*, while his *description* refers to the true *P. bicinctus*, Temm., a South African bird.

80. *Otis rhaad*, Gmelin. W.

81. *Eupodotis denhami* (Vigors). N. W.

82. *Ortyxelos meiffreni*, Vieillot. W.

83. *Œdicnemus crepitans*, Linnæus. W. This seems to me undistinguishable from *Œ. senegalensis* (Swains. Birds W. Afr. vol. ii. p. 228), the description of which agrees with the European bird.

84. *Œdicnemus affinis*, Rüppell. So exactly does this agree in size and form with *Œ. crepitans*, that I should have suspected it to be an

immature bird, did not M. Rüppell appear so convinced of its distinctness.

85. *Pluvianus ægyptius* (Linnæus).

86. *Glareola limbata,* Rüppell. Closely resembles *G. orientalis* of India, but has the external rectrices about an inch longer.

87. *Squatarola helvetica* (Linnæus). N. W.

88. *Rhinoptilus chalcopterus* (Temminck). N. W. (*Cursorius chalcopterus,* Temm.) This, with the nearly allied *M. bitorquatus,* Blyth, of India, form a very distinct group, connecting Cursorius with Charadrius. Mr. Blyth first formed it into a genus, under the name of Macrotarsus (Journ. As. Soc. Beng. vol. xvii. pt. 1. p.254); but as the name has been previously used by Lacépède for genera of mammals and of birds, and by Schönherr for a coleopterous insect, I propose the name Rhinoptilus, indicating the advanced position of the frontal feathers, which, with other characters, distinguish it from Charadrius.

89. *Chætusia gregaria* (Pallas). N.

90. *Lobivanellus albicapillus* (Vieillot). W. (*Vanellus strigilatus,* Swains.)

91. *Hoplopterus persicus* (Bonnaterre). W. (*H. spinosus,* auct. recentiorum.)

92. *Sarciophorus pileatus* (Gmelin). W.

93. *Charadrius hiaticula,* Linnæus.

94. *Charadrius alexandrinus,* Linnæus. (*C. cantianus,* Lath.)

95. *Charadrius pecuarius,* Lichtenstein.

96. *Ardeola coromanda* (Boddaert). W. (*Ardea coromandelensis,* Kuhl; *A. coromandelica,* Licht.; *A. affinis,* Horsf.; *A. russata,* Temm.; *A. bicolor,* Vieill.; *A. ruficapilla,* Vieill.; *A. bubulcus,* Audouin; *A. caboga,* Franklin; *A. verrani,* Roux; *A. lucida,* Raffl.; *Lepterodas ibis,* Ehrenb.) I could have wished that M. Rüppell had given us the diagnoses of *A. bubulcus* and *coromandelica* when he pronounced them distinct. As far as my own comparisons extend, the African and Indian birds are specifically the same.

97. *Botaurus stellaris* (Linnæus).

98. *Grus cinereus* (Linnæus).

99. *Ciconia alba* (Linnæus). W.

100. *Ibis æthiopica,* Latham.

101. *Glottis canescens* (Gmelin). W. (*G. chloropus,* Nilss.)

102. *Totanus hypoleucus* (Linnæus). W.

103. *Pelidna minuta,* Leisler. W.

104. *Pelidna subarquata* (Gmelin). W.

105. *Machetes pugnax* (Linnæus).

106. *Crex pratensis,* Bechstein.

107. *Sarkidiornis africana,* Eyton. W.

108. *Chenalopex ægyptiacus* (Linnæus).

109. *Dendrocygna viduata* (Linnæus). W. We have the authority of Jacquin, Azara, and other authors, for the occurrence of this bird in South America. If this be the case, it will form the *only known instance* of a non-marine bird being indigenous to both the African and South American continents, without occurring in Europe, Asia, or North America. Before, however, admitting this remarkable exception to the laws of geographical distribution, the absolute specific identity of the African and American specimens should be established by careful comparison, which, as far as I am aware, has not yet been done.

110. *Sterna anglica*, Montague.

111. *Hydrochelidon nigra* (Linnæus).

112. *Pelecanus rufescens*, Latham. W.

NOMENCLATURE AND CLASSIFICATION.

1. ON THE ARBITRARY ALTERATION OF ESTABLISHED TERMS IN NATURAL HISTORY.

(From Loudon's Magazine of Natural History, vol. viii. p. 36.)

[In following out the study of zoology, Mr. Strickland found great inconvenience from the want of a uniform and general system of nomenclature. Ornithology having been most attended to, that branch formed the chief basis of his criticisms and proposed alterations, and in 1835 his first ideas of reform were printed in Loudon's 'Magazine of Natural History,' with which paper this series is commenced. Subsequently, various shorter notices and replies to correspondents appeared in different periodicals, an extract from one of which we add as a note, p. 374. It gives his opinion on English and provincial nomenclature.

After his return from Asia Minor, the subject was again continued in another paper in Loudon's Magazine. This is the second paper of our series; it was published in March 1837, and in the following November of the same year he printed the first outline or draft of his proposed rules. The latter brought out much correspondence and criticism, and several controversial letters between him and Mr. Ogilvie were printed in the same and subsequent volumes of Loudon. These are mostly repetitions of the argument, and as the replies, without the letters replied to, could scarcely be understood, they are altogether omitted. The entire and matured view of the subject is found summed-up in our third paper, the Report made to the British Association at their request, and which has now been generally accepted as the guide and rule for this important subject.—Ed.]

"Trivial names ought never to be changed without the most urgent necessity." —*Fabricius.*

I would offer a few remarks on a practice which appears to me to be highly detrimental to the progress of natural history. I allude to the custom, which seems to be daily gaining ground, of altering the established generic and specific names of natural objects, without any suffifient reason for so doing. I have frequently noticed the commission of this offence by writers both in this Magazine and in other modern periodicals; but I never met with a more barefaced instance of it than in Mag. Nat. Hist. vii. 593, where an anonymous correspondent (S. D. W.) has had the temerity to alter the long-established appellations, both generic, specific and vernacular, of the common bullfinch; and that, too, on grounds the most vague and unsatisfactory. In protesting against such 'wholesale changes' as this, I am far from discarding S. D. W.'s proposal, merely 'because it is new,' and will therefore follow his advice, and endeavour to 'state my reasons' for so doing.

In any systematic work on natural history, the parts which are essential to the elucidation of each species are, *the specific name, specific character,* and the *description*; to each of which a distinct object is assigned. The *description* of a species ought to contain every character and circumstance belonging to it, except those which apply to the whole genus. All the facts that appertain to a species are thus formed into a fund, which we may apply to any purpose we have in view. The *specific character* enables us to recognize any species which we may meet with for the first time, or to compare it with others of the same genus; and, to form this character, we select from the description certain characteristics, which, singly or combined, belong to that species alone. Lastly, the object of the *specific name* is precisely the same as that of all names whatever; which have been defined to be, "*arbitrary signs adopted to represent real things or conceptions.*" Hence the use of names is, in fact, nothing more than a kind of *memoria technica* [artificial memory]; by means of which, in writing or speaking, the idea of an object is suggested, without the inconvenience of a lengthened description. The advantages of this principle are found to be so great, that mankind have in all ages applied it to every subject on which they have had occasion to discourse, as well as to every individual of their own species. It is remarkable however, that Linnæus was the first to distinguish each species of natural object by a peculiar appellation. Before his time, naturalists were obliged to resort to the singularly inconvenient method of repeating the specific character every time that they wished to designate any species. Thus, in the time of Ray, our *Lanius collurio,* for instance, was known as the '*Lanius minor rufus*;' and *Lanius rufus* of Temminck was designated '*Lanius minor cinerascens, cum maculâ in scapulis albâ.*' [See the inconvenience of this mode humorously shown, in i. 134, 135, by Miss Kent.] The *specific name* was introduced as a substitute and representative of the *specific character* in common discourse; that *character* itself being preserved, for purposes of reference, in systematic works. Now, in order that the object of the specific name may be duly performed, it is essential that it be universally adopted, and therefore never, or very rarely altered. But it is not, I think, essential that the meaning of the name should precisely designate the species; or, indeed, that it should have any *meaning* at all. [See vi. 232, vii. 636.] Proper names, in general, have either no meaning, or one not in any way referring to the persons they represent; and yet they are found fully to answer their object of defining individuals, in consequence of their being universally recognized and never altered; and the peculiar names of species ought, I think, to be in the same way regarded as proper names, the end of which is defeated by repeated alterations. And, although viewed in the light of a *memoria technica,* the recognition of species, by means of their names, is certainly facilitated, if the meaning of those names has a reference to the objects which they represent; yet this is so far from being essential, that it is in some cases prejudicial, by blinding persons to the distinction between the *specific name* and the *specific character*, and causing them to regard the objects of the latter as belonging to the former. Surely, then, the evil of changing a name, which has once become current among naturalists, is much greater than any advantage supposed to result from substituting a term which is 'more appropriate.'

Some persons even maintain that the name of every species should have a meaning applying to it *exclusively*. The above arguments show that such a system of nomenclature, being founded on a mistaken view of the object of specific names, is therefore *unnecessary*; and it is easy to prove that it would in many cases be *impossible*: the *specific character* is supposed to represent, in the most condensed form, that combination of properties which distinguishes the species in question from all others. In a few cases, one word is sufficient for this purpose; but in general a whole sentence is necessary, before the species can be accurately defined. Therefore, *if the necessity existed* to express the exclusive characters of a species every time it is mentioned, it is plain that we must give up *specific names* altogether, and recur to the *long sentences* of Willughby and Ray. Nor should we mend the matter by attempting, in defiance of all rules of language and of euphony, to melt these sentences into such sesquipedalian words as those which Mr. Thomas Hawkins has adopted for the Ichthyosauri. [vii. 478.]

In addition to the arguments above stated, there are several other strong objections to the arbitrary and unlicensed alteration of established names. If the species with which the naturalist is concerned amounted only to a few hundreds, or a few thousands, then, indeed, the supposed improvement of the nomenclature might be in some measure excusable; but, since the profuse fecundity of Nature has overwhelmed her admirers with such myriads of forms, that their number alone constitutes the chief difficulty with which they have to contend, it is surely the height of folly to increase that difficulty, by bestowing a multiplicity of names on the same object. From the excess of this practice, the rectification of synonyms has become the most laborious part of the process of compiling systematic works on natural history; which are thus vastly increased in size and price, and rendered more repulsive to the general reader. There are also many other sources of inconvenience in this practice. Can S. D. W., for instance, expect that the whole republic of science will take the trouble of relabelling their cabinets, altering their catalogues, or making notes in their works of reference, because an anonymous writer fancies that he can improve *Pyrrhula vulgaris* by changing it to *Densirostra atricapilla*? Again, if some adopt the alteration, a large number will not; and hence it is that we rarely find the same species labelled alike in two different museums. In short, if this practice be once given way to, there will soon be an end of all nomenclature, and, through it, of all science; for true it is, that,

"Nomina si pereunt, perit et cognitio rerum."
If names perish, the knowledge of things perishes with them.

The above arguments apply equally to the proper names of genera, or of larger groups, where such groups are retained unaltered, their appellations only being changed; as in the case of the genus Pyrrhula, which S. D. W. has altered to Densirostra. Where an old genus is divided into several new ones, new appellations must, of course, be found for them; but, even then, the original name should be retained for that group which is the most typical of the whole.

A complete parallel seems to exist between the proper names of species and of men. The first discoverer of a species may be regarded as its parent or godfather; who bestows on it any name he thinks fit, and publishes it to the scientific world in some standard work, as in a parish register: and as the laws of the land forbid men to change their names without due cause, so the laws of natural history ought to be equally severe against those who encumber species with a multitude of *aliases*. It would, I think, be highly desirable if an authorized body could be constituted, to frame a code of laws for naturalists, instead of the present anarchical state of things, in which every one does that which is right in his own eyes.

With respect to established species, priority seems to be the universal law for the adoption of specific names (see some excellent remarks, by Mr. Westwood, in the Zoological Journal, iv. 3–9); subject to the exception, that the same name be not repeated twice in a genus. And if persons could be brought to view specific appellations purely in the light of proper names, this law need have no other exception; for as we do not object to *William Whitehead's* name because his hair may happen to be *red*, so, if the *meaning* of a specific name be downrightly inapplicable to the object, this need not prevent its *sound* being adopted as the conventional sign of the species. But as, in practice, such a circumstance would produce confusion, we are compelled to admit the farther exception to the above law of priority, that if, as rarely happens, the specific name have a meaning contradictory to the species which it represents, that name should be changed for one that is not contradictory. For the same reason, in naming a new species, it is *desirable* that the name be as expressive as possible; but if no term that is applicable should occur, the term fixed upon should, at least, be *not contradictory*.

I hope that S. D. W. and other correspondents will regard this subject with the attention which it deserves *.

* See details of Modern History of Nomenclature in Memoir, chap. vi.

2. ON THE INEXPEDIENCY OF ALTERING ESTABLISHED TERMS IN NATURAL HISTORY.

[From Charlesworth's Mag. of Nat. Hist. vol. i. 2 ser. p. 127.]

It is now about two years since I published some remarks on this subject in this Magazine (vol. viii. p. 36), and in the 'Analyst' (vol. ii. p. 317). Those remarks were directed against a practice which certain naturalists (?) had commenced, of altering at their own good will and pleasure numerous scientific terms, many of which had been current in the republic of science for nearly a century. Absence from England has since prevented me from recurring to the subject until now. On looking over this Magazine and the 'Analyst' for the last year and a half, I find that the lovers of confusion have been hard at work, and that corresponding efforts are required on the part of the true friends of science to counteract this evil tendency. Among other papers on this subject, is one in this Magazine (vol. ix. pp. 139 and 337), by Mr. C. T. Wood, in which he attacks, in not the most courteous terms, my former communications of vol. viii. p. 36. Having a great dislike to the personalities of ordinary paper warfare, it is not my intention, at present, to expose the fallacies and misstatements in Mr. Wood's paper*. I will merely remark that, in his zeal for improved nomenclature, he might have found a more appropriate epithet for me than *anti-reformer*. No one is more desirous of the improvement of science than myself; but *reform* implies something more than *change*; and it is precisely because I do not consider that the proposed changes are *for the better*, that I enter my protest against them. On a superficial view of the case, it may certainly appear, that to change a less appropriate scientific name for one that is more so, *is* a change for the better: but what is the result? If, to take the most favourable view of the case, the scientific world should agree to adopt an 'improved' nomenclature, yet, even then, all our standard works on natural history would become, in great measure, a dead letter; every museum in the world would require to be relabelled; and the disentanglement of synonyms (already a sufficiently laborious, though necessary duty) would become almost hopeless. But if, as would most certainly be the case, these "improved nomenclatures" should be only partially adopted, the disentanglement of synonyms

* Not content with "improving" nomenclature, Mr. C. T. Wood and an anonymous colleague of his (S. D. W.) try their hand at orthography also, and insist upon writing Fasiànus, Faláropus, níctea, Cípselus, &c., in defiance of all the laws of etymology which have been acknowledged these 2000 years. Such puerilities do not require further comment.

would then become *quite* hopeless, and the curse of Babel would be entailed on the scientific world. Natural history would become divided into nations and languages; each country would remain in happy ignorance of the state of the science among its neighbours; and we should have genera and species described as new which had been known for twenty years in France or Germany. Surely there are at present sufficient impediments to scientific intercourse (such as loss of time and money, passports, quarantine, duties on books and the like), without further obliging the traveller to learn a new scientific nomenclature in each country previously to visiting its museums, or conversing with its professors.

This 'improving' system has already gone far enough to afford a fair specimen of its merits. Any one who will take the trouble to examine the various papers on ornithological nomenclature in the 'Analyst,' will find that specific names are as variable as the London fashions. Every new number of that work contains some fresh change in the nomenclature; and sage reasons are given why the names invented, three months before, by A. B., are not satisfactory to C. D. When the golden rule of *priority* is once laid aside, there seems to be no limit to these alterations, for however appropriate a nomenclature may be, dabblers in science will always be found who will prefer terms of their own coinage to those which are already established. The result will be, that, after these obscure individuals have involved the science in an inextricable maze of confusion, the real cultivators of zoology will cut the Gordian knot, and fall back upon the names originally established by the fathers of the science; names recognized in all standard works, and current among naturalists in all parts of the world. Then will the nomenclature of Linnæus, Cuvier and Temminck triumph over the crude inventions of a host of anonymous scribblers*.

That these proposed innovations will never be adopted by the highest class of naturalists is my firm belief. If I am mistaken in this point; if three-fourths of the terms now current in science *must* and *will* be called in, and a new coinage issued; then, at least, let that coinage proceed from a duly authorized mint. Let a committee be appointed; not from the 'Constant Readers' of a provincial magazine, but from the most eminent naturalists of every country where the science is cultivated, to take the subject of nomenclature into consideration. The number of persons selected from each country should be proportionate to the degree

* Sir James E. Smith, quoted by Swainson (Birds, vol. i. p. 245), says, "Those who alter names, often for the worse, according to arbitrary rules of their own, or in order to aim at consequence which they cannot otherwise attain, are best treated with silent neglect. The system should not be encumbered with such names, even as synonyms."

in which it encourages science, in such ratio as the following:—Germany, including Austria, 6; France, 5; Prussia, 4; England, 4; Holland, Switzerland, and Italy, 3 each; America and Russia, 2 each; Sweden, Denmark, and Spain, 1 each; total, 35. Let a list be prepared of such names of classes, orders, families, genera and species as are considered objectionable, and let the committee have full power to retain or condemn as they should think best. By this means, whatever changes were once made would be unalterably established, and uniformly adopted, in all parts of the world. Such a plan might not gratify the vanity of those individuals who, by altering names, "aim at consequence which they cannot otherwise attain;" but it would be the most beneficial method of reforming the language of science, if reformed it must be.

In my former paper of vol. viii. p. 36, I showed that it is of much more importance that a name should be universally adopted, than that its *meaning* should exclusively apply to the object. This indeed, it very rarely can do; for nothing short of a whole sentence will, in general, express that aggregate of characters which distinguishes a species or group from its congeners*. Nay, a name will even answer its purpose if it has no meaning at all; a doctrine admitted by some of my opponents with respect to genera, though, for some inexplicable reason, they deny it in the case of species. The *meaning* of a name is therefore a point of less importance than its *universality*; and, when the latter object has been once gained, I would never sacrifice it to the former. Yet it must not be supposed that I consider the meaning of a name a point of *no* importance. As a general principle, the meaning of an epithet may be made of great use in recalling to the mind many facts connected with the object which it represents; and, therefore, in bestowing names upon species discovered, or groups defined, *for the first time*, it is desirable to make those names as expressive as possible. I therefore fully approve of many of the rules laid down by my opponent Mr. Wood in vol. ix. p. 341; and by Mr. Swainson, in his 'Birds,' part 2. ch. 2, to which I gladly refer for support to my own views. I would not, however, go so far as Mr. Swainson in erasing many of the generic names to which he objects. Thus Tinamus, Catarrhactes, Dendrocolaptes, &c., having *once become established*, may, in my opinion, be advantageously retained, though they are not to be imitated in future.

As a proof that a name may answer the purpose of suggesting an object to the mind, even when its *meaning* implies something actually false and erroneous, I may instance the names Caprimulgus and *Paradisea apoda*. The mention of these names immediately suggests the

* Mr. Wood denies that any one maintains this. Then why, let me ask, do the innovators continually change names which are already appropriate, in the vain hope of finding others which shall be exclusively so? (See the 'Analyst,' *passim.*)

idea of certain birds to the ornithologist; but there are surely none so ignorant as to be led by the etymology of these terms into a belief that the one sucks goats, and that the other has no feet. All that is *now* implied by the term *Caprimulgus* is, that 'these are the birds once accused of sucking goats'; and by *apoda*, that 'this is the bird anciently supposed to be destitute of feet, and which to this day is so rarely obtained with those organs perfect.' Now, though these facts are of no great importance, they are at least amusing and harmless associations connected with the birds in question. Therefore, names whose very meaning is erroneous do not necessarily mislead: hence the term Caprimulgus, which has been established for 2000 years, ought, and I trust will, triumph over its ephemeral rivals Nyctichelidon, Vociferator, and Phalænivora.

Yet, in general, it cannot be denied that, in the rare cases when the derivation of a name, whether generic or specific, is likely to propagate a really false opinion, such a name should be erased. When such changes are made, some plan should be adopted to consult the opinions of foreign as well as English naturalists, and to ensure a universal adoption throughout the whole scientific world.

In what has been said above, I have had in view only the Latin and Græco-Latin names for natural objects. These, being recognized by naturalists of all nations, form the only legitimate language of science. English names, therefore, are not wanted in scientific discourse; for, when one name for a thing is sufficient, two are superfluous. There are however certain persons who find the Latin names too learned for them, and will not be satisfied without English names for genera and species. Not being able to make the vernacular names in use among our peasantry square with their ideas of systematic nomenclature, they set to work to coin terms out of their mother-tongue; and thus we are beset with a host of such names as kinglet, treeling, mufflin, &c. Such puerilities may be very well for country bird-stuffers, and for idle boys, who, instead of going to school, spend their time in bird-nesting, and call themselves 'field-naturalists;' but they are beneath the notice of men of science, whom I would gladly see discarding the superfluous English names altogether. Some concession must, however, be made to the vulgar cry; for, if science be not put in a 'popular' form, the labours of its real cultivators will not be repaid. If, then, English terms must be superadded to the Latin, I would suggest that the old terms, such as warbler, finch, &c., be as much as possible retained, and that, where a genus has no peculiar English name already, the Latin one should be used in preference to coining a new term. Thus, I would rather say, 'hedge accentor, fire-crested regulus, bearded calamophilus, yellow budytes,' &c., than burden science with such clumsy appellations as dunnock, kinglet, longtail, and willet (*alias* oatear); to say nothing of abern, surn, popin, sprigtail, and the like.

But I would strongly protest against extending the system of English nomenclature beyond the English fauna*. What possible benefit can accrue from coining fresh substantives, of Saxon derivation, for the countless genera of exotic zoology? I trust we may never see the forms of New Holland or of South America burdened with such uncouth appellations as have been invented for the birds of England.

* There is a paper on the "Nomenclature of Birds," by Mr. Neville Wood, on which I am desirous of offering a few remarks. I think Mr. Wood's plan of altering many of the English names of our common birds objectionable, because they more properly form part of our vernacular tongue than of the language of science. They are consecrated by usage as much as any other part of the English language, and consequently, when we speak of an hedge *sparrow*, we are much more likely to be understood than if we called it an hedge *dunnock*, though I willingly admit that it is *unscientific* to give the same generic name to an *Accentor* and a *Passer*. But the truth is, that the science of ornithology does not suffer by this incorrect application of English names, because those familiar appellations have no real or necessary connexion with *science*. The first and most important requisite in scientific terms is that they should be universally adopted, and hence the fathers of natural history have wisely employed the Latin language as the source of their nomenclature, being generally understood by the learned among all civilized nations. English names are useful only to denote those natural objects which are so common or remarkable in our own country as to attract the attention even of the vulgar; but as the *science* of natural history does not in the least require their assistance, I should be sorry to see them in any degree *substituted* for those *Latin* appellations which are universally current in the republic of science. I may remark that French naturalists are much more addicted to the adoption of vernacular names, to the exclusion of scientific ones, than the English. By endeavouring to coin a French term for every natural object, in addition to the Latin one which it already possesses, they exactly double the enormous labour of bearing in memory the innumerable terms with which science is unavoidably encumbered.

If, then, I am correct in regarding the *English* names of birds as belonging not to science, but to our mother-tongue, it is clearly better to let them remain as they are, than, by endeavouring to reform the English language, to make changes which are certain not to be universally adopted.—H. E. S.—*Analyst*, ii. p. 317.

3. REPORT OF A COMMITTEE APPOINTED "TO CONSIDER OF THE RULES BY WHICH THE NOMENCLATURE OF ZOOLOGY MAY BE ESTABLISHED ON A UNIFORM AND PERMANENT BASIS."

[Report Brit. Assoc. vol. xi. Reports, p. 105, 1842.]

[*Minute of Council, Feb.* 11, 1842.

"Resolved,—That (with a view of securing early attention to the following important subject) a Committee consisting of Mr. C. Darwin, Prof. Henslow, Rev. L. Jenyns, Mr. W. Ogilby, Mr. J. Phillips, Dr. Richardson, Mr. H. E. Strickland (reporter), Mr. J. O. Westwood, be appointed, to consider of the rules by which the Nomenclature of Zoology may be established on a uniform and permanent basis; the report to be presented to the Zoological Section, and submitted to its Committee, at the Manchester Meeting.

Minute of the Committee of Section D, June 29, 1842.

"Resolved,—That the Committee of the Section of Zoology and Botany have too little time during the Meeting of the Association to discuss a Report on Nomenclature, and therefore remit to the special Committee appointed to draw up the Report, to present it on their own responsibility."]

THE Committee appointed by the Council of the British Association to carry out the above object, beg leave to report, that at the meetings which they held in London the following gentlemen were added to the Committee and assisted in its labours:—Messrs. W. J. Broderip, Prof. Owen, W. E. Shuckard, G. R. Waterhouse, and W. Yarrell. An outline of the proposed code of rules having been drawn up and printed, copies of it were sent to many eminent zoologists at home and abroad, who were requested to favour the Committee with their observations and comments. Many valuable suggestions were obtained from this source, by the aid of which the Committee were enabled to introduce several important modifications into the original plan. A few copies of the plan as amended were then printed for the use of the Committee, and the total cost of printing these two editions amounts to £4 10*s*.

As the probable success of this measure must greatly depend on its obtaining a rapid and extensive circulation among foreign as well as British zoologists, the Committee beg to recommend that a small sum (say £5 10*s*.) be appropriated for printing and distributing extra copies of this report in the form which it may finally assume in our Transactions.

The plan as amended has been further considered by the Committee during the present meeting at Manchester, and the Committee having thus given their best endeavours to maturing the plan, beg now to submit it to the approval of the British Association under the title of a

Series of Propositions for rendering the Nomenclature of Zoology Uniform and Permanent.

Preface.—All persons who are conversant with the present state of zoology must be aware of the great detriment which the science sustains from the vagueness and uncertainty of its nomenclature. We do not here refer to those diversities of language which arise from the various methods of classification adopted by different authors, and which are unavoidable in the present state of our knowledge. So long as naturalists differ in the views which they are disposed to take of the natural affinities of animals there will always be diversities of classification, and the only way to arrive at the true system of nature is to allow perfect liberty to systematists in this respect. But the evil complained of is of a different character. It consists in this, that when naturalists *are* agreed as to the characters and limits of an individual group or species, they still disagree in the appellations by which they distinguish it. A genus is often designated by three or four, and a species by twice that number of precisely equivalent synonyms; and in the absence of any rule on the subject, the naturalist is wholly at a loss what nomenclature to adopt. The consequence is, that the so-called commonwealth of science is becoming daily divided into independent states, kept asunder by diversities of language as well as by geographical limits. If an English zoologist, for example, visits the museums and converses with the professors of France, he finds that their *scientific* language is almost as foreign to him as their *vernacular*. Almost every specimen which he examines is labelled by a title which is unknown to him, and he feels that nothing short of a continued residence in that country can make him conversant with her science. If he proceeds thence to Germany or Russia, he is again at a loss: bewildered everywhere amidst the confusion of nomenclature, he returns in despair to his own country and to the museums and books to which he is accustomed.

If these diversities of scientific language were as deeply rooted as the vernacular tongue of each country, it would of course be hopeless to think of remedying them; but happily this is not the case. The language of science is in the mouths of comparatively few, and these few, though scattered over distant lands, are in habits of frequent and friendly intercourse with each other. All that is wanted then is, that some plain and simple regulations, founded on justice and sound reason, should be drawn up by a competent body of persons, and then be extensively distributed throughout the zoological world.

The undivided attention of chemists, of astronomers, of anatomists, of mineralogists, has been of late years devoted to fixing their respective languages on a sound basis. Why then do zoologists hesitate in performing the same duty? at a time too when all acknowledge the evils of the present anarchical state of their science.

It is needless to inquire far into the causes of the present confusion of zoological nomenclature. It is in great measure the result of the same branch of science having been followed in distant countries by persons who were either unavoidably ignorant of each other's labours, or who neglected to inform themselves sufficiently of the state of the science in other regions. And when we remark the great obstacles which now exist to the circulation of books beyond the conventional limits of the states in which they happen to be published, it must be admitted that this ignorance of the writings of others, however unfortunate, is yet in great measure pardonable. But there is another source for this evil, which is far less excusable,—the practice of gratifying individual vanity by attempting on the most frivolous pretexts to cancel the terms established by original discoverers, and to substitute a new and unauthorized nomenclature in their place. One author lays down as a rule, that no specific names should be derived from geographical sources, and unhesitatingly proceeds to insert words of his own in all such cases; another declares war against names of exotic origin, foreign to the Greek and Latin; a third excommunicates all words which exceed a certain number of syllables; a fourth cancels all names which are complimentary of individuals, and so on, till universality and permanence, the two great essentials of scientific language, are utterly destroyed.

It is surely, then, an object well worthy the attention of the Zoological Section of the British Association for the Advancement of Science, to devise some means which may lessen the extent of this evil, if not wholly put an end to it. The best method of making the attempt seems to be, to entrust to a carefully selected committee the preparation of a series of rules, the adoption of which must be left to the sound sense of naturalists in general. By emanating from the British Association, it is hoped that the proposed rules will be invested with an authority which no individual zoologist, however eminent, could confer on them. The world of science is no longer a monarchy, obedient to the ordinances, however just, of an Aristotle or a Linnæus. She has now assumed the form of a republic, and although this revolution may have increased the vigour and zeal of her followers, yet it has destroyed much of her former order and regularity of government. The latter can only be restored by framing such laws as shall be based in reason and sanctioned by the approval of men of science; and it is to the preparation of these laws that the Zoological Section of the Association have been invited to give their aid.

In venturing to propose these rules for the guidance of all classes of zoologists in all countries, we disclaim any intention of dictating to men of science the course which they may see fit to pursue. It must of course be always at the option of authors to adhere to or depart from these principles, but we offer them to the candid consideration of zoologists, in the hope that they may lead to sufficient uniformity of method in future to rescue the science from becoming a mere chaos of words.

We now proceed to develope the details of our plan; and in order to make the reasons by which we are guided apparent to naturalists at large, it will be requisite to append to each proposition a short explanation of the circumstances which call for it.

Among the numerous rules for nomenclature which have been proposed by naturalists, there are many which, though excellent in themselves, it is not now desirable to enforce*. The cases in which those rules have been overlooked or departed from, are so numerous and of such long standing, that to carry these regulations into effect would undermine the edifice of zoological nomenclature. But while we do not adopt these propositions as authoritative laws, they may still be consulted with advantage in making such additions to the language of zoology as are required by the progress of the science. By adhering to sound principles of philology, we may avoid errors in future, even when it is too late to remedy the past, and the language of science will thus eventually assume an aspect of more classic purity than it now presents.

Our subject hence divides itself into two parts; the first consisting of *Rules* for the rectification of the present zoological nomenclature, and the second of *Recommendations* for the improvement of zoological nomenclature in future.

PART I.

Rules for rectifying the present Nomenclature.

[*Limitation of the Plan to Systematic Nomenclature.*]

In proposing a measure for the establishment of a permanent and universal zoological nomenclature, it must be premised that we refer solely to the Latin or systematic language of zoology. We have nothing to do with vernacular appellations. One great cause of the neglect and corruption which prevails in the scientific nomenclature of zoology, has been the frequent and often exclusive use of vernacular names in lieu of the Latin binomial designations, which form the only legitimate language of systematic zoology. Let us then endeavour to render perfect the Latin or Linnæan method of nomenclature, which being far removed

* See especially the admirable code proposed in the 'Philosophia Botanica' of Linnæus. If zoologists had paid more attention to the principles of that code, the present attempt at reform would perhaps have been unnecessary.

from the scope of national vanities and modern antipathies, holds out the only hope of introducing into zoology that grand desideratum, a universal language.

[*Law of Priority the only effectual and just one.*]

It being admitted on all hands that words are only the conventional signs of ideas, it is evident that language can only attain its end effectually by being permanently established and generally recognized. This consideration ought, it would seem, to have checked those who are continually attempting to subvert the established language of zoology by substituting terms of their own coinage. But, forgetting the true nature of language, they persist in confounding the *name* of a species or group with its *definition*; and because the former often falls short of the fullness of expression found in the latter, they cancel it without hesitation, and introduce some new term which appears to them more characteristic, but which is utterly unknown to the science, and is therefore devoid of all authority*. If these persons were to object to such names of men as *Long, Little, Armstrong, Golightly,* &c., in cases where they fail to apply to the individuals who bear them, or should complain of the names *Gough, Lawrence,* or *Harvey,* that they were devoid of meaning, and should hence propose to change them for more characteristic appellations, they would not act more unphilosophically or inconsiderately than they do in the case before us; for, in truth, it matters not in the least by what conventional sound we agree to designate an individual object, provided the sign to be employed be stamped with such an authority as will suffice to make it pass current. Now in zoology no one person can subsequently claim an authority equal to that possessed by the person who is the first to define a new genus or describe a new species; and hence it is that the name originally given, even though it may be inferior in point of elegance or expressiveness to those subsequently proposed, ought as a general principle to be permanently retained. To this consideration we ought to add the injustice of erasing the name originally selected by the person to whose labours we owe our first knowledge of the object; and we should reflect how much the permission of such a practice opens a door to obscure pretenders for dragging themselves into notice at the expense of original observers. Neither can an author be permitted to alter a name which he himself has once published, except in accordance with fixed and equitable laws. It is well observed by DeCandolle, "L'auteur même qui a le premier établi un nom n'a pas plus qu'un autre le droit de le changer pour simple cause d'impropriété. La pri-

* Linnæus says on this subject, "Abstinendum ab hac innovatione quæ nunquam cessaret, quin indies aptiora detegerentur ad infinitum."

orité en effet est un terme fixe, positif, qui n'admet rien, ni d'arbitraire ni de partial."

For these reasons, we have no hesitation in adopting as our fundamental maxim, the "law of priority," viz.

§ 1. The name originally given by the founder of a group or the describer of a species should be permanently retained, to the exclusion of all subsequent synonyms (with the exceptions about to be noticed).

Having laid down this principle, we must next inquire into the limitations which are found necessary in carrying it into practice.

[*Not to extend to authors older than Linnæus.*]

As our subject-matter is strictly confined to the *binomial system of nomenclature*, or that which indicates species by means of two Latin words, the one generic, the other specific, and as this invaluable method originated solely with Linnæus, it is clear that, as far as species are concerned, we ought not to attempt to carry back the principle of priority beyond the date of the 12th edition of the 'Systema Naturæ.' Previous to that period, naturalists were wont to indicate species not by a *name* comprised in one word, but by a *definition* which occupied a sentence, the extreme verbosity of which method was productive of great inconvenience. It is true that one word sometimes sufficed for the definition of a species, but these rare cases were only binomial by accident and not by principle, and ought not therefore in any instance to supersede the binomial designations imposed by Linnæus.

The same reasons apply also to generic names. Linnæus was the first to attach a definite value to genera, and to give them a systematic character by means of exact definitions; and therefore although the *names* used by previous authors may often be applied with propriety to modern genera, yet in such cases they acquire a new meaning, and should be quoted on the authority of the first person who used them in this secondary sense. It is true that several of the old authors made occasional approaches to the Linnæan exactness of generic definition, but still these were but partial attempts, and it is certain that if in our rectification of the binomial nomenclature we once trace back our authorities into the obscurity which preceded the epoch of its foundation, we shall find no resting-place or fixed boundary for our researches. The nomenclature of Ray is chiefly derived from that of Gesner and Aldrovandus, and from these authors we might proceed backward to Ælian, Pliny and Aristotle, till our zoological studies would be frittered away amid the refinements of classical learning*.

* "Quis longo ævo recepta vocabula commutaret hodie cum patrum?"—*Linnæus.*

We therefore recommend the adoption of the following proposition:—

§ 2. The binomial nomenclature having originated with Linnæus, the law of priority, in respect of that nomenclature, is not to extend to the writings of antecedent authors.

[It should be here explained, that Brisson, who was a contemporary of Linnæus, and acquainted with the 'Systema Naturæ,' defined and published certain genera of birds which are *additional* to those in the 12th edition of Linnæus's work, and which are therefore of perfectly good authority. But Brisson still adhered to the old mode of designating species by a sentence instead of a word, and therefore while we retain his defined genera, we do not extend the same indulgence to the titles of his species, even when the latter are accidentally binomial in form. For instance, the *Perdix rubra* of Brisson is the *Tetrao rufus* of Linnæus; therefore as we in this case retain the generic name of Brisson and the specific name of Linnæus, the correct title of the species would be *Perdix rufa*.]

[*Generic names not to be cancelled in subsequent subdivisions.*]

As the number of known species which form the groundwork of zoological science is always increasing, and our knowledge of their structure becomes more complete, fresh generalizations continually occur to the naturalist, and the number of genera and other groups requiring appellations is ever becoming more extensive. It thus becomes necessary to subdivide the contents of old groups and to make their definitions continually more restricted. In carrying out this process, it is an act of justice to the original author, that his generic name should never be lost sight of; and it is no less essential to the welfare of the science, that all which is sound in its nomenclature should remain unaltered amid the additions which are continually being made to it. On this ground we recommend the adoption of the following rule:—

§ 3. A generic name when once established should never be cancelled in any subsequent subdivision of the group, but retained in a restricted sense for one of the constituent portions.

[*Generic names to be retained for the typical portion of the old genus.*]

When a genus is subdivided into other genera, the original name should be retained for that portion of it which exhibits in the greatest degree its essential characters as at first defined. Authors frequently indicate this by selecting some one species as a fixed point of reference, which they term the "type of the genus." When they omit doing so, it may still in many cases be correctly inferred that the *first* species mentioned on their list, if found accurately to agree with their definition, was regarded by them as the type. A specific name or its synonyms will also often serve to point out the particular species which by implication must be regarded as the original type of a genus. In such cases we are justified in restoring the name of the old genus to its typical signification, even when later authors have done otherwise. We submit therefore that

§ 4. The generic name should always be retained for that portion of the original genus which was considered typical by the author.

Example.—The genus Picumnus was established by Temminck, and included two groups, one with four toes, the other with three, the *former* of which was regarded by the author as typical. Swainson, however, in raising these groups at a later period to the rank of genera, gave a new name, Asthenurus, to the *former* group, and retained Picumnus for the *latter*. In this case we have no choice but to restore the name Picumnus, Temm., to its correct sense, cancelling the name Asthenurus, Sw., and imposing a new name on the three-toed group which Swainson had called Picumnus.

[*When no type is indicated, then the original name is to be kept for that subsequent subdivision which* first *received it.*]

Our next proposition seems to require no explanation :—

§ 5. When the evidence as to the original type of a genus is not perfectly clear and indisputable, then the person who first subdivides the genus may affix the original name to any portion of it at his discretion, and no later author has a right to transfer that name to any other part of the original genus.

[*A later name of the same extent as an earlier to be wholly cancelled.*]

When an author infringes the law of priority by giving a new name to a genus which has been properly defined and named already, the only penalty which can be attached to this act of negligence or injustice, is to expel the name so introduced from the pale of the science. It is not right then in such cases to restrict the meaning of the later name, so that it may stand side by side with the earlier one, as has sometimes been done. For instance, the genus Monaulus, Vieill., 1816, is a precise equivalent to Lophophorus, Temm., 1813, both authors having adopted the same species as their type, and therefore when the latter genus came in the course of time to be divided into two, it was incorrect to give the condemned name Monaulus to one of the portions. To state this succinctly,

§ 6. When two authors define and name the same genus, *both making it exactly of the same extent,* the later name should be cancelled *in toto,* and not retained in a modified sense*.

This rule admits of the following exception :—

§ 7. Provided, however, that if these authors select their respective types from different sections of the genus, and these sections be afterwards raised into genera, then both these names may be retained in a restricted sense for the new genera respectively.

Example.—The names Œdemia and Melanetta were originally coextensive synonyms, but their respective types were taken from different sections which are now raised into genera, distinguished by the above titles.

[No special rule is required for the cases in which the later of two generic names is so defined as to be *less extensive* in signification than the earlier, for if the later includes the type of the earlier genus, it would be cancelled by the operation of § 4; and if it does not include that type, it is in fact a distinct genus.]

But when the later name is *more extensive* than the earlier, the following rule comes into operation :—

[*A later name equivalent to several earlier ones is to be cancelled.*]

The same principle which is involved in § 6 will apply to § 8.

§ 8. If the later name be so defined as to be equal in extent to two or more previously published genera, it must be cancelled *in toto.*

Example.—Psarocolius, Wagl., 1827, is equivalent to five or six genera previously published under other names, therefore Psarocolius should be cancelled.

If these previously published genera be *separately adopted* (as is the case with the equivalents of Psarocolius), their original names will of course prevail; but if we follow the later author in combining them into one, the following rule is necessary :—

* These discarded names may however be *tolerated,* if they have been afterwards proposed in a totally new sense, though we trust that in future no one will *knowingly* apply an old name, whether now adopted or not, to a new genus. (See proposition *q, infra.*)

[*A genus compounded of two or more previously proposed genera whose characters are now deemed insufficient, should retain the name of one of them.*]

It sometimes happens that the progress of science requires two or more genera, founded on insufficient or erroneous characters, to be combined together into one. In such cases the law of priority forbids us to cancel *all* the original names and impose a *new* one on this compound genus. We must therefore select some one species as a type or example, and give the generic name which it formerly bore to the whole group now formed. If these original generic names differ in date, the oldest one should be the one adopted.

§ 9. In compounding a genus out of several smaller ones, the earliest of them, if otherwise unobjectionable, should be selected, and its former generic name be extended over the new genus so compounded.

Example.—The genera Accentor and Prunella of Vieillot not being considered sufficiently distinct in character, are now united under the general name of Accentor, that being the earliest. So also Cerithium and Potamides, which were long considered distinct, are now united, and the latter name merges into the former.

We now proceed to point out those few cases which form exceptions to the law of priority, and in which it becomes both justifiable and necessary to alter the names originally imposed by authors.

[*A name should be changed when previously applied to another group which still retains it.*]

It being essential to the binomial method to indicate objects in natural history by means of *two words* only, without the aid of any further designation, it follows that a generic name should only have one meaning, in other words, that two genera should never bear the same name. For a similar reason no two species in the same genus should bear the same name. When these cases occur, the later of the two duplicate names should be cancelled, and a new term, or the earliest synonym, if there be any, substituted. When it is necessary to form new words for this purpose, it is desirable to make them bear some analogy to those which they are destined to supersede, as where the genus of birds, Plectorhynchus, being preoccupied in Ichthyology, is changed to Plectorhamphus. It is, we conceive, the bounden duty of an author when naming a new genus, to ascertain by careful search that the name which he proposes to employ has not been previously adopted in other departments of natural history*.

* This laborious and difficult research will in future be greatly facilitated by the very useful work of M. Agassiz, entitled 'Nomenclator Zoologicus,' 4to. 1842–1846.

By neglecting this precaution he is liable to have the name altered and his authority superseded by the first subsequent author who may detect the oversight, and for this result, however unfortunate, we fear there is no remedy, though such cases would be less frequent if the detectors of these errors would, as an act of courtesy, point them out to the author himself, if living, and leave it to him to correct his own inadvertencies. This occasional hardship appears to us to be a less evil than to permit the practice of giving the same generic name *ad libitum* to a multiplicity of genera. We submit, therefore, that

§ 10. A name should be changed which has before been proposed for some other genus in zoology or botany, or for some other species in the same genus, when still retained for such genus or species*.

---

as well as the more recent and extensive work in 8vo, 'Nomenclatoris Zoologici Index universalis,' 1848.

* Your correspondent "Viator" states as a well-known rule, "that not more than *one* genus in zoology, or *one* in botany, or *one* in any other of the like sciences, should receive the same name." It is not very clear from the above words, whether "Viator" objects to the employment of duplicate terms in *different* sciences, or *merely* to the repetition of a generic term in the *same* science; or, in other words, whether he has the same objection to the *double emploi* of the term Posidonia, *e. g.* in botany and zoology, which he justly has to that of Proteus for an amphibian and an infusorian. I hope and believe that the former was the intended meaning of "Viator," for it is now generally agreed by naturalists, that no generic term must be common both to botany and zoology. The question between the two rival claimants to the title of Proteus is one of considerable difficulty. The minute and despised infusorian might plead the undoubted right of priority, its baptism being registered, as observed by "Viator," in 1755, whereas the amphibian was not drawn from its murky caverns till 1768. Moreover, it might be argued that the term Proteus is peculiarly expressive of the versatile powers of the animalcule, while it is utterly inapplicable to the amphibian, whose essential and distinguishing character is that of undergoing no metamorphosis at all. Laurenti, therefore, was guilty of a great oversight in giving to his animal a name, which not only had already been appropriated, but which was perhaps the most unsuitable one that could have been chosen. On the other hand, it must be remembered that the amphibian has now been known by the name of *Proteus anguinus* for seventy years, and that this name has been universally adopted by the numerous naturalists who have made this extraordinary animal their study. This question of names is therefore too knotty to be decided by an anonymous author, though our thanks are due to him for having directed our attention to it. I trust the time may arrive when a Committee of nomenclature may be appointed, out of the whole republic of science, invested with power to revise the systematic terms of zoology and botany, and to establish their terminology on a sure foundation. But when changes in nomenclature are attempted by individuals, whose weight in the scientific world is not sufficient to ensure the universal adoption of their improvements (admitting them to be such), the science receives more injury by the multiplication of synonyms, than benefit by these partial ameliorations.

I have only to add, that "Viator" has been unlucky in his choice of the term Thetis as a substitute for Proteus, that name having been adopted some years since

[*A name whose meaning is glaringly false may be changed.*]

Our next proposition has no other claim for adoption than that of being a concession to human infirmity. If such proper names of places as Covent Garden, Lincoln's Inn Fields, Newcastle, Bridgewater, &c., no longer suggest the ideas of gardens, fields, castles, or bridges, but refer the mind with the quickness of thought to the particular localities which they respectively designate, there seems no reason why the proper names used in natural history should not equally perform the office of correct indication even when their etymological meaning may be wholly inapplicable to the object which they typify. But we must remember that the language of science has but a limited currency, and hence the words which compose it do not circulate with the same freedom and rapidity as those which belong to every-day life. The attention is consequently liable in scientific studies to be diverted from the contemplation of the thing signified to the etymological meaning of the sign, and hence it is necessary to provide that the latter shall not be such as to propagate actual error. Instances of this kind are indeed very rare, and in some cases, such as that of Monodon, Caprimulgus, *Paradisea apoda,* and Monoculus, they have acquired sufficient currency no longer to cause error, and are therefore retained without change. But when we find a Batrachian reptile named in violation of its true affinities, Mastodonsaurus, a Mexican species termed (through erroneous information of its habitat) *Picus cafer,* or an olive-coloured one *Muscicapa atra,* or when a name is derived from an accidental monstrosity, as in *Picus semirostris* of Linnæus, and *Helix disjuncta* of Turton, we feel justified in cancelling these names, and adopting that synonym which stands next in point of date. At the same time we think it right to remark that this privilege is very liable to abuse, and ought therefore to be applied only to extreme cases and with great caution. With these limitations we may concede that

§ 11. A name may be changed when it implies a false proposition which is likely to propagate important errors.

[*Names not clearly defined may be changed.*]

Unless a species or group is intelligibly defined when the name is given, it cannot be recognized by others, and the signification of the name is consequently lost. Two things are necessary before a zoological term can acquire any authority, viz. *definition* and *publication.* Definition properly implies a distinct exposition of essential characters, and in all cases we conceive this to be indispensable, although some authors

by Mr. Sowerby in his 'Mineral Conchology,' vol. xvi. pl. 513, for a genus of fossil shells, and is confirmed by Bronn, 'Lethæa Geognostica,' p. 704.—Strickl. in Mag. Nat. Hist. new ser. ii. p. 53.

maintain that a mere enumeration of the component species, or even of a single type, is sufficient to authenticate a genus. To constitute *publication*, nothing short of the insertion of the above particulars *in a printed book* published and for sale can be held sufficient. Many birds, for instance, in the Paris and other continental museums, shells in the British Museum (in Dr. Leach's time), and fossils in the Scarborough and other public collections, have received MS. names which will be of no authority until they are published*. Nor can any unpublished descriptions, however exact (such as those of Forster, which are still shut up in a MS. at Berlin), claim any right of priority till published, and then only from the date of their publication. The same rule applies to cases where groups or species are published, but not defined, as in some museum catalogues, and in Lesson's 'Traité d'Ornithologie,' where many species are enumerated by name, without any description or reference by which they can be identified. Therefore

§ 12. A name which has never been clearly defined in some published work should be changed for the earliest name by which the object shall have been so defined.

[*Specific names, when adopted as generic, must be changed.*]

The necessity for the following rule will be best illustrated by an example. The *Corvus pyrrhocorax*, Linn., was afterwards advanced to a genus under the name of Pyrrhocorax. Temminck adopts this generic name, and also retains the old specific one, so that he terms the species *Pyrrhocorax pyrrhocorax*. The inelegance of this method is so great as to demand a change of the specific name, and the species now stands as *Pyrrhocorax alpinus*, Vieill. We propose therefore that

§ 13. A new specific name must be given to a species when its old name has been adopted for a genus which includes that species.

N.B. It will be seen however, below, that we strongly object to the further continuance of this practice of elevating specific names into generic.

[*Latin orthography to be adhered to.*]

On the subject of orthography it is necessary to lay down one proposition,—

§ 14. In writing zoological names the rules of Latin orthography must be adhered to.

In Latinizing Greek words there are certain rules of orthography

* These MS. names are in all cases liable to create confusion, and it is therefore much to be desired that the practice of using them should be avoided in future.

known to classical scholars which must never be departed from. For instance, the names which modern authors have written Aipunemia, Zenophasia, poiocephala, must, according to the laws of etymology, be spelt Æpycnemia, Xenophasia, and pœocephala. In Latinizing modern words the rules of classic usage do not apply, and all that we can do is to give to such terms as classical an appearance as we can, consistently with the preservation of their etymology. In the case of European words whose orthography is fixed, it is best to retain the original form, even though it may include letters and combinations unknown in Latin. Such words, for instance, as Woodwardi, Knighti, Bullocki, Eschscholtzi, would be quite unintelligible if they were Latinized into Vudvardi, Cnichti, Bullocci, Essolzi, &c. But words of barbarous origin, having no fixed orthography, are more pliable, and hence, when adopted into the Latin, they should be rendered as classical in appearance as is consistent with the preservation of their original sound. Thus the words Tockus, awsuree, argoondah, kundoo, &c. should, when Latinized, have been written Toccus, ausure, argunda, cundu, &c. Such words ought, in all practicable cases, to have a Latin termination given them, especially if they are used generically.

In Latinizing proper names, the simplest rule appears to be to use the termination *-us*, genitive *-i*, when the name ends with a consonant, as in the above examples; and *-ius*, gen. *ii*, when it ends with a vowel, as Latreille, Latreillii, &c.

In converting Greek words into Latin the following rules must be attended to:—

Greek.		Latin.	Greek.		Latin.
αι	becomes	æ.	θ	becomes	th.
ει	,,	i.	φ	,,	ph.
ος	terminal,	us.	χ	,,	ch.
ον	,,	um.	κ	,,	c.
ου	becomes	u.	γχ	,,	nch.
οι	,,	œ.	γγ	,,	ng.
υ	,,	y.	ʽ	,,	h.

When a name has been erroneously written and its orthography has been afterwards amended, we conceive that the authority of the original author should still be retained for the name, and not that of the person who makes the correction.

PART II.

Recommendations for improving the Nomenclature in future.

The above propositions are all which in the present state of the science it appears practicable to invest with the character of laws. We have endeavoured to make them as few and simple as possible, in the hope that they may be the more easily comprehended and adopted by naturalists in general. We are aware that a large number of other regulations, some of which are hereafter enumerated, have been proposed and acted upon by various authors who have undertaken the difficult task of legislating on this subject; but as the enforcement of such rules would in many cases undermine the invaluable principle of priority, we do not feel justified in adopting them. At the same time we fully admit that the rules in question are, for the most part, founded on just criticism, and therefore, though we do not allow them to operate retrospectively, we are willing to retain them for future guidance. Although it is of the first importance that the principle of priority should be held paramount to all others, yet we are not blind to the desirableness of rendering our scientific language palatable to the scholar and the man of taste. Many zoological terms, which are now marked with the stamp of perpetual currency, are yet so far defective in construction, that our inability to remove them without infringing the law of priority may be a subject of regret. With these terms we cannot interfere, if we adhere to the principles above laid down; nor is there even any remedy, if authors insist on infringing the rules of good taste by introducing into the science words of the same inelegant or unclassical character in future. But that which cannot be enforced by law may, in some measure, be effected by persuasion; and with this view we submit the following propositions to naturalists, under the title of *Recommendations for the improvement of Zoological Nomenclature in future.*

[*The best names are Latin or Greek characteristic words.*]

The classical languages being selected for zoology, and words being more easily remembered in proportion as they are expressive, it is self-evident that

§ A. The *best* zoological names are those which are derived from the Latin or Greek, and express some distinguishing characteristic of the object to which they are applied.

[*Classes of objectionable names.*]

It follows from hence that the following classes of words are more or less objectionable in point of taste, though, in the case of *genera,* it is often necessary to use them, from the impossibility of finding characteristic words which have not before been employed for other genera. We will commence with those which appear the least open to objection, such as

a. Geographical names.—These words being for the most part adjectives can rarely be used for *genera.* As designations of *species* they have been so strongly objected to, that some authors (Wagler, for instance) have gone the length of substituting fresh names wherever they occur; others (*e. g.* Swainson) will only tolerate them where they apply, *exclusively*, as *Lepus hibernicus, Troglodytes europæus*, &c. We are by no means disposed to go to this length. It is not the less true that the *Hirundo javanica* is a Javanese bird, even though it may occur in other countries also, and though other species of Hirundo may occur in Java. The utmost that can be urged against such words is, that they do not tell the *whole truth*. However, as so many authors object to this class of names, it is better to avoid giving them, except where there is reason to believe that the species is chiefly confined to the country whose name it bears.

b. Barbarous names.—Some authors protest strongly against the introduction of exotic words into our Latin nomenclature, others defend the practice with equal warmth. We may remark, first, that the practice is not contrary to classical usage, for the Greeks and Romans did occasionally, though with reluctance, introduce barbarous words in a modified form into their respective languages. Secondly, the preservation of the trivial names which animals bear in their native countries is often of great use to the traveller in aiding him to discover and identify species. We do not therefore consider, if such words have a Latin termination given to them, that the occasional and judicious use of them as scientific terms can be justly objected to.

c. Technical names.—All words expressive of trades and professions have been by some writers excluded from zoology, but without sufficient reason. Words of this class, *when carefully chosen*, often express the peculiar characters and habits of animals in a metaphorical manner, which is highly elegant. We may cite the generic terms Arvicola, Lanius, Pastor, Tyrannus, Regulus, Mimus, Ploceus, &c., as favourable examples of this class of names.

d. Mythological or historical names.—When these have no perceptible reference or allusion to the characters of the object on which they are conferred, they may be properly regarded as unmeaning and in bad taste. Thus the generic names Lesbia, Leilus, Remus, Corydon, Pasiphaë, have been applied to a Humming-bird, a Butterfly, a Beetle, a Parrot, and a Crab respectively, without any perceptible association of ideas. But mythological names may sometimes be used as generic with the same propriety as technical ones, in cases where a direct allusion can be traced between the narrated actions of a personage and the observed habits or structure of an animal. Thus when the name Progne is given to a Swallow, Clotho to a Spider, Hydra to a Polyp, Athene to an Owl, Nestor to a grey-headed Parrot, &c., a pleasing and beneficial connexion is established between classical literature and physical science.

e. Comparative names.—The objections which have been raised to words of this class are not without foundation. The names, no less than the definitions of objects, should, where practicable, be drawn from positive and self-evident characters, and not from a comparison with other objects, which may be less known to the reader than the one before him. Specific names expressive of comparative size are also to be avoided, as they may be rendered inaccurate by the after-discovery of additional species. The names Picoides, Emberizoides, Pseudoluscinia, rubeculoides, maximus, minor, minimus, &c., are examples of this objectionable practice.

f. Generic names compounded from other genera.—These are in some degree open to the same imputation as comparative words; but as they often serve to express the position of a genus as intermediate to, or allied with, two other genera, they may occasionally be used with advantage. Care must be taken not to adopt such compound words as are of too great length, and not to corrupt them in trying to render them shorter. The names Gallopavo, Tetraogallus, Gypaëtos, are examples of the appropriate use of compound words.

g. Specific names derived from persons.—So long as these complimentary designations are used with moderation, and are restricted to persons of eminence as scientific zoologists, they may be employed with propriety in cases where expressive or characteristic words are not to be found. But we fully concur with those who censure the practice of naming species after persons of no scientific reputation, as curiosity-dealers (*e.g.* Caniveti, Boissoneauti), Peruvian priestesses (Cora, Amazilia), or Hottentots (Klassi).

h. Generic names derived from persons.—Words of this class have been very extensively used in botany, and therefore it would have been well to have excluded them wholly from zoology, for the sake of obtaining a *memoria technica* by which the name of a genus would at once tell us to which of the kingdoms of nature it belonged. Some few personal generic names have however crept into zoology, as Cuvieria, Mülleria, Rossia, Lessonia, &c., but they are very rare in comparison with those of botany, and it is perhaps desirable not to add to their number*.

i. Names of harsh and inelegant pronunciation.—These words are grating to the ear, either from inelegance of form, as Huhua, Yubina, Craxirex, Eschscholtzi, or from too great length, as chirostrongylostinus, Opetiorhynchus, brachypodioides, Thecodontosaurus, not to mention the *Enaliolimnosaurus crocodilocephaloides* of a German naturalist. It is needless to enlarge on the advantage of consulting euphony in the con-

* "Hoc unicum et summum præmium laboris, sancte servandum, et caste dispensandum ad incitamentum et ornamentum Botanices," was the opinion of Linnæus, Phil. Bot. p. 171.—Ed.

struction of our language. As a general rule it may be recommended to avoid introducing words of more than five syllables.

k. Ancient names of animals applied in a wrong sense.—It has been customary, in numerous cases, to apply the names of animals found in classic authors at random to exotic genera or species which were wholly unknown to the ancients. The names Cebus, Callithrix, Spiza, Kitta, Struthus, are examples. This practice ought by no means to be encouraged. The usual defence for it is, that it is impossible now to identify the species to which the name was anciently applied. But it is certain that if any traveller will take the trouble to collect the vernacular names used by the modern Greeks and Italians for the Vertebrata and Mollusca of Southern Europe, the meaning of the ancient names may in most cases be determined with the greatest precision. It has been well remarked that a Cretan fisher-boy is a far better commentator on Aristotle's 'History of Animals' than a British or German scholar. The use, however, of ancient names, *when correctly applied*, is most desirable, for "in framing scientific terms, the appropriation of old words is preferable to the formation of new ones*."

l. Adjective generic names.—The names of genera are, in all cases, essentially substantive, and hence adjective terms cannot be employed for them without doing violence to grammar. The generic names Hians, Criniger, Cursorius, Nitidula, &c. are examples of this incorrect usage.

m. Hybrid names.—Compound words, whose component parts are taken from two different languages, are great deformities in nomenclature, and naturalists should be especially guarded not to introduce any more such terms into zoology, which furnishes too many examples of them already. We have them compounded of Greek and Latin, as Dendrofalco, Gymnocorvus, Monoculus, Arborophila, flavigaster; Greek and French, as Jacamaralcyon, Jacamerops; and Greek and English, as Bullockoides, Gilbertsocrinites†.

n. Names closely resembling other names already used.—By Rule 10 it was laid down, that when a name is introduced which is *identical* with one previously used, the later one should be changed. Some authors have extended the same principle to cases where the later name, when correctly written, only approaches in form, without wholly coinciding with the earlier. We do not, however, think it advisable to make this law imperative, first, because of the vast extent of our nomenclature, which renders it highly difficult to find a name which shall not bear

* Whewell, Phil. Ind. Sc. vol. i. p. lxvii.

† This, we regret to say, has been of late indulged in to an extent which makes our nomenclature appear almost ridiculous. A few of the later coinages are Kaupifalco, Lichtensteinipicus, Blythipicus, Forbesocrinus, Cookilaria, and, among Botanists, Uroskinnera.—Ed.

more or less resemblance in sound to some other; and, secondly, because of the impossibility of fixing a limit to the degree of approximation beyond which such a law should cease to operate. We content ourselves, therefore, with putting forth this proposition merely as a recommendation to naturalists, in selecting generic names, to avoid such as too closely approximate words already adopted. So with respect to species, the judicious naturalist will aim at variety of designation, and will not, for example, call a species *virens* or *virescens* in a genus which already possesses a *viridis*.

o. Corrupted words.—In the construction of compound Latin words, there are certain grammatical rules which have been known and acted on for two thousand years, and which a naturalist is bound to acquaint himself with before he tries his skill in coining zoological terms. One of the chief of these rules is, that in compounding words all the radical or essential parts of the constituent members must be retained, and no change made except in the variable terminations. But several generic names have been lately introduced which run counter to this rule, and form most unsightly objects to all who are conversant with the spirit of the Latin language. A name made up of the first half of one word and the last half of another, is as deformed a monster in nomenclature as a Mermaid or a Centaur would be in zoology; yet we find examples in the names Corcorax (from Corvus and Pyrrhocorax), Cypsnagra (from Cypselus and Tanagra), Merulaxis (Merula and Synallaxis), Loxigilla (Loxia and Fringilla), &c. In other cases, where the *commencement* of both the simple words is retained in the compound, a fault is still committed by cutting off too much of the radical and vital portions, as is the case in Bucorvus (from Buceros and Corvus), Ninox (Nisus and Noctua), &c.

p. Nonsense names.—Some authors having found difficulty in selecting generic names which have not been used before, have adopted the plan of coining words at random without any derivation or meaning whatever. The following are examples: Viralva, Xema, Azeca, Assiminia, Quedius, Spisula. To the same class we may refer *anagrams* of other generic names, as Dacelo and Cedola of Alcedo, Zapornia of Porzana, &c. Such verbal trifling as this is in very bad taste, and is especially calculated to bring the science into contempt. It finds no precedent in the Augustan age of Latin, but can be compared only to the puerile quibblings of the middle ages. It is contrary to the genius of all languages, which appear never to produce new words by spontaneous generation, but always to derive them from some other source, however distant or obscure. And it is peculiarly annoying to the etymologist, who, after seeking in vain through the vast storehouses of human language for the parentage of such words, discovers at last that he has been pursuing an *ignis fatuus*.

q. Names previously cancelled by the operation of § 6.—Some authors consider that when a name has been reduced to a synonym by the operation of the laws of priority, they are then at liberty to apply it at pleasure to any new group which may be in want of a name. We consider, however, that when a word has once been proposed in a given sense, and has afterwards sunk into a synonym, it is far better to lay it aside for ever than to run the risk of making confusion by re-issuing it with a new meaning attached.

r. Specific names raised into generic.—It has sometimes been the practice in subdividing an old genus to give to the lesser genera so formed, the names of their respective typical species. Our Rule 13 authorizes the forming a new specific name in such cases; but we further wish to state our objections to the practice altogether. Considering as we do that the original specific names should as far as possible be held sacred, both on the grounds of justice to their authors and of practical convenience to naturalists, we would strongly dissuade from the *further continuance* of a practice which is gratuitous in itself, and which involves the necessity of altering long-established specific names.

We have now pointed out the principal rocks and shoals which lie in the path of the nomenclator; and it will be seen that the navigation through them is by no means easy. The task of constructing a language which shall supply the demands of scientific accuracy on the one hand, and of literary elegance on the other, is not to be inconsiderately undertaken by unqualified persons. Our nomenclature presents but too many flaws and inelegances already, and as the stern law of priority forbids their removal, it follows that they must remain as monuments of the bad taste or bad scholarship of their authors to the latest ages in which zoology shall be studied.

[*Families to end in* idæ, *and Subfamilies in* inæ.]

The practice suggested in the following proposition has been adopted by many recent authors, and its simplicity and convenience is so great that we strongly recommend its universal use.

§ B. It is recommended that the assemblages of genera termed *families* should be uniformly named by adding the termination *idæ* to the name of the earliest known, or most typically characterized genus in them; and that their subdivisions termed *subfamilies*, should be similarly constructed, with the termination *inæ*.

These words are formed by changing the last syllable of the genitive case into *idæ* or *inæ*, as Strix, Strigis, Strigidæ, Buceros, Bucerotis, Bucerotidæ, not Strixidæ, Buceridæ.

[*Specific names to be written with a small initial.*]

A convenient *memoria technica* may be effected by adopting our next proposition. It has been usual, when the titles of species are derived from proper names, to write them with a capital letter, and hence when the specific name is used alone it is liable to be occasionally mistaken for the title of a genus. But if the titles of *species* were *invariably* written with a *small* initial, and those of *genera* with a *capital*, the eye would at once distinguish the rank of the group referred to, and a possible source of error would be avoided. It should be further remembered that all species are *equal*, and should therefore be written all *alike*. We suggest, then, that

§ C. Specific names should *always* be written with a small initial letter, even when derived from persons or places, and generic names should be always written with a capital.

[*The authority for a species, exclusive of the genus, to be followed by a distinctive expression.*]

The systematic names of zoology being still far from that state of fixity which is the ultimate aim of the science, it is frequently necessary for correct indication to append to them the name of the person on whose authority they have been proposed. When the same person is authority both for the specific and generic name, the case is very simple; but when the specific name of one author is annexed to the generic name of another, some difficulty occurs. For example, the *Muscicapa crinita* of Linnæus belongs to the modern genus Tyrannus of Vieillot; but Swainson was the first to apply the specific name of Linnæus to the generic one of Vieillot. The question now arises, Whose authority is to be quoted for the name *Tyrannus crinitus*? The expression *Tyrannus crinitus*, Linn., would imply what is untrue, for Linnæus did not use the term *Tyrannus*; and *Tyrannus crinitus*, Vieill., is equally incorrect, for Vieillot did not adopt the name *crinitus*. If we call it *Tyrannus crinitus*, Swains., it would imply that Swainson was the first to describe the species, and Linnæus would be robbed of his due credit. If we term it *Tyrannus*, Vieill., *crinitus*, Linn., we use a form which, though expressing the facts correctly, and therefore not without advantage in particular cases where great exactness is required, is yet too lengthy and inconvenient to be used with ease and rapidity. Of the three persons concerned with the construction of a binomial title in the case before us, we conceive that the author who *first* describes and names a species which forms the groundwork of later generalizations, possesses a higher claim to have his name recorded than he who afterwards defines a genus which is found to embrace that species, or who may be the mere accidental means of bringing the generic and specific names into contact. By giving the authority for the *specific* name in preference to all others, the inquirer is referred *directly* to the original description, habitat, &c. of the species, and is at the same time reminded of the date of its discovery; while genera, being less numerous than species, may be carried in the memory, or referred to in systematic works without the necessity of perpetually quoting their authorities. The most simple mode then for ordinary use seems to be to append to the original authority for the species, when not applying to the genus also, some distinctive mark, such as (*sp.*) implying an exclusive reference to the *specific* name, as *Tyrannus crinitus,* Linn. (*sp.*), and to omit this expression when the same authority attaches to both genus and species, as *Ostrea edulis,* Linn.* Therefore,

§ D. It is recommended that the authority for a specific name, *when not applying to the generic name also,* should be followed by the distinctive expression (*sp.*).

[*New genera and species to be defined* amply *and* publicly.]

A large proportion of the complicated mass of synonyms which has now become the opprobrium of zoology, has originated either from the slovenly and imperfect manner in which species and groups have been originally defined, or from their definitions having been inserted in obscure local publications which have never obtained an extensive circulation. Therefore, although under § 12, we have conceded that mere insertion in a printed book is sufficient for *publication,* yet we would strongly advise the authors of new groups always to give in the first instance a full and accurate definition of their characters, and to insert the same in such periodical or other works as are likely to obtain an immediate and extensive circulation. To state this briefly,

§ E. It is recommended that new genera or species be *amply* defined, and *extensively* circulated in the first instance.

[*The names to be given to subdivisions of genera to agree in gender with the original genus.*]

In order to preserve specific names as far as possible in an unaltered form, whatever may be the changes which the genera to which they are referred may undergo, it is desirable, when it can be done with propriety, to make the new subdivisions of genera agree *in gender* with the old groups from which they are formed. This recommendation does not however authorize the changing the gender or termination of a genus already established. In brief,

* The expression *Tyrannus crinitus* (Linn.) would perhaps be preferable from its greater brevity.

§ F. It is recommended that in subdividing an old genus in future, the names given to the subdivisions should agree in gender with that of the original group.

[*Etymologies and types of new genera to be stated.*]

It is obvious that the names of genera would in general be far more carefully constructed, and their definitions would be rendered more exact, if authors would adopt the following suggestion :—

§ G. It is recommended that in defining new genera the etymology of the name should be always stated, and that one species should be invariably selected as a type or standard of reference.

In concluding this outline of a scheme for the rectification of zoological nomenclature, we have only to remark, that almost the whole of the propositions contained in it may be applied with equal correctness to the sister science of botany. We have preferred, however, in this essay to limit our views to zoology, both for the sake of rendering the question less complex, and because we conceive that the botanical nomenclature of the present day stands in much less need of distinct enactment than the zoological. The admirable rules laid down by Linnæus, Smith, DeCandolle, and other botanists (to which, no less than to the works of Fabricius, Illiger, Vigors, Swainson, and other zoologists, we have been much indebted in preparing the present document), have always exercised a beneficial influence over their disciples. Hence the language of botany has attained a more perfect and stable condition than that of zoology; and if this attempt at reformation may have the effect of advancing zoological nomenclature beyond its present backward and abnormal state, the wishes of its promoters will be fully attained.

(Signed)	H. E. STRICKLAND.	J. S. HENSLOW.
June 27, 1842.	JOHN PHILLIPS.	W. E. SHUCKARD.
	JOHN RICHARDSON.	G. R. WATERHOUSE.
	RICHARD OWEN.	W. YARRELL.
	LEONARD JENYNS.	C. DARWIN.
	W. J. BRODERIP.	J. O. WESTWOOD.

4. OBSERVATIONS ON CLASSIFICATION IN REFERENCE TO THE ESSAYS OF MESSRS. JENYNS, NEWMAN AND BLYTH.

[From Loudon's Magazine of Natural History, vol. vii. p. 62.]

THE two great defects of modern systems appear to be, the want of simplicity and of uniformity; and I cannot but think that both these may be in great measure attained, without any violence being offered to the 'natural system.'

Mr. Jenyns (vi. 385–390*) appears to have blended together two distinct causes of complaint: first, that in modern systems genera are founded upon characters of unequal value (*e. g.* Emberiza and Plectrophanes); and, secondly, that several genera are placed in the same family, certain of which are more nearly allied to each other than they are to the remaining ones (*e. g.* Tetrao, Lagopus, Perdix, Coturnix). In making the first complaint, Mr. Jenyns seems to have overlooked the fact, that, when a systematist separates a new genus from an old one, he must necessarily restrict the characters of the old genus, as well as establish those of the new; or else the species in question might be referred as correctly to one genus as the other. Linnæus founded the genus Emberiza on a peculiar form of the bill. He is silent concerning the other characters of the bird. Now, if we define the genus Plectrophanes to possess that peculiar bill and a long hind-claw, it is plain that we must add to our definition of Emberiza the character of having a short hind-claw. When this is done, the two genera become of equal value, be the number of species in each what it may, since they are founded on characters of equal importance, and they are therefore no longer liable to Mr. Jenyns's objection. Whether the characters of these groups are of sufficient value to constitute genera, is another question; and I fully agree with Mr. Jenyns in deprecating the practice of multiplying genera *ad infinitum* upon the most trivial and unimportant characters.

Let us now turn to Mr. Jenyns's second cause of complaint. The genera Tetrao and Perdix, if restricted as shown above, may be rendered of the same value as their offsets Lagopus and Coturnix; but still Lagopus will be more nearly related to Tetrao than to Perdix. If we place these genera, as Mr. Newman (vi. p. 484) proposes, in two distinct families of Rasores, viz. Tetraonidæ and Perdicidæ, still these two will be more nearly connected with each other than they are with Phasianidæ, or any other collateral family. Or if we follow Mr. Blyth (vi. p. 487),

* The references after the names of Jenyns, Newman and Blyth, refer to Loudon's Magazine.

in making them into subfamilies Tetraonæ and Perdicianæ, of the family Tetraonidæ, we shall diminish the simplicity of our system by multiplying our groups, and its uniformity by introducing a new kind of group, the subfamily, which it will be impossible to apply to every branch of the animal kingdom. We ought, I think, to adopt no more gradations of groups in any one class than admit of being established in every class. It therefore only remains to make these minor groups into subgenera or sections of the larger ones, Tetrao and Perdix. Subgenera are practically useful for facility of reference; and they are useful to the more philosophic naturalist, by bringing into one view those species which are nearest allied; thus giving a right direction to his comparisons and observations: but it is, I think, of the utmost importance that these subgenera should not have names imposed upon them. The needless multiplication of names is the very bane of science; loading the memory beyond its powers of endurance, and degrading the philosophic naturalist into a walking dictionary. Careful and minute observations on animals cannot be carried too far; they enrich the descriptions of species, and supply characters for subgenera; and hence Mr. Blyth deserves much praise for his careful remarks upon the habits of birds, in Rennie's 'Field Naturalist's Magazine;' but I cannot agree with him, that because distinct, and even natural, groups can be formed upon these minor characters, therefore every such group is to be made a genus, and honoured with a name. Suppose that, instead of studying one order of birds alone, Mr. Blyth was to extend the same principle to the whole animal kingdom, we should have, perhaps, 50,000 or 60,000 genera! Who, then, could be an ornithologist, much less a zoologist? Language itself would fail in finding names for such a countless multitude.

Another strong objection to naming subgenera is, that the generic and subgeneric name are continually confounded and used promiscuously. This is often the case with the French writers; and even the immortal 'Règne Animal' of Cuvier is not free from this blemish. The confusion which hence arises is evident. Mr. Jenyns's plan seems, therefore, to be the best; to distinguish subgenera, or, as I would rather call them, sections, by signs or letters. But now comes Mr. Newman's difficulty: how are we, in writing or conversation, to indicate subgenera? I think that, for common purposes, it is not necessary to do so at all. When the name of a genus is mentioned, a knowledge of the larger groups in which it is contained is presupposed. If, therefore, we carry the standard of our genera too low, it is probable that none but those few who have leisure to make themselves perfect in zoology will know the class or order of a genus which another person may casually mention. If a naturalist at Calcutta is told that the *Fringilla cœlebs* is common in England, he at once recognizes the general characters of the bird; but if it were called *Schiza cœlebs*, as Mr. Blyth would call it, it is a great chance whether he would be the wiser for the information. But, if greater accuracy be required, the informant may add that it belongs to section A or B; and, if he were describing a new species, he would, of course, either indicate the section, or describe it with sufficient accuracy to enable any one to refer it to the right one.

In offering these remarks, I am far from desiring the Linnæan genera to be retained unaltered, but merely wish the practice of forming new genera not to be carried too far.

Mr. Newman says (vi. p. 481), that the orders Feræ, Accipitres, and Coleoptera are not of the same value, because the latter contains many groups analogous to the two former, and others quite different. He seems to have overlooked the remark of DeCandolle, in Mr. Jenyns's paper (vi. p. 389, note), "that the same characters are not of equal value in different groups." Hence there can be no doubt that it is far more natural to found the orders of insects upon the structure of their wings, than to put Vespa, Libellula, and Cicindela into one order, on the ground of being rapacious; and Apis, Papilio, and Chrysomela into another, because they are herbivorous.

Mr. Newman objects to uniting the Cetacea with the other Mammalia; but if we attend not only to the number of characters which they have in common, but to the value of those characters, that is, to the high station which they hold in the scale of existence, there can be little doubt that this is a natural union. The claim of the Marsupialia, and especially the Scansores, is more doubtful; founded, as they are, on single characters only, and those not, perhaps, of very great importance.

---

5. OBSERVATIONS UPON THE AFFINITIES AND ANALOGIES OF ORGANIZED BEINGS.

[From 'Charlesworth's Magazine of Natural History,' iv. new ser. p. 219.]

I HAVE read with much interest the paper by Mr. Westwood at p. 141, on affinity and analogy. The writings of this gentleman are distinguished no less for scientific accuracy than for a spirit of sound philosophy, untainted by those visionary and theoretical views entertained by some of our modern zoologists. Instead of assuming an *à-priori* system of his own, and then twisting facts into a partial coincidence with that system, he is content to take Nature as he finds her, and not the less to admire her luxuriant variety because she refuses to marshal her irregular troops into straight lines, circles, or pentagons. This healthy tone of mind imparts a high value to all that proceeds from Mr. Westwood's pen, and it is, therefore, with much diffidence that I venture to make a few remarks on the short essay above referred to.

There is no branch of the philosophy of zoology so obscure as the subject of affinity and analogy; and although many naturalists can correctly apply these two kinds of relations to particular cases, yet few can give any clear explanation of the rules which influence their practice. Mr. Westwood's remarks go deeper into the subject than those of most of his predecessors, yet it seems to me that he has not quite set the question in its true light. Before referring to his observations, I will endeavour to explain my own views on this difficult subject.

Relations of affinity and analogy are in my opinion perfectly distinct from each other in every point of view. In order to arrive at their definitions, we must first prove the existence of a real *natural system*, a subject which involves an inquiry into the designs of creative power, one of the most awful themes which the human intellect can attempt. The most obvious and undeniable examples of design in the organized creation are seen in the adaptation of each species to the circumstances in which it lives. Now, if this were the *sole* mark of design, if each species constituted a being *per se*, adapted to its peculiar condition of existence, but not allied in physiological structure to its fellow species, there would then be no *natural system*; man might indeed classify such objects according to their accidental or fancied resemblances, but there would be none of those essential peculiarities of structure which we find to pervade vast groups of beings whose external forms are often widely dissimilar. The existence then of a comparatively few grand types of structure, or "centres of creation," from the different modifications of which the innumerable species now existing derive their characters, may be taken as a proof that species were created not absolutely, but relatively; not merely with reference to their destined mode of life, but also with reference to other species whose destination was similar, though not identical with their own. If these views be correct, it results that the *resemblances* of *different* species in essential points of structure, furnish evidences of design, less obvious, perhaps, but not less certain, than the *adaptation* of any one species to its external condition of existence; and the "natural system" thus acquires an air of truth not inferior to the ocular demonstrations of anatomy. The reality of the natural system is not affected by the difficulty experienced by man in detecting it; for it is no more to be expected that systematists should have already unravelled all the resemblances between species contemplated by the Creator, than that anatomists should have arrived at the final cause of every organ of the human body. The variety of classifications adopted by different naturalists, shows that we are still far from the true system of Nature, yet I think there can be no doubt that naturalists have already sketched out its principal features with considerable accuracy. Who, for instance, can doubt that such groups as Vertebrata, Insecta, Mammalia, Pisces, Coleoptera, &c., are not merely human generalizations, but real apartments in the edifice of the Divine Architect? It is not, however, sufficient that man should detect these natural groups, he must also give a definition of their characters,— not of the *superficial* and *arbitrary* ones, but of the *essential* and *important*, and this is often the most difficult part of his task. Although these *essential characters* form the groundwork of the natural system, yet no rule can apparently be laid down for their determination in particular cases. All that man can do is to use his best judgement in selecting such characters for a group, as seem to him the most important in their influence on the vital functions of the beings which compose it. They must, in great measure, be left to the determination of what Linnæus called a "latent instinct," which Professor Whewell defines to be "an unformed and undeveloped apprehension of physiological functions*."

When by these considerations we have arrived at the notion of a natural system, composed of natural groups arranged in a determinate order, we may proceed to define *affinity* as *the relation which subsists between two or more members of a natural group*, or in other words, *an agreement in essential characters*. After the *essential characters* of such a group have been discovered and defined, then all the objects which possess those essential characters are said to have an affinity for one another. Hence we see why the idea of a *natural system* is necessary to the definition of *affinity*, for in an *artificial system* the characters of the groups are not *essential*, but *arbitrary*, and the relation between the members of such

* History of the Inductive Sciences, vol. iii. p. 312.

a group would be, not *affinity*, but mere *resemblance* or *analogy*. Thus, if an author were to establish the characters of the class Pisces, not on the essential characters derived from the circulatory system, but on the arbitrary one of being adapted for swimming, he would then include the Cetacea and the Phocidæ among his fish. Now, on comparing a porpoise with a cod, no one could deny that they both were fish according to the assumed definition, yet no naturalist would assert the resemblance between them to be one of affinity. It is evident then, that the word *affinity* derives its meaning from a belief, acknowledged or tacit, in a *natural system*; and I do not see how a person who denies the latter, can attach any meaning to the former, as distinguished from *analogy*.

From the above definition of affinity, it follows that *the degree of affinity is inverse to the rank of the group*, in other words, that the members of the lowest group have the highest or nearest affinity, and *vice versâ*. The nearest of all affinities is that which subsists between species of the same genus, and the most remote is that between *animals* and *vegetables*, as members of the next highest group, viz. *organized bodies*. The affinity between two very distantly allied species, is merely that between the highest separate groups to which they belong. Thus, the affinity between a bat and a goatsucker (to take Mr. Westwood's illustration), is merely that which subsists between mammals and birds, as members of the group Vertebrata, and is seen quite as perfect in the whale and the humming-bird, or any other examples of the two classes. By parity of reasoning, the affinity of a goatsucker to a dragon-fly is merely that which subsists between the subkingdoms Vertebrata and Annulosa, as members of the natural group *Animals*, and is, therefore, quite as strongly exhibited in the case of a shark and a butterfly, or an elephant and a mite, &c. We thus perceive the distinction between affinity and analogy to consist, not in *degree*, but in *kind*, for there is undoubtedly a very strong *analogy* between a goatsucker and a dragon-fly, though the *affinity*, as above shown, is very remote. Analogy, in short, is nothing more than *an agreement in non-essential characters*, or a resemblance which does not constitute affinity. Hence, analogy is necessarily a very partial resemblance, existing, as Mr. Westwood remarks, in the "numerical minority" of characters, and often confined to one organ alone. Analogy originates, not in the *intentional relation* of one species to another at their first creation, but in the other instance of creative design above referred to, viz. the *adaptation* of organic beings to their destined conditions of existence. To perform any given mechanical action, there is one, and in general, only one, arrangement of mechanical structure which is better adapted to that end than all others, and hence, when any two beings whose affinities are remote, are destined to perform a similar function, we find that they are provided with more or less similar instruments for that purpose. The *resemblance*, in such a case, goes no further than the fulfilment of the required object, and may, therefore, be regarded as *unintentional*, or in common parlance, accidental. For instance, there can be no question that a lengthened form, destitute of sharp angles, and anteriorly pointed, is the best adapted for passing through the water; and accordingly, we find it to prevail, not only in fish, but in Cetacea, aquatic birds, Dyticidæ, Notonectidæ, cuttlefish, &c., and man imitates it in his naval constructions. Yet we have no evidence that such *resemblance* is *intentional*, or in other words, that whales and Dyticidæ were created *for the sake* of resembling fish, but we merely suppose that in each case the boat-shaped structure was given to adapt the animal to an aquatic life. The examples of these analogies are innumerable, and appear to me to be owing to the fact, that the real variations of circumstances which this planet affords are very few, compared with the number of organized beings destined to inhabit it, so that the performance of the same function continually recurs in different groups of the natural system, and requires, in each case, a corresponding or analogous organization. Thus, *e. g.* there are not more than *four* principal varieties of *locality*, viz. the air, the ground, shallow water, and deep water. These four variations of *habitat* have determined the structure of the four orders of birds, Insessores, Rasores, Grallatores, and Natatores. Again, the twofold division of *food* into *animal* and *vegetable*, has caused the group Raptores to be divided off from the Insessores, and we thus get the five groups under which the class Birds is commonly arranged. Now, as every other species of animal must inhabit one of the above four localities, and must feed on one of the above two kinds of food, it follows that the organs of locomotion and of nutrition are susceptible of comparatively very few grand differences of structure, and that the inhabitants of the same element, or the eaters of the same food, must present numerous points of resemblance, quite independent of their *natural* or *essential affinities*. This it is which has given to distantly allied groups an *appearance* of regularity in their analogies, whence has arisen the "theory of representation," respecting which I will take occasion to say a few words.

The theory of representation announces, that "the contents of every circular group are symbolically or analogically represented by the contents of every other circle in the animal kingdom*." This has always appeared to me one of the most unsound and unphilosophical of the doctrines maintained by the advocates of the circular system. It seems derogatory to Creative Power to suppose that the principle of *representation* had any place in the scheme of creation, or that certain organs were given to species, not with a view to the discharge of certain destined functions, but for the apparently useless object of *imitating* or *represent-*

* Swainson, 'Geog. and Classif. of Animals,' p. 230.

ing other species in a distant part of the system. The advocates of this theory would have us believe that the long tail of the horse was given it, not for the purpose of brushing off flies, but in order to represent the long "tail" [train] of the peacock*, and that both pigs and humming-birds have small eyes, *because* they are the tenuirostral types of their respective "circles†." Without wasting words upon the serious discussion of such puerilities, I will merely repeat my deliberate conviction, that relations of analogy are not to be regarded as affording any evidence of προαίρεσις, or *intention*, in the scheme of creation, but are mere coincidences of structure, incidental to the grand design of adapting a large number of organized beings to perform a comparatively limited number of functions.

It will be seen that the above view of affinity and analogy differs considerably from that of Mr. Westwood in vol. iv. p. 143 of this Magazine. Mr. Westwood seems to regard affinity and analogy as the same relation under different points of view, and as depending upon the numeral majority or minority of the points of agreement between the objects compared. Mr. Westwood's views may be explained by the following tabular arrangement, showing the number of points of agreement between four analogous genera:—

Goatsucker.	Bat.	Dragon-fly.	Dionæa.
Organized.	Organized.	Organized.	Organized.
Animal.	Animal.	Animal.	
Vertebrate.	Vertebrate.		
Fly-catching.	Fly-catching.	Fly-catching.	Fly-catching.
4	4	3	2

According to Mr. Westwood, the dragon-fly would be said to have an *affinity* to the bat or goatsucker, and an *analogy* to the Dionæa, because it agrees with the former creatures in three points, and with the latter in only two. Again, the bat has an *affinity* to the goatsucker, from agreeing with it in four points, and an *analogy* to the dragon-fly and Dionæa, from agreeing with them in only three and two points respectively. So that an *affinity* subsists between the bat and dragon-fly, when compared with the Dionæa, and an *analogy* when compared with the goatsucker. This seems to me to be a correct statement of Mr. Westwood's views, if I rightly understand them, and they certainly merit the praise of ingenuity. It seems to me, however, that they contain a fallacy, owing to Mr. Westwood not having attended to the distinction between *essential* and *non-essential* characters. Thus, the words *organized*, *animal*, and *vertebrate* in the above table, refer to characters of the highest import-

* Swainson, 'Classif. of Birds,' vol. ii. p. 159.
† Ib. vol. i. p. 43.

ance to the vital functions of the creature, and consequently, to its place in the natural system, whereas the word *fly-catching* merely relates to a point of detail in the habits of the creature, of very secondary value compared to the former characters*. I should say then, that these four creatures have *affinities* for one another, in consequence of their agreeing in the essential characters above stated, and that the degree of their affinities is proportionate to the number of the essential points in which they respectively agree, but that their *analogies* are derived solely from the one non-essential point of *fly-catching,* which applies to them all in an equal or nearly equal degree. In short, however strong may be the *analogy* which the goatsucker bears to the dragon-fly, I do not consider that it has any more *affinity* to the latter than it has to a beetle, a lobster, or any other of the Annulosa.

Since writing the above, I have referred to the very valuable remarks by Mr. Blyth on affinity and analogy, in 'Mag. Nat. Hist.,' vol. ix. p. 399, &c., to which I had not sufficiently attended at the time of their publication. His views appear to me to be more nearly correct than any others which I have seen in print. The chief point in which they differ from mine, is in the introduction of a third term, *approximation,* as distinct both from affinity and analogy. Mr. Blyth considers it to be a strong resemblance between certain members of groups really distinct, and he illustrates it by the similitude of Anthus to Alauda, of Ornithorhynchus to birds, of Myxine to Mollusca, &c. Now, it seems to me, that this *approximation* resolves itself into affinity or analogy, accordingly as we admit one or other of these two propositions; either that natural groups are quite distinct from each other in every part of their contents, or that they touch, or show a tendency to touch, each other at some particular point. Thus, if we suppose *all birds* to be equally distinct in essential structure from *all mammals, all Vertebrata* from *all Mollusca,* it is plain that the *approximation* between Ornithorhynchus and birds, and between Myxine and Mollusca, resolves itself into mere analogy. But if birds have a *tendency* to unite with *mammals* by means of Ornithorhynchus, and *Vertebrata* with *Mollusca* by means of Myxine, then this *approximation* must be regarded as an affinity. So that in either case, *approximation* is not to be considered as a distinct principle, but only as an undetermined analogy or affinity.

With regard to the above inquiry, I am inclined to believe that the larger natural groups are not only widely separated, but have no real tendency to unite,—that no mammal, for instance, is in essence any nearer a bird,—no vertebrate any nearer a mollusc than another. Be this,

* I only mean that the character of *fly-catching* is unimportant in comparing groups of such high rank, but of course it becomes an essential character when applied to smaller groups, such as *families* or *genera.*

however, as it may, we cannot assert the same complete separation of natural types, when we look to the smaller groups. There can be no doubt that the lower groups, such as families and genera, do, in numerous instances, come into contact, or *pass into* one another; and in other cases, where the contact is not complete, yet a *tendency* towards it is very evident, and in such cases the *approximation* becomes one of real affinity. Such is most probably the case with Anthus and Alauda, quoted by Mr. Blyth as examples of approximation.

6. ON THE TRUE METHOD OF DISCOVERING THE NATURAL SYSTEM IN ZOOLOGY AND BOTANY.

[From Ann. of Nat. Hist. vi. p. 184.]

[At the Meeting of the British Association in 1840, at Glasgow, the paper printed below was read, and an abstract appeared in the 'Reports' for that year. In 1843 the subject was resumed, and a large chart, showing the affinities of the genera in the order Insessores, was exhibited. The intention of this chart was to demonstrate that the true affinities of organic structures branch out irregularly in all directions, and that no symmetrical arrangement or numerical uniformity is discoverable in the system of nature when studied independently of preconceived theory. The paper was printed at length in the Annals of Natural History for 1841, and an example of the system of the chart given in Pl. X. In addition we have reduced another portion of it, which will exhibit more clearly the views the author entertained of ornithological arrangement, Pl. VI. of Part I.]

It is probable that most naturalists at the present day have an instinctive belief in the existence of a natural system in zoology and botany, but there are very few who if questioned on the subject could give any clear explanation of the grounds of their belief, of the nature of that system, or of the mode by which a knowledge of it may be attained. The uncertainty which hangs over the subject is doubtless owing to the obscure and metaphysical nature of some of the principles involved, and still more to the vague conceptions and crude theories which have been promulgated on the subject.

This essay is contributed in the hope that, even if its own arguments are of little value, it may, at least, induce others to investigate the subject on more correct principles than have hitherto been followed.

The postulate with which I commence the inquiry is, to let it be granted that there are such things as *species*, distinct in their characters and permanent in their duration. This being admitted, we define the natural system to be *the arrangement of species according to the degree of resemblance in their essential characters*. In other words, the natural system is that arrangement in which the distance from each species to every other is in exact proportion to the degree in which the essential characters of the respective species agree. Hence it follows that the whole difficulty of discovering the natural system consists in forming a right estimate of these degrees of resemblance. For the degree in which one species resembles another must not be estimated merely by the conspicuousness or numerical amount of the points of agreement, but also by the physiological importance of these characters to the existence of the species. On this point no certain rules have yet been laid down; for though naturalists in general admit, for instance, that the nervous system is superior in importance to the circulatory, and the latter superior to the digestive system, yet this subject is still in a very indeterminate state, and until our knowledge of physiology is much further advanced, disputes will always arise respecting the true position of certain species in the natural classification. Such differences of opinion, however, will continually diminish as our knowledge increases, and they are even now very few in comparison with the numerous facts in classification on which all naturalists are agreed. Much may be effected by education and habit, which impart to the naturalist a peculiar faculty (termed by Linnæus a "latent instinct") for appreciating the relative importance of physiological characters to the satisfaction of himself and others, even in cases where he is unable to explain the principles which determine his decision.

Granting, then, that by combining the *number* of points in which any two species agree, with an estimate of the *physiological importance* of those several points of agreement, the naturalist may, in practice, form a tolerably exact conception of the *degree of resemblance* between them; he will proceed in his construction of the natural system to place these species at greater or less distance from each other, in proportion to that degree of resemblance. If we suppose that by a repetition of this process every species is placed in its true position, we obtain a definition of those much-disputed terms, *affinity* and *analogy*,—the former of which consists in those *essential* and *important* resemblances which determine the place of a species in the natural system, while the latter term (analogy) expresses those *unessential* and (so to speak) *accidental* resemblances which sometimes occur between distantly allied species without influencing their position in the system. With *analogy*, therefore, we have no further concern in the present discourse, as it is a principle in no way involved in the natural system. *Affinity*, on the contrary, forms the chief element in this inquiry; and to place species in the order of their affinities is to construct the natural system*.

It appears from the above views that the natural system is an accumulation of facts which are to be arrived at only by a slow inductive process, similar to that by which a country is geographically surveyed. If this be true, it is evident how erroneous must be all those methods which commence by assuming an *à-priori* system, and then attempt to classify all created organisms in conformity with that system. This, nevertheless, is a defect which exists more or less in many modern methods of classification. The greater part of these arrangements are

* I am aware that by many naturalists analogy is considered to be as important an element in the natural system as affinity is. As the discussion of this question would lead us away from the present object, I will not enter upon it now, especially as my views respecting it are stated more at large in the 'Mag. of Nat. Hist.' for May last, iv. pp. 222 *et seq.* Collected Papers, p. 403.

based on an assumption that organic beings have been created on a regular and symmetrical plan, to which all true classifications must conform. Some naturalists have attempted to place all animal species in a straight line, descending from man to a monad. This theory assumes that each species (excepting the two extremes) has two, and only two, direct affinities; one, namely, with the species which precedes, and the other with that which follows it. Others, perceiving the existence in many cases of more than two direct affinities, have compared the natural system to a series of circles, or to the reticulations of network. Many authors have assigned the most mathematical symmetry to the different parts of the system by maintaining the prevalence throughout of a constant number, such as 2, 3, 4, 5, or 7. In applying these views to facts, they have of course found numerous exceptions to the regularity of their assumed formulæ; but by adducing the extermination of some species, and our ignorance of the existence of others, and by applying a Procrustean process to those groups which were either larger or smaller than the regulation standard, they have removed the most glaring objections to their theory, and have with wonderful ingenuity given their systems an appearance of truth*. But when the unprejudiced naturalist attempts to apply any one of these systems to nature, he soon perceives their inefficiency in expressing the real *order of affinities*. The fact is, that they all labour under the vital error of assuming that to be symmetrical, which is in an eminent degree irregular and devoid of symmetry. I will now proceed to give my reasons for taking this view of the subject.

1. *A-priori* considerations, so far from leading us to assume a *regular geometrical pattern*, or *numerical property* in the groups of organized beings, appear to indicate the direct contrary; for the analogies of external nature all indicate the utmost variety and irregularity. Beautiful as are the examples of creative design exhibited in the universe, and admirable as are the adaptations of one part of nature to another, there is no department of the creation which is tied down to mathematical laws and numerical properties further than is sufficient for the due performance of its destined functions. There are indeed certain mathe-

* As these remarks may appear somewhat severe, it is right to substantiate them by a few examples. So long as these systems are admitted by their authors to be *artificial*, it would be as unjust to object to them, as to complain of the alphabetical arrangement of an encyclopædia, that it broke the connection of the subjects. The reply would of course be, that an encyclopædia does not profess to arrange subjects in their natural order, but merely aims at convenience of reference. The remarks in the text, therefore, merely apply to those *symmetrical* methods which profess to exhibit *The Natural System*. The examples are selected from Mr. Swainson's 'Classification of Birds,' in which work the *reality* of the quinary system is insisted on throughout. See Appendix.

matical laws which regulate the motions of bodies and their chemical combinations, but these do not give to the face of nature that symmetrical and *artificial* appearance which is aimed at by the zoological systems above mentioned. For example, the relative distances of the planets, their magnitudes, and the number of their satellites conform to no known numerical law. The fixed stars exhibit no regular arrangement, either in their magnitudes, distances, or positions, but appear scattered at random across the sky. To descend to our own earth, no symmetry is traceable in the forms of islands or continents, the courses of rivers, or the directions of mountain-chains. Organic life exhibits the same irregularity,—no two plants, and no two leaves of the same plant, were ever perfectly identical in size, shape, colour and position. In the "human face divine," portrait-painters affirm that the two sides never correspond; and even when the external form of an animal exhibits an appearance of bilateral or radiate symmetry, nature departs from it in her arrangement of the internal structure. In short, variety is a great and most beautiful law of Nature; it is that which distinguishes her productions from those of art, and it is that which man often exerts his highest efforts in vain to imitate. When, therefore, we find a system of classification proposed as the natural one which departs from this universal law of variety, and fetters the organic creation down to one unalterable geometrical figure or arithmetical number, there is, I think, a strong *à-priori* presumption that such a system is the work not of nature but of art.

2. It follows from the irregularity of external nature, as seen on the surface of the earth, that the groups of organized beings *must* be irregular also, both in their magnitudes and in their affinities. In proof of this it must be granted, that the final cause of the creation of every animal and plant is the discharge of a certain definite function in nature, and not the mere occupation of a certain post in the classification; in short, that the design of creation was to form, not a cabinet of curiosities, but a living world. Few, I trust, would hesitate to admit this proposition. If, then, the different modifications of structure which constitute the characters of groups were given solely with reference to the external circumstances in which the creature is destined to live, it follows that the irregularities of the external world must be impressed upon the groups of animals and of plants which inhabit it. The supply of organic beings is exactly proportioned to the demand; and Nature does not, for the sake of producing a regular classification, go out of her way to create beings where they are not wanted, or where they could not subsist. Thus, for instance, the warm climate and varied soil of the tropics admits of the growth of a vast variety of flowers and fruits. The group of Humming-birds which feed on the former, and of Parrots

which feed on the latter, are accordingly found to be developed in a vast variety of generic and specific forms; while the family of Gulls, which seek their food in the monotonous and thinly inhabited regions of the North, are few in species and still fewer in genera. Again, the variety of plants in the tropics admits the existence of a great variety of insects, and the family of Woodpeckers is proportionately numerous; while the Oxpecker (*Buphaga*), which seems to form a group fully equivalent in value to the Woodpeckers, is limited to but one or two species, because its food is confined to a few species of insects which only infest the backs of oxen.

It follows, then, that the groups of organized beings will be great or small, and the series of affinities will be broken or continuous, solely as the variations of external circumstances admit of their existence, and not according to any rule of classification. If, indeed, we were to imagine a world laid out with the regularity of a Chinese garden, in which a certain number of islands agreeing in size, shape, soil, and form of surface, were placed at exactly equal distances on both sides of the equator, we might then conceive the possibility of a perfect symmetry in the groups of beings which inhabit them; but without some such supposition, I do not see how a class of animals or plants can be symmetrical in themselves, and yet be expressly adapted for conditions of existence which are eminently irregular.

3. To pass from syllogism to induction, it is most certainly not the case that any definite number or geometrical property runs through the animal or vegetable kingdom. I do not wish on the present occasion to enter on any criticism of individual systems, but it would be easy to show that no *symmetrical* system yet proposed is a true picture of the real series of affinities. Without referring to the numerous gaps in these systems, which are referred by their authors to species being extinct or unknown, I could point out numerous examples in which natural affinities are violated, insignificant groups promoted, or important ones reduced to the ranks, in the vain endeavour to drill the irregular troops of Nature into the square, the column, and the phalanx*. And although in some cases we do find examples of the recurrence of a certain number in the subdivisions of natural groups, yet when we remember the ease with which groups may be extended or curtailed to support a theory, the numerous exceptions which occur to these numbers, and the variety of numerical theories which have been maintained with equal firmness by different authors, we cannot, I think, regard these occasional coincidences of number as otherwise than accidental.

If, then, the diversities of organic structure, being adapted to the varying conditions of the earth's surface, are, like them, full of irregu-

* See Appendix.

larity and variety, it is plain that we can no more speculate theoretically as to what groups are likely to remain undiscovered, than we can predict the discovery of rivers, lakes or islands in any unexplored portion of the earth's surface. Both inquiries must be pursued in the same way, viz. by a careful induction of facts; and it will be found that there is much analogy between the process here recommended and that of a geographical survey. The plan proposed is to take any species, A, and ask the question, What are its nearest affinities? If, after an examination of its points of resemblance to all other known species, it should appear that there are two other species, B and C, which closely approach it in structure, and that A is intermediate between them, the question is answered, and the formula B A C would express a portion of the natural system, the survey of which is so far completed. Then take C, and ask the same question. One of its first affinities, that of C to A, is already determined; and we will suppose that D is found to form its nearest affinity on the other side. Then B A C D will represent four species, the relative affinities of which are determined. By a repetition of this process, supposing our knowledge of the structure of each species to be complete, and our rules for determining the degrees of affinity correct, the whole organized creation might be ultimately arranged in the order of its affinities, and our survey of the natural system would then be finally effected. Now, if each species never had more than two affinities, and those in opposite directions, as in the above example, the natural system would form a straight line, as some authors have assumed it to be. But we shall often find, in fact, that a species has only one direct affinity, and in other cases that it has three or more, showing the existence of lateral ramifications instead of a simple line; as shown in this example, where C, besides its affinity to A and D, has an affinity to a third species, E, which therefore forms a lateral ramification.

```
B - - A - - C - - D
            '
            '
            E
```

It was the observation of this fact which led some naturalists to adopt the circular instead of the linear theory, still adhering to the assumption of a symmetrical figure, but changing their notions of its form. Now although we find occasional ramifications in the affinities, and although these ramifications may occasionally anastomose and form a circle, yet it has been shown that the doctrine of a regular figure cannot be sustained, and therefore if ever it be permitted to man to discover what the true figure is which will express all the affinities of organic bodies, it can only be effected by constructing it piecemeal in the way above proposed. All that we can say at present is, that ramifications of affinities exist; but whether they are so simple as to admit of being correctly depicted on a plane surface, or whether, as is more probable, they assume the form of an irregular solid, it is premature to decide. They may even be of so complicated a nature that they cannot be correctly expressed by terms of space, but are like those algebraical formulæ which are beyond the powers of the geometrician to depict. Without, however, going deeper into this obscure question, let us hope that the affinities of the natural system will not be of a higher order than can be expressed by a solid figure; in which case they may be shown with tolerable accuracy on a plane surface; just as the surface of the earth, though an irregular spheroid, can be protracted on a map. The natural system may, perhaps, be most truly compared to an irregularly branching tree, or rather to an assemblage of detached trees and shrubs of various sizes and modes of growth *. And as we show the form of a tree by sketching it on paper, or by drawing its individual branches and leaves, so may the natural system be drawn on a map, and its several parts shown in greater detail on a series of maps.

In order to show that the views here maintained are not chimerical, I will here present one or two *sketch*-maps of different families of birds, though I am well aware that our knowledge of natural history is as yet far too imperfect to pretend to accuracy †. Such sketches as these can be compared only to the rude efforts at map-making made by the ancients, of which the Peutinger Table is an example; and it is probably reserved for a distant age to introduce that degree of exactness into natural history which in modern geography is attained by a trigonometrical survey. For the sake of simplicity, in making these sketches I have omitted the consideration of *species*, but assuming that the genera of modern authors consist solely of closely allied species, I have proceeded to group them in what appeared to be their true position in respect of their affinities. In order to place these groups at their true distances, it is necessary to form a *scale of degrees of affinity*, to which the intervals between each genus shall correspond. I am aware that this scale must be, in some measure, arbitrary; but for this there is no remedy. The division of the fixed stars into *seven* magnitudes is arbitrary also, yet it is found in practice to answer the purpose. It is evident, from the complex ramifications assumed by the natural system, that it is impossible, in a zoological work, to describe each genus or species in the exact order of their affinities, but that leaps must often be made from

* If this illustration should prove to be a just one, the order of affinities might be shown in museums in a pleasing manner by constructing an artificial tree, whose ramifications should correspond with those of any given family of birds, and by then placing on its branches a stuffed specimen of each genus in its true order.

† See Plate X., which exhibits one of these attempts at zoological map-making.

one part of the system to another, just as in a geographical work we cannot describe the counties of Great Britain in their exact order of position, but must continually make lateral digressions, and then return to the main line of our route. So in anatomy, we not only cannot study or describe the several parts in the order in which they join each other in the human body, but each part must even be dissected out from the rest, and removed from its natural position, before we can comprehend its characters and functions. This is an inconvenience inseparable from the nature of the case, and it is therefore no just complaint to make against a systematic work, that it frequently makes diversions which break the order of affinities. We are therefore at liberty to consult our own convenience, and consequently, whatever may be the form which the natural system, on further survey, may assume, there will be no reason for departing widely from the usual custom of commencing with Mammalia, and proceeding through Birds, Reptiles, and Fish, to the Mollusca, Annulosa, Radiata, &c. Let it not then be objected to the method here proposed, that it is subversive of the arrangements now in use. No *linear* arrangement, whether adopted in a museum, a catalogue, or a descriptive work, ever *can* express the true succession of affinities: such an arrangement, therefore, is necessarily in great measure artificial, and, if sanctioned by custom, may still be adhered to. The true order of affinities can only be exhibited (if at all) by a pictorial representation on a *surface,* and the time may come when our works on natural history may all be illustrated by a series of *maps* on the plan of those rude sketches which are here exhibited.

Those symmetrical systems which are here combated are the natural result of that instinctive love of order which is innate in man, and which produces all the noblest works of art. It would doubtless have been more *convenient* for the arrangement of our museums, and more agreeable to our love of order, if the groups of organized beings *had* resolved themselves into a symmetrical plan; but if such is not the case, we must not sacrifice truth to convenience. My object in communicating these remarks will be gained if they induce naturalists to study Nature simply as she exists,—to follow her through the wild luxuriance of her ramifications, instead of pruning and distorting the tree of organic affinities into the formal symmetry of a clipped yew-tree.

It is needless to observe, that although the above remarks have been applied chiefly to the animal kingdom, yet that the principles here announced, if true at all, may be applied with equal correctness to botanical as to zoological systems.

Appendix.

In Mr. Swainson's 'Classification of Birds,' the Procrustean process is effected in *five* different ways :—1. By transferring the members of redundant groups to fill the blanks in those which are deficient. Examples : Haliaëtus is transferred from Aquilinæ, and made a subgenus of Astur ; Myiophonus is transferred from Merulinæ to Myiotherinæ ; Cinclosoma from Turdidæ, and made a subgenus of Grallina ; Irena from Dicrurinæ, and made a subgenus of Oriolus ; Querulinæ from Ampelidæ to Muscicapidæ ; Coracinæ from Ampelidæ to Corvidæ ; Carduelis and Linaria are transferred from Fringillinæ to Coccothraustinæ ; Scythrops from Cuculidæ to Rhamphastidæ ; Tichodroma from Sittinæ to Troglodytinæ ; Orthonyx from Crateropodinæ (where it comes next Psophodes) to Buphaginæ ; Hæmatopus from Charadriadæ to Ardeadæ ; Eurypyga from Ardeadæ to Scolopacidæ ; Phaëton from Pelecanidæ to Laridæ ; and Dromas from Charadriadæ to Laridæ.

2. By uniting together groups which are naturally distinct. Examples : Harpyia is united with Morphnus ; Ibycter with Daptrius ; Corvinella, Less. (*Lanius flavirostris*, Swains.) with Lanius ; Cyclorhis with Falcunculus ; Psophodes, Sphenura, and Dasyornis with Timalia ; Mecistura and Calamophilus with Parus. The Todinæ are united with Muscicapinæ ; Corydon, Less. (*Coracias sumatranus*, Raffl.) with Eurylaimus ; Cissopis with Pitylus ; the Furnarinæ with Certhianæ ; the Phœnicophainæ with Crotophaginæ ; Dacnis with Nectarinia ; the Tamatiadæ with Halcyonidæ ; Syrrhaptes with Pterocles ; the Chionidæ with the Columbidæ ; the Cracinæ and Psophinæ with Megapodinæ ; Gallinula (*G. chloropus*) with Fulica ; Mergulus and Utamania with Mormon ; and Puffinus with Thalassidroma.

3. By dividing groups which are naturally united. Examples : the Philomelinæ are divided from the Sylvianæ, and the Agelainæ from the Icterinæ.

4. By raising subordinate groups above their natural station. Examples : Budytes, a subgenus of Motacilla, is made a genus equivalent to Lessonia, Enicurus, and Anthus ; Leptonyx and Plectrophanes, sub genera of Emberiza, are made of equal value with the genus Fringilla ; Nyctiornis, a subgenus of Merops, is put on a par with Coracias ; Lamprotila, a subgenus of Galbula, is made a genus.

5. By degrading important groups below their natural station. Examples : Circaëtus is made a subgenus of Gypogeranus ; Cossypha of Orpheus ; Pomatorhinus and Timalia of Malacocercus ; Seiurus of Accentor ; and Blechropus of Fluvicola : Rhamphopsis is made a subgenus of Tanagra ; Euphonia of Aglaia ; Crithagra and Spermophila of Pyrrhula ; Gymnophrys of Manorhina ; Pterocles of Tetrao ; Apteryx of

Struthio; Alechthelia of Gallinula; Phalaropus of Scolopax; Recurvirostra and Totanus of Himantopus; Tachydromus of Glareola; and Phaëton and Rhynchops of Sterna.

Without pretending to assert that in all the above instances my views of the affinities are right and Mr. Swainson's wrong, I will only ask any unbiassed naturalist to examine the objects themselves, without reference to books, and then say whether, in the majority of the above examples, the true order of affinities has not been violated for the sake of supporting a preconceived theory.

It may be added, that after all these efforts, the system of ornithology proposed by Mr. Swainson is very far from being a quinary one. Without referring to the very numerous instances in which his subdivisions *fall short of* the number five, there are several cases in which that number is *exceeded*. Thus the group Fringillinæ has *six* subdivisions; Pyrrhulinæ has *six*; Meliphagidæ *nine*; Tetraonidæ *six*; Ardeadæ *six*, or including Grus (which is apparently omitted through inadvertence), *seven*; and Alcadæ has six.

I feel bound to state, that, notwithstanding these objections, the 'Classification of Birds' is an exceedingly useful manual of ornithology, and it must be regretted that the mass of original observations which it contains is intermixed with so much that is of a visionary nature.

Note.—The questions which are the subject of the above paper were discussed at much length in the 'Philosophical Magazine,' in 1823 and 1825. The reader is referred to vol. lxii. pp. 192, 255, 274; vol. lxv. pp. 105, 183, 372, 428; vol. lxvi. p. 172; also to 'Phil. Mag.' and 'Annals,' New Series, 1830, vol. vii. p. 431; vol. viii. pp. 52, 134, and 200.—Ed.

7. ON THE STRUCTURE AND AFFINITIES OF UPUPA, LINNEUS, AND IRRISOR, LESSON*.

[From Ann. of Nat. Hist. xii. p. 238.]

THE African continent presents us with several species of birds constituting a well-marked genus, to which Lesson in 1831 applied the name *Irrisor*, being a translation of Lavaillant's name "*Moqueur.*" This group of birds was included by Latham in the genus Upupa, by Shaw in Promerops, by Vieillot in Falcinellus, by Cuvier in Merops, and by Temminck and Wagler in Epimachus; but as they differ essentially from the types of all these genera, it is necessary to give them a distinct appellation. Mr. Swainson, Mr. Vigors, the Baron de la Fresnaye, and Mr. G. R. Gray restrict to this group the name Promerops of Brisson; but Brisson was wholly unacquainted with the group before us, and the true type of his genus Promerops is a totally different bird, called by Vieillot Falcinellus, and by Swainson Ptiloturus. It is plain then that the right course is to supplant Falcinellus and Ptiloturus in favour of the old generic name Promerops, and to adopt for the present group the name Irrisor as proposed by M. Lesson.

Having now settled the nomenclature of this group, I will proceed to speak of its affinities, and to show first its relation to the genus Upupa; and secondly, its position in the general system of Nature.

It should be premised, that the genera Upupa and Irrisor agree in the form of the beak, but differ in many other particulars. In Upupa the plumage is ferruginous, varied with white and black; the head is crested; the tail moderate and even, composed of ten rectrices; the feet adapted for walking; the lateral toes being nearly equal, the exterior ones divided nearly to their base; the anterior claws short and blunt, and the hinder claw lengthened and approaching to straightness. In Irrisor, on the contrary, the plumage is black with rich metallic tints, varied only with a few white spots on the wings and tail; the head is not crested; the tail is long and much graduated, composed of twelve rectrices, and the feet are essentially arboreal, the outer toe being much longer than the inner, and united to the middle one for the whole length of the basal joint; the hind toe is very long, and all the claws are compressed, sharp, and much curved. It is evident, therefore, that these birds must differ greatly in their habits; and accordingly we find that the hoopoe lives chiefly on the ground, while the Irrisor is described by Levaillant as exclusively inhabiting trees. The question then arises, whether the

* Read to the Zoological Section of the British Association at Cork, August 19, 1843.

agreement in the form of their beaks is to preponderate over the disagreements of their other organs; in other words, whether this resemblance in the beaks is to be considered as indicating an affinity or only an analogy.

The majority of authors have classed the Irrisors either amongst or very near the Hoopoes. But the Baron de la Fresnaye, in the 'Proc. of the Zool. Soc.' for 1840*, p. 124, contends that the genera Upupa and Irrisor (or as he terms it, Promerops) have in reality no near affinity to each other. He argues that birds have in many cases been arranged artificially in consequence of authors being guided solely by the form of the beak without attending to the structure of the other organs. After pointing out the marked differences between the feet of Upupa and those of Irrisor, he concludes that Upupa has evident affinities with the larks (Alaudinæ), but that its true position is in a special family of the Tenuirostres, in conjunction with Upucerthia and some other allied South American genera. The genus Irrisor, on the contrary, he considers to belong to the Cinnyridæ, or, as they are more correctly called, Nectariniidæ, to which they have much resemblance in their glossy plumage.

Now it is undoubtedly true that the most unnatural classifications of birds have in many cases resulted from the beak being taken as the sole ground of arrangement, to the exclusion of the other organs. I do not however think that the juxtaposition of Upupa and Irrisor is really an instance of such a vicious arrangement, and I hope to show, that notwithstanding the disagreements in their feet, tail and plumage, these two genera are in reality very closely allied.

It will generally be found that when several genera of remote affinity have been brought together in consequence of a resemblance in the form of their beaks, that resemblance is more apparent than real, consisting in a general and superficial agreement in the form and outline, while the minor details of structure present differences which at once indicate the true affinities of the respective groups. Thus the genus Scythrops was till very lately classed by all authors among the toucans, on account of the general resemblance of the beak; while, if the slightest attention had been paid to the position of the nostrils, it would have been seen at once that its true place is among the cuckoos. A similar superficial resemblance in the beak has caused Tichodroma to be classed with Certhia instead of with Sitta, Spermophila with Pyrrhula instead of with Guiraca, Oreoïca and Falcunculus among the Laniinæ instead of the Parinæ, and numerous other cases which might be quoted.

On comparing Upupa with Irrisor, however, we find a coincidence of structure not only in the general forms, but in the minutest details of the structure of their beaks; and what is of still greater importance, the

* See Ann. Nat. Hist. vol. vii. p. 551.

beaks of these two birds present certain characters which are found in no other group of birds with which I am acquainted.

Upupa and Irrisor both present to us the remarkable combination of a very long beak with a very short tongue. The two mandibles are for three-quarters of their length perfectly solid, the surfaces of contact being smooth and flat; while in all other long-billed birds the interior of both mandibles is provided with a hollow space for the reception and action of a lengthened tongue, or for the temporary retention of their food. This very remarkable and peculiar structure has been noticed by no author (as far as I am aware) except Wagler, who in his definitions of Upupa and Epimachus, in which last genus he includes Irrisor, notices this character, but without making any comment on its singularity. It is sufficiently evident from this structure that both Upupa and Irrisor have very little affinity to the Tenuirostres, in which the tongue is remarkably lengthened and adapted to the purposes of suction, and Irrisor cannot therefore be referred to the Nectariniidæ, as supposed by the Baron de la Fresnaye. The fact is, that the beaks of these birds are not constructed for *suction* but for *probing*, i. e. for reaching into deep holes and crevices in quest of the larvæ of insects. We know that the hoopoe obtains its food by inserting its beak into the holes made in the ground by coprophagous insects, and it is probable that the Irrisor feeds in a similar manner upon the larvæ which perforate decayed trees.

The beaks of these two genera of birds present another character unnoticed by all previous authors, and, like the former one, believed to be peculiar to these two genera alone. The basal and medial portion of the

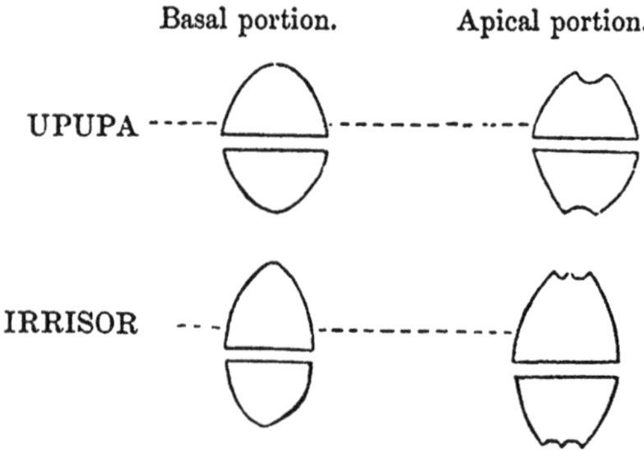

ridge of both mandibles is obtusely and roundedly carinate, but in proceeding towards the apex, the ridge first becomes flattened, then hollowed, and at last deeply grooved. In the Irrisor this flattened portion commences in both mandibles about the middle of the beak, and soon changes into a flat-bottomed groove, which towards the apex is divided into two by a fine intermediate ridge. In Upupa the flat space commences about two-thirds of the total length from the base, and wants the intermediate ridge. With these slight differences the beaks of the two birds may be considered as quite identical in structure, while they differ as before remarked from those of all other known birds. These characters are shown in the above figures, which represent magnified transverse sections of the mandibles.

This peculiar coincidence of structure must, I submit, be considered to indicate something more than mere analogy, and rather to show that Upupa and Irrisor form two subdivisions of the same superior group; or in other words, that they have more affinity to each other than either of them has to any other group which it may resemble.

Nor are the points of mutual agreement in these two genera wholly confined to the structure of the beak. Considerable as their differences undoubtedly are, yet they are not overpoweringly so. They both nidificate in hollow trees. The wings in both are similarly formed, the quills being much graduated, and the fourth and fifth longest. The differences in the style of colouring are not greater than we often meet with in genera of the same subfamily, while the large patches of white on the remiges and rectrices of Upupa have their counterparts on the same feathers of Irrisor. The differences in the form of their tails is a character admitted to be only of *generic*, and in some genera only of *specific* importance. The most weighty distinction is undoubtedly to be found in the structure of their feet, but this is not greater than will be found in the feet of many terrestrial genera when compared with the arboreal forms of the same families. If we look at the feet of ground-cuckoos, ground-woodpeckers, ground-parrots, or ground-pigeons, we shall find that in every case these members are specially modified to suit the habits of the bird, yet this modification of the feet does not blind us to the true affinities of the species which exhibit it.

It may be said, that in the present case the evidence of the feet neutralizes that of the beak, and renders it indifferent which way we decide the question. But this is not a correct view of the case, because neither the feet of Upupa nor of Irrisor present any peculiar and unique structure, such as we see in the beaks of both; they only exhibit a slight modification of the same organs adapted for special modes of life, and such as are to be met with in many other instances of genera belonging to one and the same subfamily.

I conclude, therefore, that the true and natural series of affinities will be most correctly exhibited by preserving Upupa and Irrisor in juxtaposition, and by including them both in the family Upupidæ, which may be divided into two subfamilies, Upupinæ and Irrisorinæ.

We now come to a more difficult question, viz. what is the position of the Upupidæ with respect to the other families of birds? They certainly are a very insulated group, forming what in geology would be termed a *remote outlier*, and it is not easy to say to which of the more continental masses they most nearly approximate. Guided by the elongation of the beak, the majority of authors have placed them unhesitatingly among the Tenuirostres or suctorial birds; but the latter are distinguished by the length of the tongue no less than by that of the beak, and this arrangement cannot therefore be called a natural one. Baron de la Fresnaye, while retaining both Irrisor and Upupa among the Tenuirostres, connects the latter with Upucerthia and its allied genera (Cinclodes, Geositta, Limnornis and Furnarius). But these last are merely a subfamily of the great South American family Certhiidæ, modified in accordance with their terrestrial habits, and the resemblance which they bear to Upupa is very remote, and seems to be one of analogy only.

Where then are the Upupidæ to be placed? This question cannot I think be answered satisfactorily till more facts are collected respecting the food, habits and anatomy of this group and of others with which it may be compared. It may however be conjectured that they are allied in one direction by means of Epimachus or Astrapia to the Paradiseidæ, and in another by Merops to the Alcedinidæ, as shown in 'Ann. Nat. Hist.,' vol. vi. plate 8*, and as originally suggested by Mr. Vigors in 'Linn. Trans.,' vol. xiv. p. 466. In a third direction they are perhaps connected through Lamprotornis with the Corvidæ.

* Coll. papers, p. 408, pl. x.

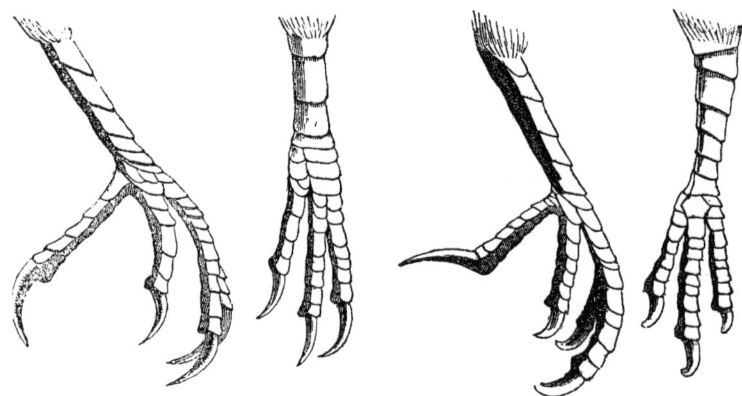

MISCELLANEOUS PAPERS.

1. ON THE MODE OF PROGRESSION OBSERVED IN THE GENUS LIMA, BRUGUIÈRE.

[From Charlesworth's Mag. of Nat. Hist. new ser. i. p. 23.]

In May 1836 I was detained at Malta to perform quarantine; the accustomed penance to which all must submit who visit the Levant. I used to beguile this tedious imprisonment with watching the amphibious occupations of the Maltese fishermen, in the harbour fronting the Lazzaretto. The barren island of Malta teems with a vast population, who live in the greatest poverty, and can with difficulty procure subsistence. Land produce failing them, they take to the sea, and devour indiscriminately all that they can find, whether vertebrated, crustaceous, molluscous, or radiated. Among others of these *frutti di mare*, as they are called, the Maltese are particularly fond of the *Lithodomus dactylus*, Cuv., a shell which perforates the limestone rocks in the same manner as the Pholas. When these shells occur in detached blocks of limestone, the latter are fished up by means of a large pair of iron forceps, attached to ropes. The fisherman sprinkles a few drops of oil on the water, and renders its surface as smooth as glass. He is then enabled, by the clearness of the water, to detect every object in a depth even of six fathoms, and to raise the blocks of stone to the surface by means of his grappling-irons. He then breaks them to pieces, and extracts the *sea-dates*, as these perforating Lithodomi are termed. Not content with those which are obtained by this means, the Maltese will dive to great depths, armed with hammers and chisels, and extract the *dates* from the solid rock.

While watching these operations, and examining the miscellaneous productions obtained by the fishermen, I was struck by the lively motions of some specimens of a Lima, about $1\frac{1}{2}$ in. in length. On placing them in a glass of sea-water, I found that this bivalve possesses the power of swimming in great perfection. The shell opens to a much greater width than any other bivalve that I am acquainted with; and, when thus filled with water, the valves are suddenly closed by a rapid contraction of the adductor muscle. The water is thus ejected with violence from the front margin, and causes the shell to move with considerable speed in an opposite direction. By repeating these contractions and expansions, the animal is enabled to swim in a straight course, and to an indefinite distance. So great is the force with which the valves are brought together, that their collision produces a snapping noise, which is distinctly audible.

I have not observed any other molluscous bivalve which possesses this singular mode of progression. If I mistake not, however, the genus Cypris (a crustaceous animal with a bivalve shell) swims in a manner somewhat analogous.

The animal of the Lima which I noticed at Malta is of a delicate pink colour; the margin of the mantle fringed with numerous slender filaments. It is described, in some works on Mollusca, as provided with a byssus; but of this I could see no traces; nor is it probable that an animal possessed of such great locomotive powers should have one. Cuvier, indeed, though aware of its power of natation, assigns it a small byssus. He further asserts that some species of Pecten are able to swim, while others are permanently attached by their byssus.

Now, it appears to me, that the great locomotive powers of the Lima betoken an organization so different from that of those Pectines which remain constantly rooted to one spot by their byssus, as to make a wide interval between these two genera. It is desirable, therefore, to learn whether this interval really exists, or whether a passage may be traced between the locomotive and the attached species of Pectinidæ. I have therefore drawn up the following queries for the consideration of your conchological readers :—

1. What are the *natatory* species of Pecten to which Cuvier alludes?
2. Do they possess a byssus? If so, can they attach themselves spontaneously?—or are they attached at any definite period of their existence?
3. Do these *natatory* Pectines exhibit any generic or subgeneric difference from the *attached* species?
4. Do the *attached* species become *detached* at certain periods of life, or can they detach themselves spontaneously?
5. Are there any species of other *byssiferous* genera, such as Arca, Pinna, &c., which are constantly or occasionally detached and locomotive?
6. Do any species of Lima possess a byssus? If so, are they either permanently or occasionally attached?

Lastly. If it should appear that the whole genus Lima is destitute of a byssus, and is permanently locomotive; and that all the other genera with which it has been hitherto classed possess a byssus, and are permanently attached;—what other organic differences characterise the genus Lima? and where ought it to be placed in a systematic classification?

2. ON THE NATURALIZATION OF *DREISSENA* (VAN BENEDEN) *POLYMORPHA* (PALLAS) IN GREAT BRITAIN.

[From Charlesworth's Mag. of Nat. Hist. new ser. ii. p. 361.]

THERE are several interesting facts connected with the history of this exotic mollusc which deserve to be recorded. The recent date at which it has been introduced, and the rapidity with which it has spread over the country, afford some valuable data for scientific inquiry. The zoologist will be led to speculate on the means by which species are transported to distant localities, and on the greater capabilities for emigration possessed by certain species and genera, in consequence of being more tenacious of life, more accommodating to circumstances, and more prolific than their congeners. And although the Dreissena has unquestionably been introduced into this country by human agency, yet it might be worth while to inquire how far the same result might have been effected by natural causes alone, such as the drifting of a piece of timber with these shells attached, down one of the European rivers, and across the German Ocean, into some brackish or freshwater estuary on our own coasts. To the geologist also this inquiry is important, as connected with the distribution of organic remains, and the sudden appearance and disappearance of particular species in a given stratum.

The *Dreissena polymorpha* is admirably figured in vol. ix. p. 573 of this Magazine. Among the various names which have been proposed for this genus, the term Dreissena is the least to be commended, being chosen in compliment to an individual, instead of expressing any fact with regard to the genus itself. Nevertheless, the term Dreissena has the clear right of priority, and, being now generally adopted, will, it is hoped, be permanently retained.

Mr. Sowerby first noticed the introduction of this species into Britain in a communication to the Linnean Society, dated November 2nd, 1824. (See Linn. Trans. vol. xiv. p. 585.) He states it to have been introduced into the Commercial Docks, Rotherhithe, whither it had probably been brought attached to Baltic timber. On visiting those docks a few months since, I observed multitudes of these shells in the depôt for bonded timber. It appeared that at least one generation had passed away since 1824, for myriads of the dead shells were scattered over the bottom of the water, and served as points of attachment for the *byssi* of the living.

This shell was next noticed in the Union Canal near Edinburgh, by Mr. Stark, who detected it in the year 1834. In November 1836 the

Rev. M. J. Berkeley communicated to this Magazine his discovery of this shell in the river Nen, in Northamptonshire.

I have further to add the following instances of its occurrence which have come to my own knowledge:—

From the years 1828 to 1834 inclusive I was in the frequent habit of conchologizing in the Avon near Evesham, during which period, if the *Dreissena polymorpha* had inhabited that river, I could hardly have failed in detecting it: not the slightest vestige of it, however, occurred to me during that time. An interval of two years elapsed, during which I was absent from England, and in the beginning of 1837 was much surprised at finding several specimens of this shell among the refuse on the river-banks. On further search I found that the Dreissena had become completely established in the beds of gravel in the river, and in the course of an hour I collected several hundred full-grown specimens. There is therefore clear evidence of the recent introduction of this mollusc into the Avon, and of the rapidity with which it has reached maturity and multiplied. I have since observed it in the canal between Warwick and Birmingham: it has also been detected in the canal at Wednesbury in Staffordshire.

In all the cases hitherto cited this shell has been found in navigable waters, where its transport has doubtless been effected by means of timber. I only know of one instance to the contrary, which is that of the Leam at Leamington, where it has been found of a large size by Dr. Lloyd of that place. But though the Leam itself is not navigable, yet it is in the immediate vicinity of the canal, from which the Dreissena has probably been introduced. I have further to add, that this shell has lately been *planted* by Mr. Stuchbury of Bristol in some of the waters near that place.

It appears desirable to record these particulars, because it may interest some of our field naturalists to watch the gradual spread of this species over the kingdom. Its propagation is so astonishingly rapid, that it will probably become in a few years one of our commonest British shells.

I lately kept several of these molluscs, of different ages, alive for some time in a basin of water. The full-grown individuals, though they had been torn from their native bed, soon secreted fresh byssus, and became anchored to the bottom of the basin. It is evident from this, that the byssus, when first secreted, must be in a highly glutenous state, which enabled it to become attached to the smooth surface of glazed earthenware. The young individuals still retained the power of locomotion, and crawled like Gasteropods over the bottom. They effected this by protruding a foot in advance of the anterior or cardinal end of the shell, and advanced by alternate expansion and contraction of this foot, dragging the shell after them. They indulged their wandering propensities for a few weeks, and then wisely followed the example of their parents, by selecting some convenient nook to which they attached themselves contentedly for life.

I further remarked that these molluscs, acephalous though they be, have still an evident perception of light. When in a quiescent state, they kept the shell partly open, with the siphuncular and branchial apertures protruded; but if any object was suddenly brought over them, they immediately receded, and partially closed their valves, although care was taken that no concussion should be given to the basin.

I have only to add, that I have examined many hundred specimens without detecting any of the crystals mentioned by Mr. Sowerby, vol. ix. p. 643, and figured p. 573.

Irregular pearly concretions are, however, of freqtient occurrence.

Evesham, June 18, 1838.

3. ON THE SATELLITARY NATURE OF SHOOTING STARS AND AËROLITES.

[From the Philosophical Magazine and Journal of Science, s. 3. xxix. p. 2.]

In the truly philosophic work 'Cosmos,' in which the profound Humboldt embodies the results of his life-long studies, he expresses some opinions on the subject of *shooting stars,* which there appears to me considerable difficulty in adopting. If we assume with him that the observations of Benzenberg and Brandes on the parallax of shooting stars are correct, it appears that these bodies have a velocity of from 17 to 36 geographical miles per second, that their elevation above the earth is from 16 to 140 geographical miles, and their diameters from 80 to 2600 feet. It may further be taken for granted, that these bodies revolve in orbits according to the laws of gravitation, that they are ordinarily invisible, but become momentarily luminous whenever they plunge into the earth's atmosphere; and that aërolites are fragments projected or swept from these asteroids (possibly by the resistance of the atmosphere), and hurled to the earth by terrestrial attraction.

Admitting these premises, the next question is to determine the nature of the orbits in which these mysterious bodies revolve, and the influences to which they are subjected in their course. Humboldt here adopts the opinion first propounded by Chladni, that shooting stars and meteors are planetary bodies revolving round the sun in elliptic orbits, and only rendered visible to us at the nodes, where the orbits of the earth and of these asteroids intersect. Their number on this view of the subject must be prodigious, as it is only those whose orbits happen to traverse the earth's orbit, and which happen to pass at the moment when the earth is crossing their node, which can ever be visible to us. How then, let me ask, is it brought about, that these innumerable planetary bodies, which are so continually entering our atmosphere and passing within a few miles of the earth, never come in contact with it? for be it remembered that aërolites are not regarded as being the shooting stars themselves, but only as fragments left behind them in their course. Can we suppose that our earth, a body of nearly 8000 miles diameter, should be incessantly forcing its way through showers of these planetary bodies, hundreds of which daily approach in their circumsolar revolutions within from 16 to 140 miles of the earth, and yet that they should never impinge upon its surface? Should we not in that case continually hear of these fiery masses, with diameters from 80 to 2600 feet, and velocities of

36 miles a second, dashing into the body of our earth like cannon-balls into an earthen rampart? If, in order to meet this objection, it be asserted that the real diameter of the shooting stars has been over-estimated, and that aërolites are not fragments of, but are identical with these bodies themselves, which accordingly really do fall upon the earth's surface, still on the doctrine of chances it would follow that the earth's disc, which presents a far larger surface than that portion of its atmosphere which surrounds and projects beyond its limb, must receive a proportionally larger number of these projectiles.

The attraction of the earth would still further increase the amount of those asteroids which would come in contact with it, as compared with those which pass through and escape from the atmospheric stratum; yet, what is the real proportion between the two classes of phænomena? We find in reality that shooting stars, that is, asteroids rendered visible by atmospheric contact, occur to the amount of scores, sometimes of hundreds, every night, while the fall of aërolites upon the earth's surface is a phænomenon of very much rarer occurrence. It seems evident therefore that there is some cause which renders the circulation of asteroids in orbits approximately parallel to the earth's surface, the normal condition, to which the fall of aërolites to the ground (whether we regard them as being the entire *nuclei*, or merely detached fragments of these meteors) forms only a casual exception.

To what then must we attribute this constant flight of asteroids in lines closely approximating, yet not impinging upon, the earth's surface? It seems evident that we cannot regard them as solar *planets*, pursuing their course through the system regardless of intervening obstacles, as they must inevitably in that case come into very frequent contact with our earth. Why then may we not suppose them to be *satellites*, revolving rapidly round the earth in orbits more or less eccentric, and occasionally plunging into the upper regions of the atmosphere? It does not follow because these bodies move with "planetary velocity," that they must therefore be planets. The satellites nearest to the bodies of Jupiter and Saturn revolve round those planets with a velocity of about ten miles per second, which is not very greatly inferior to that assigned to some shooting stars; and as the velocities of satellites increase with their proximities, we may well suppose that satellites revolving within 150 miles of their primary would have very high velocities. The alleged velocities of shooting stars accord sufficiently well (allowing for the perturbations to which the proximity of the earth may give rise) with Kepler's law, that the squares of the times are proportionate to the cubes of the distances. By applying this law to the known velocity of the moon, it results that a satellite revolving round our earth at 5000 miles from the centre, or about 1000 miles from the surface, would have a velocity of about

40 miles per second, which is even greater than that hitherto assigned to shooting stars.

We may surely then conceive these bodies to be of the nature of satellites, having all their elements so adjusted as to ensure a perpetual revolution round the earth, into whose atmosphere they occasionally dip and undergo a momentary ignition.

It appears moreover difficult to conceive, that if the motion of meteors is of a planetary nature, such small bodies could pass within a few miles of the earth, and then proceed on their course round the sun, comparatively uninfluenced by the terrestrial attraction. The perturbation produced by the earth's mass on a planet of only a few hundred feet in diameter passing within 100 or 150 miles' distance, would surely be so enormous as wholly to destroy the original orbit of the minor body, and the most probable effect would be to convert it into a satellite and to retain it permanently within the earth's attraction. So that even admitting that these asteroids may have once been in the condition of planets, and that many such bodies may still, unknown to us, be revolving in circumsolar orbits, we must yet regard all the shooting stars which ordinarily make their appearance within our atmosphere, as being at present in the condition of satellites.

The main objection, and it is certainly a very important one, to the satellitary theory of shooting stars, is founded on the fact of the nearly (though not quite regular) periodical recurrence of an increased number of these meteors at certain annual epochs. This has been explained by supposing that the earth at these periods intersects certain zones or orbital rings, in which vast numbers of asteroids are constantly revolving round the sun. On the supposition that these bodies are satellitary, and not planetary, it is certainly difficult to account for the fact of their becoming visible in greater numbers at one season of the year than at another. It is however conceivable that the luminosity of shooting stars may be caused by their coming in contact, not with our gaseous atmosphere, but with an electrical atmosphere, which may extend far beyond the limits of the gaseous one; and it may be further conjectured, that from unknown cosmical causes this electric atmosphere may at certain points in the earth's orbit receive quantitative or qualitative modifications, which may enable it at those seasons to illuminate a larger numerical proportion of the meteoric satellites.

At these annual epochs the showers of meteors are said to have apparently proceeded during several hours of observation from the same point in the heavens, viz. the constellation Leo.

But this alleged fact seems irreconcileable with either the planetary or the satellitary theory of shooting stars. Even admitting an approximate constancy in the directions in which these bodies approach our atmosphere, yet as their distance when rendered visible is considered to be not greater than about 150 miles, it is evident that their parallax (whether viewed simultaneously by two distant observers, or at successive intervals of a few hours by the same observer) would be so great as to destroy the appearance of perfect uniformity in the point of the starry heavens, where they make their first appearance. We must therefore suppose that the amount of this uniformity has been overstated; still it is possible that there may be a prevailing direction in which the majority of these bodies enter the atmosphere, and the predominance of this direction may still be in some degree apparent, notwithstanding the influence of parallax. Such a predominance of direction (if it really exist) does not however necessarily prove the shooting stars to be planets, but may be equally explained on the satellitary theory in the following manner:—

Let *a a* be an elliptical ring composed of great numbers of these satel-

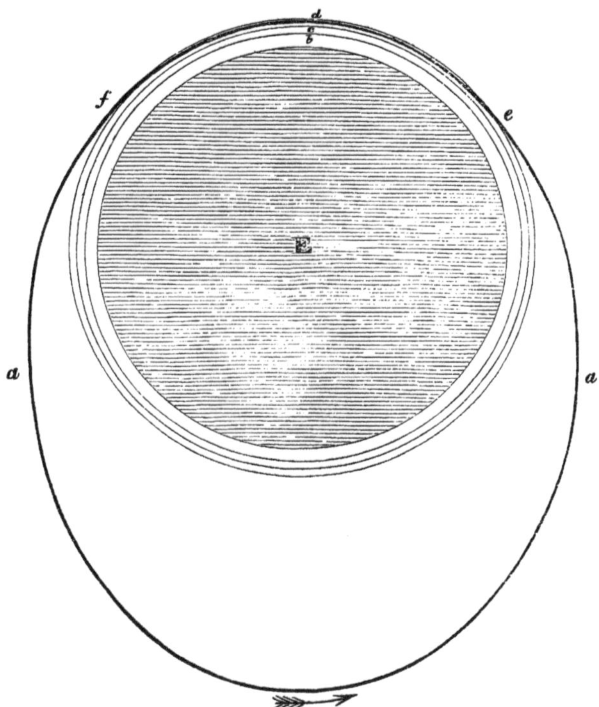

litary bodies revolving in parallel curves round E, the earth; let *b* be the limit of the gaseous, and *c* that of the electric atmosphere in its normal condition. It is evident that while this condition lasts, the meteors in the ring *a a* will be wholly external to the electric atmosphere *c*, and will be consequently invisible. The only meteors which would be seen in these circumstances, would be such stray ones as may revolve in minor orbits occasionally intersected by the circle *c*. But let the electrical atmosphere from some annually recurring cause be temporarily extended to *d*, then while this condition lasts, the meteors revolving in the ring *a a* would be rendered visible during their course from *e* to *f*, and (the ring *a a* remaining always parallel to itself) the meteors would appear (allowing for the effects of parallax) to proceed from nearly the same points of the heavens.

This explanation must however be regarded only as a rude conjecture to remove an apparent difficulty; and as this difficulty, if a real one, may be equally explained on the planetary and satellitary theory of meteors, it need not prevent us from giving the preference to the latter, if the arguments which I have adduced in its favour are of any weight.

4. ANCIENT COLOSSAL STATUE NEAR MAGNESIA.

[From the Archæologia, vol. xxx. p. 524. Read to Antiquarian Society, 17 February, 1842.]

THE curious relic of antiquity which I am about to describe appears to have escaped the notice of the more modern travellers in Asia Minor. Although this ancient sculpture is within an easy day's journey of Smyrna, yet, being situate on a cross road not leading directly to that emporium, it has attracted far less attention than it deserves. I have hence considered that a brief description, accompanied by a sketch made on the spot, may not be uninteresting to the Society of Antiquaries, and may lead others to examine this relic with more care than I was able to do during my hasty visit in April 1836*.

The road from Magnesia† to Sardis proceeds for several miles along the northern foot of Mount Sipylus. This lofty mountain, composed of compact grey cretaceous limestone, not unlike the carboniferous limestone of Derbyshire in appearance, rises abruptly from the alluvial plain of the Hermus‡. It presents in many places a precipitous cliff, occasionally perforated with sepulchral chambers which have been long since plundered, and exhibit no feature of interest. Numerous springs rise at the foot of the mountain, and form a small marsh, which Chandler conjectures to be the lake Sale, mentioned by Pausanias as the site of the ancient town of Sipylus, destroyed by an earthquake§.

One of these springs near a cafinet, about four miles east of Magnesia, is more abundant than the rest, and at the early hour of my visit felt warm to the touch, though without the aid of a thermometer I should hesitate in pronouncing it thermal. Immediately above this spring is a steep and rugged ascent, terminating at the height of about 100 feet in a vertical face of rock, in which is sculptured the colossal statue in question. It represents a figure sitting on a throne, and contained in a niche. Its height from the base of the niche to the crown of the head may be about 20 feet; but, as I had not leisure for taking measurements, this statement is only to be regarded as an approximation. The attitude of the figure is stiff and formal, devoid of grace and dignity, and indicating

* A short notice of my visit to this statue is inserted in Mr. Hamilton's 'Researches in Asia Minor,' vol. i. p. 50.

† This name is sometimes written *Manisa*, but it is more correct to retain the ancient orthography. The modern Greeks spell it Μαγνεσία, but the γ being mute, and the penultimate accented, they pronounce it as an Englishman would *Maniseeah*.

‡ For an account of the geology of Mount Sipylus, see Geol. Trans. ser. 2. vol. v. p. 399.

§ Chandler, 'Travels in Asia Minor,' ch. 79. Pausanias, Ach. ch. 24.

a period of art long anterior to those beautiful specimens of classic sculpture which abound in other parts of Asia Minor. The upper parts of the figure are quite detached from the rock behind, the niche being excavated to a considerable depth, and forming a concave roof over the head of the statue. Notwithstanding this protection, it has suffered so much from the ravages of time, that the features and the component parts of the costume are hardly traceable; and I can therefore give no further details respecting it, but must refer the reader to the following sketch.

This rude specimen of sculpture seems to be quite *sui generis*, as I am not aware of any other example of a similar statue in Asia Minor. Its vast antiquity seems evident from the rude proportions, the colossal size, the weather-beaten surface, and from the circumstance that the style is completely different from any known sculpture of the classic Greek, the Roman, or any later age. On referring to ancient authors, there seems little doubt that we here behold the very ancient statue of Cybele mentioned by Pausanias as having been sculptured by Broteas, son of Tantalus, on a rock named Coddinus, on the north side of Sipylus, in the country of the Magnesians. After mentioning that the town of Acriæ possessed the most ancient statue of this goddess in Peloponnesus, he adds, "*ἐπεὶ Μάγνησι γε, οἱ τὰ πρὸς Βοῤῥᾶν νέμονται τοῦ Σιπύλου, τούτοις ἐπὶ Κοδδίνου πέτρᾳ μητρός ἐστι θεῶν ἀρχαιότατον ἁπάντων ἄγαλμα· ποιῆσαι δὲ οἱ Μάγνητες αὐτὸ Βροτέαν λέγουσι τὸν Ταντάλου.*" (Lacon. ch. 22.)

If this conjecture be correct, we may assign a date to the sculpture of about 1250 B.C.

The only other supposition respecting it is, that this statue is the figure of Niobe lamenting for her children, as alluded to by Homer (Il. Ω, 614), and mentioned by Pausanias in the following words:—"*Ταύτην τὴν Νιόβην καὶ αὐτὸς εἶδον ἀνελθὼν ἐς τὸν Σίπυλον τὸ ὄρος· ἡ δὲ πλησίον μὲν πέτρα καὶ κρημνός ἐστιν, οὐδὲν παρόντι σχῆμα παρεχόμενος γυναικὸς, οὔτε ἄλλως οὔτι πενθούσης· εἰ δέ γε ποῤῥωτέρω γένοιο, δεδακρυμένην δόξεις ὁρᾷν καὶ κατηφῆ γυναῖκα.*" (Attic. ch. 21.)

It seems clear that this passage cannot refer to the sculpture in question; for, in the first place, Pausanias states that he saw Niobe *as he was going up Mount Sipylus*, whereas our statue can only be seen by a person going along the plain between Magnesia and Sardis, and is not visible from any of the tracks over the mountain. It is further evident that Pausanias here refers to some natural rock, which, when viewed at a distance, bore a resemblance to a human form, but lost that appearance on a near approach;—a description quite inapplicable to the unquestionably artificial sculpture above described.

No modern traveller seems to have described this statue except Chishull, who merely surveyed it from the road below, and considered it to be the figure of Niobe*. Chandler and Emerson both passed within 100 yards of it without noticing it. The former fancied that he saw the form of Niobe in the rude outline of some cliff about a mile from Magnesia†, but his account is very obscure, and, I believe, no succeeding traveller has verified his conjecture. Emerson involves the matter in confusion by connecting Chishull's description of the statue with Chandler's fanciful account of the lights and shadows on the cliff at Magnesia‡. Whether or not the latter be right in his identification of Niobe, it seems quite clear that the statue here described is not that personage. If it be not the statue of Cybele which Pausanias mentions, we may suppose that it represents some deity presiding over the spring below, or possibly some hero whose tomb may be concealed in the rock behind.

* Travels in Turkey, p. 12. † Travels in Asia Minor, ch. 79.
‡ Letters from the Ægean, vol. i. p. 225.

5. ON ACCENT AND QUANTITY, AND ON THE PRACTICABILITY OF INTRODUCING ANCIENT METRES INTO MODERN LANGUAGES, CONFORMABLY TO THE LAWS OF PROSODY.

Much has been written, and many ingenious theories have been offered, to explain the apparent discrepancies between Accentuation and the laws of Prosody, as delivered to us by the ancients. I do not propose on this occasion to discuss the views of others, but briefly to explain my own, in the hope that they may throw some farther light on what has hitherto been an obscure and difficult subject*.

The difficulty has been much increased by the diversity in the language, the ideas, and the external circumstances of the investigator and of the people whose writings he studies. The native of the north-west angle of Europe, a Teuton by extraction, with his organs of speech indurated by rigour of climate, and his utterance rendered peculiarly rapid and emphatic by the ceaseless over-excitement of an age of unparalleled social progress, endeavours to comprehend and to resuscitate the pronunciation and the rhythm of poems composed thousands of years ago, at the opposite extremity of Europe, by men of another race, another climate, and of totally different habits of thought. It is no wonder that as long as he views the subject from the recesses of his library, with no materials before him but the writings of the ancients, he should fall into countless errors, or lose himself in a maze. To be a competent judge of the matter he ought to visit the sunny climes of the South where those poems were composed, study the languages, the manners and the ideas of the modern descendants of those classic people, and the still more immutable characters of the countries where they dwell. Though I do not profess to have done this to any great extent, yet during a visit to Italy and Greece a few years ago, I gave some attention to the light which their modern languages may throw upon the ancient ones, and now offer a sketch of the results.

I need hardly say that by *Accent* is understood that peculiar emphasis or elevation of tone which in most languages is laid upon one particular syllable, and one only, of every polysyllabic word. Of modern European tongues, the English probably possesses this characteristic in the greatest, the French in the least, and other languages in an intermediate degree. It is in the excitement of oratory or of rapid vocal utterance that accent becomes most developed.

* I gave a short abstract of these views some time since in the 'Athenæum,' No 1130, p. 644; but as the subject seemed to deserve being treated at greater length than was compatible with the plan of that Journal, I am induced to publish a farther exposition of it.

Quantity, on the other hand, depends not on *elevation or depression of tone*, but on *time*; the syllables may be all uttered in one key, provided their duration be duly regulated; they are classed, not into high or low, but into long or short sounds. Quantity implies a calm state of the feelings, and a distinct enunciation of every syllable. The distinctions of accent are wholly laid aside; all the long syllables are pronounced slowly and with an equal amount of emphasis, while the short syllables, though more rapidly uttered, are still distinctly audible, and are not slurred over as in accentual discourse. A measured flow of sound is thus produced, which though quite practicable in prosaic speech, is more peculiarly applicable to versification. It follows that pronunciation according to quantity is especially adapted for the act of reading aloud, or for rehearsing from memory, such subjects as do not affect the immediate interests of the hearers; while accent is an almost indispensable adjunct to oratory, *vivâ voce* discourse, and all emphatic and energetic language.

It may hence be inferred, that no language, ancient or modern, can ever have been wholly devoid of the element of accent. In animated discourse, the most important words *must* receive a stronger emphasis than the trivial ones, and to lay this emphasis equally on *every* syllable of such words, would produce an unnatural check to the flow of speech. One syllable therefore must be selected to be the permanent recipient of such peculiar stress, and hence the origin of accent. Some languages, as the English, make this accent audible on every word of a sentence, while any word in it that demands especial emphasis, is distinguished by an extra amount of accent on its appropriate syllable: in other languages, as the French, the accent is almost wholly sunk in a monotonous flow of words, except where some peculiarly energetic expression calls it into momentary existence.

If then the accent, though less employed by some nations than by others, is yet an essential element of language, it follows that the ancient Greek and Latin must have possessed it in at least a sufficient degree for the purposes of extempore speech. It seems impossible to believe that the impassioned eloquence of the orators, or the lively wit of the comedians can have been devoid of this, the only oral means of rendering words emphatic. Moreover, the modern Greek and Italian are highly accentual languages; the Greek especially is nearly as much so as the English, and there is hence a strong *à priori* probability that their parents, the ancient Greek and Latin, were accentual languages also.

This probability becomes a certainty in the case of the Greek, for there, by a rare piece of good fortune, the ancient accents have been accurately preserved to us in writing. Greek is, I believe, the only language in existence where this is the case, the marks called 'accents' in French being merely marks of pronunciation,—*letters* in fact, resembling true accents in form, but utterly unconnected with them in essence. And it can be proved that these accentual marks, which English scholars commonly disregard as completely as they do the right pronunciation of the Greek alphabet, do really indicate accent or syllabic emphasis, and not any peculiarity of pronunciation. For, strange to say, they have been retained unaltered for two, if not three thousand years, among the valleys and rocks of the Greek peninsula and islands. Excepting perhaps the Chinese, there is probably no language of equal antiquity which has undergone so little change in the lapse of ages as the ancient Greek. A few foreign words have crept into it, and a few changes of inflexion have been introduced; but still the language of the Cretans and Athenians of the present day is substantially and essentially the same as that of Homer and of Hesiod. Centuries of barbarous despotism have made great changes in the manners and character of the Greek: his exalted taste has become debased, his literature almost extinguished; but his language still survives, like the Parthenon amid the mud hovels and plaster palaces of modern Athens, a glorious relic of the past.

It is with no little interest that the scholar of distant Britain, to whom the accentual marks of the ancient Greek have always been a stumbling-block and a mystery, finds, on visiting the shores of Greece, that every man, woman and child, not only uses many of the words of Homer and of Thucydides, but invariably pronounces them with the accent on the same syllable which bears the accentual mark in ancient writing. The words *ἄνθρωπος* and *μάλιστα*, for instance, are spoken with the accent on the *first* syllable,—not on the *second*, as in English schools; *ἡμέρα* and *ὀλίγος* are pronounced with the accent on the penultimate, and *αὐτὸς* and *ἐπειδὴ* on the ultimate, precisely as is indicated by the accentual marks. It may therefore be considered as certain, that the ancient Greek was, like the modern, an accentual language, that the marks placed over a certain syllable in every word indicated its accents, and consequently that the modern pronunciation of the words is essentially the same as the ancient one.

This being granted, we are suddenly met by a difficulty, which to many persons has appeared formidable. In all modern languages, the poetry is composed with especial reference to the accentuation, the words being so arranged that the accents only fall on those syllables where the rhythm of the verse requires them. But in the ancient Greek and Latin, we find the poetry constructed according to a series of elaborate rules, based on the *quantity* of the syllables, but wholly irrespective of the *accent*. If a word in one line be so placed that its accent correctly falls in with the metre, we shall find the same word in another line placed with its accent at utter variance therewith. In hexameter verse, the accent, *if audible at all*, must coincide with the *first* syllable of every foot, as it is always made to do in the modern imitations of hexameters:—*e.g.*

Líon | heárted | Ríchard was | thére, re|dóubtable | warrior.

But no such rule was followed by the ancients. If the word *possunt*, for example, was accented, as it probably was, on the *first* syllable, then its position would be correct, according to modern practice, in the line

Nec ve|ro ter|ræ fer|re omnes | omnia | póssunt.

But what shall we say to the verse

Carmina | vel cœ|lo pos|sunt de|ducere | lunam,

where the accent on *possunt*, if it were audible, must have been placed on the *last* syllable, not the *first*?

Hence it is quite evident, that whatever may have been the use of the accent in prosaic discourse, it formed no element in the poetry of the ancients. So long as the laws of quantity were not violated, the poetical ear was satisfied. If we read classic verse, whether Greek or Latin, conformably with the quantity, the ear perceives a beautifully melodious metre; but if we pronounce each word according to accent, it instantly degenerates into prose.

It appears then to be equally certain that the ancient Greeks and Romans must have employed the accent in their extempore speech, and that they must have entirely ignored it in their poetry. This is a very remarkable peculiarity of those tongues, and one to which we find no parallel in modern literature; for in all recent languages the accent is considered to be so far an essential and immutable element, that it is on no account permitted to be discarded in poetry. No matter what metre or what language be employed, the accent must always fall into its right place in every foot of modern verse; even the modern Greeks have learned to regard accent as so inseparable a property of their words, that all their attempts at verse take accentuation for their basis. The learned among them have so far lost the true conception of *quantity*, that in reading Homer or any other ancient poet, they invariably, as I have ascertained by experiment, pronounce according to accent; they consequently reduce the poetry to prose, and however much they may appreciate the sentiments of the original, they lose all the beauty of its rhythm. The accents of oral speech have been traditionally preserved among them, but the elaborate and highly artificial ornaments of their ancient verse, like the geometrical beauties of their architecture, have been long since forgotten.

Such appears to me to be the only satisfactory way of accounting for what is undoubtedly a great difficulty, the simultaneous employment in ancient times of accentual prose and of prosodaic verse. I conceive that however rapid and accentual may have been the utterance of everyday speech, yet that by a mutual understanding between poets and the hearers or readers of poetry, this emphatic diction was in their case deliberately laid aside, and a slow stately measure substituted. Poetry and prose must each have had their peculiar languages, agreeing in vocabulary and syntax, but greatly differing in the mode of utterance. It affords one more instance of the high degree of cultivation and refinement attained in literature and the arts by those extraordinary people.

If this idea of the nature of prosody or quantitative versification be correct, it becomes an interesting inquiry how far it would be possible, as a matter of experiment, to revive it in modern languages. Should such an attempt be successful, it will (if it have no other merit) at least serve to give a clearer idea of the nature of prosodaic verse than is attainable by means of the ancient examples alone.

Many poets in modern times, struck by the beauty of the ancient metres, and especially of the hexameter, have endeavoured to introduce them into their respective languages: they have commenced however on a wrong foundation. Adhering to the inveterate notion of the immutability of the accent, they have felt themselves bound to make the latter invariably coincide with the *first* syllable of every spondee, dactyl or trochee, and with the *last* syllable of every iambus and anapæst. By adopting a law unknown to the ancient poets, they have been compelled to depart from the laws which those poets observed. Their verses are consequently a mere spurious jingle, bearing indeed a rude resemblance in metre to their classic prototypes, but utterly unlike them in every essential point. The ear, accustomed to the melodious flow of ancient verse, is shocked and offended by these rugged and debased imitations, for which not all the genius and ideal beauty of Goethe, Schiller or Southey can atone. One or two examples, taken at random, will suffice:—

"Auf den Mauern erschienen, den Säugling im Arme, die Mütter,
Blickten dem Heerzug nach, bis ihn die Ferne verschlang."—*Schiller.*

"Aber der Himmel trübte sich bald. Um den Vortheil der Herrschaft
Stritt ein verderbtes Geschlecht, unwürdig das Gute zu schafften."—*Goethe.*

"Lift up your heads, ye Gates; and ye everlasting Portals,
Be ye lift up. For lo! a glorified Monarch approacheth,
And who in righteousness reigned and religiously governed his people."—*Southey.*

Can anything be more remote from the music of Ovid's or Virgil's verse than such imitations, where almost every syllable exhibits either a false quantity, an open vowel, or some other infraction of poetic law?*

* The defects of the modern so-called hexameters are ably exposed in the Athenæum (No. 1159, p. 40), in a review of a recent translation of Goethe's 'Hermann und Dorothea,' though I altogether demur to the writer's inference, that "in no modern language can the true prosodaic canon, on which the charm of classic versification depends, be observed."

The only way in which we moderns can really attempt to imitate the ancient poets, is simply to do as they did, to obey *their* rules, and to follow *their* practice. If the trial be fairly made, we shall soon settle the questions, whether the structure of our language renders this possible, and whether its genius renders it successful.

In writing prosodaic hexameters in the English language, some difficulties occur which did not affect the ancients. The frequent recurrence of monosyllabic words and of terminal consonants, the non-inflexion of nouns, and other peculiarities of our tongue, may present obstacles unknown to the Greek and Latin, but an occasional exercise of poetical license in the elision of terminal letters will remove this difficulty; a greater one exists in the absence of all rule or precedent as to the *quantity* of English words. We have no *Gradus ad Parnassum* to fly to, and even the orthography and derivation of words will sometimes mislead us as to their quantity. The only standard of quantity in the English language seems to be, not the *spelling* of the words, but their *pronunciation*. Thus the word *region*, e. g., would be a tribrach (rĕgĭŏn) if we regard etymology, but we pronounce it as a trochee (rēējŭn). In all cases therefore, where the rules of the ancient prosody fail to determine the quantity, we ought to be guided by the customary pronunciation.

How far prosodaic English hexameters, constructed like the ancient ones, irrespectively of accent, are to be considered pleasing to the ear and poetically successful, is a question on which tastes will probably differ. To appreciate them fairly, the reader must be well acquainted with the ancient metre, and must endeavour to forget, for a time, the accent of the words. He must read the lines deliberately, giving every syllable its due time and its distinct enunciation. He must not suppose that the accents are to be *transposed* to suit the metre, as if a line were to be read

Far, far away midst my rŏyál fŏrtrésses in Argos.

The accents must be wholly *repudiated*, and the line must be read, according to quantity alone,—

Fār, făr ăwāy mīdst my rōyāl fōrtrēssĕs ĭn Argos.

By this process only can the real beauty of the ancient metres be exhibited in a modern dress. I would gladly see some of our living poets make the attempt to cultivate this style of composition; and I am satisfied that if their efforts meet with fair play at the hands of their readers, a great improvement will be effected in the application of ancient metres to modern languages.

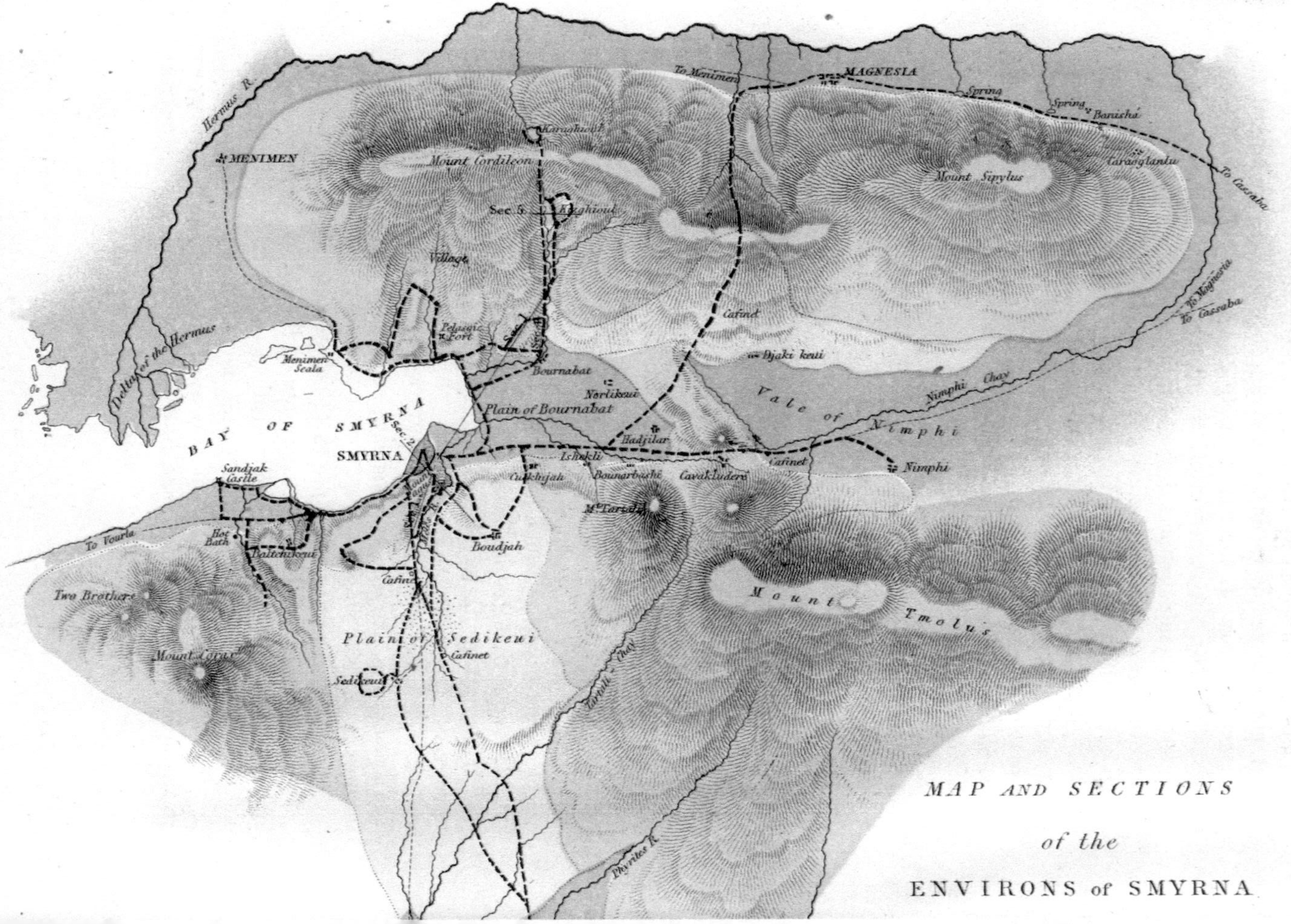

MAP AND SECTIONS of the ENVIRONS of SMYRNA

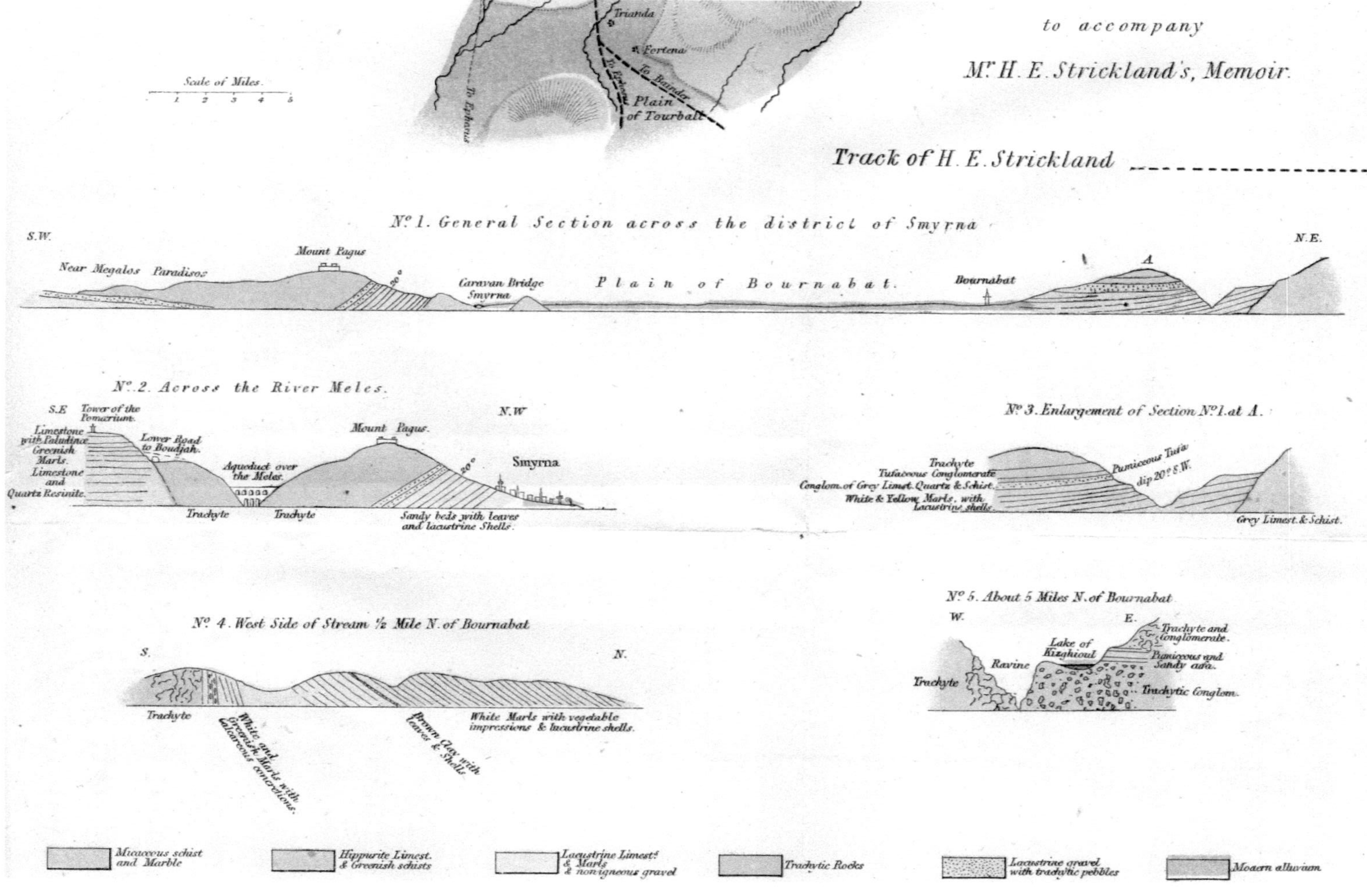
to accompany
Mr. H. E. Strickland's, Memoir.
Scale of Miles.
1 2 3 4 5
Trianda
Fortena
To Ephesus
To Bainder
Plain of Tourbali
Track of H. E. Strickland
No. 1. General Section across the district of Smyrna
S.W.
N.E.
Near Megalos Paradisos
Mount Pagus
Caravan Bridge
Smyrna
Plain of Bournabat.
Bournabat
A
No. 2. Across the River Meles.
S.E
Tower of the Pomarium.
Limestone with Paludinae
Greenish Marls.
Limestone and Quartz Resinite.
Lower Road to Boudjah.
Aqueduct over the Meles.
Mount Pagus.
N.W
Smyrna
Trachyte
Trachyte
Sandy beds with leaves and lacustrine Shells.
No. 3. Enlargement of Section No. 1. at A.
Trachyte
Tufaceous Conglomerate
Conglom. of Grey Limest. Quartz & Schist.
White & Yellow Marls, with Lacustrine shells.
Pumiceous Tufa
dip 20° S.W.
Grey Limest. & Schist.
No. 4. West Side of Stream ½ Mile N. of Bournabat
S.
N.
Trachyte
White and Greenish Marls with Calcareous concretions.
Brown Clay with leaves & Shells
White Marls with vegetable impressions & lacustrine shells.
No. 5. About 5 Miles N. of Bournabat
W.
E.
Lake of Kizghioul
Ravine
Trachyte
Trachyte and Conglomerate.
Pumiceous and Sandy tufa.
Trachytic Conglom.
Micaceous schist and Marble
Hippurite Limest. & Greenish schists
Lacustrine Limests. & Marls & non-igneous gravel
Trachytic Rocks
Lacustrine gravel with trachytic pebbles
Modern alluvium

PLATE II.

Sketch of the
ISLAND OF ZANTE.
Pt. Skinari
Mineral Spring
Catastari
Section 3
Anticlinal Axis
Section 1.
Town of Zante
Convent
Section 2
Port Cheri
Springs of Bitumen
Pt. Cheri
Scale of Miles.
2
4
6
8
Plan of the Coast near the Mineral Spring.
Sea
Small Springs
Large Spring
Sea
Fault
Downcast Side
Upcast Side
Section I. North Side of the Bay of Zante.
Citadel
S.
N.

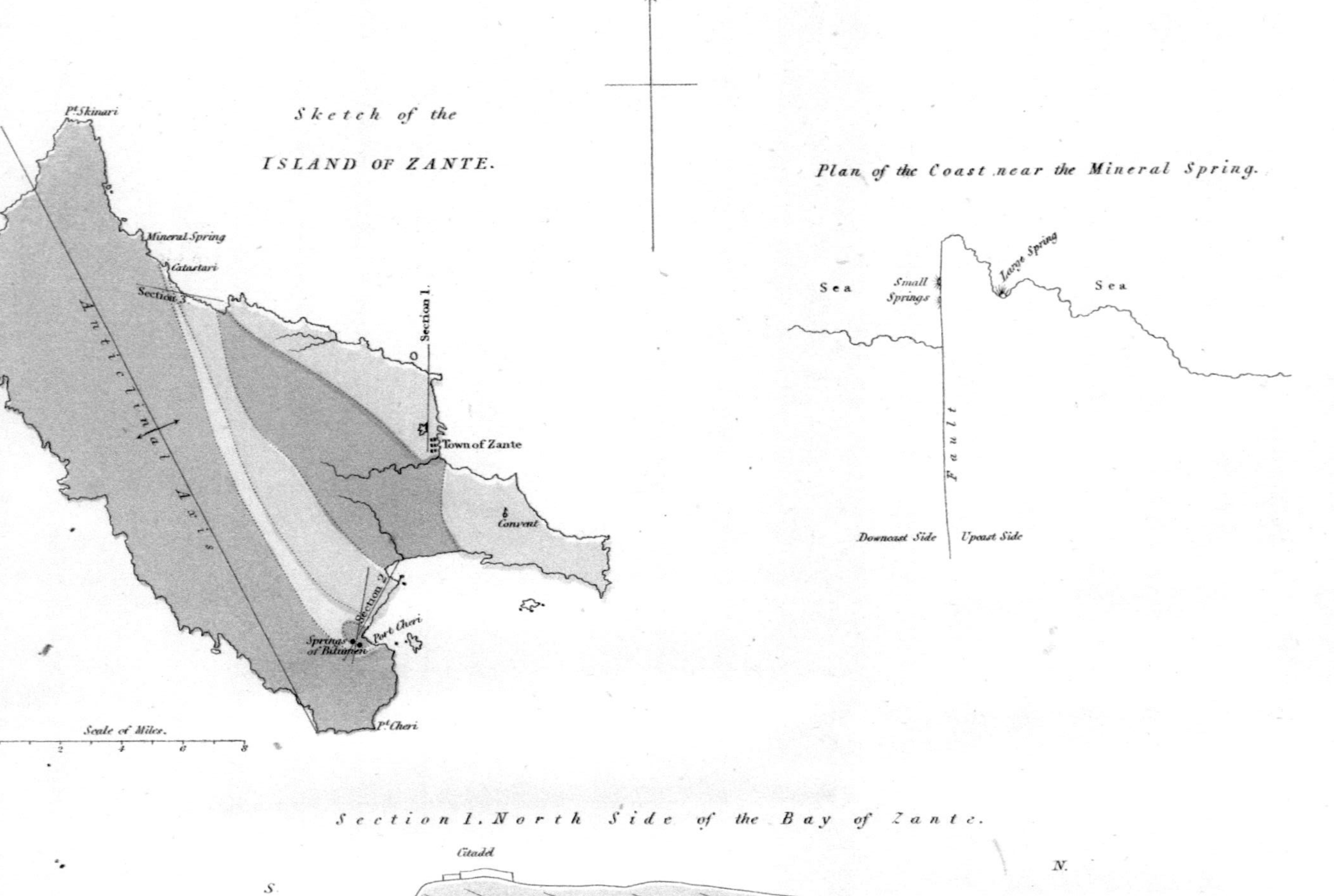

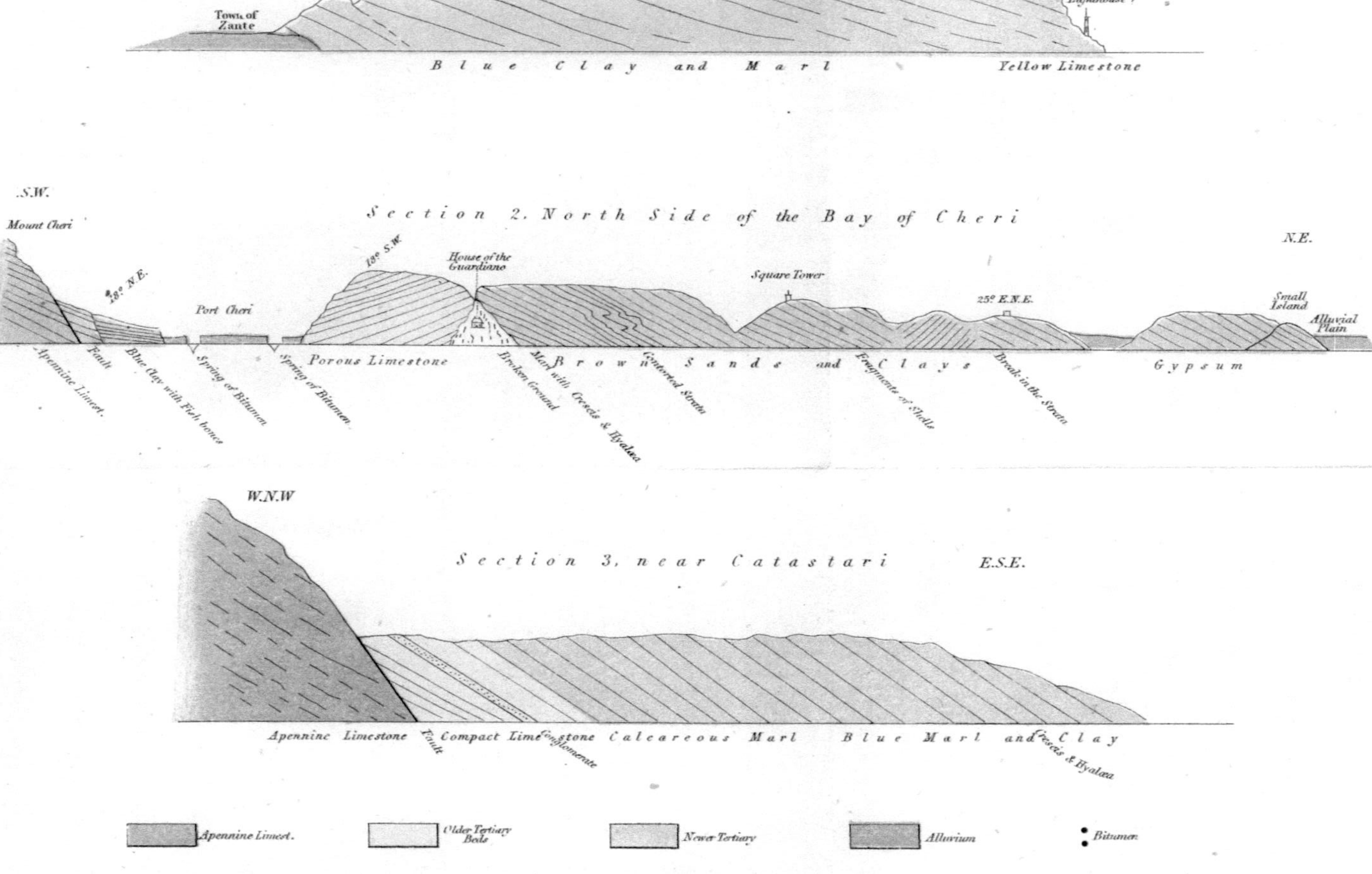

Town of Zante
Lighthouse
Blue Clay and Marl
Yellow Limestone
S.W.
Section 2. North Side of the Bay of Cheri
N.E.
Mount Cheri
18° N.E.
Port Cheri
18° S.W.
House of the Guardiano
Square Tower
25° E.N.E.
Small Island
Alluvial Plain
Apennine Limest.
Fault
Blue Clay with Fish bones
Spring of Bitumen
Spring of Bitumen
Porous Limestone
Broken Ground
Marl with Creseis & Hyalæa
Brown Sands and Clays
Contorted Strata
Fragments of Shells
Break in the Strata
Gypsum
W.N.W
Section 3, near Catastari
E.S.E.
Apennine Limestone
Fault
Compact Limestone
Conglomerate
Calcareous Marl
Blue Marl and Clay
Creseis & Hyalæa
Apennine Limest.
Older Tertiary Beds
Newer Tertiary
Alluvium
Bitumen
W.H. LIZARS, CHROMO LITHOG. EDINR

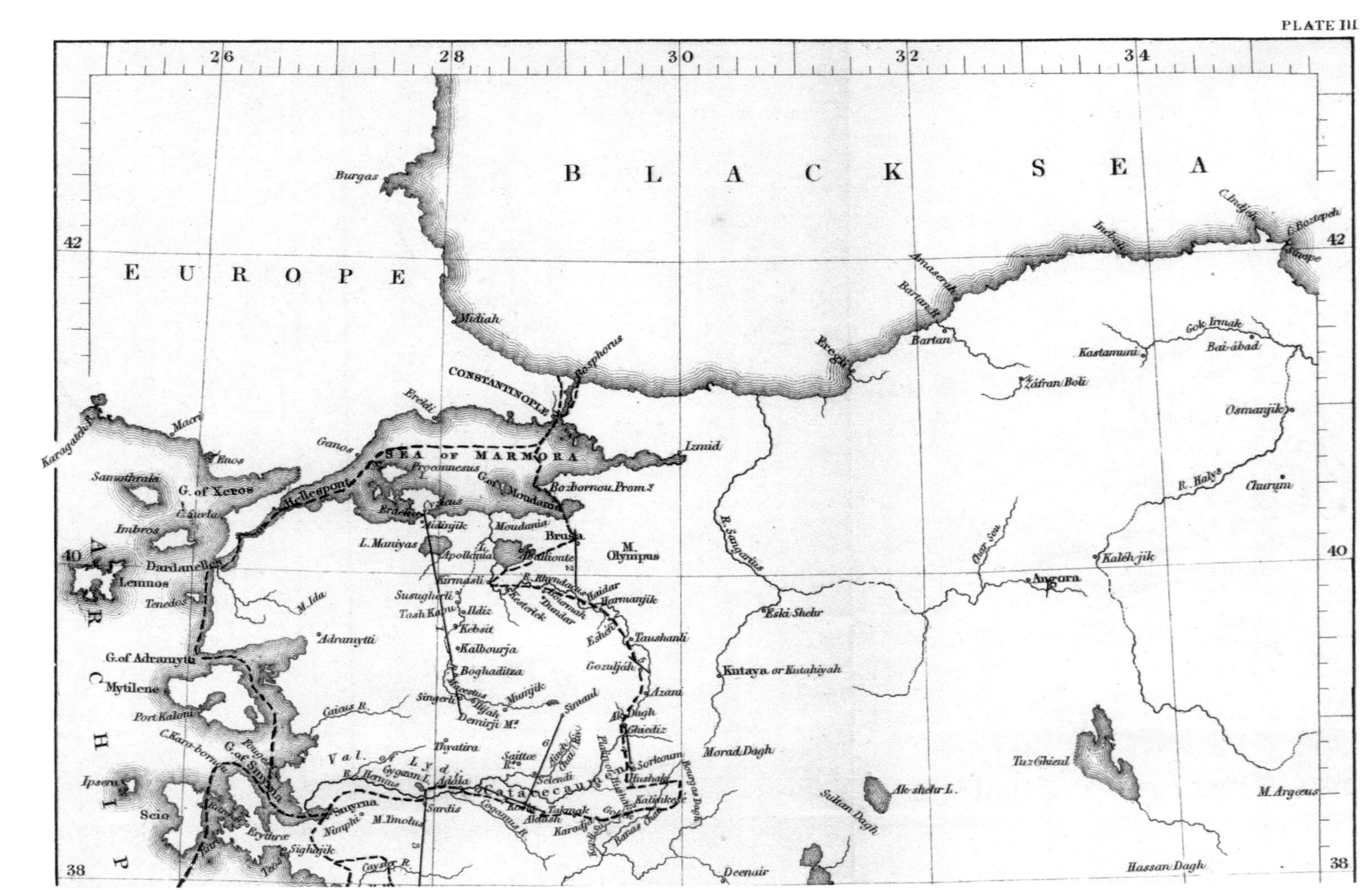
BLACK SEA
EUROPE
ARCHIP
SEA OF MARMORA
CONSTANTINOPLE
Bosphorus
Burgas
Midiah
Ereklí
Ganos
Hellespont
Dardanelles
G. of Xeros
Enos
Macri
Samothraki
Imbros
Lemnos
Tenedos
G. of Adramyti
Adramyti
M. Ida
Mytilene
Port Kaloni
G. of Smyrna
Smyrna
Scio
Ipsera
Erythræ
Sighajik
Caÿster R.
Caicus R.
Thyatira
Sardis
M. Tmolus
Val. of Lyd
Gygæan L.
R. Hermus
Cogamus R.
Katakekaumene
Proconnesus
Cyzicus
G. of Moudania
Moudania
Bozbornou Prom.
Brusa
M. Olympus
L. Maniyas
Apollonia
Kirmasli
Susugherli
Tash Kapu
Ildiz
Kebsit
Kalbourja
Boghaditza
Singerli
Demirji Mt.
Simaul
Gozuljah
Taushanli
Harmanjik
Haidar
Azani
Ak Dagh
Ghiediz
Sorkoun
Ushak
Kalinkeóe
Bourgas Dagh
Izmid
R. Sangarius
Eski Shehr
Kutaya or Kutahiyah
Morad Dagh
Sultan Dagh
Ak-shehr L.
Deenair
Erekli
Bartan
Bartan R.
Amaserah
Ineboli
C. Indjeh
C. Boztepeh
Sinope
Gok Irmak
Bai-ábad
Kastamuni
Záfran Boli
Osmanjik
R. Halys
Churum
Kaleh-jik
Angora
Char Sou
Tuz Ghieul
M. Argæus
Hassan Dagh
26
28
30
32
34
42
40
38

Map of the
WESTERN PORTION
of
ASIA MINOR
to illustrate the Memoir of
W. J. Hamilton Esqr.
and
H. E. Strickland Esqr.
P. I.
Track of H. E. Strickland
10 20 40 60 80
English Miles
26
28 Longitude East from Greenwich 30
32
34
36
34
John Arrowsmith
Samos
Nikaria
Mycori
Tino
Naxia
Paro
Nio
Sikyno
Santorin
Anaphi
Amorgo
Levita
Lero
Patino
Kalimno
Kos
G. of Kos
Bodroom (anc. Halicarnassus)
Cnidus
Stampalia
Piseopia
G. of Syme
Syme
Rhodes
Camiro
Lindo
Scarpanto
Caxo
Stan-dia
Crete
Spina-longa
CANDIA
Calderoni
Ephesus
Scala Nuova
M. Mycale
M. Messogis
Tralles
Maeander R.
M. Latmus
Latmic L.
Hierapolis
M. Cadmus
L. Beg-shehr
Koniyeh
Kara Dagh
L. Seidi-shehr
Taurus Range
Karagia-aghatch B.
Makry
G. of Makry
Patara
Port Vathy
Myra
Phineka Bay
Adalia
Esky Adalia
C. Avova
Adratchan B.
Alaya
Silinty
Karadran
C. Anamur
C. Cavaliere
Lissan-el-kahpeh
Mezetlu
C. S. Andreas
Cerina
Nicosia
R. Pedia
Famagosta
CYPRUS
C. Salizano
Baffa
Limasol
Larnica
Deenair

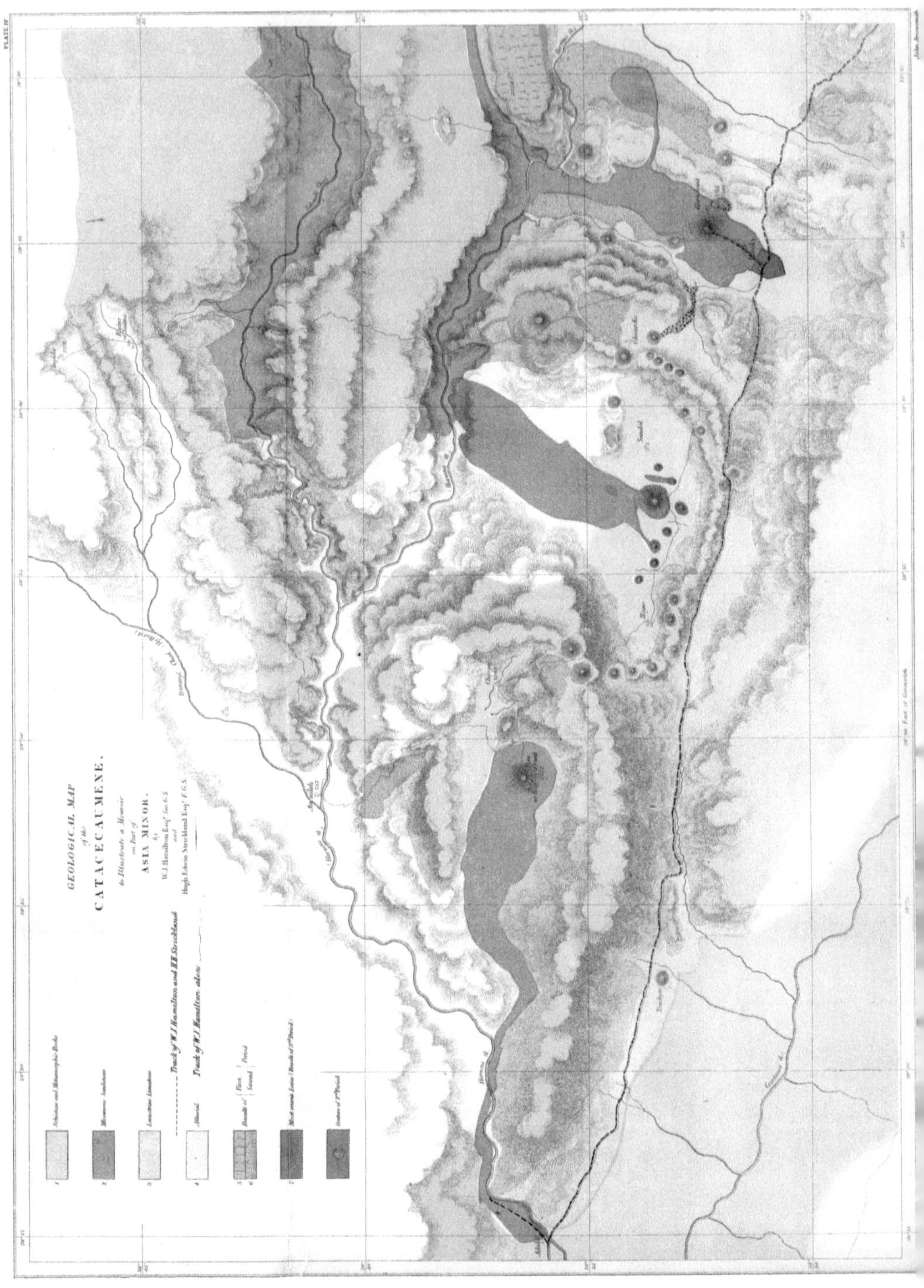

The material originally positioned here is too large for reproduction in this reissue. A PDF can be downloaded from the web address given on page iv of this book, by clicking on 'Resources Available'.

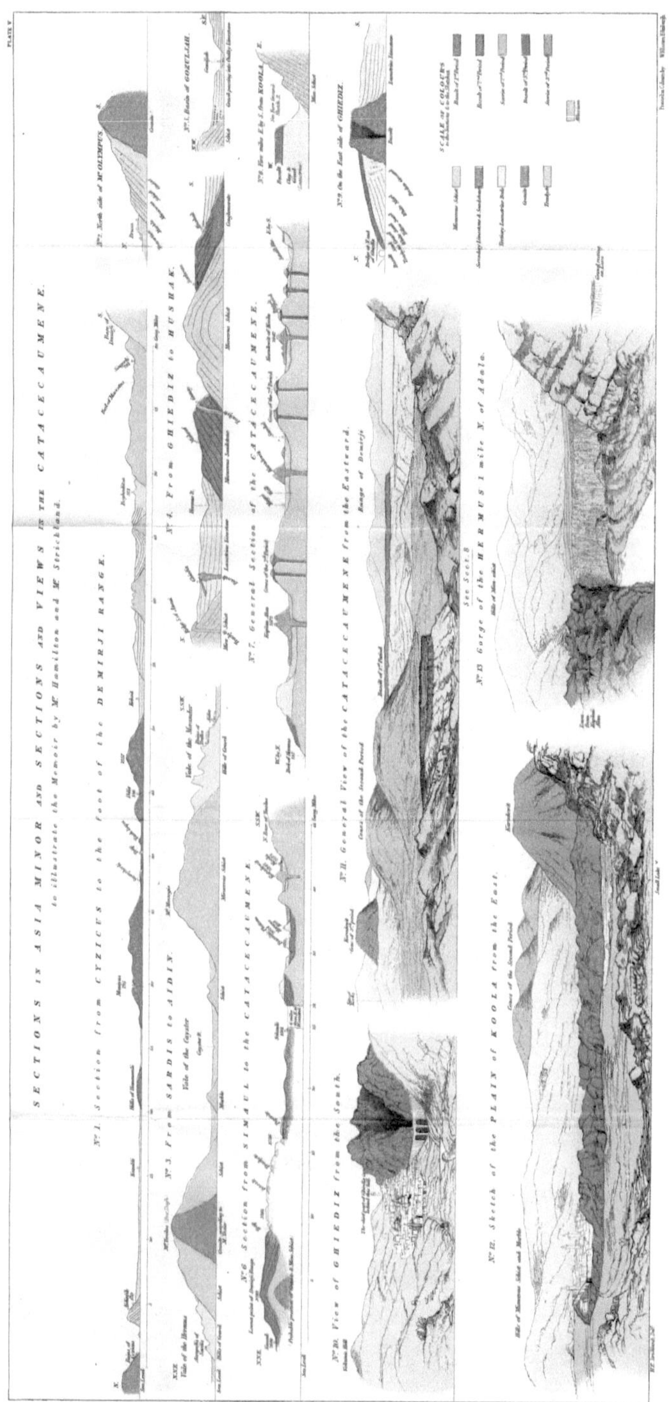

The material originally positioned here is too large for reproduction in this reissue. A PDF can be downloaded from the web address given on page iv of this book, by clicking on 'Resources Available'.

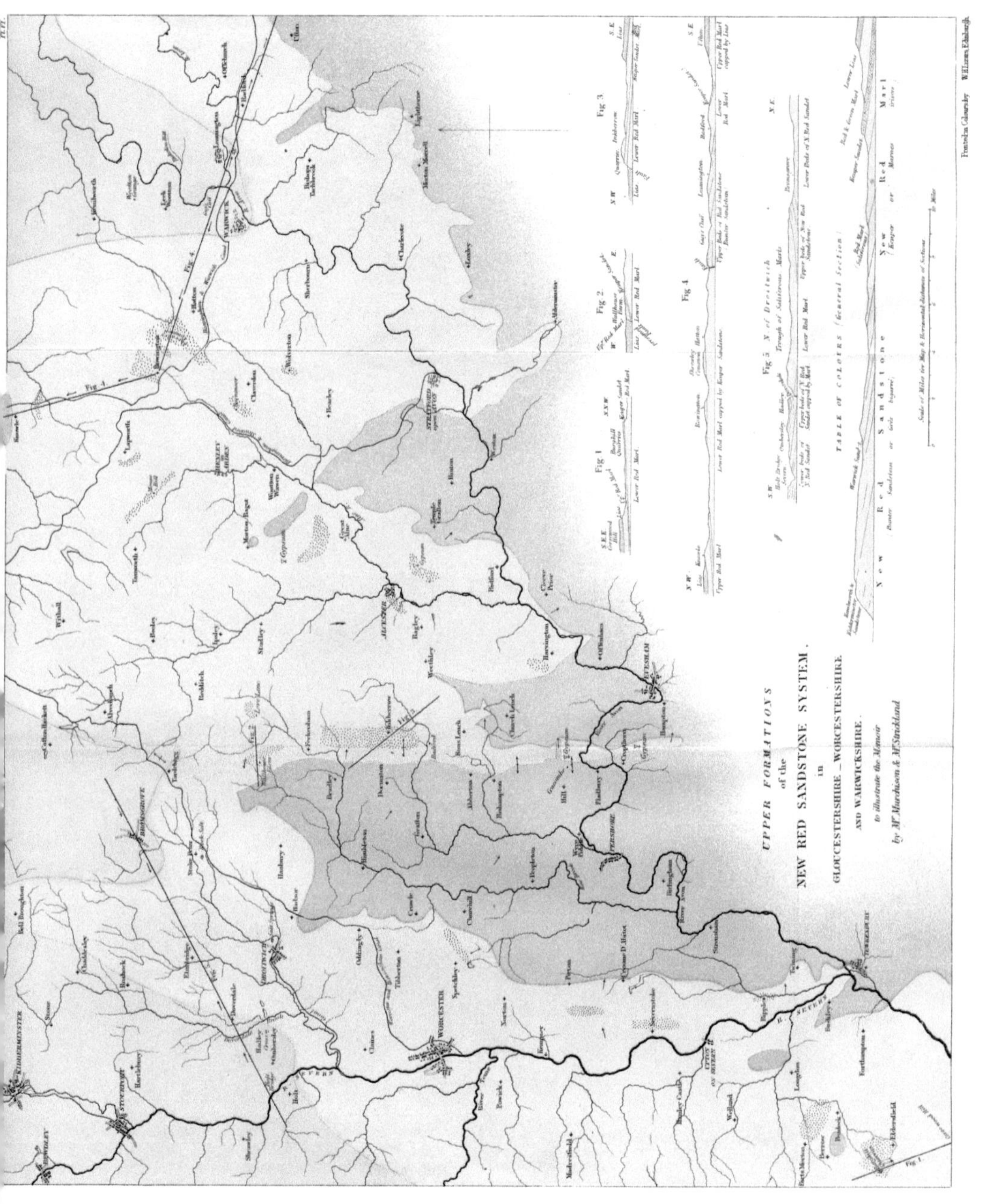

The material originally positioned here is too large for reproduction in this reissue. A PDF can be downloaded from the web address given on page iv of this book, by clicking on 'Resources Available'.

Plate VI a.

Fossils of the Keuper Sandstein.

Fig. 1. Slab of sandstone with ripple marks and impressions of footsteps.
Fig. 2. The footsteps. Fig. 3. Icthyodorulite of Hybodus keuperi.
Fig. 3. a. Fragment of the ribbed surface. Fig. 3. b. Portion of interior.
Fig. 3.* Tooth of an Hybodus. Fig. 4. Posidonomya minuta.

C.D.M.S. lith.

W. West Imp.

Plate VI b.

Fossils of the Bunter Sandstein.

Fig. 5.

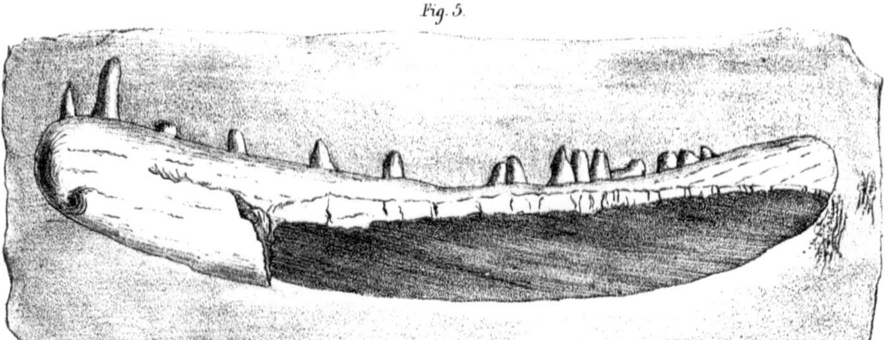

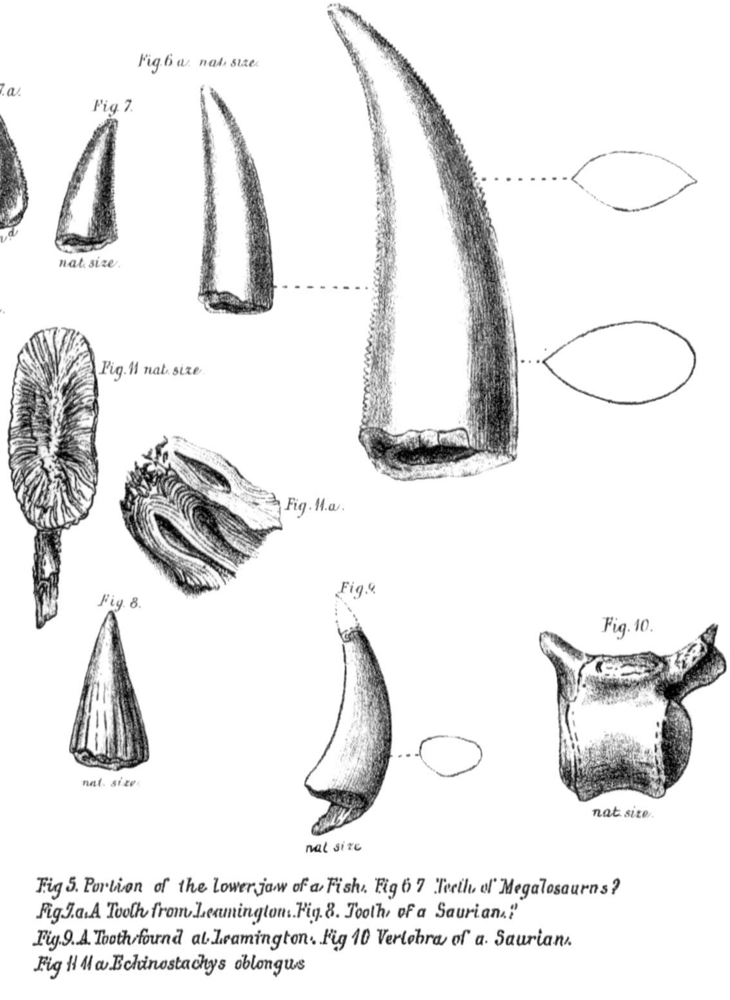

Fig. 6. b.

Fig. 6 a. nat. size.

Fig. 7. a.

Fig. 7.

3 times magn^d

nat. size.

nat. size.

Fig. 11 nat. size.

Fig. 11. a.

Fig. 8.

nat. size.

Fig. 9.

nat. size

Fig. 10.

nat. size.

Fig 5. Portion of the lower jaw of a Fish. Fig 6 7 Teeth of Megalosaurus?
Fig 7. a. A Tooth from Leamington. Fig. 8. Tooth of a Saurian.?
Fig. 9. A Tooth found at Leamington. Fig 10 Vertebra of a Saurian.
Fig 11 11 a. Echinostachys oblongus

C.D.M.S: litho. W. West Imp.

The material originally positioned here is too large for reproduction in this reissue. A PDF can be downloaded from the web address given on page iv of this book, by clicking on 'Resources Available'.

Plate VIII.

Sections of the "Bone Bed" around Tewkesbury, 1841.

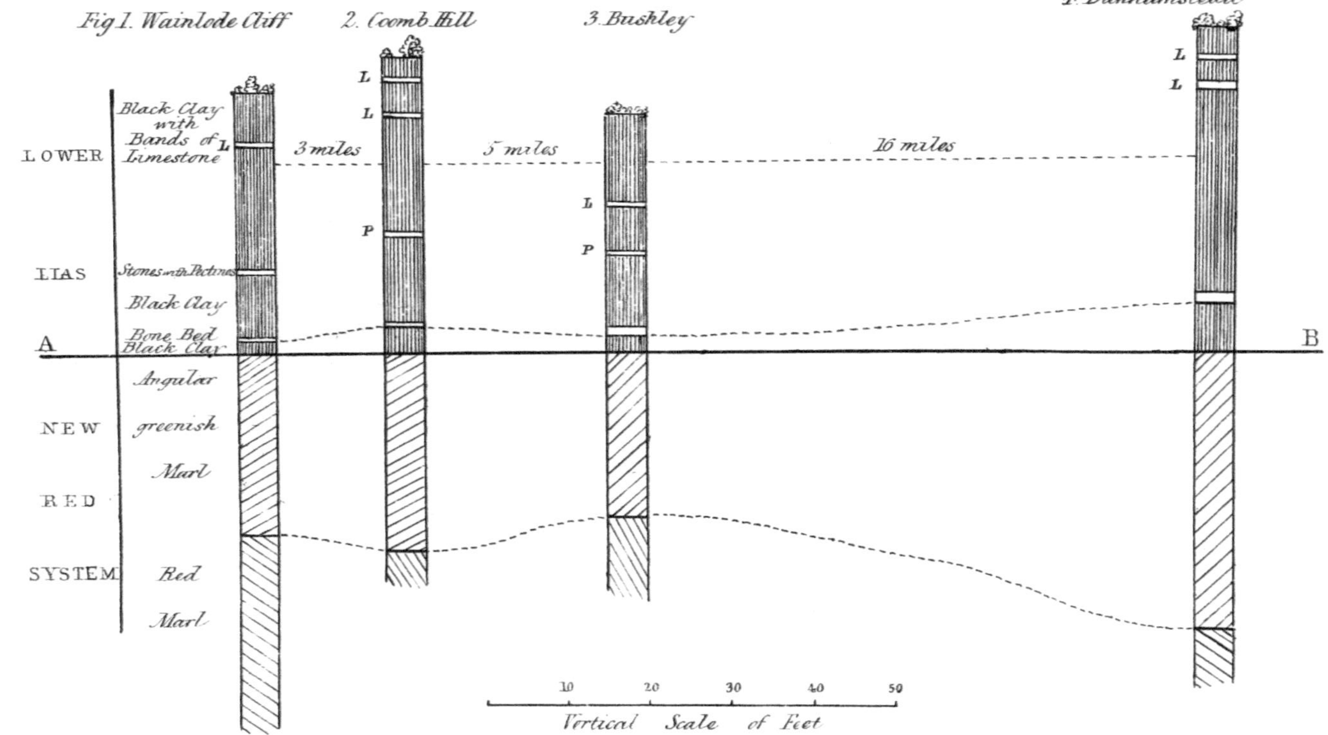

C.D.S.M. litho.

W. West Imp.

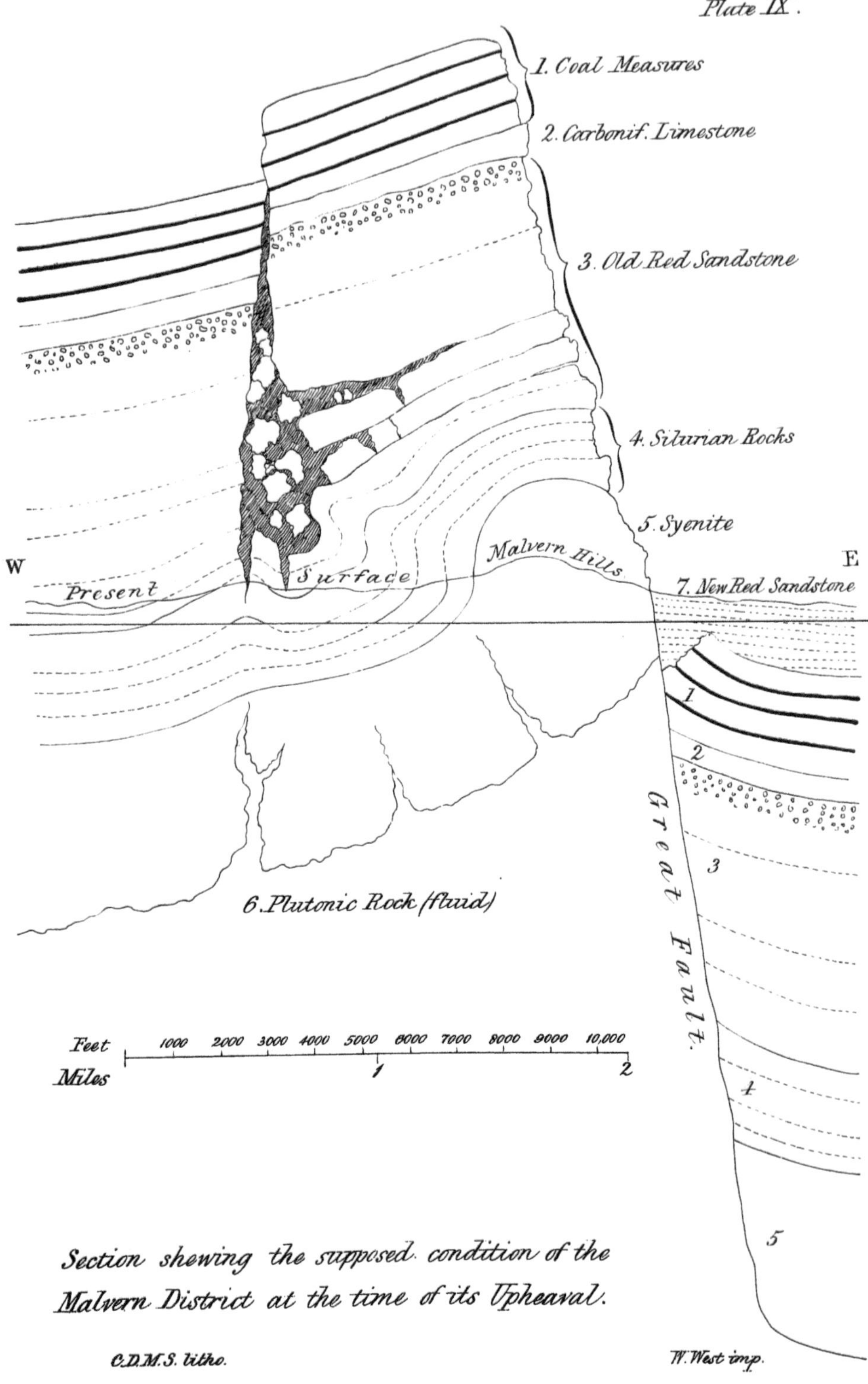

Section shewing the supposed condition of the Malvern District at the time of its Upheaval.

C.D.M.S. litho.

W. West imp.

Map of the Family Alcedinidæ.

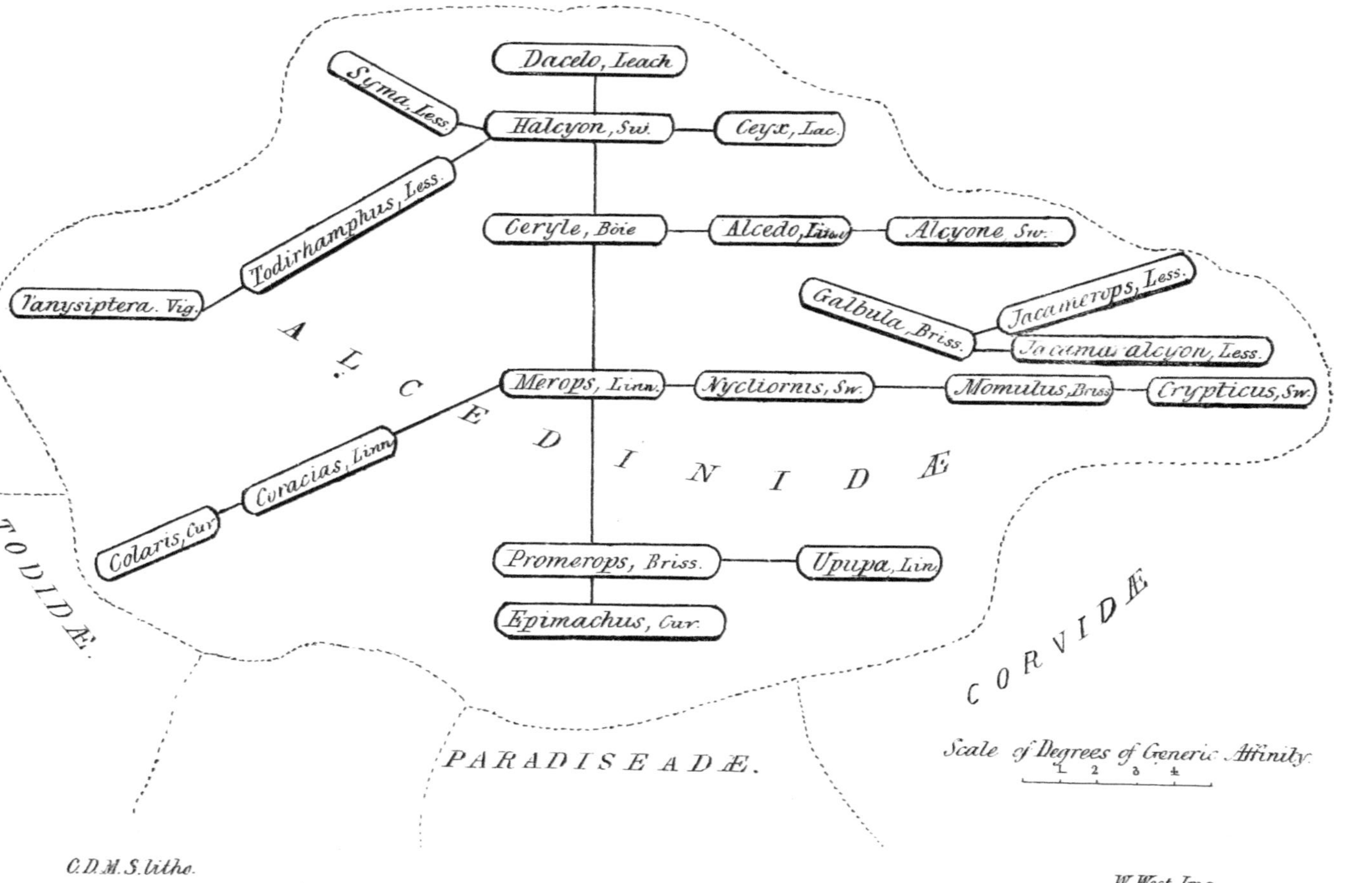

C.D.M.S. litho.

W. West Imp.

For EU product safety concerns, contact us at Calle de José Abascal, 56–1°, 28003 Madrid, Spain or eugpsr@cambridge.org.

www.ingramcontent.com/pod-product-compliance
Ingram Content Group UK Ltd.
Pitfield, Milton Keynes, MK11 3LW, UK
UKHW041900190726
13854UKWH00002B/1000

* 9 7 8 1 1 0 8 0 3 7 6 9 3 *